Embedded System Design *with the* C8051

Han-Way Huang
Minnesota State University, Mankato

CENGAGE Learning™

Australia • Brazil • Japan • Korea • Mexico • Singapore • Spain • United Kingdom • United States

Embedded System Design with the C8051
Han-Way Huang

Director, Global Engineering Program:
Chris Carson

Senior Developmental Editor:
Hilda Gowans

Editorial Assistant: Jennifer Dinsmore

Marketing Specialist: Lauren Betsos

Director, Content and Media Production:
Barbara Fuller-Jacobsen

Content Project Manager: D. Jean Buttrom

Production Service: RPK Editorial Services

Copyeditor: Shelly Gerger-Knechtl

Proofreader: Martha McMaster

Indexer: Shelly Gerger-Knechtl

Compositor: Newgen

Senior Art Director: Michelle Kunkler

Internal Designer: Juli Cook

Cover Designer: Andrew Adams

Cover Image: © Norebbo/Shutterstock

Senior First Print Buyer: Doug Wilke

Library of Congress Control Number: 2008906898

U.S. Student Edition:
ISBN-13: 978-0-495-47174-5
ISBN-10: 0-495-47174-7

Cengage Learning
200 First Stamford Place, Suite 400
Stamford, CT 06902
USA

Cengage Learning is a leading provider of customized learning solutions with office locations around the globe, including Singapore, the United Kingdom, Australia, Mexico, Brazil, and Japan. Locate your local office at: **international.cengage.com/region.**

Cengage Learning products are represented in Canada by Nelson Education Ltd.

For your course and learning solutions, visit **www.cengage.com/engineering.**
Purchase any of our products at your local college store or at our preferred online store **www.ichapters.com.**

Printed in the United States of America
1 2 3 4 5 6 7 12 11 10 09 08

CONTENTS

Chapter 1 Introduction to Microcontroller and the Intel 8051 1

Chapter 10 The SPI Function 451

Chapter 11 I^2C Bus and SMBus 513

Chapter 13 Controller Area Network (CAN) 635

PREFACE

The 8051 was introduced in 1980 and is the first and most popular 8-bit microcontroller in the market. The original 8051 has a very limited amount of peripheral functions and on-chip memory. However, the 8051 microcontroller market has experienced significant change during the last ten years. The most important developments during this period include:

- **Feature-Rich Timer System.** Most of the new 8051 variants provide input capture, output compare, timer auto-reload on match, watchdog timer, and PWM functions.
- **In-System Programming Capability.** Most new 8051 devices provide on-chip flash program memory which allows the user to upgrade the application program in the system (there is no need to replace any chip in the product).
- **JTAG Interface.** Many new 8051 devices include the JTAG interface that implements the IEEE 1149.1 boundary scan architecture. This interface supports boundary scan for production and in-system testing, flash memory read/write operation, and non-intrusive in-circuit debugging. This interface allows the user to design an inexpensive adapter that allows the software running on the PC or workstation to communicate with the 8051 MCU on the target hardware and provides source-level debugging capability. The debug adapters from SiLabs and STMicroelectronics are examples of this approach. This makes software debugging easy and inexpensive.
- **Multiple Serial Interfaces.** Many newer 8051 variants support industrial-standard UART, SPI, I^2C, CAN, and USB. UART allows the user to interface with the PC using the popular RS232, while at the same time supports

multi-processor communication. The SPI and I^2C allow the 8051 device to interface with many peripheral devices, such as the LED driver, LCD controller, A/D converter, D/A converter, Ethernet controller chip, PLL chip, real-time clock IC, and serial memory chips. The CAN bus was proposed as a communication bus for automotive applications. However, this bus also is used widely in industrial control. The USB bus has become ubiquitous these days. It enables very flexible interfacing between a host and a slave. The 8051 with a USB controller can serve either as a slave device controller or a host.

- **On-Chip Oscillator.** When frequency accuracy is not very critical, the on-chip oscillator can reduce the pin count, save the cost for external oscillator, and make the product more compact.
- **On-Chip A/D and D/A Converters.** Some 8051 variants provide A/D and D/A converters with very high resolutions, which make them very competitive in data acquisition, instrumentation, and digital audio applications.

These new features make the newer 8051 variants more competitive when compared with 8-bit microcontrollers of different architectures. These features also make the 8051 variants suitable for learning modern microcontroller interfacing and applications.

Intended Audience

This book is written for two groups of readers:

1. Students in electrical and computer engineering and technology who are taking an introductory course in microcontroller interfacing and applications. For this group of people, this book provides a broad and systematic introduction to microcontrollers.
2. Senior electrical and computer engineering students and working engineers who want to learn the 8051 and use it to design projects. For this group of readers, this book provides numerous more complicated examples to explore the functions and applications of the 8051. C language should be the choice for this group of readers.

Prerequisites

The writing of this book has assumed that the reader has taken a course on digital logic design and has been exposed to at least one high-level language programming. Knowledge of digital logic design will greatly facilitate learning the 8051. Knowledge of assembly language programming is not required, because one of the goals of this book is to teach 8051 assembly language programming.

Approach

During the last few years, we have seen more and more universities use C language in addition to assembly language to teach embedded system and microcontroller interfacing. Readers may gain more insight to the hardware working through learning assembly language. However, writing application software in C language certainly will make the designer much more productive for more complicated applications. Both assembly and C language are used in illustrating the programming of the 8051 in this text.

Organization of the Book

Chapter 1 presents basic concepts of computer hardware and software, microcontroller applications, the 8051 addressing modes, a subset of the 8051 instructions, and the program execution process. Chapter 2 introduces basic assembly programming skills, the 8051 instructions, a tutorial to the software and hardware development tools, and a few tips and examples on program debugging. Chapter 3 provides an introduction to parallel I/O ports, I/O device interfacing, time delay creation, and system clock generation. Chapter 4 covers more advanced assembly programming skills, subroutine calls, and using the UART port to perform terminal I/O. A few tips on program debugging involving subroutine calls are provided.

Chapter 5 provides a brief tutorial on the C language syntax and the use of the freeware SDCC compiler under the SiLabs IDE. Tips on program debugging in C language are provided. Chapter 6 discusses the concepts and programming of interrupts and resets. Chapter 7 introduces the basic concepts of parallel I/O. This chapter also covers the interfacing and programming of simple I/O devices, including DIP switches, keypad scanning and debouncing, LEDs, LCDs, D/A converters, and stepper motor control. Electrical compatibility issues for I/O interfacing also are discussed. Chapter 8 explores the operation and applications of the timer system, including capture mode, auto-reload in counting, toggle clock output, compare, PWM, time delay creation, frequency measurement, pulse width measurement, waveform generation, siren generation, and song playing. Chapter 9 deals with the universal asynchronous receiver and transceiver (UART) interface. Chapter 10 deals with the SPI interface and the applications of the SPI-compatible peripheral chips.

Chapter 11 introduces the SMBus and I^2C bus protocol and several peripheral chips with I^2C interface. Chapter 12 discusses A/D and D/A converters and their applications in temperature, humidity, and barometric pressure measurement and analog waveform generation. Chapter 13 presents the CAN 2.0 protocol and the C8051F040 CAN module. Several examples of the programming of the CAN module are provided. Chapter 14 describes the

8051 internal SRAM, and flash memory. This chapter also explores the erasure and programming of the flash memory.

Pedagogical Features

Each chapter starts with a list of objectives. Every subject is presented in a step-by-step manner. Background issues are presented before the specifics related to each 8051 function are discussed. Numerous examples are then presented to demonstrate the use of each 8051 I/O function. Procedural steps and flowcharts are used to help the reader to understand the program logic in most examples. Each chapter concludes with a summary, numerous exercises, and lab assignments.

Which 8051 Devices to Learn?

At the time of this writing, there are more than fifty 8051 vendors and more than 1200 different 8051 devices. These 8051 variants share the same architecture (including instruction sets and addressing modes). However, they may differ significantly in their CPU throughput and the peripheral functions that they provide. When deciding to learn the 8051, the first issue is choosing an 8051 variant to learn. The considerations for choosing the 8051 variant to learn are different from those in choosing the 8051 variant to design an embedded product. When selecting the 8051 variant to learn, we will consider at least the following factors:

- *CPU Throughput:* The 8051 uses instruction clock signals to control the timing of CPU and all peripheral operations. The original 8051 derives the instruction clock by dividing the crystal oscillator input by 12 and the highest crystal oscillator frequency is limited to 12 MHz. This speed is very slow by today's standards. However, many 8051 vendors have reduced the number of crystal oscillator cycles contained in one instruction clock cycle from 12 to 6 or 4 or 2 or even 1.
- *On-Chip Flash Memory and In-System Programming Capability:* This feature makes it easy to erase and reprogram the program memory without the need to remove MCU from the demo board. The size of flash memory must be big enough to satisfy the foreseeable need in learning the microcontroller or doing the design project.
- *On-Chip XRAM:* The on-chip XRAM is the on-chip SRAM other than the 256 bytes provided by the original 8051. Because the on-chip SRAM of the original 8051 is very small, it is necessary to add external data memory in order to handle larger data arrays. A few KBytes of on-chip XRAM makes external data memory unnecessary. Many of today's 8051 variants provide several KBytes of on-chip XRAM. The reason that the additional on-chip

SRAM is called XRAM is because the user needs to use the DPTR register to access it just like external data memory.

- *Additional Serial Interface* (such as SPI, I^2C, and CAN): The major competitors of 8051 variants such as Microchip PIC18, Freescale MC9S08, and Atmel AVR all provide these three serial interfaces. The availability of USB interface is a plus.
- *More Flexible Timer Functions:* This include extra features in addition to those provided by Timer 0, 1, and 2. The desirable timer features include input capture, timer compare (or called output compare), and flexible waveform generation functions provided by the pulse width modulation (PWM) mode.
- *A/D and D/A Converters:* A/D and D/A converters are important for data acquisition, digital audio signal handling, digital video signal processing, and instrumentation. The most popular resolution is 10-bit in the 8-bit and 16-bit microcontroller markets. However, higher resolutions may be needed in some applications.
- *Number of I/O Pins:* Without enough I/O pins, multiple functions need to share the use of the same pin. Only a subset of the available peripheral functions can be used at a time. More and more 8051 vendors provide more than 40 I/O pins, so that more peripheral functions can be used simultaneously.
- *Hardware Debug Support:* Many newer 8051 variants provide JTAG (compliant with IEEE1149.1 standard) to support boundary scan capability while at the same time support in-system programming and real-time debugging capability.

Considering these factors, devices from a small number of vendors stand out as good candidates. Silicon Laboratory (also called SiLabs), STMicroelectronics, Infineon, Analog devices, Philips, Atmel, and Dallas Semiconductors are among the top choices. However, the development tools support is just as important as the features of the MCU itself for the choice of 8051 variant to learn. The discussion of this text will be based on the original 8051 and the SiLabs C8051F040.

Software Development Tools

The ideal software development tool is an integrated development environment (IDE) that comprises a programmer's text editor, a cross assembler, a C compiler, a project manager, a linker, and a communication program. Using an IDE, the user can enter the program text, assemble/compile the program, link the object code, download the hex file into the program, and then perform program debugging without exiting any program. An approach used by a few microcontroller vendors is to provide a debug adaptor between the PC and the target hardware. The IDE runs on the PC and communicates with the target hardware via the debug adaptor to perform the debug activities.

Most 8051 vendors rely on software tool vendors to provide development tools. Silicon Laboratory is among the very few that provides its own IDE. The SiLabs IDE can work with many cross assemblers and C compilers (including the freeware SDCC C compiler) to provide source-level debugging capability. SiLabs also provides its own inexpensive debug adaptor. The freeware SDCC compiler can be downloaded from http://sdcc.sourceforge.net. IAR, Keil, and Raisonance are major 8051 C Compiler and IDE vendors. Keil has a 2-KB demo version of IDE (μVision 2 and 3) and discount price for education institutions. Both IAR and Raisonance have a 4-KB demo version of IDE available for evaluation and learning the 8051 microcontroller programming. Keil's μVision can also provide source-level debugging capability for the SiLabs 8051 variants using SiLabs's debug adaptor.

Demo Boards

Most companies provide simple and inexpensive starter kits for users to learn and experiment with their microcontrollers. However, the starter kits from 8051 vendors and many other microcontroller vendors lack the accessories (such as LCD, LEDs, DIP switches, potentiometers, peripheral chips with SPI or I^2C interfaces, digital function generator, speaker, and debounced switches) to perform experiments. These starter kits are good for doing student design projects because of their low price. SiLabs also provides starter kits for all of its MCUs. Educators can purchase these kits at a discounted price. The C8051F040TB target board could be used to learn the C8051F040 MCU. More detailed information about this kit can be found at www.silabs.com.

The demo board SSE040 made by Shuan-Shizu Enterprise is based on SiLabs's C8051F040 and includes many accessories including LEDs, LCD, potentiometer, DIP switches, CAN transceiver and connector, digital function generator, debounced switches, buzzer, real-time clock chip, serial EEPROM, digital temperature sensor, and others. More detailed information about SSE040 can be found at www.evb.com.tw. Most programs in this book are tested using the SSE040 demo board.

The following hardware and software combinations are recommended for learning and teaching the C8051 microcontroller:

Alternative 1
SiLabs IDE + compiler from Keil (demo version is OK) + debug adaptor from SiLabs + SSE040 demo board (or SiLabs target board)

Alternative 2
SiLabs IDE + compiler from Raisonance (demo version is OK) + debug adaptor from SiLabs + SSE040 demo board (or SiLabs target board)

Alternative 3
Keil's μVision + debug adaptor from SiLabs + SSE040 demo board (or SiLabs target board)

Alternative 4
SiLabs IDE + SDCC compiler (freeware without size limit) + SSE040 demo board (or SiLabs target board)

These four recommended hardware/software combinations require a PC running Windows (XP) operating system. We have confidence that the Window Vista operating system would be able to handle these combinations of tools.

To Instructors

It is unnecessary for instructors to follow strictly the order of chapters of this book in their teaching. If only assembly language programming is to be taught, then the following order is recommended:

- Chapter 1 through 4 in that order
- Chapter 6 and 7 in any order
- Chapter 7
- Chapter 8
- Chapter 9 through 14 in any order

If your microprocessor (or microcontroller) course covers only C language, then the following order is recommended:

- Section 1.6
- Subjects related to I/O ports and Timer 0/1 in Chapter 3
- Chapter 5
- Chapter 6 and 7 in any order
- Chapter 8
- Chapter 9 through 14 in any order

If both the assembly and C languages are to be taught, then the following order is recommended:

- Chapter 1 through 4 in that order
- Chapter 5
- Chapter 6 and 7 in any order
- Chapter 8
- Chapter 9 through 14 in any order

Complementary Material

The following materials are useful to the learning of the C8051 and are made available through the book website under StudentResources at **www.cengage.com/engineering**.

- Source code of all example programs in the text
- The PDF files of datasheets of the C8051F040 and peripheral chips

- The software (including freeware software SDCC, demo version of RIDE, and the demo version of μVision 3)
- Utility programs (in assembly and C languages) for time delays, LCD, UART, and SPI

Supplements

A printed Solutions Manual will be available to instructors. Solutions and PowerPoints will also be available through the book website, and professors should request a password from their Sales Representative. Professors are encouraged to modify the PowerPoint lecture notes to suit their teaching needs.

Feedback and Update

The author has tried his best to eliminate errors from this text. However, this author knows that it is impossible to eliminate all of the errors. The solutions in the examples of this book may not be the best either. Error reports and suggestions are welcomed. Please send them to hanwayh@gmail.com.

Acknowledgments

This book would not be possible without the help of a number of people. I would like to thank Minsu Choi, Missouri University of Science and Technology; Yoon Geun Kim, Virginia State University, Shahram Shirani, McMaster University, and Thomas Stuart, University of Toledo, for their valuable opinions on how to improve the quality of this book. I would also like to thank Christopher Carson, Director, Global Publishing, and Hilda Gowans, Senior Development Editor, of Cengage Learning for their enthusiastic support during the preparation of this book. I also appreciate the outstanding work of the production staff of Cengage Learning. I also would like to express my heart-felt appreciation to my students and colleagues at the Department of ECET at Minnesota State University, Mankato who allow me to test out the manuscript.

Finally, I would like to express my thanks to my wife, Su-Jane, and my sons, Craig and Derek, for their encouragement and support during the entire preparation of this book.

HAN-WAY HUANG
Mankato, Minnesota

Introduction to Microcontroller and the Intel 8051

1.1 Objectives

Upon successful completion of this chapter, you will be able to:

- Convert decimal numbers to binary
- Convert binary numbers to decimal
- Convert binary numbers to octal and vice versa
- Convert binary numbers to hexadecimal and vice versa
- Perform addition and subtraction using two's complement number addition
- Explain the structure of a computer hardware
- Explain the memory technologies
- Explain the structure of an ALU that performs ADD, SUB, AND, and OR operations
- Explain the instruction execution process
- Explain the memory organization of the 8051 microcontroller
- Explain the differences among the register inherent, direct, immediate, indirect, indexed, relative, absolute, long, bit-inherent, and bit-direct addressing modes
- Use **MOV** instructions to perform data-movement operations
- Use **ADD** instructions to perform addition operations
- Use **SUB** instructions to perform subtraction operations
- Combine the use of MOV, ADD, and SUB instructions to perform more complex operations
- Find out 8051 instruction execution time

1.2 Number System Issue

Computers were initially designed to deal with numerical data. Due to the on-and-off nature of electricity, numbers were represented in a binary base from the beginning of the electronic computer age.

However, the binary number system is not natural to us, because human-beings have been mainly using the decimal number system for thousands of years. A binary number needs to be converted to a decimal number before it can be quickly interpreted by a human being.

For convenience of discussion, we will add a subscript 2, 8, or 16 to a number to indicate the base of the given number. For example, 101_2 is a binary number; 234_8 is an octal number; and 2479_{16} is a hexadecimal number. No subscript will be used for a decimal number.

1.2.1 Converting from Binary to Decimal

A binary number is represented by two symbols: 0 and 1. Here, we refer to 0 and 1 as a *binary digit.* A binary digit is also referred to as a *bit.* In the computer, 8 bits are referred to as a *byte.* Depending on the computer, either 16 bits or 32 bits are referred to as a *word.* The values of 0 and 1 in a binary number system are identical to their counterparts in the decimal number system. To convert a binary number to decimal, we compute a weighted sum of every binary digit contained in the binary number. If a specific bit is k bits from the rightmost bit, then its weight is 2^k. For example,

$$10100100_2 = 2^7 + 2^5 + 2^2 = 128 + 32 + 4 = 164$$

$$11011001_2 = 2^7 + 2^6 + 2^4 + 2^3 + 2^0 = 128 + 64 + 16 + 8 + 1 = 217$$

$$10010010.101_2 = 2^7 + 2^4 + 2^1 + 2^{-1} + 2^{-3} = 146.625$$

1.2.2 Converting from Decimal to Binary

A decimal integer can be converted to a binary number by performing the repeated-division-by-2 operation until the quotient becomes 0. The remainder resulting from the first division is the least significant binary digit, whereas the remainder resulting from the last division is the most significant binary digit.

Example 1.1 Convert the decimal number 73 to decimal.

Solution: The procedure of conversion is shown in Figure 1.1.

$73 = 1001001_2$

Example 1.2 Convert the decimal number 95 into binary.

Solution: The repeated-division-by-2 process is shown in Figure 1.2.

$95 = 1011111_2$

If a decimal number has a *fractional part,* then it needs to be converted using a different method. The fractional part can be converted to decimal by performing the repeated-multiplication-by-2 operation to the fraction until either the fraction part becomes 0 or the *required accuracy* is achieved. The resulting integer binary digit of the first multiplication is the most significant binary digit of the fractional part, whereas the resulting integer binary digit of the last multiplication is the least significant binary digit of the fractional part.

Let m and k be the number of digits of the decimal fraction and the binary fraction, respectively. Then the desired accuracy has been achieved if and only if the following expression is true:

$$2^{-k} < 10^{-m}$$

A few pairs of k and m values that satisfy the previous relationship are shown in Table 1.1. To be more accurate, if the resultant fractional part of the last multiplication is 5 or larger, then we should round it up by adding 1 to the least significant binary digit.

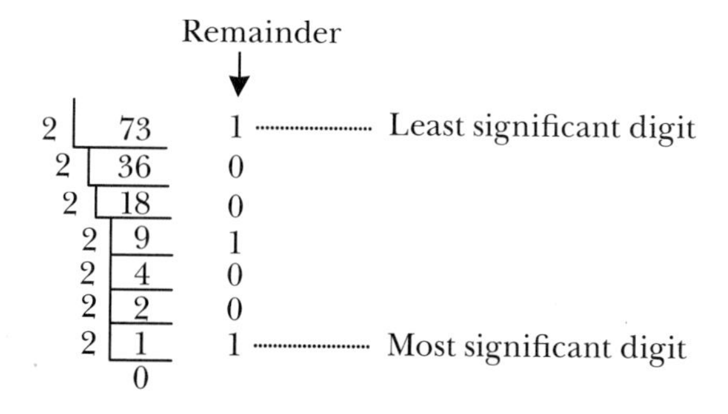

FIGURE 1.1
An example of decimal-to-binary conversion.

Remainder
2 | 95 1 Least significant digit
2 | 47 1
2 | 23 1
2 | 11 1
2 | 5 1
2 | 2 0
2 | 1 1 Most significant digit
0

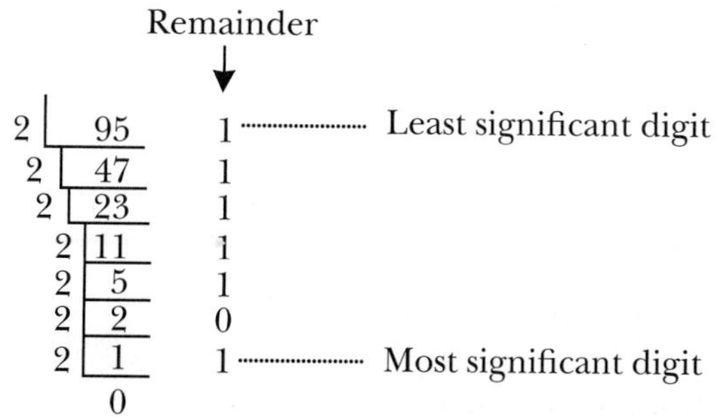

FIGURE 1.2
Another example of decimal-to-binary conversion.

TABLE 1.1 *k and m that Satisfy* $2^{-k} < 10^{-m}$

k	m
<=4	1
<=7	2
<=10	3
<=14	4

Example 1.3 Convert the decimal fraction 0.6 to binary format.

Solution: According to Table 1.1, we need to perform four repeated-multiplication-by-2 operations, as shown in Figure 1.3.

$0.6_{10} = 0.1001_2 + 0.0001_2 = 0.1010_2$

Example 1.4 Convert the decimal number 59.75 to binary representation.

Solution: Since the given number has both the integral and fractional parts, they need to be converted separately and then combined together. The conversion process is shown in Figure 1.4.

FIGURE 1.3 *Convert a decimal fraction to a binary fration.*

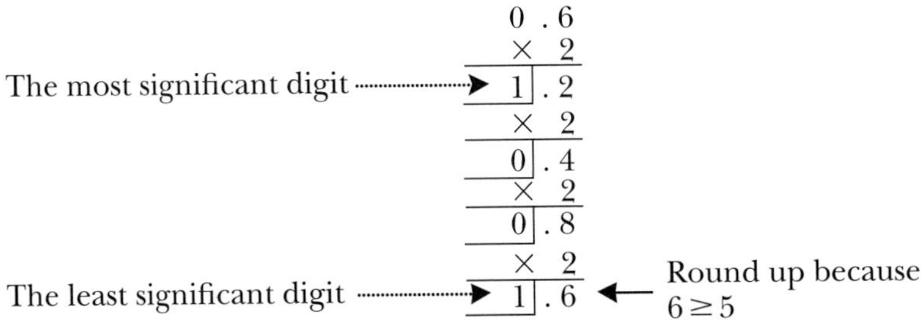

0 . 6
× 2
The most significant digit → 1 . 2
× 2
0 . 4
× 2
0 . 8
× 2
The least significant digit → 1 . 6 ← Round up because 6 ≥ 5

FIGURE 1.4 *Convert a decimal number with both integral and fractional parts to binary.*

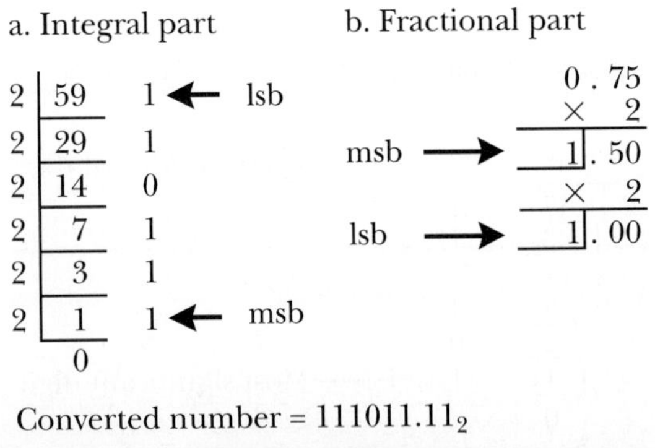

a. Integral part

2	59	1 ←	lsb
2	29	1	
2	14	0	
2	7	1	
2	3	1	
2	1	1 ←	msb
	0		

b. Fractional part

0 . 75
× 2
msb → 1. 50
× 2
lsb → 1. 00

Converted number = 111011.11_2

1.2.3 Why Octal and Hexadecimal Numbers?

After the computation, the computer needs to output the result so that the user can see and determine whether the result is correct. The result can be displayed in either binary or decimal format. To display a number in binary format is not convenient, because it takes many 0s and 1s to represent a large number. To display a number in decimal format will require the computer to perform some complicated conversion operations. A compromise is to use an octal or hexadecimal format to represent a number. Hexadecimal representation is used more often than the octal representation. The shorthand of *hexadecimal* is *hex*.

1.2.4 Convert from Binary to Octal

The octal number system uses digits 0 through 7 to represent a number. The digit 0 corresponds to 000_2, whereas the digit 7 corresponds to 111_2. There are two steps to convert a binary number to octal:

Step 1
Partition the given number (a string of 0s and 1s) from the right to the left into groups of 3 bits. Add leading 0s if the leftmost group has less than 3 bits.

Step 2
Convert each 3-bit group to the corresponding octal digit (0 to 7).

Example 1.5 Convert the following two binary numbers into octal representation.

(a) 10101001_2

(b) 1010101010_2

Solution:

(a) $10101001_2 = 10,101,001 = 010,101,001_2 = 251_8$

(b) $1010101010_2 = 1,010,101,010 = 001,010,101,010_2 = 1252_8$

1.2.5 Convert from Octal to Binary

To convert from octal to binary, simply convert each octal digit to its 3-bit binary equivalent and delete the leading 0s from the resultant binary number.

Example 1.6 Convert the following octal numbers into binary format.

(a) 532_8

(b) 246_8

Solution:

(a) $532_8 = 101{,}011{,}010_2$

(b) $246_8 = 010{,}100{,}110_2 = 10{,}100{,}110_2$

1.2.6 Convert from Binary to Hexadecimal

The hex number system uses digits 0 through 9 and letters A through F to represent a number. The digit 0 corresponds to 0000_2 and letter A corresponds to 1010, whereas the letter F corresponds to 1111_2. There are two steps to convert a binary number to hex:

Step 1
Partition the given number (a string of 0s and 1s) from the right to the left into groups of 4 bits. Add leading 0s if the leftmost group has less than 4 bits.

Step 2
Convert each 4-bit group to the corresponding hex digit (0 to F).

Example 1.7 Convert the following two binary numbers into hex representation.

(a) 100101011001_2

(b) 1011010101110_2

Solution:

(a) $100101011001_2 = 1001{,}0101{,}1001_2 = 959_{16}$

(b) $1011010101110_2 = 1{,}0110{,}1010{,}1110_2 = 0001{,}0110{,}1010{,}1110_2 = 16AE_{16}$

1.2.7 Convert from Hexadecimal to Binary

To convert from hex to binary, simply convert each hex digit into its 4-bit binary equivalent and delete the leading 0s from the resultant binary number.

Example 1.8 Convert the following hex numbers into binary format.

(a) $5CB_{16}$

(b) $2A6_{16}$

Solution:

(a) $5CB_{16} = 0101{,}1100{,}1011_2 = 101{,}1100{,}1011_2$

(b) $2A6_{16} = 0010{,}1010{,}0110_2 = 10{,}1010{,}0110_2$

1.2.8 Specifying the Number Base

To facilitate the specification of the number base used in a number representation, we will use a notation that adds a suffix to a number to indicate the base used in the number representation. The suffixes for hex, binary, and octal are H (or h), B (or b), and O (or o, Q, q), respectively. Decimal numbers do not use suffix. For examples,

10101011B and 10101011b

specify the binary number 10101011_2.

123O, 123o, 123q, and 123Q

specify the octal number 123_8.

A097H and A097h

refer to the hexadecimal number $A097_{16}$.

3467

is a decimal number.

Hexadecimal numbers are also represented by adding the prefix 0x to the number. For example, 0x1000 stands for the hex number 1000_{16}. In this text, we will mix the use of these two methods for hexadecimal numbers.

1.3 Binary Addition and Subtraction

Addition in binary representation follows the familiar rules of decimal addition. When adding two numbers, add the successive bits and any carry. You will need the following addition rules:

$$0 + 0 = 0$$

$$0 + 1 = 1$$

$$1 + 1 = 0 \text{ carry} = 1$$

$$1 + 1 + 1 = 1 \text{ carry} = 1$$

Carry generated in any bit position is added to the next higher bit.

Binary numbers are subtracted using the following rules:

$$0 - 0 = 0$$

$$0 - 1 = 1 \text{ with a borrow } 1$$

$$1 - 0 = 1$$

$$1 - 1 = 0$$

Example 1.9 Add the following pairs of binary positive numbers.

(a) 1110110_2 and 1100100_2

(b) 01101_2 and 10001_2

Solution: The addition process is shown in Figure 1.5.

Example 1.10 Perform the following binary subtractions.

(a) $11001_2 - 110_2$

(b) $110011_2 - 11001_2$

Solution: The subtraction process is shown in Figure 1.6.

1.4 Two's Complement Numbers

In a computer, the number of bits that can be used to represent a number is fixed. As a result, computers always manipulate numbers that are fixed in length.

Another restriction on number representation in a computer is that both positive and negative numbers must be represented. However, a computer does not include the plus and minus signs in a number. Instead, all modern computers use the two's complement number system to represent positive and negative numbers. In the two's complement system, all numbers that have a most significant bit (msb) set to 0 are *positive*, and all numbers with an msb set to 1 are *negative*. Positive two's complement numbers are identical to binary numbers, except that the msb must be a 0. If N is a positive number, then its two's complement N_C is given by the following expression:

$$N_C = 2^n - N \tag{1.1}$$

where n is the number of bits available for representing N.

```
a. Carry   1   1                 b. Carry        1
           1 1 1 0 1 1 0                 0 1 1 0 1
         + 1 1 0 0 1 0 0               + 1 0 0 0 1
         ---------------               -----------
         1 1 0 1 1 0 1 0                 1 1 1 1 0
```

FIGURE 1.5 *Examples of binary addition.*

```
a.      1 1          Borrow        b.  1 1            Borrow
    1 1 0 0 1        Minuend           1 1 0 0 1 1    Minuend
+       1 1 0        Subtrahend    -     1 1 0 0 1    Subtrahend
---------------------------------  ------------------------------
    1 0 0 1 1        Difference        0 1 1 0 1 0    Difference
```

FIGURE 1.6 *Examples of binary subration.*

The two's complement (2's complement) of N is used to represent $-N$. Machines that use the two's complement number system can represent integers in the range

$$-2^{n-1} \leq N \leq 2^{n-1} - 1 \quad (1.2)$$

Example 1.11 Find the range of integers that can be represented by an 8-bit two's complement number system.

Solution: The range of integers that can be represented by the 8-bit two's complement number system is.

$$-2^7 \leq N \leq 2^7 - 1$$

or

$$-128_{10} \leq N \leq 127_{10}$$

Example 1.12 Represent the negative binary number -11001_2 in 8-bit two's complement format.

Solution: Use Equation 1.1 and the subtraction method given in Section 1.3.

$$N_C = 2^8 - 11001 = 11100111_2$$

An easy way to find the two's complement of a binary number is flip every bit from 0 to 1 or 1 to 0 and then add 1 to it.

1.5 Two's Complement Subtraction

Subtraction can be performed by adding the two's complement of the subtrahend. This allows the same piece of hardware to perform addition and subtraction. All of today's computers use this method to perform subtraction.

After the addition, throw away the carry generated from the most significant bit. The result is the desired difference. If the most significant bit is 0, then the resultant difference is positive. Otherwise, it is negative.

Example 1.13 Subtract 13 from 97 using two's complement arithmetic.

Solution: The 8-bit binary representation of 13 and 97 are 00001101_2 and 01100001_2, respectively. The two's complement of 13 is 11110011_2. The value of 97 − 13 can be computed as shown in Figure 1.7.

FIGURE 1.7
Subtraction by adding two's complement of subtrahend.

```
   0 1 1 0 0 0 0 1
 + 1 1 1 1 0 0 1 1
 -----------------
 1 0 1 0 1 0 1 0 0  = 84_10
```

Throw away carry

FIGURE 1.8
Subtraction by adding two's complement of subtrahend.

```
   0 1 0 0 0 0 0 1
 + 1 0 0 1 1 1 1 0
 -----------------
   1 1 0 1 1 1 1 1  = -33_10
```

Example 1.14 Subtract 98 from 65 using two's complement arithmetic.

Solution: The 8-bit binary representation of 98 and 65 are 01100010_2 and 01000001_2, respectively. The two's complement of 98 is 10011110_2. The value of $65 - 98$ is computed in Figure 1.8.

The binary number 11011111_2 is the two's complement of 00100001_2 (33). Therefore, the resultant difference is -33 and is correct.

1.6 Overflow

Overflow can occur with either addition or subtraction in two's complement representation. During addition, overflow occurs when the sign of the sum of two numbers with like signs differs from the sign of two numbers. Overflow never occurs when adding two numbers with unlike signs. In subtraction, overflow can occur when subtracting two numbers with unlike signs. If

Negative $-$ positive $=$ positive

or

Positive $-$ negative $=$ negative

then overflow has occurred.

Overflow never occurs when subtracting two numbers with like signs.

Example 1.15 Does overflow occur in the following 8-bit operations?

(a) $01111111_2 - 00000111_2$

(b) $01100101_2 + 01100000_2$

(c) $10010001_2 - 01110000_2$

Solution:

(a) Negation of 00000111_2 is 11111001_2. $01111111_2 - 00000111_2$ is performed as shown in Figure 1.9.

The difference is $01111000_2 = 120_{10}$. There is no overflow.

(b) The sum of these two numbers is as shown in Figure 1.10.

Overflow has occurred.

(c) The two's complement of 01110000_2 is 10010000_2. $10010001_2 - 01110000_2$ is performed as shown in Figure 1.11.

Overflow has occurred.

An important benefit of using two's complement arithmetic is that the same computer hardware can perform signed and unsigned (all numbers are nonnegative) addition and subtraction without modification. It is up to the user to interpret the number to be negative.

1.7 Representing Nonnumeric Data

Computers are also used to process nonnumeric data. Nonnumeric data is often in the form of character strings. A character can be a letter, digit, or special character symbol. A unique number is assigned to each character of the

```
    0 1 1 1 1 1 1 1
  + 1 1 1 1 1 0 0 1
  -----------------
  1 0 1 1 1 1 0 0 0
  ↗
Carry out to be discarded
```

FIGURE 1.9 *Example of addition that causes no overflow.*

```
    0 1 1 0 0 1 0 1
  + 0 1 1 0 0 0 0 0
  -----------------
    1 1 0 0 0 1 0 1
  ↗
The sign has changed from positive to negative
```

FIGURE 1.10 *Example of addition that causes overflow.*

```
                              1 0 0 1 0 0 0 1
                            + 1 0 0 1 0 0 0 0
                            -----------------
Carry out to be discarded → 1 0 0 1 0 0 0 0 1
                              ↗
The sign has changed from negative to positive
```

FIGURE 1.11 *Another example of addition that causes overflow.*

character set so that they can be differentiated by the computer. To facilitate the processing of characters by the computer and the exchange of information, all the users must use the same number to represent (or encode) each character.

Two character code sets are in widespread use today: American Standard Code for Information Interchange (ASCII) and Extended Binary Coded Decimal Interchange Code (EBCDIC). Several types of mainframe computers use EBCDIC for internal storage and processing of characters. In EBCDIC, each character is represented by a unique 8-bit number, and a total of 256 different characters can be represented. Most microcomputer and minicomputer systems use the ASCII character set. A computer does not depend on any particular set, but most input/output (I/O) devices that display characters require the use of ASCII codes.

ASCII characters have a 7-bit code. They are usually stored in a fixed-length 8-bit number. The $2^7 = 128$ different codes are partitioned into ninety-five printable characters and thirty-three control characters. The control characters define communication protocols and special operations on peripheral devices. The printable characters consist of the following:

26 uppercase letters (A–Z)
26 lowercase letters (a–z)
10 digits (0–9)
1 blank space
32 special-character symbols, including !@#$%^&*()−_=+'[];:'",<.>/?{}

The complete ASCII code set is shown in Table 1.2.

1.8 Computer Hardware Organization

A computer consists of hardware and software. The hardware of a computer consists of the processor, input devices, output devices, and memory:

- *Processor.* The processor is responsible for performing all of the computational operations and the coordination of the usage of resources of a computer. A computer system may consist of one or multiple processors. A processor may perform general-purpose computations or special-purpose computations, such as graphics rendering, printing, or network processing.
- *Input Devices.* A computer is designed to execute programs that manipulate certain data. Input devices are needed to enter the program to be executed and data to be processed into the computer. There are a wide variety of input devices: keyboards, keypads, scanners, bar code readers, sensors, and so on.
- *Output Devices.* Whether the user uses the computer to perform computations or to find information from the Internet or a database, the end results must be displayed or printed on paper so that the user can see

TABLE 1.2 *Complete ASCII Code table*

Seven-Bit Hexa-decimal Code	Character	Seven-Bit Hexa-decimal Code	Character	Seven-Bit Hexa-decimal Code	Character	Seven-Bit Hexa-decimal Code	Character
0	NUL	20	SP	40	@	60	`
01	SOH	21	!	41	A	61	a
02	STX	22	"	42	B	62	b
03	ETX	23	#	43	C	63	c
04	EOT	24	$	44	D	64	d
05	ENQ	25	%	45	E	65	e
06	ACK	26	&	46	F	66	f
07	BEL	27	'	47	G	67	g
08	BS	28	(	48	H	68	h
09	HT	29	)	49	I	69	i
0A	LF	2A	*	4A	J	6A	j
0B	VT	2B	+	4B	K	6B	k
0C	FF	2C	,	4C	L	6C	l
0D	CR	2D	-	4D	M	6D	m
0E	SO	2E	.	4E	N	6E	n
0F	SI	2F	/	4F	O	6F	o
10	DLE	30	0	50	P	70	p
11	DC1	31	1	51	Q	71	q
12	DC2	32	2	52	R	72	r
13	DC3	33	3	53	S	73	s
14	DC4	34	4	54	T	74	t
15	NAK	35	5	55	U	75	u
16	SYN	36	6	56	V	76	v
17	ETB	37	7	57	W	77	w
18	CAN	38	8	58	X	78	x
19	EM	39	9	59	Y	79	y
1A	SUB	3A	:	5A	Z	7A	z
1B	ESC	3B	;	5B	[	7B	{
1C	FS	3C	<	5C	\	7C	|
1D	GS	3D	=	5D	]	7D	}
1E	RS	3E	>	5E	^	7E	~
1F	US	3F	?	5F	_	7F	DEL

them. There are many media and devices that can be used to display information: CRT displays, flat-panel LCD displays, seven-segment displays, printers, light-emitting diodes (LEDs), and so on.

- *Memory Devices.* Users write *programs* to tell computer what to do with the data at hand. Programs to be executed and data to be processed must be stored in memory devices so that the processor can readily access them.

1.8.1 The Processor

A processor is also referred to as the central processing unit (CPU). A processor consists of three major components: *arithmetic logic unit* (ALU), *control unit,* and *registers.*

The Arithmetic Logic Unit

The ALU performs arithmetic and logic operations requested by the user's program. The complexity of the ALU varies from one computer to another. If the processor designer wants to implement more operations directly in the hardware, then the ALU will get more complicated. An ALU that implements addition, subtraction, AND, and OR operations is illustrated in Figure 1.12.

The four-operation ALU operates in the following manner:

1. When **opcode = 00**, the adder selects the *n*-bit A as it's X input and the *n*-bit B as its Y input and **CIN** as its carry input (**ci**) and generate **SUM** and **Carry** operations. MUX3 selects SUM to become **Result** whereas **Carry** is connected to **Cout** directly. For this opcode, the ALU performs the **ADD** operation.
2. When **opcode = 01**, the comparator output is 1 and the inversion of B input (change 0 to 1 and 1 to 0) is selected as the Y input. Then 1 is selected as the **ci** input to the adder. The adder essentially adds the two's complement of B to A, which is equivalent to performing the **SUB** opera-

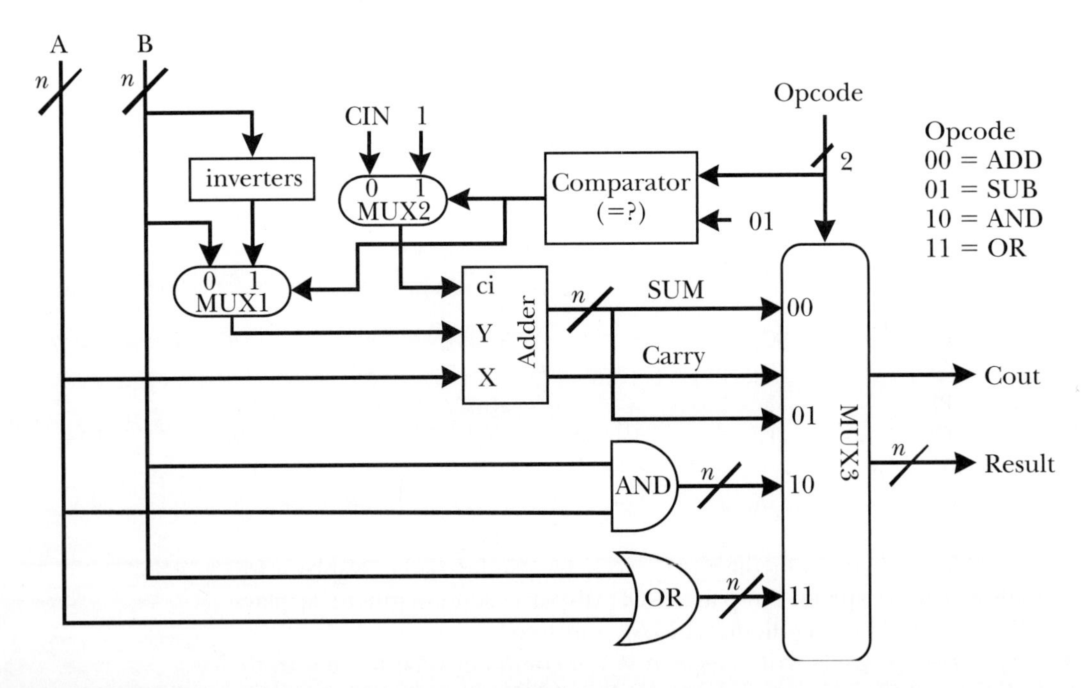

FIGURE 1.12
An ALU that implements ADD, SUB, AND, and OR operations.

tion. Since opcode is 01 and **SUM** is connected to both the 00 and 01 inputs, it will be selected and sent to **Result**. For this opcode, the ALU performs the **SUB** operation.

3. When **opcode = 10**, the MUX3 multiplexer selects the value connected to the **10** input and sends it to **Result**. Therefore, the ALU performs the **AND** operation.
4. When **opcode = 11**, the MUX3 multiplexer selects the value connected to the **11** input and sends it to **Result**. Therefore, the ALU performs the **OR** operation.

An ALU that performs more operations can be implemented by expanding the circuit shown in Figure 1.12.

The maximum number of bits that can be operated on by the ALU in one operation is referred to as the *word length* of the processor. A processor may be called 8-bit, 16-bit, or 32-bit, because it can operate on 8-bit, 16-bit, or 32-bit operands in one operation, respectively.

Control Unit

From the beginning, the electronic digital computer is designed to execute *machine instructions* only. A machine instruction is a combination of number 0s and 1s. To simplify the computer hardware design, most computers limit the instruction length to a few choices that are a multiple of 8 bits. For example, the HCS12 microcontroller from Freescale has instructions that are 8-bit, 16-bit, 24-bit, 32-bit, 40-bit, and 48-bit. You won't see instructions that are 13 bits, 29 bits, etc.

A machine instruction has several fields. A mandatory field for every instruction is *opcode*, which tells the ALU what operation to perform. Other fields are optional; when conditions exist, they specify the operand(s) to be operated on.

To make the instruction execution time predictable, a *clock* signal is used to synchronize and set the pace of instruction execution. A clock signal is also needed to control the access of registers in the processor and external memory. The clock frequencies of the 8-bit and 16-bit microcontrollers range from a few MHz to over 100 MHz.

Since a program consists of many machine instructions, there is a need to keep track of what instruction to execute next. The control unit has a register called *program counter* (PC) that serves this function. Whenever the processor fetches an instruction from memory, the program counter will be incremented by the length of that instruction so that it points to the next instruction. The fetched instruction will be placed in the *instruction register* (IR), decoded, and executed. During this process, appropriate control signals will be generated to control the hardware circuit operation.

A program is normally not sequential. The execution order of machine instructions may be changed due to the need to execute instructions based on the value of a certain condition or to repeat a group of instructions. This is called *program flow control*. The decision to change program flow is often based on certain conditions; for example, whether the previous instruction caused carry out to be 1, whether the result of the previous operation is zero, or whether the result of the previous operation is negative. These conditions are often collected in a *status register* so that they can be used to make decisions. This type of program flow change is implemented by a *conditional branch* instruction (may also be called *conditional jump* instruction). There is a limit to the distance that the processor can branch (or jump) conditionally. The *branch distance* (referred to as *branch offset*) is from −128 byte to 127 bytes for most 8-bit microcontrollers, because they use 8 bits to specify branch offset. In other situations, the programmer wants to force the processor to execute instruction in any location within the available memory space. In this situation, a jump instruction is used. The *jump target* may be specified in 16 bits, 32 bits, or 64 bits, depending on the width of the program counter.

It is easy to figure out that writing programs in machine instructions is extremely difficult. Over the years, assembly language and high-level languages such as FORTRAN, COBOL, BASIC, C, C++, JAVA, and so on have been invented. Assembly language uses mnemonic symbols to represent each machine instruction. The result is that each machine instruction is represented by an *assembly instruction.* The programmer can see the assembly instruction and figure out what operation is going to be performed quickly for most instructions. The assembly language makes programming much easier than in machine language. However, assembly language is still at a very low level. It is not very productive to write large and complicate programs in assembly language. Moreover, it needs a translator called an *assembler* to translate assembly instructions into machine instructions so that they can be executed by the computer.

High-level language is at a much higher level. Therefore, one statement written in high-level language may be translated into tens or even hundreds of machine instructions. A program written in high-level language also needs a translator to translate it into machine instructions so that it can be executed by the computer. The translator of a high-level programming language is called a *compiler.* The translation from high-level language to machine language is often not optimal. Therefore, there are some applications that require very tight performance control still written in assembly language. It is not unusual to find large programming projects that are written in both assembly and high-level languages.

Registers

A register is a storage location inside the CPU. It is used to hold data and/or a memory address during the execution of an instruction. Because the register is very close to the CPU, it can provide fast access to operands for program execution. The number of registers varies greatly from processor to processor.

A processor may add a special register called an *accumulator* and include it as one of the operands for most instructions. The Intel 8051 variants, the Microchip PIC18, and the Freescale HCS12 microcontroller use this approach. Using the dedicated accumulator as one of the operands can shorten the instruction length. Other processors, for example, Atmel AVR, may include many general-purpose data registers (16 or 32) in the CPU and allow any data register to be used as any operand of most instructions with two or three operands. This provides great freedom to the compiler during the program translation process. A processor designed with this approach is considered to be *orthogonal.*

1.8.2 MICROPROCESSOR

The earlier processors may be implemented in one or multiple printed circuit boards. With the advent of integrated circuit technology, a complete processor can be implemented in one integrated circuit (an integrated circuit is often called a *chip*). A microprocessor is a processor implemented in a single integrated circuit.

In 1968, the first microprocessors (Intel 4004 and TI TMS 1000) were introduced. Both the Intel 4004 and TI TMS 1000 are 4-bit microprocessors. In 1972, Intel introduced the 8008, which is the first 8-bit microprocessor in the world. Several other 8-bit microprocessors were introduced after the Intel 8008, including Intel 8080, Zilog Z80, Motorola 6800, etc. Microprocessors were quickly used as the controller of many products. Because of their small size (compared to discrete logic), programmability, ease-of-use, and low cost, microprocessors were well received and quickly replaced discrete logic devices.

However, the microprocessor still has a few disadvantages:

1. The microprocessor does not have on-chip memory. The designer needs to add external memory chips and other "glue" logic circuits such as decoder and buffer chips, to provide program and data storage.
2. The microprocessor cannot drive the I/O devices directly due to the fact that a microprocessor may not have enough current to drive the I/O devices or due to the incompatibility in voltage levels between the microprocessor and I/O devices. This problem is solved by adding peripheral chips as a buffer between the microcontroller and the I/O devices. The Intel *8255* parallel interface chip is one of the earliest interface chips.
3. The microprocessor does not have peripheral functions, such as timers, an A/D converter, communication interfaces, and so on. These functions must be implemented using external chips.

Because of these limitations, a product designed with microprocessors cannot be made as compact as might be desired. One of the design goals of microcontrollers is to eliminate these problems.

1.8.3 MICROCONTROLLER

A microcontroller (MCU) incorporates the processor and one or more of the following modules in one very large scale integrated circuit (VLSI):

- Memory
- Timer functions

- Serial communication interfaces, such as USART, SPI, I^2C, and CAN
- A/D converter
- D/A converter
- DMA controller
- Parallel I/O interface (equivalent to the function of Intel 8255)
- Memory-component interface circuitry
- Software-debugging support hardware

The discussion of these functions is the subject of this textbook. Since their introduction, MCUs have been used in almost every application that requires a certain amount of intelligence. They are used as controllers for displays, printers, keyboards, modems, charge-card phones, palm-top computers, and home appliances (such as refrigerators, washing machines, and microwave ovens). They are also used to control the operations of engines and machines in factories. One of the most important applications of MCUs is probably automobile control. Today, a luxurious car may use more than 100 MCUs. Today, most homes have one or more MCU-controlled electronic appliance.

1.8.4 Embedded Systems

An embedded system is a special-purpose computer system designed to perform a dedicated function. Unlike a general-purpose computer (such as a personal computer), an embedded system performs one or a few predefined tasks (usually with very specific requirements) and often includes task-specific hardware and mechanical parts not usually found in a general-purpose computer. Since the system is dedicated to specific tasks, design engineers can optimize it, reducing the size and cost of the product. Embedded systems are often mass-produced, benefiting from the economy of scale.

Physically, embedded systems range from portable devices (such as digital watches and MP3 players) to large stationary installations (like traffic lights, factory controllers, or the systems that control power plants). In terms of complexity, embedded systems run from simple (with a single microcontroller chip) to very complex (with multiple units, peripherals, and networks mounted inside a large chassis or enclosure).

Mobile phones or hand-held computers share some elements with embedded systems (such as the operating systems and microprocessors which power them) but are not truly embedded systems themselves, because they tend to be more general purpose, allowing different applications to be loaded and peripherals to be connected to them.

Characteristics of Embedded Systems

Embedded systems have the following characteristics:

- Embedded systems are designed to perform some specific task, rather than being a general-purpose computer for multiple tasks. Some also have real-time performance constraints that must be met for reasons such as safety and usability; others may have low or no performance requirements, allowing the system hardware to be simplified to reduce costs.

- An embedded system is not always a separate block—very often it is physically built into the device it is controlling.
- The software written for embedded systems is often called *firmware* and is stored in read-only memory or flash memory chips rather than a disk drive. It often runs with limited computer-hardware resources: small or no keyboard, screen, and little memory.

User Interfaces

Embedded systems range from having no interface at all—dedicated to only one task—to full user interfaces—similar to desktop operating systems in devices such as PDAs.

A simple, embedded system may use buttons for input and use LEDs or a small character-only display for output. A simple menu system may be provided for users to interface with.

A more complex system may use a full graphical screen that has touch sensing or screen-edge buttons to provide flexibility while at the same time minimize space. The meaning of the buttons can change with the screen.

Hand-held systems often have a screen with a "joystick button" for a pointing device. The rise of the World Wide Web has given embedded designers another quite different option: providing a Web page interface over a network connection. This avoids the cost of a sophisticated display yet provides complex input and display capabilities when needed on another computer. This is successful for remote, permanently installed equipment. In particular, routers take advantage of this ability.

1.9 Memory

There are three major memory technologies in use today: magnetic, optical, and semiconductor.

1.9.1 Magnetic Memory

The magnetic drum, magnetic tape, and magnetic hard disk are three major magnetic memory devices that have been invented. The magnetic drum has long been obsolete, and magnetic tape is only used for data archival. Currently, only the magnetic hard disk is still being used in almost every PC, workstation, *server*, and *mainframe* computer. Hard-drive vendors are still vigorously improving the hard disk density. It doesn't seem possible that any memory technology can totally replace the hard disk yet.

1.9.2 Optical Memory

There are two major optical memory technologies in use today: compact disc (CD) and digital video disc (DVD). The CD was introduced to the market in 1982 and has several variations. The most popular single-sided CD has a 12-cm diameter and can hold about 700 MB of data. The CD-R version of the compact disc can be recorded once, whereas the CD-RW disc can be re-recorded many times. A single-sided DVD with 12-cm diameter can hold 4.7 GB of data. There are several versions of DVD. Among them, DVD-R can be recorded only once, whereas DVD-RW can be re-recorded many times by the end user. There are several possible competing successors to the current DVD technology. They have single-sided capacities ranging from 15 GB to 25 GB.

1.9.3 Semiconductor Memory

Semiconductor memory is the dominant memory technology used in embedded systems. Memory technologies can be classified according to several criteria. Two common criteria are volatility and read-writability. Basing on volatility, semiconductor memories are divided into *volatile* and *non-volatile* memories. Basing on read-writability, semiconductor memories are divided into *random access memory* (RAM) and *read-only memory* (ROM).

1.9.4 Non-Volatile and Volatile Memory

A memory device is *non-volatile* if it does not lose the information stored in it even without the presence of power. If a memory device cannot retain its stored information in the absence of power, then it is *volatile.*

1.9.5 Random Access Memory

Random access memory allows the CPU to read from or write to any location within the chip for roughly the same amount of time. RAM can be *volatile* or *non-volatile.* RAM also is called *read/write memory,* because it allows the processor to read from and write into it. As long as the power is on, the microprocessor can write data into a location in the RAM chip and later read back the same contents. Reading memory is nondestructive. Writing memory is destructive. When the microprocessor writes data to memory, the old data is written over and destroyed.

There are four types of commercially available RAM technology: *dynamic* RAM (DRAM), *static* RAM (SRAM), *magnetoresistive random access memory* (MRAM), and *ferroelectric memory* (FRAM).

DRAMs are memory devices that require a periodic refresh of the stored information. *Refresh* is the process of restoring binary data stored in a particular memory location. The dynamic RAM uses one transistor and one capacitor to store one bit of information. The information is stored in the capacitor in the form of electric charges. The charges stored in the capacitor will leak away over time, so a periodic refresh operation is needed to maintain the contents in the DRAM. The time interval over which each memory location of a DRAM chip must be refreshed at least once in order to maintain its contents is called its *refresh period.* Refresh periods typically range from a few milliseconds to over a hundred milliseconds for today's high-density DRAMs.

SRAMs are designed to store binary information without needing periodic refreshes and require the use of more complicated circuitry for each bit. Four to six transistors are needed to store one bit of information. As long as power is stable, the information stored in the SRAM will not be degraded.

MRAMs were first developed by IBM. Several other companies also were involved in the research, development, and marketing of this technology. MRAMs use a magnetic moment to store data. A MRAM chip combines a magnetic device with standard silicon-based microelectronics to achieve the combined attributes of nonvolatility, high-speed operation and unlimited read and write endurance. The first MRAM device from Freescale is the 4-Mbit MR2A16A. This device is a parallel memory (8 or 16 bits can be accessed in one operation) and has a 35-ns access time at the time of this writing (reported in 2006).

FRAMs use the property of ferroelectric crystals to store data bits. Much of the present FRAM technology was developed by Ramtron International. The FRAM technology already has achieved high maturity. Both the serial and parallel versions of FRAM chips are available. Ramtron even incorporates FRAM in some of its 8051 microcontroller products. At the time of this writing, the fastest access time of FRAM from Ramtron is 60 ns (reported in 2006). However, the access time of FRAM may improve over time in the future.

RAM is mainly used to store *dynamic* programs or data. A computer user often wants to run different programs on the same computer, and these programs usually operate on different sets of data. The programs and data must therefore be loaded into RAM from the hard disk or other secondary storage, and for this reason, they are called *dynamic.*

1.9.6 Read-Only Memory

ROM is nonvolatile. When power is removed from ROM and then reapplied, the original data will still be there. As its name implies, ROM data can only be read. If the processor attempts to write data to a ROM location, ROM will not accept the data, and the data in the addressed ROM memory location will not be changed. However, this statement is not completely true. For some ROM technologies (EEPROM and flash memory), the user program can still write data into the memory by following a special procedure prescribed by the manufacturer.

Mask-programmed read-only memory (MROM) is a type of ROM that is programmed when it is manufactured. The semiconductor manufacturer places binary data in the memory according to the request of the customer. To be cost-effective, many thousands of MROM memory units—each consisting of a copy of the same data (or program)—must be sold. MROM is the major memory technology used to hold microcontroller application programs and constant data. Most people simply refer to MROM as ROM. The design of MROM prevents it from being written into.

The **programmable read-only memory** (PROM) was invented in 1956 by Wen Tsing Chow. It is a form of memory where the setting of each bit is locked by a fuse or antifuse. The memory can be programmed just once after manufacturing by "blowing" the fuses (using a *PROM blower*), which is an irreversible process. Blowing a fuse opens a connection while blowing an antifuse closes a connection (hence the name). Programming is done by applying high-voltage pulses which are not encountered during normal operation (typically 12 to 21 volts). Fused-based PROM technology is no longer used today.

The **erasable programmable read-only memory** (EPROM) was invented by Israeli engineer Dov Frohman in 1971. It is an array of floating-gate transistors individually programmed by an electronic device that supplies higher voltages than those normally used in electronic circuits. Programming is achieved via *hot carrier injection* onto the floating gate. Once programmed, an EPROM can be erased only by exposing it to strong ultraviolet (UV) light. That UV light usually has a wavelength of 235 nm for optimum erasure time. EPROMs are easily recognizable by the transparent fused-quartz window in the top of the package, through which the silicon chip can be seen, and which permits UV light to go through during erasing.

As the quartz window is expensive to make, one-time programmable (OTP) chips were introduced; the only difference is that the EPROM chip is packed in an opaque package, so it cannot be erased after programming. OTP versions are manufactured for both EPROMs themselves and EPROM-based microcontrollers. However, OTP EPROM (whether separate or part of a larger chip) is being increasingly replaced by EEPROM for small amounts (where the cell cost isn't too important) and flash memory for larger amounts.

A programmed EPROM retains its data for about ten to twenty years and can be read an unlimited number of times. The erasing window must be kept covered with a foil label to prevent accidental erasure by sunlight. Old PC *basic input/output system* (BIOS) chips were often EPROMs, and the erasing window was often covered with a label containing the BIOS publisher's name, the BIOS revision, and a copyright notice.

The **electrically erasable programmable read-only memory** (EEPROM) was developed in 1983 by George Perlegos at Intel. It was built on earlier EPROM technology but used a thin-gate oxide layer so that the chip could erase its own bits without requiring a UV source. EEPROM is programmed and erased using the process called field emission (more commonly known in industry as *Fowler-Nordheim tunneling*). EEPROM allows the user to selectively erase a single location, a row, or the whole chip. This feature requires a complicated programming circuitry. Because of this, the EEPROM cannot achieve the density of the EPROM technology.

Flash memory was invented by Fujio Masuoka while working for Toshiba in 1984. Flash memory incorporates the advantages and avoids the drawbacks of EPROM and EEPROM technologies. Flash memory can be erased and reprogrammed in the system without using a dedicated programmer. It achieves the density of EPROM, but it does not require a window for erasure. Like EEPROM, flash memory can be programmed and erased electrically. However, it does not allow individual locations to be erased—the user can only erase a block or the whole chip. Today, the BIOS programs of many high-performance PCs are stored in flash memory. Most microcontrollers introduced today use on-chip flash memory as their program memory.

Flash memory chips have also been used in flash disk memory, personal digital assistants (PDA), digital cameras, cell phones, and so on.

1.10 Memory System Operation

A simplified-memory system block diagram is shown in Figure 1.13. A memory system may consist of one or multiple memory chips. Both memory chips and memory systems are organized as an array of memory locations. A memory location may hold any number of bits (most common numbers are 4 bits, 8 bits, 16 bits, 32 bits, and 64 bits). The memory organization of a memory chip or a memory system is often indicated by **m × n**, where **m** specifies the number of memory locations in the memory chip or memory system and n specified the number of bits in each location. Every memory location has two components: *contents* and *address.*

The memory chip or system can only be accessed (read or write) one location at a time. This is enforced by implementing a decoder on the memory chip to select one and only one location to be accessed. There are two types of memory accesses: *read* and *write.*

1.10.1 Read Operation

Whenever the processor wants to read a memory, it sends out the address of the location it intends to read. Since the memory access can be a read or a write, the processor needs to use a control signal to inform the memory the type of access. In Figure 1.13, the RD signal from the processor indicates a read access, whereas the WR signal indicates a write operation. The memory chip also has control signals to control the read or write operation. The OE signal in Figure 1.13 means *output enable* and is connected to the RD signal from the processor, whereas the WE signal means *write enable* and is connected to the WR signal from the processor. For digital systems, there are three logic states for each signal: high, low, and high-impedance (no current flows). When the OE input to the memory chip is low, the data pins are in high-impedance state.

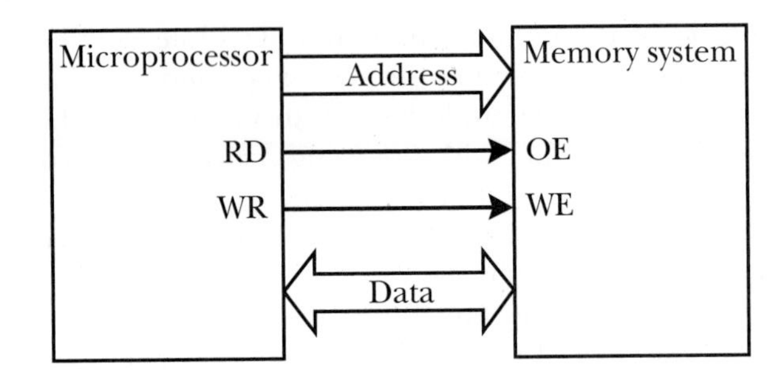

FIGURE 1.13
Block diagram of a simplified memory system.

The processor uses a set of signals (referred to as *address signals*) to specify a memory location to access. The number of address signals needed for selecting a memory location is $\log_2 m$, where **m** is the number of memory locations in the memory. The set of conductor wires that carry address signals is referred to as *address bus.*

The number of conductor wires that carry the data must be equal to the number of bits in each memory location. The set of conductors that carry the data to be accessed is called a *data bus.*

To read a memory location, the processor sends out the address of the memory location to be accessed and applies a logic 1 (high voltage) to the RD signal and a logic 0 to the WR signal (this specifies a read operation). In response, the memory system decodes the address input and enables the specified memory location to send out its contents to the data bus to be read by the processor.

1.10.2 Write Operation

To write a value to a location of the memory system in Figure 1.13, the processor places the data to be written on the data bus, places the address of the memory location on the address bus, applies a logic 1 (high voltage) to the WR signal, and a logic 0 to the RD signal (this specifies a write operation). In response, the memory system uses its address decoder to select a location and writes the value on the data bus to that location.

The actual memory system design and the signals involved may be different from that in Figure 1.13, but the concept would be the same. The semiconductor vendors may use $\overline{RD}$ instead of RD and $\overline{WR}$ instead of WR to refer to read and write signals. These types of signals are active low (i.e., when they are low they are considered to be at logic 1).

1.11 Program Execution

In order to allow the computer to execute a program immediately after the power is turned on, part of the program must be stored in non-volatile memory. Some computers place the startup program in the non-volatile memory, which will perform the system initialization. After the system initialization is completed, it loads additional programs from secondary storage, such as hard disk or optical storage into the semiconductor memory (often called *main memory*), for execution. Mainframe computers, workstations, and personal computers follow this approach. After power is turned on, the processor starts to execute the program from the BIOS, which performs the system initialization. After the system initialization is completed, the processor loads additional programs, such as Windows operating system, into the main memory for execution. Other computers (including most embedded systems) place all of their programs in the non-volatile memory. After power-up, the processor starts to execute programs from the non-volatile memory.

The following sections deal with several important issues related to program execution.

1.11.1 The Circuit of the Program Counter

The program counter consists of flip-flops and other additional logic gates. There are several types of flip-flops in use. Among them, D-type flip-flop is the most popular one. The circuit of a D-type flip-flop with set and reset capability is shown in Figure 1.14.

In Figure 1.14,

- Depending on the design, the D value may be transferred to Q on either the rising or the falling edge (but not both edges) of the CLK input.
- The CLK signal is the clock input signal of the D flip-flop.
- The Q signals of all the flip-flops of the program counter determine the address of the next instruction to be fetched.
- The $\overline{\text{set}}$ and $\overline{\text{reset}}$ inputs are active low (low voltage means logic 1) and cannot be low at the same time. When $\overline{\text{set}}$ is low, the Q signal is forced to 1. When $\overline{\text{reset}}$ is low, the Q signal is forced to 0.

As described in Section 1.8.1, a microprocessor or microcontroller has instructions to change the program flow. The design of the program counter circuit must take this into account. Figure 1.15 shows the block diagram of a

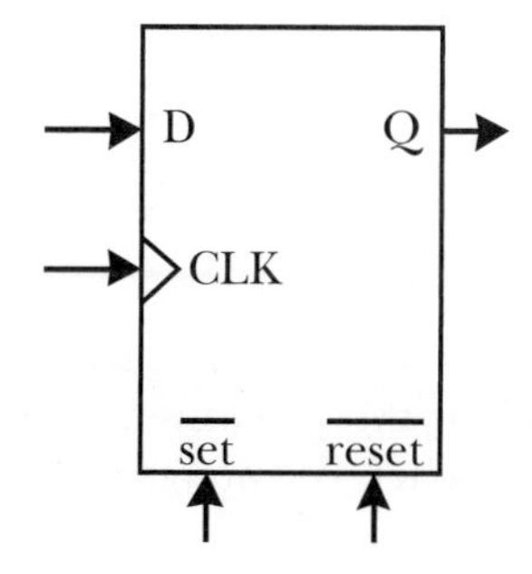

FIGURE 1.14
Block diagram of a D flip-flop with set and reset.

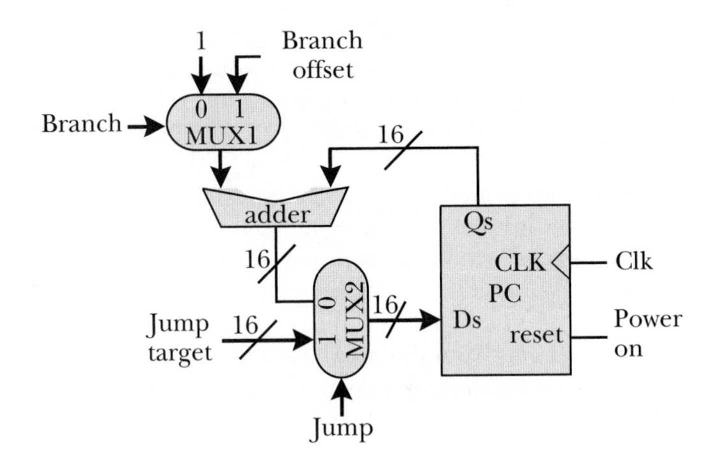

FIGURE 1.15
A simplified block diagram of the program counter (PC) of an 8-bit microcontroller.

possible program counter of an 8-bit microcontroller that allows the program counter to be

- Forced to 0
- Incremented by 1
- Incremented by a field in the IR register
- Loaded with a jump target

For the program counter circuit shown in Figure 1.15,

- Whenever power is turned on to the microcontroller, the program counter is forced to 0, and the instruction fetch will start from address 0.
- If the instruction being executed is a conditional branch instruction and the branch condition is true, then the **Branch** signal will be 1 and the sum of the current PC, **Branch offset** will be loaded into the PC, and instruction execution will continue from that address.
- If the instruction being executed is a jump instruction, then the value **Jump target** will be loaded into PC.
- If the instruction being executed is not a program flow-control instruction, then PC is simply incremented by 1 after each instruction byte is fetched.

Other microprocessors or microcontrollers may have different program flow-control schemes and may fetch more instruction bytes in one fetch. The program counter shown in Figure 1.15 will need to be modified accordingly.

1.11.2 Where does the Processor Start to Execute Program?

As it was discussed earlier in this chapter that the program counter holds the address of the next instruction to be fetched, the value of the program counter must be known when the power is turned on. One approach is to force the PC to a fixed value when the power is turned on. The circuit shown in Figure 1.15 forces the PC to 0 whenever the power is turned on. Many 8-bit microcontrollers (including Microchip PIC, all Intel 8051 variants, and Atmel AVR) use this approach, because it is easy to implement.

Another approach is to fetch the program starting address from a fixed (known) memory location whenever the power is turned on. The Freescale microcontrollers use this approach. The HCS12 microcontroller from Freescale fetches the program starting address from memory locations at 0xFFFE and 0xFFFF into the PC and then start program execution from there. This approach is slightly more complicated.

Another way to restart program execution is to apply a reset signal to the processor. All microprocessors and microcontrollers have a reset pin that allows the user to force the processor to start from scratch. The effect is identical to turning on the power.

1.11.3 Instruction Execution Process

The instruction sets of most commercial processors are irregular and complicated. The complexity of the instruction set makes it difficult to explain the instruction execution process. In the following example, we assume that there is an 8-bit processor X with an instruction set shown in Table 1.3. The opcode

TABLE 1.3 *The Instruction Set of the Processor X*

Assembly Instruction Mnemonic		Machine Code	Meaning
ld	addr, #val	75 aa xx	Load the 8-bit value (val) into memory location at **addr**.
ld	ptr, #data16	90 yyyy	Load the 16-bit value (data16) into the register **ptr**.
ld	A, @ptr	E0	Load the contents of memory location pointed by **ptr** into A.
and	A, #val	54 xx	And the 8-bit value (val) with A and leave the result in A.
bnz	addr, offset	70 zz	Branch to a location that is offset from the next instruction if the value at **addr** is zero.
inc	addr	05 aa	Increment the contents of memory location at **addr**.
dbnz	addr, offset	D5 aa zz	Decrement the contents of memory location at **addr** and branch if the result is not zero. The branch distance is offset.

Note:
aa: an 8-bit value that represent an 8-bit address.
xx: an 8-bit value.
yyyy: a 16-bit value.
zz: distance of branch from the first byte of the instruction after the **Branch** instruction.
Machine code are expressed in hex format.

of any instruction is one byte and is always the first byte of the instruction. The processor X has an 8-bit accumulator A and a 16-bit pointer register **ptr**. The data memory and program memory are separate and are both 64-kB in size. The register **ptr** is used to point to data memory and supports indirect memory addressing for data memory. The instruction set of the processor X allows the instructions to use the 8-bit address to access the lowest 256 bytes (address 0 to 255) of data memory. The processor X can use the 16-bit **ptr** register to access any location of the 2^{16} data-memory locations.

To facilitate the access of data memory, processor X includes the **MDR** register to hold the data received from data memory and data to be written to the data memory.

1.11.4 Instruction Sequence Example

Assume that the following instruction sequence is stored in the program memory starting from address 0 so that it will be executed immediately after a power on or reset.

```
        ld      0x20,#0       ; place 0 in data memory located at address 0x20
        ld      0x21,#20      ; place 20 in data memory located at address 0x21
        ld      ptr,#0x2000   ; load 0x2000 into the register ptr
loop:   ld      A,@ptr        ; load the memory contents pointed to by ptr into A
        and     A,#0x03       ; and the value 0x03 with A and leave the result in A
        bnz     next          ; branch if the result in A is not zero
        inc     0x20          ; increment the memory location at 0x20 by 1
next:   dbnz    0x21,loop     ; decrement memory location at 0x21 and branch if the
                              ; result is not 0
```

The corresponding machine code of the given instruction sequence is shown in Table 1.4.

The next section explains the process of instruction execution.

1.11.5 Instruction Execution Process

Processor X executes the instruction sequence given in Table 1.4 in the following sets of example.

Instruction ld 0x20,#0 (Machine Code 75 20 00)

When the processor comes out of a reset or a power-on process, the program counter is forced to 0, and this instruction will be fetched and executed. The execution of this instruction involves the following steps.

Step 1

The value in the PC (0x0000) is placed on the address bus of the program memory with a request to read the contents of that location.

TABLE 1.4 *The Processor X Instruction Sequence to be Executed*

Assembly Instruction Mnemonic			Address	Machine Code	Comment
	ld	0x20, #0	0x0000	75 20 00	
	ld	0x21, #20	0x0003	75 21 14	
	ld	ptr, #0x2000	0x0006	90 20 00	
loop:	ld	A, @ptr	0x0009	E0	
	and	A, #0x03	0x000A	54 03	
	bnz	next	0x000C	70 02	02 is the branch offset
	inc	0x20	0x000E	05 20	
next:	dbnz	0x21, loop	0x0010	D5 21 0A	0A is the branch offset

Note:
1. The user uses a label to specify the instruction to branch to and the assembler needs to figure out the branch offset.
2. The assembler figures out that the label "next" is 2 bytes away from the "inc 0x20" instruction.
3. The asembler figures out that the label "loop" is 10 (0A) bytes away from the first byte after the "next: dbne 0x21,loop" instruction.

Step 2

The 8-bit value at the location 0x0000 is the instruction opcode 0x75. At the end of this read cycle, the PC is incremented to 0x0001. The opcode byte 0x75 is fetched. Figure 1.16 shows the opcode read cycle.

Step 3

The control unit recognizes that this version of the **ld** instruction requires one read cycle to fetch the direct address and another cycle to read the data operand. These two bytes are stored immediately after the opcode byte. Two more read cycles to the program memory are performed to access the data memory address 0x20 (held in IR) and the value 0x00 (held in IR). After these two read cycles, the PC is incremented to 0x0003.

Step 4

The control unit places 0x0020 on the data-memory address bus and the value 0x00 on the data-memory data bus to perform a write operation. The value 0x00 is to be stored at the data-memory location at 0x0020, as shown in Figure 1.17.

Instruction **ld 0x21,#20** (Machine Code 75 21 14)

The execution of this instruction is identical to that of the previous instruction. After the execution of this instruction, the PC is incremented to 0x0006, and the data-memory location at 0x21 receives the value of 20.

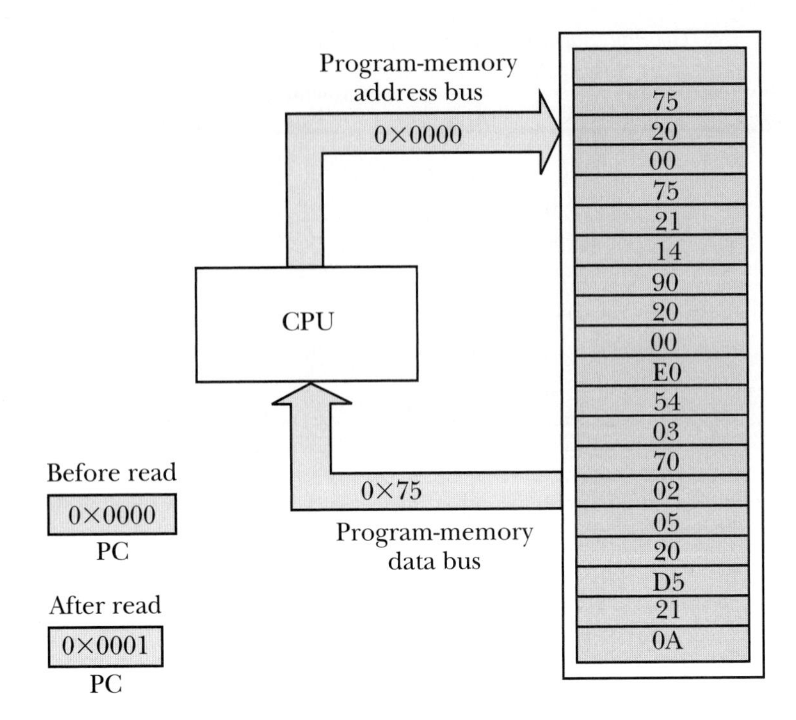

FIGURE 1.16 *Instruction 1—opcode read cycle.*

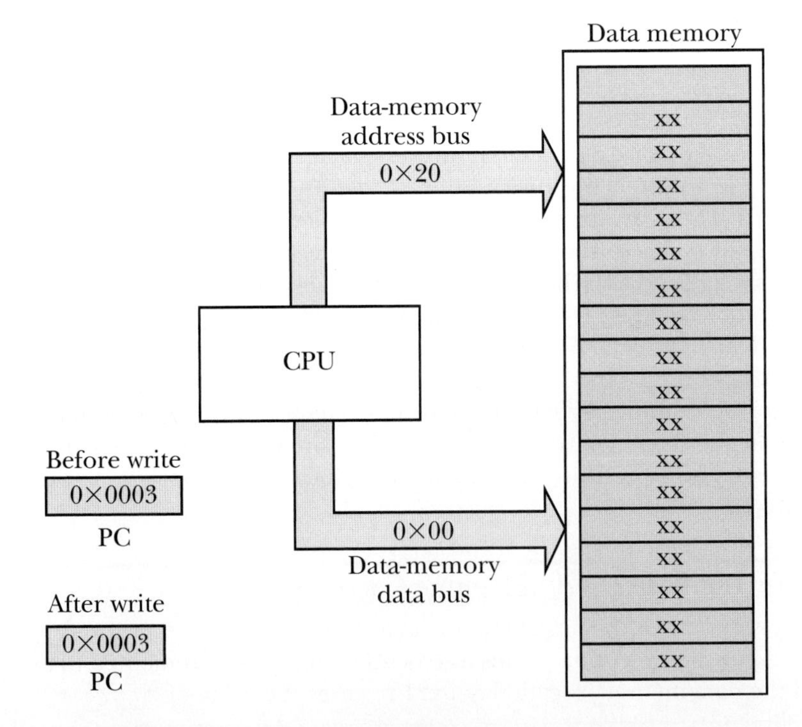

FIGURE 1.17 *Instruction 1—data-memory write cycle.*

Instruction ld ptr,#0x2000 (Machine Code 90 20 00)

Step 1
The value in the PC (0x0006) is placed on the program-memory address bus with a request to read the contents of that location.

Step 2
The 8-bit value at the location 0x0006 is the instruction opcode 0x90. At the end of this read cycle, the PC is incremented to 0x0007. The opcode byte 0x90 is fetched. Figure 1.18 shows the opcode read cycle.

Step 3
The control unit recognizes that this instruction requires two more read cycles to the program memory to fetch the 16-bit value to be placed in the **ptr** register. These two bytes are stored immediately after the opcode byte. The control unit continues to perform two more read cycles to the program memory. At the end of each read cycle, the processor X stores the received byte in the **ptr** register upper and lower bytes, respectively. After these two read operations, the PC is incremented to 0x0009.

Instruction ld A,@ptr (Machine Code E0)

Step 1
The value in the PC (0x0009) is placed on the program-memory address bus with a request to read the contents of that location.

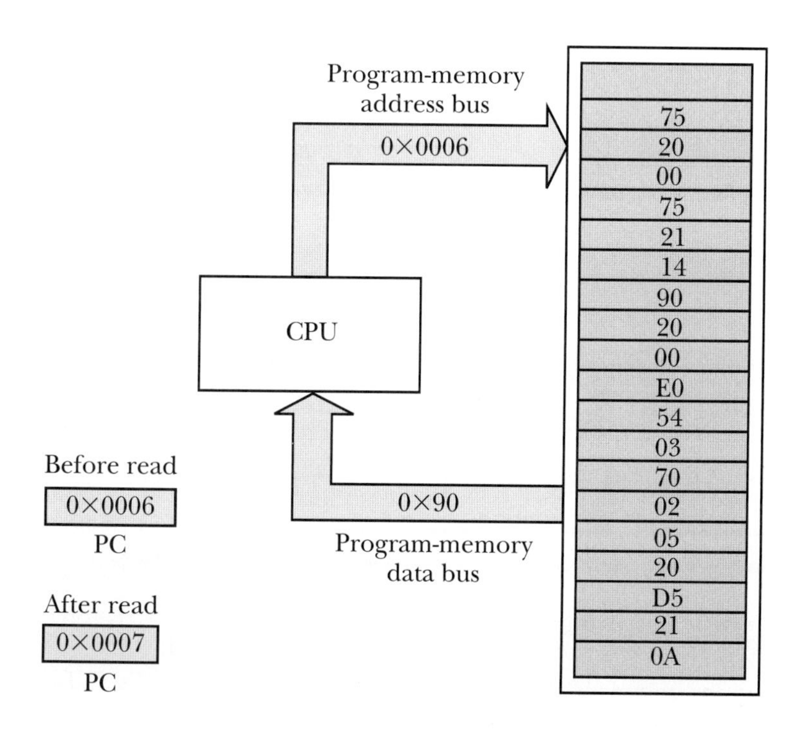

FIGURE 1.18 *Instruction 3—opcode read cycle.*

Step 2
The 8-bit value at the location 0x0009 is the instruction opcode 0xE0. At the end of this read cycle, the PC is incremented to 0x000A. The opcode byte 0xE0 is fetched.

Step 3
The control unit recognizes that the current instruction requires performing a read operation to the data memory with the address specified by the **ptr** register. The processor places the 16-bit value of the **ptr** register on the data-memory address bus and indicates this is a read operation.

Step 4
The data memory returns the contents to the processor, and the processor places it in accumulator A. This process is shown in Figure 1.19.

Instruction **and A,#0x03** (Machine Code 54 03)

Step 1
The value in the PC (0x000A) is placed on the program-memory address bus with a request to read the contents of that location.

Step 2
The 8-bit value at the location 0x000A is the instruction opcode 0x54. At the end of this read cycle, the PC is incremented to 0x000B. The program memory returns the opcode byte 0x54 to the CPU.

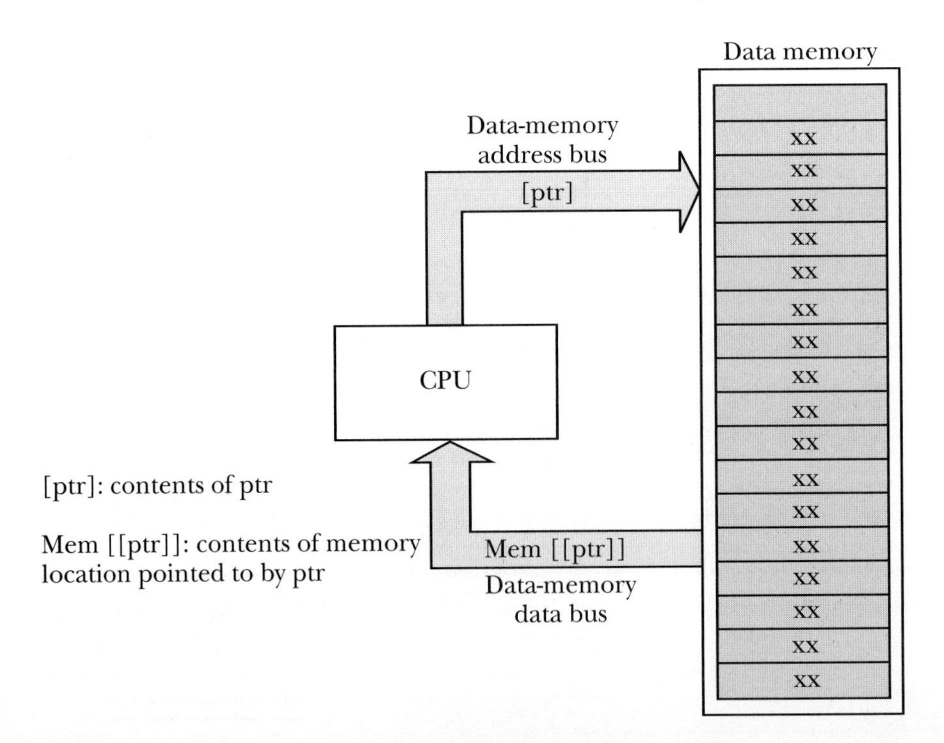

FIGURE 1.19
Instruction 4—data-memory read cycle.

Step 3
The control unit recognizes that the current instruction requires performing a read operation on the program memory to fetch the operand for the AND operation. It then places the PC value on the program-memory address bus again with a "read" request.

Step 4
The program memory returns the value 0x03 to the CPU. The PC is incremented to 0x000C.

Step 5
The CPU then performs an AND operation on the contents of accumulator A and the value 0x03 and places the result in A.

Instruction **bnz next** (Machine Code 70 02)

Step 1
The value in the PC (0x000C) is placed on the program-memory address bus with a request to read the contents of that location.

Step 2
The 8-bit value at the location 0x000C is the instruction opcode 0x70. At the end of this read cycle, the PC is incremented to 0x000D. The program memory returns the opcode byte 0x70 to the CPU.

Step 3
The processor recognizes that this is a conditional branch instruction, and it needs to fetch the branch offset from the program. So it places the PC value (0x000D) on the program-memory address bus with a read request. At the end of this read cycle, the processor increments the PC to 0x000E.

Step 4
The program memory returns the branch offset 0x02 to the CPU (held in IR). The CPU checks the contents of accumulator A to determine whether the branch should be taken. Let's assume that A contains zero and the branch is not taken. The PC remains at 0x000E. If A contains a nonzero value, the next instruction will be skipped.

Instruction **inc 0x20** (Machine Code 05 20)

Step 1
The value in the PC (0x000E) is placed on the program-memory address bus with a request to read the contents of that location.

Step 2
The 8-bit value at the location 0x000E is the instruction opcode 0x05. At the end of this read cycle, the PC is incremented to 0x000F. The program memory returns the opcode byte 0x05 to the CPU.

Step 3
The processor recognizes that it needs to increment a data-memory location which requires it to fetch an 8-bit address from the program memory.

Step 4
The processor places the value in the PC on the program-memory address bus with a read request. At the end of the read cycle, the PC is incremented to 0x0010, and the value 0x20 is returned to the CPU and is placed in the IR register.

Step 5
The processor places the value 0x20 on the data-memory address bus with a request to read the contents of that location. The data memory returns the value of that memory location at the end of the read cycle, which will be placed in MDR.

Step 6
The processor adds 1 to the MDR register.

Step 7
The processor places the contents on the data-memory data bus, places the value 0x0020 on the data-memory address bus, and indicates this is a write cycle. At the end of the cycle, the value in MDR is written into the data-memory location at 0x20.

Instruction **dbnz 0x21,loop** (Machine Code = D5 21 0A)

Step 1
The value in the PC (0x0010) is placed on the program-memory address bus with a request to read the contents of that location.

Step 2
The 8-bit value at the location 0x0010 is the instruction opcode 0xD5. At the end of this read cycle, the PC is incremented to 0x0011. The program memory returns the opcode byte 0xD5 to the CPU.

Step 3
The CPU recognizes that it needs to read a data-memory address and a branch offset from the program memory.

Step 4
Processor X performs two more read operations to the program memory. The Program memory returns 0x21 and 0x0A (both are held in IR). At the end of these two read cycles, the PC is incremented to 0x0013.

Step 5
Processor X places 0x21 on data-memory address bus with a read request. At the end of the read cycle, the value of the data-memory location at 0x21 is returned to the CPU, which will be held in the MDR register.

Step 6
Processor X decrements the contents of the MDR register. The contents of the MDR register are then placed on the data-memory data bus. Processor X also places the address 0x21 on the data-memory address bus with a write request to store the contents of MDR in data memory.

Step 7
If the value stored in the MDR is not zero, processor X adds 0x0A to the PC and places the result in the PC (this causes a branch behavior). Otherwise, the PC is not changed.

This section demonstrates the activities that may occur during the execution of a program. Overall, the operations performed by the processor are dictated by the opcode.

1.12 Overview of the 8051 Microcontroller

The **8051** is a single-chip microcontroller (μC) which was developed by Intel in 1980 for use in embedded systems. It was popular in the 1980s and early 1990s, but today it has largely been superseded by a vast range of enhanced devices with 8051-compatible processor cores that are manufactured by more than 20 independent manufacturers, including Atmel, Infineon Technologies, Maxim Integrated Products (via its Dallas Semiconductor subsidiary), NXP (formerly Philips Semiconductor), Winbond, ST Microelectronics, Silicon Laboratories (formerly Cygnal), Texas, Instruments, and Cypress Semiconductor. Intel's official designation for the 8051 family of μCs is **MCS 51**.

The original Intel 8051 family was developed using NMOS technology. However, later versions, identified by a letter 'C' in their name (e.g., 80C51) used CMOS technology and were less power-hungry than their NMOS predecessors—this made them eminently more suitable for battery-powered devices.

The original 8051 microcontroller has the following features:

- 8-bit data bus and 8-bit ALU. So it can access and operate on 8 bits of data in one operation
- 16-bit address bus which allows the 8051 to access 64 kB of memory locations in program memory and data memory
- 128 bytes of on-chip RAM (data memory)
- 4 kB of on-chip ROM (program memory)
- Four 8-bit bi-directional I/O ports
- UART (serial port)
- Two 16-bit timers
- Two-level interrupt priority
- Power saving mode

The 8051's predecessor, the 8048, was used in the keyboard of the first IBM PC, where it converted key presses into the serial data stream which is sent to the main unit of the PC. The 8031 was a cut down version of the original Intel 8051 that did not contain any internal program memory. The 8052 was an enhanced version of the 8051 that featured 256 bytes of internal SRAM instead of 128 bytes, 8 kB of ROM instead of 4 kB, and a third 16-bit timer. The 8032 had all of these features, except for the internal ROM program memory.

In this text, we will start with the coverage of the original 8051 and then extend to the C8051F040 MCU, a device manufactured by SiLabs. The features of the C8051F040 can be applied to most of the 8051 devices from SiLabs.

1.13 The 8051 Memory Space

The programs and data are stored in separate memory spaces in the 8051. The user program can read from and write into the data memory by providing an 8-bit or a 16-bit address. However, the user program can only read from but not write into the program memory.

Upon power-up, the 8051 CPU begins instruction execution from location 0 of the program memory. Depending on the 8051 variant, a portion or all of the program memory may be located inside the microcontroller chip. The internal program memory of the 8051 is mainly implemented using flash memory today. Older 8051 devices may still use ROM or EPROM as their program memory.

For most 8051 variants, only a small portion of the data memory is on the MCU chip. Internal data memory is addressed using a 1-byte address and hence is limited to a size of 256 bytes. To eliminate the need for adding external memory, many 8051 variants have added on-chip data memory, referred to as **XRAM**. The XRAM is treated as an external data memory and must be accessed using the same addressing mode as required by the external data memory.

Internal data memory is implemented using SRAM. The lowest 32 bytes are divided into four banks of eight registers. The byte 0 to byte 7 of each bank are referred to as register R0 to R7. Two bits in the *program status word* (PSW) select the register bank in use. These registers are used as the general-purpose registers of other microcontrollers. The address ranges of these four banks are as follows:

- Bank 0: 0x00 ~ 0x07
- Bank 1: 0x08 ~ 0x0F
- Bank 2: 0x10 ~ 0x17
- Bank 3: 0x18 ~ 0x1F

The 16 bytes immediately above the register bank (0x20 ~ 0x2F) form a block of bit-addressable memory space. The 8051 provides a group of bit-oriented instructions that can directly address these 128 bits. The bit addresses of this area start from 0 to 127_{10}.

1.14 The 8051 Registers

As listed in Table 1.5, the original 8051 MCU has a group of on-chip registers for supporting the ALU operations and controlling the operations of peripheral functions (such as parallel I/O ports, timers, and UART). These registers are called *special function registers* (SFRs) and occupy the data memory space from 128 to 255. The improved variants of the 8051 (such as C8051F040 from

TABLE 1.5 *Special Function Registers of the Original 8051*

Address	Symbol	Name
0x80	*P0	Port 0 data register
0x81	SP	Stack pointer
0x82	DPTR	Data pointer
0x82	DPL	Data pointer low byte
0x83	DPH	Data pointer high byte
0x87	PCON	Power control register
0x88	*TCON	Timer/Counter control register
0x89	TMOD	Timer/Counter mode control register
0x8A	TL0	Timer/Counter 0 low byte
0x8B	TL1	Timer/Counter 1 low byte
0x8C	TH0	Timer/Counter 0 high byte
0x8D	TH1	Timer/Counter 0 high byte
0x90	*P1	Port 1 data register
0x98	*SCON	Serial control register
0x99	SBUF	Serial data buffer
0xA0	*P2	Port 2 data register
0xA8	*IE	Interrupt enable register
0xB0	*P3	Port 3 data register
0xB8	*IP	Interrupt priority control register
0xC8	*+TCON2	Timer/Counter 2 control register
0xCA	+RCAP2L	Timer/Counter 2 Capture register low byte
0xCB	+RCAP2H	Timer/Counter 2 Capture register high byte
0xCC	+TL2	Timer/Counter 2 low byte
0xCD	+TH2	Timer/Counter 2 high byte
0xD0	*PSW	Program status word
0xE0	*ACC	Accumulator
0xF0	*B	Register B

* = Bit addressable.
+ = 8052 only.

SiLabs) has many more special function registers and will be discussed in later chapters.

The SFRs that are related to the ALU operations are described in the following:

ACC: ACC is used as one of the operands of most arithmetic and logic operations. ACC is also referred to as A. ACC is *bit-addressable.*

B: In addition to being used with ACC for performing multiplication and division operations, B can also be treated as a general-purpose, scratch-pad register and is bit-addressable.

SP: SP points to the top byte of the *stack* data structure. A stack is a last-in-first-out data structure and is implemented using the internal data memory. The stack operations are discussed in Chapter 4.

DPTR: The 16-bit DPTR register holds an address for accessing program memory, on-chip XRAM, and external data memory. DPTR actually consists of two 8-bit registers: DPH and SPL. The DPTR register has three major applications:

1. Access data in the program memory space by using the **movc A,@A+DPTR** instruction.
2. Perform multi-way jump operations by using the **JMP @A+DPTR** instruction.
3. Access on-chip XRAM or external data memory by using the **movx A,@DPTR** or **movx @DPTR,A** instruction.

PC: The 16-bit PC register holds the address of the next instruction to be executed. After a reset, PC is forced to 0, and hence, the CPU will start to fetch instruction from location 0. The 8051 fetches instruction one byte at a time. After each instruction byte is fetched, the PC is incremented by 1.

PSW: This 8-bit register keeps track of the program execution status and allows the CPU to make a decision on whether to change the program flow, as shown in Figure 1.20. In addition to being used as an operand, the **CY** flag may also be used in making conditional jump decisions. The **F0** flag can be used as a general purpose bit.

1.15 Methods for Addressing Instruction Operands

An 8051 instruction consists of one byte of opcode and zero to two bytes of operand information. The operation to be performed is specified by the opcode byte, whereas operand (s) to be operated on is (are) specified by one of the addressing modes.

The 8051 provides ten addressing modes: *implied register, direct, immediate value, indirect, indexed, relative, absolute, long, bit inherent,* and *bit direct.*

Both the implied register mode and the direct mode specify a register as an operand. The difference is that the implied register mode specifies the

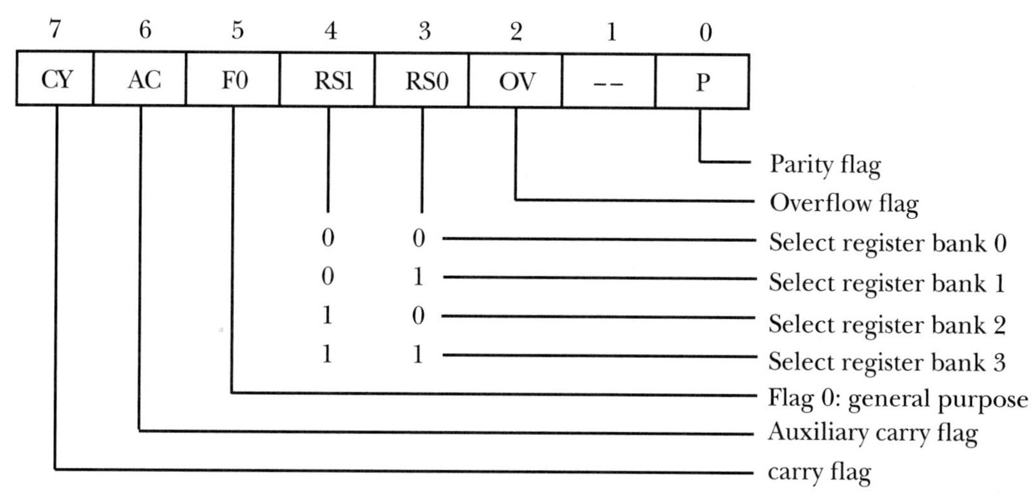

FIGURE 1.20
The 8051 program status word (PSW).

operand using the opcode byte, whereas the direct mode uses an extra byte to specify the register to be operated on.

1.15.1 Implied Register Mode

This mode is used whenever one of the following registers is used as an operand:

- R0~R7
- A
- DPTR
- PSW
- SP
- PC

The following instructions illustrate the implied register mode:

```
mov R7,#30
```

places the decimal value 30 in register R7.

```
inc R6
```

increments the contents of register R6 by 1.

```
dec A
```

decrements accumulator A by 1.

The bit 4 and bit 3 of the PSW register select the active register bank. For example, the instruction

```
mov PSW,#00011000B
```

selects bank 3 as the active register bank.

1.15.2 Direct Register Mode

The direct mode uses an 8-bit value to specify the operand. Both the internal SRAM and the peripheral registers can be specified by this mode. When the bit 7 of the direct address is 0, it selects the lower 128 bytes of the on-chip SRAM. When the bit 7 is 1, a peripheral register is specified.

The following instructions illustrate the direct addressing mode:

```
mov 10H,A
```

copies the contents of accumulator A into the memory location at 10H (hex address).

```
add A,20H
```

adds the value stored at memory location 20H (hex address) to accumulator A.

```
mov 88H,A
```

loads the contents of the accumulator A into peripheral register located at 88H.

1.15.3 Immediate Value Mode

The immediate value mode supplies a value in the instruction to be operated on by the CPU. Either an 8-bit or a 16-bit value can be specified. A 16-bit immediate value can be used only with the DPTR register. An 8-bit immediate value can be used for all other registers.

The immediate value is contained in the byte or bytes immediately following the instruction opcode. In assembly language syntax, an immediate value is indicated by a # character. The following instructions illustrate the immediate addressing mode:

```
mov A,#50
```

loads the decimal value 50 into accumulator A.

```
sub A,#30
```

subtracts the decimal value 30 and **CY** flag from accumulator A.

```
mov DPTR,#1000H
```

loads the hex value 1000 into the data pointer DPTR.

1.15.4 Indirect Addressing Mode

The indirect mode uses a register to hold the address of the memory location to be accessed. Only registers R0, R1, or DPTR can be used to hold the address. Before using the indirect mode, the address of the memory location to be accessed must be loaded into one of these registers.

Both internal and external data memory can be addressed indirectly. Both R0 and R1 can hold only 8-bit addresses, whereas DPTR can hold a 16-bit address. When being used with the mov instruction, R0 and R1 hold an address to the internal data memory. When being used with the movx instruction, R0 and R1 hold the lower 8 bits of the external data memory, and the port P0 holds the upper 8 bits of the address. In indirect mode, DPTR can be used only with the movx instruction and holds an address to the external data memory. The following instructions illustrate the use of indirect addressing mode:

```
subb A,@R0
```

subtracts the contents of the internal data-memory location pointed to by R0 from accumulator A.

```
mov @R0,A
```

copies the contents of A into the internal data-memory location pointed to by R0.

```
movx @DPTR,A
```

stores the contents of A in external data-memory location pointed to by DPTR.

1.15.5 Indexed Addressing Mode

This addressing mode uses a base register (PC or DPTR) and an offset (in accumulator A) to form the memory address for a **jmp** or a **movc** instruction. The indexed addressing mode can be used to access a jump table or a lookup table stored in the program memory.

The following instructions illustrate the use of the indexed addressing mode:

```
jmp @A+DPTR
```

causes the CPU to jump to execute the instruction with an address equal to the sum of accumulator A and the data pointer DPTR.

```
movc A,@A+DPTR
```

fetches the contents of the program memory location with an address equal to the sum of the contents of A and DPTR into A.

```
movc A,@A+PC
```

fetches the contents of the program-memory location with an address equal to the sum of the contents of A and the PC into A. The PC points to the byte immediately after the movc instruction.

1.15.6 Relative Addressing Mode

This addressing mode is used by the 8051 MCU to change program flow based on the specified condition. It is used with the *conditional jump* instruction. The direction of jump may be forward or backward. The distance of the jump is measured in reference to the first byte after the jump instruction.

The distance of the jump is specified in 8 bits and hence has a range from −128 to +127 bytes. When a jump is taken, the offset is added to the program counter to form the jump address. The jump destination is usually specified using a label, and the assembler calculates the jump offset and places it in the machine code. The following instructions illustrate the relative addressing mode:

```
jnz notequal
```

will jump to the instruction with the label **notequal** if the value of accumulator A is not zero, and

```
jc smaller
```

will jump to the instruction with the label **smaller** if the CY flag in PSW register is one.

1.15.7 Absolute Addressing Mode

The absolute addressing mode is used to specify the lowest 11 bits of the destination for the **acall** and **ajmp** instructions. The upper 5 bits of the destination address are identical to the upper 5 bits of the current program counter. This limits branching to within the current 2-KB page of the program memory. The absolute address is often specified using a label, and the assembler will be responsible for generating the absolute address. The following instructions illustrate the use of the absolute addressing mode:

```
acall subx
```

calls the subroutine that is started at the label **subx**, and

```
ajmp next
```

causes the instruction with the label **next** to be executed after the current instruction.

1.15.8 Long Address Mode

The long address mode specifies the 16-bit destination address for the **lcall** and **ljmp** destination. The user normally uses a label to specify the long address and to let the assembler figure out the actual long address. The following instructions illustrate the long address mode:

```
lcall random
```

calls the subroutine that is started at the label **random**, and

```
ljmp loop1
```

causes the CPU to execute the instruction with the label **loop1** after the current instruction.

The 8051 MCU has a group of Boolean instructions that operate on bit operands. The internal data memory ranging from 20H to 2FH and many special function registers are also bit addressable. The 8051 provides two methods for specifying bit operands.

1.15.9 Implied Bit Mode

Instructions that operate on the carry flag do not need an extra byte to specify the carry flag. The opcode bytes of these instructions have an implied carry flag as an operand. For example,

```
setb C
```

sets the **CY** flag to 1.

```
clr C
```

clears the **CY** to 0.

```
cpl C
```

complements the **CY** flag.

1.15.10 Direct Bit Mode

The bit-direct addressing mode is used to specify any bit in the internal SRAM ranging from 20H to 2FH and peripheral registers with address divisible by 8. The 8051 uses a byte to specify the direct bit address and hence can potentially specify 256 bit operands. However, not all of them are implemented in the 8051 family. The bit address assignment of the internal data SRAM space 20H to 2FH is shown in Figure 1.21. For example, the bit 7 and bit 0 of data RAM location 0x00 can be accessed by the bit addresses 0x07 and 0x00, respectively.

Special-function registers with addresses that are a multiple of 8 are bit addressable. The bit address of any bit-addressable SFR is formed by concatenating the upper five bits of the SFR byte address and the bit position (indicated by three bits) of each individual bit. The bit addresses of SFRs are in the

FIGURE 1.21
Bit address map of the data memory from 0x20 to 0x2F.

RAM Byte address	(msb)							(lsb)
2FH	7F	7E	7D	7C	7B	7A	79	78
2EH	77	76	75	74	73	72	71	70
2DH	6F	6E	6D	6C	6B	6A	69	68
2CH	67	66	65	64	63	62	61	60
2BH	5F	5E	5D	5C	5B	5A	59	58
2AH	57	56	55	54	53	52	51	50
29H	4F	4E	4D	4C	4B	4A	49	48
28H	47	46	45	44	43	42	41	40
27H	3F	3E	3D	3C	3B	3A	39	38
26H	37	36	35	34	33	32	31	30
25H	2F	2E	2D	2C	2B	2A	29	28
24H	27	26	25	24	23	22	21	20
23H	1F	1E	1D	1C	1B	1A	19	18
22H	17	16	15	14	13	12	11	10
21H	0F	0E	0D	0C	0B	0A	09	08
20H	07	06	05	04	03	02	01	00

range from 128 to 255. The bit-addressable SFRs and their bit-address map are shown in Figure 1.22. For example, the TCON register address is 0x88. The upper five bits of this address are 10001_2, whereas bit 3 is represented by 011_2. Therefore, the bit address of bit 3 of the TCON register is 10001011_2 (0x8B).

The C8051F040 has more than 128 SFRs, and hence, they are divided into pages. In any moment, only one SFR page is active, and hence, only a subset of SFRs is accessible. The C8051F040 uses the 8-bit **SFRPAGE** register to specify the active page for SFRs. Theoretically, there can be up to 256 pages. However, the C8051F040 has only four pages (page 0, 1, 2, 3, and F) of SFRs. The complete list of the C8051F040 SFRs is given in Appendix D. After a reset, the SFRPAGE is cleared to 0, and hence, the SFRs in page 0 are accessible.

A bit address can be specified in one of the following three methods in assembly language.

1. Using a number or expression that evaluates to a value in the range of 0 to 255.

(msb) Bit addresses							(lsb)	Direct-byte address	Page 0	Page 1	Page 2	Page 3	Page F
FF	FE	FD	FC	FB	FA	F9	F8	0×F8	SPI0CN	CAN0CN			P7
F7	F6	F5	F4	F3	F2	F1	F0	0×F0	B				
EF	EE	ED	EC	EB	EA	E9	E8	0×E8	ADC0CN		ADC2CN		P6
E7	E6	E5	E4	E3	E2	E1	E0	0×E0	ACC				
DF	DE	DD	DC	DB	DA	D9	D8	0×D8	CAN0 DATL				P5
D7	D6	D5	D4	D3	D2	D1	D0	0×D0	PSW				
CF	CE	CD	CC	CB	CA	C9	C8	0×C8	TMR2CN	TMR3CN	TMR4CN		P4
C7	C6	C5	C4	C3	C2	C1	C0	0×C0	SMB0CN				
BF	BE	BD	BC	BB	BA	B9	B8	0×B8	IP				
B7	B6	B5	B4	B3	B2	B1	B0	0×B0	P3				
AF	AE	AD	AC	AB	AA	A9	A8	0×A8	IE				
A7	A6	A5	A4	A3	A2	A1	A0	0×A0	P2				
9F	9E	9D	9C	9B	9A	99	98	0×98	SCON0				
97	96	95	94	93	92	91	90	0×90	P1				
8F	8E	8D	8C	8B	8A	89	88	0×88	TCON	CPT0CN	CPT1CN	CPT2CN	
87	86	85	84	83	82	81	80	0×80	P0				
7	6	5	4	3	2	1	0						

(a) bit address (b) SFR names

FIGURE 1.22
The C8051F040 bit-addressable SFRs and their bit-address map.

2. Concatenating the name or address of the register containing the bit, the period character, and the bit position in the register (7–0).
3. Using the predefined assembler symbols of the control and status registers. The symbols and meaning of these bits will be explained in later chapters.

The assembler translates the bit operand specified in these three formats into the appropriate bit-direct address. The following instructions illustrate bit-direct addressing:

```
setb PSW.3
```

sets bit 3 (carry flag) of the program status word,

```
clr IE1
```

clears bit 3 of the TCON register (see Chapter 6) to 1,

```
setb 20H
```

sets bit 0 of the internal RAM location at 20H to 1, and

```
jb P2.5,is_equ
```

will cause the CPU to jump to the instruction with the label **is_equ** if bit 5 of the P2 register is equal to 1.

1.16 Examples of 8051 Instructions

It would be beneficial to get familiar with a few groups of instructions that are often used in assembly programs before formally studying assembly programming. In the following subsections, we examine the 8051 data movement, addition, and subtraction instructions.

1.16.1 Data Movement Instructions

Data movement instructions are used to place values in a register or memory location so that a meaningful operation can be performed. A partial list of data movement instructions is given in Table 1.6. The data movement instruction can be used to place a direct value or to load the contents of a memory location into a register. For example,

```
mov A,#20H
```

loads the hex value 20_{16} into accumulator A.

```
mov A,20H
```

copies the contents of the memory location at 20_{16} into accumulator A.

A complete list of MOVE instructions is given in Appendix A.

To facilitate the expression of concepts, we will use the following notations throughout this book:

[reg]: The contents of the register **reg**. **reg** is one of the 8051 registers.
[addr]: The contents of the memory location at address **addr**. Depending on the type of supplied address, this expression may refer to on-chip data memory, XRAM, external data memory, or program memory.
SRAM[addr]: The data memory (lowest 256 bytes) location at address **addr**.
XRAM[addr]: The XRAM memory or external data memory at address **addr**.
Prog_Mem[addr]: The program memory location at address **addr**.

For example,

[R7] refers to the contents of register R7
[B] refers to the contents of register B
[0x2000] refers to the contents of the memory location at 0x2000

TABLE 1.6 *8051 A Sample of MOVE Instructions*

Instruction	Operation
mov A,Rn	A ← [Rn], n = 0, …, 7
mov A,direct	A ← [direct][1]
mov A,@Ri	A ← [[Ri]][1], i = 0 or 1
mov Rn, #data	A ← data
mov direct, A	SRAM [direct] ← [A]
mov direct1, direct2	SRAM [direct1] ← [direct2][1]
mov direct, #data	SRAM [direct] ← data
mov @Ri, A	SRAM [[Ri]] ← [A], i = 0 or 1
mov DPTR, #data	DPTR ← data
movc A,@A+DPTR	A ← [[A] + [DPTR]][2]
movx A,@DPTR	A ← [[DPTR]][3]

NOTE:
1. Internal data-memory contents.
2. Program-memory contents.
3. Internal XRAM or external data-memory contents.

[[R0]] refers to the contents of data memory with address specified in R0
[[DPTR]] refers to the contents of XRAM with address supplied in DPTR register
[[A] + [DPTR]] refers to the contents of program memory with address equals the sum of the contents of A and the contents of DPTR
A ← [R0] refers to loading the contents of R0 into A
A ← [[R0]] refers to loading the contents of internal SRAM (lowest 256 bytes) with address stored in R0 to A
Mem[2000H] ← [A] means "store the contents of accumulator A in memory location at 2000H"
SRAM[direct] ← [A] means "store the contents of A in the on-chip SRAM located at *direct*", where *direct* is a value between 0 and 127

Example 1.16 Give data movement instructions to carry out the following operations:

(a) Copy the contents of register R0 to accumulator A

(b) Load the value 30 into register R0

(c) Copy the contents of memory location at 10H to memory location at 20H

(d) Copy the internal memory location pointed to by R0 to accumulator A

(e) Place the value 50 in data memory located at 10H

Solution: The following instructions will perform the specified operations:

(a) **mov A,R0**

(b) **mov R0,#30**

(c) **mov 20H,10H**

(d) **mov A,@R0**

(e) **mov 10H,#50**

Example 1.17 There is an array with 80 8-bit integers. The array is stored in the program memory starting at the address 2000H. Write an instruction sequence to load the sixth element into accumulator A.

Solution: The first element of an array has index 0. The *k*th element has an index ***k-1***. This problem can be solved in three steps.

Step 1 Place the starting address 2000H in DPTR.

Step 2 Place 5 in accumulator A.

Step 3 Use the indexed addressing mode to copy the contents of the tenth element into A.

```
mov     DPTR,#2000H
mov     A,#5
movc    A,@A+DPTR
```

Example 1.18 Write an instruction sequence to swap the contents of the memory locations at 30H and 40H, respectively.

Solution: This problem can be solved as follows:

Step 1 Load the contents of the memory location at 30H into A.

Step 2 Store the contents of the memory location at 40H into the memory location 30H.

Step 3 Store the contents of A into the memory location at 40H.

The appropriate instructions are

```
mov     A,30H
mov     30H,40H
mov     40H,A
```

Example 1.19 Write an instruction sequence to copy the contents of the program-memory location at 2000H to the data-memory location at 80H.

Solution: The data-memory location at 80H must be accessed using the indirect mode, whereas the program-memory contents must be accessed using the indexed addressing mode. The following instruction will perform the desired operation.

```
mov     DPTR,#2000H     ; make DPTR point to 2000H
mov     R0,#80H         ; make R0 point to 80H
mov     A,#0            ; set a dummy index 0 in A
movc    A,@A+DPTR       ; load the program memory contents to A
mov     @R0,A           ; write the value into data memory location
```

Example 1.20 Write an instruction sequence to copy the contents of the XRAM location at 0x1000 to the internal data-memory location at 60H.

Solution: The XRAM must be accessed using the indirect mode with DPTR as the pointer. The following instruction sequence will perform the desired operation.

```
mov     DPTR,#1000H     ; make DPTR point to 1000H
movx    A,@DPTR         ; load XRAM contents into A
mov     60H,A           ; store contents of A in data memory at 60H
```

1.16.2 The ADD Instructions

An 8051 ADD instruction may have two or three operands. In a three-operand ADD instruction, the **C** flag of the PSW register is always included as one of the source operands. Three-operand ADD instructions are mainly used in multi-precision arithmetic, which will be discussed in Chapter 2. The ADD instruction has the following constraints.

- Accumulator is both the destination and one of the source operand
- The **CY** flag must be one of the source operands for a three-operand ADD instruction

A sample of ADD instructions is given in Table 1.7.
For example,

```
add     A,#20
```

TABLE 1.7 *A Sample of ADD Instructions*

Instruction	Operation
add A,Rn	A ← [A] + [Rn], n = 0, ..., 7
add A,direct	A ← [A] + [direct]
add A,@Ri	A ← [A] + [[Ri]], i = 0 or 1
add A,#data	A ← [A] + data
addc A,Rn	A ← [A] + [Rn] + CY, n = 0, ..., 7
addc A,direct	A ← [A] + [direct] + CY
addc A,@Ri	A ← [A] + [[Ri]] + CY, i = 0 or 1
addc A,#data	A ← [A] + data + CY

adds the decimal value 20 to the contents of A and places the sum in accumulator A.

```
add        A,@R0
```

adds the contents of the memory location with address specified in R0 to the contents of A and leaves the sum in A.

```
addc       A,B
```

adds the carry bit (in the PSW register) and the contents of register B to accumulator A and leaves the sum in A.

The ADD instruction may change the values of the **CY, AC, OV**, and **P** flags of the PSW register. A complete list of ADD instructions is given in Appendix A.

Example 1.21 Write an instruction sequence to add the contents of the memory locations at 30H and 40H and leave the sum in accumulator 50H.

Solution: This problem can be solved in two steps:

Step 1 Load the contents of the memory location at 30H into accumulator A.

Step 2 Add the contents of the memory location at 40H to accumulator A.

Step 3 Store the contents of A in the memory location at 50H

The appropriate instructions are

```
mov        A,30H
add        A,40H
mov        50H,A
```

Example 1.22 Write an instruction sequence to add 1 to data-memory locations from 30H through 32H.

Solution: For each data-memory location, perform the following three steps.

Step 1 Load the data memory location contents into A

Step 2 Add 1 to A

Step 3 Save the sum in the appropriate data memory location

The following instruction sequence will perform the desired operation.

```
mov     A,30H     ; load the contents of 30H in A
add     A,#1      ; add 1 to A
mov     30H,A     ; save the sum in 30H
mov     A,31H
add     A,#1
mov     31H,A
mov     A,32H
add     A,#1
mov     32H,A
```

1.16.3 THE SUB INSTRUCTIONS

The SUB instruction has three operands. The **CY** flag in the PSW register is included as one of the source operands. Like the ADD instruction, accumulator A is both a source and a destination operand. A summary of all SUB instruction is given in Table 1.8. After the SUB operation, the 8051 CPU updates the values of the **CY, AC, OV**, and **P** flags in the PSW register.

For example,

```
subb    A,#10
```

subtracts the decimal value 10 and the carry flag **CY** from accumulator A and stores the difference in accumulator A.

```
subb    A,B
```

subtracts the contents of the register B and the carry flag **CY** from accumulator A and leaves the difference in accumulator A.

TABLE 1.8 *A Sample of Subtract Instructions*

Instruction	Operation
subb A,Rn	A ← [A] − [Rn] − CY, n = 0, …, 7
subb A,direct	A ← [A] − [direct] − CY
subb A,@Ri	A ← [A] − [[Ri]] − CY, i = 0 or 1
subb A,#data	A ← [A] − data − CY

Example 1.23 Write an instruction sequence to subtract the value of the memory location at 50H and the carry flag from that of the memory location at 60H and store the result in 70H.

Solution: This problem can be solved in three steps:

Step 1 Load the contents of the memory location at 60H into accumulator A.

Step 2 Subtract the contents of the memory location at 50H and the carry flag **CY** from A.

Step 3 Store A in the memory location 70H.

The appropriate instructions are

```
mov     A,60H
subb    A,50H
mov     70H,A
```

Example 1.24 Write an instruction sequence to subtract 5 and **CY** flag from the data-memory location at 30H~31H.

Solution: The following four steps are needed to perform the desired operation.

Step 1 Load the contents of the data-memory location at 30H into A.

Step 2 Subtract 5 and **CY** flag from A.

Step 3 Store A back to the memory location at 30H.

Step 4 Repeat Step 1 to 3 for the memory location at 31H.

The following instruction sequence will perform the desired operation.

```
mov     A,30H       ; load the contents of 30H into A
subb    A,#5        ; subtract 5 and CY flag from A
mov     30H,A       ; store A back to memory location 30H
mov     A,31H       ; load contents at 31H to A
subb    A,#5        ; subtract 5 and CY flag from A
mov     31H,A       ; save the difference back to location 31H
```

In the previous operation, the **CY** flag is always included as one of the subtrahends. Sometimes this is undesirable. We will learn how to clear the **CY** flag in Chapter 2.

1.17 Machine Instruction Timing

The 8051 uses *machine cycle clock* to control its internal operations and external read/write cycles. Internally, the instruction execution is controlled by the *machine cycle clock* (machine cycle clock is also referred to as the *instruction cycle clock*). The original 8051 derives the machine cycle clock by dividing the external oscillator clock input by 12.

However, many 8051 variants derive their machine cycle clocks by dividing the oscillator cycle by a smaller number, for example, 6, 3, or 1. With the advancement of the semiconductor, this approach can dramatically improve the CPU throughput. SiLabs derives its machine cycle clock by dividing the oscillator clock by one. This essentially reduces the machine cycle clock period to 1/12 of the original 8051.

For the original 8051, one instruction may take one to four machine clock cycles to execute. For the C8051F040 from SiLabs, one instruction may take from one to eight machine clock cycles. The instruction execution times for the original 8051 and the SiLabs 8051 are listed in Appendix A.

Example 1.25 Suppose a member of the original 8051 and the SiLabs C8051F040 are both operating with an oscillator running at 24 MHz. Find out the execution times in machine clock cycles and in microseconds for the following instructions:

(a) subb A,35H

(b) mov DPTR,#1000H

(c) movx A,@DPTR

Solution: Solution: The machine cycle clock frequency is 2 MHz (period = 0.5 μs) and 24 MHz (period = 41.7 ns) for the original 8051 and the SiLabs 8051, respectively. The execution times of the given three instructions can be found in Appendix A, as shown in Table 1.9.

TABLE 1.9 *Instruction Execution Times of a Sample of Instructions*

	Instruction Execution Time			
	Original 8051		SiLabs 8051	
Instruction	Machine Cycle	μs	Machine Cycles	ns
subb A,40H	1	0.5	2	83.3
mov DPTR,#1000H	2	1	3	125
movx @DPTR,A	2	1	3	125

1.18 Chapter Summary

The computer uses a binary format to represent all numbers. It performs subtraction by adding the two's complement of the subtrahend to the minuend. Large binary numbers are difficult to display on output devices, and hence, hexadecimal and octal formats are used instead. Computers also need to process non-numerical information. Non-numerical data are represented in character strings. Each character is encoded using some types of "internal code." The most popular internal code is the 7-bit ASCII code.

A computer system consists of hardware and software. Computer hardware has into four major components: central processing unit (also called the processor), input unit, output unit, and memory system.

The processor consists of three major parts: arithmetic logic unit (ALU), control unit, and registers. The ALU performs all arithmetic logic operations. The control unit coordinate the use of resources in the computer system and generates appropriate control signals to control the ALU operations. Registers are used to hold data or addresses during the execution of an instruction. Because the register is very close to the ALU, it can provide fast access to operands for program execution. The number of registers varies greatly from processor to processor.

The processor uses the instruction register (IR) to hold the instruction to be executed and uses the program counter (PC) to keep track of the address of the instruction to be executed next. Clock signals are used to pace the instruction execution process while at the same time control accesses to the registers in the CPU and memory system. The CPU collects the instruction execution status (such as carry flag, negative flag, overflow flags, etc.) in the status register.

For many processors, the program counter is forced to 0 when the power is turned on or when a reset signal is asserted. After that, the processor starts to execute instruction from the address 0x0000. Other processor may fetch the starting address of the program from a fixed location. The program counter must also support the control flow change for the application program. Both the branch and jump instructions must be supported by all processors.

During the instruction execution, the processor needs to send out the PC value on the program-memory address bus and fetch the opcode from the program memory. The opcode informs the CPU what operation to perform and what operands to access during the execution process.

Introduced in 1980, the 8051 is the original member of the Intel MCS-51 family and is the core for all 8051 variants. In addition to Intel, there are many other companies producing 8051-compatible microcontrollers. These devices can execute the same instruction set and support the same addressing modes.

All 8051 devices have separate address spaces for program and data memories. The data memory can be accessed by either 8-bit or 16-bit address. Program memory can only be read, not written into. A portion of the data/program memory is located inside the MCS-51 microcontroller chip. The lowest 32 bytes of the internal data memory are grouped into four banks of eight registers. These registers are referred to as R0 through R7. Two bits in the program status word select which register bank is in use. The MCS-51 has many registers. Registers related to ALU operations include ACC, B, SP, DPTR, PSW, and PC.

An instruction uses an address to specify the memory location that it wants to access. The address information is sent to the memory on the address bus, while the data are transferred between the memory and the CPU on the data bus. The 8051 has an 8-bit data bus and a 16-bit address bus. Instructions operate on operand(s). Operands are specified by addressing modes. The 8051 implements the following addressing modes: register inherent, direct, immediate, indirect, indexed, relative, absolute, long, bit inherent, and bit direct.

The 8051 instructions can be classified into five categories: arithmetic operation, logical operation, data transfer, Boolean variable manipulation, and program branching.

The original 8051 uses 12 oscillator clock cycles to generate a machine clock to control its CPU operation. Enhanced 8051 cores may run at six, four, two, or even one oscillator clock cycle per machine cycle, have clock frequencies of up to 100 MHz, and are capable of executing an even greater number of instructions per second. Even higher speed, single-cycle 8051 cores in the range from 130 MHz to 150 MHz are now available in Internet downloadable form for use in programmable logic devices (such as FPGAs) and at many hundreds of MHz in ASICs (for example, the netlist from **e8051.com**).

1.19. Exercise Problems

E1.1 Convert the following decimal numbers to binary:
(a) 1357 (b) 560 (c) 521 (d) 683

E1.2 Convert the following binary numbers to decimal:
(a) 110011010_2 (b) 1001011101_2 (c) 11001001100_2 (d) 10101010001_2

E1.3 Convert the following binary numbers to octal:
(a) 1010111011_2 (b) 100001011010_2 (c) 10101101_2 (d) 1110011011_2

E1.4 Convert the following binary numbers to hexadecimal:
(a) 110101010_2 (b) 1001001101_2 (c) 11001001110_2 (d) 10101011001_2

E1.5 Convert the following decimal numbers to binary:
(a) 45.67 (b) 320.125 (c) 640.375 (d) 429.5625

E1.6 Expand the circuit shown in Figure 1.12 so that it can also perform XOR, and INV. The INV operation operates on the A input only. Assume opcodes 000, 001, 010, 011, 100, and 101 represent ADD, SUB, AND, OR, XOR, and INV operations, respectively.

E1.7 Explain the execution process of the instruction sequence starting at the address 0x0500 when the PC holds the value of 0x0500.

```
        ld      0x02,#0A
        ld      ptr,#1000
loop    ld      A,@ptr
        add     A,#5
        st      @ptr,A
        dbne    0x02,loop
```

The opcode of the instruction **add A,#val** is 0x39. The opcode of the instruction **st @ptr,A** is 0x49. The opcode of the other instructions are given in Table 1.3.

E1.8 What special-function registers of the original 8051 are bit-addressable? What are not? (*Hint*: Compare Table 1.5 and 1.22.)

E1.9 Write an instruction sequence to copy the contents of the XRAM location at 0x1003 into accumulator A.

E1.10 Write an instruction sequence to store the value of accumulator A at the XRAM location at 0x200.

E1.11 Give an instruction sequence to copy the contents of the program-memory location at 0x1020 into A.

E1.12 Write an instruction sequence to copy the contents of the XRAM location at 0x200 to the internal SRAM location at 0x40.

E1.13 Write an instruction sequence to copy the contents of the internal SRAM location at 0x50 to the XRAM location at 0x300.

E1.14 Write instructions to swap the contents of the data-memory locations at 50H and 60H.

E1.15 Write instructions to subtract 4 from the data-memory locations from 30H through 33H.

E1.16 Write an instruction sequence to translate the following high-level language statements into 8051 assembly instructions.

```
X := 15;
Y := 35;
Z := Y – X;
```

Assume that variables X, Y, and Z are located at the internal data-memory locations 60H, 62H, and 64H, respectively.

E1.17 What are the contents of the memory locations at 30H, 31H, and 32H after the execution of the following instruction sequence, given that [30H] = 01, [31H] = 02, and [32H] = 03?

```
mov     40H,30H
mov     30H,31H
mov     31H,32H
mov     32H,40H
```

E1.18 Write down the contents of the data-memory locations at 50H, 51H, and 52H after the execution of the following instruction sequence, given that [50H] = 3, [51H] = 5, and [52H] = 8.

```
mov     A,50H
add     A,#3
mov     R1,A
mov     A,51H
dec     A
mov     50H,A
mov     A,52H
inc     A
mov     51H,A
mov     52H,R1
```

E1.19 Find out the execution time (in machine cycles) of the following instructions by referring to Appendix A.

```
(a) mov         A,#0
(b) mov         DPTR,#1000H
(c) movx        @DPTR,A
(d) add         A,#2
(e) inc         DPTR
(f) movx        @DPTR,A
```

E1.20 Suppose that the C8051F040 is operating under the control of an external crystal oscillator running at 25 MHz.

(a) What is the period of the oscillator clock signal?

(b) How much time (in machine cycles) does it take to execute the following four instructions once?

```
loop:   nop
        nop
        nop
djnz    R2,loop
```

(c) How many times must this instruction sequence be executed in order to create a delay of 10 ms?

E1.21 Write an instruction sequence to copy the contents of the data-memory location at 40H to the data-memory location at 95H.

E1.22 Write an instruction sequence to copy the contents of the program-memory location at 2000H to the data-memory location at 40H.

E1.23 Write an instruction sequence to add 2 to the data-memory locations from 50H to 52H.

E1.24 Write an instruction sequence to add 6 to the data-memory locations from 90H to 92H.

CHAPTER 2

Introductory C8051 Assembly Programming

2.1 Objectives

Upon successful completion of this chapter, you will be able to:

- Explain the organization of an assembly language program
- Write correct assembly language statements
- Use assembler directives to define constants and symbols, allocate memory locations, etc.
- Place instruction sequence in a macro so that it can be reused
- Write pseudocode and/or use flowcharts to describe program logic flow
- Write assembly programs to perform arithmetic operations
- Use appropriate constructs to implement program loops
- Use logical instructions to set, clear, and toggle a few bits in a byte
- Use SiLabs IDE to enter, assemble, and download a program onto a demo board for execution
- Detect and eliminate syntax and logical errors from the assembly program

2.2 What is an Assembly Language Program?

An assembly language program consists of a sequence of *assembly instructions, assembler directives,* and *comments.* Assembly instructions inform the CPU what operations to perform. Assembler directives instruct the assembler how to process subsequent assembly language instructions, define program constants, reserve space for dynamic variables, and set the location counter for the active segment. The meaning of *segment* will be explained later. Comments explain what operations are performed by the related instructions or directives. Comments provide documentation to the program and make the program easier to understand.

Each line of an 8051 assembly program consists of four fields: *label, operation* or *directive, operand* or *argument,* and *comment.* Not every field needs to be present in each line of the assembly program.

Most 8051 assemblers are not case sensitive. The assembly language programmer can use uppercase or lowercase or even mix the use of cases to represent each of the four fields. Many syntax errors can be avoided when the assembler is case-insensitive.

2.2.1 The Label Field

This optional field may serve one of the following functions:

- Identify locations in the program memory
- Identify locations in the data memory
- Define constants or alternative names for registers

The label field starts from column one of a line and must begin with a letter and is followed by a mix of zero or more letters, digits, and special characters. For all assemblers, only a small number of special characters (such as an underscore and quotation mark) may be used to form the label field. There is a limit to the length of the label for most assemblers. A label must be terminated with a colon character (:) if it represents a memory location. Otherwise, it must not.

Example 2.1 Give a few examples of valid and invalid labels.

Solution: Labels in the following instructions are valid.

```
target:    subb    A,#10
output:    acall   putc2UART
max        equ     100          ; max equals 100
min        equ     20           ; min equals 10
```

Labels in the following instructions are invalid.

```
to go:     add     A,#10        ; a blank is included in the label
again      subb    A,R0         ; the label is not terminated with a colon
```

2.2.2 The Operation or Directive Field

The mnemonic names for machine instructions or assembler directives are specified in this field. If preceded by a label, this field must be separated from the label field by at least one space. Otherwise, this field must be at least one space from the left margin.

Example 2.2 Give a few examples of operation fields.

Solution:

```
            inc     A           ; inc is the operation field
            subb    A,#50       ; subb is the operation field
true        equ     1           ; the assembler directive equ is the operation field
```

2.2.3 The Operand or Argument Field

The operand field (if present) follows the operation field and is separated from the operation field by at least one space. Operands for instructions or arguments for assembler directives are provided in this field. The following examples illustrate the operand field.

```
max         equ     100         ; 100 is the argument of assembler directive equ
count       set     R0          ; R0 is the argument to the directive set
loop:       movx    @DPTR,A     ; @DPTR and A are operands
            dec     count       ; count is operand
            jnc     loop        ; loop is the operand
```

2.2.4 The Comment Field

Comments can be included anywhere in your assembler program. Comments must be preceded with a semicolon character (;). A comment may appear on a line by itself or at the end of an instruction. For example:

```
;This is a whole-line comment
nop             ;This is also a comment
```

When the assembler recognizes the semicolon character on a line, it ignores subsequent text on that line. A comment may be added for the following three purposes.

- Explain the function or purpose of a single instruction
- Explain the operation or purpose of a group of instructions
- Explain the function, operands, parameters, author, date of creation, and so on of a complete program or subroutine

Example 2.3 Point out the four fields in the following assembly language statement.

```
next:       movx    A,@DPTR     ; get the next character
```

Solution: The four fields in the above source statement are as follows:

next is a label.

movx is an instruction mnemonic.
A and *@DPTR* are the operands
; get the next character is a comment

2.3 The 8051 Memory Classes

The Intel 8051 architecture has been licensed to many semiconductor manufacturers. These companies added a wide variety of peripheral functions to make their products more competitive. Some companies even expand the memory capacity of their 8051 devices to as much as 16 MB. Those 8051 variants with 64 KB of program memory capacity are referred to as *classic 8051.* Those with more than 64 KB of program memory capacity are referred to as *extended 8051.* Only the classic 8051 microcontrollers will be discussed in this text. Because different sections of the physical memory of the 8051 may serve different purposes, they are divided into the different classes shown in Table 2.1. Bank0 . . . Bank31 are not available in a classic 8051. The description of this section is adapted from Keil's assembly language user's manual.

2.4 The 8051 Assembler Directives

Each assembler provides a different set of directives. This text describes the directives that are used in Keil Software's A51 cross assembler because of its popularity.

TABLE 2.1 *Memory Classes of the 8051*

Memory Class	Address Range	Description
DATA	00–7F	Direct addressable on-chip data memory
BIT	20–2F	Bit-addressable SRAM; accessed by bit instructions
IDATA	00–FF	Indirect addressable on-chip data memory
XDATA	0000–FFFF	64 KB data memory accessed with the **movx** instruction
CODE	0000–FFFF	64 KB program memory (read only)
BANK0	B0:0000–B0:FFFF	Code banks for expanding the program code space to 32 × 64 KB ROM
. . . BANK31	B31:0000–B31:FFFF	

2.4.1 Segments

The A51 assembler divides an assembly program into segments. A segment is a block of code or data memory. Segments are created by using segment directives, which will be described in Section 2.4.3. There are two types of segments: *absolute* and *generic.*

Segments that reside in a fixed memory location are called *absolute segments.* Generic segments have a name and a class as well as other attributes. Generic segments may be *relocatable* (i.e., when they are moved to other memory locations, the program can still run correctly). When the assembler is processing a certain segment, that segment is referred to as the *active segment.*

2.4.2 Segment Location Counter

Each program segment uses a *location counter* to point to the location where the next byte is to be assigned to the machine code or data storage. When the assembler is first started, the location counter for each segment is initialized to 0. As the program is assembled, the location counter is changed by the length of the instruction or data. By using the location counter, the assembler knows where to store an instruction or a data byte. The **org** directive forces the active location counter to a new value. The dollar sign (**$**) is used to indicate the value of the active location counter. When the **$** character is used as an operand of an instruction or directive, it represents the address of the first byte of that instruction or directive. For example, the following instruction continues to test the RI0 bit until it is set to 1.

```
jnb        RI0, $
```

2.4.3 Directives for Controlling Segments

The A51 assembler provides the following directives for controlling segments: *segment, rseg, cseg, dseg, bseg, iseg,* and *xseg.* These directives are used to create new segments. The **segment** directive declares a **generic** segment, whereas other directives declare an **absolute** segment.

Segment

The **segment** directive declares a generic segment. The declaration may include its relocation type and allocation type. The syntax and the meaning of each keyword of this directive are given in Table 2.2.

For example,

```
xds        segment XDATA
```

defines a segment with the name **xds** and the memory class XDATA.

```
newseg     segment CODE at 0x1000
```

TABLE 2.2 *The Segment Directive*

Command Syntax	*segname* **segment** *memclass* *reloc_type* *alloc_type*
Keyword	**Description**
segname	This field is the name assigned to the new segment.
memclass	This field specifies the *memory class* of the new segment and can be **bit**, **code**, **data**, **idata**, and **xdata**.
reloc_type	This field specifies the relocation operation to be performed by the **linker/locator**–the valid relocation types are listed in Table 2.3.
alloc_type	This field defines the allocation operation that may be performed by the **linker/locator**—the only valid **alloc_type** is **page**. A segment with this type starts at an address that is dividable by 256.

defines a segment with the name **newseg** and the memory class CODE that will be located at address 0x1000.

```
exmem        segment  XDATA  PAGE
```

defines a segment with the name **exmem** and the memory class XDATA. The segment starts on a 256-byte page boundary, that is, it is page-aligned.

Once a relocatable segment name is defined, it must be selected by using the **RSEG** directive. After that, it becomes the active segment that A51 uses for subsequent code and data until the active segment is changed with RSEG or with an absolute segment directive. For example, the following directive makes the **exmem** segment the current active segment.

```
rseg exmem
```

The A51 provides absolute-segment control directives **bseg**, **cseg**, **dseg**, **iseg**, and **xseg**. Each of these directives defines an absolute segment. The selected segment becomes the active segment. The formats of these directives are as follows.

```
bseg    at    addr    ; defines an absolute BIT segment
```

For example,

```
bseg    at    0x20
```

defines a bit addressable segment starting at 0x20.

```
cseg    at    addr    ; defines an absolute code segment
```

For example,

```
cseg    at    0x2000
```

defines an absolute code segment starting at 0x2000.

```
dseg    at    addr    ; defines an absolute DATA segment
```

TABLE 2.3 *Relocation Types*

Relocation Type	Purpose
AT *address*	Specifies an absolute segment to be placed at the specified address.
Bitaddressable	This type specifies a segment that will be located inside the bit addressable memory area (0x20 to 0x2F in data memory). This type is allowable only for segments with the DATA class that have size not exceeding 16 bytes.
INBLOCK	This type specifies a segment that must be contained in a 2048-byte block and is valid for segments with the class of CODE.
INPAGE	This type specifies a segment that must be contained in a 256-byte page.
OFFS offset	This type specifies an absolute segment that is placed at the starting address of memory class plus the specified *offset.* The advantage compared to the AT relocation type is that the start address can be modified with the Lx51 linker/ locator control CLASSES.
OVERLAYABLE	This type indicates that a segment can share memory with other segments declared with this relocation type.
INSEG	This type specifies a segment that must be contained in a 64-kB segment.

(Courtesy of Keil)

For example,

```
dseg    at    0x40
```

defines an absolute data segment starting at 0x40.

```
iseg    at    addr    ; defines an absolute IDATA segment
```

For example,

```
iseg    at    0x90
```

defines an absolute indirect segment starting at 0x90 of internal data memory.

```
xseg    at    addr    ; defines an absolute XDATA segment
```

For example,

```
xseg    at    0x100
```

defines an absolute segment in XRAM starting at 0x100.

2.4.4 Directives for Defining Symbols

These directives allow the user to use symbols to represent registers, numbers, and addresses. Symbols defined by these directives may not have been defined previously and may not be redefined. The only exception to this is the **SET** directive.

Using symbols to represent values or registers bring forth the following benefits:

- The readability of the program is improved. Using symbols to refer to registers or values makes the meaning and purpose of each instruction more understandable.
- The program becomes easier to maintain. The programmer needs to change the values of the constants defined by these directives in only one place for all of the statements that make reference to those symbols to be modified.

EQU

The **equ** directive assigns a numeric value or register symbol to the specified symbol name. The symbols defined by **equ** may not have been defined and may not be redefined. The format of this directive is as follows.

```
symbol    equ    expr
symbol    equ    reg
```

For example,

```
hi_limit    equ    125    ; whenever hi_limit is encountered, it will be replaced by 125
lo_limit    equ    40     ; whenever lo_limit is encountered, it will be replaced by 40
lpcnt       equ    R7     ; R7 is used as loop count
```

SET

This directive assigns a numeric value or a register symbol to the specified symbol name. The defined symbol may later be redefined. The format of this directive is as follows.

```
symbol    set    expr
symbol    set    reg
```

For example,

```
size       set    2 * N    ;
cptr       set    R0       ; use R0 as cptr
average    set    R6       ; use R6 to represent average
arCnt      set    R7       ; use R7 as arCnt (array counter)
```

BIT

This directive assigns a bit address to the specified symbol. The format of this directive is as follows.

```
symbol    bit    baddress    ; baddress is a bit address no greater than 255
```

For example,

```
valid        bit        0x50
```

assigns the bit address 0x50 to the symbol **valid**.

CODE

This directive assigns a code address to the specified symbol. The format of this directive is as follows.

```
symbol      code      caddress      ; caddress is an address in the code memory space
```

For example,

```
start        code       0x00
```

makes the code address 0x00 to be identical to the symbol **start**.

DATA

This directive assigns a data-memory address to the specified symbol. The format of this directive is as follows.

```
symbol      data      daddress      ; daddress is an address in the data memory space
```

For example,

```
P3           data       0xB0          ; a special function register
```

assigns the data-memory address 0xB0 to the symbol **P3**.

IDATA

This directive assigns an indirect data-memory address to the specified symbol. The format of this directive is as follows.

```
symbol      idata     iaddress      ; iaddress is an address in the indirect data memory
```

For example,

```
arrMax       xdata      0x1010      ; use the external data memory location 0x1010
                                    ; as variable arrMax
```

XDATA

This directive assigns an **xdata** address value to the specified symbol. Symbols defined with this directive may not be changed or redefined. The format of this directive is as follows.

```
symbol      xdata     xaddress      ; xaddress is an address in XDATA memory space
```

For example,

```
ibuf         xdata      0x70          ; use a location at XDATA as ibuf
```

2.4.5 Directives for Initializing Memory Locations

There is a need to provide data to test the program under development and to provide data to be used by the program. These directives serve this purpose. The memory initialization directives can initialize code space in either byte or word units. The values are stored starting at the point indicated by the current value of the location counter in the currently active segment.

DB

The **db** directive can be used only when the **CSEG** is the active segment. It initializes program memory with byte values. The syntax of the **db** directive is as follows.

```
label:      db      expr1, expr2, . . .
```

where

label is the symbol that is given the address of the first initialized memory location (expr1).

expr1, expr2 are byte values. **expr1** and **expr2** may be a symbol, a character string, or a numeric expression. A character string is enclosed by a pair of single or double quotes. Each character in the string is represented by its ASCII code.

Examples of the use of the **db** directive are as follows.

```
msg:        db      'Please enter your name:'   ; ASCII literal
seg7:       db      0x81,0x9F,0x92,0x86,0xCC,0xA7,0xAF,0x8F,0x80,0x84
newbie:     db      12,'good morning!',0    ; literals and numbers can be mixed
```

DW

This directive can be used only when **CSEG** is the active segment. It initializes the program memory with 16-bit values. The format of the **dw** directive is as follows.

```
label:      dw      expr1, expr2, . . .
```

The following examples illustrate the use of the **dw** directive:

```
caseTab:    dw      case1,case2,case3,case4,case5,case6,case7
extra:      dw      'H',0x1240,0x33          ; 1st byte contains 0
                                             ; 2nd byte contains 0x48 (ASCII code of letter H)
                                             ; 3rd byte contains 0x12
                                             ; 4th byte contains 0x40
                                             ; 5th byte contains 0x00
                                             ; 6th byte contains 0x33
```

The first directive stores all of the symbols that represent the targets to be jumped to when the corresponding cases occur.

2.4.6 Directives for Reserving Memory Locations

Memory reservation directives can be used to reserve space in bit, byte, or word units without setting their values. The starting point for the reserved space is indicated by the current value of the location counter in the currently active segment.

DBIT

This directive reserves bits in the **BIT** segment. The location counter of the segment is advanced by the value of the directive. The format of this directive is as follows.

```
symbol:     dbit     bit_count
```

The location counter for the bit segment is incremented by the amount of **bit_count**. The user must make sure that the location counter does not advance beyond the limit of the bit segment. Two examples of the use of this directive are.

```
            dbit     8          ; reserve 8 bits
flags:      dbit     4          ; reserve 4 bits
```

DS

This directive reserves a specified number of bytes in the currently active space. Only when **ISEG, DSEG**, or **XSEG** is the currently active segment can this directive be used. The **DS** directive has the following format.

```
symbol:     ds       byte_count
```

This directive advances the current active segment by the amount of **byte_count**. The user must make sure that the location counter does not advance beyond the limit of the segment. The following are examples of using the **DS** directive.

```
InBuf:      ds       10         ; reserve 10 bytes (label is optional)
OutBuf:     ds       50         ; reserve 50 bytes for output
```

It is possible to tell the assembler where to reserve a certain number of bytes of memory. This can be achieved by using the **AT** and **DS** directives together. For example, the following directives reserve 60 bytes starting from the address 0x30:

```
            dseg     at 0x30    ; reserve memory block starting from the address 0x30
array:      ds       60         ; reserve 60 bytes to be used as array space
```

2.4.7 Directives for Controlling Addresses

The A51 assembler provides the org and using directives for controlling the address location counter and the absolute register symbols.

ORG

The **ORG** directive sets the location counter of the currently active segment to a certain value. The user should make sure that the location counter does not advance beyond the limit of the selected segment.

The syntax of this directive is as follows.

```
org        expression
```

For example,

```
org        0x2000        ; set active location counter to 8192
```

sets the location counter of the currently active segment to 0x2000.

```
org        LookUpTab+0x100        ; arithmetic expression
```

sets the location counter of the currently active segment to **LookUpTab+ 0x100**.

USING

When a program contains subroutines, the user may have the need to save and restore registers R0 to R7 in the stack using the push and pop instructions. These instructions require the use of only absolute register addresses. The symbols AR0-AR7 are used to represent absolute-data addresses of R0 through R7 in the current register bank. When the register bank is changed, the absolute addresses for these registers also change. The **USING** directive is used to select the register bank without actually changing the active register bank. The address of any register in the four register banks can be specified by combining the use of the **USING** directive with the symbols AR0-AR7. For example,

```
mov        PSW,#0x08        ; switch to register bank 1
mov        AR4,#0x15        ; place the value 15H into the location 0x0C
using      2                ; select the register bank 2
push       AR1              ; push the memory location 11H into the stack
mov        R0,#22H          ; place the value 22H in memory location 0x08
```

This directive makes the direct addressing of a specified register bank easier.

2.4.8 Other Directives

END

The END directive terminates the assembly program and is required in every assembly source file. Any statement that appears after the **END** directive is ignored.

2.4.9 Directives for Macro Definitions

A macro is a name assigned to a group of instructions and/or directives. There are situations in which the same sequence of instructions/directives needs to be executed in several places. The macro mechanism enables the programmer to enter the same sequence of instructions only once. By passing different values to the macro, the macro may operate on different values and produce different results. The macro capability also enables software reuse. The most commonly used macros can be placed in a single file and be included in many different programs. This approach promotes software reuse.

MACRO and ENDM

The **MACRO** directive is used to define the start of a macro, whereas the **ENDM** directive terminates a macro definition. The syntax of a macro is

```
MacroName:      macro arg1,. . .,argn       ; comment
                .
                .
                .
                endm
```

Invoking the Macro

A macro is invoked by specifying the name of the macro and supplying appropriate arguments to the macro. Assume we have the macro definition:

```
sum3:       macro   arg1,arg2,arg3      ; a macro that computes the sum of arg1,
            mov     A,arg1              ; arg2, and arg3 and places the result in A
            add     A,arg2
            add     A,arg3
            endm                        ; terminate the macro definition
```

To invoke this macro, simply enter the macro name and its parameters. The statement

```
sum3    R1,R2,R3
```

will make the assembler to generate the following instructions starting from the current location counter.

```
mov     A,arg1      ; places arg1 in A
add     A,arg2
add     A,arg3
```

If a macro contains many instructions and will be invoked many times in the program, it is better to convert it into a subroutine so that the program size won't increase dramatically.

Labels within Macros

One can use labels within a macro definition. By default, labels used in a macro are global, and if the macro is used more than once in a module, the A51 assembler will generate an error.

Labels used in a macro should be local labels. Local labels are visible only within the macro and will not generate errors if the macro is used multiple times in one source file. A local label is defined with the **LOCAL** directive. Up to 16 local symbols may be defined using the **LOCAL** directive.

The following macro has a local label and initializes an array pointed to by R0 to 0:

```
clrmem      macro     addr, len
            local     loop
            mov       R1, #len          ; initialize loop count
            mov       R0, #addr         ; set up array pointer
            mov       A, #0             ; set up value for initialization
loop:       mov       @R0, A
            inc       R0
            djnz      R1, loop
            endm
```

The A51 assembler generates an internal symbol for local symbols defined in a macro. The internal symbol has the form of **??0000** (?? refers to the local label) and is incremented each time the macro is invoked. Therefore, local variables used in a macro are unique and will not generate errors.

The A51 assembler provides the ability to repeat a block of text within a macro. The **REPT**, **IRP**, and **IRPC** directives are used to specify text to repeat within macro. Each of these directives must be terminated with an **ENDM** directive.

REPT

The **REPT** directive repeats a block of text a fixed number of times. The following macro inserts 10 NOP instructions when it is invoked.

```
delay     macro
            rept      10        ; insert 10 NOP instructions
            nop
            endm                ; end REPT
          endm                  ; end of macro definition
```

IRP

The **IRP** directive repeats a block once for each argument in a specified list. A specified parameter in the text block is repeated by each argument. The following macro replaces the argument **RNUM** with R0, R1, . . . , R7.

```
initREGS    macro
              IRP        RNUM, <R0, R1, R2, R3, R4, R5, R6, R7>
                         mov   RNUM, #0
              endm                 ; end IRP
            enm                    ; end macro definition
```

The A51 assembler generates the following instruction sequence when the previous macro is invoked.

```
mov      R0, #0
mov      R1, #0
mov      R2, #0
mov      R3, #0
mov      R4, #0
mov      R5, #0
mov      R6, #0
mov      R7, #0
```

IRPC

The **IRPC** directive repeats a block once for each character in the specified argument. A specified parameter in the text block is replaced by each character. The following macro replaces the argument **CHR** with the characters T, E, S, and T.

```
debugout    macro
            IPRC    CHR, <TEST>
            jnb     TI, $           ; wait for transmitter
            clr     TI
            mov     A, #'CHR'
            mov     SBUF,A          ; transmit CHR
            endm                    ; end IRPC
            endm                    ; end macro definition
```

The A51 assembler generates the following instruction sequence when the previous macro is invoked.

```
            jnb     TI, $       ; wait for transmitter to be empty
            clr     TI
            mov     A,#'T'
            mov     SBUF,A      ; transmit letter T
            jnb     TI, $       ; wait for transmitter to be empty
            clr     TI
            mov     A,# 'E'
            mov     SBUF,A      ; transmit letter E
            jnb     TI, $       ; wait for transmitter to be empty
            clr     TI
            mov     A,# 'S'
```

```
    mov     SBUF,A      ; transmit letter S
    jnb     TI, $       ; wait for transmitter to be empty
    clr     TI
    mov     A, #'T'
    mov     SBUF,A      ; transmit letter T
```

2.5 Software Development Issue

A complete discussion of issues involved in software development is beyond the scope of this text. However, we do need to take a serious look at some of them, because embedded system designers must spend a significant amount of time on software development.

Software development starts with *problem definition*. The problem presented by the application must be understood fully before any program can be written. At the problem definition stage, the most critical thing is to get the programmer and the end user to agree upon what needs to be done. To achieve this, asking questions is very important. For complex and expensive applications, a formal, written definition of the problem is formulated and agreed upon by all parties.

Once the problem is known, the programmer can begin to lay out an overall plan of how to solve the problem. The plan is also called an *algorithm*. Informally, an algorithm is any well-defined computational procedure that takes some value or a set of values as input and produces some value or a set of values as output. An algorithm is thus a sequence of computational steps that transforms the input into the output. We can also view an algorithm as a tool for solving a well-specified computational problem. The statement of the problem specifies in general terms the desired input/output relationship. The algorithm describes a specific computational procedure for achieving that input/output relationship.

An algorithm is expressed in *pseudocode* that is very much like C or PASCAL. What separates pseudocode from "real" code is that in pseudocode, we employ whatever expressive method that is most clear and concise to specify a given algorithm. Sometimes, the clearest method is English, so do not be surprised if you come across an English phrase or sentence embedded within a section of "real" code.

An algorithm provides not only the overall plan for solving the problem but also documentation to the software to be developed. In the rest of this book, all algorithms will be presented in the following format:

Step 1

...

Step 2

...

An example of algorithm that generates a 1-kHz waveform from the Port 1 pin 0 is as follows:

Step 1
Pull Port 1 pin 0 to high.

Step 2
Wait for 0.5 ms.

Step 3
Pull Port 1 pin 0 to low.

Step 4
Wait for 0.5 ms.

Step 5
Go to Step 1.

An earlier alternative for providing the overall plan for solving software problems was the use of flowcharts. A flowchart shows the way a program operates. It illustrates the logic flow of the program. Therefore, flowcharts can be a valuable aid in visualizing programs. Flowcharts are used not only in computer programming, but also in many other fields, such as business and construction planning.

The flowchart symbols used in this book are shown in Figure 2.1. The *start/stop symbol* is used at the beginning and the end of each program. When it is used at the beginning of a program, the word **Start** is written inside it. When it is used at the end of a program, it contains the word **Stop**.

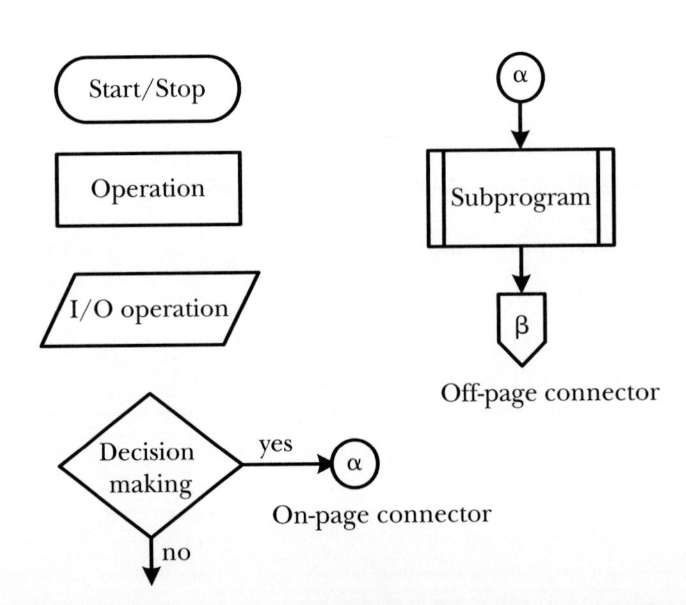

FIGURE 2.1
Flowchart symbols used in this book.

The *operation box* indicates what must be done at this point in the program execution. The operation specified by the process box could be shifting the contents of one general-purpose register to a peripheral register, setting a flag, and so on.

The *I/O operation box* is used to specify what data are to be read or displayed by the microcontroller.

The *decision box* has a question of which the answer could be yes or no. When the answer is yes, the computer will take one action. When the answer is no, the computer will take a different action.

The *subroutine* box represents the start of a sequence of instructions which may be called from many places within a program. The place that calls a subroutine should use an arrow to specify the entrance connector of the subroutine and the places to be returned from the subroutine should have a connector identical to the exit connector of the subroutine.

The *on-page connector* enables the flowchart to continue elsewhere on the same page. The place where it is continued will have the same label as the on-page connector. The *off-page connector* allows the flowchart to continue on different page. The place that the flowchart continues can be found by locating the matching off-page connector in the following pages.

The normal flow of a flowchart is from top to bottom and from left to right. Any line that does not follow this normal flow should have an arrowhead on it. When the program gets complicated, the flowchart that documents the logic flow of the program also becomes difficult to follow. This is the limitation of the flowchart. In this book, we will mix the use of flowcharts and the algorithm procedures to describe the solution to a problem.

After the programmer is satisfied with the algorithm or the flowchart, it is converted to source code in one of the assembly or high-level languages. Each statement in the algorithm (or each block of the flowchart) will be converted into one or multiple assembly instructions or high-level language statements. If an algorithmic step (or a block in the flowchart) requires many assembly instructions or high-level language statements to implement, then it might be beneficial to either

1. Convert this step (or block) into a subroutine and just call the subroutine.
2. Further divide the algorithmic step (or flowchart block) into smaller steps (or blocks) so that it can be coded with just a few assembly instructions or high-level language statements.

This process is referred to as *top-down design with hierarchical refinement* and is considered to be the most efficient software development methodology.

The next major step is *testing the program.* Testing a program means testing for anomalies. The first test is for normal inputs that are always expected. If the result is what is expected, then the borderline inputs are tested. The maximum and minimum values of the input are tested. When the program passes this test, then illegal input values are tested. If the algorithm includes several branches, then enough values must be used to exercise all of the possible branches. This is to make sure that the program will operate correctly under all possible circumstances.

In the rest of this book, most of the problems are well defined. Therefore, our focus is on how to design the algorithm that solves the specified problem as well as how to convert the algorithm into source code.

2.6 Assembly Program Template

After the power–on reset, the 8051 microcontroller starts to execute a program from address 0. Address 0 is referred to as the *reset vector.* The 8051 reserves three bytes for handling the reset. The normal approach is to place the instruction **ljmp start** in program memory locations 0 to 2 to jump to the starting point of the program. The symbol **start** is the label attached to the first instruction to be executed. The 8051 reserves a block of program-memory locations after the reset vector to handle all interrupt sources. Eight bytes are allocated to handle each interrupt source. For the SiLabs C8051F040, the address of the first program-memory location available for normal program execution is 0xAB.

When writing an 8051 assembly program, we need to add the include file of the target microcontroller. This file contains the definitions of SFRs and bits. By including the MCU include file, we can use symbolic names instead of addresses to access SFRs and bits.

After reserving space for handling interrupts, an assembly program would look like (using the Keil assembler).

```
            $nomod51                        ; disable default definition
            $include    (c8051F040.inc)
            org         000H                ; reset vector
            ljmp        start
            . . .                           ; interrupt service routines
            org         0ABH                ; starting address of normal program
start:
```

The **c8051F040.inc** file contains the SFR register and bit definitions that allow the user to use mnemonic names instead of numbers (addresses) to refer to SFRs and bits in his/her program, which will make his/her program more readable. This file is located at the directory of **c:\silabs\MCU\IDEfiles\C51\asm**, which is the default search path for the include file. The user can also use other companies' cross assemblers. However, because they are not installed in this default directory, the user will need to specify the full path in the **include** statement.

2.7 Writing Programs to Perform Computation

In Chapter 1, we have learned how to write instruction sequences to perform some simple arithmetic. In this chapter, we are going to learn how to write complete programs that can be simulated either using the software simulator or using a hardware demo board. In this section and the rest of the chapter,

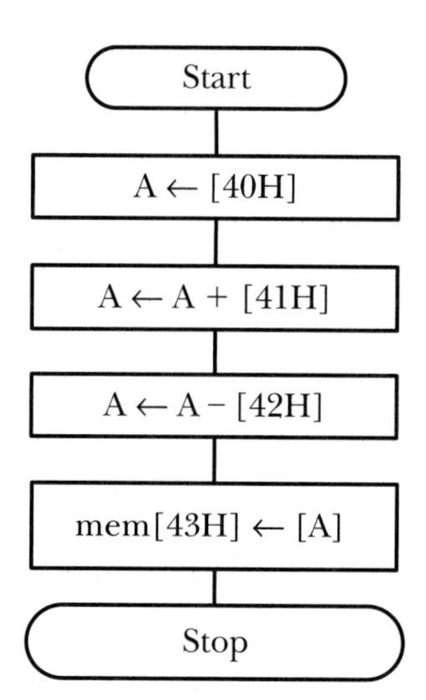

FIGURE 2.2 *Flowchart for adding the contents of three memory locations.*

we will use short programs that perform simple calculations to demonstrate how to write an assembly program.

Example 2.4 Write a program to subtract the value of the memory location at 42H from the sum of two memory locations at 41H and 42H and save the result at 43H.

Solution: The procedure for solving this problem is illustrated in the flowchart shown in Figure 2.2.

The program is

```
        $nomod51
        $include (C8051F040.inc)    ; include register definition file
        org     00H
        ljmp    start
        org     0ABH
start:  mov     A,40H               ; load the contents of the memory location at 40H into A
        add     A,41H               ; add the contents of the memory location at 41H to A
        clr     C                   ; clear the carry flag
        subb    A,42H               ; add the contents of the memory location at 42H to A
        mov     43H,A               ; save the sum at 43H
        end
```

Example 2.5 Write a program to add 9 to three 8-bit numbers stored at 0x40, 0x41, and 0x42.

Solution: The limitation of the 8051 addition instruction is that the data-memory location cannot be the destination of the operation. Therefore, three steps must be performed to carry out the addition.

Step 1. Load the memory contents into accumulator A.

Step 2. Add 9 to accumulator A

Step 3. Store the contents of accumulator A in the same memory location.

The program is

```
        $nomod51
        $include (c8051F040.inc)
        org     0x00
        ljmp    start
        org     0xAB
start:  mov     A,0x40          ; load the first number into A
        add     A,#9            ; add 9 to the first number
        mov     0x40,A          ; store the result at the same memory location
        mov     A,0x41
        add     A,#9
        mov     0x41,A
        mov     A,0x42
        add     A,#9
        mov     0x42,A
        end
```

2.7.1 The Carry/Borrow Flag

Bit 7 of the PSW register is the carry (**CY**) flag. The **CY** flag is affected by all of the 8051 addition and subtraction instructions. The sum of two 8-bit numbers may or may not be stored in an 8-bit register. For this reason, the 8051 uses the **CY** flag to hold the most significant bit of the sum of two 8-bit numbers. For example, the execution of the following two instructions will yield the sum of 131H, and the **CY** flag will be set to 1.

```
mov     A,#98H
add     A,#99H
```

When the sum of two 8-bit numbers can be held in an 8-bit register, then the **CY** flag will be cleared to 0. For example, the execution of the following two instructions will set the **CY** flag to 0 because the sum of 35H and 42H is 77H.

```
mov     A,#35H
add     A,#42H
```

In summary, the **CY** flag will be set to 1 or 0 during an addition depending on whether there is a carry out.

2.7.2 Multi-Byte Addition

When two numbers longer than 8 bits are to be added, the user must program the 8051 to add one byte at a time and proceed from least significant byte toward the most significant byte with carry out included in all except the least significant byte addition. Addition performed in this manner is referred to as *multiprecision addition.* Multiprecision addition makes use of the carry flag in the program status word (PSW). The next example illustrates the procedure of multiprecision addition.

Example 2.6 Write a program to add two four-byte numbers stored at 30H~33H and 40H~43H respectively. Save the sum at 50H~53H. Assume that the most significant to least significant bytes are stored from low address to high address.

Solution: The addition should start from the least significant byte and then proceed toward the most significant byte. The carry flag must be included in the addition in all except the least significant byte.

The program is

```
        $nomod51
        $include (c8051F040.inc)
        org     00H
        ljmp    start
        org     0ABH
start:  mov     A,33H   ; load the lowest byte of the first number into A
        add     A,43H   ; add the lowest byte of the second number to A
        mov     53H,A   ; save the lowest byte of the sum
        mov     A,32H   ; load the 2nd lowest byte of the first number into A
        addc    A,42H   ; add the 2nd lowest byte of the second number and
                        ; the carry to A
        mov     52H,A   ; save the 2nd lowest byte of the sum
        mov     A,31H   ; load the 2nd highest byte of the first number into A
        addc    A,41H   ; add the 2nd highest byte of the second number and the
                        ; carry to A
        mov     51H,A   ; save the high byte of the sum
        mov     A,30H   ; load the most significant byte of the first number to A
        addc    A,40H   ; add the most significant bytes and the carry
        mov     50H,A   ; save the most significant byte of sum
        end
```

2.7.3 The CY Flag and Subtraction

The **CY** flag enables the 8051 to borrow from the high byte to the low byte during a multiprecision subtraction. The **CY** flag operates as follows during a subtraction.

- The **CY** flag will be set to 1 if subtrahend is larger than the minuend during a subtraction.
- The **CY** flag will be cleared to 0 if subtrahend is equal to or smaller than the minuend during a subtraction.

2.7.4 Multi-Byte Subtraction

Multiprecision subtraction is the subtraction of numbers that are longer than one byte. For the 8051, multiprecision subtraction is performed one byte at a time from the least significant bytes toward the most significant bytes with the **CY** flag involved to take care of the required borrow. When subtracting the least significant bytes, the **CY** flag must be cleared to 0, because the 8051 SUB instruction always includes the borrow flag (**CY** flag) as one of the subtrahends. The borrow flag must be included in the subtraction of all higher bytes. The next example illustrates the process.

Example 2.7 Write a program to subtract the two-byte number stored at 31H~32H from the two-byte number stored at 41H~42H and save the difference at 51H~52H.

Solution: Subtraction proceeds from the least significant byte toward the most significant byte:

```
        $nomod51
        $include (c8051F040.inc)
        org     00H
        ljmp    start
        org     0ABH
start:  clr     C         ; clear the carry (borrow) flag to 0
        mov     A,42H     ; place the middle byte of the first number in A
        subb    A,32H     ; subtract the middle byte of the second number and
                          ; borrow from A
        mov     52H,A     ; save the middle byte of the difference
        mov     A,41H     ; place the high byte of the first number in A
        subb    A,31H     ; subtract the high byte of the second number and
                          ; borrow from A
        mov     51H,A     ; save the high byte of the difference
        end
```

2.7.5 Multiplication and Division

The 8051 provides one multiplication instruction and one division instruction. The **MUL AB** instruction multiplies the unsigned 8-bit integers in A and B and produces a 16-bit product. The low byte of the product is left in accumulator A, whereas the high byte is left in B. The carry flag is always cleared.

Example 2.8 Multiply the integers stored at 31H and 32H and save the product at 41H~42H. Place the upper and lower bytes in 41H and 42H, respectively.

Solution: We need to place two integers in A and B before the multiplication.

```
        $nomod51                      ; disable 8051 default definitions
        $include (c8051F040.inc)      ; include register definition file
        org     00H
        ljmp    start
        org     0ABH
start:  mov     A,31H                 ; place the first integer in A
        mov     B,32H                 ; place the second integer in B
        mul     AB                    ; perform multiplication
        mov     41H,B                 ; save the upper byte of the product
        mov     42H,A                 ; save the lower byte of the product
        end
```

When the multiplier and multiplicand are larger than 8 bits, multiprecision multiplication must be performed. The multi-byte multiplication issue is discussed in Chapter 4.

The **DIV AB** instruction performs an unsigned 8-bit division operation. It divides the value in B into the value in A. After division, the quotient is placed in A, whereas the remainder is placed in B. Both the **CY** and **OV** flags are cleared. However, if B initially contains a 0, the values placed in accumulator A and register B will be undefined, and the overflow flag will be set to 1.

Example 2.9 Write a program to convert the binary number held in A to an ASCII string that represents its decimal value. Save the string at 0x30, 0x31, and 0x32 and terminate it with a NULL character (ASCII code 0). For example, the value 0x76 (= 118_{10}) will be converted to 0x31, 0x31, and 0x38.

Solution: The given number can be converted by performing a repeated divide-by-10 operation using the following procedure.

Step 1 Initialize memory locations 0x30, 0x31, 0x32, and 0x33 with 0x20, 0x20, 0x20, and 0, respectively. The value 0x20 is the ASCII code of the space character.

Step 2 Place the value 10 in the register B and then divide accumulator A by register B.

Step 3 Swap A and B. Add 0x30 to A and store the sum in the memory location at 0x32. This saves the ASCII code of the one's digit.

Step 4 Swap A and B. Place 10 in B and perform a divide operation. Accumulator A holds the quotient that is the most significant digit of the initial value.

Step 5 Add 0x30 to A and store the sum in memory location 0x30.

Step 6 Swap A and B and add 0x30 to A. Store the sum in memory location at 0x31.

The program is

```
            $include (c8051F040.inc)
testVal     equu     195             ; value to test the program
            org      0x00
            ljmp     start
            org      0xAB
start:      mov      0x33,#0         ; store a NULL character
            mov      0x32,#0x20      ; store a space character
            mov      0x31,#0x20      ;      "
            mov      0x30,#0x20      ;      "
            mov      A,#testVal      ; place a test value in A
            mov      B,#10           ; place decimal 10 in register B
            div      AB              ; divide A by B
            xch      A,B             ; swap the remainder in A
            add      A,#0x30         ; convert to ASCII code
            mov      0x32,A          ; store the one's digit in memory
            xch      A,B             ; place the quotient back to A
            mov      B,#10           ; separate ten's and hundred's digits
            div      AB              ;      "
            add      A,#0x30         ; convert hundred's digit to ASCII code
            mov      0x30,A          ; save the most significant digit
            xch      A,B             ; place the ten's digit value in A
            add      A,#0x30         ; convert to ASCII code
            mov      0x31,A          ; save the ten's digit in memory
            nop
            end
```

2.8 Writing Program Loops

Program loops enable microprocessors to perform repeated operations. There are two types of program loops: a *finite loop* is a sequence of instructions that will be executed only a finite number of times whereas an *infinite loop* is one in which the microprocessor stays forever.

2.8.1 The Infinite Loop

The 8051 has three instructions that can be used to implement an infinite loop:

```
ajmp     addr11     ; absolute jump with 11-bit jump distance
ljmp     addr16     ; long jump with 16-bit jump distance
sjmp     rel        ; relative jump with 8-bit jump distance
```

- In the **ajmp** instruction, **addr11** specifies the lowest 11 bits of the jump target whereas the highest 5 bits come from the highest 5 bits of the current PC value.
- The **ljmp** instruction specifies the whole 16 bits of the jump target and allows the 8051 CPU to jump to any location of the 64-kB program memory space.
- The **sjmp** instruction specifies the 8-bit jump distance of the target from the first byte of the next instruction.

The jump distance can be negative or positive depending upon whether the jump direction is backward or forward. In practice, the programmer simply uses a label to specify the jump destination and let the assembler to figure out the appropriate value to be placed in the machine instruction.

An infinite loop has the format shown in Figure 2.3.

The finite loop can be implemented by using the following three constructs.

- For loop
- While loop
- Repeat until loop

2.8.2 The for Loop

The syntax of a **for-loop** is

For i = n1 to n2 do S or For i = n2 downto n1 do S

where **i** is the loop index, which runs from **n1** to **n2** or from **n2** down to **n1**, depending on the format used ($n2 \geq n1$).

The two alternatives for implementing the **for-loop** are shown in Figure 2.4. The flow-control part of a **for-loop** can be implemented by using the CJNE instruction, which has the following four variants:

```
CJNE    A,direct,rel      ; rel is the jump distance specified in relative
CJNE    A,#data,rel       ; addressing mode
CJNE    Rn,#data,rel
CJNE    @Ri,#data,rel
```

```
loop:   ...
        ...
        sjmp    loop    ; sjmp can be replaced by ajmp or
                        ; ljmp
```

FIGURE 2.3
An Instruction sequence that implements the infinite loop.

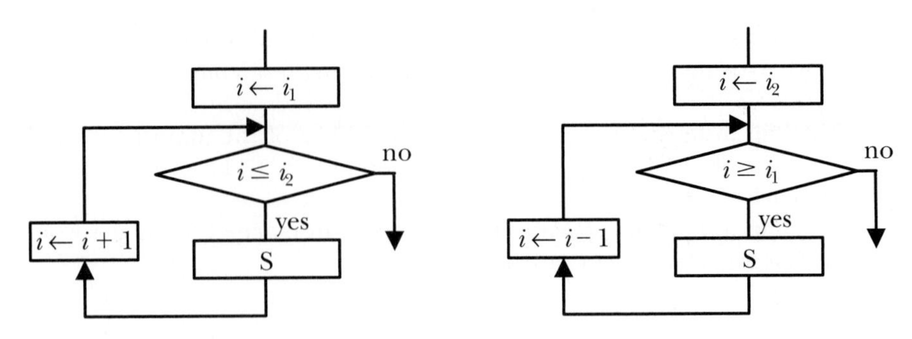

FIGURE 2.4
a) For $i = i_1$ *to* i_2 *Do S; b) For* $i = i_2$ *to* i_1 *Do S.*

The **CJNE** instruction compares the first two operands in the instruction (A with **direct** or A with **#data** or Rn with **#data** or @Ri with **#data**) and jump to the destination with distance specified by the **rel** field if the first two operands are not equal.

The **for-loop** can be implemented in the following manner:

```
n1      equ     xx              ; start loop index
n2      equ     yy              ; end loop index
ii      set     R7              ; use R7 as loop index
        ...
        mov     ii,#n1          ; set up loop index
iloop:  cjne    ii,#n2+1,next   ; has loop been executed to upper limit n2 yet?
        sjmp    done            ; if yes, get out the loop
next:   ...                     ; normal loop iteration
        ...
        ...
        inc     ii              ; increment loop index
        sjmp    iloop           ; go back to the beginning of the loop
done:   ...
```

Example 2.10 Write a program that uses the **for-loop** to add integers from *n*1 to *n*2.

Solution: The algorithm of the program is as follows.

Use **ii** as the loop index and use **sum** to hold the sum of all numbers from *n*1 to *n*2. The procedure for adding integers from *n*1 to *n*2 is as follows.

Step 1

ii ← n1, sum ← 0

Step 2

```
if (ii > n2) exit;
else
   sum ← sum + ii;
   ii ← ii + 1;
```

Step 3 Go to Step 2.

The assembly program that implements the previous algorithm is

```
        $nomod51
        $include (c8051F040.inc)
ii      set     R7              ; for-loop index
N       equ     30              ; loop
n1      equ     1               ; for-loop start index
n2      equ     30              ; for-loop end index
sumHi   set     0x30            ; high byte of sum
sumLo   set     0x31            ; low byte of sum
        org     00H
        ljmp    start
        org     0ABH
start:  mov     ii,#n1          ; initializes loop index to start index
        mov     sumHi,#0        ; initialize sum to 0
        mov     sumLo,#0        ;   "
iloop:  cjne    ii,#n2+1,next   ; loop index test
        sjmp    done            ; yes, get out of the loop
next:   mov     A,ii            ; enter normal loop iteration
        add     A,sumLo         ; add ii to sum
        mov     sumLo,A         ;   "
        mov     A,sumHi         ;   "
        addc    A,#0            ;   "
        mov     sumHi,A         ;   "
        inc     ii              ; increment the loop index
        sjmp    iloop
done:   nop
        end
```

As shown in this program, the loop-index test is done at the beginning of the loop. The last instruction of the loop is a **sjmp** instruction that causes the program flow to move to the start of the loop.

2.8.3 The while Loop

The syntax of a **while-loop** is as follows.

```
While C Do S
```

Whenever a **while-loop** construct is executed, the logical expression C is evaluated first. If it yields a false value, statement S will not be executed. The action of a **while-loop** construct is illustrated in Figure 2.5.

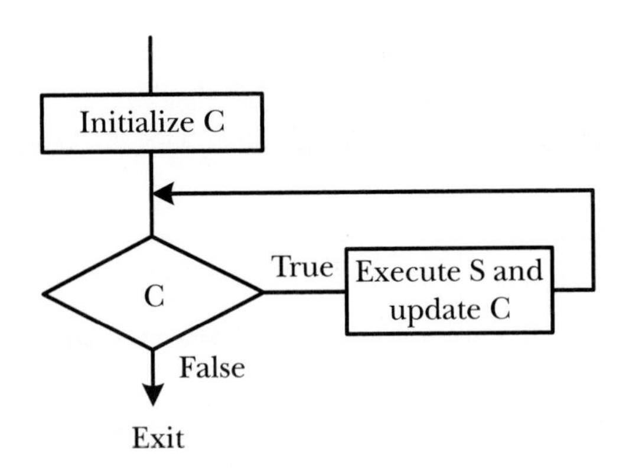

FIGURE 2.5
The while-loop construct.

TABLE 2.4 *8051 Conditional Jump Instructions*

Mnemonic		Description
jc	rel	Jump occurs when carry is set
jnc	rel	Jump occurs when carry is not set
jb	bit,rel	Jump occurs when direct bit is set
jnb	bit,rel	Jump occurs when direct bit is not set
jbc	bit,rel	Jump occurs when direct bit is set and clear bit
jz	rel	Jump occurs when accumulator A is zero
jnz	rel	Jump occurs when accumulator A is not zero
cjne	A,direct,rel	Jump occurs when direct byte and accumulator A are not equal
cjne	A,#data,rel	Jump occurs when immediate data and accumulator A are not equal
cjne	Rn,#data,rel	Jump occurs when immediate data and register are not equal
cjne	@Ri,#data,rel	Jump occurs when immediate data and indirect byte are not equal
djnz	Rn,rel	Decrement Rn and jump if [Rn] $\neq$ 0
djnz	direct,rel	Decrement direct byte and jump if [direct] $\neq$ 0

The flow-control part of a **while-loop** can be implemented by using one of the conditional jump instructions shown in Table 2.4. The jump decision may be based on the value of accumulator A, the value of the C flag, the value of a bit, or the comparison result of A with an immediate value or memory location, and so on. The next example illustrates a while-loop that uses **jz** to control the program flow.

Example 2.11 Write a program to count the number of an array of 8–bit elements that are a multiple of 8. The array is terminated by 0.

Solution: The algorithm for finding the element that is divisible by 4 is as follows.

Use **cnt** to keep track of the number of elements that are divisible by 8 and use **ii** as the array element index. A number with bit 2, bit 1, and bit 0 equals 000 is divisible by 8.

Step 1

cnt ← 0; ii ← 0;

use DPTR to point to the first element of the array.

Step 2 Fetch the array element indexed by **ii** (use **movc A,@A+DPTR**).

Step 3 If the element is 0, then exit.

Step 4 If bit 2, bit 1, and bit 0 of the element are 000, add 1 to cnt.

Step 5

ii ← ii + 1

Step 6 Go to Step 2.

The following program counts the number of array elements that are divisible by 8.

```
            $nomod51                            ; disable 8051 default definitions
            $include (c8051F040.inc)            ; include register definition file
cnt         set         R7                      ; use R7 for number count
ii          set         R6                      ; use R6 as array index
            org         0x00
            ljmp        start
            org         0xAB
start:      mov         cnt,#0                  ; initialize number count to 0
            mov         ii,#0
            mov         DPTR,#arr               ; initialize sum to 0
loopx:      mov         A,ii
            movc        A,@A+DPTR               ; get one array element
            jz          done                    ; while condition test (end of array)
doit:       anl         A,#0x07                 ; is the element divisible by 8?
            jnz         next                    ;              "
            inc         cnt                     ; yes, increment the count
next:       inc         ii                      ; move to next element
            sjmp        loopx
done:       sjmp        $
arr:        db          1,2,3,4,5,6,7,8,9,10,33,32,48,92,96
            db          11,12,13,14,15,16,17,18,19,20,78,80
            db          21,22,23,24,25,26,27,28,29,30,40,0
            end
```

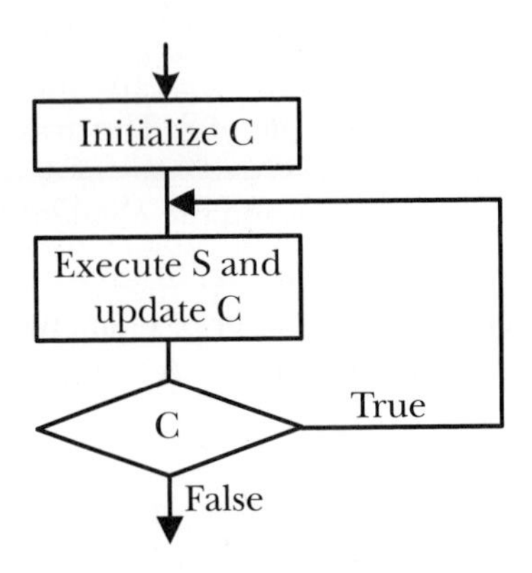

FIGURE 2.6
The repeat . . . until construct.

2.8.4 The Repeat Until Loop

The syntax of the **repeat–until** looping construct (also referred to as **do S until C**) is as follows.

Repeat S Until C

The action of this looping construct is shown in Figure 2.6. Statement S will be executed at least once.

The flow-control part of a **repeat–until** loop can be implemented by using one of the conditional jump instructions shown in Table 2.4. The next example illustrates the application of the **repeat–until** looping construct.

Example 2.12 Write a program to find the number of elements of an 8–bit array that are larger than 30 using the **repeat–until** looping construct.

Solution: The algorithm for finding the number of array elements that are larger than 30 is as follows.

Let **arcnt**, **cnt**, **ii**, and **lpcnt** represent the array count, the number of elements that are larger than 30, array index, and loop count, respectively

Step 1

cnt ← 0, ii ← 0, lpcnt ← arcnt

use DPTR to point to the first array element.

Step 2 Fetch the array element indexed by **ii** (use **movc A,@A+DPTR**).

Step 3 If the fetched array element is larger than 30, **cnt ← cnt + 1**

Step 4

ii ← ii + 1; lpcnt ← lpcnt – 1

Step 5 If **lpcnt** is 0, then exit. Otherwise, go to Step 2.

The assembly program that implements this algorithm is

```
        $nomod51                           ; disable 8051 default definitions
        $include    (c8051F040.inc)        ; include register definition file
cnt     set         R7                     ; use R7 for number count
ii      set         R6                     ; array index
lpcnt   set         R5                     ; loop count
arcnt   equ         40                     ; array count
NN      equ         30                     ; number to be compared
        org         0x00
        ljmp        start
        org         0xAB
start:  mov         cnt,#0                 ; initialize number count to 0
        mov         lpcnt,#arcnt           ; initialize loop count
        mov         DPTR,#arr              ; use DPTR to point to the array
        mov         ii,#0                  ; initialize array index to 0
loopi:  mov         A,ii                   ; fetch one array element
        movc        A,@A+DPTR              ;        "
        clr         C                      ; compare array element with NN
        subb        A,#NN+1                ;        "
        jc          next                   ; current element is smaller
        inc         cnt                    ;
next:   inc         ii                     ; increment array index
        djnz        lpcnt,loopi            ; repeat-until condition check
        nop
arr:    db          10,20,30,40,45,50,55,60,65,70,75,80,85,90
        db          11,12,14,15,16,17,23,39,41,43,91,48,53,63
        db          21,24,25,26,33,35,47,51,52,89,99,77
        end
```

2.9 Jump Table

There are situations that the program needs to perform a multi-way branch based on the value of interest. This situation is similar to the **switch** statement of the C language. The **jmp @A+DPTR** instruction can be used to implement the case jumps. The jump destination is computed as the sum of the 16-bit DPTR register and accumulator A when the instruction is executed. To implement a multi-way jump, the DPTR register is loaded with the base address of the jump table, whereas accumulator A is given the index to the table. In a four-way branch, for example, an integer 0 through 3 is loaded into accumulator A. The instruction sequence to be executed might be as follows.

```
1    mov     DPTR,#JmpTable
1    mov     A,#caseVal
1    rl      A                  ; rotate left (multiply the index by 2)
     jmp     @A+DPTR
```

The **rl A** instruction doubles the index number (0 through 3) to an even number in the range from 0 to 6, because each entry in the jump table is 2 bytes long.

```
JmpTable:   ajmp   case0
            ajmp   case1
            ajmp   case2
            ajmp   case3
```

2.10 Looping-Support Instructions

The 8051 provides a group of instructions that can be used to change the loop counter or other variables by a small quantity (such as 1). These instructions are listed in Table 2.5. The **clear** and **set** instructions listed in Table 2.6 may

TABLE 2.5 *8051 Increment and Decrement Instructions*

Instruction		Operation
dec	A	A ← [A] − 1
dec	direct	SRAM [direct] ← [direct] −1
dec	Rn	Rn ← [Rn] − 1, n = 0, . . . ,7
dec	@Ri	SRAM [[Ri]] ← [[Ri]] − 1, i = 0 or 1
inc	A	A ← [A] + 1
inc	direct	SRAM [direct] ← [direct] + 1
inc	Rn	Rn ← [Rn] + 1, n = 0, . . . ,7
inc	@Ri	SRAM [[Ri]] ← [[Ri]] + 1, i = 0 or 1
inc	DPTR	DPTR ← [DPTR] + 1

(Courtesy of Intel)

TABLE 2.6 *SETB and CLR Instructions*

Instruction		Operation
clr	A	A ← 0
clr	C	C ← 0
clr	bit	bit ← 0
setb	C	C ← 1
setb	bit	bit ← 1

(Courtesy of Intel)

also be useful in writing a program loop, because they can be used to initialize variables to either 0 or 1. The **CLR C** instruction is needed when subtracting a byte-sized number from another byte-sized number. For example, the following instruction sequence subtracts 5 from A.

```
clr     C
subb    A,#5
```

Example 2.13 Write a program to find the largest element of an array of *N* integers. The array is in the program memory. Leave the result in data memory location 0x30.

Solution: Since the array is stored in the program memory, we need to use DPTR to point to the array. The program logic flow is shown in Figure 2.7 and the program is as follow.

```
        $nomod51
        $include  (c8051F040.inc)
N       equu      20              ; array count
buf     set       R3              ; temporary storage
i       set       R0              ; loop index
        dseg      at 0x30
arMax:  ds        1               ; largest element of the array
        cseg at   0x00
        ljmp      start
        org       0xAB
start:  mov       i,#0            ; initialize i to 1
        mov       DPTR,#arrx      ; place the starting address of the array in DPTR
        mov       A,i
        movc      A,@A+DPTR       ; set A[0] as the arMax at the beginning
        mov       arMax,A         ;          "
        mov       i,#1            ; start from arr[1]
again:  mov       A,i
        movc      A,@A+DPTR       ; get the next array element
        mov       buf,A           ; save the current array element
        clr       C               ; clear the borrow before subtraction
        subb      A,arMax         ; compare arMax with A[i]
        jc        next_i          ; If A[i] is larger then check the next element
        mov       arMax,buf       ; make A[i] the current arMax
next_i: cjne      i,#N-1,next     ; if i is smaller than array count than continue
        jmp       exit            ; otherwise stop
next:   inc       i               ; increment the loop count
        jmp       again           ; and continue
exit:   jmp       $               ; forever loop
arrx:   db        4,2,3,5,1,6,7,8,49,10
        db        11,12,13,14,15,16,17,18,19,50,27
        end
```

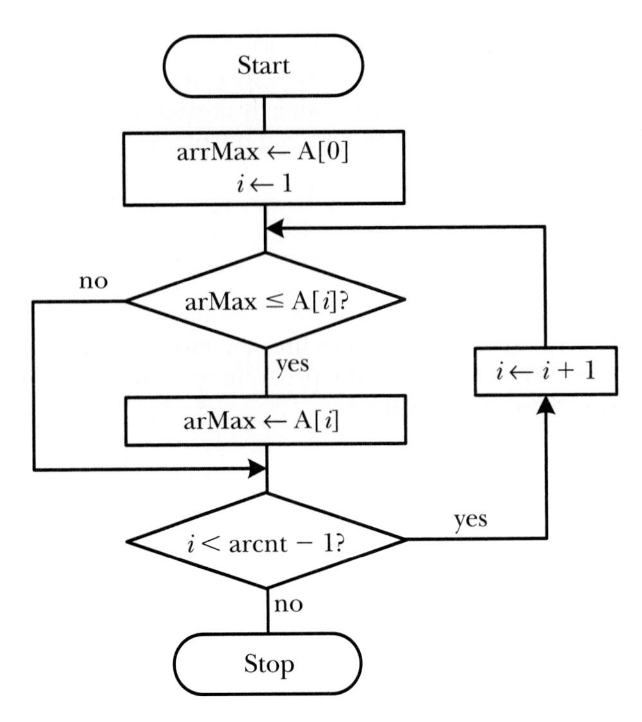

FIGURE 2.7
Program flow for finding array max.

Example 2.14 Write a program to convert a string of letters into lowercase. The string is stored in data memory and is terminated with a NULL character. The ASCII code of the NULL character is 0.

Solution: The ASCII code of an uppercase letter is from 0x41 to 0x5A. We need to provide a testing string. The string will be defined in the program memory and copied to the data memory. Therefore, the program consists of two parts. The first part is to copy the string from the program memory to the data memory. The second part is to perform the conversion. This part checks to see if a character is an uppercase letter. If it is, then convert it to lowercase letter by setting the bit 5 of the letter.

The flowchart of the program is illustrated in Figure 2.8. Register R0 is used as the string pointer. The program checks each character and takes different actions depending on whether the character is an uppercase letter or the NULL character:

- Uppercase letter: the program sets the bit 5 by ORing the letter with 0x20
- NULL character: stop

The program is as follows.

```
            $nomod51    ; disable 8051 default definitions
            $include    (c8051F040.inc)     ; include register definition file
cptr        set         R0                  ; data pointer
buf         set         R1                  ; temporary character holder
cnt         set         R2                  ; character index
            dseg at     0x30
cbuf:       ds          30                  ; buffer in data memory to hold the string
            cseg at     0x00
            ljmp        start
            org         0xAB
start:      mov         DPTR,#string        ; let DPTR points to the string in code memory
            mov         cnt,#0              ; initialize string index to 0
            mov         cptr,#cbuf          ; R0 is the pointer to the data memory buffer
loop:       mov         A,cnt               ; copy a character from program memory to
            movc        A,@A+DPTR           ; data memory buffer
            mov         @cptr,A             ; save it in buffer
            jz          exit                ; start to convert when reaching NULL
            mov         buf,A               ; save a copy of the character's ASCII code
            clr         C
            subb        A,#0x41             ; is it less than ASCII code of 'a'?
            jc          next                ; skip if it is not a uppercase letter
            mov         A,buf
            clr         C
            subb        A,#0x5B             ; is it higher than the ASCII code of 'z'?
            jnc         next                ; not a valid uppercase letter
clear:      mov         A,buf               ; get back the character
            setb        ACC.5               ; convert to lowercase
            mov         @cptr,A             ; store it back to memory
next:       inc         cnt                 ; increment the string pointer
            inc         cptr
            sjmp        loop                ; check next character
exit:       nop
string:     db          "Wyoming Yellow stone is a national park.",0
            end
```

2.11 Logical Operations

The 8051 provides logical instructions (listed in Table 2.7) for the user to perform logical operations that are mainly used to manipulate bit patterns. The user can selectively clear a few bits in a byte to 0s, set a few bits in a byte to 1s, or toggle the values of a few bits in a byte, and so on.

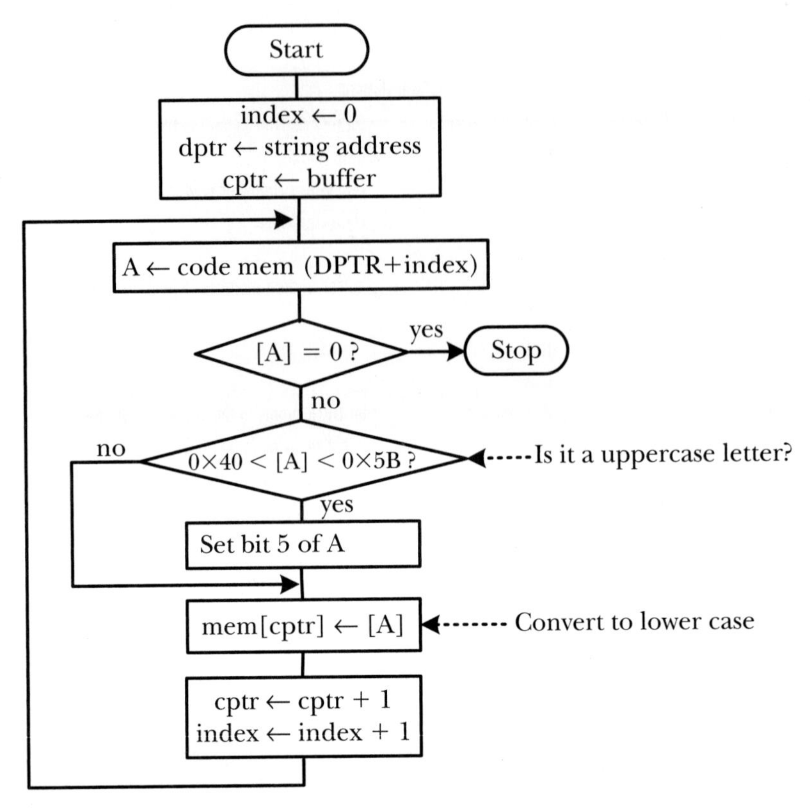

FIGURE 2.8
Program flow for converting letters of a string to lowercases.

Example 2.15 Write instruction sequences to perform the following operations:

- Clear the even bits of the memory location at P0
- Set the even bits of the memory location at P1
- Toggle the upper four bits of the memory location at P2
- Toggle every bits in A

Solution: These operations can be achieved by executing appropriate logical instructions as follows:

```
anl     P0,#0xAA
```

clears the even bits of **P0** to 0s.

```
orl     P1,#0x55
```

sets the even bits of **P1** to 1s.

```
xrl     P2,#F0H
```

toggles the upper four bits of **P2**.

```
cpl     A
```

toggles every bit in A.

2.12 Rotate Instructions

The 8051 has four rotate instructions that can be used to manipulate bit fields or accelerate multiplication or division operation by a power of 2. These four rotate instructions are listed in Table 2.8.

TABLE 2.7 *8051 Logical Instructions*

Instruction		Operation
ANL	A,Rn	A ← [A] AND [Rn], n = 0, . . . , 7
ANL	A,direct	A ← [A] AND [direct]
ANL	A,@Ri	A ← [A] AND [[Ri]], i = 0 or 1
ANL	A,#data	A ← [A] AND data
ANL	direct,A	SRAM[direct] ← [direct] AND [A]
ANL	direct,#data	SRAM[direct] ← [direct] AND data
ORL	A,Rn	A ← [A] OR [Rn], n = 0, . . . , 7
ORL	A,direct	A ← [A] OR [direct]
ORL	A,@Ri	A ← [A] OR [[Ri]], i = 0 or 1
ORL	A,#data	A ← [A] OR data
ORL	direct,A	SRAM[direct] ← [direct] OR [A]
ORL	direct,#data	SRAM[direct] ← [direct] OR data
XRL	A,Rn	A ← [A] XOR [Rn], n = 0, . . . , 7
XRL	A,direct	A ← [A] XOR [direct]
XRL	A,@Ri	A ← [A] XOR [[Ri]], i = 0 or 1
XRL	A,#data	A ← [A] XOR data
XRL	direct,A	SRAM[direct] ← [direct] XOR [A]
XRL	direct,#data	SRAM[direct] ← [direct] XOR data
CPL	A	A ← ~[A], ~ indicates 1's complement
SWAP	A	Swap the upper and lower nibbles of accumulator A

(Courtesy of Intel)

Example 2.16 Compute the new value of accumulator A after executing the four rotate instructions independently by assuming that A contains 0x73 and C = 1 before the operation.

Solution: The operation results of four rotate instructions are shown in Figure 2.9.

The multiply-A-by-2 operation can be implemented by clearing the **CY** flag to 0 before executing the RLC A instruction as follows.

```
clr     C
rlc     A
```

TABLE 2.8 *8051 Rotate Instructions*

Mnemonic	Description	
rl A	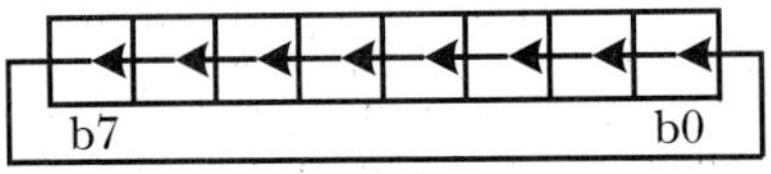	Rotate A left one place
rlc A	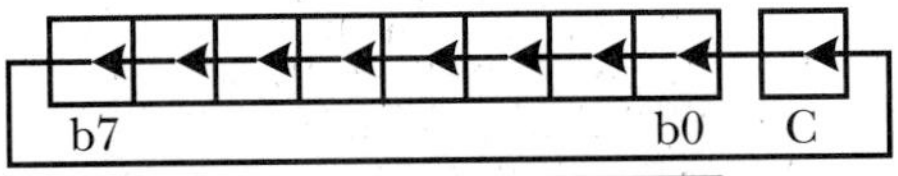	Rotate A left one place thru CY
rr A	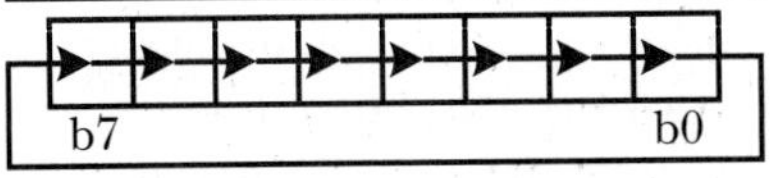	Rotate A right one place
rrc A	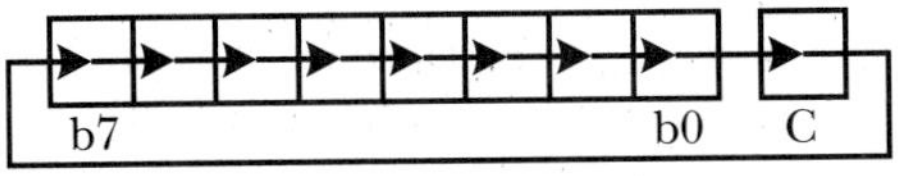	Rotate A right one place thru CY

(Courtesy of Intel)

Initial value of A	Initial C	Operation	Final value in A	Final C
0 1 1 1 0 0 1 1	1	RL A	1 1 1 0 0 1 1 0	1
0 1 1 1 0 0 1 1	1	RLC A	1 1 1 0 0 1 1 1	0
0 1 1 1 0 0 1 1	1	RR A	1 0 1 1 1 0 0 1	1
0 1 1 1 0 0 1 1	1	RRC A	1 0 1 1 1 0 0 1	1

FIGURE 2.9
Result of four rotate instructions.

The unsigned divide-A-by-2 operation can be implemented by clearing the CY flag to 0 before executing the RRC A instruction as:

```
clr     C
rrc     A
```

Example 2.17 Write an instruction sequence to multiply and divide (unsigned divide) the 16-bit number stored in data memory located at 0x30~0x31 by 2.

Solution: The multiply operation is started by first rotating the least significant byte left and continue working toward the most significant byte. Only the least significant byte requires the **CY** flag to be cleared. The instruction is as follows.

```
mov     A,0x31      ; multiply the least significant byte
clr     C           ; by 2
rlc     A           ;    "
mov     0x31,A      ;    "
mov     A,0x30      ; multiply the most significant byte
rlc     A           ; by 2 with CY bit shifted in
mov     0x30,A      ;    "
```

Multiplying a number by a power of 2 can be implemented by repeating the previous instruction sequence by an appropriate number of times.

The unsigned divide operation is started by rotating right the most significant byte first and work toward the least significant byte. Only the most significant byte requires the **CY** flag to be cleared. The instruction sequence is as follows.

```
mov     A,0x30
clr     C
rrc     A
mov     0x30,A
mov     A,0x31
rrc     A
mov     0x31,A
```

Dividing a number by a power of 2 can be implemented by repeating the previous instruction sequence by an appropriate number of times.

A signed divide-by-2 operation must maintain the sign bit of the dividend, which would require the previous instruction sequence to be modified. This will be left as an exercise problem for you.

2.13 Boolean Variable Manipulation Instructions

The 8051 provides a group of instruction that can perform move, set, clear, complement, OR, and AND operations on bit operands. The internal RAM from location 0x20 through 0x2F contains 128 addressable bits, and the SFR space supports up to 128 other addressable bits. All of the I/O port lines are bit-addressable, and each line can be treated as a separate single-bit port. The bit-oriented instructions are listed in Table 2.9.

Addressable RAM bits can be used as software flags or to store program variables. For example, in the following instruction sequence

```
mov     C,high
jc      ishigh
```

high is the name of one of the addressable bits in the SRAM locations from 0x20 to 0x2F or SFR space. The jump operation will occur if **high** is 1.

The **CY** flag in the PSW is referred to as C in bit manipulation instructions (for example, **clr C**). The carry bit also has a direct address, because it resides in the bit addressable PSW register.

TABLE 2.9 *A list of the 8051 Boolean Instructions*

Instruction		Operation
anl	C,bit	C ← C AND bit
anl	C,/bit	C ← C AND (NOT bit)
orl	C,bit	C ← C OR bit
orl	C,/bit	C ← C OR (NOT bit)
mov	C,bit	C ← bit
mov	bit,C	bit ← C
clr	C	C ← 0
clr	bit	bit ← 0
setb	C	C ← 1
setb	bit	bit ← 1
cpl	C	C ← NOT C
cpl	bit	bit ← NOT bit
jc	rel	Jump if C = 1
jnc	rel	Jump if C = 0
jb	bit,rel	Jump if bit = 1
jnb	bit,rel	Jump if bit = 0
jbc	bit,rel	Jump if bit = 1; bit ← 0

(Courtesy of Intel)

The **jbc** instruction tests the specified bit, jumps if the addressed bit is set, and also clears the bit. Thus, a flag can be tested and cleared in one operation. All of the bits of the PSW register are addressable and are available to the bit-test instructions.

2.14 Hardware and Software Development Tools

The SiLabs 8051 MCUs incorporate an on-chip JTAG debug circuitry allowing non-intrusive, full-speed, in-circuit programming and debugging using the target MCU installed in the printed circuit board. This debug system supports inspection and modification of memory and registers, breakpoints, and single-stepping through the program.

The following three components are needed to utilize the JTAG debug circuit to perform program debugging:

1. A debug adapter interfacing the PC and the target board. This debug adapter accepts debug commands from the PC software and takes corresponding actions.
2. A debug software running on the PC that allows the user to set program breakpoints, modify and display registers and memory locations, single step through a segment of instructions, reset the target MCU, and so on. In response to the user request, the debug software sends appropriate commands to the debug adapter to carry out the debug operations.
3. A cable that connects the debug adapter to the target board.

SiLabs provides many 8051 development kits that utilize this concept to perform software debugging. The hardware setup for these demo kits and any demo board that uses the JTAG debug circuit is illustrated in Figure 2.10.

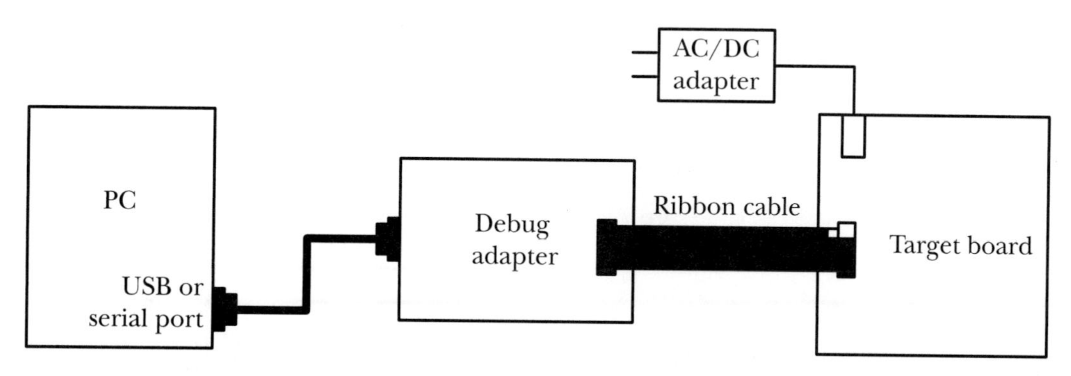

FIGURE 2.10
Hardware setup for the SiLabs MCU demo boards.

2.14.1 Debug Adapter

There are two versions of the debug adapter: the older version communicates with the PC via the serial communication port whereas the newer version uses USB interface to communicate with the PC. The debug adapter uses a ribbon cable to connect to a 10-pin connector on the target board. The signals of the 10-pin connector are defined in Table 2.10.

2.14.2 Demo Boards

The C8051F040TB target board is designed to facilitate the product development for the C8051F040 MCU. This demo board can also be used in learning the programming and interfacing of the C8051F040 MCU.

The block diagram of the C8051F040TB is shown in Figure 2.11. The RS232 connector allows the user to exchange data with a PC via the serial port (need to run a terminal program such as HyperTerminal on the PC). The CAN connector allows the C8051F040TB to communicate with other MCUs over a CAN bus. The CAN bus is discussed in Chapter 13. All I/O port signals are available via eight 10-pin connectors. A JTAG connector is provided for connecting to the SiLabs debug adapter. Information about the C8051F040TB can be found at www.silabs.com.

The SSE040 is a demo board made by Shuan-Shizu Enterprise. This demo board has the following features:

- One RS232 connector
- Two debounced push-button switches (can be used as external interrupt sources)
- One 8-bit DIP switch for digital input (connected to Port 6)
- One EEPROM (24LC08B) with an I^2C interface
- One 20 x 2 LCD (connected to Port 7)
- One JTAG connector (to be connected to SiLabs debug adapter)
- Eight LEDs (connected to Port 5)
- One 4 x 4 keypad connector for interfacing a 16-key keypad (connected to Port 3)

TABLE 2.10 *JTAG Connector Pin Description*

Pin #	Description
1	+3 VD (+3.3 VDC)
2, 3, 9	GND (Ground)
4	TCK
5	TMS
6	TDO
7	TDI
8, 10	Not connected

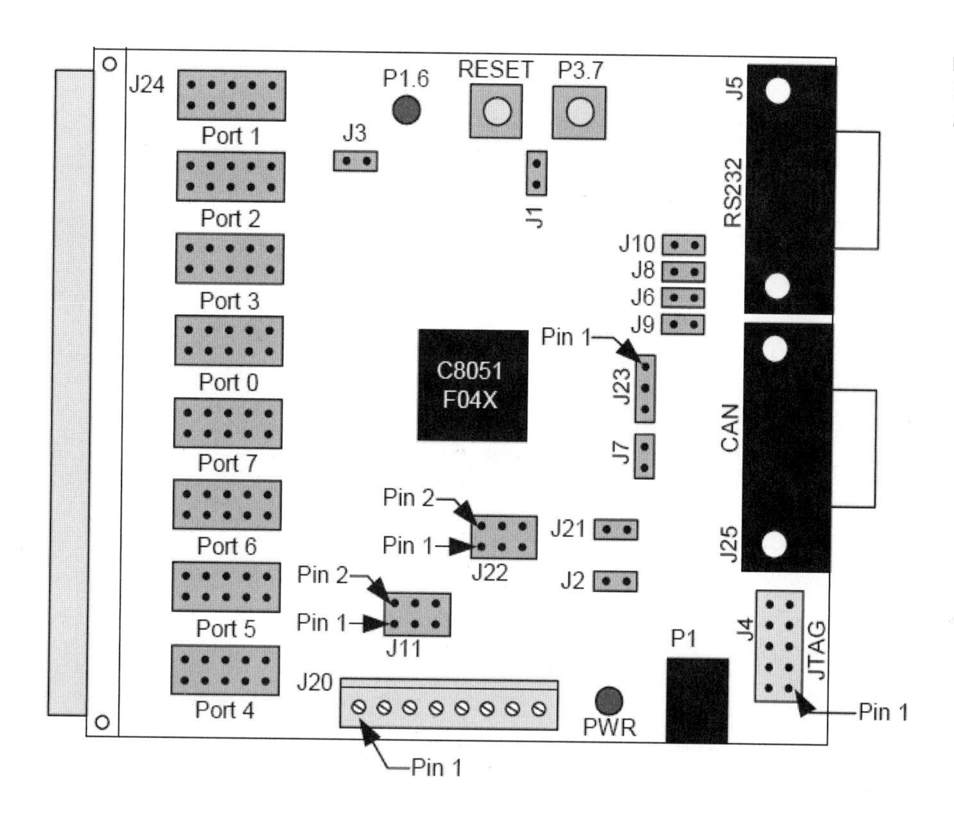

FIGURE 2.11
C8051F040 TB target board.

- One DS1337 real-time clock chip with I^2C interfacing
- One SN65HVD230 CAN transceiver with a jumpered terminating resistor and terminal block connectors for CAN bus applications
- One TC72 digital temperature sensor with SPI interface
- One potentiometer to serve as an analog input to the A/D converter
- One buzzer for sound generation
- A function generator that can generate digital waveforms with frequencies ranging from 1 Hz up to 16 MHz (located to the right of the 16-MHz crystal oscillator in Figure 2.12) with frequencies equal to a power of 2
- Eight female connectors for accessing MCU signals

The photograph of the SSE040 demo board is shown in Figure 2.12. Additional information about the SSE040 demo board can be found at www.evb.com.tw.

2.14.3 Software Tools

To develop software to be run on an 8051 demo board, we need a text editor to enter the program, an assembler and compiler to assemble and compile the program, a linker to resolve variables and subroutine cross-referencing and memory assignment, a simulator, and a debugger to debug the software being developed. A project manager is also needed to coordinate the overall

FIGURE 2.12
The SSE040 demo board.

debugging activities. These pieces of software are often integrated into a single package called Integrated Development Environment (IDE). For example, the μVision 3 from Keil and the RIDE from Raisonance are two popular IDEs for the 8051 MCU. The assemblers and C compilers of these IDEs are compatible. Programs written in one IDE can also be assembled or compiled in another IDE without modification. Demo versions of μVision (2 kB size limit) and RIDE (4 kB size limit) are available for free to the end user.

SiLabs provides a free IDE, which includes a text editor, a project manager, and a debugger. The users need to provide their own assemblers, C compilers, and linkers to work with this IDE. The SiLabs IDE does not have a simulator, and hence, all debugging activities must be performed directly on the C8051 MCU hardware. The SiLabs IDE can work with the assemblers and C compilers from Keil, Raisonance, and several other vendors.

2.15 Using the SiLabs IDE

The SiLabs IDE is the centerpiece of Silicon Laboratory's software tools strategy. With the SiLabs IDE, the user can use his or her favorite cross assembler and compiler to perform source-level debugging using the target hardware. The target hardware requires only a 10-pin JTAG connector to take advantage of this capability.

2.15.1 C8051 Program Memory Map

The C8051 starts to execute instructions from program-memory location 0 after it leaves the reset state. Three bytes are reserved for the reset. A block of memory locations after the first three bytes are reserved for handling interrupts. The 8051 reserves eight bytes for each interrupt source. The program-memory configuration for the C8051F040 is shown in Figure 2.13. The first byte that can be assigned to the application is at 0xAB. For other microcontroller with *n* interrupt sources, the first byte that can be assigned to the user application is at **8n + 3**.

The first thing to do after reset is to configure the MCU to appropriate settings. Since three bytes are not enough to perform any configuration, most programs will place an **ljmp** instruction at memory location 0**x**0000 to jump to the first instruction that performs the system configuration. One of the essential configurations is selecting the source of system clock. The user can select the internal oscillator or external crystal oscillator to generate system clock (also called instruction clock). The internal oscillator of the C8051F040 has been factory calibrated to 24.5 MHz. The system clock can be derived by dividing the internal oscillator output by 1, 2, 4, or 8. By default, the initial

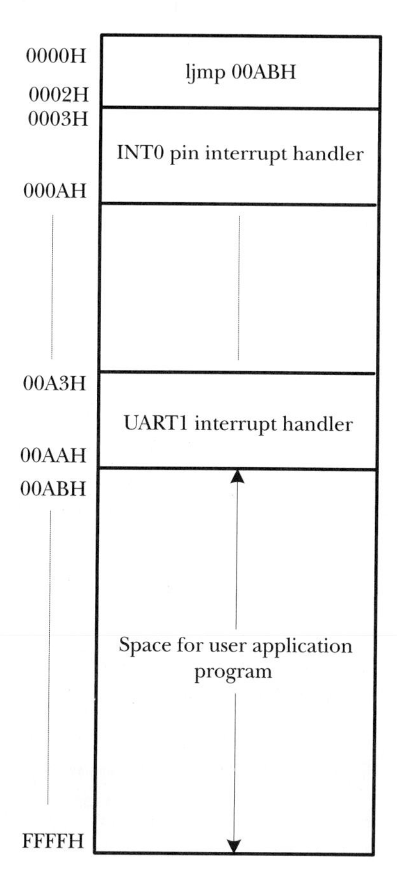

FIGURE 2.13 *Program memory configuration for the C8051F040.*

clock frequency is 24.5 Hz divided by 8 and is about 3 MHz. The following instruction sequence can be used to select the internal oscillator output divided by 1 as the system clock.

```
            mov     SFRPAGE,#0x0F     ; switch to the SFR page F
            mov     CLKSEL,#0         ; derive system clock from internal oscillator
            mov     OSCICN,#0x83      ; set prescaler to the oscillator to 1
```

If applications require a very accurate clock frequency, then an external crystal oscillator should be used to generate the system clock (machine cycle clock). The following instruction sequence selects the external crystal oscillator to generate the instruction cycle clock.

```
            mov     SFRPAGE,#0x0F     ; switch to the SFR page F
            mov     CLKSEL,#0         ; temporarily used internal oscillator to generate SYSCLK
            mov     OSCXCN,#0x67      ; configure external oscillator control
            mov     R7,#xFF
            djnz    R7,$              ; wait for about 1 ms
chkstable:  mov     A,OSCXCN          ; wait until external crystal oscillator is stable
            anl     A,#0x80           ; before using it
            jz      chkstable         ;     "
            mov     CLKSEL,#1         ; switch to use external crystal oscillator
```

The configuration of the system clock will be discussed in Chapter 6. It is desirable to use symbolic names to refer to registers and the bits of a specific register. The file **C8051F040.inc** that contains definitions for special function registers and bits allows the user to do just that. From now on, we will add the following statement to every assembly program to be run on the C8051F040 demo board:

```
$include (c8051F040.inc)
```

Example 2.18 Write a program that counts the number of elements that are a multiple of 4 from an array of 30 8-bit elements.

Solution: A number with the least significant two bits equal to 00 is divisible by 4. The logic flow of this program is shown in Figure 2.14. The assembly program that performs the desired operation is as follows.

```
            $include    (c8051F040.inc)
N           equ         45
            dseg        at 0x30          ; data segment
i:          ds          1
arr_cnt:    ds          1
count:      ds          1
            cseg        at 0             ; code segment
            ljmp        main             ; reset vector
```

```
          org     0xAB
main:     mov     SFRPAGE,#0x0F      ; switch to SFR page F
          mov     OSCICN,#0x83       ; enable internal oscillator and set divide factor to 1
          mov     CLKSEL,#0          ; derive system clock from internal oscillator
          mov     SFRPAGE,#0         ; switch to SFR page 0
          mov     DPTR,#array        ; use DPTR as a pointer to the array
          mov     i,#0               ; initialize i to 0
          mov     count,#0           ; initialize count to 0
          mov     arr_cnt,#N         ; set array count to 30
loop:     mov     A,i
          movc    A,@A+DPTR          ; get the ith array element
          anl     A,#03              ; check bit 1, and 0
          jnz     next_i
          inc     count
next_i:   inc     i
          djnz    arr_cnt,loop
          jmp     $
array:    db      1,2,3,4,5,6,7,8,9,10,11,12,13,14,15
          db      16,17,18,19,20,21,22,23,24,25,26,27,28,29,30
          db      31,32,33,34,35,36,37,38,39,40,41,42,43,44,45
          end
```

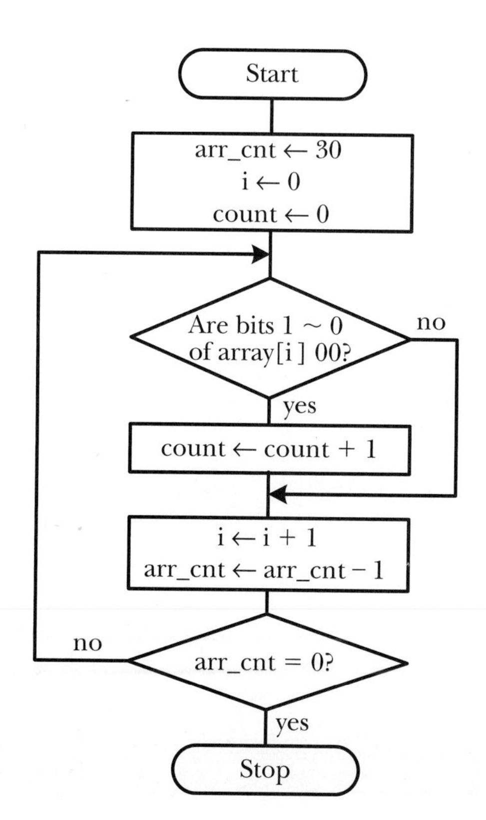

FIGURE 2.14 *Logic flow for finding the number of elements divisible by 4.*

2.15.2 Getting Started with SiLabs IDE

The SiLabs IDE can be started by double-clicking on its icon on the windows. After that, a project window similar to that in Figure 2.15 appears. If either the **Project Window** or the **Output Window** does not appear, then pull down the **View** menu and select them. The row immediately above the **Project Window** and **Editor Window** contains icons that can be clicked to perform certain debugging operations. These buttons provide short cuts for invoking the debug commands. Move the mouse cursor on top of each icon to find out the command that it can invoke. The SiLabs IDE allows you to perform the following operations:

- Using **Tool Chain Integration**
- Creating projects
- Target build configuration
- Editing and building projects
- Connecting to the hardware
- Downloading a file to the on-chip flash memory
- Using the debugger to debug the program
- Using **Watch Window** to examine variable values after program execution

2.15.3 Tool Chain Integration

The user can integrate his or her favorite cross assembler and compiler into the SiLabs IDE. This is done by pressing the **Project** menu and select **Tool Chain Integration . . .** from the SiLabs IDE window. After this, a popup dialog as shown in Figure 2.16 appears.

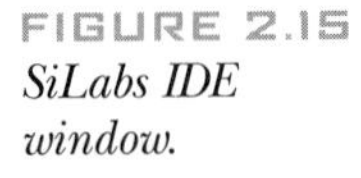
FIGURE 2.15 *SiLabs IDE window.*

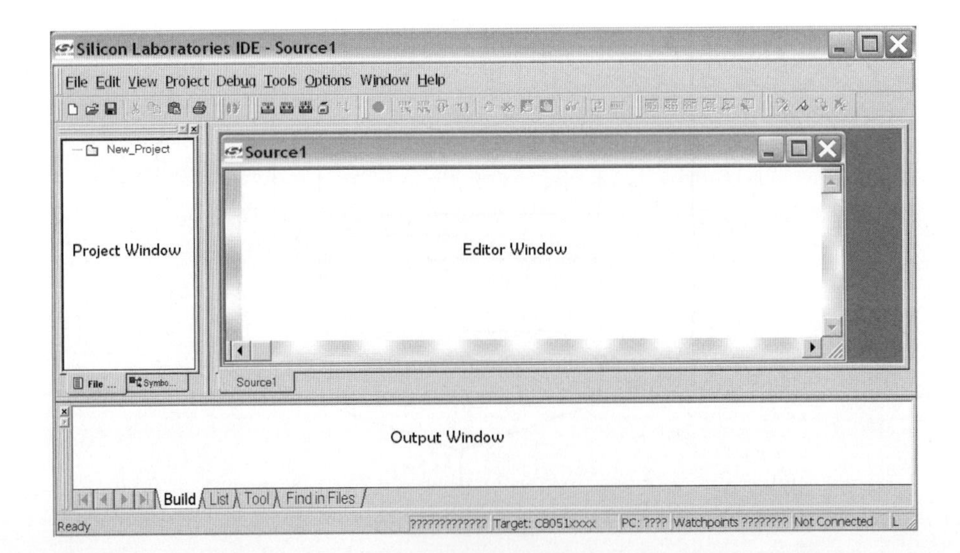

The setting in Figure 2.16 indicates that Keil's A51 is the cross assembler of choice. You need to choose a compiler and linker, as shown in Figures 2.17 and 2.18, if Keil's tool suite is your development tool of choice. These three figures select the Keil's demo version software bundled with the SiLabs IDE. If you have the full version of Keil's software, then the executables for the assembler, compiler, and linker would be **c:\keil\bin\a51.exe**, **c:\keil\bin\C51.exe**, and **c:\keil\bin\bl51.exe**, respectively. Other vendors' tools can also be integrated into the SiLabs IDE. For example, to integrate Raisonnance's tools, select **c:\RIDE\bin\ma51.exe**, **c:\RIDE\bin\ccomp51.exe**, and **c:\RIDE\bin\lx51.exe** as the assembler, compiler, and linker of choice before working on

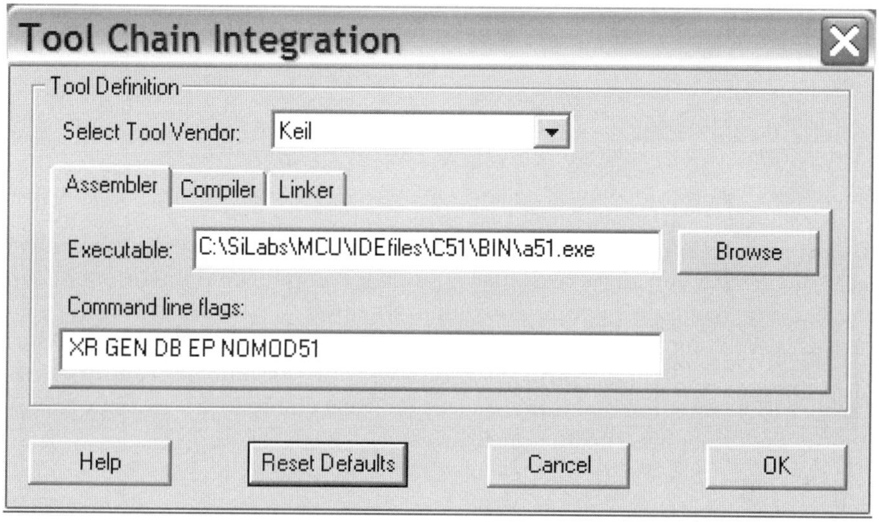

FIGURE 2.16 *Dialog for Tool Chain Integration (assembler).*

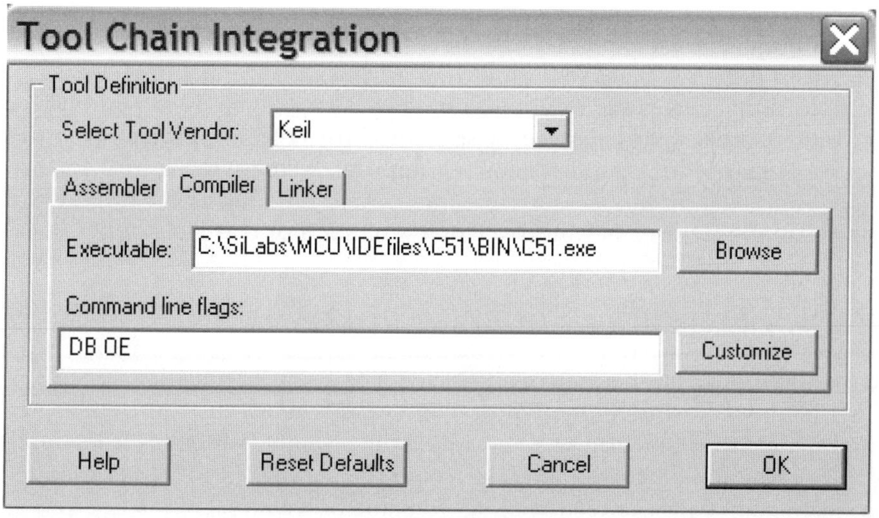

FIGURE 2.17 *Integrate Keil's C compiler in the SiLabs IDE.*

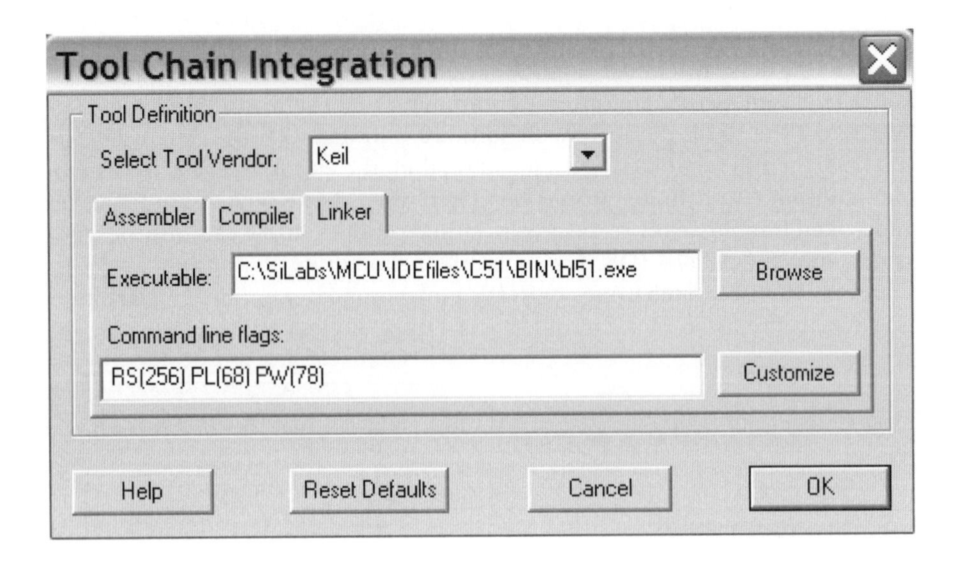

FIGURE 2.18 *Integrate Keil's linker into SiLabs IDE.*

the project. To use SDCC tools, choose **c:\sdcc\bin\asx8051**, **c:\sdcc\bin\sdcc.exe**, and **c:\sdcc\bin\sdcc.exe** as the assembler, compiler, and linker of choice, respectively. When performing tool chain integration, you first select the tool vendor. By pressing the downward ▼ you will find several tool vendors. After selecting the tool vendor, you click on the tool (**Assembler**, **Compiler**, or **Linker**) to be integrated and then select the **Executable** for the tool by browsing the file directory (clicking on **Browse**).

2.15.4 Creating a New Project

The first step in managing a programming project is to create a new project. Select **Project->New Project** from the SiLabs IDE window to open a new project. After this, the SiLabs IDE window will be changed to that in Figure 2.15. You need to give a name to the new project. This is done by saving the new project. A project can be saved by selecting **Project->Save Project** from the SiLabs IDE window. A popup dialog box appears, as shown in Figure 2.19. The user needs to enter the project name to be saved. After saving the project, the screen of the SiLabs IDE window changes to that shown in Figure 2.20.

2.15.5 Entering Source Programs

To enter a program file, select **File->New File** from the pull-down menu of the SiLabs IDE window. An **Editor Window** (as shown in Figure 2.21) will appear in the SiLabs IDE window. Enter the assembly program in Example 2.18 in the **Editor Window** and save it with the file name **eg02_18.asm**. Enter as many source files as needed using the same procedure.

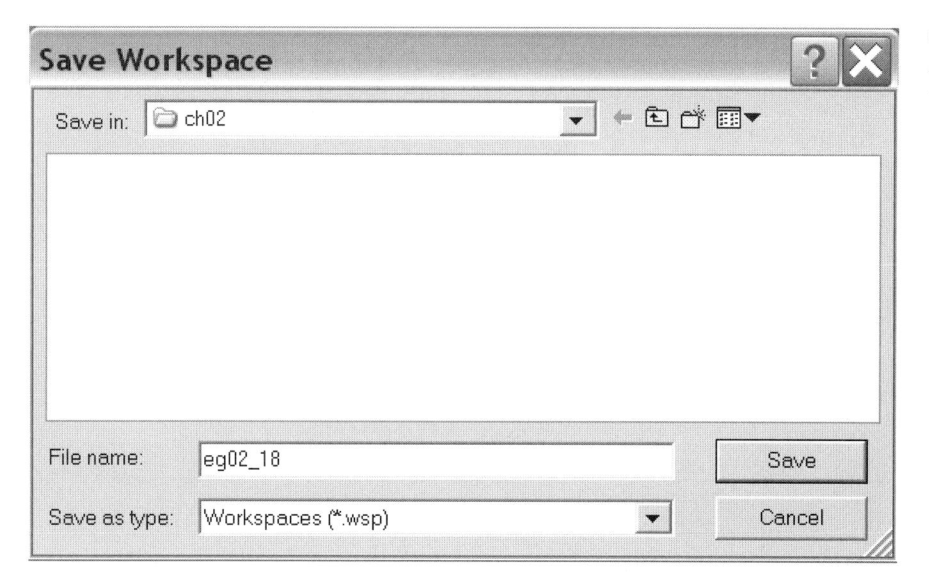

FIGURE 2.19 *Dialog for saving a new project.*

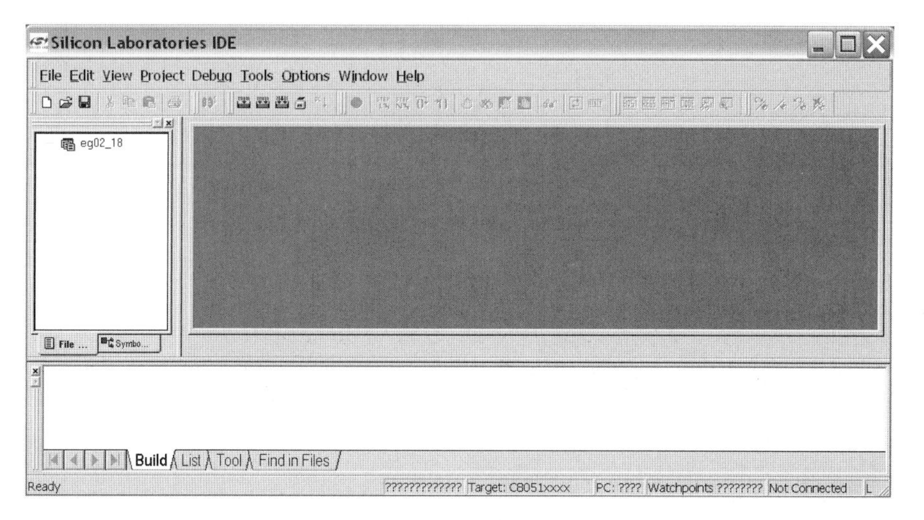

FIGURE 2.20 *Screen of SiLabs IDE after saving a Project.*

2.15.6 Adding Files to the Project

Right-click the mouse on **eg02_18** (your new project) in the **Project Window**. Select **Add file to project**. Select files using the file browser and click **Open**. Continue adding files until all project files have been added. The dialog for adding files to a project is shown in Figure 2.21. For each file in the **Project window** that you want assembled, compiled, and linked, right-click on the file name and select **Add file to build**. A popup box (as shown in Figure 2.22) will appear, informing the user to take an appropriate action. Each file will be assembled or compiled as appropriate and linked to the build of the absolute object file.

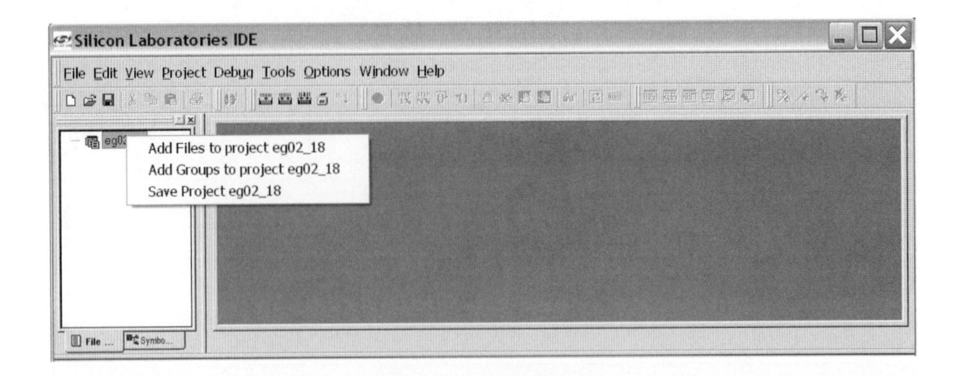

FIGURE 2.21 *Dialog for adding files to the project.*

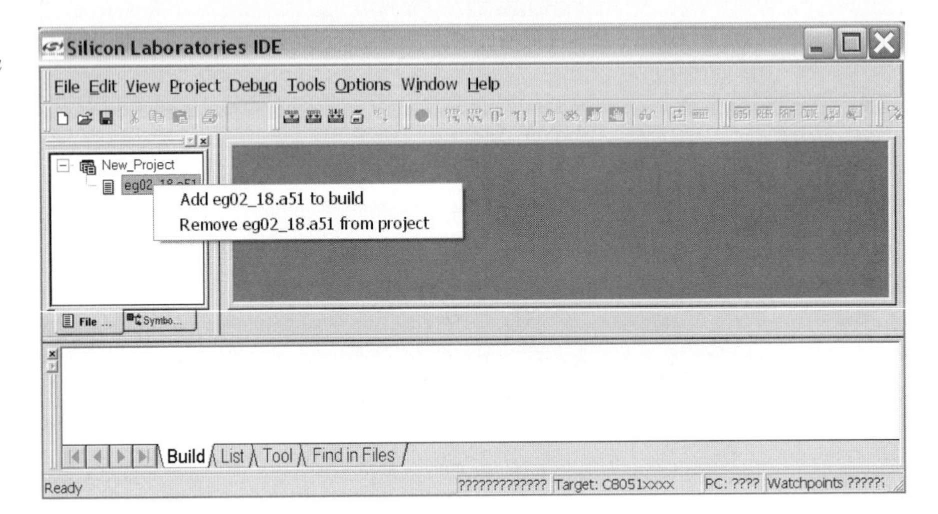

FIGURE 2.22 *Screen for adding a file to build.*

2.15.7 Building the Project

Once all source files have been added to the project, build the project by selecting **Project->Build/Make Project** from the SiLabs IDE menu. After the project has been built for the first time, the **Build/Make Project** command will build only the files that have been changed since the previous build. To rebuild all files and project dependencies, select **Project->Rebuild All** from the menu.

2.15.8 Downloading the Program for Debugging

Before one can download the project into the target MCU, one needs to select the debugging interface. This is done by selecting **Connection Options . . .** (shown in Figure 2.23) and then click on JTAG under **Debug Interface**, as

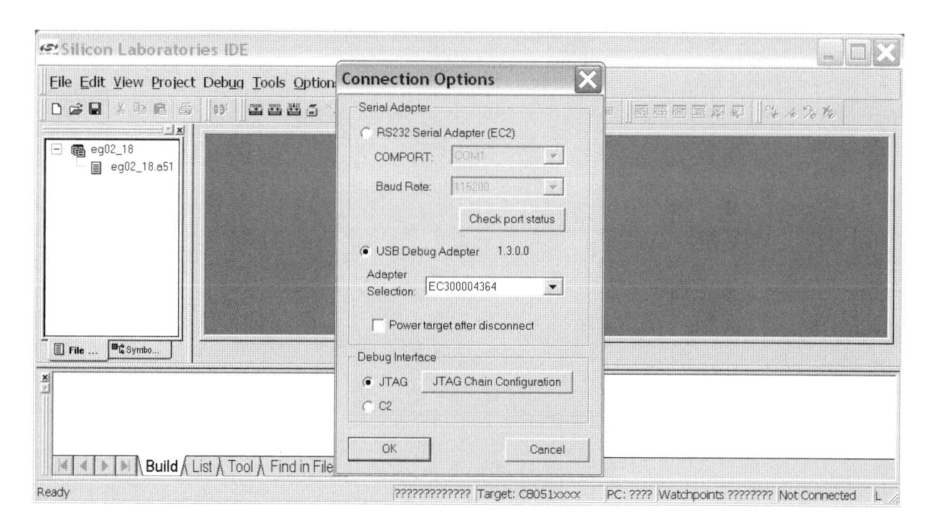

FIGURE 2.23 *Screen for selecting connection options.*

shown in Figure 2.23. Click on **OK** after selecting JTAG. This step needs to be done only during the first use of the SiLabs IDE. In Figure 2.23, the **USB Debug Adaptor** is selected. However, you should select **RS232 Serial Adapter** if this is what you are using.

After the debug interface is selected, the next step is to connect the target hardware. This is done by selecting **Debug->Connect** from the SiLabs IDE menu to connect to the target MCU.

To download machine code to the target, select **Debug->Download Object File . . .** from the SiLabs IDE menu. A popup dialog as shown in Figure 2.24 will appear asking the user to enter the name of the file to be downloaded. Use the file browser to select the file (eg02_18) to be downloaded and then click on **Download**. After the hex file is successfully downloaded, the SiLabs IDE screen will change to that in Figure 2.25. The blue flat hexagon to the left of the instruction **ljmp main** indicates that the target MCU has been reset and the program counter is set to 0. File downloads can also be started by clicking on the downward arrow with the letters DL to its left, as shown in Figure 2.25.

2.15.9 Target Build Configuration

One can choose to **Enable automatic save for project files before build**, **Enable automatic connect/download after build**, and **Run to main( )** (for C program). These options can be set by selecting **Project->Target Build Configuration** from the SiLabs IDE menu. The dialog for this configuration is shown in Figure 2.26. The default options are adequate for this tutorial. However, one can also enable the options of automatic connect and download after build.

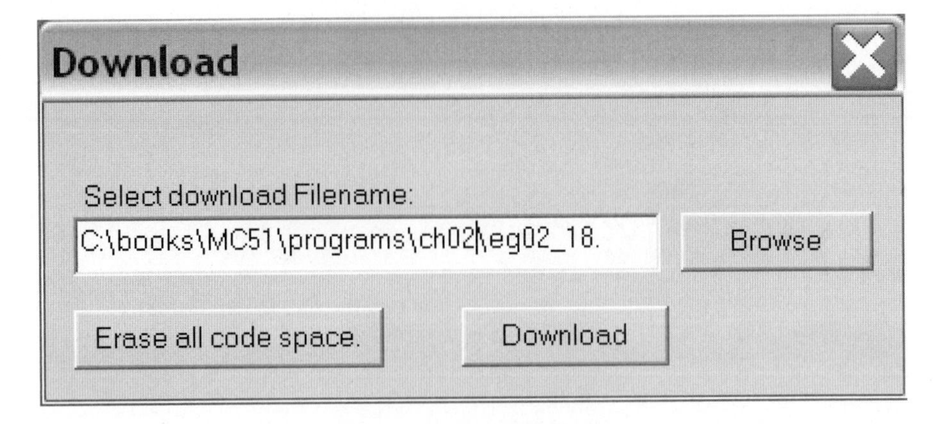

FIGURE 2.24
Dialog for selecting a file to download.

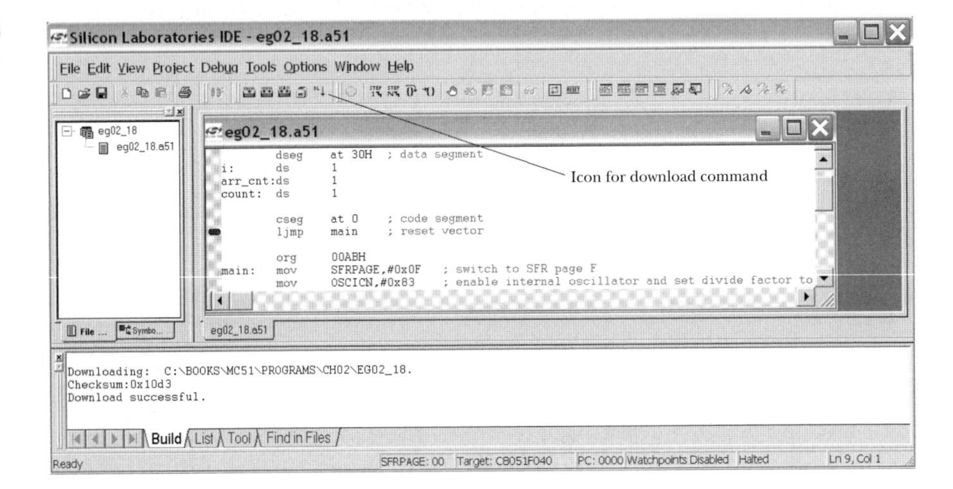

FIGURE 2.25
SiLabs IDE screen after a successful download.

2.15.10 Program Execution and Debugging

After the executable code has been downloaded successfully into the target MCU, you are ready to debug the program. This section discusses the techniques commonly used in a program debugging process.

Setting Up Debug Windows

Debug windows must be set up before a program can be debugged. There are many MCU resources that can be watched during the program execution process. These resources are shown in Figure 2.27. You can select any one or more of these resources to be displayed in a separate window. For example, the SiLabs IDE window with SFRs and registers R0 to R7 displayed in separate windows is illustrated in Figure 2.28.

The **Watch Window** is probably the most useful window among all resource windows because this window allows the user to display all program

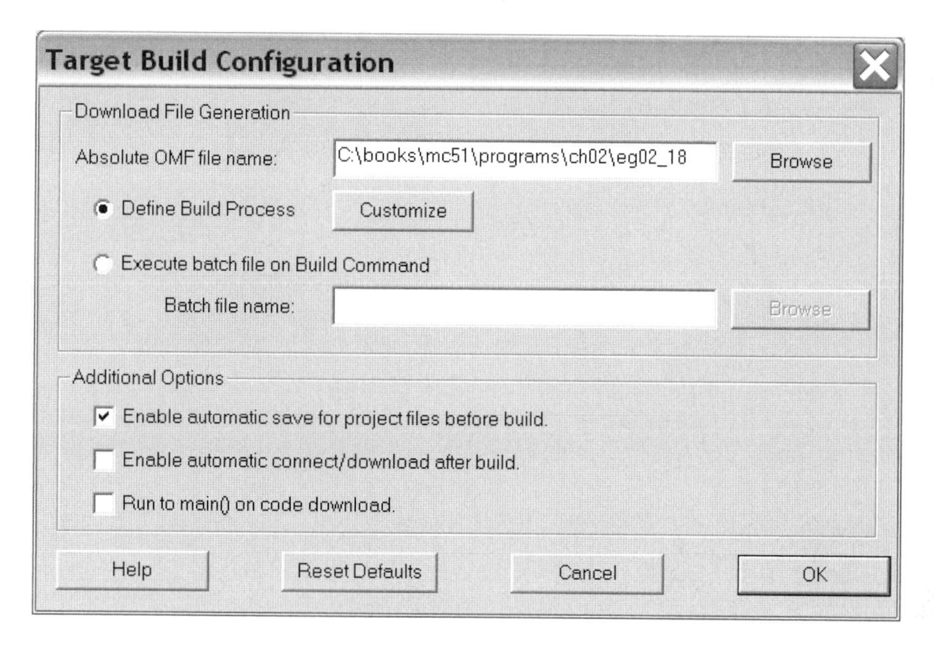

FIGURE 2.26 *Dialog for Target Build Configuration.*

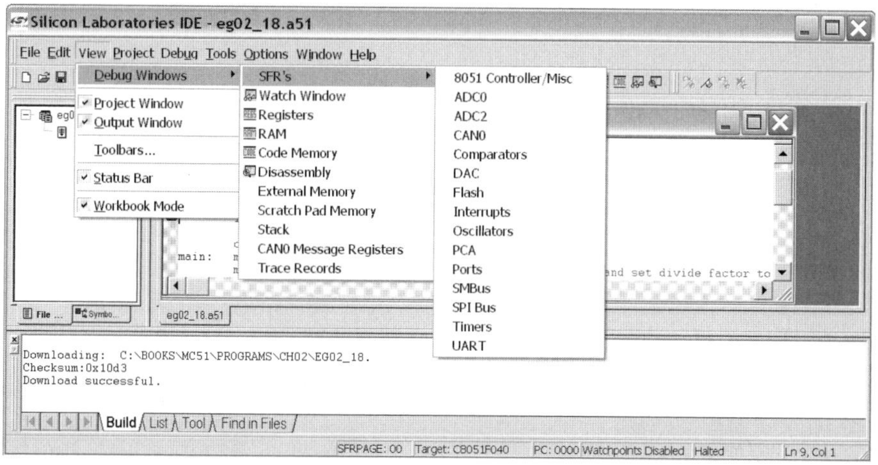

FIGURE 2.27 *C8051 MCU resources that can be watched.*

variables relevant to program execution, for example, the variables **i**, **arr_cnt**, and **count** for this tutorial.

The watch window can be brought up by selecting **View->Debug Window->Watch Window**. After bringing up the **Watch Window**, one can then add the variables of interest into the window. To add a variable to the **Watch Window**, click on the variable in the **Editor Window** (in which your program is displayed), press on the right mouse button, and bring up **Add xx to Watch**. Keep moving the mouse button to the right, as shown in Figure 2.29, to complete the insertion action. Figure 2.29 illustrates how to add the variable **i** into

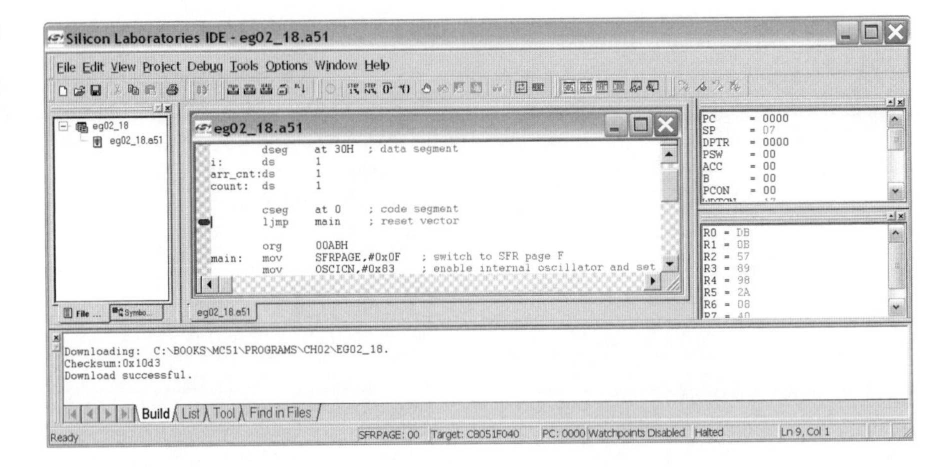

FIGURE 2.28 *SiLabs IDE window with SFRs and registers displayed in separate windows.*

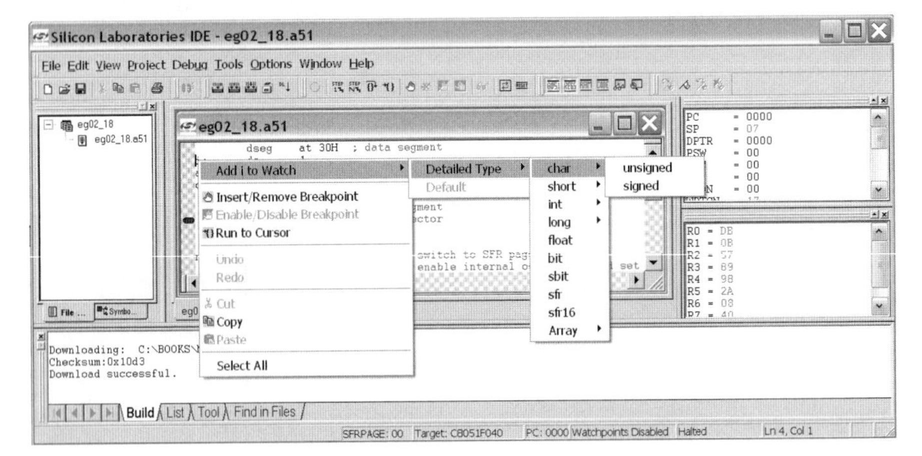

FIGURE 2.29 *Screen for adding variable i into the Watch Window.*

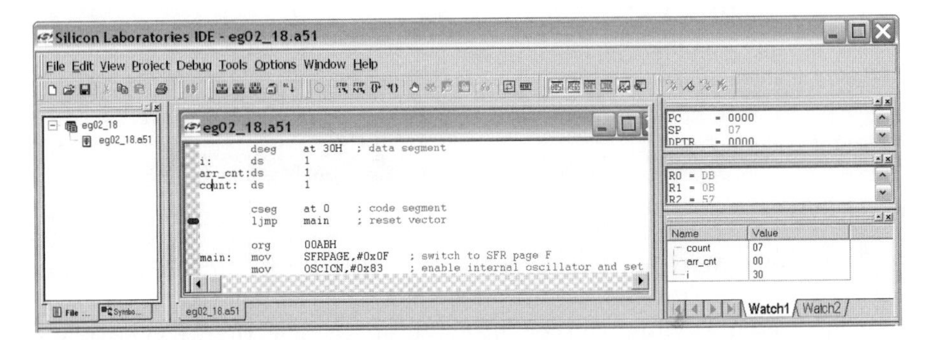

FIGURE 2.30 *SiLabs IDE screen with three variables added into the watch window.*

the **Watch Window**. The **Watch Window** with three variables added is shown in Figure 2.30. Beware that the **Watch Window** is not active if the target hardware is not connected. When the variable name has only one letter, you need to click the mouse in front of that single letter. If the variable name has multiple letters, you can click on any letter. SiLabs IDE supports two **Watch Windows**.

When the program has too many variables to be watched, they can be entered in two separate **Watch Windows**.

Resetting the Target MCU

Make sure that the target MCU is reset before running the program. The program counter of the target MCU will be forced to 0 after a reset. The target MCU can be reset by selecting the **Debug->Reset** menu in the SiLabs IDE window, as shown in Figure 2.31. You can also click on the Reset on the toolbar on the IDE window to reset the MCU. Resetting the MCU will force program counter to 0.

Setting a Breakpoint

A breakpoint is the address of an instruction where program execution will stop. After you select **Debug->go** (or press the function key F5), the MCU will execute the instructions starting from the one pointed to by the current program counter value up to the one before the breakpoint. A breakpoint is set by pressing the right mouse button on the instruction to be set as a breakpoint and selecting **Insert/Remove Breakpoint**, as shown in Figure 2.32. Repeating the same action will remove the breakpoint. After an instruction is set as a breakpoint, there will be a red circle placed to the left side of the instruction. Multiple breakpoints can be set during a debugging process.

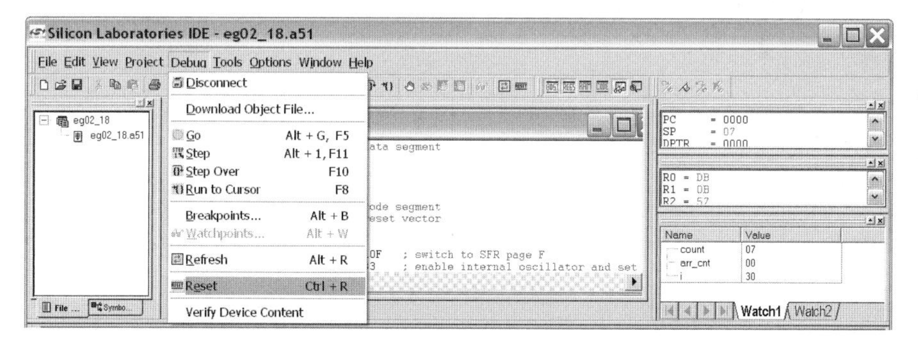

FIGURE 2.31 *SiLabs IDE screen for resetting the MCU.*

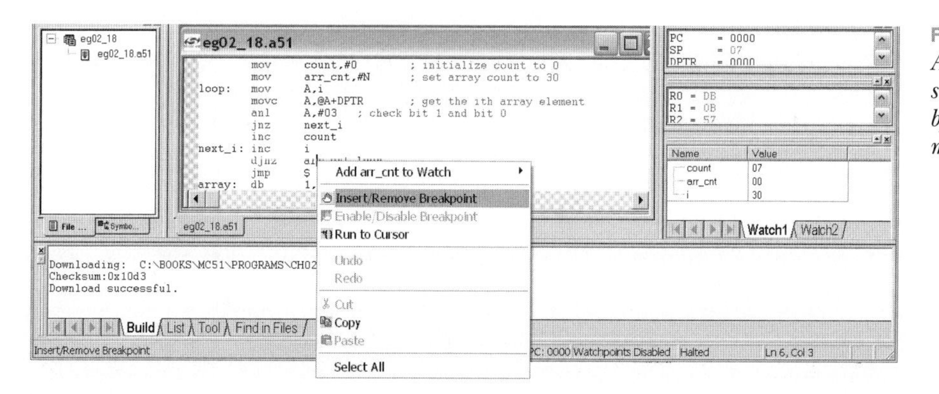

FIGURE 2.32 *An Instruction is set as a breakpoint by pressing right mouse button on it.*

When a breakpoint is not desired, it can be disabled (not deleted). A disabled breakpoint has no effect on program execution. When a disabled breakpoint is needed again, it can be reenabled. A breakpoint is disabled or enabled by pressing the right mouse button on the breakpoint instruction and selecting **Enable/Disable Breakpoint**, as shown in Figure 2.33.

Run to Cursor

The option **Run to Cursor** is a quick way for running the program up to the cursor and then stopping. When the program stops, the SiLabs IDE updates the values of the variables in the **Watch Window**. One can use this option to quickly find out if the program executes correctly up to the cursor position. This option can be exercised by pressing the right mouse button on the instruction where the program is expected to stop. Press the cursor at the **jmp $** instruction and select **Run to Cursor** and examine the **Watch Window** as shown in Figure 2.34. When the program stops, the value for the variable

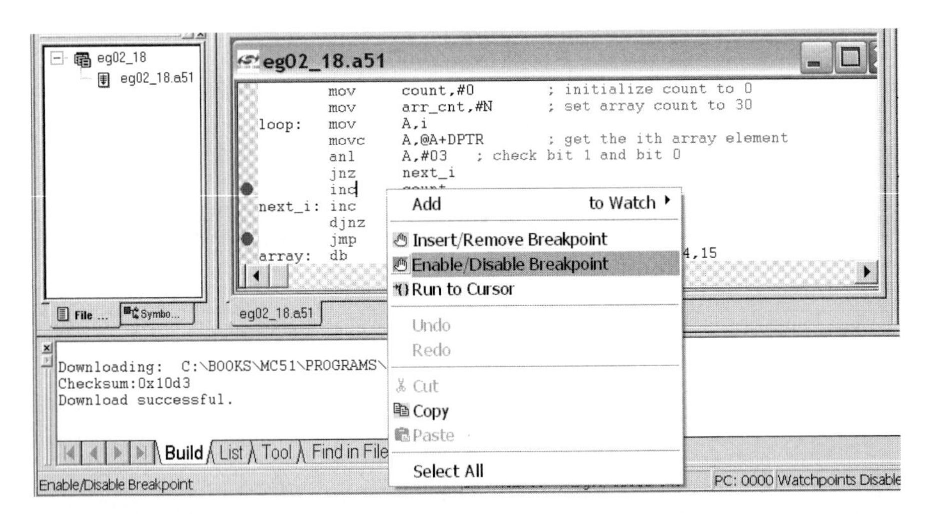

FIGURE 2.33
A breakpoint can be disabled and enabled.

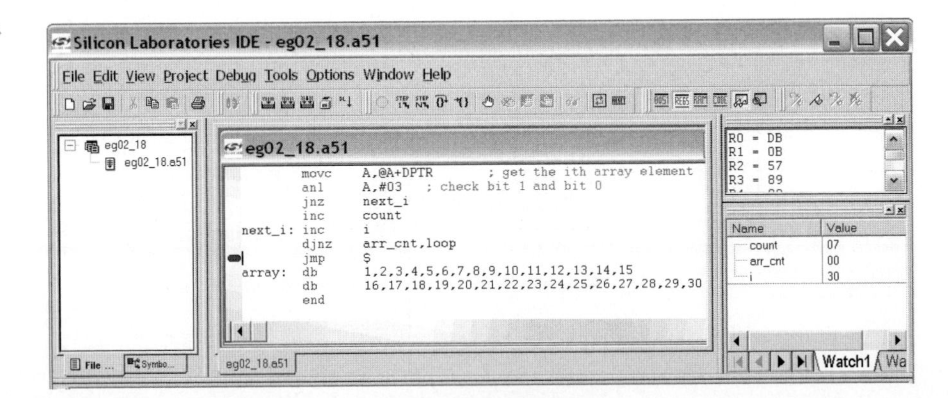

FIGURE 2.34
Screen after running program to the cursor position.

count is 7 and is correct. The example program finds the number of elements in an array that are divisible by 4.

Stepping Over Instructions

There are times that the programmer has the need to find out the effect of each individual instruction in some area in order to pinpoint the error in the program logic. This objective can be achieved by using instruction stepping. Instruction stepping is performed by selecting **Debug->Step** or **Debug->Step Over** from the SiLabs IDE menu. It will be more convenient to use a single keystroke to perform this operation. The SiLabs IDE provides these alternatives. Pressing the function key **F11** steps over an instruction, and pressing the function key **F10** steps over a subroutine call instruction. Subroutine calls will be discussed in Chapter 4. You can experiment with this command by resetting the MCU and then press F11 or F10 repeatedly and watch the change in the watch window. The effects of **Step** and **Step Over** are identical if the instruction to be stepped over is not an ACALL or LCALL instruction.

2.16 Tips for Assembly Program Debugging

Assembly program errors can be classified into two categories:

- Syntax errors
- Logical errors

2.16.1 Syntax Errors

Syntax errors are common for beginners. Syntax errors can be divided into following categories.

- ***Misspelling of instruction mnemonics.*** This type of error will be highlighted by the assembler (using a ^ character) in the output window with the line number and the type of error (syntax error). If you change the instruction mnemonic on line 12 from **mov** to **move** and reassemble the program, you will get the error message shown in Figure 2.35.
- ***A hex immediate value does not start with a 0 character.*** If you delete the leading 0 from the value of #0C3H on line 12, you will get the undefined-symbol error shown in Figure 2.36. However, a hex immediate value starting with a digit between 1 and 9 is legal.
- ***Symbol not terminated with a colon character when it is defined.*** This will cause a syntax error at the line where the symbol is defined and will cause undefined symbol errors in those lines that reference this symbol.
- ***Missing operands.*** Depending on what is missing, the error message will vary. For example, if you have the following instruction in the program, you will get two errors.

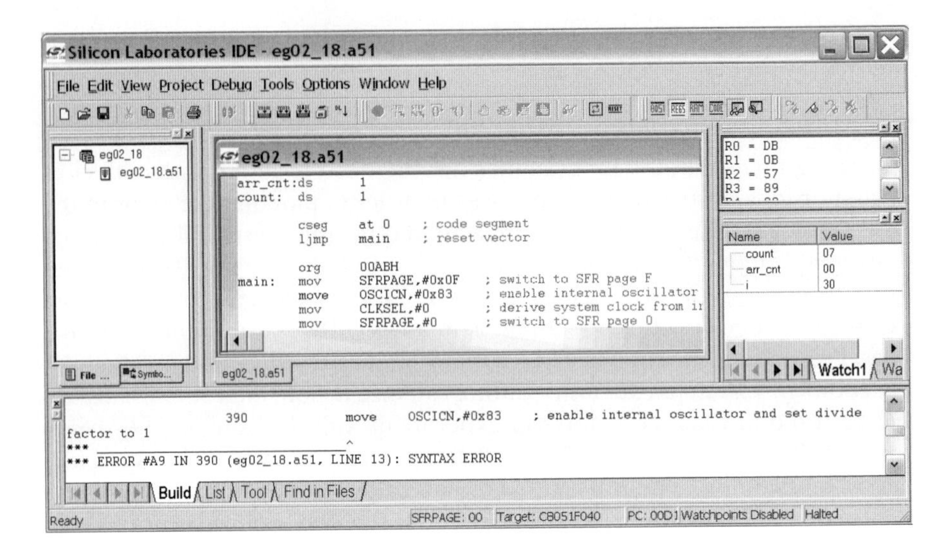

FIGURE 2.35 *SiLabs IDE syntax error report example.*

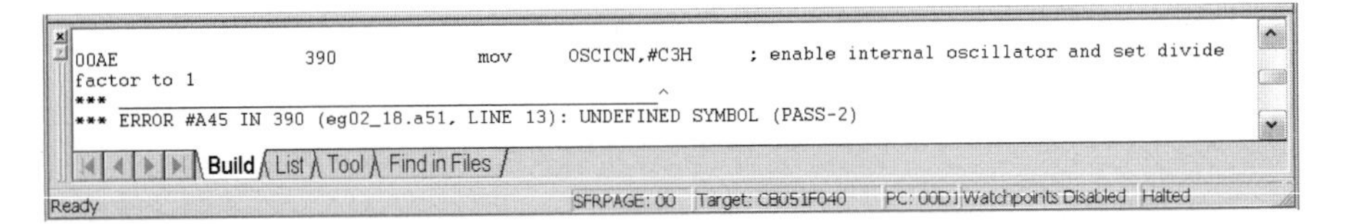

FIGURE 2.36 *Syntax error caused by missing a 0 in front of a hex number.*

As shown in Figure 2.37,

```
mov i,
```

For the instruction,

```
inc
```

the assembler outputs the message "Number of operands does not match the instruction".

As time goes on, one gains experience and also memorizes the instruction mnemonics better; this type of error will reduce and even disappear.

2.16.2 Logical Errors

Beginners make many logical errors. The most common ones are as follows.

Using Direct Addressing Mode instead of Immediate Mode

This error is very common for beginners. The following program is written to compute the sum of an array of *N* 8-bit elements.

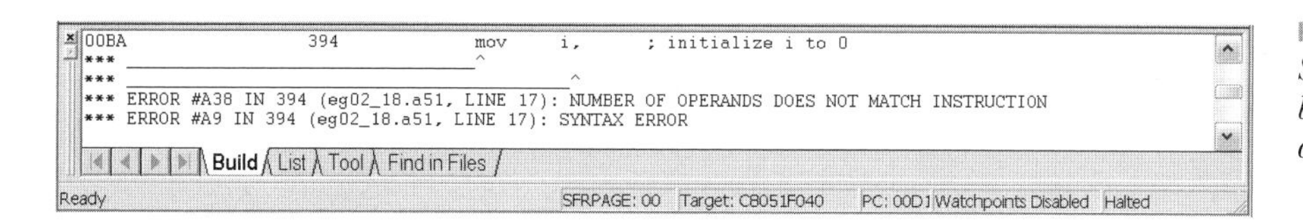

FIGURE 2.37 *Syntax error caused by missing operands.*

```
        $nomod51
        $include    (c8051F040.inc)
N       equ         20
        dseg        at 30H
sum:    ds          2               ; reserve two bytes for sum
i:      ds          1               ; loop count
arrcnt: ds          1               ; loop count
        cseg        at 00H
        ljmp        main
        org         00ABH
main:   mov         SFRPAGE,#0FH    ; switch the SFR page F
        mov         OSCICN,#0C3H    ; enable internal oscillator and set divide factor to 1
        mov         CLKSEL,#0       ; derive system clock from internal oscillator
        mov         DPTR,#array     ; use DPTR as a pointer to the array
        mov         i,0             ; initialize loop index to 0
        mov         sum,0           ; initialize sum to 0
        mov         sum+1,0         ;       "
        mov         arrcnt,N        ; initialize array count to N
sum_lp: mov         A,i
        movc        A,@A+DPTR       ; get the ith element
        add         A,sum+1         ; add the lower byte of sum
        mov         sum+1,A         ; save the lower byte of sum
        mov         A,#0
        addc        A,sum           ; add carry to the upper byte of sum
        mov         sum,A           ; save the upper byte of sum
        inc         i
        djnz        arrcnt,sum_lp
        jmp         $
array:  db          1,2,3,4,5,6,7,8,9,10
        db          11,12,13,14,15,16,17,18,19,20
        end
```

At first look, this program appears to be fine and should work. After assembling the program, we download it onto the demo board, run the program up to the last instruction (jmp $), and get the **Watch Window** output, as shown in Figure 2.38. Apparently, the result is wrong, because the sum should be 210 for the given test data. We first reset the MCU and set a breakpoint at the instruction with the label of **sum_lp** to find out if variables **i**, **sum**, and **arrcnt** are initialized correctly. The **Watch Window** output is shown in Figure 2.39.

We expect variables **i**, **sum**, and **arrcnt** to be initialized to 0, 0, and 20, respectively. But they are not. Maybe the addressing modes used in the following instructions are wrong.

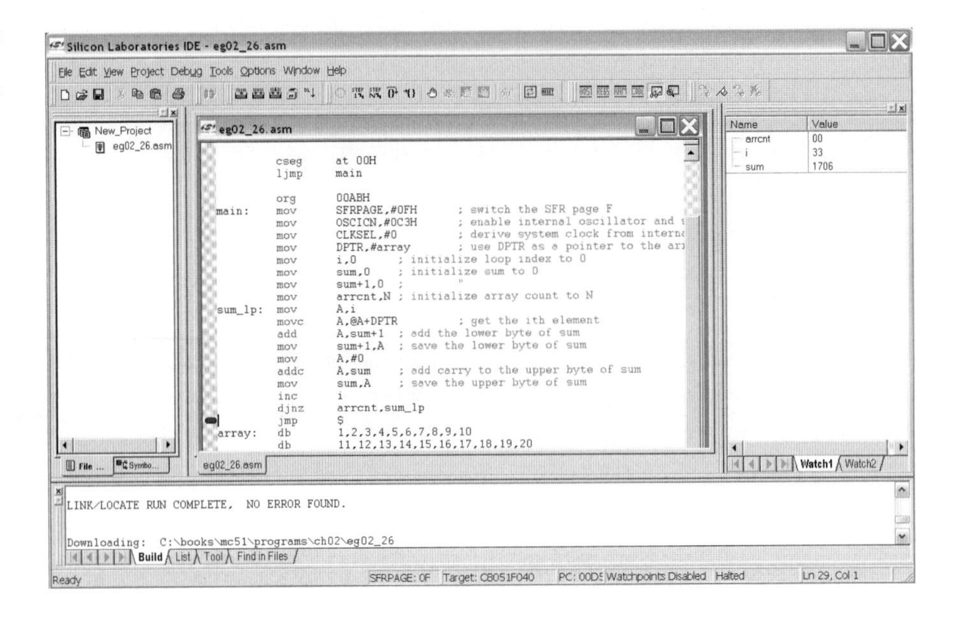

FIGURE 2.38 *The output of a program that adds 1 to 20.*

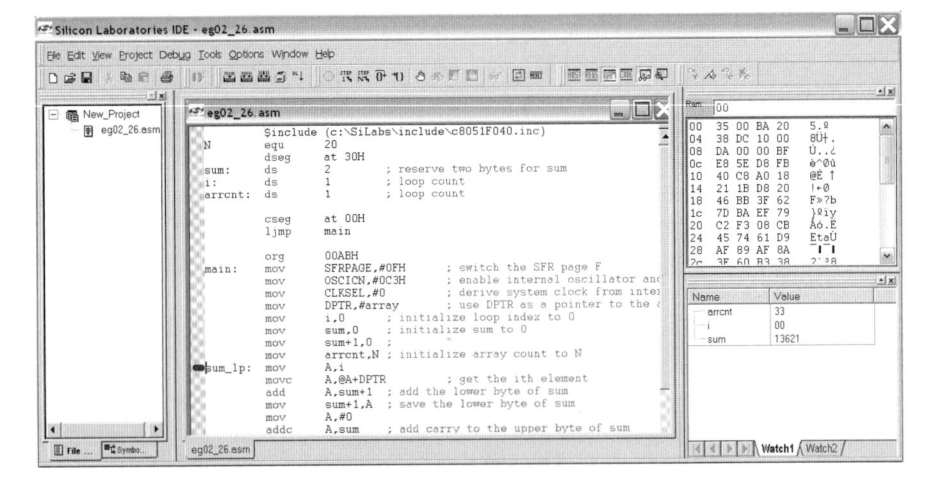

FIGURE 2.39 *Watch Window update when program executes to the* ***lable sum_lp***.

```
mov     i, 0         ; initialize loop index to 0
mov     sum,0        ; initialize sum to 0
mov     sum+1,0      ;            "
mov     arrcnt,0     ; initialize array count to 0
```

Therefore, we changed them to as follows.

```
mov     i, #0        ; initialize loop index to 0
mov     sum,#0       ; initialize sum to 0
mov     sum+1,#0     ;            "
mov     arrcnt,#0    ; initialize array count to 0
```

After rebuilding the project and running the program to the breakpoint, we find that variables **i**, **sum**, and **arrcnt** are initialized correctly. We let the program run to the last instruction (**jmp $**), check the result, and find that the program result is correct. This example demonstrates an error caused by the use of incorrect addressing mode. This is a common error for beginners.

Use of Incorrect Instructions

Using incorrect instructions can cause errors that are difficult to detect and debug. For example, assume that we change the instruction **addc A,sum** to **add A,sum** without changing anything else and use the same data to test the program. We will get the correct result after running the program. However, if we use the following array to test the same program the resultant sum will be incorrect.

```
array:      db      2,4,6,8,10,12,14,16,18,20
            db      22,24,26,28,30,32,34,36,38,40
```

For this array, the sum should be 420. However, the sum becomes 164, because we use the ADD instruction that does not include the carry.

This error is not that easy to detect for the beginner. The debugging process may proceed as follows.

Step 1
Add variables **i**, **sum**, and **arrcnt** to the **Watch Window**.

Step 2
Reset the MCU to set the PC to 0x0000.

Step 3
Press the cursor at the instruction with the label **sum_lp** and select **Run to Cursor**. The values in the **Watch Window** appear to be correct.

Step 4
We next execute the summing loop that begins at the instruction mov A,i and ends at the instruction djnz arrcnt,sum_lp. Again we use the **Run to Cursor** operation to test each summing iteration (by pressing the right mouse button at the instruction djnz arrcnt,sum_lp and select **Run to Cursor**). For the array numbers 2 to 30, the **Watch Window** reports the correct **sum** value. When the iteration reaches the value 32, we notice that *sum* is reported to be 16 instead of 272! This is caused by the missing of a carry from the lower byte of **sum** to the upper byte of **sum**. The exclusion of the carry causes the **sum** to differ by 256. Now we find the error and can fix it.

Mismatch of Operand Size

Since the user may use one byte, two bytes, or any other number of bytes to represent a number, there is a limit to the magnitude of the number that can be represented. When reserving memory bytes to hold a variable, this issue needs to be considered. Many errors are caused by the mismatch of operand size. These types of errors can also be detected by using the approach similar to those described previously.

Incorrect Algorithm

An incorrect algorithm can never result in a correct program. This type of problem cannot be fixed by tracing the program. After program tracing fails to fix the problem, one should reexamine the algorithm to see if it is incorrect.

A few more tips for debugging programs that involve subroutine calls will be discussed in Chapter 4.

2.17 Using Keil's μVision

If you choose to learn the 8051 devices from other vendors, then you cannot use the SiLabs IDE to perform program debugging. However, you can use Keil's μVision along with your favorite demo boards to debug your program. The μVision software supports source-level debugging for many 8051 variants from other vendors. In order to provide source-level debugging capability, the demo board must have external SRAM to hold application programs. Before using the Keil software, a monitor program must be programmed into the on-chip flash memory. This monitor communicates with μVision to provide source-level debugging capability. This type of demo boards will cost more to build for the same set of features. Keil provides an ULINK USB-JTAG adapter to work with the 8051 variants from many other 8051 variants (including STMicroelectronics). A tutorial on using the μVision software is given in Appendix B.

SiLabs provides μVision drivers (a version for μVision 2 and a version for μVision 3) that allow the user to use μVision and the SiLabs debug adapter to perform source-level debugging directly on the demo-board hardware. A simple setup is needed, which will be detailed in Appendix B. This is good news for those who prefer using Keil's development tool.

2.18 Using the Raisonance's RIDE

The Raisonance RIDE is another popular IDE for developing software for 8051 variants. The RIDE is similar to Keil's μVision in many areas. For almost every feature in μVision, you can find an equivalent in RIDE and vice versa. Raisonance provides an RLINK USB-Adapter, which can interface with the 8051 variants that have JTAG debug interface and supports source-level debugging activities. A tutorial on using the RIDE is given in Appendix C.

2.19 Chapter Summary

An assembly language program consists of three types of statements: assembler directives, assembly language instructions, and comments. An assembler directive tells the assembler how to process subsequent assembly language instructions. Directives also provide a way for defining program constants and reserving space for dynamic variables. A statement of an assembly program consists of four fields: label, operation code, operands, and comment.

Although the 8051 can perform only 8-bit arithmetic operations, numbers that are longer than 8 bits can still be added, subtracted, or multiplied by performing multiprecision arithmetic. The multiprecision addition can be implemented by using the **addc** instruction, and multiprecision subtraction can be performed by using the **subb** instruction. To perform multiprecision multiplication, both the multiplicand and the multiplier must be divided into 8-bit chunks. The next step is to generate partial products and align them properly before adding them together. A subroutine that performs multiprecision multiplication is described in Chapter 4. Multiprecision division is performed by repeated shifting and subtraction method and also is detailed in Chapter 4.

Performing repetitive operations is the strength of a microcontroller and microprocessor. In order for a computer to perform repetitive operations, we need to write program loops to tell the computer what instruction sequence to repeat and how many times to repeat. Two components are required in implementing a finite loop: one is the test of loop condition, whereas the second is the change of program flow. The 8051 and all microcontrollers provide instructions for both operations. The 8051 also provides instructions for initializing and updating variables and loop indices.

The 8051 instruction set includes many logical instructions, which may be useful for implementing control and I/O applications.

Rotate instructions are useful for bit field operations. They can also be used to implement multiplying and dividing a variable by a power by 2. All rotate instructions operate on accumulator A only. When required, we can write a sequence of instructions to rotate a number longer than 8 bits.

SiLabs provides a debug adapter and an IDE that can perform source-level debugging directly on the target hardware. The debug adapter requires a 10-pin connector implemented on the target hardware (or demo board) with signal assignment conformed to that in Table 2.10. The SiLabs IDE allows the users to use their favorite cross assembler and C compiler. A tutorial on using the SiLabs IDE is provided.

Keil and Raisonance provide a 2-kB and 4-kB demo version of their development tools for the user to evaluate, respectively. The target boards and development kits from SiLabs also include the demo version of Keil's development tools (4-kB size limit). The demo version of Keil's development tools distributed by SiLabs will be installed under the directory of **c:\silabs\MCU\IDEfile**.

It is important to provide test data for the program being developed. A common method is using assembler directives such as DB to define numbers or strings to be tested in program memory. If the program needs to modify the test data, then you need to copy the test data from program memory to data memory or XRAM.

2.20 Exercise Problems

Unless specified otherwise, the most significant byte of a multiprecision number is stored at the lowest address, whereas the least significant byte is stored at the highest address for all the problems.

E2.1 For the following assembly statements, write down the valid and invalid labels and explain why the invalid labels are not valid.

```
column 1
 ↓
1.  start:     clr    A            ; initialize A to 0
2.  ABC:       mov    R7,#20
3.  he_too:    dec    A
4.  hi+lo:     inc    R7
5.  CNN        djnz   R7,smaller
6.  loop       subb   A,lp_cnt     ; check loop count
```

E2.2 Specify the four fields of an assembly program statements for the following instructions:

```
column 1
 ↓
a. lp_cnt      equ    20
b.             org    100H         ; starting address of the program
c.             mov    B,#40H       ; initialize A to 30H
d. loop:       jnz    greater      ; branch if not zero yet
```

E2.3 Write assembler directives to reserve 20 bytes starting from the program-memory location 0x1200 and initialize them to 0x20.

E2.4 Write assembler directives to construct a table of the ASCII code of uppercase letters starting from the program-memory location at 0x40.

E2.5 Write assembler directives to store the following message in program-memory locations starting from 0x2000: "Welcome to Silab IDE software!"

E2.6 Write a program to add the 16-bit numbers stored at internal data-memory locations 0x34~0x35 and 0x36~0x37, respectively, and save the sum at 0x40~0x41.

E2.7 Write a program to add the 3-byte numbers stored at data-memory locations 0x40~0x42 and 0x44~0x46, respectively, store the sum at 0x4A~0x4C.

E2.8 Write a program to subtract the 16-bit number stored at internal data-memory locations 0x40~0x41 from the 16-bit number stored at 0x42~0x43 and store the difference at 0x50~0x51.

E2.9 Write a program to subtract the 32-bit number stored at internal data-memory locations 0x40~0x43 from the 32-bit number stored at 0x30~0x33 and save the difference at 0x44~0x47.

E2.10 Write a program to subtract the 16-bit number stored at data-memory locations 0x40~0x41 from the sum of two 16-bit numbers stored at 0x30~0x31 and $32~$33, respectively, and store the result at 0x36~0x37. All numbers are stored in internal data-memory space.

E2.11 Write a program to divide the 8-bit number stored at the internal data-memory at 0x36 into the 8-bit number stored at the 0x37 and store the quotient and remainder at 0x40 and 0x41, respectively.

E2.12 Write a program to find the number of elements of an array of 8-bit integers that have bit 3 and bit 1 set to 1. Use the **repeat–until** looping construct and leave the result in A.

E2.13 What would be the values stored in A and B after the execution of the **mul AB** instruction if they originally contain the following values?
(a) 0x58 and 0x37, respectively
(b) 0x29 and 0x49, respectively

E2.14 Write a program that swaps the upper 4 bits with the lower 4 bits of every array element. Provide a test array in the program memory. Your program needs to copy the array to data memory before swapping them.

E2.15 Write a program to find the number of elements in an array of 8-bit integers that are larger than 16. Use the **for-loop** looping construct.

E2.16 What will be the **CY** flag value after the execution of each of the following instructions? Assume that accumulator A contains 0x79 and the **CY** flag is 0 before the execution of each instruction.
(a) add A,#0x29
(b) addc A,#0x98
(c) subb A,#0x39
(d) subb A,#0xA7

E2.17 Write a program to find the average of the largest and the smallest elements of an array of 8-bit integers. Store the result in the memory location at 0x30. The sum of the largest and smallest elements may be larger than 255. Use a **while-loop** looping construct.

E2.18 Determine the number of times that the following loop will be executed.

```
        mov     A,#128
again:  clr     C
        rrc     A
        mov     0x20,A
        add     A,0x20
        clr     C
        rrc     A
        jnz     repeat
```

E2.19 Write a small program to shift the 32-bit number stored at internal dat-memory locations 30H~33H four places to the right. Zeros are shifted into the leftmost 4 bits of the 32-bit number.

E2.20 Write a program to copy an array of 30 8-bit elements in program memory starting from 0x2000 to XRAM starting from 0x200.

E2.21 Write a macro to shift a data-memory location at **loc** (directly addressable) to the right one place logically. A logic shift right will shift in a 0 to the bit 7 of the operand.

E2.22 Write a macro to shift a data-memory location at **loc** (directly addressable) to the right one place arithmetically. An arithmetic shift right operation will duplicate the bit 7 after the shift operation.

E2.23 Write a macro to shift a data-memory location at **loc** (directly addressable) to the left one place. A zero will be shifted into bit 0 after the operation.

E2.24 Write a macro to shift a data-memory location at **loc** (indirectly addressable) to the right one place logically. A zero will be shifted into bit 7 after the operation.

E2.25 Write a macro to shift a data-memory location at **loc** (indirectly addressable) to the right one place arithmetically. An arithmetic shift-right operation will duplicate the bit 7 after the shift operation.

E2.26 Write a macro to shift a data-memory location at **loc** (indirectly addressable) to the left one place. A zero will be shifted into bit 0 after the operation.

2.21 Laboratory Exercise Problems and Assignments

L2.1 The following program copies an array of *N* 8-bit integers from the program memory to the data memory and then swaps the first element with the last element, swaps the second element with the second last element, and so on:

```
                $nomod51
                $include    (c8051F040.inc)
N               equ         30
                dseg        at 0x30
index:          ds          1               ; array index
arcnt:          ds          1               ; array count
array:          ds          30              ; buffer to hold array
                cseg        at 0
                ljmp        main            ; reset vector
                org         0xAB
main:           mov         OSCICN,#0xC3    ; enable internal oscillator and set divide factor to 1
                mov         CLKSEL,#0       ; derive system clock from internal oscillator
```

```
        mov     DPTR,#arr1          ; use DPTR to point to the array
        mov     R0,#array           ; use R0 as a pointer to array buffer
        mov     arcnt,#N            ; array count set to N (20)
        mov     index,#0            ; array index initialized to 0
copylp: mov     A,index
        movc    A,@A+DPTR           ; get one element from program memory
        mov     @R0,A               ; copy to data memory
        inc     R0                  ; move pointer
        inc     index               ; increment loop index
        djnz    arcnt,copylp        ; done with copying of array?
        mov     R0,#array           ; R0 points to the first array element
        mov     R1,#array+N-1       ; R1 points to the last array element
        mov     arcnt,#N/2          ; set loop count to N/2
swaplp: mov     A,@R0               ; swap array[i] with array[N-i-1]
        mov     B,@R1               ;   "
        mov     @R0,B               ;   "
        mov     @R1,A               ;   "
        inc     R0                  ; move to next pair of elements
        dec     R1                  ;   "
        djnz    arcnt,swaplp        ; done with swapping?
        jmp     $
arr1:   db      1,2,3,4,5,6,7,8,9,10
        db      11,12,13,14,15,16,17,18,19,20
        db      21,22,23,24,25,26,27,28,29,30
        end
```

Perform the following operations:

(a) Create a new project and call it **lab1**.

(b) Create a new file (called it **lab1.asm**) and enter the previous program into this file.

(c) Add the program **lab1.asm** to the project **lab1**.

(d) Make the project.

(e) Connect to the C8051F040 kit or SSE040 demo board.

(f) Download the executable code to the target MCU.

(g) Open a Debug Window for registers (to view R0 to R7)

(h) Open a Debug Window for data memory and adjust the window so that the contents of locations from 0x30 to 0x4D are visible in the window.

(i) Open a Debug Window for program memory and adjust the window so that the contents of program memory from 0xD8 to 0xF7 are visible in the window.

(j) Open a Watch Window and insert the variables **index** and **arcnt** into it. What are the initial values for **index** and **arcnt**?

(k) Reset the MCU.

(l) Press the right mouse button at the instruction with the label **copylp** and select **Run to Cursor**. What are the contents of Debug Windows for data memory, program memory, registers, and Watch Window?

(**m**) Press the right mouse button at the instruction with the label **swaplp** and select **Run to Cursor**. What are the contents of Debug Windows for data memory, program memory, registers, and Watch Window?

(**n**) Press the right mouse button at the instruction **jmp $** and select **Run to Cursor**. What are the contents of Debug Windows for data memory, program memory, registers, and Watch Window? Do you see the array in data memory to be the reverse of that in the program memory?

L2.2 Write a program to count the number of elements in an array of N 8-bit integers that are greater than 30 but smaller than 50. The array is stored immediately after the program. Save the result in the data-memory location at 30H. Use the SiLabs IDE to enter and execute the program if you are using the C8051F040 kit or the SSE040 demo board.

L2.3 Write a program to compute the average of an array of 32 8-bit integers. The array is stored immediately after your program. Save the average in the data-memory location at 31H. Use the SiLabs IDE to enter and execute the program if you are using the C8051F040 kit or the SSE040 demo board. The operation of divide-by-32 can be implemented by shifting to the right by five places.

Assembly Programming and Simple I/O Operations

3.1 Objectives

Upon successful completion of this chapter, you will be able to:

- Perform input and output operation using the parallel I/O port
- Assign I/O pins to peripheral functions of the C8051F040
- Configure the C8051F040 I/O port for input and output
- Understand the C8051F040 SFRs paging issue
- Output data to light-emitting diodes and seven-segment displays
- Input data from DIP switches
- Create time delays using program loops
- Create time delays using Timer 0
- Perform digital-to-analog conversion
- Use DAC to generate waveforms

3.2 Introduction to I/O Ports

The term **port** often is used with I/O functions. An I/O port consists of a group of pins (often eight) and a set of registers including data register (or latch), data direction register (optional), and control register (optional). An I/O port is often bi-directional in the sense that it can output and input data at different times. The directions of I/O pins are established by programming the data direction register. The data register (or latch) allows the I/O pins to maintain their voltage levels if the application program does not change them. Some I/O port may have more sophisticated functions and hence require a control register for its configuration.

The I/O port allows the user to interface with I/O devices, such as DIP switches, keypads, keyboards, light-emitting diodes (LEDs), seven-segment displays, liquid crystal displays (LCDs), stepper motors, D/A converters, and so on. When performing I/O operations, the CPU deals directly with the I/O port registers. These I/O registers are most often treated as memory locations and share the address space with data memory. They can be accessed by using the general data-movement instructions just like any other data memory location. For example, the following instruction outputs the contents of accumulator A to port P2.

```
mov        P2,A      ; output contents of A to P2
```

For the 8051, the user will need to output a 1 to an I/O pin before reading from it. Writing a 1 to an I/O pin configures the pin to be an input. The following instruction sequence reads the value of P2 into accumulator A.

```
mov        P2,#0xFF        ; configure P2 for input
mov        A,P2            ; read the value of P2 and place it in A
```

In the early days before the microcontroller was invented, a general-purpose microprocessor (such as the Intel 8085) did not have I/O ports. Instead, it used interface chips (for example, i8255) to interface with I/O devices. These interface chips were treated as memory devices. The microprocessor sent out address signals to access the registers on the interface chip and the interface chip performed the desired I/O operation for the microprocessor. An address decoder was needed to select one and only one device to respond to the data-transfer request from the microprocessor. This configuration is illustrated in Figure 3.1. The interface chip i8255 is still being used

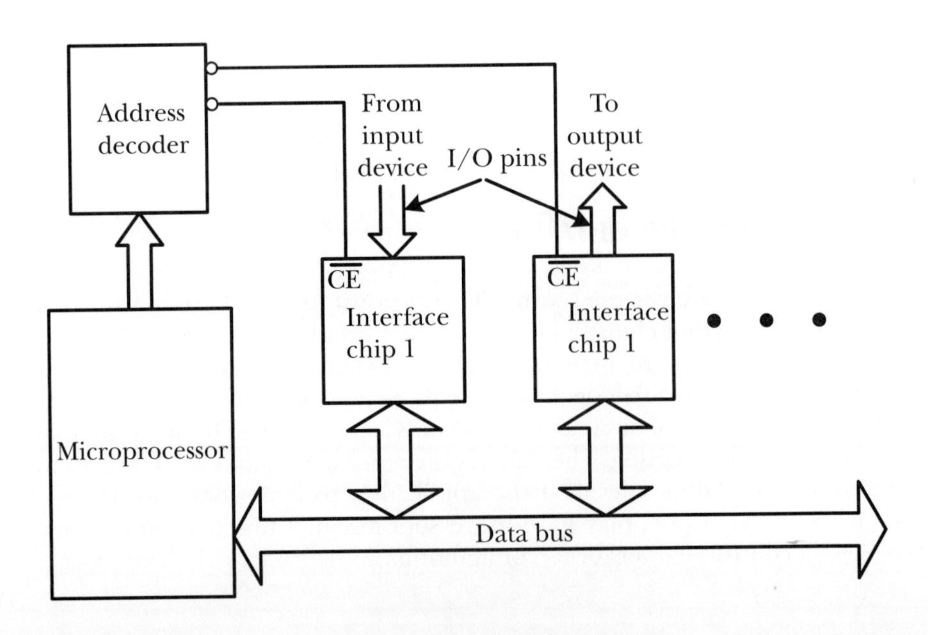

FIGURE 3.1 *Interface chip, I/O devices, and microprocessor.*

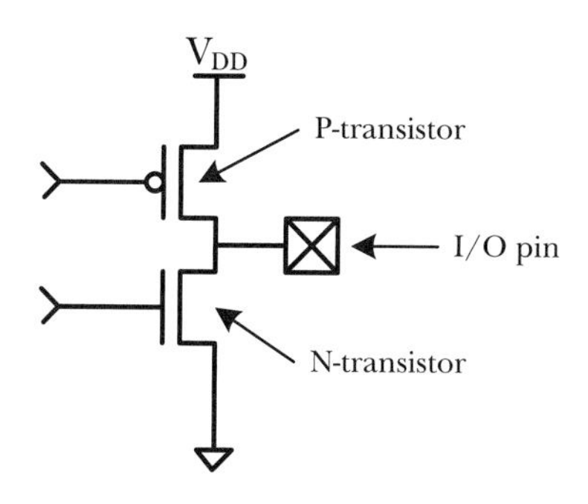

FIGURE 3.2
I/O pin driver circuit.

by some older 8051 variants to expand their I/O ports. However, this approach is not recommended for any new design due to the extra power consumption, glue logic, larger PCB size requirement, and additional cost.

Inside a CMOS microcontroller, the circuit that drives the I/O pin is usually a CMOS inverter. The CMOS inverter circuit is shown in Figure 3.2. When the P-transistor is missing, the resultant driver circuit is called an *open drain.* An open-drain I/O pin cannot pull the voltage to high. There must be an external pullup device (for example, a resistor) to pull it to high. However, multiple open-drain I/O pins can be wired together. When both the P-transistor and N-transistor are present, the driver circuit is called *push-pull* in the sense that it can push the pin voltage to low and also pull the pin voltage to high.

3.3 Original 8051 I/O Port Structures and Operations

The original 8051 has 32 I/O pins that are further divided into four ports: P0, P1, P2, and P3. All four ports are bi-directional. Each port consists of a latch, an output driver, and an input buffer.

The 8051 can access external data memory. When performing this function, P2 and P0 carry the upper 8 bits (in 16-bit address mode) and lower 8 bits of the memory address, respectively. The data to be written and read is time-multiplexed with P0. When only an 8-bit address (carried by P0) is used for external data-memory access, the P2 port pins continue to emit the contents of the P2 register. This allows the external data memory to be paged.

All of the P3 pins and two P1 pins are multifunctional. They are not only port pins, but also serve the functions of various special features, as shown in Tables 3.1 and 3.2. The alternate functions can only be activated if the corresponding bit latch in the port special-function register contains a 1. Otherwise, the port pin is stuck at 0.

There are many 8051 instructions that need to read a value from I/O port, modify it, and then write back to the I/O port. These instructions are called *read-modify-write* instructions (listed in Table 3.3). These instructions access (read and write) the latch rather than the I/O pins. These instructions access

TABLE 3.1 *Original 8051 Port 1 Pin Alternate Functions*

Port Pin	Alternate Functions
P1.0	T2 (external count input to timer/counter 2, clock out)
P1.1	T2EX (timer/counter 2 capture/reload trigger and direction control)
P1.2	NA
P1.3	NA
P1.4	NA
P1.5	NA
P1.6	NA
P1.7	NA

TABLE 3.2 *Original 8051 Port 3 Pin Alternate Functions*

Port pin	Alternate Functions
P3.0	RXD (serial input port)
P3.1	TXD (serial output port)
P3.2	INT0 (external input 0)
P3.3	INT1 (external input 1)
P3.4	T0 (timer 0 external input)
P3.5	T1 (timer 1 external input)
P3.6	WR (external data memory write strobe)
P3.7	RD (external data memory read strobe)

TABLE 3.3 *8051 Read-Modify-Write Instructions*

Instruction	Operation	Example
anl	Logical AND	anl P1,A
orl	Logical OR	orl P2,#0x55
xrl	Logical XOR	xrl P3,#0xFF
jbc	Jump if bit = 1 & clear bit	jbc P1.7,true
cpl	Complement bit	cpl P3.4
inc	Increment	inc P2
dec	Decrement	dec P3
djnz	Decrement & jump if not 0	djnz P2,loop
mov PX.Y,C	Move carry bit to bit Y of port X	mov P3.7,C
clr PX.Y	Clear bit Y of port X	clr P3.0
setb PX.Y	Set bit Y of port X	setb P1.7

the latch rather than the pin in order to avoid a possible misinterpretation of the voltage level at the pin. This situation might occur when a port pin is used to drive the base of a NPN transistor. Whenever the CPU writes a 1 to the pin, the transistor is turned on. If the CPU then reads back the same port pin rather than from the latch, it will receive the base voltage of the transistor and interpret it as a 0. This error can be avoided by reading the latch rather than the pin.

3.4 C8051F040 I/O Ports

The C8051F040 has 64 I/O pins organized as eight 8-bit ports. Like the original 8051, all ports are bit-addressable and byte-addressable. All port pins are 5-V tolerant and all support configurable open-drain and push-pull output modes and weak pull-ups. An I/O pin with open-drain needs an external pull-up device (often a resistor) to pull it high. A block diagram of the port I/O cell is shown in Figure 3.3.

The digital peripherals of the C8051F040 are sharing the use of four lower I/O port pins (Port P0, P1, P2, and P3). The association of a peripheral function and the I/O pins is not fixed. The I/O pin assignment to the peripheral function is achieved by using a *priority crossbar decoder*. The priority crossbar decoder allocates and assigns port pins on P0 through P3 to the digital peripherals (UARTs, SMBus, PCA, etc.) on the device using a priority order listed in Table 3.4, with UART0 having the highest priority and CNVSTR2 having the lowest priority.

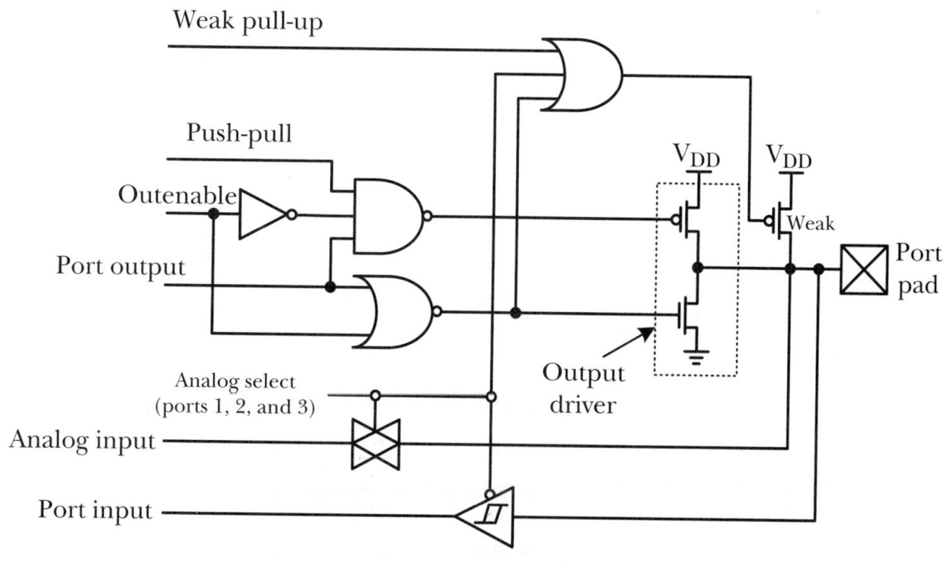

FIGURE 3.3
C8051F040 I/O cell block diagram.

TABLE 3.4 *Priority Crossbar Decode Table*

	P0								P1								P2								P3								Crossbar Register Bits
Pin I/O	0	1	2	3	4	5	6	7	0	1	2	3	4	5	6	7	0	1	2	3	4	5	6	7	0	1	2	3	4	5	6	7	
TX0	●																																UART0EN: XBR0.2
RX0		●																															
SCK	●		●																														SPI0EN: XBR0.1
MISO		●		●																													
MOSI			●		●																												
NSS				●		●			NSS is not assigned to a port pin when the SPI is placed in 3-wire mode																								
SDA	●		●	●	●	●	●																										SMB0EN: XBR0.0
SCL		●		●	●	●	●	●																									
TX1	●		●	●	●	●	●	●	●																								UART1EN: XBR2.2
RX1		●		●	●	●	●	●	●	●																							
CEX0	●		●	●	●	●	●	●	●	●	●																						PCA0ME: XBR0.[5:3]
CEX1		●		●	●	●	●	●	●	●	●	●																					
CEX2			●		●	●	●	●	●	●	●	●	●																				
CEX3				●		●	●	●	●	●	●	●	●	●																			
CEX4					●		●	●	●	●	●	●	●	●	●																		
CEX5						●		●	●	●	●	●	●	●	●	●																	
ECI	●	●	●	●	●	●	●	●	●	●	●	●	●	●	●	●	●																ECI0E: XBR0.6
CP0	●	●	●	●	●	●	●	●	●	●	●	●	●	●	●	●	●	●															CP0E: XBR0.7
CP1	●	●	●	●	●	●	●	●	●	●	●	●	●	●	●	●	●	●	●														CP1E: XBR1.0
CP2	●	●	●	●	●	●	●	●	●	●	●	●	●	●	●	●	●	●	●	●													CP2E: XBR3.3
T0	●	●	●	●	●	●	●	●	●	●	●	●	●	●	●	●	●	●	●	●	●												T0E: XBR1.1
/INT0	●	●	●	●	●	●	●	●	●	●	●	●	●	●	●	●	●	●	●	●	●	●											INT0E: XBR1.2
T1	●	●	●	●	●	●	●	●	●	●	●	●	●	●	●	●	●	●	●	●	●	●	●										T1E: XBR1.3
/INT1	●	●	●	●	●	●	●	●	●	●	●	●	●	●	●	●	●	●	●	●	●	●	●	●									INT1E: XBR1.4
T2	●	●	●	●	●	●	●	●	●	●	●	●	●	●	●	●	●	●	●	●	●	●	●	●	●								T2E: XBR1.5
T2EX	●	●	●	●	●	●	●	●	●	●	●	●	●	●	●	●	●	●	●	●	●	●	●	●	●	●							T2EXE: XBR1.6
T3	●	●	●	●	●	●	●	●	●	●	●	●	●	●	●	●	●	●	●	●	●	●	●	●	●	●	●						T3E: XBR3.0
T3EX	●	●	●	●	●	●	●	●	●	●	●	●	●	●	●	●	●	●	●	●	●	●	●	●	●	●	●	●					T3EXE: XBR3.1
T4	●	●	●	●	●	●	●	●	●	●	●	●	●	●	●	●	●	●	●	●	●	●	●	●	●	●	●	●	●				T4E: XBR2.3
T4EX	●	●	●	●	●	●	●	●	●	●	●	●	●	●	●	●	●	●	●	●	●	●	●	●	●	●	●	●	●	●			T4EXE: XBR2.4
/SYSCLK	●	●	●	●	●	●	●	●	●	●	●	●	●	●	●	●	●	●	●	●	●	●	●	●	●	●	●	●	●	●	●		SYSCKE: XBR1.7
CNVSTR0	●	●	●	●	●	●	●	●	●	●	●	●	●	●	●	●	●	●	●	●	●	●	●	●	●	●	●	●	●	●	●	●	CNVSTE0: XBR2.0
CNVSTR2	●	●	●	●	●	●	●	●	●	●	●	●	●	●	●	◐	○	○	○	○	○	○	○	○	○	○	○	○	○	○	○	○	CNVSTE2:XBR3.2
						ALE	/RD	/WE	AIN1.0/A8	AIN1.1/A9	AIN1.2/A10	AIN1.3/A11	AIN1.4/A12	AIN1.5/A13	AIN1.6/A14	AIN1.7/A15	A8m/A0	A9m/A1	A10m/A2	A11m/A3	A12m/A4	A13m/A5	A14m/A6	A15m/A7	AD0/D0	AD1/D1	AD2/D2	AD3/D3	AD4/D4	AD5/D5	AD6/D6	AD7/D7	

3.4.1 Pin Assignment and Allocation using the Crossbar Decoder

The crossbar makes pin assignment to peripheral functions by following the priority shown in Table 3.4 and the settings of crossbar configuration registers XBR0, XBR1, XBR2, and XBR3. The definitions of these four registers are shown in Figure 3.4, 3.5, 3.6, and 3.7, respectively. There are restrictions to the port pin assignment:

- The crossbar logic assigns port pins from the pin 0 of Port P0 toward pin 7 of Port P3 to the peripheral signals that are enabled according to the priority order from top to bottom of the first column in Table 3.4.
- The signals at the bottom of Table 3.4 have higher priority to be assigned to port pins than those signals in the left column.

7	6	5	4	3	2	1	0	
CP0E	ECI0E	PCA0ME			UART0EN	SPI0EN	SMB0EN	Reset value 0x00
rw	rw	rw			rw	rw	rw	

CP0E: Comparator 0 output enable bit
  0 = CP0 unavailable at port pin
  1 = CP0 available at port pin
ECI0E: PCA0 external counter input enable bit
  0 = PCA0 external counter input unavailable at port pin
  1 = PCA0 external counter input available at port pin
PCA0ME: PCA0 module I/O enable bits
  000 = All PCA0 I/O unavailable at port pins
  001 = CEX0 routed to port pin
  010 = CEX0, CEX1 routed to 2 port pins
  011 = CEX0, CEX1, and CEX2 routed to 3 port pins
  100 = CEX0, CEX1, CEX2, and CEX3 routed to 4 port pins
  101 = CEX0, CEX1, CEX2, CEX3, and CEX4 routed to 5 port pins
  110 = CEX0, CEX1, CEX2, CEX3, CEX4, and CEX5 routed to 6 port pins
UART0EN: UART0 I/O enable bit
  0 = UART0 I/O unavailable at port pins
  1 = UART0 TX routed to P0.0, and RX routed to P0.1
SPI0EN: SPI0 bus I/O enable bit
  0 = SPI0 I/O unavailable at port pin
  1 = SPI0 SCK, MISO, MOSI, and NSS routed to 4 port pins
SMB0EN: SMBus0 bus I/O enable bit
  0 = SMBus0 I/O unavailable at port pins
  1 = SMBus0 SDA and SCL routed to 2 port pins

Note. rw indicates a bit is read/writable, r is read only

FIGURE 3.4
Port I/O crossbar register 0 (XBR0).

FIGURE 3.5
Port I/O crossbar register 1 (XBR1).

7	6	5	4	3	2	1	0	
SYSCKE	T2EXE	T2E	INT1E	T1E	INT0E	T0E	CP1E	Value after reset: 00h
rw	rw	rw	rw	rw	rw	rw	rw	

SYSCKE: /SYSCLK output enable bit
 0 = /SYSCLK unavailable at port pin
 1 = /SYSCLK available at port pin
T2EXE: T2EX input enable bit
 0 = T2EX unavailable at port pin
 1 = T2EX routed to port pin
T2E: T2 input enable bit
 0 = T2 unavailable at port pins
 1 = T2 routed to port pin
INT1E: /INT1 input enable bit
 0 = /INT1 unavailable at port pin
 1 = /INT1 available at port pin
T1E: T1 input enable bit
 0 = T1 unavailable at port pin
 1 = T1 routed to port pin
INT0E: /INT0 input enable bit
 0 = /INT0 unavailable at port pins
 1 = /INT0 routed to port pin
T0E: T0 input enable bit
 0 = T0 unavailable at port pin
 1 = T0 routed to port pin
CP1E: CP1 output enable bit
 0 = CP1 unavailable at port pin
 1 = CP1 routed to port pin

- Each of the signals at the bottom of Table 3.4 can be assigned only to one port pin. For example, the signals ALE, /RD, and /WE can only be assigned to pins 5, 6, and 7 of Port P0, respectively.

As shown in Table 3.4, the pin assignment is made in the order from Port P0 pin 0 toward Port P3 pin 7. For example, if the UART0 module is enabled, the TX0 and RX0 will be assigned to pin P0.0 and P0.1, respectively.

When a port pin is used as a function specified at the bottom of the table, then it is not available for the alternate function at the left of the table. For example, if Port 1 pins are used as A/D analog inputs AIN1.0~AIN1.7, then they are not available for alternate functions.

Example 3.1 Suppose that UART0 is enabled and the SPI0 is enabled in three-wire mode. What pins will be assigned to SDA and SCL?

Solution: According to Table 3.4, pins P0.5 and P0.6 will be assigned to SDA and SCL, respectively.

7	6	5	4	3	2	1	0	
WEAKPUD	XBARE	- -	T4EXE	T4E	UART1E	EMIFLE	CNVSTOE	Value after reset: 00h
rw	rw	rw	rw	rw	rw	rw	rw	

WEAKPUD: Weak pull-up disable bit
  0 = Weak pull-ups globally enabled
  1 = Weak pull-ups globally disabled
XBARE: Crossbar enable bit
  0 = Crossbar disabled. All pins on Ports 0, 1, 2, and 3 are forced to input mode
  1 = Crossbar enabled
T4EXE: T4EX input enable bit
  0 = T4EX unavailable at port pins
  1 = T4EX routed to port pin
T4E: T4 input enable bit
  0 = T4 unavailable at port pin
  1 = T4 routed port pin
UART1E: UART1 I/O enable bit
  0 = UART1 I/O unavailable at port pin
  1 = UART1 TX and RX routed to 2 port pins
EMIFLE: External memory interface low-port enable bit
  0 = P0.7, P0.6, and P0.5 functions are determined by the crossbar or the port latches
  1 = If EMI0CF.4 = 0 (external memory multiplexed mode), P0.7 (/WR), P0.6 (/RD), and P0.5 (ALE) are skipped by the crossbar and their outputs are determined by the port latches and the external memory interface.
  1 = If EMI0CF.4 = 1 (external memory in nonmultiplexed mode), P0.7 (/WR) and P0.6 (/RD) are skipped by the crossbar and their output states are determined by the port latches and the external memory interface.
CNVSTOE: ADC0 external convert start input enable bit
  0 = CNVST0 for ADC0 unavailable at port pin
  1 = CNVST0 for ADC0 routed to port pin

FIGURE 3.6
Port I/O crossbar register 2 (XBR2).

Example 3.2 Suppose that UART0 is disabled and SPI0 is enabled in four-wire mode. What pins will be assigned to SDA and SCL?

Solution: According to Table 3.4, pins P0.4 and P0.5 will be assigned to SDA and SCL, respectively.

Example 3.3 Suppose that the UART0 is enabled, the SPI0 is enabled in three-wire mode, SMBus is enabled, UART1 is disabled, and CEX0 and CEX5 are enabled. What pins are assigned to UART0, SPI0, SMBus, and CEX0 and CEX5?

7	6	5	4	3	2	1	0	
CTXOUT	--	--	--	CP2E	CNVST2E	T3EXE	T3E	Value after reset: 00h
rw	rw	rw	rw	rw	rw	rw	rw	

CTXOUT: CAN transmit pin (CTX) output mode
0 = CTX pin output mode is configured as open-drain mode
1 = CTX pin output mode is configured as push-pull mode
CP2E: CP2 output enable bit
0 = CP2 unavailable at port pin
1 = CP2 routed to port pin
CNVST2E: ADC2 external convert start input enable bit
0 = CNVST2 for ADC2 unavailable at port pins
1 = CNVST2 for ADC2 routed to port pin
T3EXE: T3EX input enable bit
0 = T3EX unavailable at port pin
1 = T3EX routed to port pin
T3E: T3 input enable bit
0 = T3 unavailable at port pin
1 = T3 routed to port pin

FIGURE 3.7
Port I/O crossbar register 3 (XBR3).

Solution: The crossbar makes pin assignments from P0.0 and upward without leaving any unassigned pins in the middle. According to the register XBR0, if CEX5 is to be routed to I/O pin, then CEX0 through CEX4 must all be routed to I/O pins. The pin assignment is as follows.

TX0: assigned to the pin P0.0
RX0: assigned to the pin P0.1
SCK: assigned to the pin P0.2
MISO: assigned to the pin P0.3
MOSI: assigned to the pin P0.4
SDA: assigned to the pin P0.5
SCL: assigned to the pin P0.6
CEX0: assigned to the pin P0.7
CEX5: assigned to the pin P1.4

Example 3.4 Write an instruction sequence to assign peripherals UART0, SPI0 (three-wire mode), SMBus, CEX0...CEX5, CP0, CP1, /INT0, /INT1, T2, T2EX, /SYSCLK, and CTX to use I/O pins and disallow other peripherals to use I/O port pins.

Solution: To achieve the specified assignment, we need to program the crossbar register from 0 to 3 using the following instruction sequence.

```
$nomod51
$include    (C8051F040.INC)
mov         SFRPAGE,#0      ; switch to SFR page 0
mov         SPI0CN,#01H     ; enable 3-wire SPI
mov         SFRPAGE,#0FH    ; switch to SFR page F
mov         XBR0,#0B7H      ; enable CP0, CEX0...CEX5, SPI0, UART0, and SMBus
mov         XBR1,#0F5H      ; enable /SYSCLK, T2, T2EX, /INT1, /INT0 CP1
mov         XBR2,#040H      ; enable weak pull-up, enable crossbar
mov         XBR3,#080H      ; set CTX pin output to push-pull, disable T3 and ADC2
```

3.4.2 Configuring the Output Modes of the Port Pins

The output drivers on P0 through P3 remain disabled until the crossbar is enabled by setting the XBARE bit (XBR2.4) to 1. The output mode of each port pin can be configured to be either open-drain or push-pull. In the push-pull configuration, writing a 0 to the associated bit in the port data register will cause the port pin to be driven to ground, and writing a 1 will cause the port pin to be driven to V_{DD}. In the open-drain configuration, writing a 0 to the associated bit in the port data register will cause the port pin to be driven to ground, and a 1 will cause the port pin to assume a high-impedance state. The open-drain configuration is useful to prevent contention between devices in systems where the port pin participates in a shared interconnection in which multiple output devices are connected to the same physical wire.

The output modes of the port pins on Ports 0 through 3 are determined by the bits in the associated **PnMDOUT** ($n = 0, \ldots, 3$). For example, a logic 1 in P3MDOUT.7 configures the output mode of pin 7 of P3 to push-pull; a logic 0 in P3MDOUT.7 configures the output mode of the same pin to open-drain. All port pins default to open-drain output. Ports 4 through 7 also have the output-mode options, and their output modes are also configured through their associated output-mode registers. The contents of these eight registers are shown in Figure 3.8.

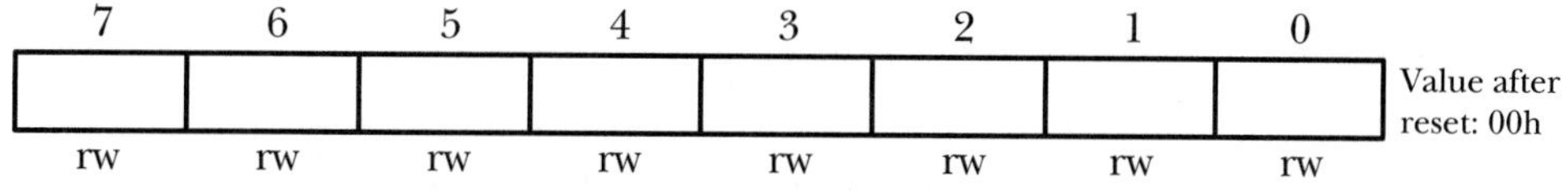

Bit 7,...,0: Port k output mode bits

0 = Port pin mode is configured as open-drain
1 = Port pin mode is configured as push-pull

FIGURE 3.8
Port 0 through 7 output mode registers (P0MDOUT, P1MDOUT, P2MDOUT, P3MDOUT, P4MDOUT, P5MDOUT, P6MDOUT, P7MDOUT).

3.4.3 Port Data Registers

After performing the appropriate configuration, we can use I/O ports to input and output data. Data Input and output from/to the I/O ports are exchanged via the port data registers P0, . . . , P7. For example, the following instruction outputs the value 0x33 to P3.

```
mov        P3,#033H
```

The following instruction sequence reads a byte from Port 3 to accumulator A.

```
mov        P3,#0FFH
mov        A,P3
```

3.4.4 Configuring Port Pins as Digital Inputs

A port pin is configured as a digital input by setting its output-mode bit of the PnMDOUT register to 0(**open-drain**) and writing a 1 to the associated bit in the port data register. For example, the pin 7 of Port 3 (indicated as P3.7) is configured as a digital input by setting the bit P3MDOUT.7 to a logic 0, which selects open-drain output mode and P3.7 to a 1, which disables the low-side output driver (in Figure 3.4). If the port pin has been assigned to a digital peripheral by the crossbar and that pin functions as an input (for example, RX0—the UART0 receive pin), then the output drivers on that pin are disabled automatically.

Example 3.5 Write an instruction sequence to configure P3 for input and read a byte from P3 into accumulator A.

Solution: The following instruction sequence will perform the specified operation.

```
mov        P3MDOUT,#0        ; configure P3 to open-drain mode
mov        P3,#0FFH          ; configure P3 for input
mov        A,P3              ; read the port P3 pin values into A
```

3.4.5 Weak Pull-ups

By default, each port pin has an internal weak pull-up device enabled which provides a resistive connection (about 100 kΩ) between the pin and V_{DD}. The weak pull-up device can be disabled globally by writing a 1 to the weak pull-up disable bit (bit 7 of XBR2). The weak pull-up is deactivated automatically on any pin that is driving a 0; that is, an output pin will not contend with its own pull-up device. The weak pull-up device also can be disabled explicitly on Ports P1, P2, and P3 pins by configuring the pin as an analog input, as described in the next section.

3.4.6 Configuring Port 1, 2, and 3 Pins as Analog Inputs

The pins on Port 1 can serve as analog inputs to the analog-to-digital converter 2 (ADC2) analog multiplexer, the pins on P2 can serve as analog inputs to the comparators, and the pins on P3 can serve as analog inputs to ADC0. A port pin is configured as an analog input by writing a 0 to the associated bit in the **PnMDIN** (n = 1, 2, or 3) registers. The contents of the **PnMDIN** registers are shown in Figure 3.9. All port pins default to the digital input mode. Configuring a port pin as an analog input causes the digital input path from the pin and the weak pull-up devices to be disabled. It also causes the crossbar to skip over the pin when allocating port pins to digital peripherals.

The output drivers on a pin configured as an analog input are not disabled explicitly. Therefore, the associated **PnMDOUT** bits configured as analog inputs should be set to 0 (open-drain output mode) explicitly, and the associated port data bits should be set to 1 (high-impedance). It is not required to configure a port pin as an analog input in order to use it as an input to the ADC or comparators; however, it is strongly recommended.

3.4.7 Port 4 through Port 7

All port pins on Ports 4 through 7 can be accessed as general-purpose I/O pins by reading and writing the associated port data registers (P4, . . . , P7). These four registers are also bit-addressable.

A read will always return the logic state present at the pin itself. An exception to this will occur during the execution of read-modify-write instruction (listed in Table 3.3) During the read cycle of the read-modify-write cycle instruction, it is the contents of the port data register—not the state of the port pins themselves—which are read.

Configuring the Output Modes of the Port Pins

Like the pins in P0 through P3, each pin of P4 through P7 has two output modes: open-drain and push-pull. The output mode of a pin is configured by programming the associated output-mode registers (PnMDOUT, n = 4, . . . , 7). The contents of these four registers are shown in Figure 3.9.

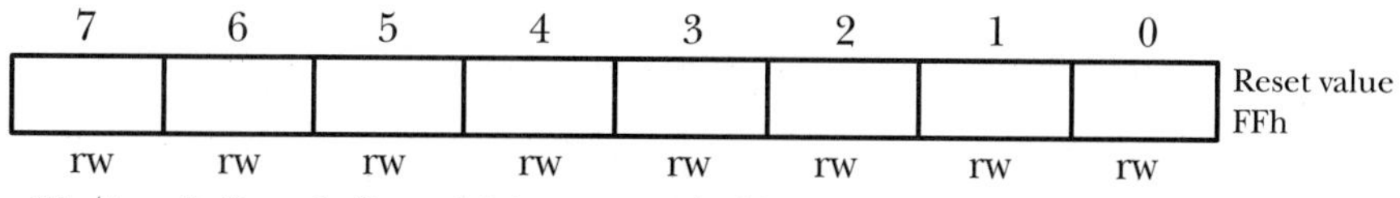

Bit 7,...,0: Port 1, 2, and 3 input mode bits

0 = Port pin is configured in analog input mode. The digital input path is disabled. The weak pull-up on the pin is disabled.

1 = Port pin mode is configured in digital input mode. A read from the port bit returns the logic level at the pin. The state of the weak pull-up is determined by the WEAKPUD bit (XBR2.7)

FIGURE 3.9 *Port 1, 2, and 3 input mode registers (P1MDIN, P2MDIN, P3MDIN).*

Configuring Port Pins as Digital Inputs

A port pin is configured as a digital input by setting its output mode bit of the PnMDOUT register to 0(**open-drain**) and writing a logic 1 to the associated port data registers. For example, pin 7 of P7 (P7.7) is configured as a digital input by setting P7MDOUT.7 to logic 0 (which selects the open-drain output mode) and setting P7.7 to a logic 1.

Weak Pull-ups

Like Ports 0 through 3, pins of Ports P4 through P7 have internal pull-up devices. By default, the pull-up of a pin is enabled after reset, which provides a resistive connection (100 kΩ) between the pins and V_{DD}. These pull-ups can be disabled globally by writing a 1 to the WEAKPUD bit of the XBR2 register.

3.5 C8051F040 Special-Function Registers Paging

The original 8051 use the data-memory locations from 0x80 to 0xFF to hold special-function registers (SFRs). This space allows no more than 128 SFRs in the MCU. However, the C8051F040 and many other devices from SiLabs have more than 128 SFRs. These SFRs are divided into pages with each page accommodating 128 SFRs. At any moment, only one SFR page is active, and this page is specified by the 8-bit SFRPAGE register. This register allows 256 SFR pages to be implemented. The C8051F040 uses SFR page 0, 1, 2, 3, and F. To access a different SFR page, you will need to write the corresponding page number into the SFRPAGE register. For example, you need to execute the following instruction before accessing the P5 data register:

```
mov     SFRPAGE,#0FH          ; switch to SFR page F
```

All SFRs and the SFR pages in which they reside are listed in Appendix D. All of the SFRs of the original 8051 are located in page 0. Some SFRs are made visible in all pages for convenience. Both Keil and Raisonance provide SFR page definitions that allow the user to use mnemonic names to refer to the SFR page in which a peripheral module is located. These definitions are listed in Appendix E.

3.5.1 Interrupts and SFR Paging

When an interrupt occurs, the SFRPAGE register will automatically switch to the SFR page containing the flag bit that caused the interrupt. The automatic SFR page switch function removes the burden of switching SFR pages from the interrupt service routine. Upon execution of the RETI instruction, the SFR page is automatically restored to the SFR page in use prior to the interrupt. This is accomplished via a three-byte SFR page stack. The top byte of the stack is SFRPAGE, the current SFR page. The second byte of the SFR page

7	6	5	4	3	2	1	0	
--	--	--	--	--	--	--	SFRPGEN	Reset value 01h
							rw	

SFRPGEN: SFR automatic page control enable
0 = SFR automatic paging disabled
1 = SFR automatic paging enabled. Upon an interrupt, the MCU will switch the SFR page to the page that contains the SFR's for the peripheral or function that is the source of the interrupt.

FIGURE 3.10
SFR page control register (SFRPGCN).

stack is SFRNEXT. The third or the bottom byte of the SFR page stack is SFRLAST. On interrupt, the current SFRPAGE value is pushed to the SFRNEXT byte, and the value of SFRNEXT is pushed to SFRLAST. Hardware then loads SFRPAGE with the SFR page containing the flag bit associated with the interrupt. On a return from interrupt, the SFR page stack is popped, causing the value of SFRNEXT returning to the SFRPAGE register, thereby restoring the SFR page context without the software intervention. The value in SFRLAST (0x00 if there is no SFR page value in the bottom of the stack) of the stack is placed in the SFRNEXT register.

Automatic hardware switching of the SFR page on interrupts may be enabled or disabled as desired using the SFR automatic page-control enable bit of the SFRPGCN register. This function is enabled by default after reset. Interrupts and resets will be discussed in Chapter 6. The contents of the SFRPGCN register are shown in Figure 3.10.

3.6 Simple Output Devices

A microcontroller is often connected to several different kinds of output devices. These output devices include light-emitting diodes (LED), seven-segment displays constructed from LEDs, liquid-crystal displays (LCDs), motors, digital-to-analog converters, and vacuum tube devices, such as fluorescent displays, printers, and so on.

3.6.1 Interfacing with LEDs

The LED is often used to indicate the system operation mode, whether power is on, whether system operation is normal, and so forth. An LED can illuminate when it is forward biased and has sufficient current flowing through it. The current required to light an LED may range from a few to more than 10 mA. The forward voltage drop across the LED can range from about 1.6 V to more than 2.2 V. The larger the current flows through the LED, the higher the voltage drop across the LED.

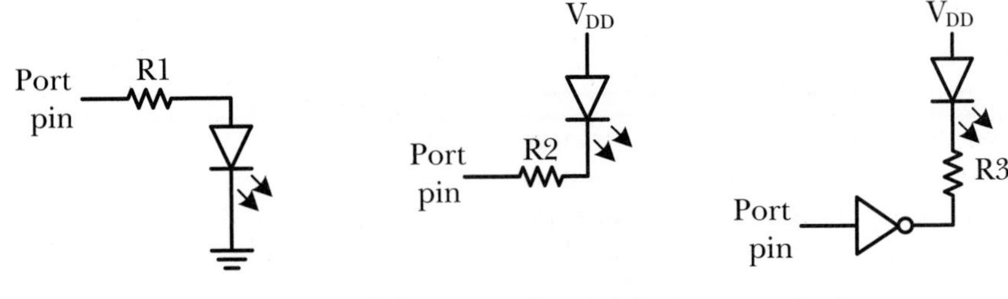

FIGURE 3.11 *Circuit for driving an LED.*

Figure 3.11 suggests three methods for driving LEDs. Methods (a) and (b) are recommended for use with LEDs that need only a few mA of current to produce enough brightness. A resistor is required in the LED circuit to limit the current drawn from the MCU. This resistor is often referred to as a **current-limiting** resistor. The circuit in part (c) is recommended for use with LEDs that need larger current to light. The current-limiting resistance for the circuits in parts (a) and (b) should be larger (between 0.5 kΩ to 1.5 kΩ for V_{DD} = 3.3V to 5V). The circuit in Figure 3.11a is being used in the SSE040 demo board.

Example 3.6 Use the C8051F040 Port P5 to drive the LED circuit and output the value 0x39 to P5. Show the circuit connection and give an instruction sequence to output the value 0x39 to P5.

Solution: The circuit for driving eight LEDs is shown in Figure 3.12.

The program that outputs the value 0x39 to P5 is as follows.

```
        $include   (c8051F040.inc)
        org        0x00
        ljmp       start
        org        0xAB
start:  mov        SFRPAGE,#CONFIG_PAGE   ; switch to configuration page
        mov        WDTCN,#0xDE            ; disable watchdog timer
        mov        WDTCN,#0xAD            ;    "
        mov        P5MDOUT,#0xFF          ; configure P5 for output
        mov        P5,#0x39               ; output the value 0x39 to P5
        sjmp       $
        end
```

3.6.2 Interfacing with Seven-Segment Displays

Seven-segment displays are often used when the embedded product needs to display only a few digits. Seven-segment displays are mainly used to display decimal digits and a small subset of letters.

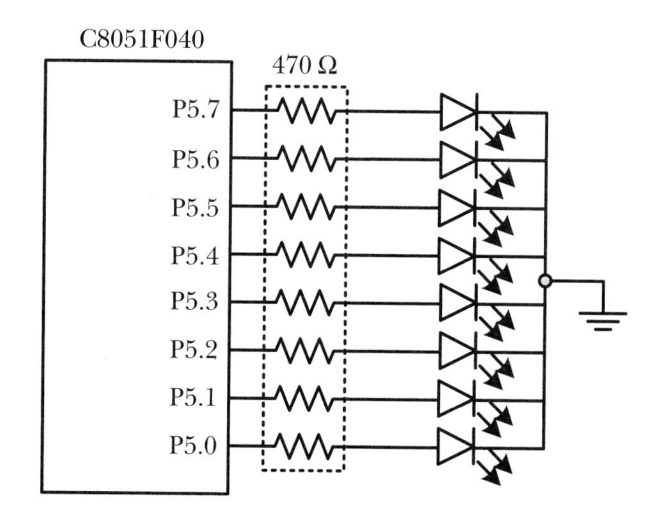

FIGURE 3.12
Circuit connection for Example 3.6.

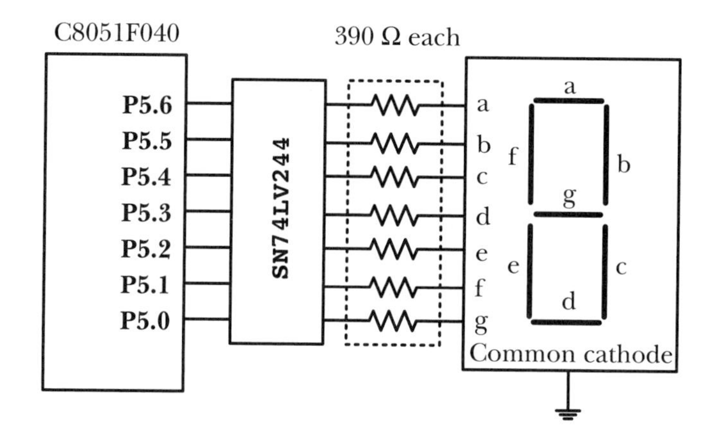

FIGURE 3.13
Driving a single seven-segment display.

Although an 8051 device has enough current to drive a seven-segment display, it is not advisable to do so when an 8051-based embedded product needs to drive many other I/O devices. In Figure 3.13, the P5 port of the C8051F040 drives a common-cathode seven-segment display through the buffer chip SN74LV244. The V_{OH} (output high voltage) value of the SN74LV244 is about 3.1 V at a small load (~ 2 mA) with a 3.3-V power supply. Adding a 390-Ω resistor will set the display segment current to about 3.3 mA (assume that the voltage drop across the LED is about 1.8 V), which should be sufficient to light an LED segment. The light patterns corresponding to 10 decimal digits are shown in Table 3.5. The numbers in Table 3.5 require that segments a ~ g be connected from the second most significant pin to the least significant pin of the output port.

When an application needs to display multiple decimal digits, the time-multiplexing technique is often used. An example of the circuit that displays six decimal digits is shown in Figure 3.14, where both the segment patterns

TABLE 3.5 *Decimal to Seven-Segment Decoder*

Decimal Digit	Segments a	b	c	d	e	f	g	Corresponding Hex Number
0	1	1	1	1	1	1	0	0x7E
1	0	1	1	0	0	0	0	0x30
2	1	1	0	1	1	0	1	0x6D
3	1	1	1	1	0	0	1	0x79
4	0	1	1	0	0	1	1	0x33
5	1	0	1	1	0	1	1	0x5B
6	1	0	1	1	1	1	1	0x5F
7	1	1	1	0	0	0	0	0x70
8	1	1	1	1	1	1	1	0x7F
9	1	1	1	1	0	1	1	0x7B

Note:
Segments **a** through **g** are connected to most significant through least significant I/O pins

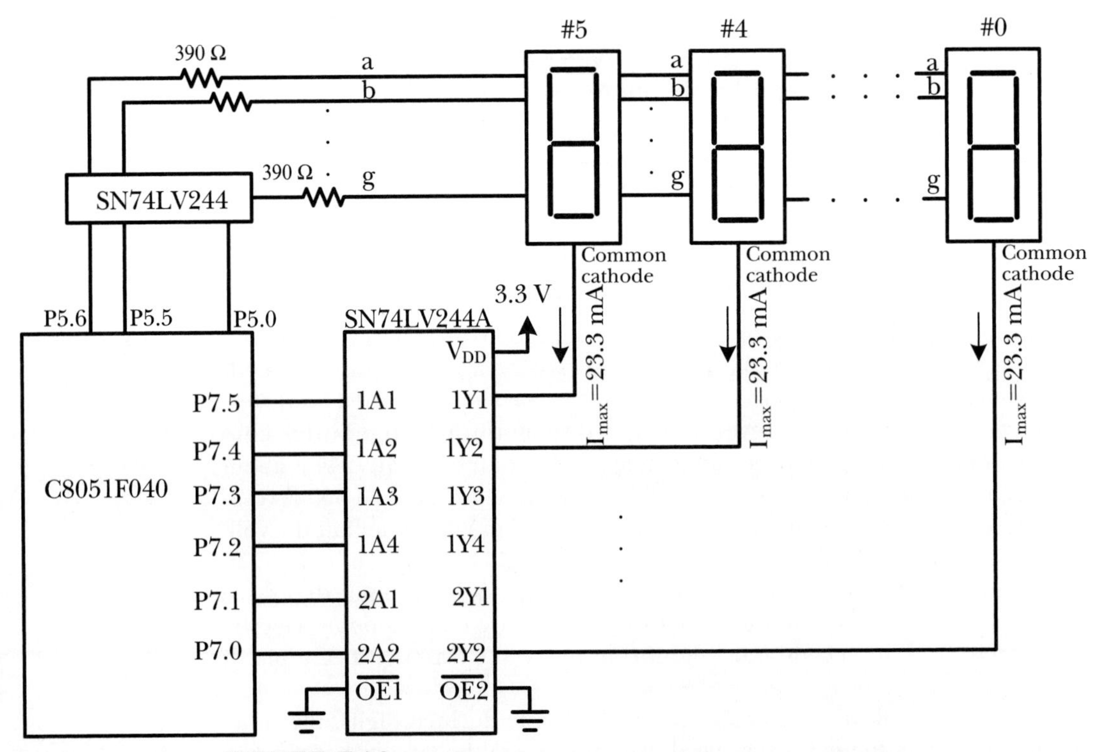

FIGURE 3.14
Port 5 and Port 6 together drive six seven-segment displays (C8051F040).

and digit-select signals are driven by the SN74LV244 buffer chip. The SN74LV244 can source and sink up to 35 mA continuously. The maximum current that needs to be sourced and sunk by the SN74LV244A is 23.3 mA and hence is within its capacity. To turn on a display, we need to drive its common cathode pin to low.

Example 3.7 Write a sequence of instructions to display **4** on the seven-segment display **#4** in Figure 3.14.

Solution: To display **4** on the display **#4**, we need to

1. Output the hex value 0x33 to P5
2. Clear the P7.4 pin to 0
3. Set pins P7.5 and P7.3 through P7.0 to 1

The instruction sequence is

```
mov     WDTCN,#0DEH         ; disable watchdog timer
mov     WDTCN,#0ADH         ;     "
mov     SFRPAGE,#0FH        ; switch to page F of SFR space
mov     P5MDOUT,#0FFH       ; enable port 5 push-pull
mov     P6MDOUT,#0FFH       ; enable port 6 push-pull
mov     P5,#0x33            ; output the pattern of 4
mov     P7,#0xEF            ; turn on digit 4
```

The circuit in Figure 3.14 can display six digits simultaneously by using the time-multiplexing technique, in which each seven-segment display is lighted in turn briefly and then turned off. When one display is lighted, all other displays are turned off. Within one second, each seven-segment display is lighted and then turned off many times. Because of the *persistence of vision*, the six displays will appear to be lighted simultaneously.

To implement the time-multiplexing technique, we need a way to create time delays. An easy way to create a time delay is to use program loops. We can choose an instruction sequence of which the execution time can be multiplied conveniently. Assume that the demo board uses a 24-MHz external crystal to create its system clock, then the following instruction would take 1 μs to execute.

```
inlp:   push    ACC     ; 2 machine cycles
        pop     ACC     ; 2 machine cycles
        push    ACC
        pop     ACC
        push    ACC
        pop     ACC
        push    ACC
        pop     ACC
        push    ACC
```

```
            pop     ACC
            nop                 ; one machine cycle
            djnz    R0,inlp     ; three machine cycles when jump is taken
```

By repeating the previous instruction sequence 250 times, a delay of 250 μs can be created as follows.

```
            mov     R0,#250     ; fSYS = 24 MHz
inlp:       push    ACC         ; 2 machine cycles
            pop     ACC         ; 2 machine cycles
            push    ACC
            pop     ACC
            push    ACC
            pop     ACC
            push    ACC
            pop     ACC
            push    ACC
            pop     ACC
            nop                 ; one machine cycle
            djnz    R0,inlp     ; three machine cycles when jump is taken
```

A longer delay can be created by repeating the above instruction sequence. For example, a 50-ms delay can be created as follows.

```
            mov     R1,#200
exlp:       mov     R0,#250     ; fSYS = 24 MHz
inlp:       push    ACC         ; 2 machine cycles
            pop     ACC         ; 2 machine cycles
            push    ACC
            pop     ACC
            push    ACC
            pop     ACC
            push    ACC
            pop     ACC
            push    ACC
            pop     ACC
            nop                 ; one machine cycle
            djnz    R0,inlp     ; three machine cycles when jump is taken
            djnz    R1,exlp
```

Example 3.8 Write a program to display 123456 on the six seven-segment displays shown in Figure 3.14.

Solution: The digits 1, 2, 3, 4, 5, and 6 are displayed on display #5, #4, . . . , and #0, respectively. The values to be output to P5 and P6 to display one digit at a time are shown in Table 3.6. This table can be created by the following assembler directives.

TABLE 3.6 *Table of Display Patterns for Example 3.8*

Seven-Segment Display	Displayed BCD Digit	Port 5	Port 6
#5	1	30H	DFH
#4	2	6DH	EFH
#3	3	79H	F7H
#2	4	33H	FBH
#1	5	5BH	FDH
#0	6	5FH	FEH

```
disptab:   db    0x30,0xDF    ; display 1 on seven-segment display #5
           db    0x6D,0xEF    ; display 2 on seven-segment display #4
           db    0x79,0xF7    ; display 3 on seven-segment display #3
           db    0x33,0xFB    ; display 4 on seven-segment display #2
           db    0x5B,0xFD    ; display 5 on seven-segment display #1
           db    0x5F,0xFE    ; display 6 on seven-segment display #0
```

The program logic of this example is shown in Figure 3.15.

The program that implements this logic flow is as follows.

```
           $nomod51
           $include   (c8051F040.inc)
cnt        set        R6
           org        00H
           ljmp       start
           org        0ABH
start:     mov        SFRPAGE,#0FH       ; switch to page F of SFR space
           mov        WDTCN,#0DEH        ; disable watchdog timer
           mov        WDTCN,#0ADH        ;    "
;*******************************************************************************************
; prepare to use external oscillator as SYSCLK
;*******************************************************************************************
           mov        CLKSEL,#0          ; used internal oscillator to generate SYSCLK
           mov        OSCXCN,#067H       ; configure external oscillator control
           mov        R7,#255
           djnz       R7,$               ; wait for about 1 ms
chkstable: mov        A,OSCXCN           ; wait until external crystal oscillator is stable
           anl        A,#80H             ; before using it
           jz         chkstable          ;    "
           mov        CLKSEL,#1          ; switch to use external crystal oscillator
           mov        P5MDOUT,#0xFF      ; enable port 5 push-pull
           mov        P7MDOUT,#0xFF      ; enable port 7 push-pull
           mov        DPTR,#disptab      ; use DPTR as the pointer to the LED table
forever:   mov        cnt,#0
```

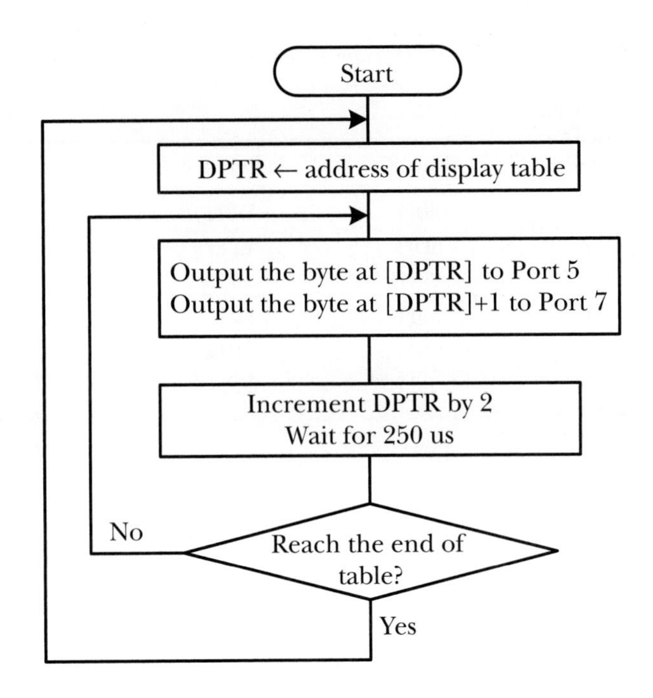

FIGURE 3.15 *Time-multiplexed seven-segment display algorithm.*

```
loop:       mov     A,cnt
            movc    A,@A+DPTR       ; output the digit pattern and increment the counter
            mov     P5,A            ;       "
            inc     cnt             ;       "
            mov     A,cnt
            movc    A,@A+DPTR       ; output the digit select value and increment the counter
            mov     P7,A            ;       "
            inc     cnt             ;       "
;**************************************************************************************************
; The following instruction sequence creates a delay of 250 us at 24 MHz SYSCLK.
;**************************************************************************************************
dly250us:   mov     R0,#250
inlp:       push    ACC             ; 2 machine cycles
            pop     ACC             ; 2 machine cycles
            push    ACC
            pop     ACC
            push    ACC
            pop     ACC
            push    ACC
            pop     ACC
            push    ACC
            pop     ACC
            nop                     ; 1 machine cycle
            djnz    R0,inlp         ; 2/3 cycles to execute (branch takes 3 cycles)
;**************************************************************************************************
```

```
          cjne     cnt,#12,loop        ; reach the end of the table?
          ajmp     forever
disptab:  db       0x30,0xDF           ; display 1 on seven-segment display #5
          db       0x6D,0xEF           ; display 2 on seven-segment display #4
          db       0x79,0xF7           ; display 3 on seven-segment display #3
          db       0x33,0xFB           ; display 4 on seven-segment display #2
          db       0x5B,0xFD           ; display 5 on seven-segment display #1
          db       0x5F,0xFE           ; display 6 on seven-segment display #0
          end
```

Example 3.9 For the circuit shown in Figure 3.12, write a program to light one LED for 200 ms in turn and repeat, assuming that the C8051F040 system clock is 24 MHz.

Solution: To light the LED driven by P5.7 through P5.0 one at a time, one should output the value 0x80, 0x40, . . . , 0x01, 0x01, 0x02, . . . , 0x80 to Port P5 in turn and repeat. The procedure to render the desired LED display pattern is as follows.

Step 1 Place the values 0x80, 0x40, . . . , 0x01, 0x01, 0x02, . . . , 0x80 in a table. Use the register DPTR to point to the start of this table.

Step 2 Output the value pointed to by DPTR to P5. Increment the pointer DPTR.

Step 3 Wait for 200 ms

Step 4 If DPTR points to the end of the table, reset DPTR to the start of the table. Go to Step 2.

The assembly program that implements this algorithm is as follows.

```
          $nomod51
          $include (c8051F040.inc)
cnt       set      R6                  ; use R6 as a loop counter
          cseg     at 00H
          ljmp     main
          org      0xAB
main:     mov      DPTR,#led_tab       ; use DPTR as a pointer to the light pattern
          mov      SFRPAGE,#0x0F       ; switch to page F of SFR space`
          mov      WDTCN,#0DEH         ; disable watchdog timer
          mov      WDTCN,#0ADH         ;      "
;*********************************************************************************************
; switch to external oscillator as SYSCLK
;*********************************************************************************************
          mov      CLKSEL,#0           ; used internal oscillator to generate SYSCLK
          mov      OSCXCN,#067H        ; prepare to use external oscillator control as SYSCLK
          mov      R7,#255
          djnz     R7,$                ; wait for about 1 ms
```

```
chkstable:  mov     A,OSCXCN          ; wait until external crystal oscillator is stable
            anl     A,#80H            ; before using it
            jz      chkstable         ;         "
            mov     CLKSEL,#1         ; use external crystal oscillator to generate SYSCLK
;*****************************************************************************************
            mov     P5MDOUT,#0xFF     ; enable port P5 push-pull
forever:    mov     cnt,#0
loopi:      mov     A,cnt             ; place the index to LED pattern in A
            movc    A,@A+DPTR
            mov     P5,A              ; send out LED pattern
;*****************************************************************************************
; The following instruction sequence creates a delay of 200 ms.
;*****************************************************************************************
            mov     R2,#4             ; wait for 200 ms
outlp:      mov     R1,#200
extlp:      mov     R0,#250
inlp:       push    ACC               ; 2 machine cycles
            pop     ACC               ; 2 machine cycles
            push    ACC
            pop     ACC
            push    ACC
            pop     ACC
            push    ACC
            pop     ACC
            push    ACC
            pop     ACC
            nop                       ; 1 machine cycle
            djnz    R0,inlp           ; 2/3 cycles to execute (branch takes 3)
            djnz    R1,extlp
            djnz    R2,outlp
;*****************************************************************************************
            inc     cnt               ; reach the end of the table?
            cjne    cnt,#16,loopi     ;         "
            ljmp    forever           ; go to the start of the pattern table
led_tab:    db      80H,40H,20H,10H,08H,04H,02H,01H
            db      01H,02H,04H,08H,10H,20H,40H,80H
            end
```

It is quite easy to use program loops to create time delays. However, there is an overhead in this approach. For example, the following two statements in the program of the previous example add an additional $4 \times 200 \times (2 + 3)$ machine cycles ($= 166.7\ \mu s$) to the desired delay:

```
extlp:      mov     R0,#250       ; 2 machine cycles (execute 4 × 200 times)
            djnz    R1,extlp      ; 2/3 machine cycles (execute 4 × 200 times)
```

The overhead in the program loop makes the time delay inaccurate. We need to take this overhead into account when setting up the loop count in order to make the time delay more accurate. An easier approach is to use timer function to create time delays.

The C8051F040 has several timers (Timer 0, 1, 2, 3, and 4) and a programmable counter array (PCA). These timer modules have many applications. We will discuss only how to use Timer 0 to create time delays in this Chapter. Other timer modules and applications will be discussed in Chapter 8.

3.7 Using Timer 0 to Create Time Delays

Timer 0 is implemented as a 16-bit register accessed as two separate bytes: a low byte (TL0) and a high byte (TH0). Timer 0 has four operation modes: mode 0, 1, 2, and 3. Timer 0 operates as a 16-bit timer or counter in Mode 1. The block diagram of mode 1 is shown in Figure 3.16. Timer 0 can operate as a timer or a counter depending upon whether the clock source is the system clock or external clock input (T0 pin).

The operation of Timer 0 is controlled by three registers: timer control register (TCON), timer mode register (TMOD), and clock control register. The TCON register enables and disables Timer 0 to count. The TMOD register selects Timer 0s operation mode. The CKCON register selects the clock source of Timer 0. The contents of TCON, TMOD, and CKCON are shown in Figure 3.17, 3.18, and 3.19, respectively.

3.7.1 Mode 1 of Timer 0

In Mode 1, TH0 holds the upper eight bits of the 16-bit counter/timer register, whereas TL0 holds the lower eight bits. As Timer 0 overflows from 0xFFFF to 0x0000, the Timer 0 overflow flag TF0 is set, and an interrupt will occur if

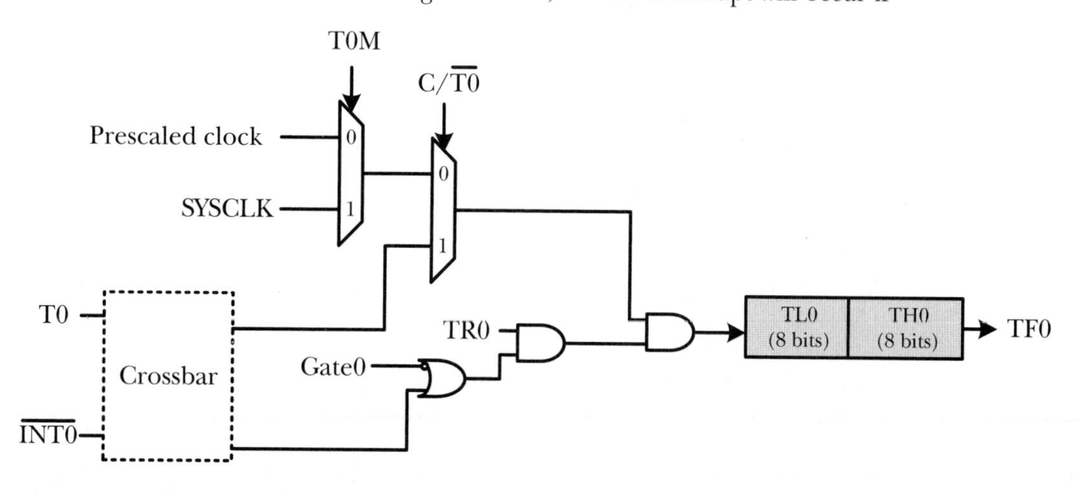

FIGURE 3.16
Timer 0 mode 1 block diagram (C8051F040).

FIGURE 3.17
Timer control register (TCON).

7	6	5	4	3	2	1	0	
TF1	TR1	TF0	TR0	IE1	IT1	IE0	IT0	Value after reset: 00h
rw	rw	rw	rw	rw	rw	rw	rw	

TF1: Timer 1 overflow flag
  0 = No Timer 1 overflow
  1 = Timer 1 has overflowed
  This flag is automatically cleared when the CPU vectors to Timer 1 interrupt service routine.
TR1: Timer 1 run control
  0 = Timer 1 disabled
  1 = Timer 1 enabled
TF0: Timer 0 overflow flag
  0 = No Timer 0 overflow
  1 = Timer 0 has overflowed
  This flag is automatically cleared when the CPU vectors to Timer 0 interrupt service routine.
TR0: Timer 0 run control
  0 = Timer 0 disabled
  1 = Timer 0 enabled
IE1: External interrupt 1 flag
  This flag is set when an edge/level of type defined by IT1 is detected. It is automatically cleared when the CPU is vectored to the external interrupt 1 service routine. It can also be cleared by software.
IT1: Interrupt 1 type select
  0 = /INT1 is level triggered, active low
  1 = /INT1 is falling-edge triggered
IE0: External interrupt 0 flag
  This flag is set when an edge/level of type defined by IT0 is detected. It is automatically cleared when the CPU is vectored to the external interrupt 1 service routine. It can also be cleared by software.
IT0: Interrupt 0 type select
  0 = /INT0 is level triggered, active low
  1 = /INT0 is falling-edge triggered

it is enabled. Another way to detect the timer overflow is by polling the TF0 flag.

The C/T0 bit of the TMOD register selects the counter/timer's clock source. When C/T0 is set to 1, the high-to-low transition on the T0 pin increments the timer register. When the C/T0 bit is set to 0, Timer 0 operates in timer mode. When in timer mode and the T0M bit of the CKCON register is set, Timer 0 is clocked by the system clock. If C/T0 is 0 and the T0M bit is also 0, then the clock source of Timer 0 is selected by the clock scale bits in the CKCON register.

7	6	5	4	3	2	1	0	
GATE1	C/T1	T1M1	T1M0	GATE0	C/T0	T0M1	T0M0	Value after reset: 00h
rw	rw	rw	rw	rw	rw	rw	rw	

GATE1: Timer 1 gate control
  0 = Timer 1 is enabled when TR1 = 1 irrespective of/INT1 logic level
  1 = Timer 1 is enabled only when TR1 = 1 and/INT1 = logic 1
C/T1: Counter/Timer 1 select
  0 = Timer function: Timer 1 is incremented by clock defined by T1M bit
  1 = Counter function: Timer 1 is incremented by high-to-low transitions on external input pin (T1)
T1M1-T1M0: Timer 1 mode select
  00 = Mode 0: 13-bit counter/timer
  01 = Mode 1: 16-bit counter/timer
  10 = Mode 2: 8-bit counter/timer with auto-reload
  11 = Mode 3: Timer 1 inactive.
GATE0: Timer 0 gate control
  0 = Timer 0 is enabled when TR0 = 1 irrespective of/INT0 logic level
  1 = Timer 0 is enabled only when TR0 = 1 and/INT0 = logic 1
C/T0: Counter/Timer select
  0 = Timer function: Timer 0 is incremented by clock defined by T0M bit
  1 = Counter function: Timer 0 is incremented by high-to-low transitions on external input pin (T0)
T0M1-T0M0: Timer 0 mode select
  00 = Mode 0: 13-bit counter/timer
  01 = Mode 1: 16-bit counter/timer
  10 = Mode 2: 8-bit counter/timer with auto-reload
  11 = Mode 3: Two 8-bit counter/timers.

FIGURE 3.18 *Timer mode register (TMOD).*

7	6	5	4	3	2	1	0	
--	--	--	T1M	T0M	--	SCA	SCA0	Value after reset: 00h
rw	rw	rw	rw	rw	rw	rw	rw	

T1M: Timer 1 clock select (this bit is ignored when C/T1 is set to 1)
  0 = Timer 1 uses the clock defined by the prescale bits SCA1-SCA0
  1 = Timer 1 uses the system clock
T0M: Timer 0 clock select (this bit is ignored when C/T0 is set to 1)
  0 = Timer 0 uses the clock defined by the prescale bits SCA1-SCA0
  1 = Timer 0 uses system clock
SCA1-SCA0: Timer 0/1 prescale bits
  00 = system clock divided by 12
  01 = system clock divided by 4
  10 = system clock divided by 48
  11 = external clock divided by 8

FIGURE 3.19 *The clock control register (CKCON) of C8051F040.*

Timer 0 is started by setting the TR0 bit of the TCON register. Timer 0 is not reset by the setting of TR0 bit.

3.7.2 Create Time Delay Using Mode 1

To operate the C8051F040 at a high frequency, one needs to choose the on-chip oscillator (frequency is 24.5 MHz) or use an external crystal oscillator to derive the system clock. The first step in using the timer function to create a time delay is to convert the time delay into the timer clock count. To create a longer time delay, it would be more convenient to use a prescaler larger than 1 (e.g., 4, 12, or 48) for the clock source to Timer 0 or Timer 1. Assume that the frequency of the system clock is f_{SYS}, then the timer count required to create the time delay of ***k*** seconds is given by the expression:

$$\textbf{timer count} = k \times f_{SYS} \div \text{prescaler} \tag{3.1}$$

Depending on the duration of the time delay to be created, this number may or may not be larger than $2^{16} - 1$. If this number is smaller than 2^{16}, then the desired time delay can be created by placing the value of **65536 − timer count** into the Timer 1 or Timer 0 registers and choose Mode 1. By waiting until Timer 0 overflows, the desired time delay is created. Otherwise, the calculated timer count must be factored into a product of two integers as

$$\textbf{timer count} = k \times f_{SYS} \div \text{prescaler} = n \times Y \tag{3.2}$$

where, Y is a number smaller than 65536 and n is the number of times that the count value Y should be repeated.

Example 3.10 Write an instruction sequence to create a time delay of 1 ms using Timer 0 of the C8051F040. The C8051F040 runs with a 24-MHz external crystal oscillator as its system clock.

Solution: The timer count corresponding to 1 ms with prescaler set to 48 is 500. By placing the value of 65036 (65536 − 500) into Timer 0 registers (TH0 and TL0), Timer 0 will overflow in 1 ms. The procedure for creating a time delay of 1 ms is as follows.

Step 1. Configure Timer 0 to operate at mode 1 and choose **system clock ÷ 48** as the clock input.

Step 2. Place the value 0xFE0C (65036_{10}) into Timer 0 registers and wait until the overflow flag (TF0) is set to 1.

The following instruction sequence will create a delay of 1 ms.

```
dly1ms:     push    SFRPAGE
            mov     SFRPAGE,#0
            mov     TMOD,#0x01        ; set up Timer 0 in mode 1
            clr     TF0
```

```
    mov     CKCON,#02       ; use SYSCLK/48 as Timer 0 clock source
    mov     TH0,#0xFE       ; place 65036 (=0xFE0C) in TMR0
    mov     TL0,#0x0C       ;    "
    setb    TR0             ; start TMR0
    jnb     TF0,$           ; wait for 1 ms (timer overflow)
    pop     SFRPAGE
```

Other delays can be created by modifying this instruction sequence. This instruction sequence can be modified into a macro so that it can be invoked in several places in the program.

3.8 Simple Input Device

A switch is probably the simplest input device we can find. To make input more efficient, a set of eight switches organized as a dual inline package (DIP) is often used. A DIP package can be connected to any input port with eight pins, as shown in Figure 3.20. When a switch is closed, the associated P6 pin has a value 0. Otherwise, the associated P6 pin has a value of 1. Each P6 pin is pulled up to high via a 10-KΩ resistor when the associated switch is open.

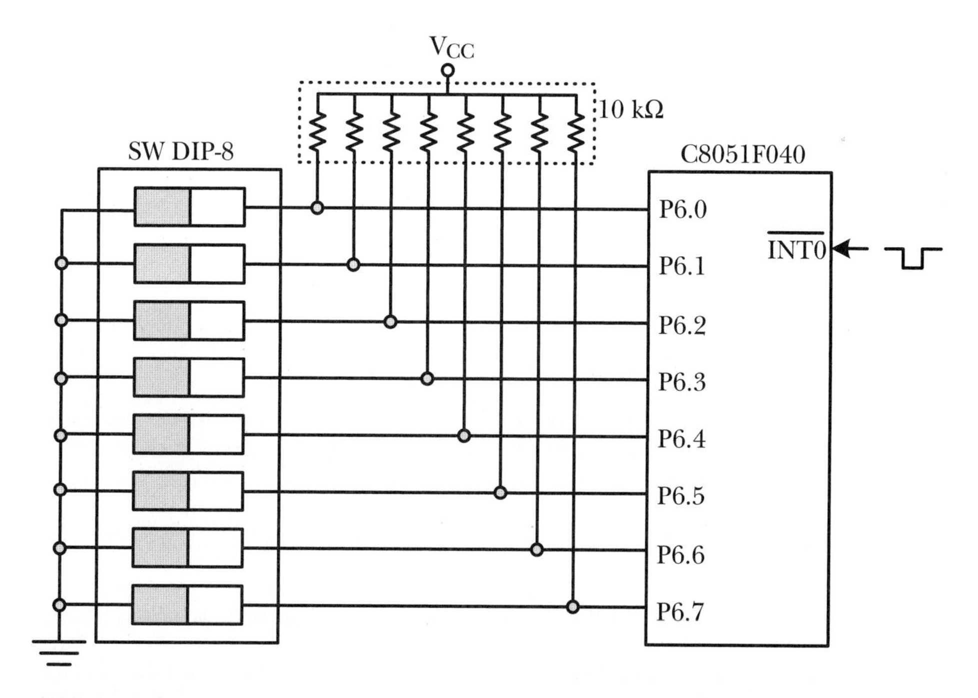

FIGURE 3.20
Connecting a set of eight DIP switches to Port 6 of the C8051F040.

Inputting data to the computer usually requires interaction with the user. The procedure is as follows.

Step 1
Output a message to prompt the user to enter a value.

Step 2
Set a new value to the eight DIP switches.

Step 3
Use interrupt (e.g., applied a negative pulse to the $\overline{\text{INT0}}$ pin) or some other method to inform the user program to read the value.

Repeat Steps 1 through 3 as many times as needed.

Example 3.11 Write an instruction sequence to read the value from the DIP switches and output it to the LEDs driven by Port P5.

Solution: The version for the C8051F040 is as follows.

```
push    SFRPAGE             ; save the current SFRPAGE
mov     SFRPAGE,#0x0F       ; switch to page F
mov     P6MDOUT,#0          ; configure port P6 to open-drain
mov     P6,#0xFF            ; configure Port 6 for input
mov     P5MDOUT,#0xFF       ; configure Port P5 to push-pull
mov     P5,P6               ; read a byte from Port P6 and send to Port P5
pop     SFRPAGE             ; restore the previous SFRPAGE value
```

3.9 Using the D/A Converter

A D/A converter converts a digital value into analog voltage. A few important DAC characteristics include:

- ***Resolution:*** The resolution of a DAC refers to the number of bits used to represent a value to be converted. In general, the more bits the better. The resolution of a commercial DAC may range from 8 bits up to 24 bits.
- ***Conversion Time.*** DAC conversion time is the amount of time required to convert a digital value to an analog voltage or current. It is desirable to have short conversion time for a DAC. The DAC conversion time has a wide range. It could be as short as 10 ns and longer than 10 μs.
- ***Number of Channels.*** This parameter refers to the number of outputs that a DAC has.
- ***Input Format.*** A DAC may require data to be input in parallel format or serial format (using the SPI or I²C interface). Parallel format is more suitable for high-speed DAC. Serial input format is more suitable for low pin count MCUs and those applications that do not require very high throughput.

3.9.1 The AD7302 DAC

The AD7302 is a dual-channel, 8-bit, DAC chip from analog devices that has a parallel interface with the microcontroller. The AD7302 converts an 8-bit digital value into an analog voltage. The block diagram of the AD7302 is shown in Figure 3.21. The AD7302 is designed to be a memory-mapped device. In order to send data to the AD7302, the $\overline{CS}$ signal must be pulled to low. On the rising edge of the $\overline{WR}$ or $\overline{CS}$ signal, the values on pins D7 through D0 will be latched into the input register. When the signal $\overline{LDAC}$ is low, the data in the input register will be transferred to the DAC register, and a new D/A conversion is started. If the $\overline{LDAC}$ signal is tied low, then the data in the input register will be transferred to the DAC register on the rising edge of the $\overline{WR}$ signal. The AD7302 needs a reference voltage to perform D/A conversion. The reference voltage can come from either the external REFIN input or the internal V_{DD}. The $\overline{A}/B$ signal selects the channel (A or B) to perform the D/A conversion. Channel A is selected when this signal is low. Otherwise, Channel B is selected. When the $\overline{PD}$ pin is low, the AD7302 is placed in the power-down mode and reduces the power consumption to 1 μW. The **conversion time** is 1.2 μs.

The AD7302 operates from a single +2.7- to 5.5-V supply and typically consumes 15 mW at 5 V, making it suitable for battery-powered applications. Each digital sample takes about 1.2 μs to convert. The output voltage ($V_{OUT}A$ or $V_{OUT}B$) from either DAC is given by

$$V_{OUT}\ A/B = 2 \times V_{REF} \times (N/255)$$

where V_{REF} is derived internally from the voltage applied at the REFIN pin or V_{DD}. If the voltage applied to the REFIN pin is within 1 V of the V_{DD}, $V_{DD}/2$

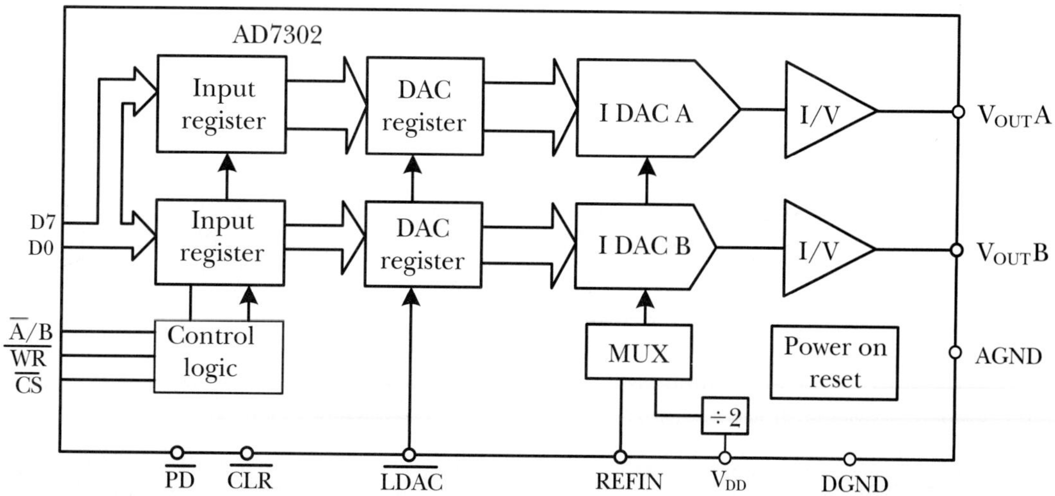

FIGURE 3.21
Functional block diagram of the AD7302.

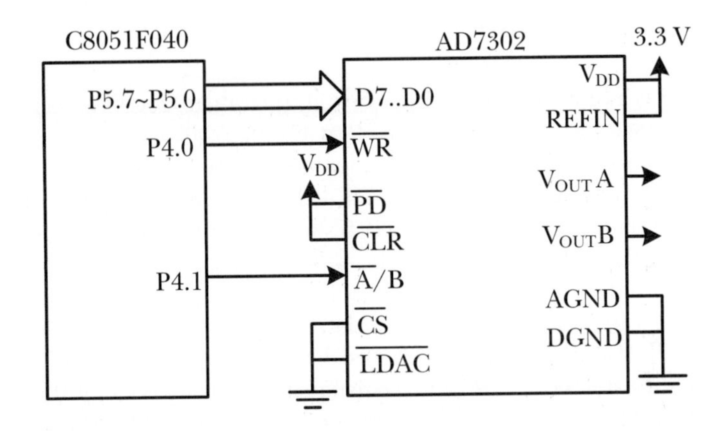

FIGURE 3.22 *Circuit connection between the AD7302 and the C8051F040.*

is used as the reference voltage automatically. Otherwise, the voltage applied at the REFIN pin is used as the reference voltage. The range of V_{REF} is from 1 V to $V_{DD}/2$.

N is the decimal equivalent of the code loaded to the DAC register, ranging from 0 to 255.

3.9.2 Interfacing the AD7302 with the 8051 Variants

The data to be sent to the AD7302 for conversion can be sent via a parallel port. Both the $\overline{CS}$ and $\overline{LDAC}$ signals can be tied to ground permanently. The value to be converted must be sent to the AD7302 via a parallel port (connect to pins D7~D0). An output pin can be used (as the $\overline{WR}$ signal) to control the transferring of data to the input register. The circuit connection between the C8051F040 and the AD7302 is illustrated in Figure 3.22.

Example 3.12 Write a program to generate a sawtooth waveform from the $V_{OUT}A$ pin, assuming that the C8051F040 is running with a 24-MHz crystal oscillator.

Solution: A sawtooth waveform can be generated by outputting the digital value from 0 to 255 and repeats. For each value, pull the P4.0 pin to low and then to high, so that the value on pins P5.7 through P5.0 can be latched into the input register and be converted to voltage. Pull the P4.1 pin to low to select channel A.

The assembly program that generates the sawtooth waveform is as follows.

```
$nomod51
$include        (C8051F040.inc)
org             0x00
```

```
            ljmp    start
            org     0xAB
start:      mov     SFRPAGE,#0x0F       ; switch to page F
            mov     P5MDOUT,#0xFF       ; configure port P6 to push-pull
            mov     P4MDOUT,#0xFF       ; configure Port P4 to push-pull
            mov     WDTCN,#0DEH         ; disable watchdog timer
            mov     WDTCN,#0ADH         ;    "
;*************************************************************************************
; switch to external oscillator as SYSCLK
;*************************************************************************************
            mov     CLKSEL,#0           ; used internal oscillator to generate SYSCLK
            mov     OSCXCN,#067H        ; configure external oscillator control
            mov     R7,#255
            djnz    R7,$                ; wait for about 1 ms
chkstable:  mov     A,OSCXCN            ; wait until external crystal oscillator is stable
            anl     A,#80H              ; before using it
            jz      chkstable           ;    "
            mov     CLKSEL,#1           ; use external crystal oscillator to generate SYSCLK
;*************************************************************************************
; start to generate the sawtooth waveform.
;*************************************************************************************
            clr     P4.1                ; select channel A output
loop:       clr     P4.0                ; latch P5 into DAC input register to be converted
            inc     P5                  ; increment the output by one step
            setb    P4.0                ; latch the input data and start conversion
            mov     R7,#7               ; create a delay of 30 machine cycles
loopdw:     nop                         ;    "
            djnz    R7,loopdw           ;    "
            ajmp    loop
```

Example 3.13 Write a program to generate a two-tone siren that alternates between 250 Hz and 500 Hz with each tone lasting for half of a second from the circuit in Figure 3.22 assuming the system clock frequency is 24 MHz.

Solution: The half period of a 250 Hz waveform is 2 ms, whereas the half period of a 500 Hz waveform is 1 ms. The procedure for generating the specified two-tone siren is as follows.

Step 1 loopLO ← 125 (low tone repetition count)

Step 2 Pull the P4.0 pin (drives R/W signal) to low. Output 255 to P5. Pull P4.0 pin to high.

Step 3 Wait for 2 ms.

Step 4 Pull the P4.0 pin to low. Output 0 to P5. Pull P4.0 pin to high.

Step 5 Decrement loopLO by 1. If loopLO ≠ 0, go to Step 2.

Step 6 loopHI ← 250 (high tone repetition count)

Step 7 Pull the P4.0 pin to low. Output 255 to P5. Pull the P4.0 pin to high.

Step 8 Wait for 1 ms.

Step 9 Pull the P4.0 pin to low. Output 0 to P5. Pull the P4.0 pin to high.

Step 10 Wait for 1 ms.

Step 11 Decrement loopHI by 1. If loopHI $\neq$ 0, go to Step 7. Otherwise, go to Step 1.

This siren can be heard by connecting the V_{OUT}A pin of AD7302 to a speaker (or buzzer on the SSE040 demo board).

The assembly program that generates the specified two-tone siren is as follows.

```
                $nomod51
                $include    (C8051F040.inc)
dly1ms          macro                           ; this macro creates a delay of 1 ms
                push        SFRPAGE
                mov         SFRPAGE,#0
                mov         TMOD,#0x01          ; set up Timer 0 in mode 1
                clr         TF0
                mov         CKCON,#02           ; use SYSCLK/48 as Timer 0 clock source
                mov         TH0,#0xFE           ; place 65036 (=0xFE0C) in TMR0
                mov         TL0,#0x0C           ;       "
                setb        TR0                 ; start TMR0
                jnb         TF0,$               ; wait for 1 ms (timer overflow)
                pop         SFRPAGE
                endm
                org         0x00
                ljmp        start
                org         0xAB
start:          mov         SP,#0x7F
                mov         SFRPAGE,#CONFIG_PAGE; switch to page F
                mov         P5MDOUT,#0xFF       ; configure port P5 to push-pull
                mov         P4MDOUT,#0xFF       ; configure Port P4 to push-pull
                mov         WDTCN,#0DEH         ; disable watchdog timer
                mov         WDTCN,#0ADH         ;       "
;*******************************************************************************************
; use internal oscillator divide by 1 as SYSCLK
;*******************************************************************************************
                mov         CLKSEL,#0           ; used internal oscillator to generate SYSCLK
                mov         OSCICN,#0x83        ; set frequency to 24.5 MHz
```

```
; ******************************************************************************************
; start to generate the two-tone siren that alternate between 250 Hz and 500 Hz with
; each tone lasts for 0.5 seconds.
; ******************************************************************************************
        mov     SFRPAGE,#0x0F
        clr     P4.1            ; select channel A output
loop:   mov     R1,#125         ; make 250 Hz lasts for 0.5 seconds
next1:  clr     P4.0            ; latch P5 into DAC input register to be converted
        mov     P5,#255         ; output a high voltage
        setb    P4.0            ; voltage
        dly1ms                  ; high voltage lasts for 2 ms
        dly1ms                  ;       "
        clr     P4.0
        mov     P5,#0           ; output a low voltage
        setb    P4.0
        dly1ms                  ; low voltage also lasts for 2 ms
        dly1ms                  ;       "
        djnz    R1,next1
        mov     R1,#250         ; make 250 Hz to last for 0.5 seconds
next2:  clr     P4.0            ; latch P5 into DAC input register to be converted
        mov     P5,#255         ; output a high voltage
        setb    P4.0
        dly1ms                  ; high voltage lasts for 1 ms
        clr     P4.0
        mov     P5,#0           ; output a low voltage
        setb    P4.0
        dly1ms                  ; low voltage also lasts for 1 ms
        djnz    R1,next2
        ajmp    loop
        end
```

3.10 Clock Generation and Control

All microcontrollers and microprocessors require a clock signal to operate. The clock signal can be generated internally or derived from an external crystal oscillator or other circuits.

3.10.1 Internal Oscillator

The C8051F040 has an internal oscillator that can be selected as the source of system clock. If the internal oscillator is chosen, then the machine cycle clock (system clock) can be derived by dividing this internal oscillator by a factor of 1, 2, 4, or 8. This oscillator is enabled after reset. The internal oscillator is

7	6	5	4	3	2	1	0
IOSCEN	IFRDY	--	--	--	--	IFCN1	IFCN0
rw	rw					rw	rw

Reset value 0xC0

IOSCEN: Internal oscillator enable bit
 0 = Internal oscillator disabled
 1 = Internal oscillator enabled
IFRDY: Internal oscillator frequency ready flag
 0 = Internal oscillator is not running at programmed frequency
 1 = Internal oscillator is running at programmed frequency
IFCN1-0: Internal oscillator frequency control bits
 00 = SYSCLK derived from internal oscillator divided by 8
 01 = SYSCLK derived from internal oscillator divided by 4
 10 = SYSCLK derived from internal oscillator divided by 2
 11 = SYSCLK derived from internal oscillator divided by 1

FIGURE 3.23
Internal oscillator control register OSCICN(C8051F040).

factory calibrated to 24.5 MHz (with a maximal 2 percent variation) and is configured by the OSCICN register. The contents of this register are shown in Figure 3.23.

3.10.2 Using an External Oscillator as the System Clock

If the application requires a clock frequency that cannot be derived by the internal oscillator or requires the frequency to have very small variation, then the user will need to use one of the following methods to provide the clock signal to the C8051F040.

1. *External Clock Source:* The user can supply a CMOS compatible clock signal to drive the XTAL1 directly.
2. *External RC Oscillator:* The circuit connection for this approach is shown in Figure 3.24. The frequency (in MHz) of the oscillator is given by the equation:

$$f = 1.23(10^3) / (R \times C) \tag{3.3}$$

 where **C** is in pF and **R** is in kΩ.
3. *Capacitor Oscillator* (work with on-chip resistor): The circuit connection for this approach is illustrated in Figure 3.25. The frequency (in MHz) of the generated clock signal is given by the equation:

$$f = KF / (C \times V_{DD}) \tag{3.4}$$

 where, C is in pF and *KF* is the *K* factor given in Figure 3.29.
4. *External Crystal Oscillator:* The circuit connection for this approach is shown in Figure 3.26.

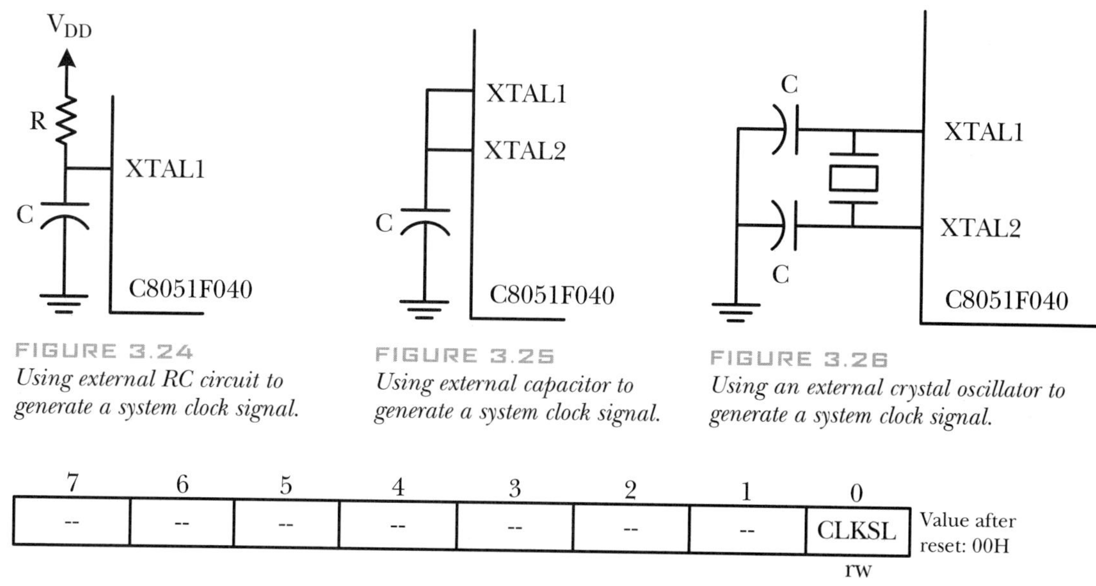

FIGURE 3.24
Using external RC circuit to generate a system clock signal.

FIGURE 3.25
Using external capacitor to generate a system clock signal.

FIGURE 3.26
Using an external crystal oscillator to generate a system clock signal.

CLKSL: System clock source select bit
0 = SYSCLK derived from the internal oscillator, and scaled as per the IFCN bits in the OSCICN register
1 = SYSCLK derived from the external oscillator circuit

FIGURE 3.27
Oscillator clock selection register CLKSEL(C8051F040).

An external RC circuit or capacitor is used mainly when the accuracy of the system clock frequency doesn't matter and the frequency is low. If the clock frequency needs to be very accurate, then either the external clock source, a crystal oscillator, or a ceramic oscillator should be used. As shown in Figure 3.27, the least significant bit of the CLKSEL register selects the internal or external oscillator to generate the system clock **(SYSCLK)**. The oscillator control is done by programming the OSCXCN register. The contents of OSCXCN are shown in Figure 3.28.

Example 3.14 Assume that an external RC oscillator is used to generate a 100 kHz system clock. Choose appropriate values for R and C.

Solution:

$$f = 1.23\ (10^3)\ /\ RC$$

therefore

$$RC = 1.23\ (10^3)\ /\ 100 \times 10^3 = 1.23\ /\ 100$$

Choose C = 50 pF, then R = 246 kΩ. The XFCN value should be set to 010.

7	6	5	4	3	2	1	0	
XTLVLD	XOSCMD2	XOSCMD1	XOSCMD0	--	XFCN2	XFCN1	XFCN0	Reset value 0x00
rw	rw	rw	rw	rw	rw	rw	rw	

XTLVLD: Crystal oscillator valid flag
  0 = Crystal oscillator is unused or not yet stable
  1 = Crystal is running and stable
XOSCMD2-XOSCMD0: External oscillator mode bits
  00x = External oscillator circuit off
  010 = External CMOS clock mode (external CMOS clock input on XTAL1 pin)
  011 = External CMOS clock mode with divide by 2 stage (external clock on XTAL1 pin)
  10x = RC/C oscillator mode with divide by 2 stage
  110 = Crystal oscillator mode
  111 = Crystal oscillator mode with divide by 2 stage
XFCN2-XFCN0: External oscillator frequency control bits

XFCN	Crystal (XOSCMD = 11x)	RC (XOSCMD = 10x)	C (XOSCMD = 10x)
000	f ≤ 32 kHz	f ≤ 25 kHz	K factor = 0.87
001	32 kHz < f ≤ 84 kHz	25 kHz < f ≤ 50 kHz	K factor = 2.6
010	84 kHz < f ≤ 225 kHz	50 kHz < f ≤ 100 kHz	K factor = 7.7
011	225 kHz < f ≤ 590 kHz	100 kHz < f ≤ 200 kHz	K factor = 22
100	590 kHz < f ≤ 1.5 MHz	200 kHz < f ≤ 400 kHz	K factor = 65
101	1.5 MHz < f ≤ 4 MHz	400 kHz < f ≤ 800 kHz	K factor = 180
110	4 MHz < f ≤ 10 MHz	800 kHz < f ≤ 1.6 MHz	K factor = 664
111	10 MHz < f ≤ 30 MHz	1.6 MHz < f ≤ 3.2 MHz	K factor = 1590

FIGURE 3.28
External oscillator control register OSCXCN (C8051F040).

Example 3.15 Assume that an external capacitor C is used to generate a 50 kHz system clock. Choose a C value that can achieve this goal.

Solution: Assume $V_{DD} = 3V$.

$$f = KF / (C \times V_{DD}) = KF / (C \times 3) = 0.05 \text{ MHz}$$

Set *KF* to 7.7, then C is computed to be 51.3 pF, and the XFCN value should be set to 010.

3.11 Chapter Summary

This chapter provides a brief introduction to the parallel I/O ports of the C8051F040 and their applications. An I/O port consists of a set of pins (usually 8 pins) and a group of registers. For the original 8051, output operation is performed by writing data to the port data register (P0, P1, P2, or P3) using a data movement instruction. Input operation requires writing the value 0xFF to the port data register before reading from the same register. For the C8051F040, the user needs to configure the port for input or output before the I/O operation can be performed. By writing the value 0xFF into the PnMDOUT register, Port Pn is configured to be an output port. By writing the value 0x00 into the same register, Port Pn is configured to be an input port.

Port P0 to Port P3 are shared by peripheral functions and general I/O pins. A peripheral signal can be assigned dynamically to an I/O pin in P0, P1, P2, or P3 by using the priority crossbar decoder of the C8051F040. If a pin is not assigned to any peripheral signal, it can then be used as an I/O pin.

Light-emitting diodes and seven-segment displays can be driven by parallel I/O ports. To make the output function interesting, appropriate time delays are often needed. The user can use either the program loops or the timer function to create time delays. When using the program loops, the user needs to pay attention to the overhead (extra instructions) incurred in setting up loops if he or she wants the delay time to be more accurate.

The instruction sequence for creating a time delay should be converted into a subroutine when the program needs to invoke time delays in more than one place. Subroutines and subroutine calls are discussed in Chapter 4.

Switches packaged in a dual-in-line package (DIP) are the simplest input devices in an embedded system.

A digital-to-analog converter can convert a value into a voltage. It is often used in generating waveforms. When choosing a DAC, the user needs to consider the resolution, number of channels, input format, conversion time, and the cost.

3.12 Exercise Problems

E3.1 Write an instruction sequence to output the value 0x48 to the I/O Port P5 of the C8051F040.

E3.2 Write an instruction sequence to read the value of input Port P4 and write the value to P5 of the C8051F040.

E3.3 Suppose that UART0, SPI, and UART1 are disabled, and SMBus and all channels of PCA are assigned to I/O pins. What pins are assigned to CEX3 and CEX4?

E3.4 Suppose that UART0 and UART1 are assigned to I/O pins, SPI, SMBus, CEX0, CEX1, and ECI are routed to I/O pins. What I/O pin is assigned to ECI signal?

E3.5 Write an instruction sequence to assign I/O pins to UART0, UART1, CP1, CP2, INT0, INT1 but not any other peripheral functions.

E3.6 Write an instruction sequence to assign I/O pins to UART0, SPI in three-wire mode, with all PCA pins including ECI, T3, and T3E signals but not any other peripheral functions.

E3.7 Suppose that UART0, UART1, SPI (three-wire mode), SMBus, CEX0, CEX1, CEX2, AIN1.0, and AIN.1 are assigned to I/O pins and all other peripheral functions are not. What I/O pins are assigned to these signals?

E3.8 If we reverse the connection in Figure 3.14 so that segments **a** to **g** are connected to pins P5.0 to P5.6, what values should be sent to P5 in order to display decimal values from 0 to 9?

E3.9 Write an instruction sequence to create a 10-ms time delay using Timer 0 in Mode 1, assuming that f_{SYS} = 24 MHz.

E3.10 Write an instruction sequence to create a 1-s time delay using Timer 0 in Mode 1 assuming that f_{SYS} = 24 MHz.

E3.11 Write an instruction sequence to create a 100-ms time delay using Timer 0 in Mode 1 assuming that f_{SYS} = 24 MHz.

E3.12 Write a program to flash the LEDs in Figure 3.12 in the following manner:

Step 1. Turn on all LEDs for 200 ms and then turn them off for 200 ms.

Step 2. Repeat Step 1 for three more times.

Step 3. Turn on one LED at a time from P5.7 toward P5.0 with each LED turned on for 200 ms.

Step 4. Repeat Step 3 for three more times.

Step 5. Turn on one LED at a time from P5.0 toward P5.7 with each LED turned on for 200 ms.

Step 6. Repeat Step 5 for three more times.

Step 7. Turn on LEDs driven by pins P5.7 and P5.0 on and off four times. The on-time and off-time are both 200 ms.

Step 8. Turn on LEDs driven by pins P5.6 and P5.1 on and off four times. The on-time and off-time are both 200 ms.

Step 9. Turn on LEDs driven by pins P5.5 and P5.2 on and off four times. The on-time and off-time are both 200 ms.

Step 10. Turn on LEDs driven by pins P5.4 and P5.3 on and off four times. The on-time and off-time are both 200 ms.

Step 11. Turn on LEDs driven by pins P5.3 and P5.4 on and off four times. The on-time and off-time are both 200 ms.

Step 12. Turn on LEDs driven by pins P5.2 and P5.5 on and off four times. The on-time and off-time are both 200 ms.

Step 13. Turn on LEDs driven by pins P5.1 and P5.6 on and off four times. The on-time and off-time are both 200 ms.

Step 14. Turn on LEDs driven by pins P5.0 and P5.7 on and off four times. The on-time and off-time are both 200 ms.

Step 15. Go to Step 1 and repeat.

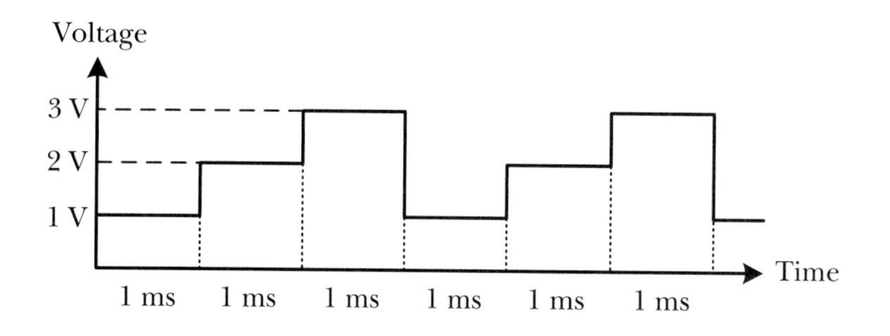

FIGURE E3.15 *Waveform to be generated by AD7302.*

E3.13 Change the program in Example 3.8 so that one digit is displayed at a time for 100 ms instead of simultaneously.

E3.14 Change the program in Example 3.8 so that one digit is displayed at a time for 500 ms instead of simultaneously.

E3.15 Write a program to use the circuit in Figure 3.22 to generate a waveform shown in Figure E3.15.

E3.16 Write a program to use the circuit in Figure 3.22 to generate a triangular waveform that increases from 0 V toward 3.3 V and then decreases from 3.3 V toward 0 V. The delay between two adjacent steps must be minimal.

E3.17 Configure the OSCICN register of the C8051F040 so that the system clock is derived by dividing the internal oscillator clock by 1.

E3.18 Configure the OSCICN register of the C8051F040 so that the system clock is derived by dividing the internal oscillator clock by 4.

E3.19 Write an instruction sequence to select external clock signal as the system clock for the C8051F040.

E3.20 Write an instruction sequence to select external clock signal divided by 2 as the system clock for the C8051F040.

3.13 Laboratory Exercise Problems and Assignments

L3.1 Use the Port P5 to drive two 74LS48s. The 74LS48 is a common cathode seven-segment display driver. It accepts a 4-bit binary number and converts it into the corresponding seven-segment pattern. Write a program to display the value from 00 to 99 on these two seven-segment displays three times and stop. Display each value for 200 ms.

You will need to convert the value to BCD format before sending to Port P5 for display. If you are using the SSE040 demo board, remove the jumper that enables LEDs. The 74LS48 requires a 5-V power supply.

L3.2 Write a program as described in Problem E3.13 and run the program on the SSE040 demo board.

L3.3 Connect the AD7302 to the C8051F040 MCU, as shown in Figure 3.22. Write a program to generate a 2-kHz waveform. Display the waveform on the oscilloscope.

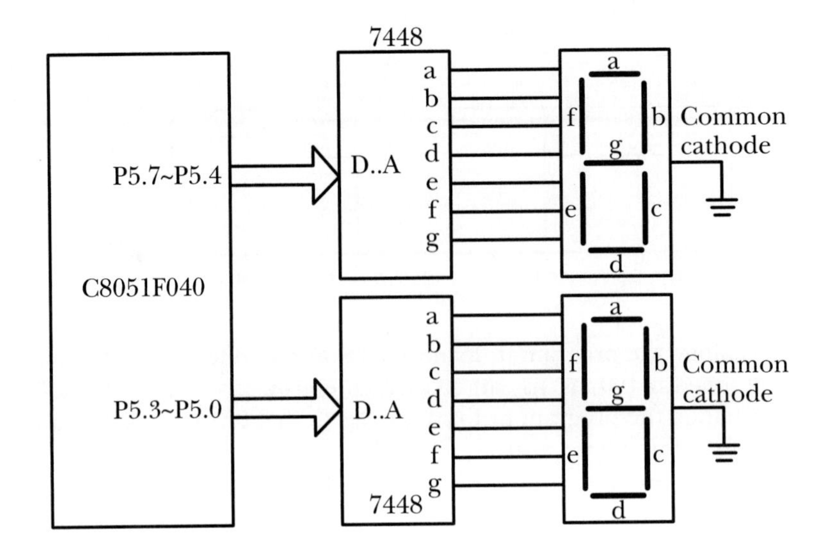

FIGURE L3.1
Two-digit seven-segment displays circuit.

L3.4 Connect the AD7302 to the C8051F040 MCU, as shown in Figure 3.22 and then connect the $V_{OUT}A$ to the buzzer on the SSE040 demo board or a speaker. Write a program to generate a two-tone siren. The frequencies of these two tones are 440 Hz and 880 Hz, respectively.

Advanced Assembly Programming and Subroutines

4.1 Objectives

Upon successful completion of this chapter, you will be able to:

- Understand the issues related to subroutine calls
- Initialize the stack pointer and perform stack operations
- Access XRAM
- Write subroutines to create time delays
- Explain and apply top-down-design-with-hierarchical-refinement software development methodology
- Write subroutines to multiply and divide signed and unsigned 16-bit or longer numbers
- Call UART functions to output data to a PC monitor screen and input data from a keyboard
- Perform program debugging that involves subroutines

4.2 Introduction

It was demonstrated in Example 3.13 that the same instruction sequence may need to be executed in several places of the program. Macros have been used so far as the solution to this issue. However, using the macro call will duplicate the same sequence of instructions in places where the macro is invoked and cause the program size to be bloated. Fortunately, all microprocessors and microcontrollers provide the *subroutine mechanism* in which the program can invoke the same sequence of instructions in several places of the program without duplicating it.

A subroutine is often written to perform operations based upon the inputs provided by the caller. The inputs provided by the caller are called *incoming parameters*. The caller of the subroutine often expects the subroutine to return certain results to it.

Subroutine calls involve the change of program control flow. In general, the processor executes a *subroutine call instruction* to change the program control flow. When calling a subroutine, the processor loads the starting address of the subroutine into the program counter, and then the CPU starts to execute subroutine instructions. When the processor finishes execution of the subroutine, it should return to the instruction immediately after the instruction that makes the subroutine call. This is achieved by executing a *return* instruction.

The subroutine call and return instructions work together to make the subroutine mechanism to work. The subroutine call instruction saves the *return address* in the stack data structure and at the same time changes the program control flow to the start of the subroutine. The return instruction fetches the return address from the stack, places it in the program counter, and hence returns the program control back to the instruction immediately after the subroutine call instruction.

The subroutine mechanism has a great impact on the program development methodology. It was mentioned in Chapter 2 that the most popular software development methodology is "*top-down design with hierarchical refinement.*" Subroutine mechanism makes this approach possible. Example 3.13 will be restructured to illustrate this process.

Reusable macros and subroutines should be made into files and included in programs that need them. This approach can increase programmers' productivity. Subroutines are especially convenient for this approach. One of the objectives of this text is to promote software reuse and writing reusable software.

4.3 The Stack Data Structure

A *stack* is a data structure from which elements can be accessed from only its top. The processor can add a new element to the stack by performing a **push** operation. To remove an element, the processor performs a **pull** (or **pop**) operation. Physically, a stack can grow from a high address toward lower addresses or from a low address toward higher addresses. Depending upon the processor, the *stack pointer* can point to the top element of the stack or point to the byte immediately above the top element of the stack. The 8051 stack grows from a low address toward higher addresses and has an 8-bit stack pointer (SP) that points to the top byte of the stack. The memory space available for use by the stack is limited in a computer system. There is always a danger of *stack overflow* and *stack underflow*. Stack overflow is a situation in which the processor pushes data into the stack too many times so that the SP points to a location outside the area allocated to the stack. Stack underflow is a situ-

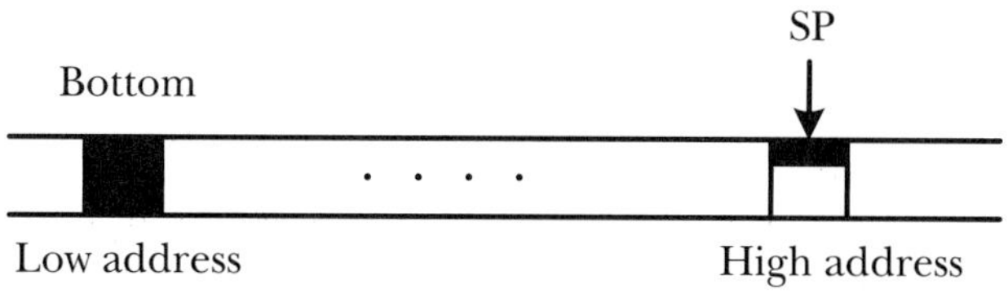

FIGURE 4.1
The 8051 stack.

ation in which the processor pulls data from the stack too many times so that the SP points to an area below the stack bottom.

4.3.1 Initializing the Stack Pointer

The SP register of the 8051 is designed to point to the internal data memory in the range from 0x00 to 0xFF. Before the stack can be used, the user must initialize SP to an appropriate value. The SP register is initialized to 0x07 after the 8051 is reset. This address is also the direct address of the register R7 in bank 0. If the user program does not assign another value to the SP, the first **Push** operation will store data in R0 of bank 1.

The SP register should be initialized to point to an internal RAM address above the highest address that will be used by the program. For example, the following instruction initializes SP to 0x7F and hence the bottom byte of the stack is located at 0x80.

```
mov         SP,#0x7F
```

4.3.2 Instructions for Stack Operation

The 8051 provides the **push direct** and **pop direct** instructions to support the stack data structure. No other addressing modes can be used with these two instructions. For example, the following two instructions cause syntax error.

```
push        A         ; incorrect syntax for representing accumulator A
push        R0        ; R0 is not direct address
```

The symbol that represents the direct address of accumulator A is ACC. To push accumulator A, using the instruction:

```
push        ACC       ; 8051 adds 1 to SP and then stores the contents of A
                      ; in the location pointed to by SP
```

Similarly, this instruction pulls the top byte of the stack into accumulator A:

```
pop         ACC       ; pop the top byte into A and then decrement SP by 1
```

Since there are four banks of registers, the Keil (and Raisonance) assembler requires two steps for pushing an **Rm** register:

```
using       k         ; specifying the bank, k = 0, 1, 2, or 3
push        ARm       ; specify the direct address of register Rm, m = 0, . . . , 7
```

For example, the following instruction pushes R0 of bank 3 into the stack.

```
using   3       ; select bank 3 without making bank 3 active
push    AR0
```

The assembler generates the direct address of the register R0 of bank 3 (0x18) and uses it to replace the symbol AR0 before translating it into machine code.

Similarly, the following instruction sequence pops out the top byte of the stack and places it in register R0 of bank 1.

```
using   1
pop     AR0
```

4.4 The XRAM of the C8051F040

The C8051F040 has 4096 bytes of on-chip XRAM which is treated as a part of the external data memory. The external data-memory interface (EMIF) can be configured to operate in one of the four modes (as shown in Figure 4.2) based on the EMIF mode bits in the EMI0CF register. The contents of the EMI0CF register are shown in Figure 4.3. The actual external data-memory expansion and timing consideration will be explored in Chapter 14.

The C8051F040 kit and the SSE040 demo board do not have external data memory. One can set bit 3 and bit 2 of the EMI0CF register to 00 or 01 or 10. Other bits are irrelevant to the operation of the demo board and can simply be set to 0. We will use the value of 03H to configure the EMI0CF register in this book.

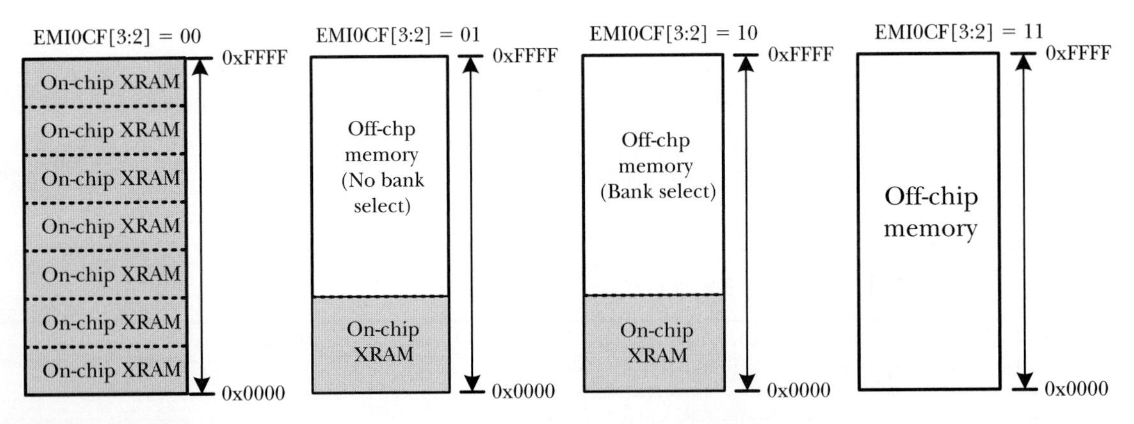

FIGURE 4.2
External data-memory operation modes.

7	6	5	4	3	2	1	0
--	--	PRTSEL	EMD2	EMD1	EMD0	EALE1	EALE0

Reset value 00000011

PRTSEL: EMIF port select
0 = EMIF active on P0-P3
1 = EMIF active on P4-P7
EMD2: EMIF multiplex mode select
0 = external memory operates in multiplexed address/data mode
1 = external memory operates in non-multiplexed mode (separate address and data pins)
EMD1~EMD0: External memory interface operating mode select
00 = internal only. The movx instruction accesses on-chip XRAM only.
01 = split mode without bank select. Accesses below the 4k boundary are directed on-chip. Accesses above the 4k boundary are directed off-chip.
10 = split mode with bank select. Accesses below the 4k boundary are directed on-chip. Accesses above the 4k boundary are directed off-chip. Eight-bit off-chip movx operations use the contents of the EMI0CN register to determine the high-byte of the address.
11 = external only. movx accesses off-chip XRAM only. on-chip XRAM is invisible to the CPU.
EALE1~EALE0: ALE pulse-width select bits
00 = ALE high and ALE low pulse width = 1 SYSCLK cycle.
01 = ALE high and ALE low pulse width = 2 SYSCLK cycle.
10 = ALE high and ALE low pulse width = 3 SYSCLK cycle.
11 = ALE high and ALE low pulse width = 4 SYSCLK cycle.

FIGURE 4.3 *EMIOCF external memory configuration.*

4.4.1 Two Miscellaneous Operators

There are situations that we need to place the upper half and lower half of a label or a value into two separate 8-bit registers. The Keil A51 assembler (also applies to Raisonance's assembler) provides two operators for this purpose: **high** and **low**. To use these operators, place one of these operators in front of the value that you want to separate. For example, this instruction places the upper half of the hex value 0x1234 in R2.

```
mov        R2,#high 0x1234        ; place 0x12 in R2
```

This instruction places the lower half of the value 0x1234 in R3 as

```
mov        R3,#low 0x1234         ; place 0x34 in R3
```

A few more examples are

```
mov        R2,#high done          ; place the upper half value of the label done in R2
mov        R3,#low done           ; place the lower half value of the label done in R3
```

4.5 An Example of Subroutine

We have learned how to write an instruction sequence to create a time delay in Chapter 3. Such an instruction sequence can be converted into a subroutine by adding a ret instruction to its end. For example, the subroutine that can create a 500-μs time delay is as follows.

```
delay500us:
            push    SFRPAGE
            mov     SFRPAGE,#0
            mov     TMOD,#0x01          ; set up Timer 0 in mode 1
            clr     TF0
            mov     CKCON,#0            ; use SYSCLK/12 as Timer 0 clock source
            mov     TH0,#high 64536     ; place 64536 (1000 clock cycles) in TMR0
            mov     TL0,#low 64536      ; so that it overflows in 1000 clock cycles (0.5 ms)
            setb    TR0                 ; start TMR0
            jnb     TF0,$               ; wait for 0.5 ms (timer overflow)
            pop     SFRPAGE
            ret
```

This subroutine would be much more useful if it could create a time delay that is a multiple of 500 μs and allow the caller to specify the multiple. The following version of the subroutine can achieve just that, and the multiple is passed in register R0.

```
delayby500us:
            push    SFRPAGE
            mov     SFRPAGE,#0
            mov     TMOD,#0x01          ; set up Timer 0 in mode 1
            clr     TF0
            mov     CKCON,#0            ; use SYSCLK/12 as Timer 0 clock source
loopw:      mov     TH0,#high 64536     ; place 64536 (1000 clock cycles) in TMR0
            mov     TL0,#low 64536      ; so that it overflows in 1000 clock cycles (0.5 ms)
            setb    TR0                 ; start TMR0
            jnb     TF0,$               ; wait for 0.5 ms (timer overflow)
            clr     TR0                 ; stop TMR0
            djnz    R0,loopw
            pop     SFRPAGE
            ret
```

The user needs to place the multiple in register R0 before calling this subroutine. For example, the following instruction sequence calls the **delayby500μs** subroutine to create a delay of 4 ms.

```
mov         R0,#8
lcall       delayby500us
```

4.6 An Example of Top-Down Design with Hierarchical Refinement

The subroutine mechanism makes the top-down design easier to implement. Suppose we want to generate a three-tone siren using the I/O pin P5.0 and a buzzer. The frequencies of these three tones are 250 Hz, 500 Hz, and 1000 Hz, respectively. Each tone lasts for half of a second. The P5.0 pin is connected to one terminal of the buzzer, whereas the other terminal of the buzzer is connected to the ground. (If the SSE040 demo board is used, you need only to run a wire between the P5.0 pin and the buzzer).

The top-down design with a hierarchical refinement approach will go through several iterations and may work as given in the following examples.

Iteration 1

This iteration lays out the major steps of the program as follows.

Step 1
Generate the 250-Hz tone for half of a second.

Step 2
Generate the 500-Hz tone for half of a second.

Step 3
Generate the 1000-Hz tone for half of a second.

Step 4
Go to Step 1.

Iteration 2

Steps 1, 2, and 3 are refined, as show in the following steps.
The 250-Hz tone is generated as

Step 2.11
Use the variable **lpcnt** to keep track of the iterations remained to be performed and initialize it to 125.

Step 2.21
Pull the P5.0 pin to high.

Step 2.31
Wait for 2 ms.

Step 2.41
Pull the P5.0 pin to low.

Step 2.51
Wait for 2 ms.

Step 2.61
lpcnt ← lpcnt − 1

Step 2.71
If **lpcnt** is not zero, go to Step 2.21.
The 500-Hz tone is generated as

Step 2.12
Use the variable **lpcnt** to keep track of the iterations remained to be performed and initialize it to 250.

Step 2.22
Pull the P5.0 pin to high.

Step 2.32
Wait for 1 ms.

Step 2.42
Pull the P5.0 pin to low.

Step 2.52
Wait for 1 ms.

Step 2.62
lpcnt ← lpcnt − 1

Step 2.72
If **lpcnt** is not zero, go to Step 2.22.
The 1000-Hz tone can be generated as

Step 2.13
Use the variable **lpcnt** to keep track of the iterations remained to be performed and initialize it to 500.

Step 2.23
Pull the P5.0 pin to high.

Step 2.33
Wait for 500 μs.

Step 2.43
Pull the P5.0 pin to low.

Step 2.53
Wait for 500 μs.

Step 2.63
lpcnt ← lpcnt − 1

Step 2.73
If **lpcnt** is not zero, go to Step 2.23.

Iteration 3

The only unresolved issue is how to generate 2-ms, 1-ms, and 500-μs time delays. They can be resolved as follows.
The 2-ms time delay can be generated as

Step 3.11
Place 4 in R0.

Step 3.21
Call the **delayby500us** subroutine.
The 1-ms time delay can be created as

Step 3.12
Place 2 in R0.

Step 3.22
Call the **delayby500us** subroutine.
The 500-μs time delay can be created as

Step 3.13
Place 1 in R0.

Step 3.23
Call the **delayby500us** subroutine.
After Iteration 3, all of the program details have been figured out, and hence, we can start to write the program to generate the three-tone siren. The program is as follows.

```
            #include    <C8051F040.inc>
lpcnt       set         R7
            org         0x00
            ajmp        start
            org         0xAB
start:      acall       sysinit
            mov         SFRPAGE,#0x0F
forever:    mov         B,#125          ; generate 250-Hz tone for half a second
loop2:      setb        P5.0            ;  "
            mov         R0,#4           ;  "
            lcall       delayby500us    ;  "
            clr         P5.0            ;  "
            mov         R0,#4           ;  "
            lcall       delayby500us    ;  "
            djnz        B,loop2         ;  "

            mov         B,#250          ; generate 500-Hz tone for half a second
loop1:      setb        P5.0            ;  "
            mov         R0,#2           ;  "
            lcall       delayby500us    ;  "
            clr         P5.0            ;  "
            mov         R0,#2           ;  "
            lcall       delayby500us    ;  "
            djnz        B,loop1         ;  "

            mov         lpcnt,#2        ; generate 1-kHz tone for half a second
loop0e:     mov         B,250           ;  "
loop0:      setb        P5.0            ;  "
            mov         R0,#1           ;  "
```

```
            lcall       delayby500us            ;   "
            clr         P5.0                    ;   "
            mov         R0,#1                   ;   "
            lcall       delayby500us            ;   "
            djnz        B,loop0                 ;   "
            djnz        lpcnt,loop0e            ;   "
            ajmp        forever
;**************************************************************************************************
; The following subroutine selects external crystal oscillator as SYSCLK, enable
; crossbar decoder, configure port P5 for output, 3-wire SPI.
;**************************************************************************************************
sysinit:    mov         SFRPAGE,#CONFIG_PAGE    ; switch to the SFR page that contains
                                                ; registers that configures the system
            mov         WDTCN,#0xDE             ; disable watchdog timer
            mov         WDTCN,#0xAD             ;   "
            mov         CLKSEL,#0               ; used internal oscillator to generate SYSCLK
            mov         OSCXCN,#0x67            ; configure external oscillator control
            mov         R7,#255
            djnz        R7,$                    ; wait for about 1 ms
chkstable:  mov         A,OSCXCN                ; wait until external crystal oscillator is stable
            anl         A,#80H                  ; before using it
            jz          chkstable               ;   "
            mov         CLKSEL,#1               ; use external crystal oscillator to generate SYSCLK
            mov         XBR0,#0xF7              ; assign I/O pins to all peripheral functions
            mov         XBR1,#0xFF              ; and enable crossbar
            mov         XBR2,#0x5D              ;   "
            mov         XBR3,#0x8F              ;   "
            mov         P5MDOUT,#0xFF           ; configure Port P5 for push-pull
            mov         SFRPAGE,#SPI0_PAGE      ;
            mov         SPI0CN,#01H             ; select 3-wire SPI mode
            ret
;**************************************************************************************************
; The following subroutine creates a time delay that is a multiple of 500 us at fOSC = 24 MHz.
;**************************************************************************************************
delayby500us:
            push        SFRPAGE
            mov         SFRPAGE,#0              ; switch to page 0
            mov         TMOD,#0x11              ; configure Timer 0 and 1 as Mode 1 timer
            mov         CKCON,#0                ; Timer 0 use system clock divided by 12 to count
repw01:     mov         TH0,#high 64536         ; place 64536 in Timer 0 so that it overflows in
            mov         TL0,#low 64536          ; 1000 clock cycles (in 500 µs)
            setb        TR0                     ; enable Timer 0
            clr         TF0
            jnb         TF0,$                   ; wait until TF0 is set again
            clr         TR0
            djnz        R0,repw01
            pop         SFRPAGE
            ret
            end
```

4.7 Issues Related to Subroutine Calls

There are three major issues related to subroutine calls: parameter passing, result returning, and local variable allocation.

4.7.1 Parameter Passing

The caller of the subroutine may pass parameters using registers, stack, or global memory. For the C8051, we can use registers R0~R7, B, DPTR, and A to pass parameters to the subroutine. The stack, internal SRAM, and XRAM can also be used to pass parameters to the subroutine.

4.7.2 Result Returning

A subroutine may return its computation result in registers, stack, or global memory. For the C8051, a subroutine can return its computation results using registers R0~R7, B, DPTR, and A. It can also use the stack, internal SRAM, and XRAM to return the computation results.

4.7.3 Local Variable Allocation and Deallocation

In addition to parameters passed to it, a subroutine may also need memory space to hold temporary variables and results. These variables are useful only during the execution of the subroutine and are called *local variables* because they are local to the subroutine.

Theoretically, it is best to allocate local variables in the stack, because local variables come into being when the subroutine is entered and disappear when the subroutine is exited—as long as the subroutine allocates local variables at its entrance and deallocates local variables before it returns to the caller. This can avoid the interference among subroutines when one subroutine needs to call other subroutines. Allocating local variables in stack has an additional benefit: It allows the creation of *re-entrant* subroutines. A *re-entrant* subroutine can call itself.

The 8051 programmer can use one of the following two methods to allocate local variables in the stack.

1. Use as many **inc SP** instructions as needed. This method takes less time than the next method if no more than 3 bytes are needed for local variables.
2. Use this instruction sequence if more than three bytes are needed for local variables.

```
mov     A, SP
add     A,#k        ; allocate k bytes
mov     SP,A        ; move the stack pointer up by k bytes
```

To deallocate local variables from the stack, the 8051 programmer can use one of the following methods.

1. Use as many **dec SP** instructions as needed. This method takes less time than the next method if there are no more than 4 bytes to be deallocated.
2. Use this instruction sequence if there are more than four bytes to be deallocated.

```
mov     A,SP
clr     C
subb    A,#k
mov     SP,A
```

4.7.4 Accessing Incoming Parameters and Local Variables in the Stack

Assume that we have a stack frame (as shown in Figure 4.4) after the subroutine is entered. Example 4.1 illustrates how to access variables in the stack.

Example 4.1 You are given the stack frame shown in Figure 4.4. Write an instruction sequence to

(a) Load the value in slot **loc3** into R6

(b) Load the value in slot **loc4** into DPTR

Solution:

(a) To load the value in slot **loc3** into R6, invoke

```
mov     A,SP
clr     C
subb    A,#2
mov     R0,A
mov     A,@R0
mov     R6,A
```

(b) To load the value in the slot **loc4** into DPTR, invoke

```
mov     A,SP
clr     C
subb    A,#3
mov     R1,A
mov     DPL,@R1
dec     R1
mov     DPH,@R1
```

It is obvious that it involves very high overhead in order to access variables stored in the stack frame. Therefore, unless we are required to write reentrant subroutines, we will not use the stack to pass parameters nor use the stack to hold local variables.

# of bytes	
1	loc1 ← SP
1	loc2
1	loc3
2	loc4
1	loc5
2	return address
1	parm4
1	parm3
2	parm2
1	parm1

FIGURE 4.4
A stack frame example.

Whenever a register or a data memory is used as a local variable, the subroutine should save its original value by pushing it into the stack. Before returning to the caller, the subroutine should restore the saved registers by popping them out from the stack. The popping order should be the reverse of the pushing order. Registers and memory locations that are used to pass parameters or return results need not be saved.

4.8 Writing Subroutines to Perform Multiprecision Arithmetic

The 8051 has no instructions that can perform 16-bit or longer arithmetic. In this section, we will write subroutines that can perform 16-bit unsigned and signed multiplication and division. An unsigned number is non-negative, whereas a signed number may be positive, zero, and negative. The most significant bit of a signed number is the signed bit.

4.8.1 Writing Subroutines to Perform 16-bit Unsigned Multiplication

To multiply two 16-bit unsigned numbers, the multiplier and the multiplicand must be broken down into 8-bit chunks, and four 8-bit by 8-bit multiplications are performed. Suppose that S and R are two 16-bit unsigned numbers to be multiplied; they are broken down as

$$R = R_H R_L = R_H \times 2^8 + R_L$$

$$S = S_H S_L = S_H \times 2^8 + S_L$$

where R_H, R_L, S_H, and S_L are the upper and lower bytes of R and S, respectively. Four byte-to-byte multiplications are performed, and then the partial

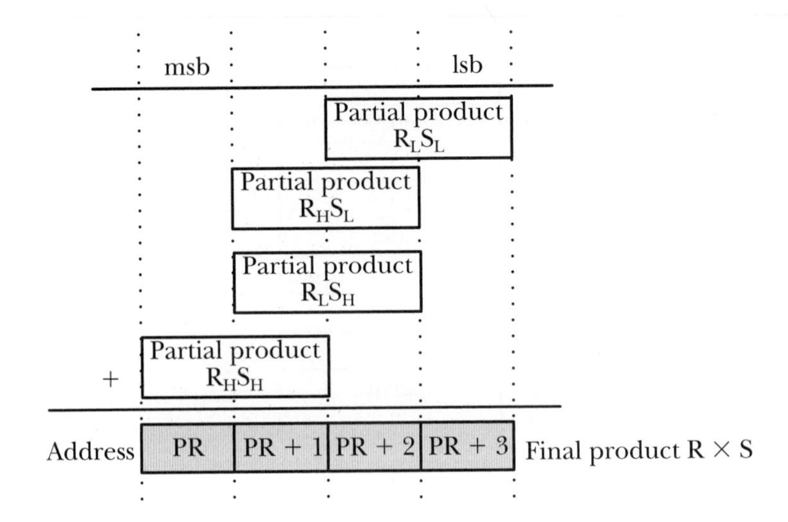

FIGURE 4.5 *16-bit by 16-bit multiplication.*

products are added together, as shown in Figure 4.5. The procedure for adding these four partial products is as follows.

Step 1
Allocate four bytes to hold the product. Refer to these four bytes using the symbols PR, PR + 1, PR + 2, and PR + 3, respectively.

Step 2
Compute the partial product $R_L S_L$ and save it at locations PR + 2 and PR + 3. Store the high byte of the product at PR + 2 and low byte of the product at PR + 3.

Step 3
Compute the partial product $R_H S_H$ and save it at locations PR and PR + 1. Store the high byte of the product at PR and the low byte of the product at PR + 1.

Step 4
Compute the partial product $R_H S_L$ and add it to the memory locations PR + 1 and PR + 2. Add the carry flag to the memory location P.

Step 5
Compute the partial product $R_L S_H$ and add it to the memory locations PR + 1 and PR + 2. Add the carry flag to the memory location PR.

Example 4.2 Write a subroutine to multiply two unsigned 16-bit integers passed in R0:R1 and R2:R3 and return the product in R4, . . . , R7. R0 and R2 hold the upper bytes of two numbers to be multiplied and R4 to R7 hold the most significant to the least significant bytes of the product.

Solution: The assembly subroutine that implements the previous algorithm and its test program are as follows.

```
         $nomod51
         $include    (c8051F040.inc)
         org         0x00
         ljmp        start
         org         0xAB
start:   mov         R0,#0x41        ; pass test numbers in R0:R1 and R2:R3
         mov         R1,#0x09        ;   "
         mov         R2,#0x41        ;   "
         mov         R3,#0x79        ;   "
         acall       mul16u
         ajmp        $
;--------------------------------------------------------------------------
; The following subroutine multiplies two 16-bit unsigned integers passed in
; R0-R1 and R2-R3 and returns the product in R4-R7 (msb to lsb).
;--------------------------------------------------------------------------
mul16u:  mov         A,R1            ; place RL in A
         mov         B,R3            ; place SL in B
         mul         AB              ; compute RL * SL
         mov         R7,A            ; save the lower byte of RL * SL
         mov         R6,B            ; save the upper byte of RL * SL
         mov         A,R0            ; place RH in A
         mov         B,R2            ; place SH in B
         mul         AB              ; compute RH * SH
         mov         R5,A            ; save the lower byte of RH * SH
         mov         R4,B            ; save the high byte of RH * SH
         mov         A,R1            ; compute RL * SH
         mov         B,R2            ;   "
         mul         AB              ;   "
         add         A,R6            ; add the lower byte of product to PR + 2
         mov         R6,A            ;   "
         mov         A,B             ; add the carry and upper byte of RL* SH to location
         addc        A,R5            ; PR + 1
         mov         R5,A            ;   "
         mov         A,R4            ; add the carry to memory location PR
         addc        A,#0            ;   "
         mov         R4,A            ;   "
         mov         A,R0            ; compute RH * SL
         mov         B,R3            ;   "
         mul         AB              ;   "
         add         A,R6            ; add the lower byte of RH * SL to location PR + 2
         mov         R6,A            ;   "
         mov         A,B             ; add the upper byte of RH * SL and carry to location
         addc        A,R5            ; PR + 1
         mov         R5,A            ;   "
         mov         A,R4            ; add carry to location PR
         addc        A,#0            ;   "
         mov         R4,A            ;   "
         ret
         end
```

The multiplication of two numbers longer than 16 bits or two numbers of different length can be performed by using the same method and hence will be left as an exercise problem.

4.8.2 Writing Subroutines to Perform 16-bit Signed Multiplication

Because the computer's word length is limited, it essentially is performing *modulus* arithmetic when performing an arithmetic operation. For example, an 8-bit MCU will drop (discard) the part of the result that is equal to or larger than 2^8. A 16-bit MCU will drop the part of the result that is equal to or larger than 2^{16}. Adding 2^8 to the result of the arithmetic operation performed by an 8-bit MCU has no effect on the result, because this value (2^8) cannot be kept in the arithmetic hardware of an 8-bit MCU. In summary, in an n-bit MCU, the result of any arithmetic operation is equal to the remainder of the initial result divided by 2^n (i.e., it is performing a modulo-2^n operation).

The multiplication of signed numbers requires the programmer to consider the signs of the multiplier and the multiplicand. Let M and N represent the magnitudes of two numbers. There are four possible situations.

Case 1. Both operands are positive (op1 = M, op2 = N). The product of these two operands can be computed by using either the **mul AB** instruction for 8-bit operands or the procedure described in Section 4.8.1 for 16-bit and longer operands.

Case 2. The first operand is negative (op1 = $-M$), whereas the second operand is positive (op2 = N). The first operand **op1** will be represented in two's complement of M ($2^8 - M$ if M is 8-bit, $2^{16} - M$ if M is 16-bit, and so on). The product of two n-bit operands $-M$ and N can be rewritten as follows.

$$-M \times N = (2^n - M) \times N = (2^n \times N) - M \times N = \underbrace{2^{2n} - M \times N}_{①} + \underbrace{2^n \times N}_{②}$$

The value 2^{2n} is added to the expression. Since the MCU is performing a modulo-2^{2n} arithmetic in this case, adding the value 2^{2n} makes no difference to the result. Item 1 is the two's complement of the number $-M \times N$ and is the correct product. The second part of the product is an extra term and should be eliminated from the product. This extra term can be eliminated by subtracting **op2** from the upper half of the product of $-M$ and N.

Case 3. The first operand is positive (op1 = M) but the second operand is negative (op2 = $-N$). Similar to Case 2, the product of two n-bit operands M and $-N$ can be rewritten as follows.

$$M \times (-N) = M \times (2^n - N) = (2^n \times M) - M \times N = \underbrace{2^{2n} - M \times N}_{①} + \underbrace{2^n \times M}_{②}$$

The value of 2^{2n} is added to the expression, and it makes no difference to the result for the same reason as in Case 2. Again, the first term is the correct product, which is represented as the two's complement of $-M \times N$. The second term of the product is an extra term and can be eliminated by subtracting **op1** from the upper half of the product of M and $-N$.

Case 4. Both operands are negative (op1 = $-M$, op2 = $-N$). The product of two n-bit operands $-M$ and $-N$ can be rewritten as

$$\begin{aligned}(-M) \times (-N) &= (2^n - M) \times (2^n - N) = 2^{2n} - 2^n \times M - 2^n \times N + M \times N \\ &= M \times N + 2^{2n} - 2^n \times M + 2^{2n} - 2^n \times N \\ &= \underbrace{M \times N}_{①} + \underbrace{2^n \times (2^n - M)}_{②} + \underbrace{2^n \times (2^n - N)}_{③}\end{aligned}$$

The value of 2^{2n} is added to the expression. It makes no difference to the result for the same reason as in Case 2. The first term is the product of $-M$ and $-N$. The second term and the third term are extra terms and must be eliminated. The second term can be eliminated by subtracting **op1** from the upper half of the product, whereas the third term can be eliminated by subtracting **op2** from the upper half of the product.

By combining these four situations, we conclude that the signed n-bit multiplication can be implemented by the following algorithm.

Step 1
Multiply two operands (that is, compute **op1** × **op2**) disregarding their signs.

Step 2
If **op1** is negative, then subtract **op2** from the upper half of the product.

Step 3
If **op2** is negative, then subtract **op1** from the upper half of the product.

Example 4.3 Write a subroutine that can multiply two 16-bit signed integers. The multiplier and multiplicand are passed to this subroutine in R0:R1 and R2:R3, respectively. The product is returned in registers R4~R7 with the most significant byte stored in R4 and the least significant byte stored in R7.

Solution: This subroutine may call the **mul16u** subroutine to perform unsigned 16-bit multiplication. The subroutine and its test program are as follows.

```
            $nomod51
            #include <C8051F040.inc>
            org     0x00
            ajmp    start
            org     0xAB
start:      mov     R0,#0x87        ; load the test numbers
            mov     R1,#0x56        ;  "
            mov     R2,#0x91        ;  "
            mov     R3,#0x52        ;  "
            acall   mul16s
            ajmp    $
; -------------------------------------------------------------------------
; The following subroutine multiplies two 16-bit signed integers and returns the product
; in R4..R7. The multiplicand and multiplier are passed in R0-R1 and R2-R3.
; -------------------------------------------------------------------------
mul16s:     acall   mul16u
chk_M:      mov     A,R0            ; test the sign bit of M
            jnb     ACC.7,chk_N     ; is M negative?
            mov     A,R5            ; M is negative, so subtract N from upper half
            clr     C               ; of the product M × N
            subb    A,R3            ;  "
            mov     R5,A            ;  "
            mov     A,R4            ;  "
            subb    A,R2            ;  "
            mov     R4,A            ;  "
chk_N:      mov     A,R2            ; test the sign bit of N
            jnb     ACC.7,done      ; if positive, then return
            mov     A,R5            ; N is negative, so subtract M from upper half
            clr     C               ; of the product M × N
            subb    A,R1            ;  "
            mov     R5,A            ;  "
            mov     A,R4            ;  "
            subb    A,R0            ;  "
            mov     R4,A            ;  "
done:       ret
            #include "mul16u.a51"
            end
```

4.8.3 Writing Subroutines to Perform Unsigned Multiprecision Division

The 8051 provides only an instruction that divides the 8-bit value in register B into the 8-bit value in accumulator A. The division operation for larger numbers must be implemented in macros or subroutines. One of the most widely used division methods is the **repeated-shift-and-subtract** method. This method assumes the use of three n-bit registers R, Q, and S.

The repeated-shift-and-subtract algorithm consists of two phases.

Initialization Phase

Place 0, dividend, and divisor in register R, Q, and S, respectively.

4.8.4 Shift-and-Subtract Phase

Repeat for n times.

Step 1
Shift the register pair R and Q to the left one place.

Step 2
Subtract S from R and place the difference back to R if the difference is non-negative.

Step 3
Set the least significant bit of Q to 1 if the result in Step 2 is non-negative. Otherwise, set the least significant bit of Q to 0

Example 4.4 Write a subroutine that can divide an unsigned 16-bit integer into another 16-bit unsigned integer. The dividend and divisor are passed in R4:R5 and R6:R7, respectively. Remainder and quotient are to be returned in register pairs R2:R3 and R4:R5, respectively.

Solution: The assembly subroutine that implements the 16-bit unsigned division and its test program are as follows.

```
            $nomod51
            $include  (c8051F040.inc)
lpcnt       set       R0              ; divide loop index
tmp         set       R1              ; temporary storage
Rhi         set       R2              ; high byte of register R
Rlo         set       R3              ; low byte of register R
Qhi         set       R4              ; high byte of register Q
Qlo         set       R5              ; low byte of register Q
Shi         set       R6              ; high byte of register S
Slo         set       R7              ; high byte of register S

rotate      macro     arg             ; a macro that rotate arg to the left through carry
            mov       A,arg           ;   "
            rlc       A               ;   "
            mov       arg,A           ;   "
            endm                      ;   "

            org       0x00
            ajmp      start
            org       0xAB
```

```
start:      mov     Qhi,#0xDA       ; pass the dividend
            mov     Qlo,#0xC0       ;   "
            mov     Shi,#0x02       ; pass the divisor
            mov     Slo,#0xBC       ;   "
            lcall   div16u          ; call the subroutine to divide
            jmp     $

div16u:     mov     lpcnt,#16       ; initialize the loop count to 0
            mov     Rhi,#0
            mov     Rlo,#0
; The following instructions shift registers R and Q to the left one place
dloop:      clr     C
            rotate  Qlo
            rotate  Qhi
            rotate  Rlo
            rotate  Rhi

; The following instructions perform a division step by performing subtract operations
            clr     C
            mov     A,Rlo
            subb    A,Slo           ; subtract the lower bytes of two numbers
            mov     tmp,A           ; save the difference in the temporary register
            mov     A,Rhi
            subb    A,Shi           ; subtract the upper bytes of the 16-bit numbers
            jc      less            ; the minuend (R) is smaller
            mov     Rhi,A           ; store the difference back to the minuend
            mov     A,tmp           ; also store the lower byte of the difference back
            mov     Rlo,A           ; to the minuend
            mov     A,Qlo           ; set the lsb of Q to 1
            orl     A,#0x01         ;   "
            mov     Qlo,A           ;   "
            ajmp    chkend
less:       mov     A,Qlo
            anl     A,#0xFE         ; set the lsb of Q to 0
            mov     Qlo,A           ;   "
chkend:     djnz    lpcnt,dloop
exit:       ret
            end
```

4.8.5 Converting an Internal Binary Number into a BCD String

A binary number in the computer memory cannot be output directly to an output device. It must be converted into a BCD string before being output. A BCD string uses ASCII code to represent each decimal digit. For example, the binary number 11111010_2 is converted to a three-byte string **0x32 0x35 0x30** before being output. After being output to a monitor screen or an LCD, the string appears as the decimal value 250. For easy processing, a BCD string is usually terminated by a NULL character (ASCII code is 0).

Example 4.5 Write a subroutine to convert a 16-bit signed integer into a string of BCD digits and save the conversion result in a memory buffer in XRAM pointed to by DPTR. The string is terminated by a NULL character. The high and low bytes of the number to be converted are passed in R4 and R5, respectively.

Solution: An unsigned binary number can be converted into a BCD string by executing an algorithm that performs repeated division by 10 until the quotient becomes zero. Adding 0x30 to the remainder converts the remainder to its ASCII code. This process generates a string that is reversed. It can be converted back to the right order by pushing it into the stack and then popping it out. Let *xx*, *quo*, *rem*, and *ptr* represent the number to be converted, quotient of the division process, remainder of the divide process, and the pointer to the buffer to hold the resultant string, respectively. The procedure is as follows.

Step 1 quo ← xx / 10; rem ← xx mod 10

Step 2 mem[ptr] ← rem + 0x30; ptr++

Step 3 If quo $\neq$ 0, xx = xx/10 and go to Step 1, otherwise **mem[ptr] ← 0.**

Step 4 Push 0 and then the string into the stack.

Step 5 Pop the string out of the stack and store it in the buffer.

A negative number should be converted into its two's complement representation before being processed by the previous algorithm. After conversion, a negative number will be preceded by a minus sign.

The assembly program that implements this algorithm is as follows.

```
        $nomod51
        $include  (c8051F040.inc)
tstHi   equ       0xE6          ; test data (-6601)
tstLo   equ       0x37
minus   equ       0x2D          ; ASCII code of minus sign

        xseg      at 0x00
buf:    ds        10            ; buffer to hold BCD string
        cseg at   0x00
        ajmp      start
        org       0xAB
start:  mov       SP,#0x7F      ; set up the stack pointer
        mov       R4,#tstHi
        mov       R5,#tstLo
```

```
                mov         DPTR,#buf
                lcall       bin2bcd
                ajmp        $                   ; spin here
; ************************************************************************************************
; The following subroutine converts a 16-bit signed integer into a BCD string. R4 and R5 hold
; the number to be converted whereas DPTR points to the buffer to store the resultant string.
; ************************************************************************************************
sign            set         B                   ; sign indicator
ptrHi           set         0x30                ; buffer pointer holder
ptrLo           set         0x31                ;    "
bin2bcd:        mov         A,sign
                push        ACC
                mov         A,ptrLo             ; save memory location contents used by
                push        ACC                 ; ptrHi:ptrLo
                mov         A,ptrHi             ;    "
                push        ACC                 ;    "
                mov         ptrLo,DPL           ; save a copy of the buffer pointer
                mov         ptrHi,DPH           ;    "
                mov         sign,#0             ; initialize sign to nonnegative
                mov         A,R4                ; check sign bit of the number to be converted
                anl         A,#0x80             ;    "
                jz          normal              ; jump if the number is positive
                mov         A,R5                ; calculate the 2's complement of the given number
                cpl         A                   ;    "
                add         A,#1                ;    "
                mov         R5,A                ;    "
                mov         A,R4                ;    "
                cpl         A                   ;    "
                addc        A,#0                ;    "
                mov         R4,A                ;    "
                mov         sign,#1             ; set sign to minus
normal:         mov         R6,#0               ; load 10 as the divisor
                mov         R7,#10              ;    "
next:           lcall       div16u              ; perform repeated division by 10
                mov         A,R3                ; get the remainder (low byte only)
                add         A,#0x30             ; convert to ASCII code
                movx        @DPTR,A             ; store the ASCII code in buffer
                inc         DPTR                ; move the buffer pointer
                mov         A,R4                ; check if the quotient is zero, continue if nonzero
                jnz         next                ;    "
                mov         A,R5                ;    "
                jnz         next                ;    "
                mov         A,#0                ; terminate the string with a NULL
                movx        @DPTR,A             ;    "
                mov         DPL,ptrLo           ; reset the buffer pointer
                mov         DPH,ptrHi           ;    "
                mov         A,#0                ; push a NULL character into the stack
                push        ACC                 ;    "
pushloop:       movx        A,@DPTR             ; get a character from the BCD string
                jz          poploop             ; reach the end of the string?
```

```
            push    ACC             ;   "
            inc     DPTR            ;   "
            ajmp    pushloop        ;   "
poploop:    mov     DPL,ptrLo       ; reset buffer pointer
            mov     DPH,ptrHi       ;   "
            mov     A,sign
            jz      notMinus        ; the number is non-negative
            mov     A,#0x2D         ; place the minus sign as the first character
            movx    @DPTR,A         ;   "
            inc     DPTR            ; also move the pointer
notMinus:   pop     ACC
            jz      donePop         ; reach the end of the string?
            movx    @DPTR,A         ; save in the buffer
            inc     DPTR
            ajmp    notMinus
donePop:    mov     A,#0            ; terminate the buffer with a NULL
            movx    @DPTR,A         ;   "
            pop     ACC             ; restore ptrHi:ptr:Lo, and B
            mov     ptrHi,A         ;   "
            pop     ACC             ;   "
            mov     ptrLo,A         ;   "
            pop     ACC             ;   "
            mov     sign,A          ;   "
            ret
;*****************************************************************************************************
; The following subroutine divides a 16-bit unsigned integer into another unsigned 16-bit integer.
; The divided is passed in R4-R5 whereas the divisor is passed in R6-R7. The quotient is
; returned in R4-R5 and the remainder is returned in R2-R3. The subroutine file div16u.a51 is
; in the project directory.
;*****************************************************************************************************
            $include    (div16u.a51)
            end
```

4.8.6 Signed Division Operation

The one complication for signed division is that we must also set the sign of the remainder. This equation must always hold for division.

Dividend = Quotient × Divisor + Remainder

Our common sense requires that the magnitude of the quotient to be the same as long as the magnitudes of the dividends are the same and the magnitudes of the divisors are the same. We can determine the sign of the remainder on the basis of this principle. To illustrate, let's use $(\pm 41) \div (\pm 7)$ as an example. The first situation is simple.

$41 \div 7$: Quotient = +5, Remainder = +6

If we change the sign of the dividend, the quotient must be changed as well.

$-41 \div 7$: Quotient $= -5$

Rewriting our basic formula to find the remainder; we have

$$\begin{aligned} \text{Remainder} &= \text{Dividend} - \text{Quotient} \times \text{Divisor} \\ &= -41 - (-5 \times 7) = -41 + 35 = -6 \end{aligned}$$

If we change the sign of the divisor and keep the sign of dividend unchanged,

$41 \div (-7)$: Quotient $= -5$

$$\text{Remainder} = 41 - (-5 \times -7) = 41 - 35 = 6$$

If we change the signs of both the dividend and the divisor,

$-41 \div -7$: Quotient $= 5$

$$\text{Remainder} = -41 - (-5 \times -7) = -41 + 34 = -6$$

From this discussion, we conclude that the correctly signed division algorithm negates the quotient if the signs of the operands are opposite and makes the sign of the nonzero remainder match that of the dividend.

The signed division subroutine is straightforward and hence will be left as an exercise.

4.9 Using the UART0 Module to Perform I/O

The UART0 module of the 8051 uses the TX and RX pin to shift data in and out of the MCU. It allows the 8051 to communicate with a PC or a workstation through the COM port. The UART0 port also allows the 8051 to communicate with other 8051 or even an MCU that has different CPU architecture.

The UART0 module has four operation Modes: Mode 0, 1, 2, and 3. Mode 1 allows the MCU of a demo board to talk to a PC. We will not discuss the operation and configuration of the UART0 module here. The UART0 function will be discussed in Chapter 9. However, we will provide a set of functions for you to call to display information on the PC monitor and input data from the PC keyboard. These utility functions include:

- **openUART0:** This subroutine configures the UART0 module to transfer data at 19,200 bits/s, assuming that you use a 24 MHz crystal oscillator or 24.5 MHz internal oscillator to generate the system clock. You need to call this function to configure the baud rate and select the UART0 operation mode (Mode 1) before performing any input and output operation.
- **putch:** This subroutine outputs the contents of accumulator A to the UART0 port. The character to be output must be placed in accumulator A before this function is called.

- **newline:** This function outputs a carriage return character followed by a linefeed character to the UART0 port. When the demo board is communicating with a PC, this function moves the screen cursor to the first column of the next line.
- **putsc:** This function outputs a string in program memory to the UART0 port. When the demo board is communicating with a PC, this function displays the string on the monitor screen. The string to be output is pointed to by DPTR.
- **putsx:** This function outputs a string in XRAM to the UART0 port. When the demo board is communicating with a PC, this function displays the string on the monitor screen. The string to be output is pointed to by DPTR.
- **getch:** This function reads a character from the UART0 port. When the demo board is communicating with a PC, this function allows the user to enter a character from the keyboard. The character is returned in accumulator A.
- **gets:** This function reads a string from the UART0 port and saves the string in a buffer in XRAM pointed to by DPTR. The string will be terminated by a NULL character in the buffer. When the demo board is communicating with a PC, this function allows the user to enter a string from the keyboard. The string is terminated when the user presses the enter key. The characters entered by the user are echoed back to the screen for the user to verify. With this subroutine, the user can backspace and retype the string from the PC keyboard.

The file **uartUtil.a51** in the complementary CD contains these subroutines. Before you can use the UART port to communicate with a PC, you need to connect the demo board to the PC using a serial cable. The serial cable connects the demo board and the COM port of the PC using the DB9 connectors.

In addition, the user needs to run a terminal program such as the HyperTerminal bundled with the Windows. The HyperTerminal can be started by selecting **Start=>All Programs=>Accessories=>Communications=> HyperTerminal**. The baud rate of the HyperTerminal and that of the demo board must be set to the same value. The procedure for setting up the HyperTerminal is detailed in Appendix F.

Example 4.6 Write a program to test the UART subroutines by outputting a message to remind the user to enter his/her name and then echoing the entered name back to the monitor screen.

Solution: The program and subroutines required to test the UART subroutines are as follows.

```
$nomod51
$include        (c8051F040.inc)
xseg            at 0x00
```

```
ibuf:     ds          20
          cseg        at 0x00
          ljmp        start
          org         0xAB
start:    mov         SP,#0x7F          ; set up stack pointer
          lcall       sysinit           ; select system clock and disable watchdog timer
          lcall       openUART0         ; configure UART operation mode, baud rate
          lcall       newline           ; move screen cursor to the start of the next line
          mov         DPTR,#msg1        ; output a message to remind the user to enter his/
          lcall       putsc             ; her name
          mov         DPTR,#ibuf        ; read the name entered by the user from the keyboard
          acall       gets              ;   "
          acall       newline           ; move cursor to the next line
          mov         DPTR,#msg2        ; output "My name is "
          acall       putsc             ;   "
          mov         DPTR,#ibuf        ; output the entered name to the monitor screen
          acall       putsx             ;   "
          acall       newline
          jmp         $
msg1:     db          "Please enter your name: ",0
msg2:     db          "My name is ",0
          $include    (systeminit.a51)  ; contains the subroutine sysinit
          $include    (UartUtil.a51)
          end
```

The file **systeminit.a51** uses an external crystal oscillator to generate system clock, enables crossbar decoder, assign I/O pins to all peripheral functions, select three-wire SPI mode, and disable the watchdog timer. This file is also included in the complementary CD.

After running the program, you will be asked to enter your name. Supposing you enter **Albert Einstein** as your name, the resultant Hyper Ternimal screen will be as shown in Figure 4.6.

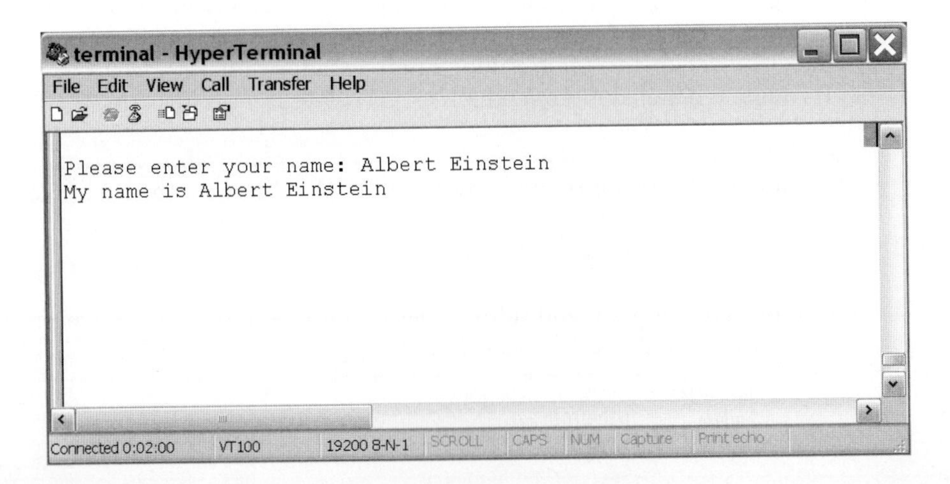

FIGURE 4.6
UART I/O session example.

The user can also enter numbers from the keyboard. A number entered from the keyboard is represented in the form of ASCII code. Before it can be processed by the MCU, the numeric string must be converted to the binary format. The following example describes a subroutine that performs the conversion.

Example 4.7 Write a subroutine that converts a string which represents a signed 16-bit integer into a binary number. The string is stored in XRAM, and the pointer to the string is passed in DPTR to the subroutine. The resultant binary number is returned in R0:R1. R0 holds the upper byte.

Solution: The algorithm of the desired conversion is based on the following representation of a *k*-digit decimal number.

$$d_{k-1}d_{k-2}\ldots d_0 = (\ldots((d_{k-1} \times 10) + d_{k-2}) \times 10 + \ldots) \times 10 + d_0$$

where, $d_{k-1}, \ldots, d_0$ are characters entered by the user from the keyboard and are represented in ASCII code. They can be converted to decimal digits by subtracting 0x30 from them. The algorithm for the conversion subroutine is as follows.

Step 0 Use the variable **num**, **neg**, **char**, and **dval** to represent the converted value, sign of the number, current character, and the value of the current character being processed, respectively. Initialize variables **num** and **neg** to 0.

Step 1 Read one character from the string. If the character is the minus character (ASCII code is 0x2D), set **neg** to 1. Otherwise, **dval ← char − 0x30**. Go to Step 3.

Step 2 Read the next character from the string. If the character is a NULL character, then exit. Otherwise, **dval ← char − 0x30**.

Step 3
num ← num × 10 + dval

Step 4 Go to Step 2.

The operation of multiplying by 10 is implemented in a subroutine called **mulby10**. The assembly subroutine that implements this algorithm and the program that tests the subroutine is as follows.

```
        $nomod51
        $include    (c8051F040.inc)
        xseg at     0x0000
ibuf:   ds          20
```

```
obuf:       ds          20
            cseg at     0x00
            ljmp        start
            org         0xAB
start:      mov         SP,#0x7F            ; set up stack pointer
            lcall       sysinit             ; subroutine sysinit is contained in systeminit.a51
                                            ; use external crystal oscillator as system clock,
                                            ; enable crossbar decoder, assign I/O pins to
                                            ; peripheral signals, 3-wire SPI
            lcall       openUART0           ; set UART0 baud rate to 19200 bits/s, mode 1
            lcall       newline             ; move cursor to the next line
            mov         DPTR,#prompt        ; remind the user to type a number on keyboard
            lcall       putsc               ;    "
            mov         DPTR,#ibuf
            lcall       gets                ; read in the number
            mov         DPTR,#ibuf          ; convert the string to binary and leave it in R0:R1
            lcall       bcd2bin             ;    "
            jmp         $                   ; stay here forever
prompt:     db          "Please enter a signed integer: ",0
;*****************************************************************************************************
; The following subroutine converts a ASCII string into a binary number and returns the result
; in R0:R1. R0 holds the most significant byte. The string is in XRAM and the pointer to this
; string is passed in DPTR.
;*****************************************************************************************************
signbd      set         R7                  ; indicate the sign of the given number
neg         equ         0x2D                ; ASCII code of minus sign
bcd2bin:    mov         A,signbd            ; save R7 (used as signbd)
            push        ACC                 ;    "
            mov         signbd,#0           ; initialize sign to positive
            mov         R0,#0               ; R0:R1 are used to accumulate the converted number
            mov         R1,#0               ;    "
            movx        A,@DPTR             ; get the first character
            cjne        A,#neg,doit         ; is it a minus character?
isneg:      mov         signbd,#1           ; record minus sign
            inc         DPTR
            movx        A,@DPTR             ; get the second character of the decimal string
doit:       push        ACC                 ; save the character in stack
            lcall       mulby10             ; compute num × 10
            mov         A,R4                ; place the product in R0:R1
            mov         R1,A
            mov         A,R3                ;    "
            mov         R0,A
            pop         ACC                 ; get back the character from stack
            clr         C
            subb        A,#0x30             ; convert to the digit value from ASCII
            clr         C                   ; add to the accumulated sum R0:R1
            add         A,R1                ;    "
            mov         R1,A                ;    "
```

4.10 A Few More Tips on Assembly Program Debugging

Program debugging becomes much more difficult when a program calls subroutines. To make debugging simpler, one should test each individual subroutine thoroughly to make sure each subroutine works correctly and returns to its caller.

4.10.1 What to Do When the Program Gets Stuck?

When a program contains one or multiple subroutines, it is common that the program gets stuck in one of the subroutines and cannot return to the caller during the program development process. The procedure for debugging in this situation is as follows.

Step 1
Write an instruction sequence to call each subroutine to find out if the program gets stuck in the subroutine. This can be done by setting a breakpoint at the instruction immediately after the **acall** or **lcall** instruction.

Step 2
Find out why the program gets stuck in the subroutine. There are at least the following four possible causes.

1. Forgetting to restore registers pushed onto the stack before returning to the caller
2. Forgetting to deallocate local variables before returning to the caller
3. Some infinite loops in the subroutine
4. Calling other subroutines that do not return

The first two causes can be identified by simply looking at the program. To determine if the program gets stuck because of forgetting to restore registers, one checks to see if the subroutine has several push instructions at its entrance. If it does, then it should have the same number of pop instructions in the reverse order before it returns (using the **ret** instruction). For the second cause, one checks to see if the subroutine has an instruction sequence that allocates space to local variables near the entry point of the subroutine. If it does, then the subroutine should have an instruction sequence to deallocate the stack space used by local variables before returning to the caller. The first two causes result in the incorrect return address to be popped out from the stack. Sometimes a subroutine has several returning points. One needs to make sure that registers saved in the stack are restored and local variables are deallocated before each **ret** instruction.

If the first two causes are not present, it is still possible that the subroutine gets stuck in some loop. To find out if the program gets stuck in a certain loop,

```
                mov     A,R0                ;   "
                addc    A,#0                ;   "
                mov     R0,A                ;   "
                inc     DPTR                ; get the next character
                movx    A,@DPTR             ;   "
                jz      doneD2B             ; if the character is NULL, prepare to return
                ajmp    doit
doneD2B:        cjne    signbd,#1,done3B
                mov     A,R1                ; The number is negative, compute it's
                cpl     A                   ; 2's complement
                add     A,#1                ;   "
                mov     R1,A                ;   "
                mov     A,R0                ;   "
                cpl     A                   ;   "
                addc    A,#0                ;   "
                mov     R0,A                ;   "
done3B:         pop     ACC
                mov     signbd,A
                ret
; ***************************************************************************************************************
; The following subroutine multiplies a number by 10. The number to be multiplied is passed in
; R0:R1 and the product is returned in R2:R4. R0 and R2 hold the more significant byte.
; ***************************************************************************************************************
mulby10:        push    B
                push    ACC
                mov     R2,#0               ; generate the first partial product
                mov     A,R1                ;   "
                mov     B,#10               ;   "
                mul     AB                  ;   "
                mov     R4,A                ; save the first product in R3:R4
                mov     R3,B                ;   "
                mov     A,R0                ; compute the second partial product
                mov     B,#10               ;   "
                mul     AB                  ;   "
                add     A,R3                ; add low byte of the product to R3
                mov     R3,A                ;   "
                mov     A,R2                ; add carry to R2
                addc    A,#0                ;   "
                mov     R2,A                ;   "
                mov     A,B                 ; add the high byte of the second partial product to R2
                add     A,R2                ;   "
                mov     R2,A                ;   "
                pop     ACC
                pop     B
                ret
                $include    (systeminit.a51)
                $include    (uartUtil.a51)
                end
```

one can insert a breakpoint after the last instruction of the suspicious loop. If the breakpoint is never reached, then one knows that the loop has some problems. Once the infinite loop is identified, it shouldn't be too hard to figure out what's wrong.

To make sure that the subroutine does not get stuck after calling other subroutines, one should make sure that the subroutines called by the current subroutine do not have the first three problems described in this step.

4.10.2 General Debugging Strategy

Subroutines can be classified into two categories: *intermediate* and *leaf* subroutines. An intermediate subroutine may call other subroutines, whereas a leaf subroutine does not call any other subroutines. Making sure that a subroutine returns to its caller does not guarantee that it produces correct results. One needs to use the methods described in Section 2.14 to debug each leaf subroutine and make sure it works correctly. After making sure that each leaf subroutine works correctly, one can start to debug the intermediate subroutines. One needs to make sure that each intermediate subroutine does not get stuck and works correctly using the methods discussed in Sections 4.10.1 and 2.14. After each intermediate subroutine has been debugged, one can then perform the top-level program debugging. Again, the method discussed in Section 2.14 can be used.

4.11 Chapter Summary

It happens very often that the same sequence of instructions needs to be executed in several places in a program. The user can place this sequence of instructions in a macro or a subroutine. Both the subroutine and the macro can be invoked in many places within a program. So, what is the difference between a macro and a subroutine? Their similarity and differences are as follows.

Similarities:

- Both the macro and subroutine can be invoked from many different places within a program.
- Parameters may be passed when a macro or a subroutine is invoked.
- Both create reusable code. A macro can be defined once and be included in many programs that can use it, so can a subroutine.
- Both the macro and subroutine can reduce the amount of typing when they are invoked more than once in the program.

Differences:

- A macro is invoked by entering the macro name and parameters, whereas a subroutine is invoked by executing a call instruction.
- A macro call does not need the saving and restoring of return address and hence will run faster than its equivalent in the subroutine format.

- Each time a macro is invoked, a copy of the instructions contained in the macro definition is inserted into the program. However, no such code duplication is made in a subroutine call.

The subroutine mechanism makes the "top-down design with hierarchical refinement" methodology easy to implement. In addition to creating reusable code, the subroutine mechanism enables the user to breakdown a difficult problem into smaller and manageable program units. A complicated problem will then become easier to solve. To implement the subroutine mechanism, the 8051 provides subroutine call and return instructions **acall**, **lcall**, and **ret**. The subroutine mechanism also requires the return address to be saved and restored. This is achieved by providing the stack data structure. The 8051 MCU allows the user to implement a stack using on-chip data memory. The 8-bit stack pointer (SP) points to the top byte of this stack. The user can push data into the stack and pop data out of the stack.

When making subroutine calls, the caller may need to pass certain values required by the subroutine for performing the intended operation. After the computation is completed, the subroutine may return the result back to the caller. Incoming parameters may be passed in registers, data memory, or the stack. The computation results from the subroutine may be returned in registers, data memory, or the stack. The subroutine may also need temporary variables for computation purpose. These temporary variables are called local variables and should be allocated in the stack to avoid interference among subroutines. However, the overhead in accessing the stack for the 8051 is so high that we avoid using the stack to allocate local variables. Local variables will be allocated in registers and data memory. The subroutine should save the contents of registers and data-memory locations that are used as local variables to preserve their values.

Although the 8051 provides instructions that can multiply and divide only 8-bit operands, the user can write subroutines to implement multiplication and division that operate on 16-bit or longer operands. These subroutines can operate on both signed and unsigned numbers.

When performing program development using a demo board together with a PC, UART is often a good vehicle for the demo board to communicate with the PC. The program running on the demo board can output information on the monitor screen of the PC and can read data from the keyboard. The data to be output to the UART port needs to be converted into an ASCII string. The data read from the keyboard must be converted to binary format before it can be processed by the MCU. UART I/O subroutines can be very useful in program debugging.

A program that contains one or multiple subroutines can be tricky to debug. One of the common program bugs is that program gets stuck in subroutine calls. Possible causes for a program to get stuck include:

- Forgetting to restore registers pushed onto the stack before returning to the caller
- Forgetting to deallocate local variables before returning to the caller

- Infinite loops in the subroutine
- Calling other subroutines that do not return

4.12 Exercise Problems

E4.1 Write a subroutine that can create a time delay that is a multiple of 1 ms. The multiple is passed in register R0.

E4.2 Write a subroutine that can create a time delay that is a multiple of 10 ms. The multiple is passed in register R0.

E4.3 Write a subroutine that can create a time delay that is a multiple of 100 ms. The multiple is passed in register R0.

E4.4 Write a subroutine that can divide a 16-bit signed integer into another 16-bit signed integer. The dividend and divisor are passed in R4:R5 and R6:R7, respectively. Remainder and quotient are to be returned in register pairs R2:R3 and R4:R5, respectively.

E4.5 Rewrite the program in Example 3.9 to use the subroutine you created in Problem E4.2 to create the desired delay.

E4.6 Write a program to generate a 100-Hz square waveform with a 50 percent duty cycle from the P5.0 pin.

E4.7 Write a program to generate a 200-Hz square waveform with a 40 percent duty cycle from the P5.0 pin.

E4.8 Write a program to generate a 10-kHz square waveform with a 50 percent duty cycle from the P5.0 pin.

E4.9 Write a program to generate a square waveform with a 50 percent duty cycle and a frequency that alternates between 1 kHz and 2 kHz every four seconds.

E4.10 Write a subroutine to find the number of elements that are a multiple of 5. The array is stored in program memory and has 16-bit elements. The array base and array count are passed to this subroutine in DPTR and B. The result is returned in B.

E4.11 Write a subroutine to implement **bubble sort**. This subroutine is designed to sort an array of n 8-bit elements. Assume that the array to be sorted is stored in XRAM. The array base and array count are passed to this subroutine in DPTR and B, respectively. The array has no more than 255 elements. The bubble sort algorithm can be found in Internet or any book in data structure.

E4.12 Write a subroutine that can divide an unsigned 32-bit integer into another unsigned 32-bit integer. Use internal data memory to pass parameters.

E4.13 The successive approximation method is an efficient method for finding the square root of an integer. The logic flow of this algorithm is shown in Figure 5.1. Write a subroutine to implement the algorithm.

E4.14 The most efficient method for finding out whether an integer is prime is to divide that number by all the prime numbers from 2 to the square root of the number to be tested. An approximation to this algorithm is to divide the number to be tested by 2 and all the odd integers from 2 to the square root of this number. Write a subroutine to implement this method for 16-bit integers.

E4.15 Write a subroutine to compute the greatest common divisor of two 16-bit unsigned integers by using the **Euclidian** method:

Step 1

```
If m = n then
   gcd ← m;
   return;
```

Step 2
If n < m then swap m and n.

Step 3
gcd ← 1

If m = 1 or n = 1 then return.

Step 4
p = n % m;

Step 5
if (p == 0) then m is the gcd.
else

```
n ← m;
m ← p;
```

Go to Step 4.

E4.16 Write a subroutine to compute the average of an array of *n* 8-bit elements. The array is stored in program memory. The starting address of the array and the array count are passed in DPTR and R7, respectively. The array average is to be returned in register B.

E4.17 Write a program to call the prime routine that you wrote in Problem E4.14 to find out all of the prime numbers between 8000 and 9999 and print them on the PC monitor screen in the format of five numbers in a row.

E4.18 Write a subroutine that can display from one to eight digits on seven-segment displays using the time multiplexing technique. The pointer to the segment pattern is passed in DPTR, whereas the number of digits is passed in B. Assume that the circuit in Figure 3.14 is used.

E4.19 Write a subroutine to implement ***selection sort*** and write a main program to test it. The selection sort method works like this:

You are given an array of integer values. Search through the array, find the largest value, and exchange it with the value stored in the first array location. Next, find the second largest value in the array, and exchange it with the value stored in the second array location. Repeat the same process until the end of the array is reached.

Assume that the array is stored in XRAM. The main program passes the starting address of the array and array count in DPTR and R7, respectively.

E4.20 Write a function to test whether a four-digit decimal number has the property:

The square of the sum of the upper half and the lower half of the given number is equal to the original number.

The number to be tested is passed in R6:R7. The function returns a 1 in A if the given number has the specified property. Otherwise, a 0 is returned in A.

E4.21 Write a program to find out all the four-digit decimal numbers that have the property described in Problem E4.20. Display the value on the PC monitor screen.

E4.22 Write a subroutine to test whether a 3-digit decimal integer has the property:

Let d2, d1, and d0 be the hundreds, tens, and ones digit. $d2^3 + d1^3 + d0^3$ = original number.

Write a program to call this function and find out all the 3-digit numbers that have this property.

4.13 Laboratory Exercise Problems and Assignments

L4.1 **Two-Tone Siren Generation.** Write a program to generate a two-tone siren from the P4.7 pin. The frequencies of these two tones are 440 Hz and 880 Hz, respectively. Each tone lasts for half of a second. Connect the P4.7 pin to a buzzer before testing the program.

L4.2 **Digital Waveform Generation.** A periodic digital waveform can be generated by toggling an I/O pin voltage level at an appropriate frequency. Use this method and the time delay subroutine to generate a digital waveform from the P4.7 pin with a frequency alternating between 1 kHz and 2 kHz every four seconds. Display the waveform using the oscilloscope.

L4.3 **Time-of-Day Display Using the PC Monitor Screen.** This lab assignment consists of the following operations.

Step 1. Output a message to ask the user to enter the current *time-of-day* from the keyboard in the format of **hh:mm:ss.**

Step 2. Invoke the **gets** subroutine to read in the time.

Step 3. Invoke the time-delay function to create a 1-second delay to update the current time-of-day.

Step 4. Display and update the current time-of-day on the monitor screen in the format of **Current time is hh:mm:ss** on the monitor screen.

Step 5. Backspace eight places (output the value 0x08 eight times) before outputting the new time on the screen.

C Language Programming

5.1 Objectives

Upon successful completion of this chapter, you will be able to:

- Explain the overall structure of a C language program
- Use the appropriate operators to perform desired operations in the C language
- Understand the basic data types and expressions of the C language
- Write program loops in the C language
- Write functions and make function calls in the C language
- Use arrays and pointers for data manipulation
- Perform basic I/O operations in the C language
- Use the SiLabs IDE to enter, compile, and debug C programs

5.2 Introduction to C

This chapter is not intended to provide a complete coverage of the C language. Instead, it provides a summary of those C language constructs that will be used in this book. You will be able to deal with the basic 8051 interfacing programming if you fully understand the contents of this chapter. In addition to providing a tutorial to the C language, this chapter will also provide a tutorial on using the SiLabs IDE to enter, compile, and debug C programs.

The C language is gradually replacing the assembly language in many embedded applications because it has several advantages over the assembly language. The most important one is that it allows the user to work on program logic at a level higher than assembly language, and thus programming productivity is improved greatly.

A C program, whatever its size, consists of functions and variables. A function contains statements that specify the operations to be performed. The types of statements in a function could be a *declaration, assignment, function call, control,* or *NULL*. A variable stores a value to be used during the computation. The **main( )** function is required in every C program and is the one to which control is passed when the program is executed. A simple C program is as follows.

```
(1)  #include <c8051F040.h>                  -- include C8051F040 header file
(2)  /* this is where program execution begins */
(3)  void main (void)                       -- defines a function named main that receives
                                            -- no argument values and returns no values
(4)  {                                      -- statements of main are enclosed in braces
(5)        SFRPAGE = 0x0F;                  -- switch to SFR page F
(6)        WDTCN = 0xDE;                    -- disable watchdog timer
(7)        WDTCN = 0xAD;                    --        "
(8)        OSCICN = 0xC3;                   -- enable internal oscillator, set prescaler to 1
(9)        CLKSEL = 0;                      -- use internal oscillator to generate SYSCLK
(10)       P5MDOUT = 0xFF;                  -- configure Port P5 to be push pull
(11)       P5 = 0x59;                       -- output 0x59 to port P5
(12) }                                      -- the end of main function
```

The first line of the program

```
#include <C8051F040.h>
```

causes the file C8051F040.h to be included in the program. This line appears at the beginning of most C programs. The header file C8051F040.h contains the definitions (association of symbolic names and their addresses) of all special function registers (SFRs) and addressable bits so that users can use symbolic names to refer to SFRs and bits.

The second line is a comment. A comment explains what will be performed and will be ignored by the compiler. A comment in C language starts with /* and ends with */. Everything in between /* and */ is ignored. Comments provide documentation to the program and enhance readability. Comments affect only the size of the text file and do not increase the size of the executable code. Many commercial C compilers also allow the user to use two slashes (//) for commenting out a single line.

The third line main() is where program execution begins. The opening brace on the fourth line marks the start of the main() function's code. Every C program must have one, and only one, main() function. Program execution is also ended with the **main()** function. The fifth line switches to the SFR page F in order to access SFRs that are residing in this page.

The sixth and seventh lines disable the watchdog timer function. The eighth line enables internal oscillator and sets its prescaler to 1. The ninth line selects the internal oscillator to generate SYSCLK.

The tenth line configures port P5 to be a push-pull item, so that it can be used as an output. The eleventh line outputs the value 0x59 to port P5. The closing brace in the twelveth line ends the main() function.

5.3 Types, Operators, and Expressions

Variables and constants are the basic objects manipulated in a program. Variables must be declared before they can be used. A variable declaration must include the name and type of the variable and may optionally provide its initial value. A variable name may start with a letter (**A** through **Z** or **a** through **z**) or underscore character followed by zero or more letters, digits, or underscore characters. Variable names cannot contain arithmetic signs, dots, apostrophes, C keywords or special symbols, such as **@**, **#**, **?**, and so on. Adding the underscore character (_) may sometimes improve the readability of long variables. Don't begin variable names with an underscore, however, since library routines often use such names. C language is case sensitive. Uppercase and lowercase letters are distinct.

5.3.1 Data Types

There are only a few basic data types in C: **void**, **char**, **int**, **float**, and **double**. A variable of type **void** represents nothing. The type **void** is used most commonly with functions, which can indicate that the function does not return any value or does not have incoming parameters. A variable of type **char** can hold a single byte of data (signed). A variable of type **int** is an integer (signed), which is normally the natural size (word length) for a particular machine. The type **float** refers to a 32-bit, single-precision, floating-point number. The type **double** represents a 64-bit, double-precision, floating-point number. In addition, there are a number of qualifiers that can be applied to these basic types. **short** and **long** apply to integers. These two qualifiers will modify the lengths of integers. An integer variable is 16-bit by default for many C compilers (including SDCC and Keil's C compiler). The modifier **short** does not change the length of an integer. The modifier **long** doubles a 16-bit integer to 32-bit. The keyword **unsigned** should be used if the variables are never negative to improve the efficiency of the generated code.

5.3.2 Variable Declarations

All variables must be declared before their use. A declaration specifies a type and contains a list of one or more variables of that type, as in

```
int     i, j, k;
char    cx, cy;
```

A variable may also be initialized when it is declared, as in

```
int     i = 0;
char    echo = 'y';     // the ASCII code of letter y is assigned to variable echo.
```

5.3.3 Constants

There are four kinds of constants: *characters, integers, floating-point numbers,* and *strings.* A character constant is an integer written as one character within single quotes, such as **x**. A character constant is represented by the ASCII code of the character. A string constant is a sequence of zero or more characters surrounded by double quotes, as in

```
"C8051F040 is a microcontroller made by Silicon Laboratory."
```

or

```
""                /* an empty string */
```

Each individual character in the string is represented by its ASCII code.

An integer constant such as 3241 is an **int**. A long constant is written with a terminal *l* (el) or *L*, as in **44332211L**. The following constant characters are predefined in C language (can be embedded in a string):

\a	alert (bell) character	\\	backslash
\b	backspace	\?	question mark
\f	formfeed	\'	single quote
\n	newline	\"	double quote
\r	carriage return	**\ooo**	octal number
\t	horizontal tab	**\xhh**	hexadecimal number
\v	vertical tab		

As in assembly language, a number in C can be specified in different bases. The method to specify the base of a number is to add a prefix to the number. The prefixes for different bases are

Base	Prefix	Example	
decimal	none	1357	
octal	0	04723	;preceded by a zero
hezadecimal	0x	0x2A	

5.3.4 Arithmetic Operators

There are seven arithmetic operators:

+	add and unary plus	%	modulus (or remainder)
–	subtract and unary minus	++	increment
*	multiply	– –	decrement
/	divide		

The expression

```
a % b
```

produces the remainder when a is divided by b. The % operator cannot be applied to **float** or **double**. The ++ operator adds 1 to the operand, and the −− operator subtracts 1 from the operand. The / operator performs a division and truncates the quotient to an integer when both operands are integers.

Example 5.1 What value will be assigned to *ax* for the following statement?

```
ax = 150 / 7;
```

Solution: The integral part of 150/7 is 21. Therefore, *ax* receives 21 after the previous statement is executed.

Example 5.2 What value will be assigned to *bx* for the following statement?

```
bx = 235 % 20;
```

Solution: The remainder of 235/20 is 15. Therefore, *bx* receives 15 after the previous statement is executed.

Example 5.3 Assume that *ak* is a four-digit (decimal) integer. Write a few C statements to separate *ak* into two parts and assign the upper two digits to the variable *ax* and the lower two digits to the variable *bx*.

Solution: A four-digit value $d_3d_2d_1d_0$ can be written as

$$d_3d_2d_1d_0 = d_3d_2 \times 100 + d_1d_0$$

where d_3, d_2, d_1, and d_0 are the thousand's, hundred's, ten's, and one's digit of the number, respectively.

Therefore, we can divide a four-digit integer by dividing 100 into that number. The following two statements will achieve the desired operation.

```
ax = ak / 100;
bx = ak % 100;
```

5.3.5 Bitwise Operators

C provides six operators for bit manipulations; these operators may be applied only to integral operands, that is, **char**, **short**, **int**, and **long**, whether they are *signed* or *unsigned*:

&	AND	~	NOT
|	OR	>>	right shift
^	XOR	<<	left shift

The **&** operator is often used to clear one or more bits to zero. For example, the statement

```
P1 = P1 & 0xBD;          /* variable P1 is 8 bits */
```

clears bits 6 and 1 of P1 to 0.

The | operator is often used to set one or more bits to 1. For example, the statement

```
P2 = P2 | 0x40;          /* variable P2 is 8 bits */
```

sets the bit 6 of P2 to 1.

The XOR operator can be used to toggle a bit. For example, the statement

```
abc = abc ^ 0xF0;        /* abc is of type char */
```

toggles the upper four bits of the variable *abc*.

The >> operator shifts the involved operand to the right for the specified number of places. For example,

```
xyz = abc >> 3;
```

shifts the variable *abc* to the right three places and assigns the result to the variable *xyz*.

The << operator shifts the involved operand to the left for the specified number of places. For example,

```
xyz = xyz << 4;
```

shifts the variable *xyz* to the left four places.

The assignment operator = is often combined with the operator when the destination operand is the same as one of the source operand. For example,

```
PTP = PTP & 0xBD;
```

can be rewritten as

```
PTP & = 0xBD;
```

The statement

```
P2 = P2 | 0x40;
```

can be rewritten as

```
P2 | = 0x40;
```

5.3.6 Relational and Logical Operators

Relational operators are used in expressions to compare the values of two operands. If the result of the comparison is true, then the value of the expression is 1. Otherwise, the value of the expression is zero. The relational and logical operators are

==	equal to (two = characters)	<=	less than or equal to
!=	not equal to	&&	and
>	greater than	||	or
>=	greater than or equal to	!	not (one's complement)
<	less than		

Some examples of relational and logical operators are

```
if (!(ADCTL & 0x80))
        statement1;        /* if bit 7 is 0, then execute statement1 */

if (i > 0 && i < 10)
        statement2;        /* if 0 < i < 10 then execute statement2 */

if (a1 == a2)
        statement3;        /* if a1 equals a2 then execute statement3 */
```

5.3.7 Precedence of Operators

Precedence refers to the order in which operators are processed. The C language maintains a precedence for all operators, as shown in Table 5.1. Operators at the same level are evaluated from left to right. A few examples that illustrate the precedence of operators are listed in Table 5.2.

5.4 Control Flow

The control-flow statements specify the order in which computations are performed. In the C language, the semicolon is a statement terminator. Braces { and } are used to group declarations and statements together into a *compound statement* or *block*, so that they are syntactically equivalent to a single statement.

5.4.1 if Statement

The **if** statement is a conditional statement. The statement associated with the **if** *statement* is executed based upon the outcome of a condition. If the condition evaluates to nonzero, the statement is executed. Otherwise, it is skipped.

TABLE 5.1 *Table of Precedence of Operators*

<table>
<tr><th>Precedence</th><th>Operator</th><th>Associativity</th></tr>
<tr><td>Highest</td><td>( ) [ ] → .</td><td>Left to right</td></tr>
<tr><td>↑</td><td>! ~ ++ -- - * & (type) sizeof</td><td>Right to left</td></tr>
<tr><td></td><td>* / %</td><td>Left to right</td></tr>
<tr><td></td><td>+ -</td><td>Left to right</td></tr>
<tr><td></td><td><< >></td><td>Left to right</td></tr>
<tr><td></td><td>< <= > >=</td><td>Left to right</td></tr>
<tr><td></td><td>== !=</td><td>Left to right</td></tr>
<tr><td></td><td>&</td><td>Left to right</td></tr>
<tr><td></td><td>^</td><td>Left to right</td></tr>
<tr><td></td><td>|</td><td>Left to right</td></tr>
<tr><td></td><td>&&</td><td>Left to right</td></tr>
<tr><td></td><td>||</td><td>Left to right</td></tr>
<tr><td></td><td>?:</td><td>Right to left</td></tr>
<tr><td></td><td>= += -= *= /= %= &= ^= |= <<= >>=</td><td>Right to left</td></tr>
<tr><td></td><td>,</td><td>Left to right</td></tr>
<tr><td>Lowest</td><td></td><td></td></tr>
</table>

TABLE 5.2 *Examples of Operator Precedence*

<table>
<tr><th>Expression</th><th>Result</th><th>Note</th></tr>
<tr><td>15 - 2 * 7</td><td>1</td><td>* has higher precedence than -</td></tr>
<tr><td>(13 - 4) * 5</td><td>45</td><td></td></tr>
<tr><td>(0x20 | 0x01) != 0x01</td><td>1</td><td></td></tr>
<tr><td>0x20 | 0x01 != 0x01</td><td>0x20</td><td>!= has higher precedence than |</td></tr>
<tr><td>1 << 3 + 1</td><td>16</td><td>+ has higher precedence than <<</td></tr>
<tr><td>(1 << 3) + 1</td><td>9</td><td></td></tr>
</table>

The syntax of the **if** *statement* is

```
if (expression)
    statement;
```

An example of an **if** statement is

```
if (a > b)
    sum += 2;
```

The value of **sum** will be incremented by 2 if the variable *a* is greater than the variable *b*.

5.4.2 IF ELSE STATEMENT

The **if-else** statement handles conditions where a program requires one statement to be executed if a condition is nonzero and a different statement if the condition is zero.

The syntax of an **if-else** statement is

```
if (expression)
        statement1
else
        statement2
```

The *expression* is evaluated. If it is true (nonzero), $statement_1$ is executed. If it is false, $statement_2$ is executed. Here is an example of the **if-else** statement:

```
if (a != 0)
        r = b;
else
        r = c;
```

The **if-else** statement can be replaced by the **?:** operator. The statement

```
r = (a != 0)? b : c;
```

is equivalent to the previous **if-else** statement.

5.4.3 MULTIWAY CONDITIONAL STATEMENT

A multiway decision can be expressed as a cascaded series of **if-else** statements. Such a series looks like

```
if (expression1)
        statement1
else if (expression2)
        statement2
else if (expression3)
        statement3
. . .
else
        statementn
```

An example of a three-way decision is

```
if (abc > 0) return 5;
else if (abc == 0) return 0;
else return -5;
```

5.4.4 SWITCH STATEMENT

The **switch** statement is a multiway decision based on the value of a control expression. The syntax of the **switch** statement is

```
switch (expression) {
    case const_expr1:
        statement1;
        break;
    case const_expr2:
        statement2;
        break;
    ...
    default:
        statementn;
}
```

As an example, consider the following program fragment.

```
switch (i) {
    case 1:
        duty = 10;
        break;
    case 2:
        duty = 20;
        break;
    case 3:
        duty = 30;
        break;
    case 4:
        duty = 40;
        break;
    case 5:
        duty = 50;
        break;
    default:
        duty = 0;
}
```

The variable **duty** receives a value that is equal to $i \times 10$. The keyword **break** forces the program flow to drop out of the **switch** statement so that only the statements under the corresponding case-labels are executed. If any **break** statement is missing, then all of the statements from that case-label until the next **break** statement within the same **switch** statement will be executed.

5.4.5 FOR-LOOP STATEMENT

The syntax of a **for-loop** statement is

```
for (expr1; expr2; expr3)
    statement;
```

where **expr1** and **expr3** are assignments or function calls, and **expr2** is a relational expression. For example, the following **for loop** computes the sum of the squares of integers from 1 to 9.

```
sum = 0;
for (i = 1; i < 10; i++)
    sum = sum + i * i;
```

The following **for loop** prints out the first 10 odd integers.

```
for (i = 1; i < 20; i++)
    if (i % 2) printf("%d ", i);
```

5.4.6 WHILE STATEMENT

While an *expression* is nonzero, the **while** loop repeats a statement or block of code. The value of the expression is checked prior to each execution of the statement. The syntax of a **while** statement is

```
while (expression)
    statement;
```

The *expression* is evaluated. If it is nonzero (true), the *statement* is executed and *expression* is reevaluated. This cycle continues until the *expression* becomes zero (false), at which point execution resumes after the *statement*. The *statement* may be a NULL statement. A NULL statement does nothing and is represented by a semicolon. Consider the program fragment:

```
intCnt = 5;
while (intCnt);
```

The CPU will stay in the **while-loop** until the variable **intCnt** is equal to zero. In microcontroller applications, the decrement of **intCnt** is often triggered by external events, such as interrupts.

5.4.7 DO-WHILE STATEMENT

The **while** and **for** loops test the termination condition at the beginning. By contrast, the **do-while** statement tests the termination condition at the end of the statement; the body of the statement is executed at least once.

The syntax of the statement is

```
do
    statement
while (expression);
```

The following **do-while** statement displays the integers 9 down to 1.

```
int digit = 9;
do
    printf("%d ", digit--);
while (digit >= 1);
```

5.4.8 GOTO STATEMENT

Execution of a **goto** statement causes control to be transferred directly to the labeled statement, which must be located in the same function as the **goto** statement. The use of the **goto** statement interrupts the normal sequential flow of a program and thus makes it harder to follow and decipher. For this reason, the use of **goto** is not considered good programming style, so it is recommended that you do not use it in your program.

The syntax of the **goto** statement is

```
goto label
```

An example of the use of a **goto** statement is

```
if (x > 100)
    goto fatal_error;
...

fatal_error:
    printf("Variable x is out of bound!\n);
```

5.5 Input and Output

Input and output facilities are not part of the C language itself. However, input and output are very important in applications. The ANSI standard defines a set of library functions that must be included so that they can exist in a compatible form on any system where C exists. Some of the functions deal with file input and output. Others deal with text input and output. In this section, we will look at four input and output functions.

int *getchar* ( ). This function returns a character when it is called. The following program fragment returns a character and assigns it to the variable *xch*.

```
char xch;
xch = getchar ();
```

int *putchar* (int). This function outputs a character on the standard output device. The following statement outputs the letter *a* from the standard output device.

```
putchar ('a');
```

int *puts* (const char *s). This function outputs the string pointed to by **s** on the standard output device. The following statement outputs the string "Learning microcontroller is fun!" from the standard output device.

```
puts ("Learning microcontroller is fun! \n");
```

int *printf* (*formatting string*, $\mathbf{arg_1}$, $\mathbf{arg_2}$, . . . , $\mathbf{arg_n}$). This function converts, formats, and prints its arguments on the standard output under the control of a *formatting string*. $\mathbf{arg_1}$, $\mathbf{arg_2}$, . . . , $\mathbf{arg_n}$ are arguments that represent the individual output data items. The arguments can be written as constants, single variable or array names, or more complex expressions. The formatting string is composed of individual groups of characters, with one character group associated with each output data item. The character group corresponding to a data item must start with %. In its simplest form, an individual character group will consist of the percent sign followed by a *conversion character*, indicating the type of the corresponding data item.

Multiple character groups can be contiguous or separated by other characters, including white-space characters. These "other" characters are simply transferred directly to the output device where they are displayed. A subset of the more frequently used conversion characters is listed in Table 5.3. Between the % character and the conversion character there may be, in order:

- A minus sign, which specifies *left adjustment* of the converted argument.
- A number that specifies the minimum *field width*. The converted argument will be printed in a field at least this wide. If necessary, it will be padded on the left (or right, if left adjustment is called for) to make up the field width.
- A period, which separates the field width from precision.
- A number for the precision, which specifies the maximum number of characters to be printed from a string, the number of digits after the decimal point of a floating-point value, or the minimum number of digits for an integer.
- An ***h*** if the integer is to be printed as **short**, or ***l*** (letter el) if as **long**.

Several valid **printf** calls are given next.

```
printf ("this is a challenging course! \n");      /* outputs only a string */
printf ("%d %d %d", x1, x2, x3);                  /* outputs variables x1, x2, x3 using a minimal number
                                                     of digits with one space separating each value */
printf("Today's temperature is %4.1d \n", temp);  /* display the string Today's temperature is followed by
                                                     the value of temp. Display one fractional digit and use at
                                                     least four digits for the value. */
```

TABLE 5.3 *Commonly used Conversion Characters for Data Output*

Conversion Character	Meaning
c	Data item is displayed as a single character.
d	Data item is displayed as a signed decimal number.
e	Data item is displayed as a floating-point value with an exponent.
f	Data item is displayed as a floating-point value without an exponent.
g	Data item is displayed as a floating-point value using either e-type or f-type conversion, depending on value; trailing zeros, trailing decimal point will not be displayed.
i	Data item is displayed as a signed decimal integer.
o	Data item is displayed as an octal integer, without a leading zero.
s	Data item is displayed as a string.
u	Data item is displayed as an unsigned decimal integer.
x	Data item is displayed as a hexadecimal integer, without the leading 0x.

By default, functions **putchar**, **puts**, and **printf** output data to the computer monitor via the UART port, whereas the **getchar** function receives data entered from the computer keyboard via the UART port. The UART port is discussed in Chapter 9. Most embedded systems are not connected to the PC, and hence, users cannot use these functions to perform I/O. However, these functions can be very helpful during the program development stage.

5.6 Functions and Program Structure

Every C program consists of one or more functions. If a program consists of multiple functions, their definitions cannot be embedded within another. The same function can be called from several different places within a program. Generally, a function will process information passed to it from the calling portion of the program and return a single value.

Information is passed to the function via special identifiers called *arguments* (also called *parameters*) and returned via the **return** statement. Some functions, however, accept information but do not return anything (for example, the library function **printf**). There are also functions that do not receive or return any values. Their purpose is to modularize the program, i.e., make it more readable and easier to develop.

The syntax of a function definition is

```
return_type function_name (declarations of arguments)
{
    declarations and statements
}
```

The declaration of an argument in the function definition consists of two parts: the *type* and the *name* of the variable. The return type of a function is *void* if it does not return any value to the caller. An example of a function that converts a lowercase letter to an uppercase letter is

```
char lower2upper (char cx)
{
    if (cx >= 'a' && cx <= 'z') return (cx - ('a' - 'A'));
    else return cx;
}
```

A character is represented by its ASCII code. A letter is in lowercase if its ASCII code is between 97 (0x61) and 122 (0x7A). To convert a letter from lowercase to uppercase, subtract its ASCII code by the difference of the ASCII codes of letters *a* and *A*.

Example 5.4 Write a function to use Timer 0 to generate a time delay that is a multiple of 100 ms, assuming that a 24-MHz external crystal oscillator is selected to generate the system clock. The multiple is passed to this function as a parameter.

Solution: As discussed in Chapter 3, Timer 0 registers are in SFR page 0. By setting the prescale factor of the clock source of Timer 0 to 48, the number of clock cycles to create 100 ms delay is 50,000. We should save the current SFR page value and then switch to page 0 before we can access Timer 0 registers. The SFRPAGE value should be restored before the function returns to the caller. The following C function creates the desired delay.

```
//****************************************************************************************************
// The following function creates a time delay that is a multiple of 100 ms.
//****************************************************************************************************
void dlyby100ms (char k)
{
    char temp, i;
    temp     = SFRPAGE;
    SFRPAGE  = TIMER01_PAGE;   // TMR0 SFRPAGE
    TMOD     = 0x01;           // configure Timer 0 to Mode 1
    TF0      = 0;
    CKCON    = 0x02;           // use SYSCLK / 48 as Timer 0 clock source
    for (i = 0; i < k; i++) {
        TH0  = 0x3C;           // place 15536 in Timer 0 so it overflows in 100 ms (50,000 cycles)
        TL0  = 0xB0;           //    "
        TR0  = 1;              // enable Timer 0 to count
        while (!TF0);          // wait for 100 ms
        TF0  = 0;
        TR0  = 0;              // stop timer 0
    }
    SFRPAGE = temp;            // restore the SFRPAGE
}
```

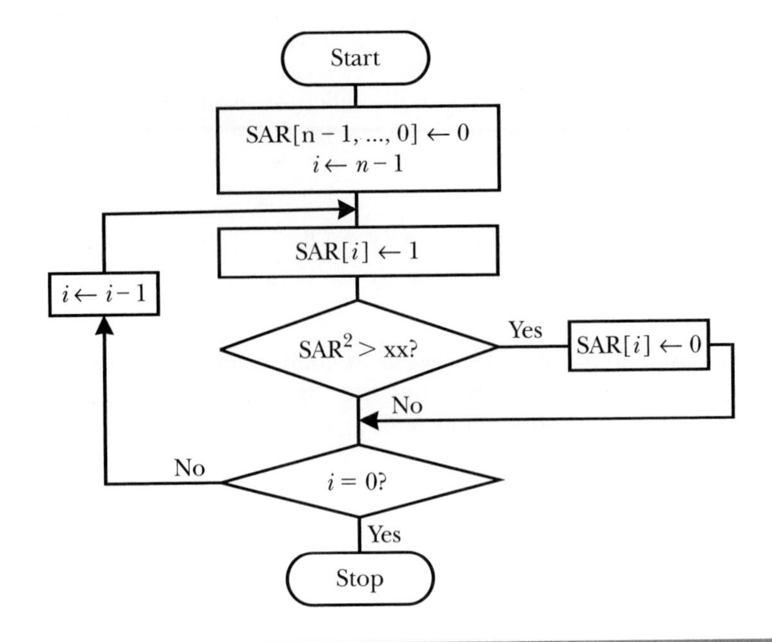

FIGURE 5.1 *Finding square root of the integer xx using successive-approximation method.*

Example 5.5 Write a function to find the integer square root of a 16-bit unsigned integer.

Solution: The *successive approximation method* is a very efficient method for computing the square root of an integer. This method is also used in performing *analog-to-digital* conversion. The logic flow of this method is illustrated in Figure 5.1. The SAR register is used to accumulate the result.

The square root found in Figure 5.1 tends to be smaller than the true square root. The integer closest to the true square root could be [**SAR**] (the contents of the SAR register) or [**SAR**] + 1. To find the closest approximation to the true square root, we need to compare the values of $(\mathbf{y} - [\mathbf{SAR}]^2)$ and $(([\mathbf{SAR}] + 1)^2 - \mathbf{y})$ where y is the integer of which the square root is to be found.

The C function that finds the square root of a 16-bit integer is as follows.

```
unsigned char sqroot (unsigned int y)
{
    unsigned int i, k, mask, test;
    k = 0;
    mask = 0x80;
    for (i = 0; i < 8; i++) {
        test = k | mask;
        if ((test * test) < y)
            k = test;
        mask = mask >> 1;
    }
```

```
    if ((y – k * k) < ((k+1) * (k+1) – y))
        return k;
    else return (k+1);
}
```

Example 5.6 Write two functions that will select the internal and external crystal oscillator to generate system clock, respectively.

Solution: The functions that select the internal oscillator and external crystal oscillator as system clock are as follows.

```
void sysclkUseIosc (void)
{
    SFRPAGE = CONFIG_PAGE;
    CLKSEL  = 0;                // select internal oscillator as system clock
    OSCICN  = 0x83;             // set frequency to 24.5 MHz (prescaler = 1)
}

void sysclkUseEosc (void)
{
    int n;                          // local variable used in delay FOR loop.
    SFRPAGE = CONFIG_PAGE;          // switch to config page
    OSCXCN  = 0x67;                 // start external oscillator; 24 MHz Crystal
                                    // system clock is 24 MHz
    for (n = 0; n < 255; n++);      // delay about 1 ms
    while ((OSCXCN & 0x80) == 0);   // wait for oscillator to stabilize
    CLKSEL  |= 0x01;                // switch to external oscillator
}
```

5.6.1 Function Prototype

A function cannot be called before it has been defined. This dilemma is solved by using the *function prototype* statement. The syntax for a function prototype statement is

return_type **function_name** (declarations of arguments);

For example, the statement

```
char test_prime (int a);
```

before **main( )** is a function prototype statement.

5.6.2 Writing a C Program with Multiple Functions

To call a function, simply write the name of the function and replace the argument declarations by actual arguments or values and terminate it with a semicolon.

Example 5.7 Write a program to drive the LED circuit in Figure 3.13, display one LED at a time from the one driven by pin 7 toward the one driven by pin 0, and then reverse. Repeat this operation forever. Each LED is lighted for 200 ms.

Solution: The values to drive Port P5 to turn on one LED at a time should be placed in a lookup table. The program reads one value at a time from the table, outputs it to P5, and then waits for 200 ms. The C program that performs the desired operation is as follows.

```
#include <C8051F040.h>
void dlyby100ms (char k);
void sysclkUseEosc (void);

void main (void)
{
    unsigned char led_tab[]=  {0x80,0x40,0x20,0x10,0x08,0x04,0x02,0x01,
                               0x01,0x02,0x04,0x08,0x10,0x20,0x40,0x80};
    char i;
    external_osc();            // use external crystal oscillator to generate SYSCLK
    WDTCN  = 0xDE;             // disable watchdog timer
    WDTCN  = 0xAD;
    while (1) {
        for (i = 0;i < 16; i++){
            P5 = led_tab[i]; /* output a new LED pattern */
            dlyby100ms(2);
        }
    }
}
// include void sysclkUseEosc (void) function here
// include void dlyby100ms (char k) function here
```

Example 5.8 Write a program to find out the number of prime numbers between 100 and 600.

Solution: The most efficient method to find out if a number is prime is to divide the given number by all the prime numbers between 2 and its square root. If none of them can divide the given number, then it is a prime number. However, we don't have all the prime numbers at hand; we will be satisfied by dividing the given number by all the numbers from 2 to its square root.

```
#include <C8051F040.h>
char test_prime (int k);
unsigned char sqroot (unsigned int y);
void main (void)
{
    int i, prime_count;
```

```
    prime_count = 0;
    for (i = 100; i <= 600; i++) {
        if (test_prime(i))
            prime_count ++;
    }
    SFRPAGE  = 0x0F;
    P5MDOUT = 0xFF;
    P5        = prime_count;      // use LEDs to display the number of prime numbers
}
//************************************************************************************************
// The following function tests if k is a prime. It returns a 1 if it is; otherwise, it returns a 0
//************************************************************************************************
char test_prime (int k)
{
    int i;
    char limit;
    if (k == 1) return 0;
    limit = sqroot(k);              // use the square root as prime test limit
    for (i = 2; i < limit; i++)
        if ((k % i) == 0) return 0; // divisible?
    return 1;
}
//************************************************************************************************
// include sqroot (unsigned int y) here
//************************************************************************************************
```

5.7 Pointers, Arrays, Structures, and Unions

A pointer holds the address of a variable. It is related closely to arrays, structures, and unions. An array is a collection of data of the same type. A structure is a collection of data which may have the same or different data types. A union is a variable that may hold (at different times) objects of different types and sizes. A pointer can be used to access an array, a structure, and a union.

5.7.1 Pointers and Addresses

A *pointer* is a variable that holds the address of a variable. Pointers are used frequently in C, as they have a number of useful applications. For example, pointers can be used to pass information back and forth between a function and its reference (calling) point. In particular, pointers provide a way to return multiple data items from a function via function arguments. Pointers also permit references to other functions to be specified as arguments to a given function. This has the effect of passing functions as arguments to the given function.

Pointers also are associated closely with arrays and therefore provide an alternative way to access individual array elements. The syntax for declaring a pointer type is

```
type_name *pointer_name;
```

For example,

```
int *ax;
```

declares that the variable *ax* is a pointer to an integer, and

```
char *cp;
```

declares that the variable *cp* is a pointer to a character.

To access the value pointed to by a pointer, use the *dereferencing* operator *. For example,

```
int     a, *b;        /* b is a pointer to int */
...
a = *b;
```

assigns the value pointed to by *b* to variable *a*.

We can assign the address of a variable to a pointer by using the unary operator **&**. The following example shows how to declare a pointer and how to use **&** and *.

```
int        x, y;
int        *ip;         /* ip is a pointer to an integer */

ip         = &x;        /* assigns the address of the variable x to ip */
y          = *ip;       /* y gets the value of x */
```

5.7.2 ARRAYS

Many applications require the processing of multiple data items that have common characteristics (e.g., a set of numerical data represented by x_1, x_2, . . . , x_n). In such situations, it is more convenient to place data items into an *array*, where they will all share the same name. The individual data items can be characters, integers, floating-point numbers, and so on. They must all be of the same type and the same storage class.

Each array element is referred to by specifying the array name followed by one or more *subscripts* with each subscript enclosed in brackets. Each subscript must be expressed as a nonnegative integer. Thus, the elements of an *n*-element array x are x[0], x[1], . . . , x[n − 1]. The number of subscripts determines the dimensionality of the array. For example, x[i] refers to an element of a one-dimensional array. Similarly, y[i][j] refers to an element of a two-dimensional array. Higher-dimensional arrays can be formed by adding additional subscripts in the same manner. However, higher-dimensional arrays are not used very often in 8- and 16-bit microcontroller applications. In general, a one-dimensional array can be expressed as

```
data-type array_name[expression];
```

A two-dimensional array is defined as

```
data-type array_name[expr1][expr2];
```

An array can be initialized when it is defined. This is a technique used in table lookup, which can speed up the computation process. For example, a song-playing program in Chapter 8 places the pitch and duration information of every note in a table and reads them sequentially during the execution process.

Example 5.9 Write the bubble sort function to sort an array of integers.

Solution: The basic idea of bubble sort is to go through the array sequentially several iterations with each iteration placing one element in its right position. An iteration consists of comparing each element in the array with its successor (**x[i]** with **x[i+1]**) and interchanging them if they are not in proper order (either ascending or descending).

For an array with N elements, $N-1$ comparisons are performed during the first iteration. As more and more iterations are performed, more and more elements would be moved to their right positions. Fewer comparisons are needed. In the worse case, $N-1$ iterations are needed, and only one comparison is made during the last iteration.

The bubble sort program can be made more efficient by keeping track of whether *swap* operations have been performed. If no swap is made in an iteration, then the array is already sorted, and the process should be stopped. The function that implements this idea is as follows.

```
void   swap (char *px, char *py);
void  bubble (char a[ ], char n)  // n is the array count
{
    char i, j;
    char inorder;     // array in order flag
    for (i = 0; i < n − 1; i++){
            inorder = 1;        // assume array is in order at the start of a new iteration
            for (j = 0; j < n − i − 1; j++)
                    if (a[j] > a[j+1]){ // are two adjacent elements not in order?
                            swap (&a[j], &a[j+1]);
                            inorder = 0;    // area not in order
            }
    if (inorder) // array is in order, there is no need to sort.
            return;
    }
}
```

```
void swap (char *px, char *py)
{
    char temp;
    temp = *px;
    *px = *py;
    *py = temp;
}
```

Example 5.10 Write a function that can use a time-multiplexing technique to display one to six digits simultaneously with the circuit shown in Figure 3.15. These digits are to be displayed for 600 ms.

Solution: The pattern and digit-select information of k ($k = 1 \ldots 6$) digits are stored as

digit-pattern$_1$, digit-select$_1$, digit-pattern$_2$, digit-select$_2$, . . . , digit-pattern$_k$, digit-select$_k$

Since there are k digits to be displayed for 600 ms, the function needs to go through the pattern array $600/k$ times. The function that can perform the desired operation is as follows.

```
//*******************************************************************************************
// The following function uses time-multiplexing technique to display k digits
// simultaneously for 600 ms.
//*******************************************************************************************
void seg7_mux(char code *ptr, char k)
{
        char ix, *cptr;
        int     jx;
        cptr    = ptr;
        for (jx = 0; jx < 600 / k; jx++) {
                ptr     = cptr;
                for (ix = 0; ix < k; ix++){
                        P5      = *ptr++;       // output pattern value
                        P7      = *ptr++;       // output digit select value
                        dlyby1ms(1);            // wait for 1 ms
                }
        }
}
//*******************************************************************************************
// The following function creates a time delay that is n ms. n is passed to this function.
//*******************************************************************************************
void dlyby1ms (char n)
{
        char temp, ix;
        temp    = SFRPAGE; // save the current SFRPAGE value
        SFRPAGE = TIMER01_PAGE; // switch to the SFR page where TMR0 resides
        TMOD    = 0x01;         // Timer 0 operate in mode 1
        CKCON   = 0x00;         // use SYSCLK / 12 as Timer 0 clock source
```

```
        TF0     = 0;            // clear TF0 flag
        for (ix = 0; ix < n; ix++){
            TH0  = 0xF8;        // place 63536 in Timer 0 so it overflows in 1 ms (2,000 cycles)
            TL0  = 0x30;        //                    "
            TR0  = 1;           // enable Timer 0 to count
            while(!TF0);        // wait for 1 ms
            TF0  = 0;
            TR0  = 0;           // stop timer 0
        }
        SFRPAGE = temp;
}
```

5.7.3 Pointers and Arrays

In C, there is a strong relationship between pointers and arrays. Any operation that can be achieved by array subscripting can also be done with pointers. The pointer version in general will be faster but somewhat harder to understand. For example,

```
int ax[20];
```

defines an array *ax* of 20 integral numbers. The notation ax[i] refers to the *i*th element of the array. If *ip* is a pointer to an integer, declared as

```
int *ip;
```

then the assignment

```
ip = &ax[0];
```

makes *ip* contain the address of ax[0]. Now the statement

```
x = *ip;
```

will copy the contents of ax[0] into *x*. If *ip* points to ax[0], then *ip* + 1 points to ax[1], and *ip* + i points to ax[i], etc.

5.7.4 Passing Arrays to a Function

An array name can be used as an argument to a function, thus permitting the entire array to be passed to the function. To pass an array to a function, the array name must appear by itself, without brackets or subscripts, as an actual argument within the function call. When declaring a one-dimensional array as a formal argument, the array name is written with a pair of empty square brackets. The size of the array is not specified within the formal argument declaration. If the array is two-dimensional, then there should be two pairs of brackets following the array name with the first pair of brackets empty and the second pair of brackets containing the column size.

The following program outline illustrates the passing of an array from the main portion of the program to a function.

```
int  average (int n, int arr[]);
void main ( )
{
        int n, avg;                   /* variable declaration */
        int arr[50];                  /* array definition */
        . . .
        avg = average(n, arr);        /* function call */
        . . .
}
int average (int k, int brr[])        /* function definition */
{
        . . .
}
```

Within **main**() we see a call to the function *average*. This function call contains two actual arguments—the integer variable *n* and the one-dimensional integer array **arr**. Note that **arr** appears as an ordinary variable within the function call. In the first line of the function definition, we see two formal arguments, *k* and **brr**. The formal argument declarations establish *k* as an integer variable and **brr** as a one-dimensional integer array. Note that the size of **brr** is not defined in the function definition. As formal parameters in a function definition,

```
int brr[];
```

and

```
int *brr;
```

are equivalent.

5.7.5 Initializing Arrays

C allows initialization of arrays. Standard data-type arrays may be initialized in a straightforward manner. The syntax for initializing an array is

```
array_declarator = { value-list }
```

The following statement shows a five-element integer array initialization.

```
int i[5] = {10, 20, 30, 40, 50};
```

The element i[0] has the value of 10 and the element i[4] has the value of 50.

A string (character array) can be initialized in two ways. One method is to make a list of each individual character, such as

```
char strgx[5] = {'w', 'x', 'y', 'z', 0};
```

The second method is to use a string constant

```
char myname [6] = "huang";
```

A NULL character is appended automatically at the end of "huang". When initializing an entire array, the array size (which is one more than the actual length) must be included, for example

```
char prompt [24] = "Please enter an integer:";
```

Example 5.11 Write a program to invoke the function created in Example 5.10 to display the following pattern forever.

```
1            (1)
2 1          (2)
3 2 1        (3)
4 3 2 1      (4)
5 4 3 2 1    (5)
6 5 4 3 2 1  (6)
7 6 5 4 3 2  (7)
8 7 6 5 4 3  (8)
9 8 7 6 5 4  (9)
```

Solution: A possible solution to this problem is as follows.

Step 1 Place the pattern in an array that contains the digit pattern and digit select values for the display patterns specified in rows 1 to 9.

Step 2 Create a loop of nine iterations to be repeated forever. The **seg7_mux** function is called in each iteration to display the digits simultaneously.

```
#include   <c8051F040.h>
#define    d1      0x30  // seven-segment pattern of 1
#define    d2      0x6D  // seven-segment pattern of 2
#define    d3      0x79  // seven-segment pattern of 3
#define    d4      0x33  // seven-segment pattern of 4
#define    d5      0x5B  // seven-segment pattern of 5
#define    d6      0x5F  // seven-segment pattern of 6
#define    d7      0x70  // seven-segment pattern of 7
#define    d8      0x7F  // seven-segment pattern of 8
#define    d9      0x7B  // seven-segment pattern of 9
#define    sel5    0xDF  // value to select display #5
#define    sel4    0xEF  // value to select display #4
#define    sel3    0xF7  // value to select display #3
#define    sel2    0xFB  // value to select display #2
#define    sel1    0xFD  // value to select display #1
#define    sel0    0xFE  // value to select display #0
```

```
void dlyby1ms(char n);
void seg7_mux(char code *ptr, char k);
void sysinit(void);
char code pattern[] = {d1,sel5,                     // table is in program memory
                       d2,sel5,d1,sel4,
                       d3,sel5,d2,sel4,d1,sel3,
                       d4,sel5,d3,sel4,d2,sel3,d1,sel2,
                       d5,sel5,d4,sel4,d3,sel3,d2,sel2,d1,sel1,
                       d6,sel5,d5,sel4,d4,sel3,d3,sel2,d2,sel1,d1,sel0,
                       d7,sel5,d6,sel4,d5,sel3,d4,sel2,d3,sel1,d2,sel0,
                       d8,sel5,d7,sel4,d6,sel3,d5,sel2,d4,sel1,d3,sel0,
                       d9,sel5,d8,sel4,d7,sel3,d6,sel2,d5,sel1,d4,sel0};
void main (void)
{
        char code *ptr;
        char ix;
        sysinit();
        while(1) {
                ptr    = &pattern[0];          // ptr points to the start of the table
                for (ix = 1; ix <= 6; ix++){
                        seg7_mux(ptr,ix);      // display ix digits simultaneously for 600 ms
                        ptr   += 2 * ix;       // move to the next row in pattern table
                }
                for (ix = 7; ix <= 9; ix++){
                        seg7_mux(ptr,6);
                        ptr   += 12;           // row 7 to 9 has 6 patterns (12 bytes)
                }
        }
}
void sysinit(void)
{
        int n;
        SFRPAGE  = CONFIG_PAGE;
        WDTCN    = 0xDE;            // disable watchdog timer
        WDTCN    = 0xAD;            //  "
        CLKSEL   = 0;               // use internal oscillator as SYSCLK temporarily
        OSCXCN   = 0x67;            // start external oscillator; 24 MHz Crystal
        for (n=0;n<255;n++);  // delay about 1 ms
        while ((OSCXCN & 0x80) == 0); // wait for oscillator to stabilize
        CLKSEL |= 0x01; // switch to external oscillator

        P5MDOUT= 0xFF;              // configure P5 for output
        P7MDOUT= 0xFF;              // configure P7 for output
}
// ************************************************************************************************************
// include seg7_mux () function here
// ************************************************************************************************************
// ************************************************************************************************************
// include the dlyby1ms () function here.
// ************************************************************************************************************
```

5.7.6 Structures

A structure is a group of related variables that can be accessed through a common name. Each item within a structure has its own data type, which can be different from the other data items. The syntax of a structure declaration is

```
struct struct_name {            /* struct_name is optional */
    type1    member1;
    type2    member2;
    ...
};
```

The **struct_name** is optional and, if it exists, defines a *structure tag*. A **struct** declaration defines a type. The right brace that terminates the list of members may be followed by a list of variables, just as for any basic type. The following example is for a card catalog in a library.

```
struct catalog_tag {
    char  author [40];
    char  title [40];
    char  pub [40];
    unsigned int date;
    unsigned char rev;
} card;
```

Here, the variable *card* is of type **catalog_tag**.

A structure definition that is not followed by a list of variables reserves no storage; it merely describes a template or the shape of a structure. If the declaration is tagged (i.e., has a name), however, the tag can be used later in definitions of instances of the structure. For example, suppose we have the following structure declaration.

```
struct point {
    int x;
    int y;
};
```

We can then define a variable *pt* of type *point* as follows.

```
struct point pt;
```

A member of a particular structure is referred to in an expression by a construction of the form

```
structure-name.member
```

or

```
structure-pointer → member
```

The structure member operator (.) connects the structure name and the member name. As an example, the square of the distance of a point to the origin can be computed as follows.

```
long integer sq_distance;
...
sq_distance = pt.x * pt.x + pt.y * pt.y;
```

Structures can be nested. One representation of a circle consists of the center and radius, as shown in Figure 5.2.

This circle can be defined as

```
struct circle {
    struct      point    center;
    unsigned    int      radius;
};
```

5.7.7 Unions

A *union* is a variable that may hold (at different times) objects of different types and sizes with the compiler keeping track of size and alignment requirements. Unions provide a way to manipulate different kinds of data in a single area of storage without embedding any machine-dependent information in the program. The syntax of the union is as follows.

```
union union_name {
    type-name1    element1;
    type-name2    element2;
    ...
    type-namen    elementn;
};
```

The field **union_name** is optional. When it exists, it is called a *union-tag*. One can declare a union variable at the same time one declares a union type. The union variable name should be placed after the right brace }. In order to

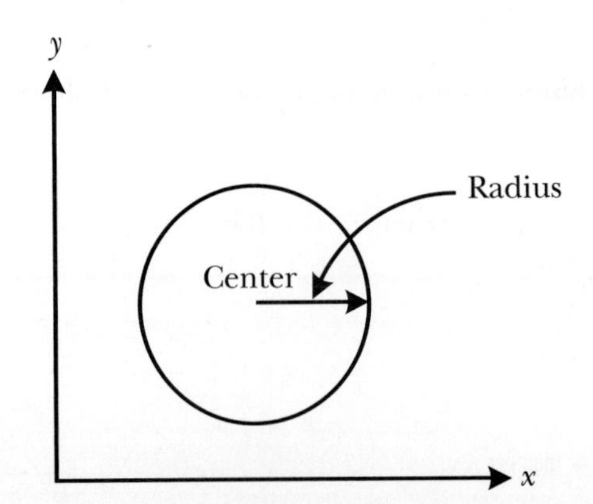

FIGURE 5.2
A circle.

represent the current temperature using both the integer and the string, one can use the following declaration.

```
union u_tag {
    int i;
    char c[4];
} temp;
```

Four characters must be allocated to accommodate the larger of the two types. Integer type is good for internal computation, whereas string type is suitable for output. Of course, some conversion may be needed before making a certain kind of interpretation. Using this method, the variable *temp* can be interpreted as an integer or a string, depending on the purpose. Syntactically, members of a union are accessed as

```
union-name.member
```

or

```
union-pointer → member
```

just as for structures.

5.8 Miscellaneous Items

5.8.1 Automatic/External/Static/Volatile

A variable defined inside a function is an *internal variable* of that function. These variables are called *automatic*, because they come into existence when the function is entered and disappear when it is left. Internal variables are equivalent to local variables in assembly language.

External variables are defined outside of any function and thus are available potentially to many functions. Because external variables are globally accessible, they provide an alternative to function arguments and return values for communicating data between functions. Any function may access an external variable by referring to it by name (if the name has been declared somewhere). External variables are also useful when two functions must share some data, yet neither calls the other.

The use of **static** with a local variable declaration inside a block or a function causes the variable to maintain its value between entrances to the block or function. Internal **static** variables are local to a particular function just as automatic variables are, but unlike automatic variables, they remain in existence rather than coming and going each time the function is activated. When a variable is declared **static** outside of all functions, its scope is limited to the file that contains the definition. A function can also be declared as **static**. When a function is declared as **static**, it becomes invisible outside of the file that defines the function.

A *volatile* variable has a value that can be changed by something other than user code. A typical example is an input port or a timer register. These variables must be declared as **volatile** so the compiler makes no assumptions on their values while performing optimizations. The keyword **volatile** prevents the compiler from removing apparently redundant references through the pointer.

5.8.2 Scope Rules

The functions and external variables that make up a C program need not all be compiled at the same time; the source text of the program may be kept in several files, and previously compiled routines may be loaded from libraries.

The scope of a name is the part of the program within which the name can be used. For a variable declared at the beginning of a function, the scope is the function in which the name is declared. Local (internal) variables of the same name in different functions are unrelated.

The scope of an external variable or a function lasts from the point at which it is declared to the end of the file being compiled. Consider the following program segment.

```
...
void f1 (...)
{
    ...
}
int a, b, c;
void f2 (...)
{
    ...
}
```

Variables *a*, *b*, and *c* are accessible to function f2 but not to f1.

When a C program is split into several files, it is convenient to put all global variables into one file so that they can be accessed by functions in different files. Functions residing in different files that need to access global variables must declare them as external variables. In addition, the prototypes of certain functions can be placed in one file so that they can be called by functions in other files. The following example is a skeletal outline of a two-file C program that makes use of external variables.

In File1:

```
extern int  xy;
extern long arr[];
void main (void)
{
    ...
}
void foo (int abc) {...}
long soo (void) {...}
```

In File2:

```
int xy;
long arr[100];
```

5.8.3 Type Casting

Type casting causes the program to treat a variable of one type as though it contains data of another type. The format for type casting is

```
(type) variable
```

For example, the following expression converts the variable *kk* to a long integer.

```
int kk;
. . .
(long) kk
```

Type casting can avoid many errors caused by size mismatch among operands. For example, in the program segment

```
long result;
int     y1, y2;
. . .
result = y1 * y2;
```

if the product of **x1** and **x2** is larger than $2^{16} - 1$, it will be truncated to 16 bits. Then the variable **result** will receive an incorrect value. To fix the error, use type casting to force **y1** and **y2** to **long**, as

```
result = ((long) y1) * ((long) y2);
```

This technique is used in several examples in this text.

Another example of the use of type casting is in the pointer type. Sometimes one needs to treat the contents of a structure type variable as a string. The most convenient way to do it is to recast the pointer to a structure-type variable into a pointer to a string (character type). For the declarations

```
struct personal {
    char name [10];
    char addr [20];
    char sub1[5];
    char sub2[5];
    char sub3[5];
    char sub4[5];
} ptr1;
char *cp;
```

one can use the following statement to treat the variable ptr1 as a string.

```
cp = (char *) &ptr1;
```

5.9 Using the C Compiler

There are many commercial C compilers that support the 8051 MCU. The freeware **Small Device C Compiler** (SDCC), **Keil C**, and **Raisonance C** compilers will be used to compile the C programs in this book. These C compilers can be integrated into the SiLabs IDE to provide source-level debugging capability in C language for target boards containing Silicon Laboratory's 8051 devices. Keil's μVision IDE and the Raisonance RIDE perform source-level debugging by utilizing the 8051 simulator. It can also work with a monitor resident on the target board to provide this capability. However, the target board must have external program memory to hold the application program. A tutorial for using the SiLabs in entering, compiling, and running the C program on the demo board will be given in Section 5.12.

Both compilers can also be run from the command line. To do this, select **start–>programs–>Accessories–>Command Prompt** from the desktop screen to open a command prompt window. The appropriate commands should be entered in this window to compile the C program. A tutorial for compiling the source code and download program onto the demo board will be given in Section 5.13.

5.10 C Language Extensions for the 8051 MCU

Language extensions are added to take advantage of the architecture features of the 8051 MCU. The statements in this section apply to both the SDCC, Keil C, and Raisonance C compilers unless stated otherwise.

5.10.1 Memory Model

The memory model determines the default memory type to use for function arguments, automatic variables, and declarations with no explicit memory type specifier. Keil and Raisonance C compilers support three memory models: small, compact, and large. SDCC supports only small and large memory models. The default memory type imposed by the memory model can be overridden by using a memory type specifier when declaring a variable.

Small Model

In this model, all variables (by default) reside in the internal data memory of the 8051 system. In this model, variable access is very efficient. However, all objects as well as the stack must fit into the internal RAM. Typically, if the linker/locator is configured to overlay variables in the internal data memory, the small model is the best model to use.

Compact Model

Using the compact model, all variables (by default) reside in one page of external data memory. (This is as if they were explicitly declared using the **pdata** memory type specifier.) This memory model can accommodate a maximum of 256 bytes of variables. This limitation is due to the addressing scheme used, which is indirect through registers R0 and R1. The compact memory model is not as efficient as the small model; therefore, access is not as fast. However, the compact model is faster than the large model. If the compact model is used with more than 256 bytes of external memory, the high-order address byte (or page) is provided by Port 2 on the 8051. In this case, you must initialize Port 2 with the proper external memory page to use. This can be done in the start code. You must also specify the starting address for **pdata** to the linker. Interested readers should refer to the Keil Cx51 or Raisonance C compiler user's guide for more details.

Large Model

In the large model, all variables (by default) reside in external data memory (up to 64 KB).

The data pointer (DPTR) is used for addressing. Memory access through this data pointer is inefficient, because it will generate longer code than the small or compact models.

5.10.2 Storage Class Language Extensions

In addition to the ANSI storage class, SDCC, Keil, and Raisonance C compilers also support the following 8051 specific storage classes.

data/near

This is the default storage class for the small memory model (**data** and **near** can be used synonymously). Variables declared with this storage class will be allocated in the directly addressable portion of the internal RAM. For example, the following statement declares an unsigned character variable in data memory.

```
data unsigned char test_data;
```

With this declaration, the C statement **test_data = 0x01;** will be translated into the assembly instruction

```
mov        _test_data,#0x01
```

All variables in the C program will be preceded by an underscore character (_) when translated into assembly program.

xdata/far

Variables declared with this storage class will be placed in the XRAM. This is the default storage class for the large memory model. The following statement declares an unsigned character variable in XRAM.

```
xdata unsigned char test_xdata;
```

With this declaration, the C statement **test_xdata = 0x02;** will be translated into the following assembly instruction sequence.

```
mov       DPTR,#_test_xdata
mov       A,#0x02
movx      @DPTR,A
```

Idata

Variables declared with this storage class will be allocated into the indirectly addressable portion of the internal data RAM. For example, this statement declares an unsigned integer variable:

```
idata  unsigned int test_idata;
```

With this declaration, the C statement **test_idata = 0x1234;** will be translated into the assembly instruction sequence

```
mov       R0,#test_idata
mov       @R0,#0x12
inc       R0
mov       @R0,#0x34
```

Please note that the first 128 bytes of **idata** physically access the same RAM as the data memory. The original 8051 has 128 bytes of idata memory; nowadays most devices have 256 bytes of idata memory. The stack is located in idata memory.

pdata

pdata stands for paged **xdata**. It typically is located at the start of **xdata** and has a maximum size of 256 bytes. The following example assigns the value 0x10 to a **pdata** variable.

```
pdata  unsigned char  test_pdata;
test_pdata = 0x10;
```

The previous statement will be translated into the following assembly instruction sequence as

```
mov       R0,#_test_pdata
mov       A,#0x10
movx      @R0,A
```

Code

Variables declared with this storage class will be placed in the program memory. Variables declared in code memory are read only. For example, the statement declared an initialized character array in program memory is given as

```
code char test_array[8] = {1, 2, 3, 4, 5, 6, 7, 8};
```

Read access to this array using an 8-bit index **ix** will be translated into the instruction sequence

```
mov     A,_ix
mov     DPTR,#_test_array
movc    A,@A+DPTR
```

Bit

This is a data type and a storage class specifier. When a variable is declared with the bit storage class, it is allocated into the bit-addressable memory of the 8051. For example, the following statement declares a bit variable.

```
bit test_bit;
```

With this declaration, the C statement **test_bit = 1;** will be translated into the assembly instruction

```
setb    _test_bit
```

sfr/sbit

Like the **bit** keyword, sfr/sbit signifies both a data type and a storage class. They are used to describe the special function registers and special bit variables of the 8051. The header file for the 8051 MCU contains statements of this type. For example, the I/O port P0 is declared as

```
sfr     at 0x80 P0;
```

The carry flag **CY** is declared as

```
sbit    at 0xD7 CY;
```

Special function registers that are located on an address dividable by 8 are bit-addressable; an **sbit** statement addresses a specific bit within one of these SFRs.

5.10.3 Pointers

Keil, Raisonance C, and SDCC compilers support the declaration of variable pointers using the * character. The pointer can be used to perform all operations in standard C language. However, because of the unique architecture of the 8051 and its derivatives, these C compilers provide two different types of pointers: generic and memory-specific pointers.

5.10.4 Generic Pointers

Generic pointers are declared in the same fashion as standard C pointers. For example,

```
char    *s;          /* string pointer */
int     *number;     /* int pointer */
long    *state;      /* a long pointer */
```

Generic pointers always are stored using three bytes. The first byte is the memory type, the second is the high-order byte of the offset, and the third is the low-order byte of the offset. Assembler support routines are called whenever data is stored or retrieved using the *generic* pointers. These pointers are useful for developing reusable library routines. Explicitly specifying the pointer type will generate the most efficient code.

You may specify the memory area in which a generic pointer is stored by using a memory type specifier. For example,

```
char    * xdata    strptr;    /* generic pointer stored in xdata space */
int     * data     numptr;    /* generic pointer stored in data RAM */
long    * idata    varptr;    /* generic pointer stored in idata space */
```

5.10.5 Memory-Specific Pointers

Memory-specific pointers always include a memory-type specification in the pointer declaration and always refer to a specific memory data. For example,

```
char    data     *str;       /* pointer to a string in data memory */
int     xdata    *numtab;    /* pointer to an integer in xdata memory */
long    code     *powtab;    /* pointer to a long variable in code memory */
```

Because the memory type is specified at compile time, the memory-type byte required by generic pointers is not needed by memory-specific pointers. Memory-specific pointers can be stored using only one byte (**idata**, **data**, **bdata**, and **pdata** pointers) or two bytes (code and **xdata** pointers).

Like generic pointers, you may specify the memory area in which a memory-specific pointer is stored. To do so, prefix the pointer declaration with a memory-type specifier. For example,

```
char    data     * xdata str;       /* pointer in xdata space to data character */
int     xdata    * data  numtab;    /* pointer in data space to xdata int */
long    code     * idata powtab;    /* pointer in idata space to code long */
```

Memory-specific pointers may be used to access variables in the declared 8051 memory area only. Memory-specific pointers provide the most efficient method of accessing data objects, but at the cost of reduced flexibility.

5.10.6 Absolute Variable Location

Data items can be assigned an absolute address with the **at <address>** keyword in addition to a storage class. Keil, Raisonance, and SDCC differ in the format for absolute addressing. For example, the following statement assigns the variable **chksum** to address 0x7FFE:

```
xdata   at 0x7FFE unsigned int chksum;     /* absolute addressing in SDCC */
xdata   at 0xFE00 unsigned int chksum;     /* absolute addressing in Raisonance */
xdata   unsigned int chksum _at_ 0x7FFE;   /* absolute address in Keil's C */
```

Memory space (**xdata** in the examples) is optional. When memory space is not specified, the default space is data memory.

The C compiler does not reserve any space for variables declared in this way (they are implemented with an EQU directive in assembler). Thus, it is left to the programmer to make sure there are no overlaps with other variables that are declared without the absolute address. The assembler listing (with file name extension **.lst**) and the linker output file (with file name extension **.rst**) and the map file (with file name extension **.map**) are good places to look for such overlaps. Variables with an absolute address are not initialized.

5.10.7 Parameters and Local Variables

Automatic (local) variables and parameters to functions can either be placed in the stack or in data space. The default action is to place these variables in the internal RAM (for small model) or external RAM (for large model). This in fact makes them similar to **static**, so by default, functions are non-reentrant. They can be placed in the stack by the keyword **reentrant** to the function definition. An example is.

```
unsigned char xyz (char i) reentrant
{
    ...
}
```

Since stack space on the 8051 is quite limited, the **reentrant** keyword option should be used sparingly. Note that the **reentrant** keyword just means that the parameters and local variables will be allocated to the stack, it does not mean that the function is register-bank independent.

5.10.8 Passing Parameters in Registers

The C compilers discussed previously allow up to three parameters to be passed in CPU registers. The assignment for registers is shown in Table 5.4. The column 4 of Table 5.4 indicates that only one long parameter (either parameter 1 or 2 but not both) will be placed in registers R4–R7. Column 5

TABLE 5.4 *Passing Parameters in 8051 Registers*

Argument Number	char, 1-byte ptr	int, 2-byte ptr	long, float	Generic ptr
1	R7	R6 & R7	R4–R7	R1–R3
2	R5	R4 & R5	R4–R7	R1–R3
3	R3	R2 & R3		R1–R3

TABLE 5.5 *Function Return Value Type and Registers Used*

Return Type	Register	Description
bit	**CY** flag	
char, unsigned char, 1-byte ptr	R7	
int, unsigned int, 2-byte ptr	R6 & R7	MSB in R6, LSB in R7
long, unsigned long	R4–R7	MSB in R4, LSB in R7
float	R4–R7	32-bit IEEE format
generic **ptr**	R1–R3	Memory type in R3, MSB in R2, LSB in R1

indicates that only one generic pointer parameter (must be one of the first three) will be placed in registers R1–R3.

5.10.9 Function Return Values

MCU registers are always used to return values in C programs. Table 5.5 lists the return types and the registers used for each.

5.10.10 Function Declaration

The C compiler provides a number of extensions for standard C function declarations. These extensions allow you to

- Specify a function as an interrupt service routine
- Choose the register bank used
- Select the memory model
- Specify reentrancy (a function is reentrant if it can call itself)

The format for function declaration is

```
[return_type] funcname ([args]) [{small | comact | large}][reentrant] [interrupt n] [using n]
```

where

return_type	is the type of the value returned from the function. If no type is specified, **int** is assumed
funcname	is the name of the function
args	is the argument list for the function
small, compact, or large	is the explicit memory model for the function
reentrant	indicates that the function is recursive or reentrant
interrupt	indicates that the function is an interrupt function
using	specifies which register bank the function uses

5.10.11 Specifying Register Bank for a Function

The lowest 32 bytes of data memory of all members of the 8051 MCU are grouped into four banks of eight registers each. Programs can access these registers as R0 through R7. The register bank is selected by two bits of the program status word (PSW). Register banks are useful when processing interrupts or when using a real-time operating system. Rather than saving these eight registers, the MCU can switch to a different register bank for the duration of the interrupt service routine. Interrupts are discussed in Chapter 6.

The bank can be selected by adding the **using** function attribute. The following function fragment illustrates the specification of register bank.

```
void xy_function (void) using 3
{
      .
      .
      .
}
```

The **using** attribute is not allowed in function prototype. The **using** attribute affects the object code of the function as.

- The current PSW is saved on the stack at function entry.
- The specified register bank is set (by modifying the PSW).
- The former PSW is restored before the function is exited.

5.10.12 In-Line Assembly Instructions

Keil, Raisonance, and SDCC C compilers allow the user to add assembly instructions into his/her C programs. In-line assembly instructions are enclosed by **_asm** (or **#pragma asm** in Keil C) and **_endasm** (or **#pragma endasm** in Keil C). The inline assembly code can contain any valid code understood by the assembler, this includes any assembler directives and comment lines. The compiler does not do any validation of the code within the **_asm . . . _endasm**; keyword pair. It is recommended that each assembly instruction (including labels) be placed in a separate line.

An **in_line** assembly instruction can access variables defined in C language. However, the name of the C variable must be preceded with an underscore (_) character. The following program fragment illustrates the use of **in_line** assembly instructions.

```
... unsigned char count, sum;
...
_asm
    mov    A,_count
    add    A,_sum
    mov    _sum,A
    inc    _count
_endasm
```

5.10.13 Header Files

An embedded system programmer may include header files to facilitate the program development. A header file may contain definitions for special function registers and bits contained in SFRs. Including appropriate header files in the program, a programmer may use symbolic names instead of addresses to refer to those registers and bits. A header file may also contain macros, constant definitions, and prototype declarations of library functions. By including these header files, the programmer may invoke appropriate constants, library functions, and macros in their programs and hence shorten the software development time.

In general, an 8051 C compiler would provide a header file (call it *MCU header file*) that contains the definitions of SFRs and bits for each specific 8051 variant. However, it is common that several MCUs have the same number of SFRs and bit definitions. In this situation, several MCUs may share the same header file. For example, the C8051F040, 041, 042, 043, 044, 045, 046, and 047 MCUs share the same header file (c8051F040.h).

The three C compilers differ in the formats used in their MCU header files. For example, the definitions for I/O Port 0 in Keil, Raisonance, and SDCC C compilers are.

```
__sfr __at 0x80 P0;     // SDCC compiler format
sfr  P0    = 0x80;      // Keil C compiler format
sfr  P0    = 0x80;      // Raisonance C compiler format
```

The bit definitions for bit 0 of the TCON register in three C compilers are.

```
__sbit __at 0x88 IT0    // SDCC compiler format
sbit IT0  =  TCON ^ 0;  // Keil C compiler format
sbit IT0  =  TCON ^ 0;  // Raisonance C compiler format
```

These three C compilers do not provide MCU header files for all 8051 variants. However, the MCU header files for those devices can be created by editing an existing MCU header file. The C compiler is written in a way that allows the user to include an MCU header file by enclosing the header file name between < and >. For example,

```
#include <c8051F040.h>
```

5.11 C Library Functions

Both the Keil C and Raisonance C compilers provide many library functions that can be invoked by the user. The library functions of the Keil C compiler are described clearly in its user manual. The SDCC also provides a group of library functions that can be invoked by the user. However, the user's manual of SDCC does not provide much description about library functions. Interested users may want to look at the source code of library which is provided under the installation directory of SDCC (for example, **c:\sdcc\lib\src**).

The library functions from the Keil's C compiler are listed in Appendix G. Most of these functions are also provided by the SDCC and the Raisonance C compilers. The prototype declarations and operations performed by library functions are listed in Appendix H.

Example 5.12 Write a program to invoke the C library functions **srand( )** and **rand( )** to generate the first 30 pseudorandom numbers.

Solution: The program is as follows.

```
#include <c8051F040.h>
#include <math.h>
#include <stdlib.h>
xdata unsigned rand_arr[30] _at_ 0x00; /* array to hold 30 random numbers */

int main (void)
{
    unsigned int ix;
    srand(rand_arr[0]);      /* set seed to rand_arr[0] */
    for (ix = 0; ix < 30; ix++)
        rand_arr[ix] = rand();
    return 0;
}
```

The function **srand( )** sets the starting seed for generating the random number. As long as the starting seed is the same, the generated random number sequence will be the same.

5.12 Using SiLabs IDE to Develop C Programs

The procedure for developing C programs in SiLabs IDE is similar to that for developing assembly programs. The user has the freedom to choose whatever 8051 C compiler available to him/her. When choosing compilers, the user may have different experiences because of the different features supported by different compilers. The following example illustrates the process of compiling a C program by using Keil's C compiler.

Example 5.13 Write a C program to find six prime numbers closest to 10,000 using the SiLabs IDE by following the procedure described in Section 3.6.2.

Solution: Three of these prime numbers are larger than 10,000, whereas the other three are smaller than 10,000. This program starts from 10,001, tests each odd number larger than 10,000 until three prime numbers are found, and then starts from 9999 and tests each odd number smaller than 10,000 until three prime numbers are found. The C program that finds six prime numbers closest to 10,000 is as follows.

```
#include <C8051F040.h>
data char notfound;
#define     yes     1
#define     no      0
int isprime (unsigned int x);
unsigned char sqroot (unsigned int y);
int main (void)
{
    xdata unsigned int prim_num[6];     // array to hold the prime numbers
    unsigned int testnum;     // number to be tested
    char i;
    testnum = 10001;
    for (i = 0; i < 3; i++) {
        notfound = yes;
        while (notfound) {// find the prime number larger than and closest to 10000
            if(isprime(testnum)){
                notfound = no;
                prim_num[i+3] = testnum;
            }
            testnum += 2;
        }
    }
    testnum = 9999;
    for (i = 2; i >= 0; i--) {
        notfound = yes;
        while (notfound) {//find the prime number smaller but closest to 10000
            if(isprime(testnum)){
```

```
                    notfound = no;
                    prim_num[i] = testnum;
                }
                testnum -= 2;
            }
        }
        return 0;
}
/**********************************************************************************************************/
/* include functions isprime() and sqroot() here.                                                        */
/**********************************************************************************************************/
```

The procedure for developing this program is as follows.

Step 1 Start the SiLabs IDE by clicking on its icon. The resultant SiLabs IDE window is similar to that in Figure 2.15.

Step 2 Select your favorite C compiler and the associated assembler and linker by performing the Tool Chain Integration step as shown in Figure 2.16 through 2.18. This step needs only be performed once if the user doesn't change the C compiler. SiLabs will remember the C compiler that the user chose until the user changes it.

Step 3 Create a new project and save it as **eg05_13**. This is done by pressing the right mouse button on the keyword **New Project** in the Project Window and select **Save Project New_Project**. A dialog box will appear and ask you to specify the project name. Enter **eg05_13** as the project name and click on **OK**.

Step 4 Select **File->New File** from the pull-down menu of the SiLabs IDE window and enter the C program in the Editor Window. Save the program with the file name **eg05_13.c**.

Step 5 Add the program **eg05_13.c** into the project **eg05_13**. This is done by pressing the right mouse button on the project name **eg05_13** in the Project Window (need to **Add file to project** and **Add file to build**). A dialog window will appear to ask you to select the files to be included in the project. The dialog is similar to that in Figure 2.21 and 2.22.

Step 6 Build the project by selecting **Project–>Build/Make Project** from the SiLabs IDE menu. To make sure all files and project dependencies are resolved, choose **Project–>Rebuild** instead. Syntax errors will be identified and error messages will be displayed in this step. You need to correct all of the syntax errors before the program can be built successfully. No hex file will be generated if there is one or multiple syntax errors.

Step 7 Download the program for debugging by following the procedure described in Section 2.15.8. SiLabs IDE runs the program on the target hardware during the debugging process.

Step 8 Execute and debug the program. Open a Watch Window and enter the variables **notfound** and **testnum** into this window. Open a window to display external memory contents. The resultant SiLabs IDE window looks like Figure 5.3.

There are two **for-loops** in the program. The first **for-loop** finds prime numbers that are larger than 10,000 whereas the second **for-loop** finds prime numbers that are smaller than 10,000. There are two **notfound = yes**; statements in the program. Place the mouse pointer at the first **notfound = yes**; statement and press the right mouse button and select **Run to Cursor**. The values of the variables **testnum** and **notfound** in the watch window will be changed to 10001 and 00, respectively. Press the right mouse button on the same **notfound = yes** statement again; and select **Run to Cursor**. The values of the variables **testnum** and **notfound** will be changed to 10,009 and 0. This tells us that the first prime number larger than 10,000 is 10,007 (**0x2717**). Press the right mouse button and select **Run to Cursor** again. The values of the variables **testnum** and **notfound** will be changed to 10011 and 0. This tells us that the second prime larger than 10,000 is 10,009. Now place the cursor at the second **notfound = yes**; statement, press the right mouse button, and select **Run to Cursor**. Continue this process until all six prime numbers are found. These six prime numbers (26DD

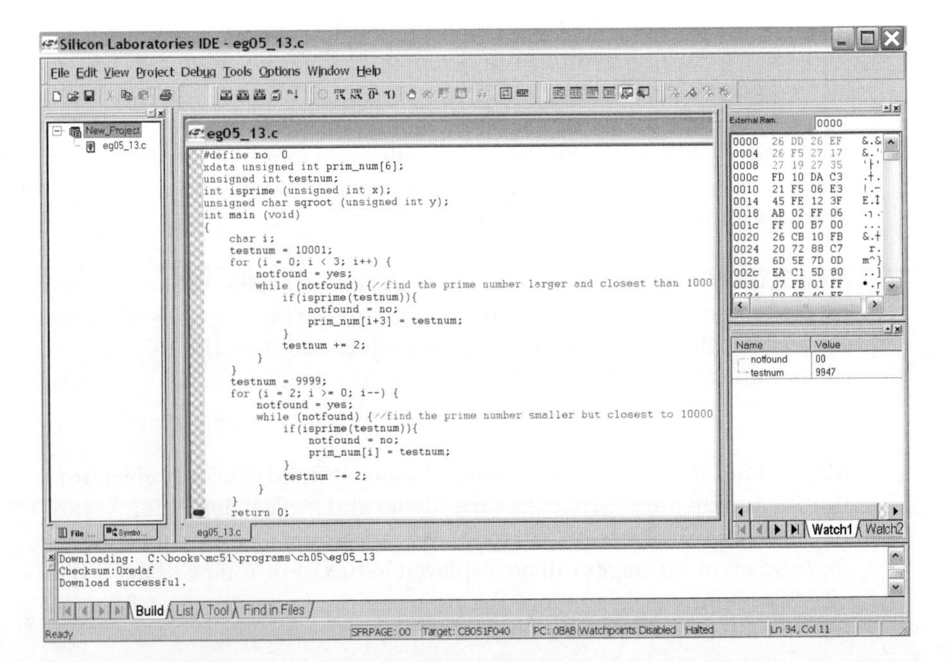

FIGURE 5.3 *SiLabs IDE window after adding watch list and external RAM window.*

26EF 26F5 2717 2719 2735) will be stored in XRAM (external RAM) starting from address **00**. Each prime number occupies two bytes and is stored with the high byte at the lower address and the low byte at the next higher address. All numbers are represented in the hex format.

The tutorial for using Keil's μVision and Raisonance's RIDE in developing C programs are given in Appendices B and C, respectively.

5.13 Building Multiple File Projects

It is normal to create projects that contain multiple files. The procedure to create a multi-file project is as follows.

Step 1
Start the SiLabs IDE and enter project files one at a time. Add each C file to the project and also **add** each C file **to build**.

Step 2
Create a header file that contains the prototypes of the functions that will be called by the main program. Include this header file to the main program.

Step 3
Build the project and run the program. The debugging process for a multi-file project is similar to a single-file project.

In the following, an example is used to illustrate how to create a multi-file project. The following functions have been created (included in the complementary CD) for illustration purpose.

- **void openUART (void):** This function configures the UART baud rate to 19200.
- **void putch (char xc):** This function outputs the character **xc** to the UART port.
- **void newline (void):** This function outputs a carriage return and linefeed character to the UART port. The effect is to move the cursor to the start of next line on a monitor screen.
- **void putsc (char *ptr):** This function calls the putcUART function repeatedly to output a string pointed to by **ptr**. The string is terminated by a NULL character.
- **void putsx (char xdata *ptr):** This function calls the putcUART function repeatedly to output the string stored in XRAM memory (pointed to by ptr) to the UART port. The string is terminated by a NULL character.
- **char getch (void):** This function reads a character from the UART port (keyboard).
- **void getsx (char xdata *ptr):** This function reads a string from the UART port by calling the getcUART function repeatedly and saves it in a buffer in XRAM pointed by **ptr**.

These functions are placed in the **uartUtil.c file**. Their prototypes are stored in the **uartUtil.h file** (to be included in the file that contains main () function). The main program that tests these functions are created and stored in the **uartTest.c file**. This program is as follows.

```
#include <c8051F040.h>
#include <uartUtil.h>    // contains the prototypes of UART functions

void systemInit(void);
void main (void)
{
    char code *ptr0 = "Enter your name: ";
    char code *ptr1 = "My name is ";
    char xdata cptr[20];
    WDTCN  = 0xDE;        // disable watchdog timer
    WDTCN  = 0xAD;        //     "
    systemInit();
    SFRPAGE = UART0_PAGE;
    openUART();
    newline();
    putsc(ptr0);              // output "Enter your name: "
    getsx(cptr);              // read the name from keyboard
    newline();
    putsc(ptr1);              // output "My name is"
    putsx(cptr);              // output my name
    while(1);
}
void systemInit(void)
{
            int n;
    SFRPAGE = CONFIG_PAGE;
    WDTCN  = 0xDE;        // disable watchdog timer
    WDTCN  = 0xAD;        //                  "
    OSCXCN  = 0x67;       // start external oscillator; 24 MHz Crystal
                          // system clock is 24 MHz
    for (n = 0;n < 255; n++);      // delay about 1 ms
    while ((OSCXCN & 0x80) == 0); // wait for oscillator to stabilize
    CLKSEL |= 0x01;       // switch to external oscillator
    XBR2 = 0x5D;          // enable crossbar and assign I/O pins to all
    XBR0 = 0xF7;          // peripheral signals,
    XBR1 = 0xFF;          //     "
    XBR3 = 0x8F;          //     "
    SFRPAGE = SPI0_PAGE;
    SPI0CN  = 0x01;       // enable 3-wire SPI (make sure SPI uses 3 pins
}
```

Add these two files to the new project and also add them to build. Start the HyperTerminal program before running the program and you should see a screen similar to that in Figure 5.4.

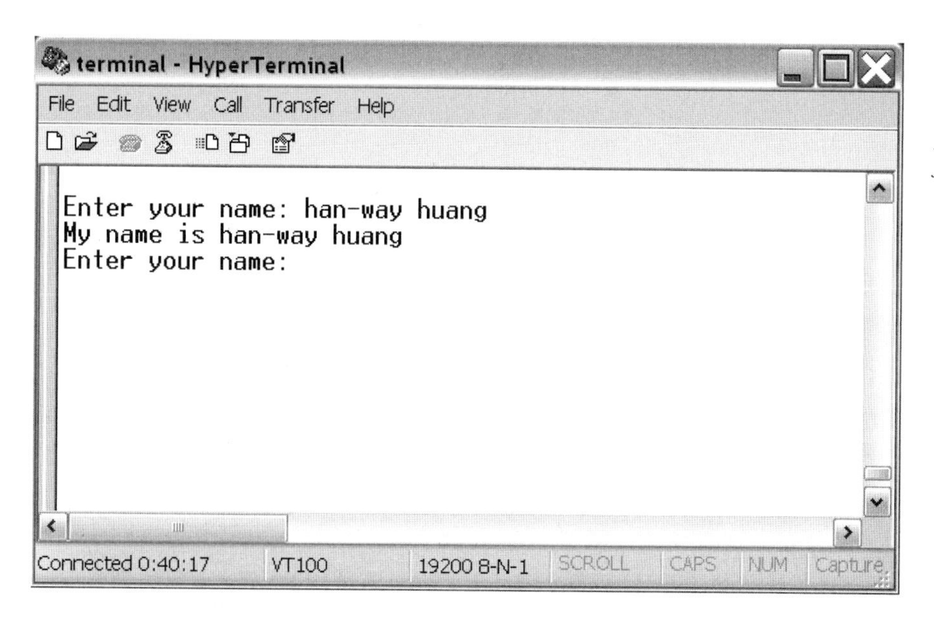

FIGURE 5.4 *Screen shot for running the program in a multi-file project.*

5.14 Using UART C functions

Among the I/O functions provided in the Keil library, the following are most useful.

- **char getchar (void):** This functions reads a character from the UART port.
- **char putchar (char):** This function outputs a character to the UART port.
- **int printf (const char *, . . .):** This function implements formatted output via the UART port.
- **int puts (const char *ptr):** This function outputs a string to the UART port.
- **char *gets (char *, int n):** This function reads a string of up to *n* characters from the UART port. A string is terminated by the carriage return (enter key).

You must configure the system clock to the appropriate clock source, enable crossbar decoder, assign I/O pins to UART signals (TX and RX), and initialize the UART module properly before calling these functions.

Example 5.14 Modify the program in Example 5.13 to print out the six prime numbers closest to 10,000 in the following format by calling the library I/O function(s)

```
Six prime numbers closest to 10000 are
xxxx xxxx xxxx yyyyy yyyyy yyyyy
```

Solution: The main program can be modified as follows.

```
#include    <C8051F040.h>
#include    <stdio.h>
#define     yes   1
#define     no    0

void systemInit(void);
void openUART0(void);
int isprime (unsigned int x);
unsigned char sqroot (unsigned int y);

int main (void)
{
char i;
char        data     notfound;
unsigned    int      xdata prim_num[6];
unsigned    int      testnum;

systemInit();
openUART0();
printf("\n");           // move cursor to the start of next row
testnum = 9999;      // prepare to find three prime numbers smaller than 10000
for (i = 2; i >= 0; i--) { // find three prime numbers that are smaller than 10000
     notfound = yes;
     while (notfound) {
          if(isprime(testnum)){ // is testnum a prime number?
               notfound = no;
               prim_num[i] = testnum; // save the found prime number
          }
          testnum -= 2;
     }
}
testnum = 10001; // prepare to find three prime numbers larger than 10000
for (i = 0; i < 3; i++) {
     notfound = yes;
     while (notfound) {
          if(isprime(testnum)){
               notfound = no;
               prim_num[i+3] = testnum;
          }
          testnum += 2;
          }
     }
     printf("Six prime numbers closest to 10000 are \n");
     for (i = 0; i < 6; i ++) // display all six prime numbers on screen
          printf("%d ",prim_num[i]);
     return 0;
}
/**********************************************************************************************/
/* Add isprime(), sqroot(), systemInit(), and openUART0() here. */
/**********************************************************************************************/
```

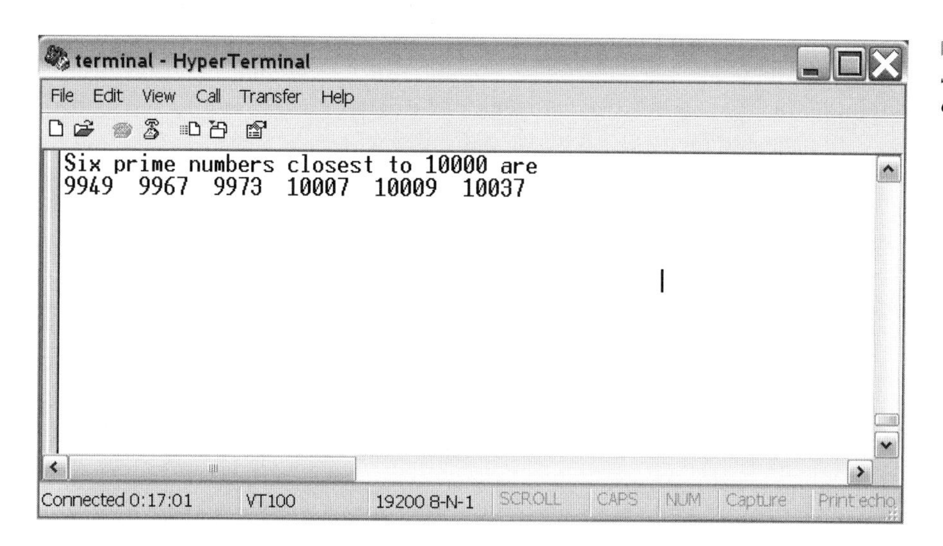

FIGURE 5.5
Six prime numbers closest to 10,000.

Set a breakpoint at the statement **return 0** and run the program. The screen of the HyperTerminal should look like Figure 5.5.

Using C language to generate waveforms using the AD7302 D/A converter described in Chapter 3 is also straightforward, and hence, it will be left for you as an exercise problem.

5.15 Tips for C Program Debugging

As in assembly language, C program errors can be divided into two categories:

- Syntax/semantics errors
- Logical errors

Many syntax/semantics errors can be discovered if the programmer follows the rules of C language:

- ***Undeclared Variables.*** A variable must be declared before it can be used. Most errors of this type can be discovered easily.
- ***Variable and Symbol Names are Case Sensitive.*** Sometimes, the programmer has declared a variable but still has an undeclared variable error with the same variable. This error could occur because one or more characters of the variable name are in the wrong case. For example, in the program segment

```
int A1, A2, xyClk;
. . .
A1 = A1 * xycLk;
```

the compiler would output the error message **undefined variable xycLk.**

- ***Missing Parenthesis (Brace or Bracket).*** A missing parenthesis can cause many other errors. This type of error can be avoided by entering the matching parenthesis whenever the left parenthesis is entered before entering statements between the parentheses.
- Mismatch of the function name in the function prototype declaration and the function definition.
- Mismatch of argument data types between the function prototype and the function definition.
- ***Declaring Variables within Code Body.*** For example, the following *for-loop* declares *i* implicitly (i.e., *i* is not declared in any earlier statement.) and is an error.

```
for (i = 0; i < n; i++) {
```

Logical errors are harder to debug than syntax and semantics errors. Whenever the program behavior is not what one expects, the first step is to read the program carefully to find out where the most likely problem spot is. After identifying the potential problem spot, the user can set a breakpoint and examine the program execution result.

If the user has a source-level debugger (for example, SiLabs IDE) to debug the program, the following actions can be taken to solve the problem.

- ***Set Breakpoints.*** Breakpoints allow the user to examine program execution results at the suspicious point.
- ***Set Up a Watch List.*** A Watch List is used together with the breakpoints. A Watch List consists of pairs of program variables of interest and their values at a certain time. It allows the user to find out program execution results quickly at any breakpoint.
- ***Trace Program Execution.*** Users can find out the execution result of a few statements by using the **run to cursor** feature available in many source-level debuggers (including SiLabs IDE). Without a source-level debugger, you will need to output some information (using LCD or LEDs) from program at the suspicious statement.

The freeware SiLabs IDE provides great support for debugging. A commercial source-level debugger for the C language can cost thousands of dollars. Without a source-level debugger, many debugging activities cannot be performed easily.

There are several purposes for setting breakpoints. Amoung these are:

- To determine whether a segment of code has ever been entered by the CPU. This can determine whether the MCU gets stuck at a certain statement. After locating the statement at which the MCU gets stuck, you can figure out what was wrong.
- To determine whether the execution result is correct up to the breakpoint. Without a source-level debugger, the programmer can output the program execution result to the LCD or terminal monitor to find out if the program executes correctly up to the breakpoint. This also serves as a watch list for the program execution.

The following guidelines can help reduce and identify logic errors.

- Make sure the precedence of operators has been observed. If you are not sure about the precedence, then use parentheses to enforce the intended precedence.
- Match the size of the source operands and that of the destination variables. Use type casting when appropriate. Type casting has been used in several examples in this text.
- Walk through the program algorithm carefully before converting it into the program code. An incorrect algorithm is often the cause of program bugs.
- Use enough data to test the program. The program must be tested with normal inputs, maximum and minimum inputs, and also illegal inputs to make sure that it operates correctly under all circumstances. When the problem to be solved gets complicated, a structured programming approach should be used to organize the program. The guidelines described in Section 2.4 should be followed to develop the program algorithm and convert the algorithm into program code. Each individual function should be tested thoroughly before the whole program is tested. A comprehensive discussion of structured programming and testing is beyond the scope of this text but can be found in many textbooks on software engineering.

5.16 Chapter Summary

A C program consists of one or more functions and variables. The **main ( )** function is required in every C program. It is the entry point of a C program. A function contains statements that specify the operations to be performed. The types of statements in a function could be *declaration, assignment, function call, control,* and *NULL.*

A *variable* stores a value to be used during the computation. A variable must be declared before it can be used. The declaration of a variable consists of the name and the type of the variable. There are four basic data types in C: **int**, **char**, **float**, and **double**. Several qualifiers can be added to the variable declarations. They are **short**, **long**, **signed**, and **unsigned**.

Constants are often needed in forming a statement. There are four types of constants: integers, characters, floating-point numbers, and strings.

There are seven *arithmetic operators*: +, −, *, /, %, ++, and −−. There are six *bitwise operators*: &, |, ^, ~, >>, and <<. Bitwise operators can be applied only to integers. *Relational operators* are used in control statements. They are ==, !=, >, >=, <, <=, &&, ||, and !.

The *control-flow statements* specify the order in which computations are performed. Control-flow statements include if-else statement, multi-way conditional statement, switch statement, for-loop statement, while statement, and do-while statement.

Every C program consists of one or more functions. If a program consists of multiple functions, their definitions cannot be embedded within another.

The same function can be called from several different places within a program. Generally, a function will process information passed to it from the calling statement of the program and return a single value. Information is passed to a function via special identifiers called *arguments* (also called *parameters*) and returned via the **return** statement. Some functions, however, accept information but do not return anything (for example, the library function **printf**).

A *pointer* holds the address of a variable. Pointers can be used to pass information back and forth between a function and its reference (calling) point. In particular, pointers provide a way to return multiple data items from a function via function arguments. Pointers also permit references to other functions to be specified as arguments to a given function. Two operators are related with pointers: * and **&**. The * operator returns the value of the variable pointed to by the pointer. The **&** operator returns the address of a variable.

Data items that have common characteristics are placed in an *array*. An array may be one-dimensional or multidimensional. The dimension of an array is specified by the number of square bracket pairs [] following the array name. An array name can be used as an argument to a function, thus permitting the entire array to be passed to the function. To pass an array to a function, the array name must appear by itself, without brackets or subscripts. An alternative way to pass arrays to a function is to use pointers.

A variable defined inside a function is an *internal variable* of that function. *External variables* are defined outside of any function and thus are available potentially to many functions. The *scope* of a name is the part of the program within which the name can be used. The scope of an external variable or a function lasts from the point at which it is declared to the end of the file being compiled.

In C language, parallel I/O is performed by an assignment statement. An input operation is performed by making the port data register as one of the source operands, whereas an output operation is performed by making the port data register as the destination of a statement.

The header file in C language is used mainly to hold constant definitions (registers and bits are in this category), macros, functional prototype declarations, and type definition. It is not meant to hold program files.

SiLabs IDE, Keil's μVision, and Raisonance's RIDE allow the user to build multi-file projects. In a multi-file project, the designer enters functions into several files. The prototype definitions of the functions in the same file are placed in a separate **header file**. Any other file should include this header file if the functions in the file need to call them. All of the program files are then added into the project. In SiLabs IDE, all of the files in the project must also be added *for build.*

Invoking I/O library functions judicially can make your programming task in C language much easier. The library functions provided by Keil and Raisonance are listed in Appendices G and H. The C programs in this Chapter are mainly tested using SiLabs IDE and Keil's development tools.

5.17 Exercise Problems

E5.1 Assume that *ax* = 83 and *bx* = 11, what is the value of *ax* / *bx*?

E5.2 Assume that *ax* = 97 and *bx* = *ax* % 23. What is the value of *bx*?

E5.3 Assume that *ax* = 0x39 and *bx* = *ax* ^ 0x79. What is the value of *bx*?

E5.4 Assume that *ax* = 0x9D and *bx* = *ax* << 2. What is the value of *bx*?

E5.5 Assume that *ax* = 0x6B and *bx* = *ax* & DE. What is the value of *bx*?

E5.6 Write a C program to find the *median* and *mode* of an array of integers. When the array has an even number of elements, the median is defined as the average of the middle two elements. Otherwise, it is defined as the middle element of the array. The mode is the element that occurs most frequently. You need to sort the array in order to find the median.

E5.7 Write a function that tests if a given number is a multiple of 8. A 1 is returned if the given number is a multiple of 8. Otherwise, a 0 is returned. The number to be tested is an integer and is an argument to this function.

E5.8 Write a function that computes the greatest common divisor (GCD) of two integers *m* and *n*.

E5.9 What is a function prototype? What is the difference between a function prototype and function definition?

E5.10 Write a **switch** statement that will examine the value of an integer variable *xx* and assign one of the following values to the variable *cy*, depending on the value assigned to *xx*:

(a) 10, if *xx* == 1
(b) 20, if *xx* == 2
(c) 30, if *xx* == 3
(d) 40, if *xx* == 4

E5.11 Write a C program to output the message "Microcontroller is fun to use!".

E5.12 Write a function that will convert an uppercase letter to lowercase.

E5.13 Write a C function to convert from miles to kilometers. The number of miles to be converted is the argument to the function.

E5.14 Write a C program that swaps the first column of a matrix with the last column, swaps the second column of the matrix with the second-to-last column, and so on.

E5.15 Write a loop to compute the sum of the squares of the first 100 odd integers.

E5.16 An *Armstrong number* is a number of *n* digits that is equal to the sum of each digit raised to the *n*th power. For example, 153 (which has three digits) equals $1^3 + 5^3 + 3^3$. Write a function to store all three-digit Armstrong numbers in an array.

E5.17 Write a program to find the first five numbers that when divided by 2, 3, 4, 5, or 6 leave a remainder of 1 and when divided by 7 have no remainder.

E5.18 Take a four-digit number. Add the first two digits to the last two digits. Now, square the sum. Surprise, you've got the original number again! Of course, not all four-digit numbers have this property. Write a C program to find all the four-digit numbers that have this special property and display them on the monitor screen.

E5.19 Write a C function to create a time delay that is a multiple of 10 ms.

E5.20 Write a C function to create a time delay that is a multiple of 1 s.

E5.21 Write a C function to create a time delay that is a multiple of 50 μs.

E5.22 Write a C program to generate a 1-kHz digital waveform using the circuit shown in Figure 3.22.

E5.23 Write a C program to generate a sawtooth waveform using the circuit shown in Figure 3.22.

E5.24 Write a C program to generate a triangular waveform using the circuit shown in Figure 3.22.

E5.25 Write a C program to generate a digital waveform that alternates between 1 kHz and 2 kHz every three seconds using the circuit shown in Figure 3.22.

E5.26 Write a C function that converts the Celsius temperature into Fahrenheit temperature. The temperature is represented as an integer part and a fractional part. Each part is represented by one byte. The fractional part is between 0 and 9. Both parts are passed to this function as parameters.

5.18 Laboratory Exercise Problems and Assignments

L5.1. Enter, compile, and download the following C program onto the SSE040 demo board for execution.

```
#include <C8051F040.h>
void systemInit(void);
void delayby100ms(char cx);

code unsigned char led_tab[] = {0x00,0xFF,0x00,0xFF,0x00,0xFF,0x00,0xFF,
                        0x80,0x40,0x20,0x10,0x08,0x04,0x02,0x01,
                        0x01,0x02,0x04,0x08,0x10,0x20,0x40,0x80,
                        0x80,0x40,0x20,0x10,0x08,0x04,0x02,0x01,
                        0x01,0x02,0x04,0x08,0x10,0x20,0x40,0x80,
                        0x80,0x40,0x20,0x10,0x08,0x04,0x02,0x01,
                        0x01,0x02,0x04,0x08,0x10,0x20,0x40,0x80,
                        0x80,0x40,0x20,0x10,0x08,0x04,0x02,0x01,
                        0x01,0x02,0x04,0x08,0x10,0x20,0x40,0x80,
                        0x81,0x42,0x24,0x18,0x18,0x24,0x42,0x81,
                        0x81,0x00,0x81,0x00,0x81,0x00,0x81,0x00,
                        0x42,0x00,0x42,0x00,0x42,0x00,0x42,0x00,
                        0x24,0x00,0x24,0x00,0x24,0x00,0x24,0x00,
                        0x18,0x00,0x18,0x00,0x18,0x00,0x18,0x00,
                        0x18,0x00,0x18,0x00,0x18,0x00,0x18,0x00,
                        0x24,0x00,0x24,0x00,0x24,0x00,0x24,0x00,
```

```
                                        0x42,0x00,0x42,0x00,0x42,0x00,0x42,0x00,
                                        0x81,0x00,0x81,0x00,0x81,0x00,0x81,0x00};

void main(void)
{
    int ix;
    systemInit();
    while(1) {
        for (ix = 0; ix < 144; ix++) {
            P5 = led_tab[ix];
            delayby100ms(2);
        }
    }
}
void systemInit(void)
{
    int n;
    SFRPAGE = CONFIG_PAGE;
    WDTCN   = 0xDE;                      // disable watchdog timer
    WDTCN   = 0xAD;                      //      "
    OSCXCN  = 0x67;                      // start external oscillator; 24 MHz Crystal
                                         // system clock is 24 MHz
    for (n=0;n<255;n++);                 // delay about 1 ms
    while ((OSCXCN & 0x80) == 0); // wait for oscillator to stabilize
    CLKSEL |  = 0x01;                    // switch to external oscillator
    XBR2      = 0x5D;                    // enable crossbar and assign I/O pins to all
    XBR0      = 0xF7;                    // peripheral signals,
    XBR1      = 0xFF;                    //      "
    XBR3      = 0x8F;                    //      "
    P5MDOUT = 0xFF;                      // configure Port P5 as output
}
void delayby100ms(char cx)
{
    char temp, i;

    temp    = SFRPAGE;
    SFRPAGE = 0;          // TMR0 is in SFRPAGE 0
    TMOD    = 0x01;       // configure Timer 0 to mode 1
    TF0     = 0;
    CKCON   = 0x02;       // use SYSCLK/48 as Timer 0 clock source
    for (i = 0; i < cx; i++) {
        TH0 = 0x3C;       // place 15536 in Timer 0 so it overflows in 100 ms
        TL0 = 0xB0;       //      "
        TR0 = 1;          // enable Timer 0 to count
        while(!TF0);      // while for 100 ms
        TF0 = 0;
        TR0 = 0;          // stop timer 0
    }
    SFRPAGE = temp;
}
```

If you are using SiLabs' C8051F040TB to perform this experiment, you might want to consider using the output test board (shown in Figure L5.1) made by Futurlec (www.futurlec.com).

L5.2. Write a C program that finds all the prime numbers between 100 and 1000 and print eight numbers in one row on the PC monitor screen.

L5.3. Write a C program to find all the four-digit numbers that have the property that the sum of the square of the upper half and the square of the lower half is equal to the original number. Display the numbers on the PC monitor screen.

L5.4. Modify the program in Example 5.8 to display the following pattern.

```
1              (1)
2 1            (2)
3 2 1          (3)
4 3 2 1        (4)
5 4 3 2 1      (5)
6 5 4 3 2 1    (6)
7 6 5 4 3 2    (7)
8 7 6 5 4 3    (8)
9 8 7 6 5 4    (9)
0 9 8 7 6 5    (10)
1 0 9 8 7 6    (11)
2 1 0 9 8 7    (12)
3 2 1 0 9 8    (13)
4 3 2 1 0 9    (14)
5 4 3 2 1 0    (15)
```

Display the pattern (1) to (5) only once but repeat the pattern (6) to (15) forever.

L5.5 Use the AD7302 ADC and the circuit shown in Figure 3.23 to generate a digital waveform that alternates between 1 kHz and 2 kHz every three seconds.

Output Test Board
Ideal for testing output ports and creating LED sequencing codes.

Features
- IDCC Cable Connector Provided
- Mounting Supports Fitted
- Circuit Diagram Included
- Cable Length 200mm

FIGURE L5.1
Futurlec output test board.

Interrupts and Resets

6.1 Objectives

Upon successful completion of this chapter, you will be able to:

- Explain the nature of interrupts and resets
- Explain how the CPU handles interrupts and resets
- Enable and disable interrupts
- Write interrupt handling service routines
- Program any interrupt to high or low priority
- Use a watchdog timer (WDT) reset to detect software bugs

6.2 Fundamental Concepts of Interrupts

Interrupts and resets are among the most useful mechanisms that a computer system provides. With interrupts and resets, I/O operations are performed more efficiently, errors are handled more smoothly, and CPU utilization is improved. This chapter will begin with a general discussion of interrupts and resets and then focus on the specific features of the 8051 interrupts and resets.

6.2.1 What is an Interrupt?

When the CPU is executing a program, the instruction execution sequence is determined by the program logic. However, this instruction execution sequence may be changed by a special event called an *interrupt.* An interrupt is an event that requires the CPU to stop normal program execution and perform some service related to the event. Most processors are designed to allow the user to use an external signal to get the attention of the CPU. The external signal used in this manner is referred to as an *external* (or *hardware*) *interrupt.* All microcontrollers have on-chip peripheral devices such as timers, A/D converters, serial peripheral interface, and so on. These devices can assert a signal to get the attention of the CPU. These signals are called *internal interrupt.* An internal interrupt can also be generated by *software* errors such as illegal opcodes, overflow, divide-by-zero, and underflow. These interrupts are called *software interrupts.* Software interrupts are also referred to as *traps* or *exceptions.*

A good analogy for interrupt is how you act when you are sitting in front of a desk to read this book and the phone rings. You probably will act like this:

1. Remember the page number or place a bookmark on the page that you are reading, close the book, and put it aside.
2. Pick up the phone and say, "Hello, this is so and so."
3. Listen to the voice over the phone to find out who is calling or ask who is calling if the voice is not familiar.
4. Talk to that person.
5. Hang up the phone when you finish talking.
6. Open the book and turn to the page where you placed the bookmark and resume reading this book.

The phone call example spells out a few things that are similar to how the microprocessor handles the interrupt.

1. As a student, you spend most of your time studying. Answering a phone call happens only occasionally. Similarly, the microprocessor is executing application programs most of the time. Interrupts will only force the microprocessor to stop executing the application program briefly and take some necessary actions.
2. Before picking up the phone, you finish reading the sentence and then place a bookmark to remind you of the page number that you are reading so that you can resume reading after finishing the conversation over the phone. Most microprocessors will finish the instruction they are executing and save the address of the next instruction in memory (usually in the stack) so that they can resume the program execution later.
3. You find out who the person is by listening to the voice over the phone, or you ask questions so that you can decide what to say. Similarly, the microprocessor needs to identify the cause of the interrupt before it can take appropriate actions. This is built into the microprocessor hardware.
4. After identifying the person who called you, you start the phone conversation with that person on some appropriate subjects. Similarly, the microprocessor will take some actions appropriate to the interrupt source by executing a short routine.
5. When finishing the phone conversation, you hang up the phone, open the book to the page where you placed the bookmark, and resume reading. Similarly, after taking some actions appropriate to the interrupt, the

microprocessor will jump back to the instruction next to the one that was being executed when the interrupt occurred and resume program execution. This can be achieved easily, because the address of the instruction to be resumed was saved in memory (stack). Most microprocessors do this by executing a *return-from-interrupt* (RETI) instruction.

6.2.2 Why are Interrupts Used?

Interrupts are useful in many applications, such as

- ***I/O Handling.*** Before reading data from an input device, the CPU needs to know if the I/O device has new data. If the input device cannot interrupt the CPU, the CPU will need to keep checking (also called *polling*) the status of the I/O device until it discovers that the input device has new data before reading it. With the interrupt capability, the CPU can continue to perform other operations and will read the data only when it is interrupted. To output data, the CPU needs to make sure that the output device can accept new data before it sends out data. Without interrupt capability, the CPU must keep polling the output device status until it discovers that the output device is ready for more data and then outputs data. If the output device can interrupt the CPU, the microcontroller does not need to wait. CPU time thus can be utilized more efficiently because of the interrupt mechanism.
- ***Real-Time Response.*** Emergent events (such as power-failure in a computer system and overheat and overpressure in a control system) require the CPU to take action immediately. The interrupt mechanism can force the CPU to divert from normal program execution and take actions immediately. Severe data loss or damage to the system thus can be avoided.
- ***Reminding the CPU to Perform Routine Tasks.*** An embedded system often needs to perform certain operations periodically. This can be achieved by using timer interrupts to remind the CPU.

6.2.3 Enabling and Disabling Interrupts

Although interrupt is a very useful mechanism, it must be used carefully in order for an embedded system to achieve the highest performance. An interrupt may be useful in some situations but not desirable in other situations. Modern microcontrollers are designed to have the capability to selectively pay attention to interrupts. Whenever an interrupt is needed, the MCU *enables* it. Whenever an interrupt is not needed, the MCU *disables* it. An interrupt that can be enabled selectively and disabled is called *maskable interrupt.* There are other types of interrupts that the CPU cannot disable and must take immediate actions for; these are called *nonmaskable interrupts.* A program can request the CPU to service or ignore a maskable interrupt by setting or clearing an *enable bit.* When an interrupt is enabled, the CPU will respond to it. When an interrupt is disabled, the CPU will ignore it. An interrupt is said to be *pending* when it is active but not yet serviced by the CPU. A pending interrupt may or may not be serviced by the CPU, depending on whether or not it is enabled.

To make the interrupt system more flexible, a computer system normally provides a global and local interrupt masking capability. When none of the interrupts are desirable, the processor can disable all of the interrupts by clearing the global interrupt enable bit (or setting the global interrupt mask bit for some other processors). In other situations, the processor can enable certain interrupts selectively, while at the same time disable other undesirable interrupts. This is achieved by providing each interrupt source an enable bit in addition to the global interrupt mask. Whenever any interrupt is undesirable, it can be disabled, while at the same time allowing other interrupt sources to be serviced (attended) by the processor. Today, almost all commercial processors are designed to provide this two-level (or even three-level) interrupt-enabling capability.

6.2.4 Handling Multiple Interrupts by Prioritizing

If a computer system is supporting multiple interrupt sources, then it is possible that several interrupts would be pending at about the same time. The CPU has to decide which interrupt should receive service first in this situation. The common solution is to prioritize all interrupt sources. In this approach, the CPU provides service to the interrupt with the highest priority among all pending interrupts. Most microcontrollers (including the 8051) prioritize interrupts in hardware. For those microcontrollers that do not prioritize interrupts in hardware, the software can be written to handle certain interrupts before others. By doing this, interrupts are essentially prioritized.

6.2.5 Servicing the Interrupt

When an enabled interrupt occurs, the CPU is forced to provide service to it by executing an *interrupt service routine.* After executing the interrupt service routine, the CPU must resume the execution of the interrupted program. In order to be able to resume the interrupted program, the CPU saves the program counter before executing the interrupt service routine. After saving the program counter, the CPU needs to identify the source of interrupt and locate the starting address of the interrupt service routine. It then starts to execute the interrupt service routine. The interrupt service routine is often very short. When the CPU finishes the execution of the interrupt service routine, it restores the saved program counter and returns to the interrupted program for execution.

6.2.6 What is an Interrupt Vector?

The starting address of the interrupt service routine is referred to as an *interrupt vector.* To provide service to an interrupt, the microcontroller must find out the starting address of its service routine. For some microcontrollers, each interrupt source has its own predefined interrupt vector (i.e., at fixed location) whereas others store all of the interrupt vectors in a table. Each interrupt vector is stored in the fixed location in the table. This table is called an *interrupt vector table.*

When the first approach is used, the CPU will jump to a predefined location to service the pending interrupt. However, the CPU will need to fetch the interrupt vector from the interrupt vector table in the second approach. The 8051 uses the first approach to locate the interrupt vector. When the first approach is used, each interrupt source is allocated with the same number of bytes to hold its service routine. This approach is easy to implement but cannot satisfy the needs of interrupt handling, because some interrupts may need much more memory space to hold their service routines. This problem can be resolved by placing the actual interrupt service routine at other location and using a jump (or **goto**) instruction to jump to it. For example, the Timer 0 interrupt vector is at 0x0B and the Timer 0 interrupt service routine can be written in the following manner.

```
            org     0x0B
            ljmp    TMR0_ISR
            ...
TMR0_ISR:                       ; actual TMR0 service routine is longer than 8 bytes
            clr     TF0         ;
            ...
            reti                ; return from interrupt
```

An interrupt service routine is not much different from a subroutine, except for the last instruction. The last instruction of an interrupt service routine must be the *return from interrupt* instruction. An interrupt service routine is not to be called. Instead, the execution of an interrupt service routine is triggered by the occurrence of the associated interrupt.

6.2.7 How to Write an Interrupt-Driven Program?

Interrupt programming deals with how to provide service to the interrupt. Interrupt programming consists of three major steps:

Step 1

Initialize the Interrupt Vector Table. The purpose of this step is to place interrupt vectors in the interrupt vector table. This can be achieved by using the assembler directives as follows.

```
ORG     0xYYYY     ; starting address of the interrupt vector table
DW      ISR1       ; the first interrupt vector
DW      ISR2
.
.
.
DW      ISRn       ; the last interrupt vector
```

This step is not needed for the 8051 microcontroller.

Step 2

Write the Service Routine. The interrupt service routine must be made to be as short as possible. An interrupt service routine for INT0 may be as simple as the following example.

```
cnt         set     R7          ; cnt is a counter value
            .
            .
            .
int0ISR:    inc     cnt         ; increment cnt by 1
            mov     P3,cnt      ; output cnt to P3
            reti                ; return from interrupt
```

Step 3

Enable Interrupts. It is important to enable interrupts. If the user forgets to enable interrupts, interrupts will never occur. This is a common mistake in writing interrupt-driven programs.

6.3 Resets

When a computer is first powered up, the initial values of program counter, CPU registers, flip-flops, and I/O control registers are unknown. The computer cannot execute programs correctly under these conditions. In order for computers to execute a program correctly after they are powered up, computers are designed to have a reset mechanism. All critical registers, the program counter, and flip-flops are forced to a default value after a power-up reset so that the computer can start to execute a program correctly. For example, the program counter of the 8051 is forced to 0x0000 after reset, and hence, it always starts program execution from the address 0x0000 after power-on reset.

In addition to *power-on reset,* a computer may provide several other resets to improve the performance of the computer system. These resets include manual reset (applied to a pin), brown-out reset, watchdog reset, clock monitor reset, and so on. A manual reset without power-down has the same effect as power-on reset and can force the program counter and critical registers to default values and allows the computer to *reboot* itself after a reset. Other reset sources are discussed later.

6.4 The 8051 Interrupts

The original 8051 has six interrupt sources. They are listed in Table 6.1. Most 8051 variants have more than six interrupt sources. For example, the Silicon Laboratory C8051F040 has 20 interrupt sources (listed in Table 6.2).

All 8051 variants have two external interrupt pins INT0 and INT1. The user can choose to allow these two pins to interrupt the CPU when they are low (level-triggered) or on their falling edge (edge-triggered). The selection of interrupting on the falling edge or low level is controlled by the setting of the IT0 and IT1 bits of the timer control register (TCON). When the ITx bit ($x = 0$ or 1) is 0, the INTx ($x = 0$ or 1) signal interrupts when it is low. Otherwise, the CPU is interrupted on the falling edge. The contents of the TCON register are shown in Figure 6.1.

When an INTx pin interrupt occurs, the corresponding flag (IE0 or IE1) will set to 1. If the INTx pin interrupt is programmed to be edge-triggered, the associated interrupt flag will be cleared when the service routine is entered. If these interrupts are programmed to be level-triggered, their flags must be cleared by the software.

In Tables 6.1 and 6.2, every interrupt source has a pending flag and an enable flag. In order for an interrupt to be requested to the CPU, both the enable flag and the pending flag of an interrupt source must be set to 1.

The interrupt enabling of the original 805/MCU is controlled by the IE register. The C8051F040 uses registers IE, EIE1, and EIE2 to control the enabling of its 20 interrupt sources. The contents of registers IE, EIE1, and EIE2 are shown in Figures 6.2, 6.3, and 6.4, respectively. The 8051 adopts a two-level interrupt enabling mechanism and has a global interrupt enable bit

TABLE 6.1 *Summary of Original 8051 Interrupt Sources*

Source	Interrupt Vector	Priority Order	Pending Flag	Bit-Addressable	Cleared by Hardware	Enable Flag	Priority Control
Reset	0x0000	Top	None	N/A	N/A	Always enabled	Always highest
/INT0	0x0003	0	IE0 (TCON.1)	Y	Y	EX0 (IE.0)	PX0 (IP.0)
Timer 0 overflow	0x000B	1	TF0 (TCON.5)	Y	Y	ET0 (IE.1)	PT0 (IP.1)
/INT1	0x0013	2	IE1 (TCON.3)	Y	Y	EX1 (IE.2)	PX1 (IP.2)
Timer 1 overflow	0x001B	3	TF1 (TCON.7)	Y	Y	ET1 (IE.3)	PT1 (IP.3)
UART	0x0023	4	RI (SCON.0)	Y		ES (IE.4)	PS (IP.4)
			TI (SCON.1)	Y		ES (IE.4)	PS (IP.4)
Timer 2	0x002B	5	TF2 (T2CON.7)	Y		ET2 (IE.5)	PT2 (IP.5)

TABLE 6.2 *C8051F040 Interrupt Sources*

Source	Interrupt Vector	Priority Order	Pending Flag	Bit-Addressable	Clear by HW	Enable Flag	Priority Control
Reset	0x0000	Top	None	N/A	N/A	always	highest
/INT0	0x0003	0	IE0 (TCON.1)	Y	Y(1)	EX0(IE.0)	PX0(IP.0)
Timer 0 overflow	0x000B	1	TF0 (TCON.5)	Y	Y	ET0(IE.1)	PT0(IP.1)
/INT1	0x0013	2	IE1 (TCON.3)	Y	Y(1)	EX1(IE.2)	PX1(IP.2)
Timer 1 overflow	0x001B	3	TF1 (TCON.7)	Y	Y	ET1(IE.3)	PT1(IP.3)
UART0	0x0023	4	RI0 (SCON0.0) TI0 (SCON0.1)	Y		ES0(IE.4)	PS0(IP.4)
Timer 2	0x002B	5	TF2 (T2CON.7)	Y		ET2(IE.5)	PT2(IP.5)
SPI interface	0x0033	6	SPIF (SPI0CN.7) WCOL (SPI0CN.6) MODF(SPI0CN.5) RXOVRN(SPI0CN.4)	Y		ESPI0 (EIE1.0)	PSPI0 (EIP1.0)
SMBus interface	0x003B	7	SI (SMB0CN.3)	Y		ESMB0 (EIE1.1)	PSMB0 (EIP1.1)
ADC0 window comparator	0x0043	8	ADWINT(ADC0CN.2)	Y		EWADC0 (EIE1.2)	PWADC0 (EIP1.2)
PCA array	0x004B	9	CF(PCA0CN.7) CCFn(PCA0CN.n)	Y		EPCA0(EIE1.3)	PPCA0 (EIP1.3)
Comparator 0	0x0053	10	CP0FIF(CPT0CN.4) CP0RIF(CPT0CN.5)			CP0IE(EIE1.4)	PCP0 (EIP1.4)
Comparator 1	0x005B	11	CP1FIF(CPT1CN.4) CP1RIF(CPT1CN.5)			CP1IE(EIE1.5)	PCP1 (EIP1.5)
Comparator 2	0x0063	12	CP2FIF(CPT2CN.4) CP2RIF(CPT2CN.5)			CP2IE(EIE1.6)	PCP2 (EIP1.6)
Timer 3	0x0073	14	TF3(TMR3CN.7)			ET3(EIE2.0)	PT3 (EIP2.0)
ADC0 end of conversion	0x007B	15	ADC0INT(ADC0CN.5)	Y		EADC0(EIE2.1)	PADC0 (EIP2.1)
Timer 4	0x0083	16	TF4(T4CON.7)			ET4(EIE2.2)	PT4 (EIP2.2)
ADC2 window comparator	0x0093	17	AD2WINT(ADC2CN.0)			EWADC2 (EIE2.3)	PWADC2 (EIP2.3)
ADC2 end of conversion	0x008B	18	ADC2INT(ADC2CN.5)			EADC2 (EIE2.4)	PADC2 (EIP2.4)
CAN interrupt	0x009B	19	CAN0CN.7		Y	ECAN0 (EIE2.5)	PCAN0 (EIP2.5)
UART1	0x00A3	20	RI1 (SCON1.0) TI1 (SCON1.1)			ES1 (EIE2.6)	PS1 (EIP2.6)

Note 1:
Will be cleared by hardware only when configured to be edge-triggered.

7	6	5	4	3	2	1	0	
TF1	TR1	TF0	TR0	IE1	IT1	IE0	IT0	Reset value = 0x00
rw	rw	rw	rw	rw	rw	rw	rw	

TF1: Timer 1 overflow flag
  0 = no timer 1 overflow
  1 = Timer 1 overflows.
  This flag will be cleared when processor vectors to its interrupt service routine
TR1: Timer 1 run control bit
  0 = Timer 1 is not allowed to run
  1 = Timer 1 is enabled to run
TF0: Timer 0 overflow flag
  0 = Timer 0 does not overflow
  1 = Timer 0 has overflowed
TR0: Timer 0 run control bit
  0 = Timer 0 is not allowed to run
  1 = Timer 0 is enabled to run
IE1: INT1 interrupt edge flag
  0 = no interrupt edge (INT1) detected
  1 = interrupt edge (INT1) has been detected
IT1: INT1 interrupt type control bit
  0 = INT1 interrupt is low-level triggered
  1 = INT1 interrupt is falling edge triggered
IE0: INT0 interrupt edge flag
  0 = no interrupt edge (INT0) detected
  1 = interrupt edge (INT0) detected
IT0: INT0 interrupt type control bit
  0 = INT0 interrupt is low-level triggered
  1 = INT0 interrupt is falling edge triggered

FIGURE 6.1 *The timer/counter control register TCON.*

(EA, bit 7 of the IE register) which must be set to 1 in order for each individual interrupt to be enabled. For example, in order to enable the Timer 0 interrupt, the user needs to set both the EA bit and the ET0 bit to 1.

6.4.1 The 8051 Interrupt Priority Structure

Both the original 8051 and the C8051F040 allow each individual to be programmed to either high or low priority. The original 8051 uses the IP register to program the interrupt priority. The SiLabs C8051F040 uses registers IP, EIP1, and EIP2 to program the priority of interrupts. The contents of these registers are shown in Figure 6.5 to 6.7. A low-priority interrupt can be interrupted by a high-priority interrupt, but not by another low-priority interrupt.

7	6	5	4	3	2	1	0	
EA	--[1]	ET2	ES[2]	ET1	EX1	ET0	EX0	Reset value = 0x00
rw	rw	rw	rw	rw	rw	rw	rw	

EA: Enable all interrupts
  0 = disable all interrupts
  1 = enable each interrupt according to its individual enable bit setting
EC: PCA interrupt enable
ET2: Timer 2 interrupt enable
ES: UART interrupt enable
ET1: Timer 1 interrupt enable
EX1: INT1 interrupt enable
ET0: Timer 0 interrupt enable
EX0: INT0 interrupt enable
  When set to 1, bit 5 to bit 0 enables the associated interrupt source
  When cleared to 0, bit 5 to bit 0 disables the associated interrupt source

Note 1: Bit 6 is used as a general-purpose flag in the original 8051 and the C8051F040.
Note 2: Bit 4 is used to enable UART0 for the C8051F040

FIGURE 6.2
The interrupt enable register IE.

7	6	5	4	3	2	1	0	
--	CP2IE	CP1IE	CP0IE	EPCA0	EWADC0	ESMB0	ESPI0	Reset value = 0x00
rw	rw	rw	rw	rw	rw	rw	rw	

CP2IE: Comparator 2 (CP2) interrupt enable bit
CP1IE: Comparator 1 (CP1) interrupt enable bit
CP0IE: Comparator 0 (CP0) interrupt enable bit
EPCA0: Programmable counter array PCA0 interrupt enable bit
EWADC0: Window comparison ADC0 interrupt enable bit
ESMB0: System management bus (SMBus) interrupt enable bit
ESPI0: Serial peripheral interface interrupt enable bit
  When a bit is set to 1, its associated interrupt request is enabled
  When a bit is cleared to 0, its associated interrupt is disabled

FIGURE 6.3
The interrupt enable register EIE1 (SiLabs C8051F040).

A high-priority interrupt can not be interrupted by any other interrupt sources.

The 8051 CPU provides service to the interrupt with the higher priority when there are two or more interrupts occurring at the same time. If two or more interrupts at the same priority level occur at the same time, an internal polling sequence determines which request is serviced. The internal polling sequence is shown in Tables 6.1 and 6.2. The interrupt that has a smaller priority number is at the higher priority.

7	6	5	4	3	2	1	0	
--	ES1	ECAN0	EADC2	EWADC2	ET4	EADC0	ET3	Reset value = 0x00
rw	rw	rw	rw	rw	rw	rw	rw	

ES1: UART1 interrupt enable bit
ECAN0: CAN controller interrupt enable bit
EADC2: ADC2 end of conversion interrupt enable bit
EWADC2: Window comparison ADC2 interrupt enable bit
ET4: Timer 4 interrupt enable bit
EADC0: ADC0 end of conversion interrupt enable bit
ET3: Timer 3 interrupt enable bit
  When a bit is set to 1, its associated interrupt request is enabled
  When a bit is cleared to 0, its associated interrupt is disabled

FIGURE 6.4
The interrupt enable register EIE2 (SiLabs C8051F040).

7	6	5	4	3	2	1	0	
--	--	PT2	PS[1]	PT1	PX1	PT0	PX0	Reset value = 0x00
rw	rw	rw	rw	rw	rw	rw	rw	

PT2: Timer 2 interrupt priority control bit
PS: UART interrupt priority control bit
PT1: Timer 1 interrupt priority control bit
PX1: INT1 pin interrupt priority control bit
PT0: Timer 0 interrupt priority control bit
PX0: INT0 pin interrupt priority control bit
  0 = low priority
  1 = high priority

Note 1: This bit is PS0 (UART0) for the C8051F040

FIGURE 6.5
The interrupt priority register IP (all 8051 variants).

6.4.2 INT0 and INT1 Pins Interrupts

If the INT0 or INT1 pin interrupt is programmed to be level-triggered, it needs to be held low until the requested interrupt is actually generated. Then it has to be pulled high before the interrupt service routine is completed or else another interrupt (from the same source) will be generated. Programming the INT0 or INT1 pin interrupt to be edge-triggered can eliminate this requirement. However, the setting of edge-trigging is not suitable for a noisy environment, because every falling edge will cause an interrupt.

7	6	5	4	3	2	1	0	
--	PCP2	PCP1	PCP0	PPCA0	PWADC0	PSMB0	PSPI0	Reset value = 0x00
rw	rw	rw	rw	rw	rw	rw	rw	

PCP2: Comparator 2 interrupt priority control bit
PCP1: Comparator 1 interrupt priority control bit
PCP0: Comparator 0 interrupt priority control bit
PPCA0: Programmable counter array (PCA0) interrupt priority control bit
PWADC0: ADC0 windowcomparator interrupt priority control bit
PSMB0: System management bus interrupt priority control bit
PSPI0: Serial peripheral interface interrupt priority control bit
  0 = low priority
  1 = high priority

FIGURE 6.6
The interrupt priority register EIP1 (SiLabs C8051F040).

7	6	5	4	3	2	1	0	
--	EP1	PCAN0	PADC2	PWADC2	PT4	PADC0	PT3	Reset value = 0x00
rw	rw	rw	rw	rw	rw	rw	rw	

EP1: UART1 interrupt priority control bit
PCAN0: CAN interrupt priority control bit
PADC2: ADC2 end of conversion interrupt priority control bit
PWADC2: ADC2 window comparator interrupt priority control bit
PT4: Timer 4 interrupt priority control bit
PADC0: ADC0 end of conversion interrupt priority control bit
PT3: Timer 3 interrupt priority control bit
  0 = low priority
  1 = high priority

FIGURE 6.7
The interrupt priority register EIP2 (SiLabs C8051F040).

Example 6.1 Write an instruction sequence to enable INT0 and INT1 interrupts and also configure them to be high priority and falling-edge triggered.

Solution: The following instruction sequence will achieve the desired configuration for the original 8051 and C8051F040.

```
            $include  (c8051F040.inc)
            cseg      at 0x00
            org       0x00
            ljmp      start
            org       0xAB
start:      setb      IT0                ; configure INT0 to be edge triggered
```

```
        setb    IT1             ; configure INT1 to be edge triggered
        setb    EX0             ; enable INT0 interrupt locally
        setb    EX1             ; enable INT1 interrupt locally
        setb    PX0             ; set INT0 interrupt at high priority level
        setb    PX1             ; set INT1 interrupt at high priority level
        setb    EA              ; enable interrupt globally
        . . .
```

When interrupts are not needed, they can be disabled by clearing the EA bit. For the C8051F040, any instruction that clears the EA bit should be immediately followed by an instruction that has two or more opcode bytes. For example,

```
// in C:
EA  = 0;        // clear EA bit
EA  = 0;        //. . . followed by another 2-byte opcode
; in assembly:
clr  EA         ; clear EA bit
clr  EA         ;. . . followed by another 2-byte opcode
```

If an interrupt is posted during the execution phase of a "clr EA" opcode, and the instruction is followed by a single-cycle instruction, the interrupt may be taken. However, a read of the EA bit will return a 0 inside the interrupt service routine. When the "clr EA" instruction is followed by a multi-cycle instruction, the interrupt will not be taken.

Example 6.2 Write a program and the service routine for the INT0 pin interrupt. The main program configures the INT0 interrupt to be edge-sensitive and then enables it. The program also initializes a counter to zero and then stays in an infinite loop while waiting for interrupt. The service routine for the INT0 interrupt increments the counter and then outputs it to the P5 port. P5 drives eight LEDs.

Solution: We need to enable the crossbar decoder and assign the INT0 signal to an I/O pin. We will assign I/O pins to all peripheral signals with priority higher than INT0 signal and select three-wire SPI mode (this is done simply to fix the pin assignment). With this setting, the INT0 signal is assigned to the P2.4 pin.

The assembly program that implements the requirement is as follows.

```
        $include (c8051F040.inc)
        dseg    at 0x30
cnt:    ds      1               ; counter
        cseg    at 0x00
        ljmp    start
        org     0x03
        ljmp    int0ISR
        org     0xAB
```

```
start:      mov     SP,#0x7F            ; establish stack pointer
            acall   sysinit
            setb    IT0                 ; INT0 interrupt is falling edge triggered
            setb    EX0                 ; enable INT0 interrupt locally
            setb    PX0                 ; place INT0 interrupt at high priority
            mov     cnt,#0              ; initialize count value to 0
            setb    EA                  ; enable interrupt globally
            ajmp    $                   ; wait for INT0 interrupt to come
; ***************************************************************************************
; Service routine for the INT0 interrupt.
; ***************************************************************************************
int0ISR:    mov     SFRPAGE,#0FH
            inc     cnt
            mov     P5,cnt
            reti
; ***************************************************************************************
; The following subroutine selects external crystal oscillator as SYSCLK, enable
; crossbar decoder,
; ***************************************************************************************
sysinit:    mov     SFRPAGE,#CONFIG_PAGE ; switch to the SFR page that contains
                                         ; registers that configures the system
            mov     WDTCN,#0xDE         ; disable watchdog timer
            mov     WDTCN,#0xAD         ;          "
            mov     CLKSEL,#0           ; used internal oscillator to generate SYSCLK
            mov     OSCICN,#0x83        ; configure internal oscillator control
            mov     XBR0,#0F7H          ; assign I/O pins to all peripheral functions
            mov     XBR1,#0FFH          ; and enable crossbar
            mov     XBR2,#5DH           ;    "
            mov     XBR3,#8FH           ;    "
            mov     P5MDOUT,#0xFF       ; configure Port P5 for push-pull
            mov     P5,#0
            mov     SFRPAGE,#SPI0_PAGE  ;
            mov     SPI0CN,#01H         ; select 3-wire SPI mode
            ret
            end
```

This program can be tested on the SSE040 demo board by connecting a wire between the INT0 (J3) pin and the P2.4 pin. After downloading the program onto the MCU, you can start to press the S1 red button, and the LEDs will show the incrementing of the count value.

The C language version of the program is as follows.

```
#include <c8051F040.h>
void int0_ISR (void); // INT0 service routine prototype
void sysinit(void);
char cnt;
void main (void)
```

```
{
    cnt  = 0;
    sysinit ();
    IT0  = 1;       // INT0 falling edge triggered
    EX0  = 1;       // enable INT0 interrupt locally
    PX0  = 1;       // place INT0 interrupt at high priority
    EA   = 1;       // enable interrupt globally
    while (1);      // wait for INT0 interrupt
}
void int0_ISR (void) interrupt 0    // 0 is the interrupt number of INT0
{
    cnt++;
    SFRPAGE = 0x0F;
    P5      = cnt;              // display count value on LEDs
}
void sysinit(void)
{
    int     n;
    SFRPAGE = CONFIG_PAGE;
    WDTCN  = 0xDE;              // disable watchdog timer
    WDTCN  = 0xAD;              //              "
    OSCXCN = 0x67;              // start external oscillator; 24 MHz Crystal
                                // system clock is 24 MHz
    for (n=0;n<255; n++);       // delay about 1 ms
    while ((OSCXCN & 0x80) == 0); // wait for oscillator to stabilize
    CLKSEL  |= 0x01;            // switch to external oscillator
    XBR2    = 0x5D;             // enable crossbar and assign I/O pins to all
    XBR0    = 0xF7;             // peripheral signals,
    XBR1    = 0xFF;             //    "
    XBR3    = 0x8F;             //    "
    P5MDOUT = 0xFF;             // configure Port P5 to push-pull for output
    P5      = 0;                // turn off all LEDs
    SFRPAGE = SPI0_PAGE;
    SPI0CN  = 0x01;             // enable 3-wire SPI (make sure SPI uses 3 pins
}
```

For other compilers (including SDCC and Raisonance C), the prototype declaration of the INT0 interrupt service routine should be

```
void int0ISR (void) interrupt 0;
```

Example 6.3 Write a main program and an interrupt service routine for Timer 0. The main program clears a count variable to 0, configures Timer 0 to overflow in every 100 ms, enables Timer 0 interrupt, places Timer 0 interrupt in high priority, and then stays in an infinite loop waiting for Timer 0 interrupt to occur. The service routine for Timer 0 interrupt clears the TF0 flag, checks the P4.7 pin, increments the count variable if it is low, and then outputs the count value to P5. If the P4.7 pin is high, the count variable remains unchanged.

Solution: The assembly program that implements the specified operation is as follows.

```
        $include  (c8051F040.inc)
        dseg      at 0x30
cnt:    ds        1               ; counter
        cseg      at 0x00
        ljmp      start
        org       0x0B            ; TMR0 interrupt vector
        ljmp      TMR0ISR
        org       0xAB
start:  mov       SP,#0x7F        ; establish stack pointer
        acall     sysinit
        acall     initTMR0        ; set up Timer 0 to mode 1, sysclk/48 as clock source
        setb      ET0             ; enable TMR0 interrupt locally
        setb      PT0             ; place TMR0 interrupt at high priority
        mov       cnt,#0          ; initialize count value to 0
        setb      EA              ; enable interrupt globally
        setb      EA
        ajmp      $               ; wait for TMR0 interrupt to come
;*****************************************************************************************
; Service routine for the TMR0 interrupt.
;*****************************************************************************************
TMR0ISR: mov      SFRPAGE,#0
        clr       TF0
        clr       TR0             ; stop Timer 0
        mov       TH0,#0x3C       ; reset TMR0 to count up from 15536 (0x3CB0)
        mov       TL0,#0xB0       ;       "
        setb      TR0             ; start TMR0 again
        mov       SFRPAGE,#0x0F
        mov       P4,0xFF         ; configure P4 for input
        mov       A,P4            ; if P4.7 is low, increment cnt. Otherwise, cnt
        anl       A,#0x80         ; is unchanged.
        jnz       exitT0          ;       "
        inc       cnt             ;       "
exitT0: mov       P5,cnt          ; display cnt value on LEDs
        reti
;*****************************************************************************************
; Set up timer 0 to use SYSCLK/48 as clock source and count up from 15536. This
; setting makes TMR0 overflow in 100 ms.
;*****************************************************************************************
initTMR0: mov     TMOD,#0x01      ; configure Timer 0 to mode 1
        mov       CKCON,#02       ; Timer 0 use SYSCLK/48 as clock source
        mov       TH0,#0x3C       ; let TMR0 count up from 15536 (0x3CB0)
        mov       TL0,#0xB0       ;       "
        setb      TR0             ; let TMR0 to count
        ret
```

```
;*****************************************************************************************************
; The following subroutine selects external crystal oscillator as SYSCLK, enable
; crossbar decoder,
;*****************************************************************************************************
sysinit:    mov     SFRPAGE,#CONFIG_PAGE    ; switch to the SFR page that contains
                                            ; registers to configure the system
            mov     WDTCN,#0DEH             ; disable watchdog timer
            mov     WDTCN,#0ADH             ;       "
            mov     CLKSEL,#0               ; used internal oscillator to generate SYSCLK
            mov     OSCXCN,#067H            ; configure external oscillator control
            mov     R7,#255
            djnz    R7,$                    ; wait for about 1 ms
chkstable:  mov     A,OSCXCN                ; wait until external crystal oscillator is stable
            anl     A,#80H                  ; before using it
            jz      chkstable               ;       "
            mov     CLKSEL,#1               ; use external crystal oscillator to generate SYSCLK
            mov     XBR0,#0F7H              ; assign I/O pins to all peripheral functions
            mov     XBR1,#0FFH              ; and enable crossbar
            mov     XBR2,#5DH               ;       "
            mov     XBR3,#8FH               ;       "
            mov     P5MDOUT,#0FFH           ; configure Port P5 for push-pull
            mov     P4MDOUT,#0              ; configure P4 for input
            mov     SFRPAGE,#SPI0_PAGE      ;
            mov     SPI0CN,#01H             ; select 3-wire SPI mode
            ret
            end
```

This program can be tested by connecting the P4.7 pin to V_{CC} and GND. When it is connected to V_{CC}, the LEDs won't change. Otherwise, the LEDs will keep changing.

If you pay attention, you will notice that many applications (including automobile) use this technique to set time. The C language version of the program is straightforward and hence will be left for you as an exercise problem.

Example 6.4 Interrupt-Driven Input. Write a main program and an interrupt service routine of $\overline{\text{INT0}}$ to read data from the DIP switches shown in Figure 3.21. The main program performs the required initialization, enables $\overline{\text{INT0}}$ interrupt, and then outputs a message to remind the user to set up a value and then to press the $\overline{\text{INT0}}$ switch to generate an interrupt. The service routine of the $\overline{\text{INT0}}$ interrupt reads the DIP switch's value (connected to P6) and outputs it to LEDS driven by P5. This program can be run directly on the SSE040 demo board. The $\overline{\text{INT0}}$ signal is assigned to P2.4 pin.

Solution: The main program calls the **sysinit** subroutine to select the system clock source, enables crossbar decoder, configures the $\overline{\text{INT0}}$ signal to be edge-sensitive, enables $\overline{\text{INT0}}$ interrupt, calls the **openUART0**

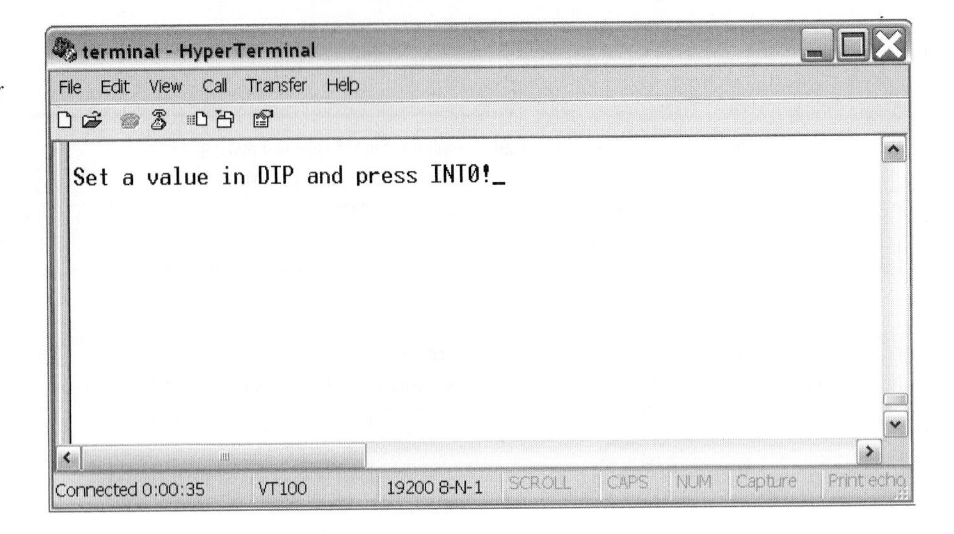

FIGURE 6.8 *A message to remind the user for input in an interrupt-driven input.*

subroutine to initialize UART0, outputs the message "Set a value in DIP and press INT0!", and then waits for the user to take action. The service routine of the $\overline{\text{INT0}}$ interrupt simply reads P6, outputs it to P5, and return from interrupt.

When the program runs, you would see the message shown in Figure 6.8. To continue the program execution, you should set a value on the DIP switches and then press the $\overline{\text{INT0}}$ switch. After that, the value that you set on the DIP switches will appear on the LEDS driven by P5.

The assembly program that implements the required operation is as follows.

```
        $include  (c8051F040.inc)
        cseg      at 0x00
        ljmp      start
        org       0x03              ; INT0 interrupt vector
        ljmp      int0ISR
        org       0xAB
start:  mov       SP,#0x7F          ; establish stack pointer
        acall     sysinit
        acall     openUART0
        setb      IT0               ; INT0 interrupt is falling edge triggered
        setb      EX0               ; enable INT0 interrupt locally
        setb      PX0               ; place INT0 interrupt at high priority
        setb      EA                ; enable interrupt globally
        acall     newline           ; move to the next line on the monitor screen
        mov       DPTR,#prompt      ; remind user to set up a value on DIP switches
        acall     putsc             ;       "
        ajmp      $                 ; wait for INT0 interrupt to come
prompt: db        "Set a value in DIP and press INT0!",0
```

```
; ********************************************************************************************************
; Service routine for the INT0 interrupt.
; ********************************************************************************************************
int0ISR:        mov         SFRPAGE,#0FH
                mov         P5,P6
                reti
; ********************************************************************************************************
; The following subroutine selects external crystal oscillator as SYSCLK, enable
; crossbar decoder, selects external crystal oscillator as the system clock, assign
; all peripheral signals to I/O pins, select 3-wire SPI (4-wire is fine too).
; ********************************************************************************************************
sysinit:        mov         SFRPAGE,#SPI0_PAGE      ;
                mov         SPI0CN,#01H             ; select 3-wire SPI mode

mov             SFRPAGE,#CONFIG_PAGE ; switch to the SFR page that contains
                                                    ; registers that configures the system
                mov         WDTCN,#0DEH             ; disable watchdog timer
                mov         WDTCN,#0ADH             ;       "
                mov         CLKSEL,#0               ; used internal oscillator to generate SYSCLK
                mov         OSCXCN,#067H            ; configure external oscillator control
                mov         R7,#255
                djnz        R7,$                    ; wait for about 1 ms
chkstable:      mov         A,OSCXCN                ; wait until external crystal oscillator is stable
                anl         A,#80H                  ; before using it
                jz          chkstable               ;       "
                mov         CLKSEL,#1               ; use external crystal oscillator to generate SYSCLK
                mov         XBR0,#0F7H              ; assign I/O pins to all peripheral functions
                mov         XBR1,#0FFH              ; and enable crossbar
                mov         XBR2,#5DH               ;       "
                mov         XBR3,#8FH               ;       "
                mov         P5MDOUT,#0FFH           ; configure Port P5 for push-pull
                mov         P6MDOUT,#0              ; configure Port P6 as input
                mov         P6,#0xFF                ;       "
                ret
                $include    (UartUtil.a51)
                end
```

The C language version of the program is as follows.

```
#include <c8051F040.h>
#include <stdio.h>
void    int0_ISR (void);        // INT0 service routine prototype
void    sysinit(void);
char    cnt;
void main (void)
{
    cnt  = 0;
    sysinit();
    IT0  = 1;                   // INT0 falling edge triggered
    EX0 = 1;                    // enable INT0 interrupt locally
```

```
    PX0 = 1;  // place INT0 interrupt at high priority
    EA  = 1;  // enable interrupt globally
    printf ("Set up a value in DIP and press INT0!");
    while (1);  // wait for INT0 interrupt
}
void int0_ISR (void) interrupt 0   // 0 is the interrupt number of INT0
{
    SFRPAGE = 0x0F;
    P5      = P6;              // reads DIP switches and outputs to LEDs
}
void sysinit(void)
{
    int       n;
    SFRPAGE  = CONFIG_PAGE;
    WDTCN    = 0xDE;           // disable watchdog timer
    WDTCN    = 0xAD;           //      "
    OSCXCN   = 0x67;           // start external oscillator; 24 MHz Crystal
                               // system clock is 24 MHz
    for (n = 0; n < 255; n++);     // delay about 1 ms
    while ((OSCXCN & 0x80) == 0); // wait for oscillator to stabilize
    CLKSEL   |= 0x01;          // switch to external oscillator
    XBR2     = 0x5D;           // enable crossbar and assign I/O pins to all
    XBR0     = 0xF7;           // peripheral signals,
    XBR1     = 0xFF;           //      "
    XBR3     = 0x8F;           //      "
    P5MDOUT = 0xFF;            // configure Port P5 to push-pull for output
    P6MDOUT = 0;               // configure Port P6 for input (open drain)
    P6       = 0xFF;           //      "
    SFRPAGE  = SPI0_PAGE;
    SPI0CN   = 0x01;           // enable 3-wire SPI (make sure SPI uses 3 pins)
}
```

6.5 The 8051 Reset

The original 8051 has two reset sources: **RST pin** and **power-on** resets. The C8051F040 adds five additional reset sources: **power-failure**, **missing-clock detector**, **comparator 0**, **CNVSTR0 pin**, and **watchdog timer resets**. The C8051F040 uses the RSTSRC register to indicate the cause of reset. The contents of the RSTSRC register are shown in Figure 6.9.

Bits 6, 5, 4, 2, and 0 are read and writable. Bits 6, 5, and 2 allow the user to enable the associated reset source to reset the MCU when it occurs. After the reset, the reset service routine can read this register to identify the cause of reset. For example, setting bit 2 to 1 enables the missing clock detector. When

7	6	5	4	3	2	1	0	
--	CNVRSEF	C0RSEF	SWRSF	WDTRSF	MCDRSF	PORSF	PINRSF	Reset value = 0x00
R	R/W	R/W	R/W	R	R/W	R	R/W	

CNVRSEF: Convert start reset source enable and flag
Write: 0 = CNVSTR0 is not a reset source
1 = CNVSTR0 is a reset source (active low)
Read: 0 = source of prior reset was not CNVSTR0
1 = source of prior reset was CNVSTR0
C0RSEF: Comparator 0 reset enable and flag
Write: 0 = Comparator 0 is not a reset source
1 = Comparator 0 is a reset source
Read: 0 = source of last reset was not Comparator 0
1 = source of last reset was Comparator 0
SWRSF: Software reset force and flag
Write: 0 = No effect
1 = forces an internal reset /RST pin is not affected
Read: 0 = source of last reset was not a write to the SWRSF bit
1 = source of last reset was a write to the SWRSF bit
WDTRSF: Watchdog timer reset flag
0 = source of last reset was not WDT timeout
1 = source of last reset was WDT timeout
MCDRSF: Missing clock detector flag
Write: 0 = missing clock detector disabled
1 = missing clock detector enabled; trigger a reset if a missing clock condition is detected
Read: 0 = source of last reset was not a missing clock detector timeout
1 = source of last reset was a missing clock detector timeout
PORSF: Power-on reset flag
Write: 0 = deselect the VDD monitor as a reset source
1 = select the VDD monitor as a reset source (also requires the MONEN pin to be tied high)
Read: 0 = source of last reset was not a power-on or VDD monitor reset
1 = source of last reset was a power-on or VDD monitor reset
PINRSF: HW pin reset flag
Write: 0 = no effect
1 = forces a power-on reset /RST is driven low
Read: 0 = source of prior reset was not /RST pin
1 = source of prior reset was /RST pin

FIGURE 6.9
The C8051F040 reset source register RSTSRC.

the missing clock condition occurs, the associated detector circuit resets and also sets this flag (bit 2).

Setting the SWRSF bit forces an MCU internal reset and also leaves this flag to be set. Setting the PINRSF bit will force the $\overline{\text{RST}}$ pin to go low to reset the peripheral devices connected to it.

The internal reset algorithm writes 0s to all the SFRs except the port latches, the stack pointer, and SBUF. The port latches are initialized to 0xFF, the stack pointer to 0x07, and SBUF is indeterminate. The internal RAM is not affected by reset. On power up, the RAM contents are indeterminate.

6.5.1 The RST Pin Reset

The RST pin is the reset input. A reset is accomplished by holding the RST pin high (holding low for the C8051F040) for at least two machine cycles while the oscillator is running. The external reset signal is asynchronous with the internal clock. In the original 8051 and many variants, the RST pin has a pull-down resistor allowing power-on reset by simply connecting an external capacitor to V_{DD}. The $\overline{RST}$ pin to the C8051F040 is active-low, therefore it is recommended to connect a pull-up resistor to this pin to avoid erroneous noise-induced resets. The MCU will remain in reset until at least 12 clock cycles after the active-low $\overline{RST}$ signal is removed. An example circuit for the manual reset signal is illustrated in Figure 6.10. This circuit can also perform a power-on reset to the MCU, as described in the next section.

6.5.2 Power-On Reset

For the original 8051, when V_{DD} is turned on an automatic reset can be obtained by connecting the RST pin to V_{DD} through an appropriate capacitor (about 10 μF) and to V_{SS} through an appropriate resistor (about 8.2 KΩ), as shown in Figure 6.10a. The push button across the capacitor allows the user to manually reset the CPU.

The circuit shown in Figure 6.10b will apply a power-on reset to the C8051F040 because the charge across the capacitors leaked away when the

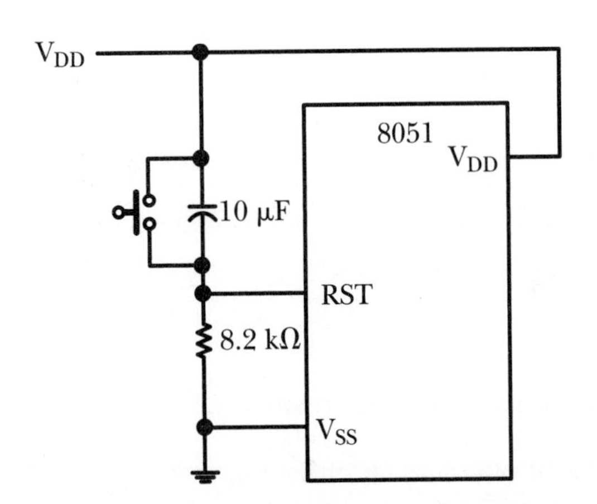

(a) Reset circuit for the original 8051

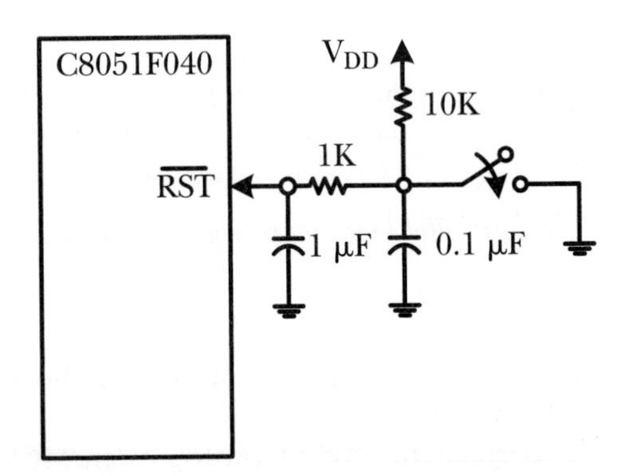

(b) Reset circuit for the C8051F040

FIGURE 6.10
Power-On reset circuit.

power was off. The power-on reset condition is recoded in the **PORSF** flag of the RSTSRC register for the C8051F040.

6.5.3 Power-Failure Reset

The power-failure reset is also referred to as **low-voltage-detect** (LVD) reset. The 8051 can operate normally only if the power supply voltage exceeds a threshold value. During a power irregularity, the power-supply voltage may drop below that threshold level and cause the 8051 MCU to operate incorrectly. To prevent the MCU from operating during this interval, most 8051 variants implement a power-failure detection circuit. Once the V_{DD} level drops below that threshold level, the power-failure detection circuit resets the MCU until V_{DD} restores to the normal level.

6.5.4 JTAG Debug Reset

JTAG is the acronym of **Joint Test Action Group** and is the usual name used for the IEEE 1149.1 standard entitled **Standard Test Access Port and Boundary-Scan Architecture** for test access ports used for testing printed circuit boards using boundary scan. JTAG was standardized in 1990 as the IEEE Standard. 1149.1–1990. In 1994, a supplement that contains a description of the boundary-scan description language (BSDL) was added. Since then, this standard has been adopted by electronics companies all over the world. Boundary scan nowadays is synonymous with JTAG.

While designed for printed circuit boards, today it primarily is used for testing sub-blocks of integrated circuits and is also useful as a mechanism for debugging embedded systems, providing a convenient "back door" into the system. When used as a debugging tool, an in-circuit emulator (which in turn uses JTAG as the transport mechanism) enables a programmer to access an on-chip debugging module which is integrated into the CPU via JTAG. The JTAG circuit has the capability to reset the MCU and force the CPU to start instruction execution from the address 0. The debugging module enables the programmer to debug the software of an embedded system.

6.5.5 Missing Clock Detector (MCD) Reset

This feature is available in the C8051F040. The missing clock detector is essentially a one-shot circuit that is triggered by the MCU system clock. If the system clock goes away for more than 100 μs, the one-shot circuit will timeout and generates a reset if it is enabled. The MCD circuit is enabled by setting bit 2 of the RSTSRC register. After a missing clock detector reset, the MCDRSF bit (bit 2 of RSTSRC) will be set to 1, signifying the MCD as the reset source; otherwise, this bit reads 0. The state of the $\overline{RST}$ pin is not affected by this reset.

6.5.6 Comparator 0 Reset

The Comparator 0 of the C8051F040 can be configured as a reset source by setting the **C0RSEF** flag of the RSTSRC register to 1. The user must set bit 7 of the CPT0CN register before writing a 1 to the **C0RSEF** flag to prevent any turn-on chatter from generating an unwanted reset. Once enabled, Comparator 0 will place the MCU in a reset state if the non-inverting input voltage (CP0+ pin) to Comparator 0 is less than the inverting input voltage (CP0− pin) to Comparator 0. After a Comparator 0 reset, the **C0RSEF** flag will read 1, signifying Comparator 0 as the reset source; otherwise, this bit reads 0. This reset does not affect the level of the $\overline{\text{RST}}$ pin.

6.5.7 External CNVSTR0 Pin Reset

The external CNVSTR0 signal of the C8051F040 can be configured as a reset input by writing a 1 to the CNVRSEF bit of the RSTSRC register. Depending on the crossbar decoder setting, this signal can appear on any of the P0, P1, P2, or P3 pins. When configured as a reset source, this signal is active-low and level-sensitive. After a reset, the CNVRSEF flag will read 1, signifying CNVSTR0 as the reset source; otherwise, this bit reads 0.

6.5.8 Watchdog Timer Reset

The C8051F040 has a dedicated programmable watchdog timer (WDT) running off the system clock. When enabled, the WDT will generate a reset whenever it overflows. To prevent the reset, the WDT must be restarted (cleared to 0) by application software before it overflows. Application software that is behaving correctly will periodically clear the WDT before it overflows. If the system experiences a software or hardware malfunction that prevents the software from restarting the WDT, the WDT will overflow and cause a reset. This should prevent the system from running out of control. By checking bit 3 of the RSTSRC register after reset, the user can discover this problem and try to fix the software error.

The WDT consists of a 21-bit timer running from the programmed system clock. This timer measures the period between specific writes to its control register. If this period exceeds the programmed limit, a WDT reset is generated. The WDT can be enabled and disabled as needed in software or can be enabled permanently if desired. The features of the WDT are controlled via the watchdog-timer control register (WDTCN). The contents of this register are shown in Figure 6.11.

The WDT is enabled and restarted by writing **0xA5** into the WDTCN register. The user's application software should include periodic writes of 0xA5 into the WDTCN register to prevent a watchdog-timer overflow. The WDT is enabled and restarted as a result of any system reset.

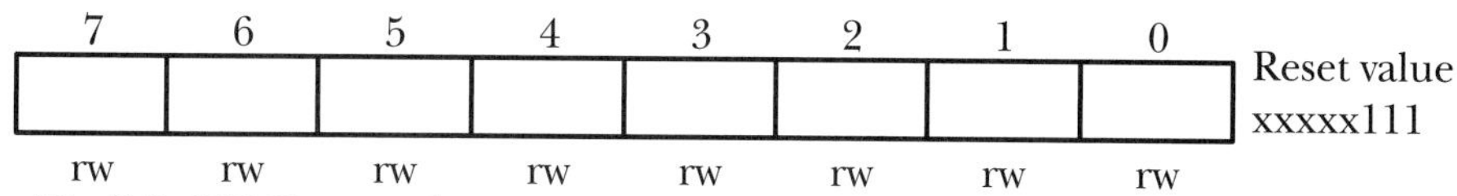

Bits 7-0: WDT control
  Writing 0xA5 enables and reloads the WDT
  Writing 0xDE followed within 4 system clocks by 0xAD disables the WDT
  Writing 0xFF locks out the disable feature
Bit 4: Watchdog status bit (when read)
  0: WDT is inactive
  1: WDT is active
Bits 2-0: Watchdog timeout interval bits
  These 3 bits set the watchdog timer interval. When writing these bits, the bit 7 of this register must be cleared to 0

FIGURE 6.11 *Watchdog timer control register (WDTCN).*

Writing 0xDE followed by 0xAD to the WDTCN register disables the WDT. The following instruction sequence disables the WDT.

```
clr     EA              ; disable all interrupts
mov     WDTCN,#0DEh     ; disable WDT
mov     WDTCN,#0ADh     ;    "
setb    EA              ; reenable interrupts
```

The writes of 0xDE and 0xAD must occur within four clock cycles of each other, or the disable operation is ignored. Interrupts should be disabled during this procedure to avoid delay between the two writes.

Writing 0xFF to WDTCN locks out the disable feature. Once locked out, the disable operation is ignored until the next system reset. Writing 0xFF does not enable or reset the watchdog timer. Applications intending to use the watchdog timer should write 0xFF to WDTCN in the initialization code. The lowest 3 bits of the WDTCN register control the watchdog timeout interval. The timeout interval is given by

$$T_{\text{timeout}} = 4^{3+\text{WDTCN}[2-0]} \times T_{\text{sysclk}} \tag{6.1}$$

where, T_{sysclk} is the system clock period.

The maximum WDT timeout interval at 24 MHz is 43.69 ms.

Using the WDT Timer

An outline of using the watchdog timer is illustrated in Figure 6.12. Most embedded applications are written as an infinite loop. The application software performs system initialization after the hardware is powered on. The WDT is disabled in this step. After the system initialization, the application is waiting for the user to make requests to perform a certain operation. If there is no user request for application, the software will constantly reset the WDT. Once the user requests to perform a certain operation, the application

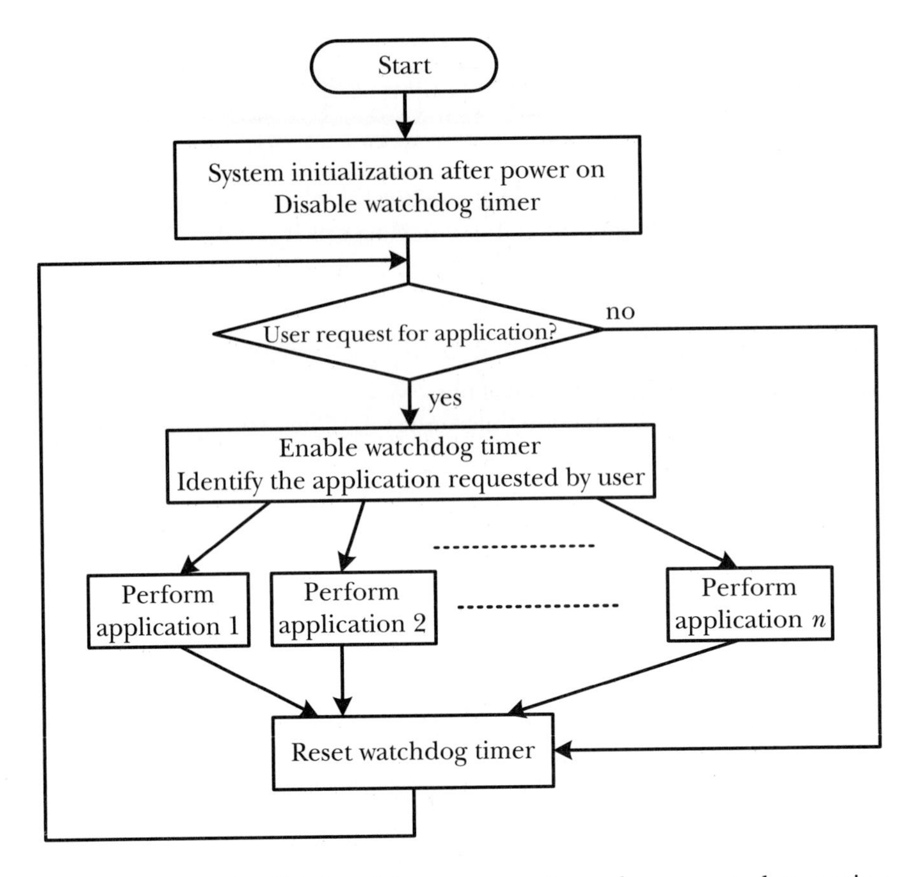

FIGURE 6.12
Outline for using the watchdog timer.

software enables the WDT and jumps to perform the requested operation. After finishing the operation, the application software resets the WDT and jumps back to wait for the next user request.

The timeout period of the WDT should be set to be slightly longer than the execution time of the longest application so that the WDT won't reset the MCU before the application is completed. However, it is possible that one or more of the applications will take much longer than the longest timeout period of the WDT. This issue can be resolved by using a regular timer along with the WDT and **this** problem will be left as an exercise.

6.6 Power-Saving Modes of Operation

Power consumption is a critical issue for many embedded applications. To save power, the C8051F040 provides two power-saving modes: *idle* and *power-down* (also called *stop mode*). In the idle mode, the CPU is halted but the peripheral functions and clocks are still running. In stop mode, the CPU is halted, all interrupts and timers (except the missing clock detector) are inactive, and the internal oscillator is stopped. Since clocks are running in idle mode, power consumption is dependent upon the system clock frequency and the number of peripheral functions left in active mode before entering the idle mode. Stop mode consumes much less power than the idle mode.

Although the 8051 MCU has idle and stop modes built in (as with any 8051 variant), power management of the entire MCU is better accomplished by enabling/disabling individual peripheral functions as needed. Each analog peripheral can be disabled when not in use and put into low-power mode. Digital peripherals, such as timers or serial buses, draw little power whenever they are not in use.

6.6.1 The IDLE Mode

The idle mode can be entered by setting the bit 0 of the PCON register. The contents of the PCON registers are shown in Figure 6.13. When the MCU is in this mode, power consumption is reduced. The special-function registers and the on-chip RAM retain their values during the idle mode, but the processor stops executing instructions. The idle mode will be exited if the chip is reset or if an enabled interrupt occurs. If the idle mode is terminated by a pending interrupt, the interrupt will be serviced and the next instruction to be executed after the return-from-interrupt instruction (RTI) will be the instruction immediately following the one that sets the idle-mode select bit. If the idle mode is terminated by an internal or external reset, the 8051 performs a normal reset sequence and begins instruction execution at address 0x0000.

For the C8051F040, an instruction that sets the IDLE bit should be immediately followed by an instruction that has two or more opcode bytes. For example,

```
// in C language:
        PCON |= 0x01;           // set IDLE bit
        PCON = PCON;            // . . . followed by a 3-cycle dummy instruction
; in assembly language
        ORL     PCON,#01h       ; set IDLE bit
        MOV     PCON,PCON       ; . . . followed by a 3-byte dummy instruction
```

7	6	5	4	3	2	1	0	
SMOD[1]	--	--	--	--	--	STOP	IDLE	Reset value = xxxxxx00b
rw	rw	rw	rw	rw	rw	rw	rw	

SMOD: Baud rate doubler enable bit
  0 = UART baud rate not doubled
  1 = UART baud rate is doubled that given in the baud rate equation
STOP: Stop mode
  This bit is set to activate the power down mode. Once set, this bit will be cleared by hardware when an interrupt or reset occurs.
IDLE: Idle mode bit
  Setting this bit will activate the idle mode. Once set, this bit will be cleared by hardware when an interrupt or reset occurs. If STOP and IDLE are both set, the STOP bit takes precedence.

Note 1: This bit is available only in the original 8051

FIGURE 6.13 *The PCON register.*

An outline of using the IDLE mode together with the WDT is shown in Figure 6.14. When the application completes system initialization or completes an application, it starts an idle timer (timeout period is settable) and waits for the next user request. If there is no user request before the idle timer times out, the software disables WDT and then enters the IDLE mode to save power.

6.6.2 The STOP Mode

To save even more power, software can invoke the stop mode. In this mode, the oscillator is stopped and the instruction that invoked the stop mode is the last instruction executed. The on-chip RAM and special-function registers retain their values until the stop mode is terminated.

The **stop mode** can be exited by an external or internal reset (excluding the WDT reset). Because the WDT is not functioning in the stop mode, the clock signal already has stopped. The method of using the STOP mode is similar to that for IDLE mode.

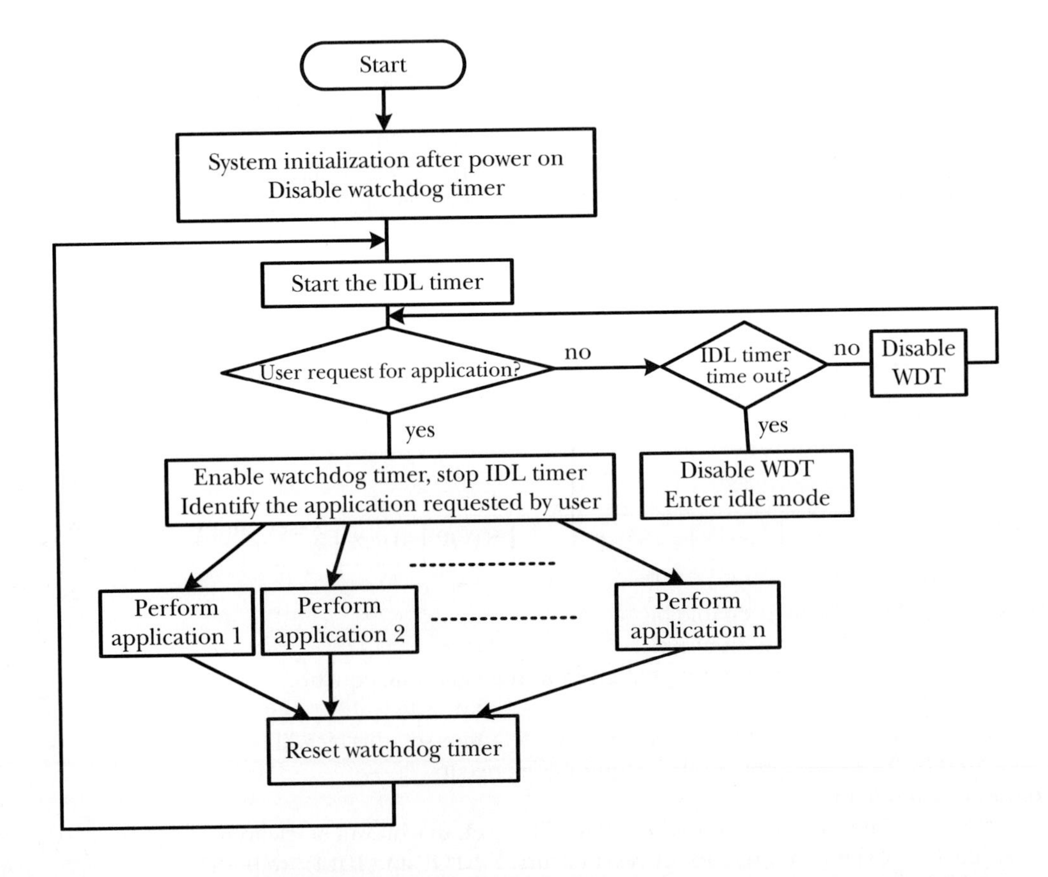

FIGURE 6.14
Outline for using the watchdog timer together with idle mode.

6.7 Chapter Summary

Interrupt is a special event that can change the normal instruction–execution sequence. Interrupt makes I/O activity more efficient, because the CPU no long needs to poll the input device to find out if it has new data nor to poll the output device to find out if it can accept more data.

In addition, interrupts also allow emergency events to be handled in time. The handling of routine works can also be facilitated by using the timer interrupt. Most microcontrollers support multiple interrupt sources. The microcontroller prioritizes interrupt sources so that whenever there are two or more interrupts occur at the same time, the pending interrupt at the highest priority will receive service first.

Interrupts are not always desirable. The CPU should ignore certain interrupts when they are not needed. This is implemented by using an enable bit to allow or disallow the interrupt to occur. An interrupt that can be enabled or disabled is called *maskable interrupt.* Otherwise, it is referred to as a *nonmaskable interrupt.*

To handle the interrupt request, the CPU needs to identify the source of interrupt and also to locate the starting address of the service routine that handles the interrupt. The starting address of the interrupt service routine is referred to as an *interrupt vector.* Interrupt vectors may be predefined or stored in a table. The original 8051 has six interrupt sources. Many 8051 variants have many more interrupt sources because they have added many peripheral functions. The C8051F040 has 20 interrupt sources. Each of these interrupts can be programmed to one of two priority levels by setting or clearing a bit in the IP register. The polling sequence within each priority determines the priority structure within each level.

There are two steps in the 8051 interrupt programming:

Step 1
Writing the interrupt service routine

Step 2
Setting the appropriate interrupt enable bits

Most microprocessor and microcontrollers have *power-on reset* and *manual reset.* A power-on reset provides the default values to registers and flip-flops and initializes all I/O interface chips so that the processor can execute a program correctly. A manual reset (when power is on) has a similar effect. The computer restarts itself after a reset. The reset service routine has a fixed starting address and is stored in ROM. The C8051F040 has seven reset sources and uses a reset source register (RSTSRC) to record the cause of the reset so that it can be identified. The watchdog-timer time-out reset often is used to check for software errors.

The 8051 has two power-saving modes: idle and power-down modes. The oscillator is not stopped in the idle mode but is stopped in the power-down mode. Power consumption is reduced significantly in these two modes. The power-down mode saves more power than the idle mode.

6.8 Exercise Problems

E6.1 What is interrupt?

E6.2 Why it is more efficient to use interrupts to handle inputs and outputs?

E6.3 Explain why the CPU can execute a program correctly (a) with interrupt occurring from time to time and (b) by changing the instruction execution sequence?

E6.4 Can the 8051 interrupt priority be programmed?

E6.5 Write an instruction sequence to (a) enable INT0 and INT1 interrupts to be high priority and edge-triggered and (b) enable TMR1 interrupt to be a low priority.

E6.6 Write a sequence of C statements to (a) enable INT0 and INT1 interrupts to be high priority and edge-triggered and (b) enable TMR1 interrupt to be a low priority.

E6.7 Write a sequence of C statements to enable all except serial-port interrupts for the C8051F040.

E6.8 Write an instruction sequence and a set of C statements (a) to set Timer 0 and Timer 1 to be high (highest) priority and (b) to set other interrupt sources to low priority for the C8051F040.

E6.9 Write an instruction sequence to enable all three comparator interrupts and place them in high priority, enable PCA interrupt but place it at low priority, disable window interrupt for ADC0, and enable SMB and SPI interrupts and place them at low priority for the C8051F040.

E6.10 Write an instruction sequence to enable UART1 interrupt and place it at a high priority, enable CAN and ADC2 interrupts and place them at a high priority, disable window interrupt for ADC2, disable Timer 4 interrupt, and enable ADC0 and Timer 3 interrupts and place them at high priority for the C8051F040.

E6.11 Write a sequence of C statements to enable Timers 0, 1, and 2 interrupts and place them in high priority, enable PCA interrupt but place it at lowest priority, disable UART and ADC interrupts, and enable INT0 and INT1 interrupts and place them at high priority for the C8051F040.

E6.12 Write an instruction sequence to enable the WDT timer and set the timeout interval to 10.49 ms for the C8051F040 at a 25-MHz clock rate.

E6.13 What would be the WDT timeout interval, assuming that the lowest 3 bits of the WDTCN register for the C8051F040 are 111 and the system clock frequency is 20 MHz?

E6.14 What would be the WDT timeout interval, assuming that the lowest 3 bits of the WDTCN register for the C8051F040 are 110 and the system clock frequency is 24 MHz?

E6.15 Write a main program and the service routine for the INT1 interrupt. The main program:

1. Performs system initialization and also initializes the UART module
2. Initializes the variable **ICount** to 0
3. Sets $\overline{\text{INT1}}$ to a high-priority level while other interrupts are set to low priority
4. Enables $\overline{\text{INT1}}$ interrupt
5. Stays in a loop while the variable **ICount** is less than 10
6. Disables $\overline{\text{INT1}}$ interrupt, outputs the message "Interrupt is over!" on the monitor screen, and stays in an infinite loop

The interrupt service routine does the following:

1. Increments the variable **ICount** by 1
2. Returns from interrupt

6.9 Laboratory Exercise Problems and Assignments

L6.1 **INT0 and INT1** ***Interrupt Experiments.*** Connect debounced pulse sources to INT0 and INT1 pins, respectively. Write a main program that enables INT0 and INT1 interrupts, places the INT0 interrupt at a high priority (whereas INT1 interrupt at low priority), initializes the variable **count0** and **count1** to 0, and then stays in an infinite loop to wait for interrupts to occur. Write the INT0 interrupt service routine that clears the **IE0** flag to zero, increments **count0** by 1, and then outputs it to P5. Write the INT1 interrupt service routine that clears the **IE1** flag to zero, increments count1 by 1, and then outputs it to P5. Assemble/Compile the assembly/C program and download it to the demo board for execution. The SSE040 demo board has provided two debounced key switches for your use. You can just connect to the I/O pins assigned to INT0 and INT1.

L6.2 Write a main program and a service routine for the Timer 0 interrupt. The main program performs the required initialization (as shown in Example 6.3) and also initializes a counter to 0. The Timer 0 interrupt service routine increments the counter, outputs it to P5, and returns from interrupt. Timer 0 overflows in 100 ms.

L6.3 Use a 555 timer to generate a 2-Hz periodical signal to drive the INT0 pin. The circuit connection is shown in Figure L6.2. Use an I/O port of your choice to drive eight LEDs (similar to Figure 3.12). Write a main program that initializes a byte counter to 0 and enables the INT0 interrupt. Write the INT0 interrupt service routine that increments the counter by 1 and outputs it to the I/O port that drives the LEDs. Enter, assemble, and download the program to the program memory for execution. If you have an SSE040 demo board, you can take the 2-Hz signal from its function generator instead of building the 555 circuit.

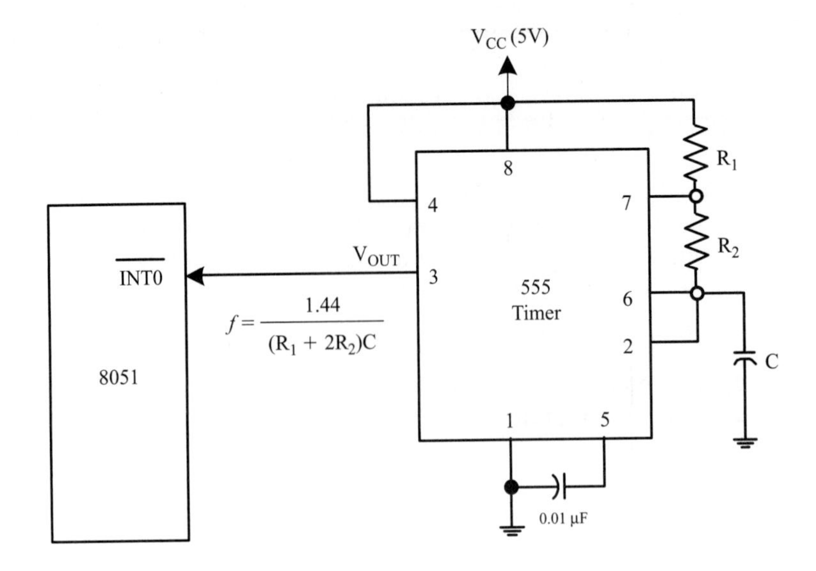

FIGURE L6.2
8051 INT0 pin interrupt circuit.

L6.4 Write a program and the service routines for INT0 and INT1 interrupts. The main program performs the following operations.

1. Disable watchdog timer, and select internal oscillator to generate SYSCLK.
2. Clear the variable **intflag** to 0. This variable will be set to 1 when either INT0 or INT1 is pressed.
3. Configure INT0 and INT1 to be high priority and edge-triggered.
4. Stay in an infinite loop to check if the **intflag** variable is set to 1. If yes, then check the value of the variable **source**. If **source** is 0, then generate a 1-kHz waveform for 5 seconds. If **source** is 1, then generates a 2-kHz waveform from pin P4.7 also for 5 seconds.
5. Clear the variable **intflag** to 0 after generating the waveform.
6. Connect the P4.7 pin to a buzzer or speaker so that you can hear the sound.

The INT0 service routine performs the following operations.

1. Set the variable **intflag** to 1
2. Set the variable **source** to 0
3. Return from interrupt

The INT1 service routine performs the following operations.

1. Set the variable **intflag** to 1
2. Set the variable **source** to 1
3. Return from interrupt

Advanced Parallel I/O

7.1 Objectives

Upon successful completion of this chapter, you will be able to:

- Understand the importance of electrical compatibility
- Verify electrical compatibility when interfacing a microcontroller with peripheral devices
- Understand the operation of the LCD controller
- Use LCD to display information
- Understand issues involved in keypad and keyboard input
- Use keypad and keyboard to input data to the MCU
- Explain the operation of a stepper motor
- Program the HCS12 parallel port to rotate the stepper motor

7.2 Electrical Characteristic Considerations for I/O Interfacing

Most embedded systems require the use of logic chips and peripheral devices in addition to the microcontroller to perform their function. Because these chips may use different types of integrated circuit (IC) technologies, there is a concern that the resultant embedded system may not function properly.

The major concern in interfacing IC chips that are made with different technologies is whether they are electrically compatible. There are two issues involved in electrical compatibility:

- ***Voltage-Level Compatibility.*** Is the high-output level of an IC chip high enough to be considered as a high for the input of another IC chip? Is the low-output level of an IC chip low enough to be considered as a low for the input of another IC chip?

- ***Current Drive Capability.*** Does the output of an IC chip have enough current to drive its load? Can the output circuit of an IC chip sink the currents of its load?

The signal timing is also an important factor for making sure that the digital circuit functions correctly. The main concern about timing is whether the signal from one chip becomes valid early enough to be used by other chips. This is a *timing compatibility* issue. When the operating frequency becomes very high, the *transmission line effect* and *ground bounce effect* will need to be considered too. However, we are not dealing with high-frequency systems in this text. These two issues will not be discussed.

7.2.1 Voltage-Level Compatibility

There are many IC technologies in use today. Some of them are bipolar; others are unipolar. A bipolar IC technology has both the electron and hole currents in any moment.

However, in a unipolar IC technology, the current in any part of the circuit is either the electron current or the hole current. The unipolar CMOS technology is the dominant IC technology in use today.

The voltage-level compatibility issue arises because IC technologies differ in the following four voltages.

- ***Input High Voltage*** (V_{IH}). This is the lowest voltage that will be treated as logic 1 when applied to the input of a digital circuit.
- ***Input Low Voltage*** (V_{IL}). This is the highest voltage that will be treated as logic 0 when applied to the input of a digital circuit.
- ***Output High Voltage*** (V_{OH}). This is the voltage level when a digital circuit outputs logic 1.
- ***Output Low Voltage*** (V_{OL}). This is the voltage level when a digital circuit outputs logic 0.

In order for the digital circuit X to be able to drive circuit Y correctly, the following conditions must be satisfied.

- $V_{OHX} \geq V_{IHY}$ (the output high voltage of circuit X must be higher than the input high voltage of circuit Y). The difference between V_{OH} and V_{IH} of the same technology is referred to as the *noise margin high* (NMH).
- $V_{OLX} \leq V_{ILY}$ (the output low voltage of circuit X must be lower than the input low voltage of circuit Y). The difference between V_{IL} and V_{OL} of the same technology is referred to as the *noise margin low* (NML).

The input and output voltage levels of a few popular logic families are listed in Table 7.1, from which one can draw the following conclusions.

- There is no problem in using CMOS logic chips to drive bipolar logic chips at the same power supply level.

TABLE 7.1 *Input and Output Voltage Levels of Common Logic Families*

Logic Family	V_{DD}	V_{IH}	V_{OH}	V_{IL}	V_{OL}
C8051F040[6]	3.3 V	2.31 V	2.5 V ~ 3.2 V	0.99 V	0.6 V–1 V
S[4]	5 V	2 V	3.0~3.4 V[1]	0.8 V	0.4~0.5 V[2]
LS[4]	5 V	2 V	3.0~3.4 V[1]	0.8 V	0.4~0.5 V[2]
AS[4]	5 V	2 V	3.0~3.4 V[1]	0.8 V	0.35 V
F[4]	5 V	2 V	3.4 V	0.8 V	0.3 V
HC[3]	5 V	3.5 V	4.9 V	1.5 V	0.1 V
HCT[3]	5 V	3.5 V	4.9 V	1.5 V	0.1 V
ACT[3]	5 V	2 V	4.9 V	0.8 V	0.1 V
ABT[5]	5 V	2 V	3 V	0.8 V	0.55 V
BCT[5]	5 V	2 V	3.3 V	0.8 V	0.42 V
FCT[5]	5 V	2 V	2.4 V	0.8 V	0.55 V

Notes:
1. V_{OH} value will get lower when output current is larger.
2. V_{OL} value will get higher when output current is larger. The V_{OL} values of different logic gates are slightly different.
3. HC, HCT, and ACT are based on the CMOS technology.
4. S, LS, AS, and F logic families are based on the bipolar technology.
5. ABT, BCT, and FCT use the BiCMOS technology.
6. The power supply for C8051F040 is from 2.7 to 3.6 V. $V_{IL} = 0.3\ V_{DD}$. $V_{IH} = 0.7\ V_{DD}$. $V_{OH} = V_{DD} - 0.7$ V for $I_{OH} = 3$ mA and $V_{OH} = V_{DD} - 0.8$ V for $I_{OH} = 10$ mA. VOL = 0.6 V for $I_{OL} = 8.5$ mA, $V_{OL} = 1$ V for $I_{OL} = 25$ mA.

- The C8051F040 is implemented in CMOS technology and has no problem in driving the CMOS logic chips and being driven by the CMOS logic chips at the same power supply level.
- The BiCMOS logic is not suitable for driving the C8051F040 and other CMOS logic chips at the same power supply level.
- The bipolar logic ICs are not suitable for driving the CMOS logic ICs or the C8051F040 at the same power supply level.

7.2.2 Current Drive Capability

A microcontroller needs to drive other peripheral I/O devices in an embedded system. The second electrical compatibility issue is whether the microcontroller can source (when the output voltage is high) or sink (when the output voltage is low) the current needed by the I/O devices that it interfaces with. The designer must make sure that the following two requirements are satisfied.

1. Each I/O pin can supply and sink the current needed by the I/O devices that it interfaces with.
2. The total current required to drive I/O devices does not exceed the maximum current rating of the microcontroller.

Each logic chip has the following four currents that are involved in the current drive calculation.

- ***Input High Current*** (I_{IH}). This is the input current (flowing into the input pin) when the input voltage is high.
- ***Input Low Current*** (I_{IL}). This is the input current (flowing out of the input pin) when the input voltage is low.
- ***Output High Current*** (I_{OH}). This is the output current (flowing out of the output pin) when the output voltage is high.
- ***Output Low Current*** (I_{OL}). This is the output current (flowing into the output pin) when the output voltage is low.

The current capabilities (each I/O pin) of several common logic families and the C8051F040 are listed in Table 7.2. In CMOS technology, the gate terminal (one of the three terminals in an N− or P−transistor) draws a significant current only when they are charged up toward V_{DD} or pulled down toward GND level. After that, the gate terminal draws only leakage currents. Bipolar technology is different from CMOS technology in that a DC current always flows into or out of the base terminal of the transistor of a bipolar logic chip.

To determine whether a pin can supply (also called source) and sink currents to all the peripheral pins that it drives directly, the designer needs to check the following two requirements.

1. The I_{OH} of an I/O pin of the microcontroller is equal to or greater than the sum of currents flowing into all peripheral pins that are connected directly to the microcontroller I/O pins.

TABLE 7.2 *Current Capabilities of Common Logic Families*[1]

Logic Family	V_{DD}	I_{IH}	I_{IL}	I_{OH}	I_{OL}
C8051F040	3.3 V	10 μA	10 μA	10mA	25mA
S	5 V	50 μA	1.0 mA	1 mA	24 mA
LS	5 V	20 μA	0.2 mA	15 mA	64 mA
AS	5 V	20 μA	0.5 mA	15 mA	20 mA
F	5 V	20 μA	0.5 mA	1 mA	25 mA
HC[2]	5 V	1 μA	1 μA	25 mA	25 mA
HCT[2]	5 V	1 μA	1 μA	25 mA	24 mA
ACT[2]	5 V	1 μA	1 μA	24 mA	64 MA
ABT[2]	5 V	1 μA	1 μA	32 mA	64 mA
BCT	5 V	20 μA	1 mA	15 mA	64 mA
FCT[2]	5 V	1 μA	1 μA	15 mA	64 mA

Notes:
1. Values are based on the 74xx244 of Texas Instrument (xx is the technology name).
2. The values for I_{IH} and I_{IL} are input leakage currents.
3. V_{OH} will drop when I_{OH} is higher.

2. The I_{OL} of an I/O pin of the microcontroller is equal to or greater than the sum of currents flowing out of all peripheral pins that are connected directly to the microcontroller I/O pins.

In addition, the designer also must make sure that the total current needed to drive the peripheral signal pins does not exceed the total current that the microcontroller can supply.

One question that arises here is what should be done if an I/O pin cannot supply (or sink) the current needed to drive the peripheral pins? A simple solution is to add buffer chips (for example, 74ABT244) that can supply enough current to the microcontroller and the peripheral chips. This technique is widely used in microcontroller applications.

The current capability of the C8051F040 is shown in Table 7.3. The "any other I/O pin" in Table 7.3 refers to non-I/O port pins. For example, DAC0, DAC1, AIN0, AIN1, and those pins that are dedicated to a certain function rather than to general input and output.

Example 7.1 For the circuit shown in Figure 3.14, do an analysis to determine whether we can use the port P6 pins to sink the current flowing out of the common cathode of the seven-segment displays without using the buffer chip.

Solution: For the circuit shown in Figure 3.14, the maximum current flowing out of the common cathode of a seven-segment display is 23.3 mA. According to Table 7.3, a port pin can sink up to 100 mA of current. Therefore, we don't need to use the SN74LV244A chip to sink the current flowing out of seven-segment displays.

7.2.3 Timing Compatibility

If an I/O pin is driving a peripheral pin that does not contain latches or flip-flops, then timing is not an issue. A latch or flip-flop usually has a control signal or clock signal to control the latching of an input signal. As illustrated in Figure 7.1, the D input to the D flip-flop must be valid t_{su} ns before the rising

TABLE 7.3 *C8051F040 Current Capability*

Parameter	Max Value
Maximum output current through VDD, AV+, DGND & AGND	800 mA
Maximum output current sunk by any port pin	100 mA
Maximum output current sunk by any other I/O pin	50 mA
Maximum output current sourced by any port pin	100 mA
Maximum output current sourced by any other I/O pin	50 mA

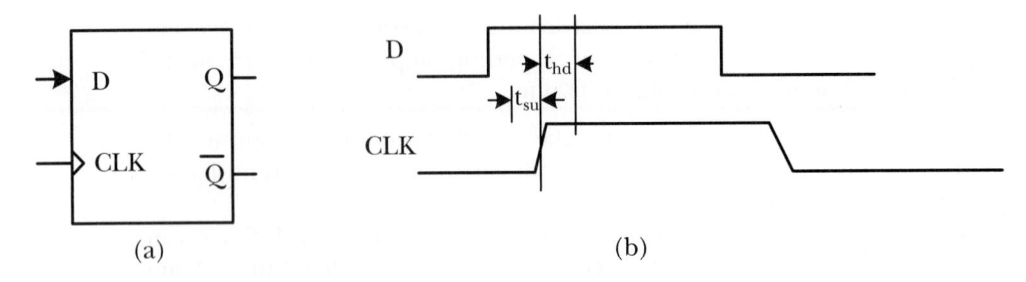

FIGURE 7.1
D flip-flop and its latching timing requirement.

edge of the CLK signal and remain valid for at least t_{hd} ns after the rising edge of the CLK signal in order for its value to be correctly copied to the output signal Q. The timing parameters t_{su} and t_{hd} are referred to as the *setup* and *hold* time requirements of the D flip-flop. The main timing consideration is that the setup- and hold-time requirements for all latches and flip-flops in a digital system must be satisfied in order for the system to work correctly. A signal may pass through several intermediate chips before it is used by the final latch or flip-flop. The time delays of all intermediate devices must be added when considering the timing analysis. Timing requirement analysis can be very complicated and is best illustrated using examples.

7.3 Liquid Crystal Displays (LCDs)

Although seven-segment displays are easy to use, they are bulky and quite limited in the set of characters that they can display. When more than a few letters and digits are to be displayed, seven-segment displays become inadequate. Liquid crystal displays (LCDs) come in handy when the application requires the display of many characters.

A liquid crystal display has the following advantages.

- High contrast
- Low power consumption
- Small footprint
- Ability to display both characters and graphics

In recent years, the price of LCD displays has dropped to such an acceptable level that most PC vendors bundle LCD displays instead of CRT displays with their PC systems. Notebook computers used LCD as displays right from the beginning. Because of the price reduction of LCDs, the prices of notebook computers also have dropped sharply, and more and more computer users have switched from desktop to notebook computers.

Although LCDs can display graphics and characters, only character-based LCDs are discussed in this text. LCDs are often sold in a module that consists of the LCD panel and its controller. The Hitachi HD44780 (with two slightly different versions: HD44780U and HD44780S) is one of the most popular LCD display controllers in use today. Because of its popularity, many semiconductor vendors are producing HD44780-compatible LCD driver chips. The following section examines the operation and programming of this controller.

7.4 The HD44780U LCD Controller

The block diagram of an LCD kit that incorporates the HD44780U controller is shown in Figure 7.2. The pin assignment shown in Table 7.4 is the industry standard for character-based LCD modules with a maximum of 80 characters. The pin assignment shown in Table 7.5 is the industry standard for character-based LCD modules with more than 80 characters.

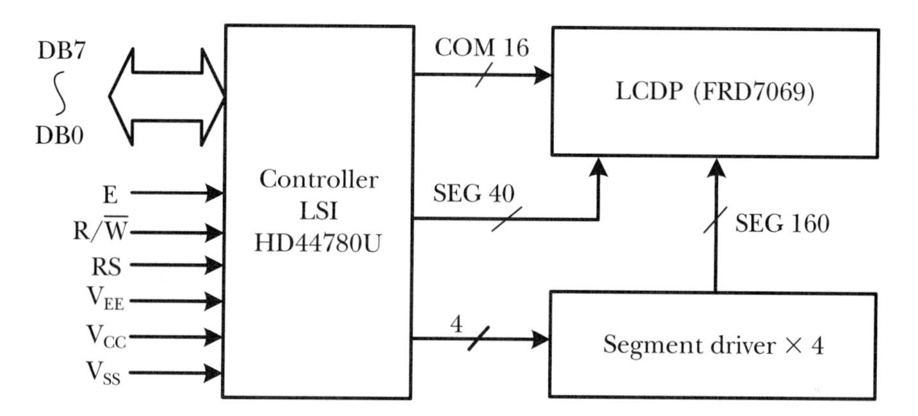

FIGURE 7.2 *Block diagram of a HD44780U-based LCD kit.*

TABLE 7.4 *Pin Assignment for Displays with Less than 80 Characters*

Pin No.	Symbol	I/O	Function
1	V_{SS}	-	Power supply (GND)
2	V_{CC}	-	Power supply (+5 V)
3	V_{EE}	-	Contrast adjust
4	RS	I	0 = instruction input, 1 = data input
5	$R/\overline{W}$	I	0 = write to LCD, 1 = read from LCD
6	E	I	Enable signal
7	DB0	I/O	Data bus line 0
8	DB1	I/O	Data bus line 1
9	DB2	I/O	Data bus line 2
10	DB3	I/O	Data bus line 3
11	DB4	I/O	Data bus line 4
12	DB5	I/O	Data bus line 5
13	DB6	I/O	Data bus line 6
14	DB7	I/O	Data bus line 7

TABLE 7.5 *Pin Assignment for Displays with More than 80 Characters*

Pin No.	Symbol	I/O	Function
1	DB7	I/O	Data bus line 7
2	DB6	I/O	Data bus line 6
3	DB5	I/O	Data bus line 5
4	DB4	I/O	Data bus line 4
5	DB3	I/O	Data bus line 3
6	DB2	I/O	Data bus line 2
7	DB1	I/O	Data bus line 1
8	DB0	I/O	Data bus line 0
9	E1	I	Enable signal row 0 and 1
10	$R/\overline{W}$	I	0 = write to LCD, 1 = read from LCD
11	RS	I	0 = instruction input, 1 = data input
12	V_{EE}	-	Contrast adjust
13	V_{SS}	-	Power supply (GND)
14	V_{CC}	-	Power supply (+5 V)
15	E2	I	Enable signal row 2 and 3
16	N.C	-	

The DB7~DB0 pins are used to exchange data with the microcontroller. The E pin is an enable signal to the kit. The $R/\overline{W}$ signal controls the direction of data transfer. The RS signal selects the register to be accessed. When the RS signal is high, the *data register* is selected. Otherwise, the *instruction register* is selected. The V_{EE} pin is used to control the brightness of the display and is often connected to a potentiometer. The V_{EE} input should not be set to the maximum value ($=V_{CC}$) for an extended period of time to avoid burning the LCD.

An LCD module can be used as a memory-mapped device and be enabled by an address decoder. The E signal is normally connected to the address decoder output qualified by other control signals to meet the timing requirement. The $R/\overline{W}$ pin can be connected to the same pin of the microcontroller. The RS pin can be connected to the least significant bit of the address pin (A0) from the microcontroller. This approach is applicable only for those microcontrollers that support external memory. The LCD programming in this approach is generally easier and straightforward.

An LCD module can also be interfaced directly with an I/O port. In this configuration, the user needs to use I/O pins to control the signals E, $R/\overline{W}$, and RS. Programming will be slightly more cumbersome than the memory-mapped approach due to the need to manipulate these three signals. The HD44780U provides a set of instructions for the user to set up the LCD parameters. The operations performed by these instructions are summarized in Table 7.6. The meanings of certain bits in these instructions are explained in Table 7.7.

TABLE 7.6 *HD44780U Instruction Set*

Instruction	Code											Execution Time
	RS	R/$\overline{W}$	B7	B6	B5	B4	B3	B2	B1	B0		
Clear display	0	0	0	0	0	0	0	0	0	1	Clears display and returns cursor to the home position (address 0)	1.64 ms
Cursor home	0	0	0	0	0	0	0	0	1	*	Returns cursor to home position without changing DDRAM contents. Also returns display being shifted to the original position.	1.64 ms
Entry mode set	0	0	0	0	0	0	0	1	I/D	S	Sets cursor move direction (I/D); specifies to shift the display (S). These operations are performed during data read/write.	40 μs
Display on/off control	0	0	0	0	0	0	1	D	C	B	Sets on/off of all display(D), cursor on/off (c), and blink of cursor position character (B).	40 μs
Cursor/display shift	0	0	0	0	0	1	S/C	R/L	*	*	Sets cursor-move or display-shift (S/C), shift direction (R/L). DDRAM contents remain unchanged.	40 μs
Function set	0	0	0	0	1	DL	N	F	*	*	Sets interface data length (DL), number of display line (N), and character font (F).	40 μs
Set CGRAM address	0	0	0	1	CGRAM address						Sets the CGRAM address. CGRAM data are sent and received after this setting.	40 μs
Set CGRAM address	0	0	1	DDRAM address							Sets the DDRAM address. DDRAM data are sent and received after this setting.	40 μs
Read busy flag and address counter	0	1	BF	CGRAM/DDRAM address							Reads busy flag (BF) indicating internal operation being performed and reads CGRAM or DDRAM address counter contents (depending on previous operation).	0 μs
Write CGRAM or DDRAM	1	0	write data								Writes data to CGRAM or DDRAM	40 μs
Read from CGRAM or DDRAM	1	1	read data								Reads data from CGRAM or DDRAM	40 μs

TABLE 7.7 *LCD Instruction Bit Names*

Bit Name	Settings	
I/D	0 = decrement cursor position	1 = increment cursor position
S	0 = no display shift	1 = display shift
D	0 = display off	1 = display on
C	0 = cursor off	1 = cursor on
B	0 = cursor blink off	1 = cursor blink on
S/C	0 = move cursor	1 = shift display
R/L	0 = shift left	1 = shift right
DL	0 = 4-bit interface	1 = 8-bit interface
N	0 = 1/8 or 1/11 duty (1 line)	1 = 1/16 duty (2 lines)
F	0 = 5 × 8 dots	1 = 5 × 10 dots
BF	0 = can accept instruction	1 = internal operation in progress

The HD44780U can be configured to control 1-line, 2-line, and 4-line LCDs. The mappings of the character positions on the LCD screen and the DDRAM addresses are not sequential and are shown in Table 7.8.

7.4.1 Display Data RAM

Display data RAM (DDRAM) stores display data represented in 8-bit character codes. Its extended capacity is 80 × 8 bits or 80 characters. The area in DDRAM that is not used for display can be used as general data RAM.

7.4.2 Character Generator ROM (CGROM)

The character generator ROM generates 5 × 8 or 5 × 10 dot character patterns from 8-bit character codes. It can generate 208 5 × 8 dot character patterns and 32 5 × 10 dot character patterns.

7.4.3 Character Generator RAM (CGRAM)

The user can rewrite character patterns into the CGRAM. For 5 × 8 fonts, eight character patterns can be written, and for 5 × 10 fonts, four character patterns can be written.

7.4.4 Registers

The HD44780U has two 8-bit registers, an **instruction register** (IR) and a **data register** (DR). The IR register stores **instruction codes**, such as **display clear** and **cursor move**, and **address information** for display data RAM (DDRAM)

TABLE 7.8A *DDRAM Address Usage for a 1-Line LCD*

	Visible	
Display Size	**Character Positions**	**DDRAM Addresses**
1 * 8	00..07	0x00..0x07
1 * 16	00..15	0x00..0x0F
1 * 20	00..19	0x00..0x13
1 * 24	00..23	0x00..0x17
1 * 32	00..31	0x00..0x1F
1 * 40	00..39	0x00..0x27

TABLE 7.8B *DDRAM Address Usage for a 2-Line LCD*

	Visible	
Display Size	**Character Positions**	**DDRAM Addresses**
2 * 16	00..15	0x00..0x0F + 0x40..0x4F
2 * 20	00..19	0x00..0x13 + 0x40..0x53
2 * 24	00..23	0x00..0x17 + 0x40..0x57
2 * 32	00..31	0x00..0x1F + 0x40..0x5F
2 * 40	00..39	0x00..0x27 + 0x40..0x67

TABLE 7.8C *DDRAM Address Usage for a 4-Line LCD*

	Visible	
Display Size	**Character Positions**	**DDRAM Addresses**
4 * 16	00..15	0x00..0x0F + 0x40..0x4F + 0x14..0x23 + 0x54..0x63
4 * 20	00..19	0x00..0x13 + 0x40..0x53 + 0x14..0x27 + 0x54..0x67
4 * 40	00..39 on 1st controller and 00..39 on 2nd controller	0x00..0x27 + 0x40..0x67 on 1st controller and 0x00..0x27 + 0x40..0x67 on 2nd controller

and character generator RAM (CGRAM). The microcontroller writes commands into this register to set up the LCD operation parameters. To write data into the DDRAM or CGRAM, the microcontroller writes data into the DR register. Data written into the DR register will be automatically written into DDRAM or CGRAM by an internal operation. The DR register is also used for data storage when reading data from DDRAM or CGRAM. When address information is written into the IR register, data is read and then stored in the

DR register from DDRAM or CGRAM by an internal operation. The microcontroller can then read the data from the DR register. After a read operation, data in DDRAM or CGRAM at the next address is sent to the DR register, and the microcontroller does not need to send another address. The IR and DR registers are distinguished by the RS signal. The IR register is selected when the RS input is low. The DR register is selected when the RS input is high. Register selection is illustrated in Table 7.9.

Busy Flag (BF)

The HD44780U has a busy flag (BF) to indicate whether the current internal operation is complete. When BF is 1, the HD44780U is still busy with an internal operation. When RS = 0 and $R/\overline{W}$ is 1, the busy flag is output to the DB7 pin. The microcontroller can read this pin to find out if the HD44780U is still busy.

Address Counter (AC)

The HD44780U uses a 7-bit address counter to keep track of the address of the next DDRAM or CGRAM location to be accessed. When an instruction is written into the IR register, the address information contained in the instruction is transferred to the AC register. The selection of DDRAM or CGRAM is determined by the instruction. After writing into (reading from) DDRAM or CGRAM, the content of the AC register is automatically incremented (decremented) by 1. The content of the AC register is output to the DB6..DB0 pins when the RS signal is low and the $R/\overline{W}$ signal is high.

7.4.5 Instruction Description

The functions of LCD instructions are discussed in this section.

Clear Display

This instruction writes the space code 0x20 into all DDRAM locations. It then sets 0 into the address counter and returns the display to its original status if it was shifted. In other words, the display disappears, and the cursor or blinking goes to the upper left corner of the display. It also sets the I/D bit to 1 (increment mode) in entry mode.

TABLE 7.9 *Register Selection*

RS	$R/\overline{W}$	Operation
0	0	IR write as an internal operation (display clear, etc.).
0	1	Read busy flag (DB7) and address counter (DB0 to DB6).
1	0	DR write as an internal operation (DR to DDRAM or CGRAM).
1	1	DR read as an internal operation (DDRAM or CGRAM to DR).

Return Home

This instruction sets DDRAM address 0 into the address counter and returns to its original status if it was shifted. The DDRAM contents are not changed. The cursor or blinking goes to the upper left corner of the display.

Entry Mode Set

The **I/D** bit of this instruction controls the incrementing (I/D = 1) or decrementing (I/D = 0) of the DDRAM address. The cursor or blinking will be moved to the right or left depending on whether this bit is set to 1 or 0. The same applies to writing and reading of CGRAM.

The **S** bit of this instruction controls the shifting of the LCD display. The display shifts if S = 1. Otherwise, the display does not shift. If S is 1, it will seem as if the cursor does not move but the display does. The display does not shift when reading from DDRAM. Also, writing into or reading from CGRAM does not shift the display.

Display On/Off Control

This instruction has three bit parameters: **D**, **C**, and **B**. When the D bit is set to 1, the display is turned on. Otherwise, it is turned off. The cursor is turned on when the C bit is set to 1. The character indicated by the cursor will blink when the B bit is set to 1.

Cursor or Display Shift

This instruction shifts the cursor position to the right or left without writing or reading display data. The shifting is controlled by two bits, as shown in Table 7.10. This function is used to correct or search the display. When the cursor has moved to the end of a line, it will be moved to the beginning of the next line.

When the displayed data is shifted repeatedly, each line moves only horizontally. The second line of the display does not shift into the first row. The contents of the address counter will not change if the only action performed is a display shift.

Function Set

This instruction allows the user to set the interface data length, select the number of display lines, and select the character fonts. There are three bit variables in this instruction.

TABLE 7.10 *LCD Shift Function*

S/C	R/L	Operation
0	0	Shifts the cursor position to the left. (AC is decremented by 1)
0	1	Shifts the cursor position to the right. (AC is incremented by 1)
1	0	Shifts the entire display to the left. The cursor follows the display shift.
1	1	Shifts the entire display to the right. The cursor follows the display shift.

DL: Data is sent or received in an 8-bit length (DB7 to DB0) when DL is set to 1, and in a 4-bit length (DB7 to DB4) when DL is 0. When the pin count is at a premium for the application, the 4-bit data length should be chosen, even though it is cumbersome to perform the programming.

N: This bit sets the number of display lines. When set to 0, one-line display is selected. When set to 1, two-line display is selected.

F: When set to 0, the 5 × 8 font is selected. When set to 1, the 5 × 10 font is selected.

Set CGRAM Address

This instruction contains the CGRAM address to be set into the address counter.

Set DDRAM Address

This instruction allows the user to set the address of the DDRAM (in address counter).

Read Busy Flag and Address

This instruction reads the busy flag (BF) and the address counter. The BF flag indicates whether the LCD controller is still executing the previously received instruction.

7.4.6 Interfacing the HD44780U to the 8051 Microcontroller

The data transfer between the HD44780U and the MCU can be done in 4 bits or 8 bits at a time. When in 4-bit mode, data are carried on the upper four data pins (DB7~DB4). The upper four bits are sent over DB7~DB4 first and followed immediately by the lower four bits.

We have the choice of using I/O ports to interface with the LCD module or treating the LCD as a memory device. Unless our application requires external data memory, there is no reason to interface LCD kit by treating it as a memory device. This chapter will treat the LCD only as an I/O device.

The LCD circuit connections for the SSE040 (4-bit data bus) is illustrated in Figure 7.3. The R/$\overline{W}$ signal to the LCD kit in the SSE040 demo board is grounded, which prevents the user from polling the BF flag to determine whether the LCD internal operation has been completed. The C8051F040 and the HD44780U use different power supplies.

Certain timing parameters must be satisfied in order to access the LCD successfully. The read and write timing diagrams are shown in Figures 7.4 and 7.5, respectively. The values of timing parameters depend on the frequency of the operation. HD44780U-based LCDs can operate at either 1 MHz (cycle time of E signal) or 2 MHz. The values of timing parameters at these two frequencies are shown in Tables 7.11 and 7.12, respectively.

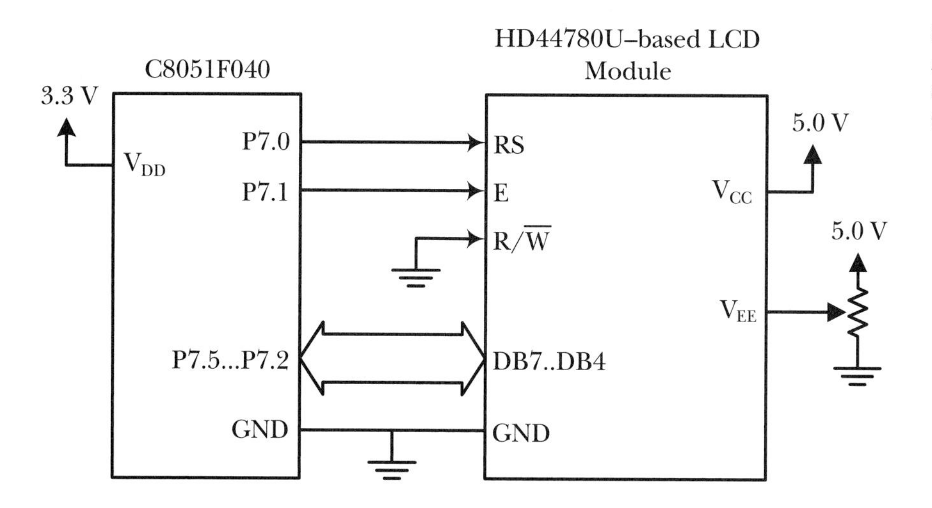

FIGURE 7.3
LCD interface example (4-bit bus, used in SSE040).

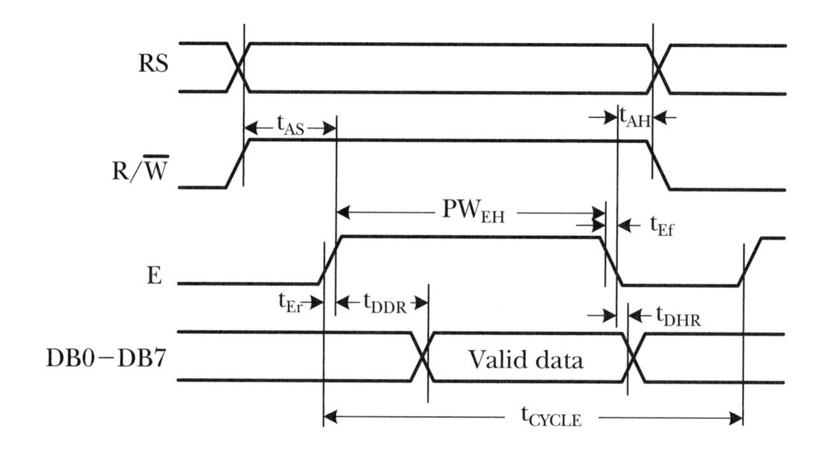

FIGURE 7.4
HD44780U LCD controller read timing diagram.

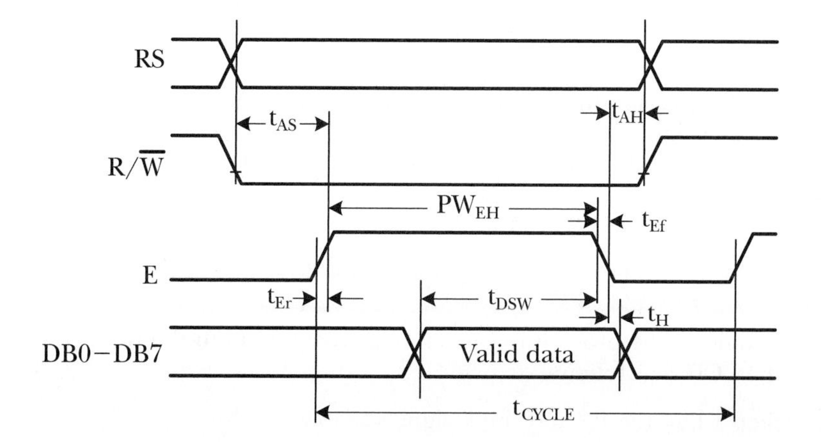

FIGURE 7.5
HD44780U LCD controller write timing diagram.

TABLE 7.11 *HD44780U Bus Timing Parameters (1 MHz operation)*

Symbol	Meaning	Min.	Max.	Unit
t_{CYCLE}	Enable cycle time	1000	-	ns
PW_{EH}	Enable pulse width (high level)	450	-	ns
t_{Er}, t_{Ef}	Enable rise and decay time	-	25	ns
t_{AS}	Address setup time, RS, R/$\overline{W}$, E	60	-	ns
t_{DDR}	Data delay time	-	360	ns
t_{DSW}	Data setup time (write)	195	-	ns
t_H	Data hold time (write)	10	-	ns
t_{DHR}	Data hold time (read)	5	-	ns
t_{AH}	Address hold time	20	-	ns

TABLE 7.12 *HD44780U Bus Timing Parameters (2 MHz Operation)*

Symbol	Meaning	Min.	Max.	Unit
t_{CYCLE}	Enable cycle time	500	-	ns
PW_{EH}	Enable pulse width (high level)	230	-	ns
t_{Er}, t_{Ef}	Enable rise and decay time	-	20	ns
t_{AS}	Address setup time, RS, R/$\overline{W}$, E	40	-	ns
t_{DDR}	Data delay time	-	160	ns
t_{DSW}	Data setup time (write)	80	-	ns
t_H	Data hold time (write)	10	-	ns
t_{DHR}	Data hold time (read)	5	-	ns
t_{AH}	Address hold time	10	-	ns

Example 7.2 Write a set of LCD routines that perform the following functions for the SSE040 demo board:

cmd2lcd (char cmd). This function sends the command **cmd** to the LCD kit.
openlcd (void). This function initializes the LCD.
putc2lcd (char cx). This function outputs the character **cx** to the LCD kit.
puts2lcd (char *ptr). This function outputs a NULL-terminated string pointed to by **ptr** to the LCD kit.

Solution: The procedure for sending a command to the IR register of the LCD is as follows.

Step 1 Pull the RS and the E signals to low.

Step 2 Pull the R/$\overline{W}$ signal to low.

Step 3 Pull the E signal to high.

Step 4 Output data to the output port attached to the LCD data bus and wait for a while to provide enough data setup time (t_{DSW}).

Step 5 Pull the E signal to low and make sure that the internal operation is complete.

The procedure for writing a byte to the LCD data register is as follows.

Step 1 Pull the RS signal to high.

Step 2 Pull the R/$\overline{W}$ signal to low.

Step 3 Pull the E signal to high.

Step 4 Output data to the I/O port attached to the LCD data bus and wait for a while to provide enough data setup time (t_{DSW}).

Step 5 Pull the E signal to low and make sure that the internal operation is complete.

The previous procedures need to be repeated once for an LCD kit that has a 4-bit interface.

The following constant definitions will be used in the specified functions.

```
lcdPort    equ    P7      ; LCD data pins (P7.5~P7.2)
lcdE       equ    02H     ; E signal pin
lcdRS      equ    01H     ; RS signal pin
```

All of the LCD commands take a much longer time to complete than an MCU instruction does. There is always a possibility that the LCD is still busy with its internal operation when one intends to send a new command to the LCD. There are two methods for solving this problem.

1. The **cmd2lcd** function calls a subroutine to make sure that the LCD is idle before proceeding with the new command. After performing the desired operation, the **cmd2lcd** function simply returns without waiting for the internal LCD operation to complete. This approach cannot work with the SSE040 demo board, because the LCD kit on the SSE040 is designed to be **write-only**.
2. The **cmd2lcd** function performs the desired operation, waits for 40 μs (or slightly longer), and then returns to the caller. By waiting for 40 μs, all except two commands (clear display and cursor home) will be completed. For these two instructions, the caller needs to call a delay subroutine that can create a longer delay (at least 1.64 ms).

The function that sends a command to the LCD kit using the second approach is as follows.

```
cmd2lcd:    push    SFRPAGE
            push    ACC             ; save a copy of the command
            mov     SFRPAGE,#0FH    ; switch to the SFR page of LCD port
            clr     lcdRS           ; select the instruction register
            setb    lcdE            ; pull E to high
            anl     A,#0F0H         ; clear the lower 4 bits
            clr     C               ; match the upper 4 bits with the LCD
            rrc     A               ; data pins
            rrc     A               ;     "
            orl     A,#02H          ; maintain E signal value
            mov     lcdPort,A       ; send the command, along with RS, and E signals
            nop                     ; insert required data setup time
            nop                     ;     "
            nop                     ;     "
            clr     lcdE            ; pull E signal low
            pop     ACC             ; get back the command
            anl     A,#0FH          ; clear the upper four bits
            clr     C               ; align the data with the LCD data pins
            rlc     A               ;     "
            rlc     A               ;     "
            setb    lcdE
            orl     A,#02H          ; maintain the value of E and RS
            mov     lcdPort,A       ; send the lower 4 command bits, E, RS to LCD
            nop                     ; insert enough data setup time
            nop                     ;     "
            nop                     ;     "
            clr     lcdE
            lcall   delay50us       ; wait for 50 μs
            pop     SFRPAGE
            ret
```

Before using the LCD, the user must configure it properly. The configuration of the LCD involves at least the following four LCD instructions.

1. **Entry mode set.** The common setting for this instruction is to move the cursor to the right after reading or writing a character from/to the LCD.
2. **Display on/off.** The common setting for this instruction is to turn on the display and cursor, and enable cursor blinking.
3. **Function set.** This instruction sets the number of rows for display, the font size, and the width of the interface data (4 or 8 bits).
4. **Clear display.** Before outputting any data, it is always a good idea to clear the LCD screen and move the cursor to the home position (upper-left corner).

The following function performs the LCD configuration.

```
OpenLCD:    mov     R0,#3
            acall   delayby100ms    ; wait for LCD to complete internal initialization
            mov     A,#33H          ; The following two commands are required by the
            acall   writelcd        ; manufacturer
            mov     A,#32H          ;   "
            acall   cmd2lcd         ;   "
            mov     A,#28H          ; set 4-bit data, 2-line display, 5 × 8 font
            acall   cmd2lcd         ;   "
            mov     A,#08H          ; turn off display, cursor, and cursor blinking
            acall   cmd2lcd         ;   "
            mov     A,#01H          ; clear display and return to home position
            acall   cmd2lcd         ;   "
            mov     R0,#2           ; wait until "clear display" command is completed
            acall   delayby1ms      ;   "
            mov     A,#06H          ; move cursor right
            acall   cmd2lcd         ;   "
            mov     A,#0FH          ; turn on display, cursor, and cursor blinking
            acall   cmd2lcd         ;   "
            ret
;*****************************************************************************************************************
; The following performs power on LCD initialization write operation.
;*****************************************************************************************************************
writelcd:   push    SFRPAGE
            push    ACC
            mov     SFRPAGE,#0FH
            clr     lcdRS           ; select the instruction register
            setb    lcdE            ; pull E to high
            anl     A,#0F0H         ; clear the lower 4 bits
            clr     C               ; match the upper 4 bits with the LCD
            rrc     A               ; data pins
            rrc     A               ;   "
            orl     A,#02H          ; maintain E signal value
            mov     lcdPort,A       ; send the command, along with RS, and E signals
            nop
            nop
            nop
            clr     lcdE            ; pull E signal low
            mov     R0,#5
            acall   delayby1ms
            pop     ACC
            anl     A,#0FH          ; clear the upper four bits
            clr     C
            rlc     A
            rlc     A
            setb    lcdE
            orl     A,#02H          ; maintain the value of E and RS
            mov     lcdPort,A       ; send the lower 4 command bits, E, RS to LCD
            nop
```

```
        nop
        nop
        clr     lcdE
        lcall   delay50us
        lcall   delay50us
        pop     SFRPAGE
        ret
```

The function that outputs a character to the LCD and makes sure that the write operation is complete is as follows.

```
putc2lcd:   push    SFRPAGE
            push    ACC
            mov     SFRPAGE,#0FH
            setb    lcdRS           ; select the instruction register
            setb    lcdE            ; pull E to high
            anl     A,#0F0H         ; clear the lower 4 bits
            clr     C               ; match the upper 4 bits with the LCD
            rrc     A               ; data pins
            rrc     A               ;    "
            orl     A,#03H          ; maintain E and RS signal values
            mov     lcdPort,A       ; send the upper 4 data bits, E, and RS to LCD
            nop
            nop
            nop
            clr     lcdE            ; pull E signal low
            pop     ACC             ; retrieve the data byte to be outputted
            anl     A,#0FH          ; clear the upper four bits
            clr     C               ; shift left two bits to match LCD data pins
            rlc     A               ;    "
            rlc     A               ;    "
            setb    lcdE
            orl     A,#03H          ; maintain the value of E and RS
            mov     lcdPort,A       ; send the lower 4 data bits, E, and RS to LCD
            nop                     ; insert enough data setup time
            nop                     ;    "
            nop                     ;    "
            clr     lcdE            ; pull E signal to low
            lcall   delay50us       ; wait for internal operation to complete
            pop     SFRPAGE
            ret
```

The function that outputs a NULL-terminated string in program memory pointed to by DPTR is as follows.

```
puts2lcd:   push    B               ; use B as an index to the string
            mov     B,#0            ; B starts with 0 and increments
loops:      mov     A,B             ; place the ith character in A to be sent to LCD
            movc    A,@A+DPTR       ;    "
```

```
            jz      dones               ; reach NULL character?
            acall   putc2lcd            ; not yet, output it to LCD.
            inc     B                   ; increment the string index
            ajmp    loops               ; continue
dones:      pop     B
            ret
```

Example 7.3 Write an assembly program to test the previous four subroutines by displaying the following messages on two lines:

I am very happy!
The LCD is working!

Solution: The program is as follows.

```
            $NOMOD51
            $include (c8051F040.inc)    ; for Keil
lcdPort     equ     P7
lcdRS       equ     P7.0
lcdE        equ     P7.1
            cseg    at 00H
            ljmp    main
            org     0ABH
main:       lcall   sysInit             ; configure system clock and crossbar
            mov     P7MDOUT,#0FFH       ; enable port 7 push-pull
            mov     SP,#0x7F            ; set stack pointer to 0x7F
            acall   OpenLCD             ; configure LCD properly
            mov     A,#80H              ; set LCD cursor to row 1, column 1
            acall   cmd2lcd             ;     "
            mov     DPTR,#msg1          ; output the first message
            acall   puts2lcd            ;     "
            mov     A,#0C0H             ; set LCD cursor to row 2, column 1
            acall   cmd2lcd             ;     "
            mov     DPTR,#msg2          ; output the 2nd message
            acall   puts2lcd            ;     "
forever:    nop
            ajmp    forever
msg1:       db      "I am very happy!",0
msg2:       db      "The LCD is working!",0

sysinit:    mov     SFRPAGE,#CONFIG_PAGE ; switch to the SFR page that contains
                                        ; registers that configures the system
            mov     WDTCN,#0DEH         ; disable watchdog timer
            mov     WDTCN,#0ADH         ;     "
            mov     CLKSEL,#0           ; used internal oscillator to generate SYSCLK
            mov     OSCXCN,#067H        ; configure external oscillator control
            mov     R7,#255
            djnz    R7,$                ; wait for about 1 ms
```

```
chkstable:  mov   A,OSCXCN        ; wait until external crystal oscillator is stable
            anl   A,#80H          ; before using it
            jz    chkstable       ;    "
            mov   CLKSEL,#1       ; use external crystal oscillator to generate SYSCLK
            mov   XBR0,#0F7H      ; assign I/O pins to all peripheral functions
            mov   XBR1,#0FFH      ; and enable crossbar
            mov   XBR2,#5DH       ;    "
            mov   XBR3,#8FH       ;    "
            ret
;--------------------------------------------------------------------------------------------
; include the previous four LCD subroutines here.
;--------------------------------------------------------------------------------------------
;********************************************************************************************
; The following function creates a time delay of 50 us. fOSC = 24 MHz
;********************************************************************************************
delay50us:  push  SFRPAGE
            mov   SFRPAGE,#0      ; switch to page 0
            mov   TMOD,#11H       ; configure Timer 0 and 1 as mode 1 timer
            mov   CKCON,#00H      ; Timer 0 use system clock divided by 12 to count
            mov   TH0,#0FFH       ; place 65436 in Timer 0 so that it overflows in
            mov   TL0,#09CH       ; 100 clock cycles (in 50 us)
            setb  TR0             ; enable Timer 0
            clr   TF0
loopw0:     jnb   TF0,loopw0      ; wait until TF0 is set again
            clr   TR0
            pop   SFRPAGE
            ret
;********************************************************************************************
; The following function creates a delay that is a multiple of 1 ms. The multiple is
; passed in R0. The system clock is 24 MHz.
;********************************************************************************************
delayby1ms:
            push  SFRPAGE
            mov   SFRPAGE,#0      ; switch to page 0
            mov   TMOD,#011H      ; configure Timer 0 and 1 as mode 1 timer
            mov   CKCON,#00H      ; Timer 0 use system clock divided by 12 to count
repw1:      mov   TH0,#0F8H       ; place 63536 in Timer 0 so that it overflows in
            mov   TL0,#030H       ; 2000 clock cycles (in 1 ms)
            setb  TR0             ; enable Timer 0
            clr   TF0
loopw1:     jnb   TF0,loopw1      ; wait until TF0 is set again
            clr   TR0
            djnz  R0,repw1
            pop   SFRPAGE
            ret
;********************************************************************************************
; The following function creates a delay that is a multiple of 100 ms. The multiple is
; passed in R0.
;********************************************************************************************
```

```
delayby100ms:
          push     SFRPAGE
          mov      SFRPAGE,#0         ; switch to page 0
          mov      TMOD,#011H         ; configure Timer 0 and 1 as mode 1 timer
          mov      CKCON,#02H         ; Timer 0 use system clock divided by 48 to count
repw3:    mov      TH0,#03CH          ; place 15536 in Timer 0 so that it overflows in
          mov      TL0,#0B0H          ; 50000 clock cycles (in 100 ms)
          setb     TR0                ; enable Timer 0
          clr      TF0
loopw3:   jnb      TF0,loopw3         ; wait until TF0 is set again
          djnz     R0,repw3
          pop      SFRPAGE
          ret
          end
```

Example 7.4 Write the C language versions of the previous four LCD functions and a test program to test them.

Solution: The C functions for the LCD kit on the SSE040 demo board and their test program are as follows:

```
#pragma src                           // required for adding in-line assembly instructions (Keil)
#include <c8051F040.h>                // use Keil C to compile this program
#include <delays.h>
#define lcdPort    P7                 // Port P7 drives LCD data pins, E, and RS
#define lcdE       0x02               // signal E
#define lcdRS      0x01               // RS signal
#define lcdE_RS    0x03               // assert both E and RS signal
void cmd2lcd (char cmd);
void openlcd (void);
void putc2lcd (char cx);
void putsc2lcd (char code *ptr);
void sysInit (void);
int main (void)
{
    char code *msg1 = "Hello World!";
    char code *msg2 = "The LCD is working!";
    sysInit();            // configure system clock, assign peripheral function pins
    openlcd();
    putsc2lcd(msg1);      // output message "Hello World!"
    cmd2lcd(0xC0);        // move cursor to 2nd row, 1st column
    putsc2lcd(msg2);      // output message "The LCD is working!"
    return 0;
}
void sysInit(void)
{
    int n;
    SFRPAGE = CONFIG_PAGE;
```

```
    WDTCN  = 0xDE;                  // disable watchdog timer
    WDTCN  = 0xAD;                  //          "
    OSCXCN = 0x67;                  // start external oscillator; 24 MHz Crystal
                                    // system clock is 24 MHz
    for (n=0;n<255;n++);            // delay about 1 ms
    while ((OSCXCN & 0x80) == 0);   // wait for oscillator to stabilize
    CLKSEL  |= 0x01;                // switch to external oscillator
    XBR2   = 0x5D;                  // enable crossbar and assign I/O pins to all
    XBR0   = 0xF7;                  // peripheral signals,
    XBR1   = 0xFF;                  //    "
    XBR3   = 0x8F;                  //    "
}
void cmd2lcd(char cmd)
{
    char  temp, temp1;
    temp    = cmd;                  // save a copy of the command
    temp1   = SFRPAGE;
    SFRPAGE = 0x0F;
    cmd     &=0xF0;                 // clear out the lower four bits
    lcdPort &= (~lcdRS);            // select LCD instruction register
    lcdPort |= lcdE;                // pull E signal to high
    cmd     >>= 2;                  // shift to match LCD data pins
    lcdPort = cmd | lcdE;           // output upper four bits, E, and RS
    #pragma asm                     // provide enough pulse width for E
    nop                             //    "
    nop                             //    "
    nop                             //    "
    #pragma endasm
    lcdPort &= (~lcdE);             // pull E signal to low
    cmd     = temp & 0x0F;          // extract the lower four bits
    lcdPort |= lcdE;                // pull E to high
    cmd     <<= 2;                  // shift to match LCD data pins
    lcdPort = cmd | lcdE;           // output lower four bits, E, and RS
    #pragma asm                     // provide enough pulse width for E
    nop                             //    "
    nop                             //    "
    nop                             //    "
    #pragma endasm
    lcdPort &= (~lcdE);             // pull E clock to low
    delayby50us(1);                 // wait until the command is complete
    SFRPAGE = temp1;
}
void openlcd(void)
{
    SFRPAGE  = 0x0F;
    P7MDOUT = 0xFF;                 // enable LCD port
    delayby100ms(3);                // wait for the LCD to be ready
    cmd2lcd(0x28);                  // set 4-bit data, 2-line display, 5 × 8 font
    cmd2lcd(0x0F);                  // turn on display, cursor, and cursor blinking
    cmd2lcd(0x06);                  // move cursor right
```

```
        cmd2lcd(0x01);                  // clear screen, move cursor to home
        delayby1ms(2);                  // wait until "clear display" command is complete
   }
   void putc2lcd(char cx)
   {
        char      temp, temp1;
        temp      = cx;
        temp1     = SFRPAGE;
        SFRPAGE = 0x0F;
        lcdPort   |= lcdRS;             // select LCD data register
        lcdPort   |= lcdE;              // pull E signal to high
        cx        &= 0xF0;              // clear the lower four bits
        cx        >>= 2;                // shift to match the LCD data pins
        lcdPort   = cx | lcdE_RS;       // output upper four bits, E, and RS
        #pragma asm                     // provide enough pulse width for E
        nop                             //     "
        nop                             //     "
        nop                             //     "
        #pragma endasm
        lcdPort   &= (~lcdE);           // pull E to low
        cx        = temp & 0x0F;        // get the lower four bits
        lcdPort   |= lcdE;              // pull E to high
        cx        <<= 2;                // shift to match the LCD data pins
        lcdPort   = cx | lcdE_RS;       // output lower four bits, E, and RS
        #pragma asm                     // provide enough pulse width for E
        nop                             //     "
        nop                             //     "
        nop                             //     "
        #pragma endasm
        lcdPort   &= (~lcdE);           // pull E to low
        delayby50us(1);
        SFRPAGE = temp1;
   }
   void putsc2lcd (char code *ptr)
   {
        while (*ptr) {
             putc2lcd(*ptr);
             ptr++;
        }
   }
```

This example is also a multiple-file project. Both the **eg07_03.c** and **delays.c** must be included into the project. When first building the project, the compiler will generate the **eg07_03.src** file and also indicate that the project has error. Add the **eg07_03.src** file to the same project and rebuild the project and the error will go away and the program will run correctly.

7.5 Using Keypad as an Input Device

A keypad is another commonly used input device. Like a keyboard, a keypad is arranged as a matrix of switches. The key switch can be mechanical, membrane, capacitors, or Hall-effect in construction. A mechanical switch relies on two metal contacts to be brought together to complete an electrical circuit. The membrane switch consists of two thin plastic or rubber membranes that can be pressed to contact each other. A capacitive switch comprises of two parallel plates that form a capacitor. When being pressed, the capacitance between these two plates is increased. A special circuitry is needed to detect this change in capacitance. A Hall-effect key switch consists of a Hall-effect crystal and a permanent magnet. Whenever the magnet moves in a direction that is perpendicular to the Hall-effect crystal, a voltage is induced between the two faces of the Hall-effect crystal. The presence of this voltage is recognized as a switch closure.

Mechanical keypads and keyboards are most popular due to their low cost and strength of construction. However, mechanical switches have a common problem called *contact bounce.* Instead of producing a single, clean pulse output, pressing a mechanical switch generates a series of pulses because the switch contacts do not come to rest immediately. This phenomenon is shown in Figure 7.6.

When the key is not pressed, the voltage output to the computer is V_{CC}. In order to detect which key has been pressed, the microcontroller needs to scan every key switch of the keypad. A human being cannot press and release a key switch in less than 20 ms. During this interval, the microcontroller can scan the same key-switch closure tens or even hundreds of thousand times, interpreting each low signal as a new input when in fact only one input should be sent.

Because of the contact bounce and the disparity in speed between the microcontroller and human key pressing, a *debouncing* process is needed. A keypad input program is normally divided into three steps:

Step 1
Scan the keypad to find out which key was pressed.

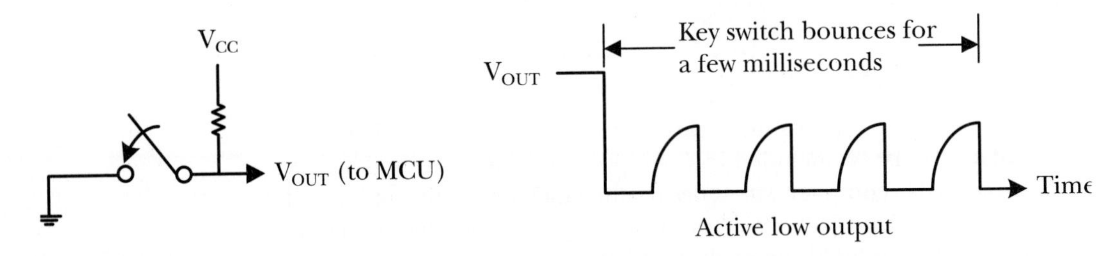

FIGURE 7.6
Key switch contact bounce.

Step 2

Debounce the key switch to make sure a key was indeed pressed. This step also guarantees that one key press is only recognized once.

Step 3

Look up the code of the key that was pressed. ASCII code is used most often.

7.5.1 How to Perform Keypad Scanning

Keypad scanning is usually performed row-by-row and column-by-column. A 16-key keypad can easily be interfaced with any available I/O port. Figure 7.7 shows a 16-key keypad organized into four rows with each row driving four switches.

For the keypad input application, the upper four pins (P3.7~P3.4) of Port 3 should be configured for output, whereas the lower four pins (P3.3~P3.0) of Port 3 should be configured for input.

The rows and columns of a keypad are simply conductors. In Figure 7.7, Port 3 pins P3.3~P3.0 are pulled up to high by pull-up resistors. Whenever a key switch is pressed, the corresponding row and column are shorted together. In order to distinguish the row being scanned and those not being scanned, the row being scanned is driven low, whereas the other rows are driven high. The P3.4 pin selects the keys 0 to 3, The P3.5 pin selects the keys 4 to 7, and so on.

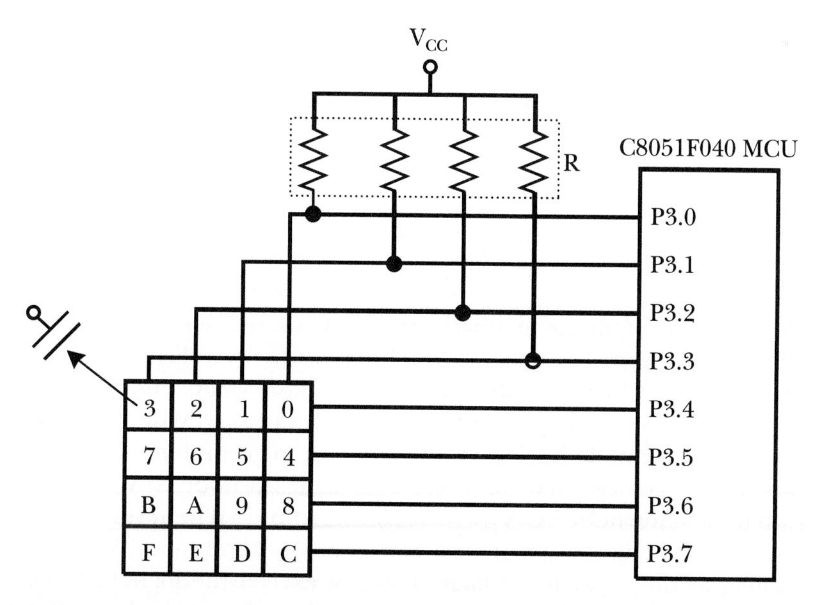

FIGURE 7.7
Sixteen-key keypad connected to the C8051F040.

7.5.2 How to Perform Keypad Debouncing

Before the key switch is closed totally when being pressed, it will bounce and cause the signal to rise and fall a few times within a period of a few milliseconds. A human being cannot press and release a switch in less than 20 ms. A keyswitch debouncer will accept that the key switch is closed after the voltage is low for about 10 ms and will recognize that the key switch is open after the voltage is high for about 10 ms.

Both hardware and software solutions to the key-bounce problem are available. Hardware solutions to contact bounce include an analog circuit that uses a resistor and a capacitor to smooth the voltage and two digital solutions that use set–reset latches or CMOS buffers and double-throw switches. Dedicated hardware scan-and-debounce chips are also available.

Hardware Debouncing Techniques

There are several well-known hardware debouncing methods. The objectives of these methods are to produce a single pulse for each key press. They are suitable for debouncing a few keys only. They can be used to generate interrupts to the CPU as long as the interrupt is edge-triggered. None of them have been used to debounce the keypad or keyboard. The following are hardware debouncing techniques.

- ***Set–Reset Latches.*** A key switch can be debounced by using the set-reset latch shown in Figure 7.8a. Before being pressed, the key is touching the set input, and the Q voltage is high. When pressed, the key moves toward the reset position. When the key touches the reset position, the Q voltage will go low. When the key is bouncing and touching neither the set nor the reset input, both set and reset inputs are pulled low by the pull-down resistors. Since both set and reset are low, the Q voltage will remain low and the key will be recognized as pressed.
- ***Noninverting CMOS Buffer Gate.*** The CMOS buffer output is identical to its input. When the switch is pressed, the input of the buffer chip 4050B is grounded, and hence, V_{OUT} is forced to low. When the key switch is bouncing (not touching the input), the resistor R keeps the output voltage low. This is due to the high-input impedance of 4050B, which causes a negligible voltage drop on the feedback resistor. Thus, the output changes value only once for each key press. This solution is shown in Figure 7.8b.
- ***Integrated RC Circuit.*** As shown in Figure 7.8c, when the switch is not pressed, the voltage V_{OUT} will be charged toward V_{DD}. When the switch is pressed toward node X, V_{OUT} is pulled down to ground immediately. The switch may bounce a few times before it settles at node X. Whenever the switch touches node X, V_{OUT} drops to ground. Whenever the switch bounces off node X, capacitor starts to charge up and causes V_{OUT} to rise. As long as the capacitor voltage does not exceed the logic 0, threshold value, the V_{OUT} signal will be recognized as logic 0, and the key switch will be recognized as pressed.

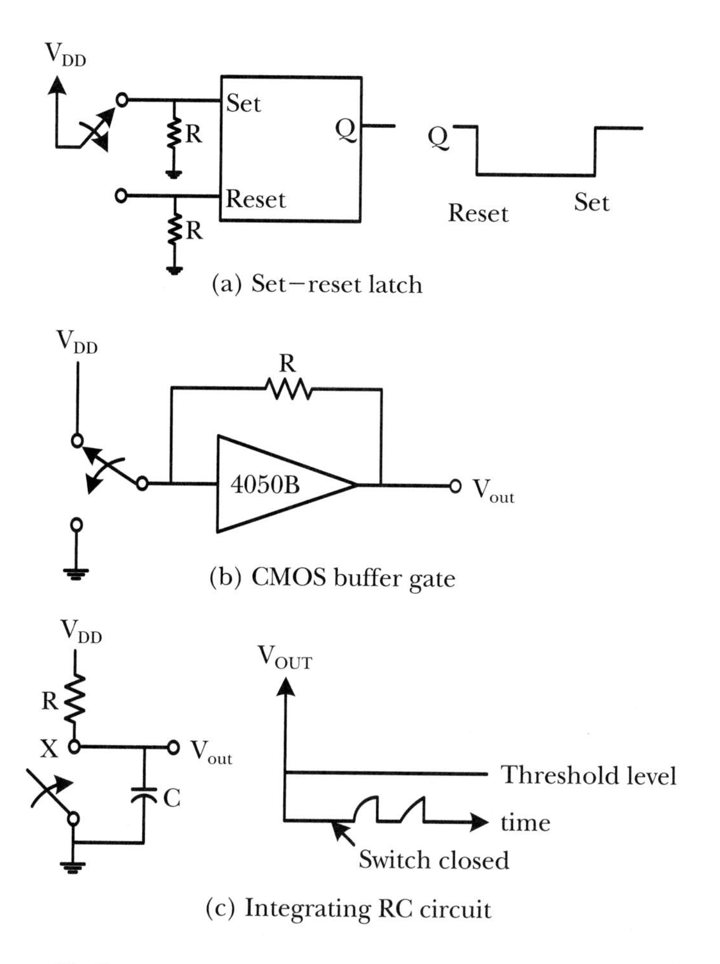

FIGURE 7.8 *Hardware debouncing techniques.*

Software Debouncing Techniques

Most keyboards use the software technique to perform key switch debouncing. The most widely used software technique for debouncing is based on the *wait-and-see* method. Once a key switch has been detected being pressed, the program waits for about 10 ms and reexamines the same key to see if it is still pressed. If the key switch output voltage is still low after 10 ms, then the key switch is considered being pressed. If the key switch output voltage is high, the program will consider the key switch is not being pressed and go on to scan the next key.

7.5.3 ASCII Code Lookup

For an application that needs a keyboard, the easiest way to find out the ASCII code of the pressed key is to perform a table lookup. However, table lookup is not necessary for the keypad because ASCII code lookup can be embedded in the program that performs scanning and debouncing.

Example 7.5 Write a subroutine to perform keypad scanning and debouncing and return the ASCII code of the pressed key to the caller.

Solution: The following assembly subroutine will perform keypad scanning and debouncing and will return the ASCII code of the pressed key in accumulator A to the caller.

```
maskr       set     R4                  ; the mask to scan a row
maskc       set     R5                  ; the mask to scan a column
column      set     R6                  ; the column being scanned
row         set     R7                  ; the row being scanned
keypad      set     P3                  ; this port drives the keypad
;****************************************************************************************************
; The following subroutine performs keypad scanning, debouncing, and returns the
; ASCII code of the pressed key.
;****************************************************************************************************
getkey:     push    SFRPAGE
            mov     SFRPAGE,#0FH        ; make SFR page F active
            mov     P3MDOUT,#0F0H       ; configure P3 upper 4 pins to push-pull,
            mov     SFRPAGE,#0          ; lower 4 pins to open drain
gkloope:    mov     row,#0              ; start from row 0
            mov     maskr,#0EFH         ; mask for selecting row 0
nextrow:    mov     column,#0           ; start from column 0
            mov     maskc,#01H          ; mask for scanning column 0
            mov     keypad,#0FFH        ; prepare to scan
            mov     A,maskr             ; select a row to scan
            anl     keypad,A            ;     "
gkloopi:    mov     A,maskc             ; select a column to check
            anl     A,keypad            ;     "
            jz      debounce            ; if low, then a key press has been detected
scan_next:  cjne    column,#3,inc_col
            cjne    row,#3,inc_row
            ajmp    gkloope             ; restart from row 0, column 0
inc_col:    inc     column              ; move to next column
            mov     A,maskc             ; update the column mask to check the
            rl      A                   ; next column
            mov     maskc,A             ;     "
            ajmp    gkloopi             ; go to scan next column
inc_row:    inc     row                 ; move to next row
            mov     A,maskr             ; update the mask for scanning
            rl      A                   ; the next row
            mov     maskr,A             ;     "
            ajmp    nextrow
debounce:   call    delay10ms
            mov     A,maskc             ; reexamine the same key
            anl     A,keypad            ;     "
            jz      getcode             ; if still low, get the ASCII code
            ajmp    scan_next           ; not low, continue to scan
getcode:    mov     A,row               ; find out the key number that
```

```
          rl      A                  ; has been pressed
          rl      A                  ;     "
          add     A,column           ;     "
          mov     R4,A               ; save a copy of A
          clr     C
          subb    A,#10              ; A <- [A] - 10
          jc      isdeci             ; jump if A < 10
          mov     A,R4               ; the number in A is greater than 9
          add     A,#37H             ; compute the ASCII code of the pressed key
          pop     SFRPAGE
          ret
isdeci:   mov     A,R4
          add     A,#30H             ; compute the ASCII code
          pop     SFRPAGE
          ret
; ******************************************************************************************
; The following routine uses TMR4 to create a time delay of 10 ms.
; ******************************************************************************************
delay10ms: push   SFRPAGE
          mov     SFRPAGE,#02H       ; switch to page 2
          mov     TMR4CN,#04H        ; enable TMR4 in timer mode
          mov     TMR4CF,#0H         ; select SYSCLK/12 as TMR4 clock source
          mov     TMR4H,#0B1H        ; load 45536 into TMR4H:TMR4L so that
          mov     TMR4L,#0E0H        ; it overflows in 10 ms
waitlp2:  mov     A,TMR4CN           ;
          anl     A,#080H            ; check bit 7 (TF4)
          jz      waitlp2            ; if TF5 bit is not set, then continue to wait
          pop     SFRPAGE
          ret
```

The C language version of the **getkey** function is as follows.

```
// rmask is row mask, cmask is column mask, row is the row being scanned, col is the
// column being scanned
char getkey (void)
{
    char rmask, cmask, row, col;
    char temp, keycode;
    temp      = SFRPAGE;
    SFRPAGE   = 0x0F
    P3MDOUT   = 0xF0;
    SFRPAGE   = 0;
    keypad    &= 0x0F; // configure lower four pins for input
    while (1) {
        rmask = 0xEF;
        for (row = 0; row < 4; row++){
            cmask       = 0x01;
            keypad      &= rmask; // select the current row
            for (col = 0; col < 4; col++){
                if (!(keypad & cmask)){          // key switch detected pressed
```

```
                    delayby10ms(1);
                    if(!(keypad & cmask)){
                        keycode = row * 4 + col;
                        if (keycode < 10)
                            return (0x30 + keycode);
                        else
                            return (0x37 + keycode);
                    }
                }
                cmask = cmask << 1;
            }
            rmask = (rmask << 1) | 0x0F;
        }
    }
}
```

7.6 Driving Stepper Motor

Stepper motors are digital motors. They are convenient for applications where a high degree of positional control is required. Printers, tape drives, disk drives, and robot joints, for example, are typical applications of stepper motors. In its simplest form, a stepper motor has a permanent magnet rotor and a stator consisting of two coils. The rotor aligns with the stator coil that is energized. By changing which coil is energized, as illustrated in Figure 7.9a through d, the rotor is turned.

In Figure 7.9a through d, the permanent magnet rotor lines up with the coil pair that is energized. The direction of the current determines the polarity of the magnetic field, and thus the angular position of the rotor. Energizing coil pair C3 and C4 causes the rotor to rotate 90°. Again, the direction of the current determines the magnetic polarity and thus the angular position of the rotor. In this example, the direction of the current causes the rotor to rotate in a clockwise direction, as shown in Figure 7.9b.

Next, coils C1 and C2 are energized again, but with a current opposite to that in Step 1. The rotor moves 90° in a clockwise direction, as shown in Figure 7.9c. The last full step moves the rotor another 90° in a clockwise direction. Note that again the coil pair C3 and C4 is energized but with a current opposite to that in Step 2.

We can also rotate the stepper motor in the counterclockwise direction. This can be done by reversing the polarities of coils C3 and C4 in Figure 7.9b and d. Figure 7.10 shows the counterclockwise sequence.

The stepper motor may also be operated with half steps. A *half step* occurs when the rotor (in a four-pole step) is moved to eight discrete positions (45°). To operate the stepper motor in half steps, half of the time both coils must be on at the same time. When both coils are energized, the resultant rotor will be on the 45°, 135°, 225°, and 315° positions relative to coil C4.

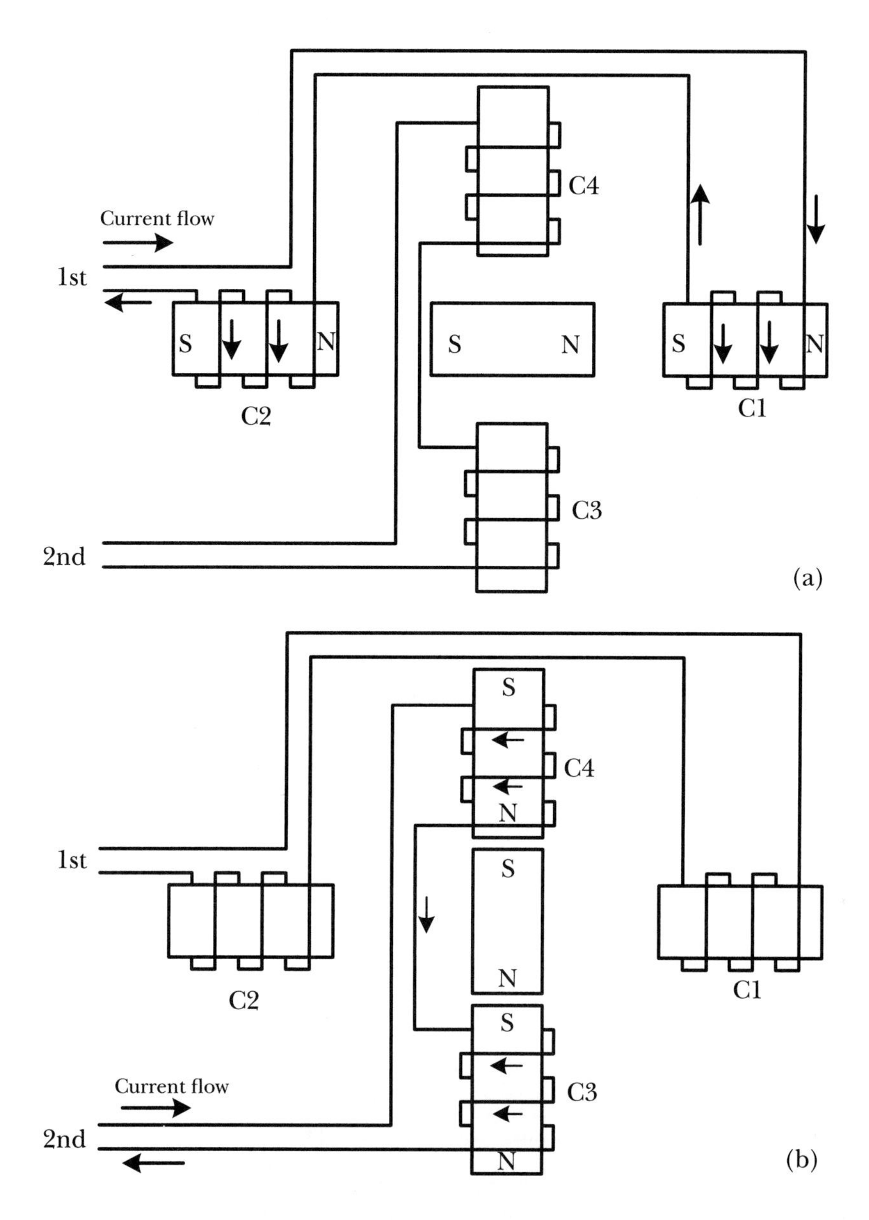

FIGURE 7.9
Stepper motor.

The step size of stepper motors may range from 0.72° to 90°. Among them, step sizes of 1.8°, 7.5°, and 15° are most common. The steps of 45° or 90° are not fine enough for most applications.

7.6.1 Stepper Motor Drivers

A stepper motor is driven by applying a series of voltages to the coils of the motor. Each time a subset of coils is energized to make the motor rotate one step. The coils energizing patterns must be followed exactly for the motor to

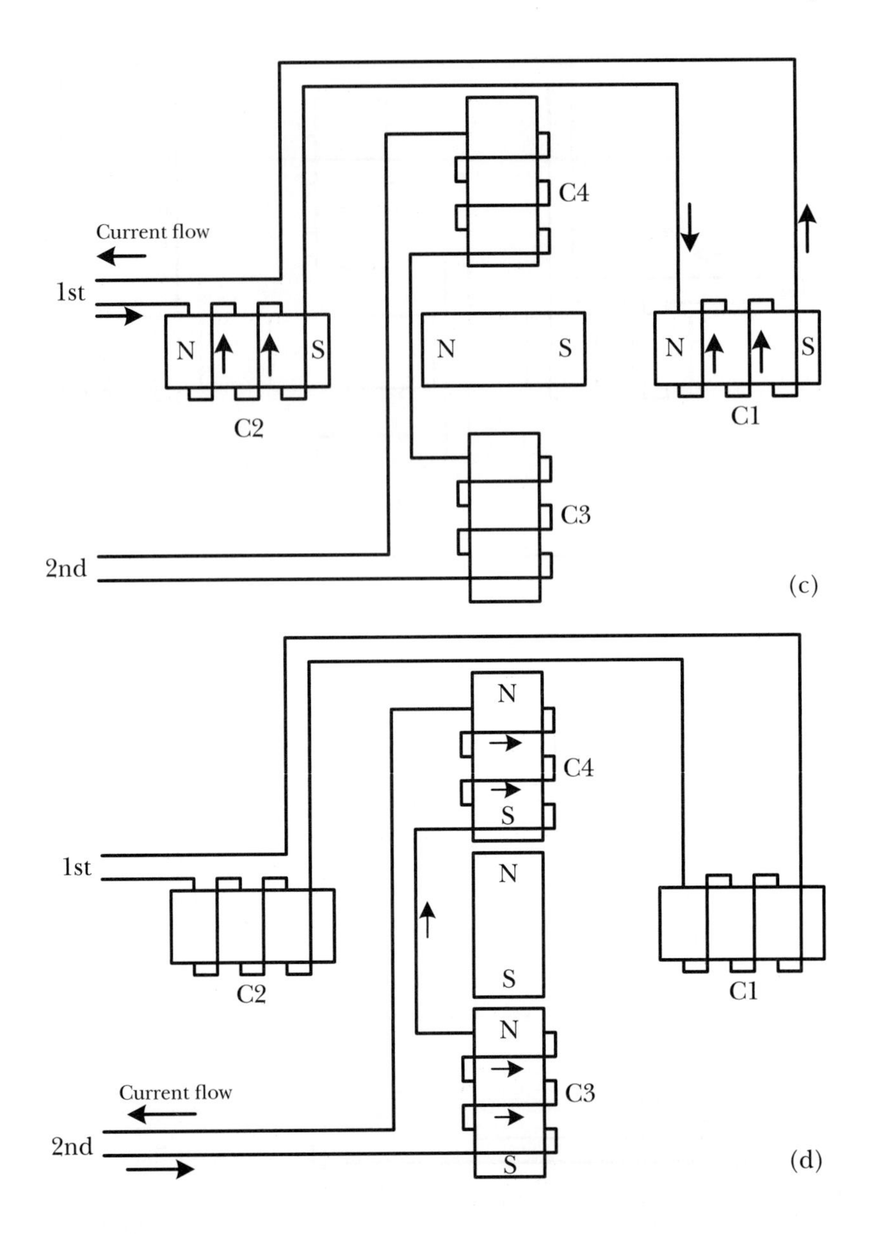

FIGURE 7.9
Stepper motor (continued).

rotate. The half-step mode and full-step mode require different energizing patterns. A delay of few milliseconds (from 5 ms to 15 ms) between two steps is needed for the motor to react to the newly applied pulse pattern due to its mechanical inertia. A microcontroller can easily provide this time delay for the coil to be energized and, hence, control the speed of the stepper motor in a precise manner.

A hobby stepper motor kit is shown in Figure 7.11. Here, four NPN transistors are used to switch the current to each of the four coils of the stepper motor. The windings in the stepper motor are the loads of these transistors.

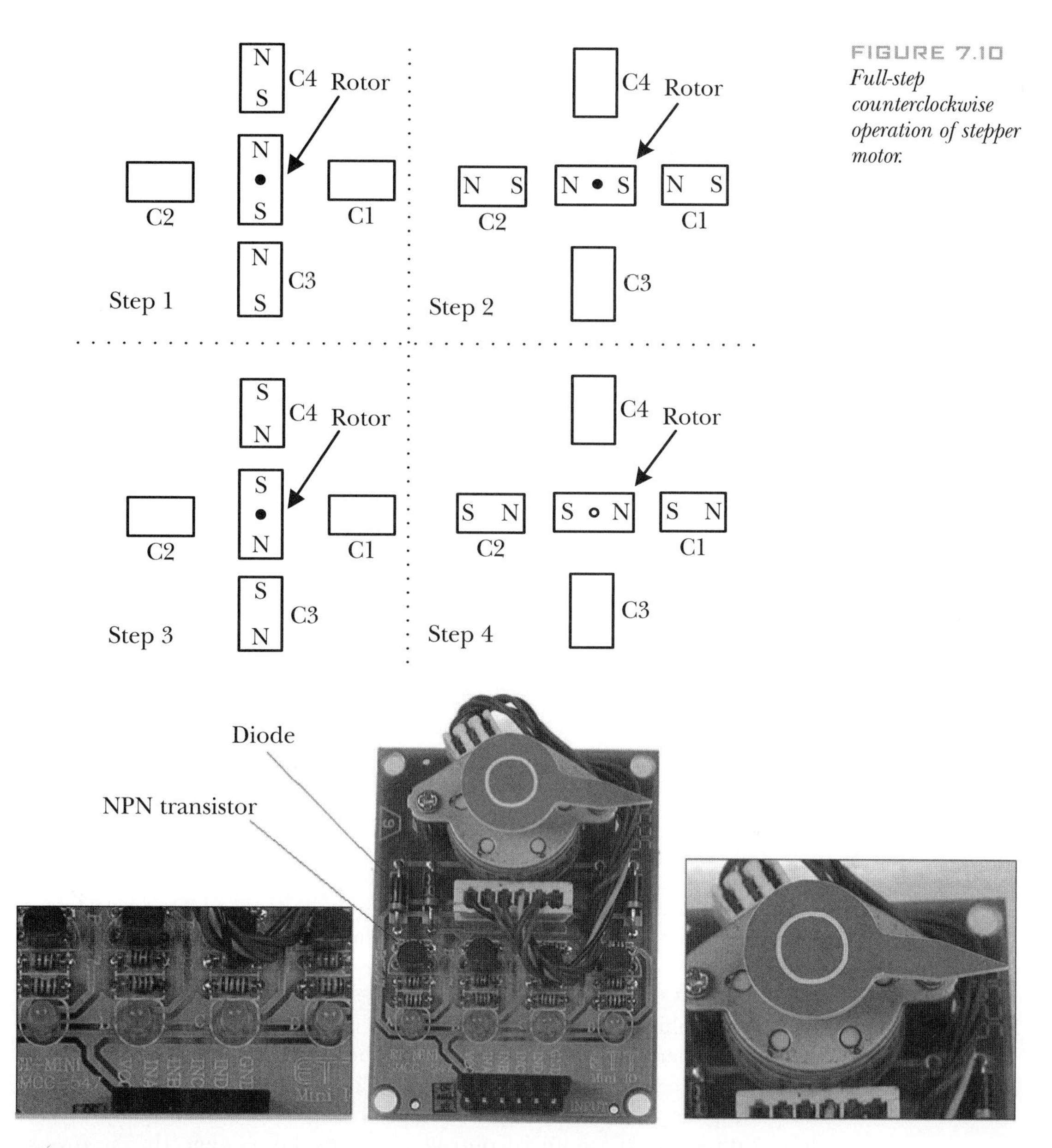

FIGURE 7.10
Full-step counterclockwise operation of stepper motor.

FIGURE 7.11
A hobby stepper motor kit
(Courtesy of Futurlec).

The windings are inductors, storing energy in a magnetic field. When the current is cut off, the inductor releases its stored energy in the form of an electric current. This current attempts to flow through the transistor from collector to emitter. Four diodes are placed between the collector and ground to prevent this from occuring.

TABLE 7.13 *Full-Step Sequence for Clockwise Rotation*

Step	Q4 P4.3	Q3 P4.2	Q2 P4.1	Q1 P4.0	Value
1	On	On	Off	Off	1100
2	Off	On	On	Off	0110
3	Off	Off	On	On	0011
4	On	Off	Off	On	1001

TABLE 7.14 *Half-Step Sequence for Clockwise Rotation*

Step	Q4 P4.3	Q3 P4.2	Q2 P4.1	Q1 P4.0	Value
1	On	Off	Off	Off	1000
2	On	On	Off	Off	1100
3	Off	On	Off	Off	0100
4	Off	On	On	Off	0110
5	Off	Off	On	Off	0010
6	Off	Off	On	On	0011
7	Off	Off	Off	On	0001
8	On	Off	Off	On	1001

Stepper motors can produce stronger torque with the full-step sequence shown in Table 7.13. However, the rotation is not very smooth with the full-step sequence. The half-step sequence is shown in Table 7.14. As shown in these tables, the lower 4 pins of Port P4 are used to drive the stepper motor. Stepper motor rotation is smoother using the half-step sequence. However, the half-sequence produces weaker torque. The full-step sequence will require a longer delay between two steps due to the higher mechanical inertia that the motor needs to overcome. At the bottom of Figure 7.11, pins IND through INA are connected to Q4 through Q1. To control the motor, the microcontroller must output the values in the table in the sequence shown. To drive the stepper motor, the microcontroller needs to send the sequence shown in Tables 7.13 and 7.14 in circular order and repeat them after reaching the last step. When outputting the values from Step 1 to Step 4 in Table 7.13, the motor will rotate in one direction. If the microcontroller outputs the sequence in the reverse order, the motor will then rotate in the reverse direction. It is essential that the order be preserved, even if the motor is stopped briefly. The next step to restart the motor must be the next sequential step following the last step used.

Example 7.6 Suppose that the pins P4.3, . . . , P4.0 are used to drive the four transistors, as shown in Figure 7.11. Write a subroutine to rotate the stepper motor clockwise one cycle using the half-step sequence.

Solution: The assembly language subroutine is as follows.

```
            $include    (c8051F040.inc)
step1       equ         00001000b
step2       equ         00001100b
step3       equ         00000100b
step4       equ         00000110b
step5       equ         00000010b
step6       equ         00000011b
step7       equ         00000001b
step8       equ         00001001b
HalfStep:
            push        SFRPAGE
            mov         SFRPAGE,#0FH        ; switch SFR page F
            mov         P4,#step1
            acall       delay10ms
            mov         P4,#step2
            acall       delay10ms
            mov         P4,#step3
            acall       delay10ms
            mov         P4,#step4
            acall       delay10ms
            mov         P4,#step5
            acall       delay10ms
            mov         P4,#step6
            acall       delay10ms
            mov         P4,#step7
            acall       delay10ms
            mov         P4,#step8
            acall       delay10ms
            mov         P4,#step1
            acall       delay10ms
            pop         SFRPAGE             ; restore the caller's SFRPAGE
            ret
```

The C language version of the routine is straightforward and hence is left as an exercise.

7.7 Chapter Summary

When interfacing with peripheral devices, we need to consider electrical compatibility issues. We need to consider if the output voltage of the microcontroller is high enough to be recognized as high or low enough to be recognized as low by the peripheral device to be driven by the MCU. In addition, we also need to consider if the microcontroller has enough capability to source or sink current to or from the peripheral devices.

Seven-segment displays are simple output devices that are suitable to display a small amount of information in the decimal or hex format. However, seven-segment displays are not suitable for displaying large amounts of information due to their high power consumption and wiring requirements. Liquid-crystal displays are more suitable for this type of application due to their lower power consumption and capability of displaying large amounts of information. An LCD needs a controller to generate the required control signals during the exchange of data between the MCU and the LCD. The most popular character-based LCD controller is the Hitachi HD44780U. Microcontrollers send commands and data to the LCD controller in order to display information in an appropriate format.

Keypads and keyboards are two of the most widely used input devices. When using the keypad to input data, one often needs to output a message to ask the user to type the keypad. An input operation is often a two-way interactive process. Like keyboards, keypads also have the key bouncing problem. Because of this reason, keypad input is divided into three steps: key-switch scanning, key debouncing, and ASCII code lookup. The purpose of key-switch scanning is to detect the key that has been pressed. The purpose of key debouncing is to make sure that the key has indeed been pressed and not to read the same key press more than once. The software *wait-and-see* method is the most widely used debouncing method.

A stepper motor is a digital motor in the sense that each step of the rotation rotates a fixed number of degrees. It is most suitable for applications that require a high degree of positional control, such as plotters, disk drives, magnetic tape drives, robot joints, and so on. The resolution of one step of a stepper motor can be as small as 0.72° and as large as 90°. The simplest stepper motor has two pairs of coils.

Driving a stepper motor involves applying a series of voltages to the coils of the motor. A subset of coils is energized at a time to cause the motor to rotate one step. The pattern of coils energized must be followed exactly for the motor to work correctly. The pattern will vary depending on the mode used on the motor. A microcontroller can easily time the duration during which the coil is energized and, hence, control the speed of the stepper motor in a precise manner.

7.8 Exercise Problems

E7.1 Write an assembly and C program to display the following information in two rows in the LCD connected to the demo board:
Date: 10 02 1952
Time: 10:20:10

E7.2 Write a program that performs the following operations.
(a) Configures the MCU to use external oscillator as the SYSCLK and initialize the LCD.

(b) Outputs a message (**Enter a value in DIP:**) on first row of the LCD to remind the user to enter data using the DIP switches connected to Port P6.

(c) The user sets up a value using the DIP switches and then presses on the switch connected to the INT0 pin.

(d) Enables INT0 interrupt writes a service routine that reads the value of DIP switches and displays it on the LCD on the second row.

E7.3 Look up the electrical characteristic table of the LCD controller NT3881D (used in SSE040 demo board) and verify the electrical compatibility between the NT3881D and the C8051F040.

E7.4 Write an instruction sequence to rotate the stepper motor shown in Figure 7.12 clockwise for one cycle using the full-step sequence.

E7.5 Write an instruction sequence to rotate the stepper motor shown in Figure 7.12 counterclockwise one cycle using the half-step sequence.

E7.6 Write an instruction sequence to rotate the stepper motor shown in Figure 7.12 counterclockwise one cycle using the full-step sequence.

E7.7 Write a sequence of C statements to rotate the stepper motor shown in Figure 7.12 clockwise one cycle using the full-step sequence.

E7.8 Write a sequence of C statements to rotate the stepper motor shown in Figure 7.12 counterclockwise one cycle using the full-step sequence.

E7.9 Suppose common-anode seven-segment displays are used instead of common-cathode seven-segment displays. Show the changes that must be made in the circuit shown in Figure 3.14. What needs to be modified in Table 3.5 for the seven-segment pattern?

E7.10 The Fairchild Semiconductor (www.fairchildsemi.com) 74C922 is a 16-key encoder that performs keypad scanning and debouncing. The pin assignment is shown in Figure 7E.10a. The data out value for each binary coded decimal (BCD) digit is its own value. That is, the data out value for 0 is 0, for 1 is 1, and so on. It has a *data available* output that can remind the microcontroller to read the keypad inputs. This signal is often connected to one of the interrupt inputs of the microcontroller. A circuit connection of the 74C922 to the C8051F040 is shown in Figure 7E.10b. Write a main program to enable keypad interrupt and initialize the DPTR to the starting address of a buffer **key_buf**. The main program then stays in a loop until the key (in accumulator A) pressed is an '**F**'. Then the main program disables the keypad interrupt and exits the loop. Whenever the keypad has been pressed, the 74C922 asserts the data-available pin, which in turn interrupts the microcontroller. Write an interrupt service routine to read the keypad and look up the ASCII code and leave it in accumulator A, set a data available flag, and return. The main program keeps checking the flag to find out if a new character has been entered. When the flag is set, the main program reads the character, saves it in the buffer, and then clears the flag and waits for the next character or the end of keypad input.

FIGURE E7.10

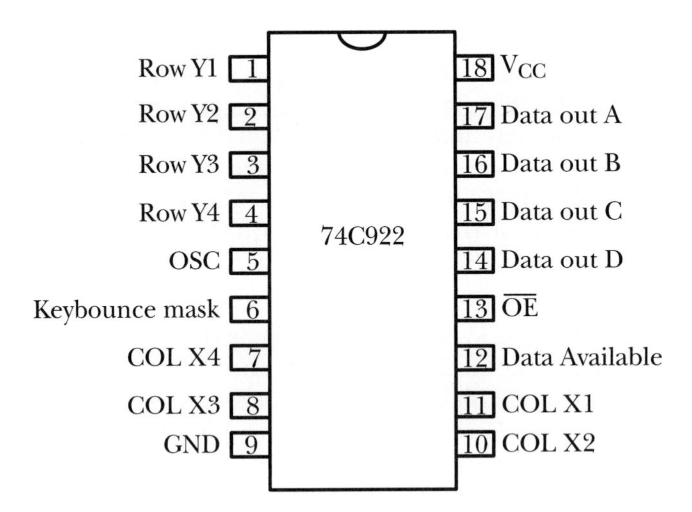

(a) Pin assignment of 74C922

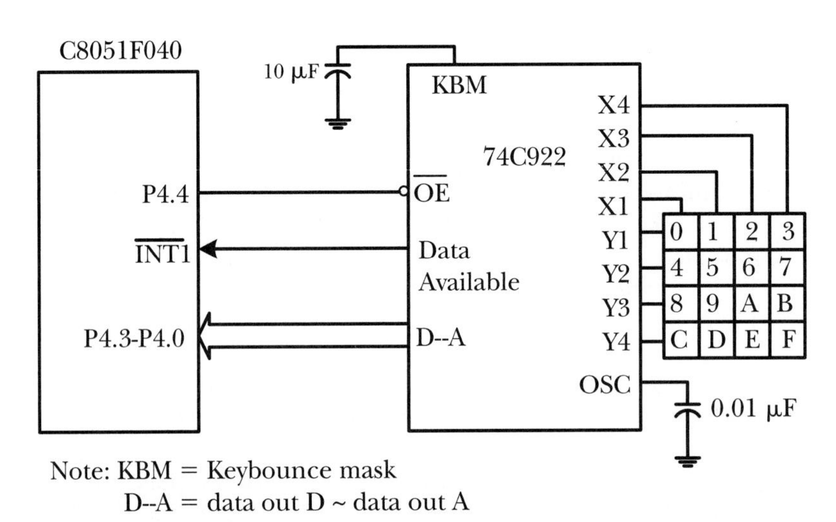

(b) The circuit connection between 74C922 and the C8051F040

E7.11 Write a program to generate a waveform shown in Figure E7.11 using the circuit shown in Figure 3.23.

E7.12 If a stepper motor takes 48 steps to complete a revolution, what is the step angle for this motor?

E7.13 Calculate the number of steps per revolution for a step angle of 1.8°.

E7.14 Finish the normal four-step sequence clockwise if the first step is 1001_2.

E7.15 Finish the normal four-step sequence clockwise if the first step is 0110_2.

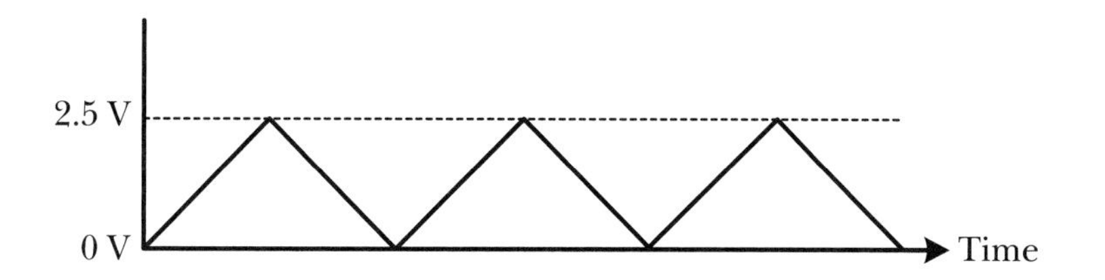

FIGURE E7.11
Waveform to be generated from $V_{OUT}A$.

E7.16 Finish the normal four-step sequence counterclockwise if the first step is 0110_2.

E7.17 Finish the normal four-step sequence counterclockwise if the first step is 1001_2.

7.9 Laboratory Exercise Problems and Assignments

L7.1 **Interactive I/O Practice.** This lab practice requires a 20 × 2 (or 16 × 2) LCD kit. Write a main program that outputs a prompt to remind the user to enter a value using the DIP switches. The user sets up a new value on the DIP switches and then presses a debounced switch that is connected to the $\overline{\text{INT0}}$ pin. The service routine for the $\overline{\text{INT1}}$ interrupt reads the value and then outputs it to the LCD kit. The lab procedure is as follows.

Step 1
The main program outputs a message **Wants to continue?** on the LCD. The user sets a value using the DIP switches and presses the $\overline{\text{INT0}}$ pin. The interrupt service routine reads the value and sets a flag to inform the main program to read the new value. If the entered value is 0, then stay in an infinite loop. Otherwise, continue to the next step.

Step 2
The main program outputs the prompt **Enter your age:** (in the first row) to remind the user to enter his/her age. The user sets up the DIP that represents his/her age and interrupts the MCU. The interrupt service routine reads the value and sets a flag to inform the main program. The main program then displays the age **I am xx years old.** on the second row.

Step 3
Repeat the question in Step 1, and if the answer is yes, then continue to this step. The main program outputs the message **Enter your height:** The user sets up his/her height in feet and inches (each uses four bits) and interrupts the MCU. The MCU reads the height and then displays **I am x feet y inches** on the LCD.

Step 4
Repeat the question in Step 1, and if the answer is yes, then continue to this step. The main program outputs the message **Enter your weight:** The user sets up his/her weight in pounds and interrupts the MCU. The MCU reads the weight and then displays it on the LCD.

After this step, the program stays in an infinite loop.

L7.2 **Keypad Input.** Use the circuit shown in Figure 7.7 to perform keypad input practice. The procedure is as follows.

Step 1
Configure the upper four pins of Port P3 for output and the lower four pins of P3 for input.

Step 2
Initialize the LCD properly.

Step 3
Output the message **Enter an integer:** on the first row of the LCD. After seeing this message, you enter a number on the keypad. Use the F key to terminate the number. Your program reads in the number, converts it into a binary, and saves it in a buffer.

Step 4
Output the message **Enter another integer:** on the first row of the LCD. After seeing this message, you enter a number on the keypad. Use the F key to terminate the number. Your program reads in the number, converts it into a binary, and saves it in a buffer.

Step 5
Compute the gcd of these two numbers and display it on the LCD screen as follows.

```
The gcd of xxxx and
yyyy is xxxx.
```

where, xxxx and yyyy are the numbers that you entered from the keypad and zzzz is the **gcd** of these two numbers.

L7.3 **Stepper Motor Experiment.** Use the stepper motor mini-board from Futurlec (www.futurlec.com) to perform the stepper motor experiment. The stepper motor mini-board implements the circuit shown in Figure 7.11. The IND, INC, INB, and INA signals must be connected from most significant to least significant pins of Table 7.13 and 7.14 are followed.

Perform the following procedure.

Step 1
Configure Port P4 for output.

Step 2
Use the lower four pins of P4 to drive the bases of four transistors.

Step 3
Write a sequence of instructions or C statements to rotate the stepper motor clockwise for 5 seconds using the full-step sequence. Use 10 ms as the delay between two steps.

Step 4
Repeat Step 3 but use the half-step sequence.

Step 5
Write instructions or C statements to rotate the stepper motor counterclockwise for 5 seconds using the full-step sequence.

Step 6
Repeat Step 3 but use the half-step sequence.

Step 7
Repeat Steps 3 through 6 but use different delays between the steps, for example, 5 ms, 15 ms, and 20 ms.

Timers and Programmer Counter Array

8.1 Objectives

Upon successful completion of this chapter, you will be able to:

- Explain the functioning of the 8051 timer subsystem
- Use timer functions to measure signal parameters such as the pulse width, period, frequency, and duty cycle
- Use timer functions to trigger certain operations periodically
- Use a PCA module capture mode to capture the arrival time of an event and/or use it as a time reference
- Use a PCA module to generate digital waveforms
- Use timer functions to make sound and play songs
- Use timer functions to drive DC motors, fans, and lamps

8.2 Introduction

The circuit of a timer is identical to that of a counter. When the clock input to a counter is a periodic signal, it is referred to as a timer. A timer is normally edge-triggered (i.e., it increments or decrements on the clock edge). There may be many variations to the operation of an *n*-bit timer:

1. The timer may count up from 0 to $2^n - 1$, roll over to 0, and count up again.
2. The timer may be preloaded with a value, count up to $2^n - 1$, and roll over to 0.
3. The timer may count up from 0 to a value less than $2^n - 1$, then reset to 0, and count up again.

4. The timer may count from $2^n - 1$ down to 0 and then repeat.
5. The timer may count from a value less than $2^n - 1$ down to 0 and then repeat.

An extra register is needed to implement variations 3 and 5, because the user will want the count-limit value to be programmable. The moment that the timer reaches the count limit, other actions may be triggered. For example, a signal pin may be pulled to high, pulled to low, or toggled when the timer reaches the count limit.

Contemporary microcontrollers implement much more complicated timer functions than we described previously. The most popular ones are *capture, compare,* and *PWM.*

The **capture function** requires an additional register (called the *capture register*) to function. The capture operation normally is triggered by an *event* that is represented by a signal edge (rising or falling). When the selected edge is detected, the count value of the timer is copied into the capture register. The capture function may be used to measure the pulse width, period, or duty cycle of a signal. The moment that an edge is captured also can be used as a time reference for triggering other actions. The pulse width of a signal can be measured by capturing one rising edge and the following falling edge and taking the difference of these two edges. The period of a signal can be measured by capturing two consecutive rising or falling edges and taking the difference of them.

The **compare function** is implemented by comparing the timer value in every clock cycle with the value stored in another register (called the *compare register*). The moment that the timer value equals that of the compare register may cause a pin to be pulled up to high, pulled down to low, or simply toggled. This capability may be used to generate digital waveforms. However, the CPU needs to start each subsequent compare operation.

The **PWM function** is used to generate periodic square waveforms with a specified frequency and duty cycle. Once the frequency and duty cycle are set, the PWM circuit will continue to repeat the same waveform without the CPU intervention. The frequency and duty cycle can be changed any time. There are applications that are controlled by their average input DC voltage. The average DC voltage of a periodic square waveform can be changed by changing its duty cycle. This capability is used mainly to control the DC motor, lamp dimming, and any other applications that are controlled by the average DC voltage level.

The original 8051 microcontrollers provide two 16-bit timers (Timers 0 and 1). Timer 2 is added to the 8052 series microcontrollers. The C8051F040 adds Timers 3 and 4 that are identical to Timer 2. The C8051F040 also adds a six-channel programmable counter array (PCA). The capabilities of Timers 0, 1, and 2 and the PCA are listed in Table 8.1.

Both Timers 0 and 1 can be used as timers or counters. A timer can be used to create a time delay, whereas a counter can be used to count external events that occurred within a time interval. Both timers can generate interrupts to the CPU.

TABLE 8.1 *Capability of the C8051F040 Timer System*

Timer Subsystem	Capabilities
Timer 0	Mode 0: 13-bit counter/timer
	Mode 1: 16-bit counter/timer
	Mode 2: 8-bit auto-reload counter/timer
	Mode 3: two separate 8-bit counters
Timer 1	Mode 0-2: identical to those in Timer 0
	Mode 3: holds its contents
Timer 2, 3, and 4	16-bit counter/timer with auto-reload
	16-bit counter/timer with capture
	Toggle clock out
PCA	Edge-triggered capture
	Software timer
	High-speed output
	Frequency output
	8-bit PWM
	16-bit PWM

In addition to being used as a timer/counter, Timers 2, 3, and 4 can be programmed to capture the arrival time of an event (represented by a falling edge only). Timers 2, 3, and 4 also can be programmed to count up or count down in the auto-reload mode.

The PCA of the C8051F040 consists of a free-running timer and six capture/compare modules. Each of these modules supports the capture, compare, and PWM functions described previously.

Signal pins related to timer functions are listed in Table 8.2. The timer pins of the C8051F040 are assigned to port pins by programming the crossbar decoder.

8.3 Timer 0 and Timer 1

Each timer is implemented as a 16-bit register accessed as two separate bytes: a low byte (TL0 and TL1) and a high byte (TH0 and TH1). The timer control register (TCON) is used to enable Timers 0 and 1 as well as indicate their status. Timer 0 interrupts can be enabled by setting the ET0 bit in the IE register. Timer 1 interrupts can be enabled by setting the ET1 bit in the IE register. Both counter/timers operate in one of four modes selected by setting the mode select bits T1M1 through T0M0 in the timer/counter mode register (TMOD). Each timer can be configured independently. The contents of the TCON and TMOD registers are shown in Figures 3.18 and 3.19, respectively. The description in the following subsections applies to both Timers 0 and 1.

There is a need to load a value into the timer registers in many timer applications. Many 8051 variants require the TR0 (or TR1) bit be cleared before

TABLE 8.2 *The Functions of Timer Pins of the C8051F040*

Pin Name	Description
T0	Timer 0 external input
T1	Timer 1 external input
T2	Timer 2 external input
T2EX	Timer/counter 2 capture/reload trigger and direction control
T3	Timer 3 external input
T3EX	Timer/counter 3 capture/reload trigger and direction control
T4	Timer 4 external input
T4EX	Timer/counter 3 capture/reload trigger and direction control
ECI	External clock input to PCA0
CEX0	External I/O for compare/capture module 0
CEX1	External I/O for compare/capture module 1
CEX2	External I/O for compare/capture module 2
CEX3	External I/O for compare/capture module 3
CEX4	External I/O for compare/capture module 4
CEX5	External I/O for compare/capture module 5
ECI1	External clock input to PCA1

loading a value into the timer registers or changing the operation mode of the timer. Otherwise, the behavior of the timer/counter is unpredictable.

8.3.1 Mode 0: 13-bit Counter/Timer

Timers 0 and 1 operate as 13-bit counter/timers in Mode 0. The TH0 register holds the upper 8 bits of the 13-bit counter/register, whereas TL0 holds the lower 5 bits in bit positions TL0.4 through TL0.0. The highest 3 bits of TL0 are indeterminate and should be masked out or ignored when reading. As the 13-bit timer register increments and overflows from 0x1FFF (all ones) to 0x0000, the timer overflow flag TF0 is set, and an interrupt will occur if it is enabled.

The **C/T0** bit of the TMOD register selects the counter/timer's clock source. When C/T0 is set to 1, the high-to-low transition on the T0 pin increments the timer register. In timer mode, the original 8051 uses $f_{OSC}/\mathbf{12}$ as the timer clock input, whereas the C8051F040 allows the user to choose from the system clock and four prescaled clock sources. The C8051F040 uses the CKCON register to choose the desired clock input for Timers 0 and 1.

The clock defined by the T0M (CKCON.3) bit is selected if the C/T0 bit is set to 0. When the T0M bit is set, Timer 0 is clocked by the system clock. When T0M is cleared, Timer 0 is clocked by the source selected by the clock scale bits in the CKCON register. The contents of the CKCON register of the C8051F040 are shown in Figure 3.20.

Setting the TR0 (TCON.4) bit enables the timer when either GATE0 is logic 0 or the input signal $\overline{\text{INT0}}$ is at logic-level 1. Setting GATE0 to 1 allows the timer to be controlled by the external input signal $\overline{\text{INT0}}$, facilitating pulse-width measurements.

Setting the TR0 bit (TCON.4) does not reset the timer. The timer registers should be loaded with the desired initial value before the timer is enabled.

TL1 and TH1 form the 13-bit register for Timer 1 in the same manner as described previously for TL0 and TH0. Timer 1 is configured and controlled using the relevant TCON and TMOD bits just as with Timer 0. The block diagram of Timer 0 Mode 0 is shown in Figure 8.1a and b.

8.3.2 Mode 1: 16-bit Counter/Timer

In this mode, THx (x = 0 or 1) and TLx are cascaded into a single 16-bit timer. The counters/timers are enabled and configured in Mode 1 in the same manner as for Mode 0. The block diagram of Mode 1 for Timers 0 and 1 are identical to those in Figure 8.1.

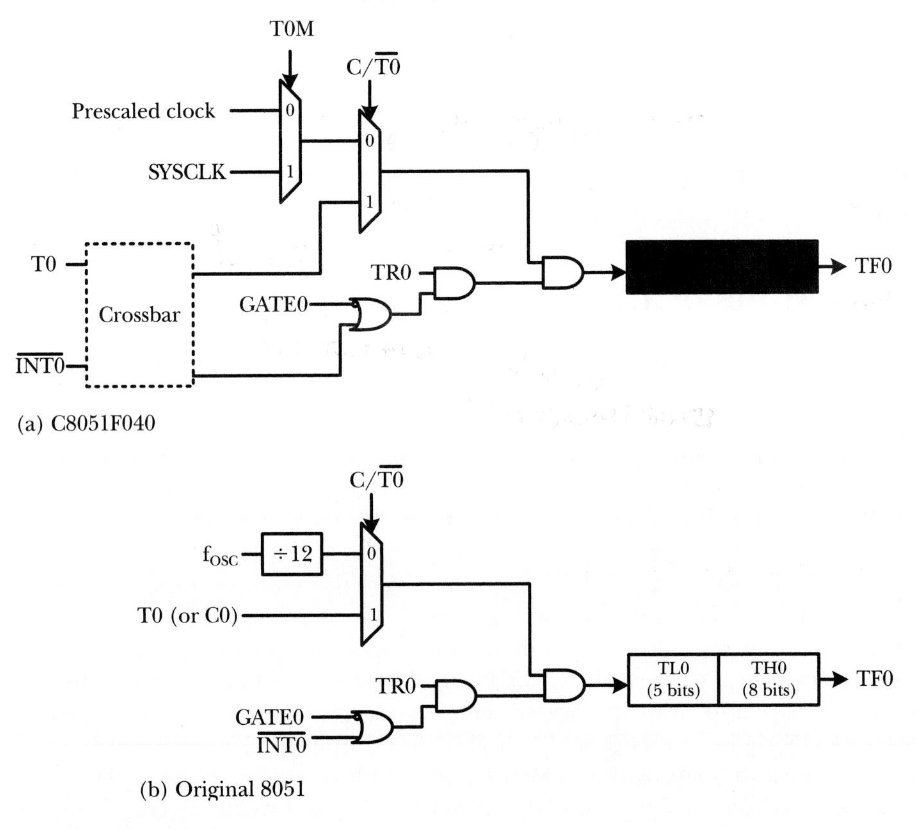

FIGURE 8.1
Timer 0 Mode 0 block diagram.

8.3.3 Mode 2: 8-bit Counter/ Timer with Auto-Reload

As shown in Figure 8.2a and b, Timers 0 and 1 are configured as 8-bit counter/timers (TLx, $x = 0$, or 1) with automatic reload in Mode 2. An overflow from TLx not only sets TFx, but also reloads TLx with the contents of THx. The reload leaves THx unchanged. If Timer 0 or Timer 1 interrupts are enabled, an interrupt will be requested when the counter TLx is reloaded from THx.

8.3.4 Mode 3: Two 8-bit Counter/ Timers (Timer 0 only)

In Mode 3, Timer 1 holds its contents, whereas Timer 0 is configured as two separate 8-bit counters/timers held in TL0 and TH0. TL0 is controlled using the Timer 0 control/status bits in TCON and TMOD: TR0, C/$\overline{\text{T0}}$, GATE0,

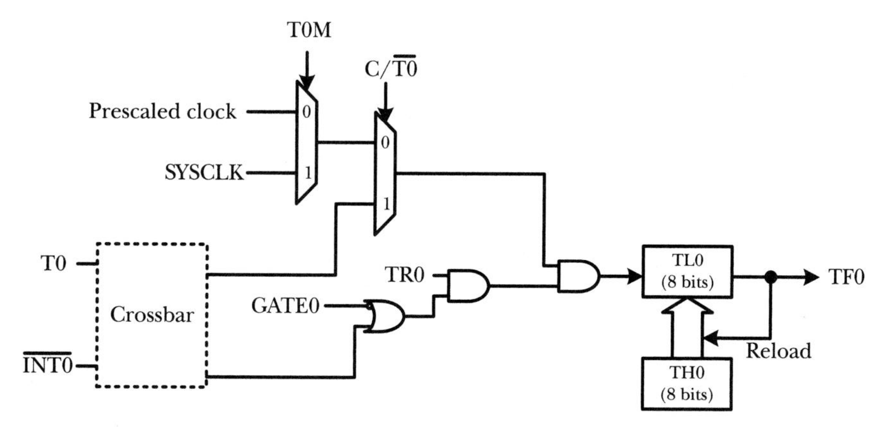

(a) C8051F040

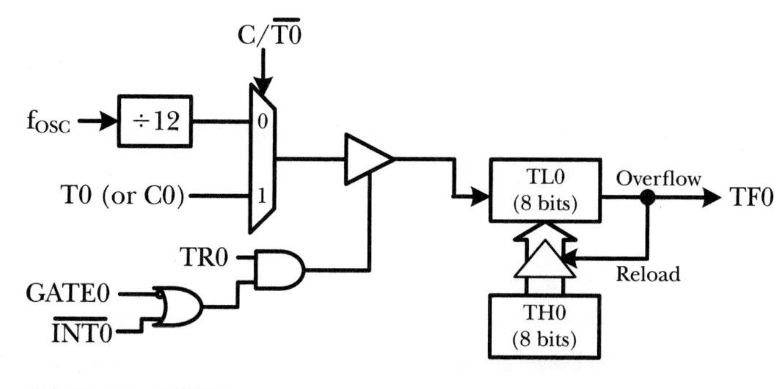

(b) Original 8051

FIGURE 8.2
Timer 0 Mode 2 block diagram.

and TF0. TL0 can use either the system clock or an external input signal as its clock source. The TH0 is restricted to a timer function clocked by the system clock or prescaled clock. TH0 is enabled using the Timer 1 control bit TR1. TH0 sets the TF1 flag on overflow and thus controls the Timer 1 interrupt.

8.3.5 Using Timer 0 and Timer 1 in Measuring Signal Parameters

Using Timer 0 (or 1) to create time delays has been discussed in Chapter 3. Timers 0 and 1 also can be used to measure signal pulse width and frequency.

Pulse-Width Measurement

The signal pulse width can be measured by using Mode 1 of Timer 0 or Timer 1. The unknown signal should be connected to the $\overline{\text{INT0}}$ or $\overline{\text{INT1}}$ pin, depending on which timer is unused. Timer registers should be cleared to 0 before the measurement is started. Timer 0 or Timer 1 is allowed to count when the signal is high. If the pulse duration is shorter than 2^{16} clock cycles, then the pulse width (in timer clock cycles) is equal to the timer final count. Otherwise, the number of timer overflows must be kept track of. The pulse width of the unknown signal is given by the expression

$$\textbf{Pulse width} = (\text{timer overflower count} \times 2^{16} + \text{timer count}) \text{ timer cycles} \quad (8.1)$$

Since the falling edge of the $\overline{\text{INT0}}$ or $\overline{\text{INT1}}$ may interrupt the CPU, it can be used to terminate the pulse-width measurement. Immediately before the pulse-width measurement is started, the main program sets the interrupt count to 1 and enables either the INT0 or INT1 interrupt. The main program simply waits for the interrupt count to be decremented to 0 by the $\overline{\text{INT0}}$ or INT1 interrupt service routine and computes the signal pulse width. This idea is implemented in Example 1.

Example 8.1 Write a program to measure the pulse width (use the clock count as its unit) of the signal connected to the INT1 pin, assuming that the C8051F040 microcontroller is controlled by an oscillator running at 24 MHz.

Solution: The overflow count of Timer 1 is recorded in the R6:R7 register pair in this example.

```
            $include    (c8051F040.inc)
intCnt      set         R3              ; INT1 interrupt count
pwLo        set         R4              ; low byte of pulse width
pwHi        set         R5              ; high byte of pulse width
t1ovCntLo   set         R6              ; use R6 & R7 as the Timer 1 overflow count
t1ovCntHi   set         R7              ;   "
```

```
            org     00H
            ljmp    start
            org     0x13                ; INT1 interrupt vector
int1ISR:    dec     intCnt              ; decrement the INT1 interrupt count
            reti
            org     0x1B                ; Timer 1 overflow interrupt vector
t1ISR:      ljmp    incOvcnt            ; jump to increment Timer 1 overflow count

            org     0xAB                ; starting address of the program
start:      lcall   sysinit
            mov     TL1,#0              ; initialize timer 1 register to 0
            mov     TH1,#0              ;     "
            mov     t1ovCntLo,#0        ; initialize Timer 1 overflow count to 0
            mov     t1ovCntHi,#0        ;
            mov     intCnt,#1           ; initialize INT1 interrupt count to 1
            mov     TMOD,#0x90          ; configure Timer 1 to gated, timer, mode 1
            mov     CKCON,#0            ; select system clock divided by 12 as clock source
            setb    IT1                 ; choose falling edge interrupt for INT1
            orl     IE,#0x8C            ; enable INT1 and Timer 1 overflow interrupts
            clr     TF1                 ; clear Timer 1 overflow flag
            setb    TR1                 ; enable Timer 1 to run
measure:    mov     A,intCnt            ; check if the INT1 falling edge arrived
            jnz     measure
            mov     pwLo,TL1            ; save lower byte of pulse width
            mov     pwHi,TH1            ; save upper byte of pulse width
            nop
            clr     TR1                 ; disable Timer 1
forever:    jmp     forever             ; stay in infinite loop
;*************************************************************************************************************
; The following routine performs system initialization that includes disabling watchdog timer,
; select external oscillator as SYSCLK, assign peripheral signals to I/O pins, and select
; 3-wire SPI mode.
;*************************************************************************************************************
sysinit:    mov     SFRPAGE,#CONFIG_PAGE ; switch to the SFR page that contains
                                        ; registers that configures the system
            mov     WDTCN,#0xDE         ; disable watchdog timer
            mov     WDTCN,#0xAD         ;     "
            mov     CLKSEL,#0           ; used internal oscillator to generate SYSCLK
            mov     OSCXCN,#0x67        ; configure external oscillator control
            mov     R7,#255
            djnz    R7,$                ; wait for about 1 ms
chkstable:  mov     A,OSCXCN            ; wait until external crystal oscillator is stable
            anl     A,#80H              ; before using it
            jz      chkstable           ;     "
            mov     CLKSEL,#1           ; use external crystal oscillator to generate SYSCLK
            mov     XBR0,#0xF7          ; assign I/O pins to all peripheral functions
            mov     XBR1,#0xFF          ; and enable crossbar
            mov     XBR2,#0x5D          ;     "
            mov     XBR3,#0x8F          ;     "
            mov     SFRPAGE,#SPI0_PAGE  ;
```

```
                mov     SPI0CN,#1           ; select 3-wire SPI mode
                ret
; ***********************************************************************************************
; Timer 1 overflow interrupt service routine.
; ***********************************************************************************************
incOvcnt:       clr     TF1                 ; clear the TF1 flag to avoid repeated interrupt
                mov     A,t1ovCntLo         ; increment Timer overflow count by 1
                add     A,#1                ;     "
                mov     t1ovCntLo,A         ;     "
                mov     A,t1ovCntHi         ;     "
                addc    A,#0                ;     "
                mov     t1ovCntHi,A         ;     "
                reti
                end
```

The C language version of the program is as follows.

```
#include <C8051F040.h>
void int1ISR(void);
void T1ISR(void);
char intCnt;
unsigned long pw;
unsigned long int T1OvCnt;
void sysinit(void);
void main(void)
{
    sysinit();      // use external oscillator to generate SYSCLK, fSYSCLK = 24 MHz
    TL1     = 0;
    TH1     = 0;
    T1OvCnt = 0;
    intCnt  = 1;                    // there is only one falling edge of INT1 pin to wait
    TMOD    = 0x90;                 // Timer 1 gated, mode 1 operation
    CKCON   = 0;                    // use SYSCLK / 12 as clock source
    IT1     = 1;                    // choose falling edge interrupt for INT1
    IE      |= 0x8C;                // enable INT1 and Timer 1 overflow interrupts
    TF1     = 0;
    TR1     = 1;                    // start Timer 1
    while (intCnt);                 // wait until the falling edge of INT1 arrives
    TR1     = 0;                    // stop the timer
    pw      = 65536*T1OvCnt + 256*(unsigned long)TH1 + (unsigned long)TL1;
    while(1);
}
void sysinit(void)
{
    int n;
    SFRPAGE = CONFIG_PAGE;
    WDTCN = 0xDE;                   // disable watchdog timer
    WDTCN = 0xAD;                   //    "
    OSCXCN = 0x67;                  // start external oscillator; 24 MHz Crystal
                                    // system clock is 24 MHz
```

```
        for (n = 0;n < 255; n++);           // delay about 1 ms
        while ((OSCXCN & 0x80) == 0); // wait for oscillator to stabilize
        CLKSEL  |= 0x01;                    // switch to external oscillator
        XBR2    = 0x5D;                     // enable crossbar and assign I/O pins to all
        XBR0    = 0xF7;                     // peripheral signals,
        XBR1    = 0xFF;                     //     "
        XBR3    = 0x8F;                     //     "
        SFRPAGE = SPI0_PAGE;
        SPI0CN  = 0x01;                     // enable 3-wire SPI (make sure SPI uses 3 pins
}
void int1ISR(void) interrupt 2
{
        intCnt--;                           // IE1 is cleared when this service routine is entered (edge-triggered)
}
void T1ISR(void) interrupt 3
{
        T1OvCnt++;                          // T1F is cleared automatically when this service is entered
}
```

Frequency Measurement. The frequency of an unknown signal can be measured by using Timer 0 (or 1) to create a delay of one second and use Timer 1 (or 0) to count the number of rising (or falling) edges arrived at during this interval.

Example 8.2 Write a program to measure the frequency of a signal by using Timers 1 and 0. Make the program general so that it can measure the signal with frequency higher than 65536. The program is to be run on a **C8051F040** demo board with a 24-MHz crystal oscillator.

Solution: Connect the unknown signal to the T1 pin and configure Timer 1 as follows:

- Select T1 pin as the clock input to Timer 1
- Configure Timer 1 to operate as a counter in Mode 1 without an external gating signal
- Clear Timer 1 register to 0
- Configure Timer 0 to operate as a timer in Mode 1 without an external gating signal
- Use the oscillator output divided by 48 as the clock source to Timer 0

The one-second delay can be created by letting Timer 0 to count up from 15536 and overflow 10 times. Each overflow creates a delay of 100 ms. The frequency of the signal is given by the following expression.

$$\text{Frequency} = \text{Timer1 overflow count} \times 2^{16} + \text{timer count} \quad (8.2)$$

The program is as follows.

```
            $include  (C8051F040.inc)
t0ovCnt     set       R4                   ; Timer 0 overflow count
t1ovCnt     set       R5                   ; Timer 1 overflow count
freqHi      set       R6                   ; lower byte of frequency
freqLo      set       R7                   ; upper byte of frequency
            org       00H
            ljmp      start
            org       0x1B
T1_ISR:     clr       TF1                  ; Timer 1 interrupt service routine
            inc       t1ovCnt
            reti
            org       0xAB                 ; starting address of the main program
start:      mov       SP,#0x80             ; set up stack pointer
            lcall     sysinit
            mov       SFRPAGE,#TIMER01_PAGE
            mov       CKCON,#02            ; Timer 0 clock = SYSCLK/48
            mov       TMOD,#0x51           ; configure Timer 1, Timer 0 to counter and timer mode 1
            mov       t0ovCnt,#10          ; initialize Timer 0 overflow count to 10
            mov       TH1,#0               ; initialize Timer 1 register to 0
            mov       TL1,#0               ;          "
            clr       ET1                  ; disable Timer 1 interrupt
            clr       ET0                  ; disable Timer 0 interrupt
            setb      TR1                  ; enable Timer 1 to count
            mov       IE,#0x88             ; enable Timer 1 interrupt
again:      clr       TR0                  ; stop Timer 0
            mov       TH0,#high 15536      ; place 15536 in Timer 0 so it overflows in 100 ms
            mov       TL0,#low 15536       ;      "
            clr       TF0                  ; clear overflow flag
            setb      TR0                  ; enable Timer 0 to operate
wait:       jnb       TF0,wait             ; wait until TF0 is set to 1
            djnz      t0ovCnt,again        ; wait until 100 ms delay is repeated 10 times
            clr       EA                   ; disable interrupt
            clr       EA
            mov       freqLo,TL1           ; save the frequency
            mov       freqHi,TH1           ;      "
            nop
forever:    jmp       forever
;-----------------------------------------------------------------------------------------
; Include the sysinit subroutine here.
;-----------------------------------------------------------------------------------------
```

The C language version of the program is as follows.

```
#include        <C8051F040.h>
void sysinit(void);
void t1_ISR (void);
unsigned char t1ovCnt;
void main (void)
```

```
{
    char i;
    long int freq;
    sysinit();
    SFRPAGE = TIMER01_PAGE;
    TMOD    = 0x51;          // configure Timer 1, Timer 0 to counter and timer mode 1
    CKCON   = 0x02;          // use SYSCLK/48 as the clock source of Timer 0
    TH1     = 0x00;          // Timer 1 count up from 0
    TL1     = 0x00;
    IE      = 0x88;          // enable Timer 1 interrupt
    ET0     = 0;             // disable Timer 0 interrupt
    for (i = 0; i < 10; i++){
        TF0 = 0;
        TR0 = 0;             // stop Timer 0
        TH0 = 0x3C;          // let Timer 0 count up from 15536 so it overflows in 100 ms
        TL0 = 0xB0;          //    *
        TR0 = 1;             // start Timer 0
        while (!TF0);        // wait for 100 ms
    }
    freq = (long)t1ovCnt * 65536 + 256 * (long)TH1 + (long)TL1;
}
void t1_ISR (void) interrupt 3
{
    TF1 = 0;
    t1ovCnt++;
}
//-----------------------------------------------------------------
// include the sysinit() function here.
//-----------------------------------------------------------------
```

8.4 Timer 2, Timer 3, and Timer 4

Timer 2 is a 16-bit timer and is provided in the original 8052 and all the 8051 variants. The 8051 variants from SiLabs also add Timers 3 and 4, which are identical to Timer 2. These three timers can operate either as a timer or as an event counter. In addition, they provide auto-reload, capture, and toggle output modes with the capability to count up or down. For the C8051F040, capture mode and auto-reload mode are selected using bits in the Timer n control registers (TMRnCN, n = 2, 3, or 4). Toggle output mode is selected using the Timer n configuration registers (TMRnCF, n = 2, 3, or 4). These timers also may generate a square wave at an external pin. For the original 8051 and some 8051 variants, these two registers are referred as T2CON and T2MOD. The contents of TMRnCR and T2CON are shown in Figure 8.3, whereas the contents of TMRnCF and T2MOD are shown in 8.4.

7	6	5	4	3	2	1	0	
TFn	EXFn	–	–	EXENn	TRn	C/Tn	CP/RLn	TMRnCN reset:00h
TF2	EXF2	RCLK	TCLK	EXEN2	TR2	C/T2	CP/RL2	T2CON reset:00h

TFn (n = 2, 3, or 4): Timer n overflow/underflow flag
  0 = no overflow or underflow occurred
  1 = Timer n overflow or underflow (timer/counter count from 0x0000 to 0xFFFF) ocurred
EXFn (n = 2, 3, or 4): Timer 2, 3, or 4 external flag
  This flag is set when either a capture or reload is caused by a high-to-low transition on the TnEX pin and EXENn is logic one. This flag must be cleared by software
RCLK: Receive clock bit
  0 = use Timer 1 overflow as receive clock for serial port in Mode 1 or 3
  1 = use Timer 2 overflow as receive clock for serial port in Mode 1 or 3
TCLK: Transmit clock bit
  0 = use Timer 1 overflow as transmit clock for serial port in Mode 1 or 3
  1 = use Timer 2 overflow as transmit clock for serial port in Mode 1 or 3
EXENn (n = 2, 3, or 4): Timer n external enable
  0 = transitions on the TnEX pin are ignored
  1 = transitions on the TnEX pin cause capture, reload, or control the direction of timer count (up or down) as follows:
    Capture mode: 1-to-0 transition on TnEX pin causes RCAPnH:RCAPnL to capture count value
    Auto-reload mode:
      DCEN = 0: 1-to-0 transition of TnEX pin causes reload of timer and sets ExFn flag
      DCEN = 1: TnEX logic level controls direction of timer (up or down)
TRn (n = 2, 3, or 4): Timer n run control
  0 = Timer n disabled
  1 = Timer n enabled
C/Tn (n = 2, 3, or 4): Counter/timer select
  0 = timer function
  1 = counter function: timer incremented by the falling edge of Tn pin
CP/RLn (n = 2, 3, or 4): Capture/reload select
  0 = timer is in reload mode
  1 = timer is in capture mode

FIGURE 8.3
Contents of the TMRnCN and T2CON registers.

Timers 2, 3, and 4 can use either the system clock (divided by 1, 2, or 12), external clock (divided by 8) or transitions on an external input pin as its clock source. The counter/timer select bit **C/Tn** (TMRnCN.1) configures the peripheral as a counter or timer. When **C/Tn** is set to 1, a high-to-low transition at the **Tn** input pin increments the counter/timer register.

7	6	5	4	3	2	1	0	
–	–	–	TnM1	TnM0	T0Gn	Tn0E	DCEN	TMRnCF reset:00h
–	–	–	–	–	–	T20E	DCEN	T2MOD reset:00h

TnM1-TnM0 (n = 2, 3, or 4): Timer clock mode select bits
00: select SYSCLK/12 as the Timer n clock source
01: select SYSCLK as the Timer n clock source
10: select External Clock/8 as the Timer n clock source
11: select SYSCLK/2 as Timer n clock source
TOGn (n = 2, 3, or 4): Toggle output state bit
When timer is used to toggle a port pin, this bit can be used to read the state of the output, or can be written to in order to force the state of the output
TnOE (n = 2, 3, or 4): Timer output enable bit
0 = output of toggle mode not available at timer's assigned port pin
1 = output of toggle mode available at timer's assigned port pin
DCEN: Decrement enable bit
0 = timer will count up, regardless of the state of TnEX
1 = timer will count up or down depending on the state of TnEX as follows:
if TnEX = 0, the timer counts down
if TnEX = 1, the timer counts up

FIGURE 8.4
Contentsof the TMRnCF and T2MOD registers.

8.4.1 Configuring Timers 2, 3, and 4 to Count Down

Timers 2, 3, and 4 have the ability to count down. Setting the DCEN bit of the TMRnCF (or T2MOD) register allows the timer to count up or count down with the counting direction determined by the logic level of the TnEX pin. When TnEX = 1, the counter/timer will count up; when TnEX = 0, the counter/timer will count down. To use this feature, the TnEX pin must be enabled in the digital crossbar and configured as a digital input (there is no need to do this in the original 8052). When DCEN = 1, other functions of the TnEX input (i.e., capture and auto-reload) are not available. TnEX will control only the direction of the timer when DCEN = 1.

8.4.2 Timers 2, 3, and 4 Capture Mode

The block diagram of Timers 2, 3, and 4 in capture mode of the original 8052 and the C8051F040 are shown in Figure 8.5 and 8.6, respectively. In capture mode, Timer n will operate as a 16-bit counter/timer with capture capability. The capture mode is entered when the timer external-enable bit (EXENn) is

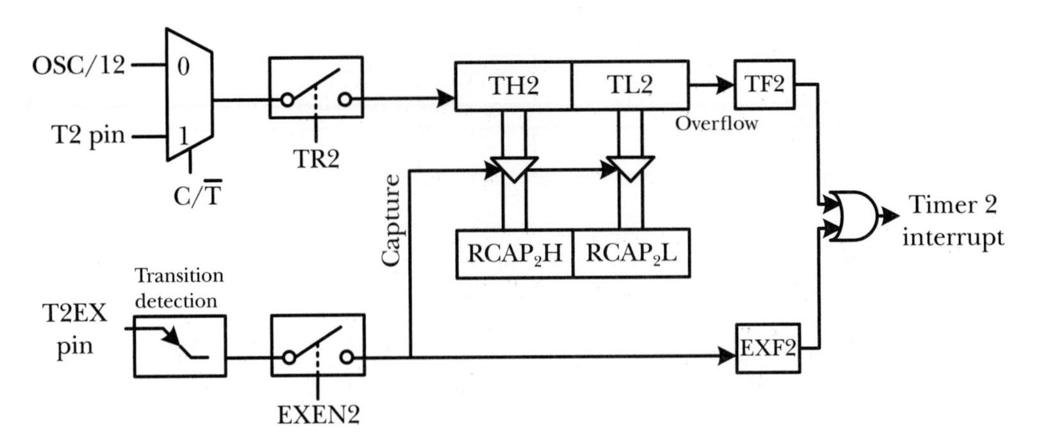

FIGURE 8.5
Timer 2 in capture mode (original 8052 with courtesy of Intel).

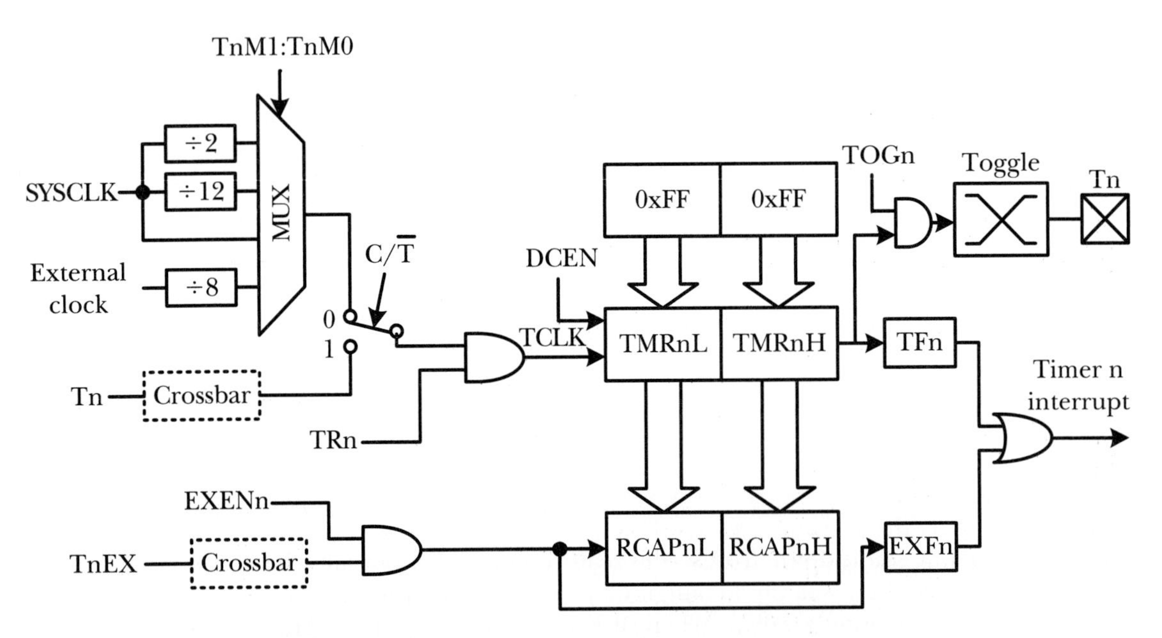

FIGURE 8.6
Timers 2, 3, or 4 in capture mode block diagram (C8051F040).

set to 1. A falling edge on the TnEX input pin causes the 16-bit value in the associated timer (TMRnH, TMRnL) to be copied into the capture registers (RCAPnH, RCAPnL) and the timer external flag (**EXFn**) to be set to 1. An interrupt will occur if it is enabled. If the EXENn bit is 0, then Timer 2 (or 3 or 4) operates as a simple 16-bit count up timer. The **TFn** (2, 3, or 4) flag will be set to 1 when timer overflows.

The **TFn** flag will be set to 1 whenever the counter/timer rollovers from 0xFFFF to 0x0000 (in count mode) or from 0x0000 to 0xFFFF (in count down mode). An interrupt will be requested if it is enabled.

Capture mode is selected by setting the CP/RLn, TRn, and EXENn bits of the TMRnCF (or T2MOD) register to 1.

8.4.3 Timers 2, 3, and 4 Auto-Reload Mode (Up or Down Counter)

In auto-reload mode, the counter/timer of Timers 2, 3, and 4 of the C8051F040 and many 8051 variants can be configured to count up or count down and cause an interrupt when the timer overflows or underflows. When counting up and the timer overflows, the values in the reload/capture register pair (RCAPnH:RCAPnL) will be reloaded into the timer, and counting is resumed. When the EXENn bit is set to 1 and the DCEN bit is 0, a falling edge on the TnEX pin will cause a timer reload (in addition to timer overflows causing auto-reload). When the DCEN bit is set to 1, the state of the TnEX pin controls whether the counter/timer counts up or down and will not cause an auto-reload or interrupt event.

When in count down mode, the counter/timer will set its **TFn** flag when the timer value equals the 16-bit value of the reload/capture registers. This will cause the value 0xFFFF to be reloaded into the timer.

The auto-reload mode is selected by clearing the CP/RLn bit. Setting the TRn bit to 1 starts the timer.

In auto-reload mode, the external flag **EXFn** toggles upon every overflow or underflow and does not cause an interrupt. The **EXFn** flag can be thought of as the most significant bit of a 17-bit counter. The block diagram of Timers 2, 3, and 4 in auto-reload mode is shown in Figure 8.7a, b, and c.

8.4.4 Toggle (Programmable) Clock-Out Mode

Timer 2 of the original 8052 provides a programmable clock-out signal from the T2 pin. To use this mode, the C/T2 bit must be cleared; the TR2 bit and the T2OE bit must be set. In this mode, Timer 2 counts up from the value stored in RCAP2H:RCAP2L. When Timer 2 overflows, the T2 level will be toggled, and Timer 2 will be reloaded with the value in RCAP2H:RCAP2L. In this mode, the rollovers of Timer 2 will not generate interrupts. The frequency of the T2 output is given by the expression

$$\textbf{Clock-out frequency} = f_{OSC} \div (4 \times (65536 - \text{RCAP2H:RCAP2L})) \quad (8.2)$$

For the C8051F040, Timers 2, 3, and 4 also have the capability to toggle the state of their associated output port pins (T2, T3, or T4) to produce a 50 percent duty-cycle waveform output. The state of the timer pin toggles

FIGURE 8.7 *Timers in auto-reload mode.*

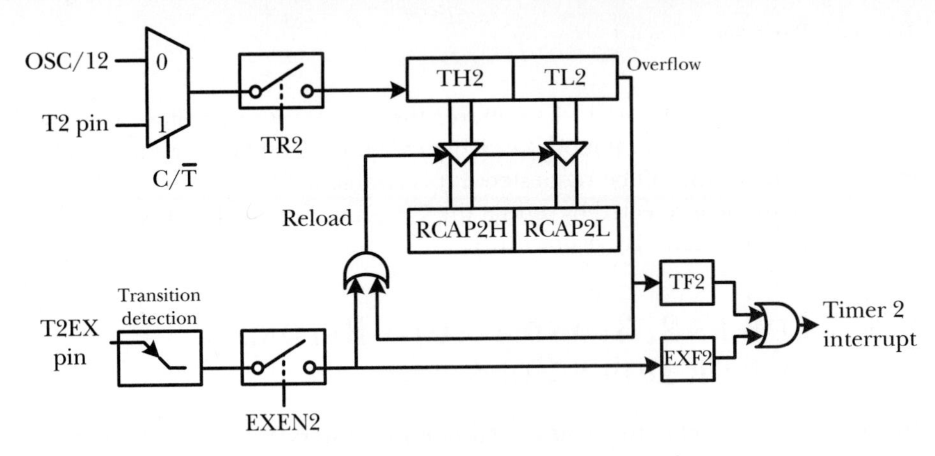

(a) Timer 2 in auto−reload mode (DCEN = 0, original 8052) (Courtesy of Intel)

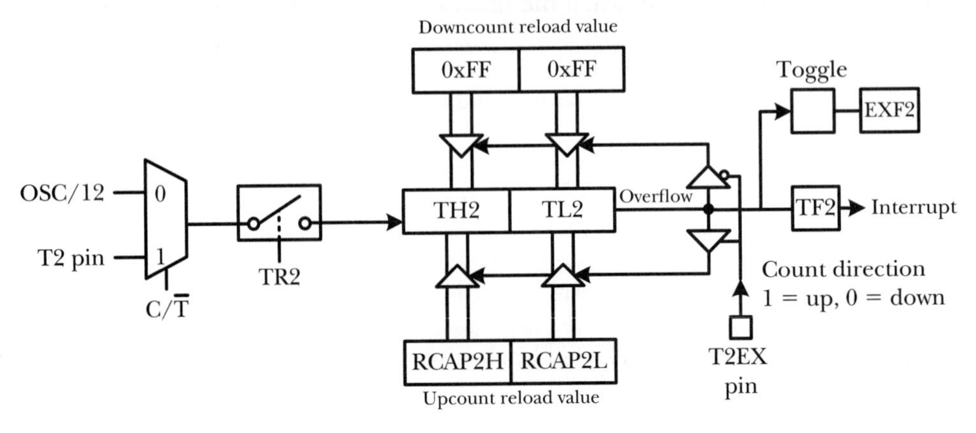

(b) Timer 2 in auto−reload mode (DCEN = 1, original 8052) (Courtesy of Intel)

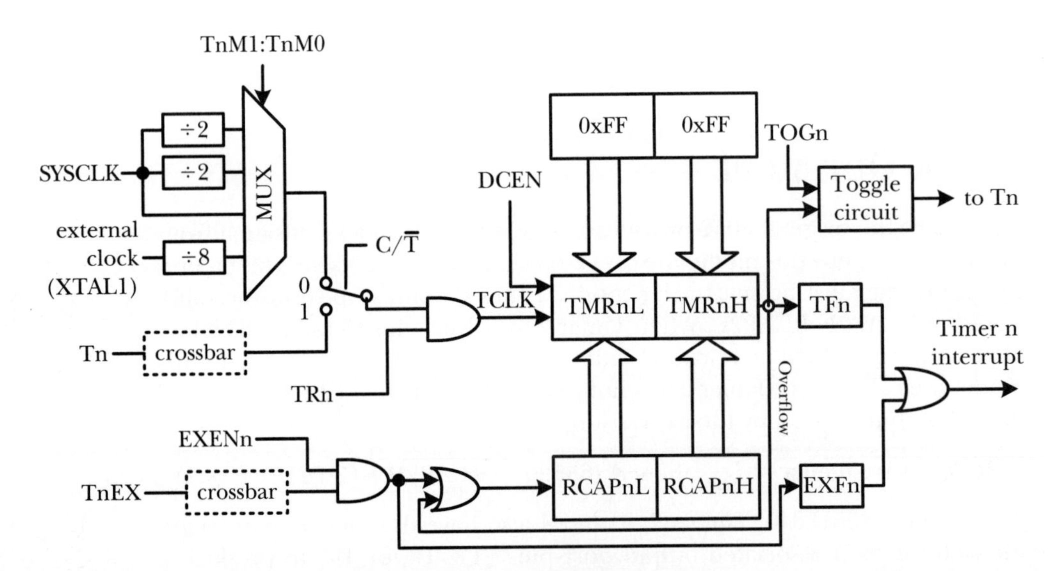

(c) Timer 2,3, or 4 in auto−reload mode updown counter diagram (C8051F040)

whenever the associated timer overflows (count up) or underflows (count down). The toggling frequency is determined by the clock source of the timer and the value loaded into registers RCAPnH and RCAPnL. When counting up, the auto-reload value for the timer is from RCAPnH:RCAPnL, and auto-reload will occur when the value in the timer reaches 0xFFFF. When counting down, the auto-reload value is 0xFFFF, and the underflow will occur when the value in the timer matches the value stored in RCAPnH:RCAPnL.

The circuit block diagram of Timers 2, 3, and 4 in toggle-output mode for the C8051F040 is shown in Figure 8.7c, which is controlled by the timer overflow or underflow. The frequency of the square waveform output is given by the expression

$$f_{SQ} = f_{TCLK} \div (2 \times (2^{16} - \mathbf{RCAPnH{:}RCAPnL})) \quad (8.3)$$

where f_{TCLK} is the frequency of the Timer 2 clock source (shown in Figure 8.6). The value to be loaded into RCAPnH:RCAPnL is given by

$$\mathbf{RCAPnH{:}RCAPnL} = 2^{16} - f_{TCLK} \div (2 \times f_{SQ}) \quad (8.4)$$

To use this mode to generate square wave, clear the **CP/RLn** and **C/Tn** bits, sets the **TnOE** bit and load an appropriate value into the RCAPnH: RCAPnL register pair. The counter/timer overflow/underflow rate should be set to half of that of the waveform frequency. The state of the timer-output pin can be read back or forced by reading or writing the TOGn bit (bit 2) of the TMRnCF register.

8.4.5 Timer 2 Baud Rate Generation Mode

In the original 8051, Timer 2 is selected as the baud rate generator for the UART by setting the RCLK and/or TCLK bits in the T2CON register. When RCLK = 1, the serial port uses Timer 2 overflow pulses the receive clock in serial port Modes 1 and 3. RCLK = 0 causes Timer 1 overflow pulses to be used for the receive clock. When TCLK = 1, the serial port uses Timer 2 overflow pulses to be used for the transmit clock. This subject will be discussed in more detail in Chapter 9.

8.4.6 Applications of Timer 2

A few application examples of Timer 2 are given in this section.

Period Measurement

In addition to creating time delays and measuring frequencies, Timer 2 also can be used to measure the period of an unknown signal. The period of a signal can be measured by capturing two consecutive falling edges of a signal, as shown in Figure 8.8.

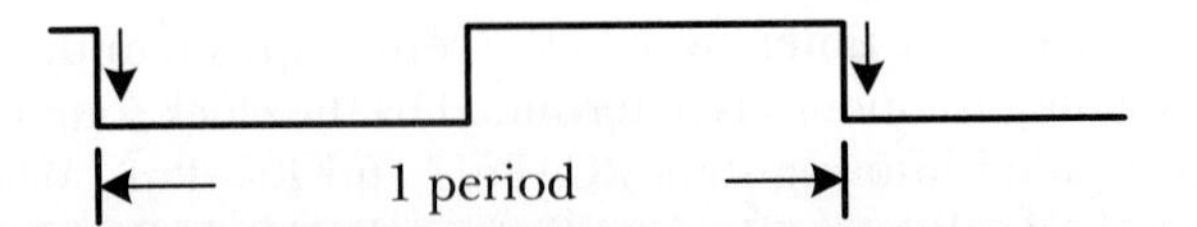

FIGURE 8.8
Period measurement.

The difference of the two captured values gives the period in number of timer clock cycles. For low-frequency signals, the timer may have overflowed many times before the second falling edge is captured. Timer overflows must be kept track of in order to obtain a correct measurement of the period.

Every timer overflow corresponds to 2^{16} timer counts. Let **ovcnt**, **1stedge**, **2ndedge** be the timer overflow count, first captured edge, and second captured edge, respectively, then

$$\text{Period} = 2^{16} \times \text{ovcnt} + \text{2ndedge} - \text{1stedge} \quad (\text{2ndedge} > \text{1stedge}) \tag{8.5}$$

Otherwise,

$$\text{Period} = 2^{16} \times (\text{ovcnt} - 1) + \text{2ndedge} - \text{1stedge} \quad (\text{1stedge} > \text{2ndedge}) \tag{8.6}$$

Example 8.3 Use Timer 2 of the C8051F040 to measure the period of an unknown signal that is connected to the T2EX pin. Assume that the C8051F040 is using a 24-MHz external crystal oscillator to generate the system clock.

Solution: For this application, Timer 2 must be enabled to capture mode. Write the value of 0x0D into the TMR2CN register. The procedure is as follows.

Step 1 ovcnt ← 0. Wait for the first edge to arrive.

Step 2 TH2 ← 0, TL2 ← 0, enable Timer 2 interrupt, and enable T2EX pin interrupt.

Step 3 When Timer 2 interrupt occurs, clear TF2 to 0 and increment **ovcnt** by 1.

Step 4 When the second edge of the T2EX pin arrives, stop period measurement and calculate the period.

The program for measuring the period is as follows.

```
            $nomod51                    ; disable default SFR definition to avoid redefinition
            $include (c8051F040.inc)    ; in uVision
periodH     set     R1                  ; period high byte
periodL     set     R2                  ; period low byte
tov2Cnt     set     R0                  ; Timer 2 overflow count
```

```
edgeCnt     set     R3                   ; T2EX edge count
            org     0x00
            ajmp    start
            org     0x2B                 ; Timer 2 interrupt vector
            ljmp    t2ISR
            org     0x00AB
start:      mov     SP,#0x80             ; establish stack pointer
            lcall   sysinit
            mov     tov2Cnt,#0           ; initialize Timer 2 overflow count to 0
            mov     IE,#0                ; disable all interrupts
            mov     TMR2CN,#0x0D         ; enable TMR2 in capture mode
edge1:      jnb     EXF2,edge1           ; wait for the first falling edge to arrive
            mov     periodL,RCAP2L       ; save the first edge
            mov     periodH,RCAP2H       ;     "
            clr     EXF2                 ; clear the flags
            clr     TF2                  ;
            mov     IE,#0xA0             ; enable Timer 2 overflow interrupt
edge2:      cjne    edgeCnt,#0,edge2     ; wait for the 2nd falling edge
            mov     A,RCAP2L
            clr     C
            subb    A,periodL            ; compute edge2 - edge1
            mov     periodL,A            ; save the difference
            mov     A,RCAP2H             ;     "
            subb    A,periodH            ;     "
            mov     periodH,A            ;     "
            jnc     done                 ; edge2 is larger
            dec     tov2Cnt              ; second edge is smaller
done:       jmp     $                    ; infinite loop
;*************************************************************************************************
; The following function performs system initialization.
;*************************************************************************************************
sysinit:    mov     SFRPAGE,#0x0F
            mov     WDTCN,#0xDE          ; disable watchdog timer
            mov     WDTCN,#0xAD          ;     "
            mov     CLKSEL,#0            ; used internal oscillator to generate SYSCLK
            mov     OSCXCN,#0x67         ; configure external oscillator control
            mov     R7,#255
againosc:   djnz    R7,againosc          ; wait for about 1 ms
chkstable:  mov     A,OSCXCN             ; wait until external crystal oscillator is stable
            anl     A,#80H               ; before using it
            jz      chkstable            ;     "
            mov     CLKSEL,#1            ; use external crystal oscillator to generate SYSCLK
            mov     XBR0,#0xF7           ; assign I/O pins to all peripheral
            mov     XBR1,#0xFF           ; signals and enable crossbar
            mov     XBR2,#0x5D           ; decoder
            mov     XBR3,#0x8F           ;     "
            mov     SFRPAGE,#0           ; switch to SFR page 0
            mov     SPI0CN,#01           ; enable 3-wire SPI
            ret
```

```
; ****************************************************************************************
; Timer 2 interrupt service routine.
; ****************************************************************************************
t2ISR:      jnb     TF2,nextF           ; check EXF2 flag
            clr     TF2
            inc     tov2Cnt
            reti
nextF:      clr     EXF2
            dec     edgeCnt
            reti
            end
```

The C language version of the program is as follows.

```
#include <c8051F040.h>  // compiled using the SDCC compiler
unsigned char tov2Cnt;      // timer 2 overflow count
unsigned char periodH, periodL, edgeCnt;
unsigned long period;
void t2ISR (void) interrupt 5
{
    if (TF2) {
        tov2Cnt ++;
        TF2 = 0;
    }
    else {
        edgeCnt --;
        EX2F = 0;
    }
}
void sysInit (void)
{
    unsigned char n;
    SFRPAGE = 0x0F;
    WDTCN  = 0xDE;                     // disable watchdog timer
    WDTCN  = 0xAD;
    OSCXCN = 0x67;                     // start external oscillator; 24 MHz Crystal
                                       // system clock is 24 MHz
    for (n = 0; n < 255; n ++);        // delay about 1 ms
    while ((OSCXCN & 0x80) == 0);      // wait for oscillator to stabilize
    CLKSEL  |= 0x01;                   // switch to external oscillator
    XBR0    = 0xF7;                    // assign all peripheral function signals to port pins
    XBR1    = 0xFF;                    //      "
    XBR2    = 0x5D;                    //      "
    XBR3    = 0x8F;                    //      "
    SFRPAGE = 0;                       // switch to SFR page 0
    SPI0CN  = 0x01;                    // enable and configure SPI to 3-wire mode
}
void main (void)
{
```

```
    sysInit();
    edgeCnt  = 1;
    tov2Cnt  = 0;                   // initialize timer 2 overflow count to 0
    TMR2CN = 0x0D;                  // configure Timer 2
    IE        = 0;                  // disable all interrupt
    while(!EXF2);                   // wait for the first falling edge on T2EX pin
    periodL  = RCAP2L; // save the first edge
    periodH  = RCAP2H; //      "
    EXF2     = 0;
    TF2      = 0;
    IE        = 0xA0;               // enable Timer 2 interrupt
    while(edgeCnt);                 // wait for the second falling edge on T2EX pin
    IE        = 0;                  // disable TMR2 interrupt
    if (RCAP2H < periodH)
        tov2Cnt--;
    else if ((RCAP2H == periodH) && (RCAP2L < periodL))
        tov2Cnt--;
    period = (unsigned long) tov2Cnt * 65536 + (unsigned long) periodH * 256 + (unsigned long) periodL;
    while(1);
}
```

Time Delay Creation

Timers 2, 3, and 4 also can be used to create delays. Their auto-reload mode makes the creation of time delay longer than $2^{16} - 1$ timer clock cycles easier. This mode also can be used to generate periodic interrupts.

Example 8.4 Use Timer 2 to create a time delay of 1 second and write a program to test it on a **C8051F040** demo board. Use the internal oscillator (24.5 MHz) as SYSCLK.

Solution: A long delay can be created by repeating a short delay multiple times. For example, a 1-s delay can be created by repeating forty 25 ms delays.

By choosing 12 as the timer clock prescaler and loading 14,494 into the Timer 2 register, a 25-ms delay (f_{SYSCLK} = 24.5 MHz) can be created when the Timer 2 register overflows. The auto-reload mode eliminates the need of manual reload of the Timer 2 register. Loading the value 0x04 into the TMR2CN register and loading 0 into the TMR2CF register will enable Timer 2 in auto-reload mode, set the prescaler to 12, and clear other unneeded flags and functions.

The subroutine that uses the auto-reload mode of Timer 2 to create a one-second delay can be tested by incrementing a counter after calling this function and then outputting the counter value to Port P5, which drives eight LEDs. The assembly program and its test program that implements the specified operation are as follows.

```
            $nomod51
            $include    (C8051F040.inc)
cnt         set         R0
lp          set         R1
            org         0x00
            ljmp        start
            org         0x2B
            dec         lp                  ; decrement repetition count
            clr         TF2
            reti
            org         0xAB
start:      mov         SFRPAGE,#0x0F
            mov         WDTCN,#0xDE         ; disable watchdog timer
            mov         WDTCN,#0xAD
            mov         CLKSEL,#0           ; select internal oscillator as SYSCLK
            mov         OSCICN,#x83         ;      "
            mov         XBR2,#0x40          ; enable crossbar
            mov         P5MDOUT,#0xFF       ; enable port 5 push pull
            mov         P5,#0               ; turn off all LEDs
            mov         SP,#0x7F
            mov         cnt,#0              ; initialize cnt to 0
loopf:      lcall       dly1s               ; call dly1s to create one second delay
            inc         cnt
            mov         P5,cnt              ; output the complement of cnt to LEDs
            ajmp        loopf               ; repeat forever
dly1s:      push        SFRPAGE
            mov         SFRPAGE,#0          ; switch to page 0
            mov         lp,#40              ; prepare to perform 40 count up operations
            clr         TF2
            mov         TMR2H,#0x38         ; let Timer 2 count up from 14494 so that it overflows
            mov         TMR2L,#0x9E         ; in 25 ms
            mov         RCAP2H,#0x38        ; Timer 2 reload value
            mov         RCAP2L,#0x9E        ;      "
            mov         TMR2CN,#0x04        ; enable Timer 2 auto-reload mode
            mov         TMR2CF,#0           ; use SYSCLK/12 as clock source to Timer 2
            mov         IE,#0xA0            ; enable Timer 2 overflow interrupt
again:      mov         A,lp
            jnz         again
            pop         SFRPAGE
            mov         IE,#0               ; disable interrupt
            ret
            end
```

The C program that performs the specified operation is as follows.

```
#include <c8051F040.h>
volatile unsigned char lp;
unsigned char cnt;
void t2ISR(void) interrupt 5
```

```
{
    TF2 = 0;
    lp--;                   /* decrement repetition count */
}
void delay1s(void)
{
    char sfrtemp;
    sfrtemp   = SFRPAGE;
    SFRPAGE   = 0x0;
    lp        = 40;
    TMR2H     = 0x38;       // count up from this number and overflow in 25 ms
    TMR2L     = 0x9E;
    RCAP2H    = 0x38;       // Timer 2 reload value
    RCAP2L    = 0x9E;
    TMR2CF    = 0;          // use SYSCLK/12 as clock source to Timer 2
    TMR2CN    = 0x04;       // enable Timer 2 auto-reload mode
    IE        = 0xA0;
    while(lp);              // wait until lp is decremented to 0
    SFRPAGE   = sfrtemp;    // restore SFRPAGE
    return;
}
void main (void)
{
    SFRPAGE   = 0x0F;       // switch to SFR page F
    WDTCN     = 0xDE;       // disable watchdog timer
    WDTCN     = 0xAD;
    CLKSEL    = 0;          // select internal oscillator output as
    OSCICN    = 0x83;       // SYSCLK
    XBR2      = 0x40;       // enable crossbar
    P5MDOUT = 0xFF;         // enable Port 5 push pull
    P5        = 0;          // turn off all LEDs
    cnt       = 0;
    while(1){
        delay1s();          // wait for one second
        cnt++;
        P5    = cnt;        // output cnt on LEDs
    }
}
```

Waveform Generation

The toggle output mode of Timers 2, 3, and 4 can simplify the generation of waveforms. After configuring Timer n (2, 3, or 4) properly, the waveform will be repeated continuously until the user change it.

Example 8.5 Write an instruction sequence to generate a 2-KHz square waveform using Timer 2 of the C8051F040, which runs with the internal 24.5-MHz internal-oscillator clock source.

Solution: Using Equation 8.4 and setting TCLK to SYSCLK (24.5 MHz), the value to be loaded into the RCAP2H:RCAP2L registers is

$$\textbf{RCAP2H:RCAP2L} = 2^{16} - 24.5 \times 10^6 \div (2 \times 2000) = 65536 - 6125 = \text{0xE813}$$

The assembly program for generating the desired waveform (using the T2 pin) is as follows.

```
            $nomod51
            $include (c:\keil\c51\asm\c8051F040.inc)
            org         00H
            ljmp        start
            org         0ABH
start:      acall       sysInit
            mov         TMR2H,#high 59411       ; count up from 59411
            mov         TMR2L,#low 59411        ;     "
            mov         RCAP2H,# high 59411     ; set reload value to 59411
            mov         RCAP2L,# low 59411      ;     "
            mov         TMR2CF,#0x0A            ; select SYSCLK as timer clock, enable T2 output
            mov         TMR2CN,#0x04            ; enable timer 2 in auto-reload timer mode
forever:    nop
            ajmp        forever
; ************************************************************************************************
; The following subroutine enables all peripheral functions and assigns port pins
; to peripheral functions. It configures SPI to 3-wire mode and switch to SFR page 0.
; ************************************************************************************************
sysInit:    mov         SFRPAGE,#0x0F
            mov         WDTCN,#0xDE             ; disable watchdog timer (must be done)
            mov         WDTCN,$0xAD             ;     "
            mov         CLKSEL,#0               ; use internal oscillator as SYSCLK
            mov         OSCICN,#0x83            ;     "
            mov         XBR0,#0xF7
            mov         XBR1,#0xFF              ; route T2 pin to port pin (P2.7)
            mov         XBR2,#0x5D
            mov         XBR3,#0x8F
            mov         P2MDOUT,#0x80           ; enable P2.7 for output
            mov         SFRPAGE,#0              ; switch SFR page 0
            mov         SPI0CN,#0x01            ; enable and configure SPI to 3-wire mode
            ret
            end
```

Siren Generation

Making sound is easy. Connecting a waveform with an appropriate frequency (between 20 Hz and 20 kHz) to a speaker or a buzzer can make a sound. A two-tone siren can be generated by connecting the T2 (or T3, T4) pin to a speaker or a buzzer and switching between two different frequencies. The circuit connection between the T2 pin and a speaker is shown in Figure 8.9.

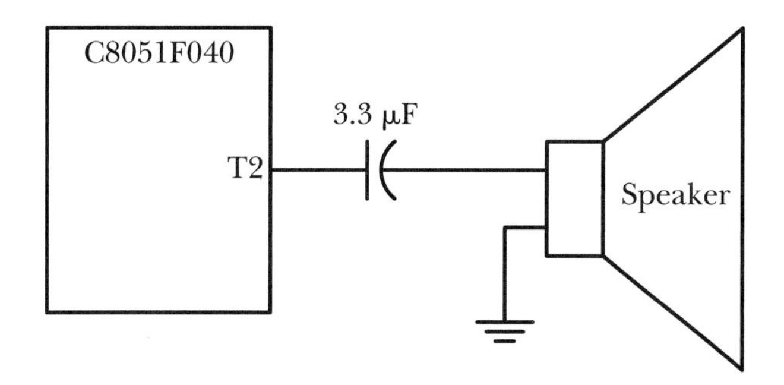

FIGURE 8.9 *Circuit connection for a speaker.*

Example 8.6 Describe how to use Timer 2 of the C8051F040 to generate a two-tone siren that oscillates between 440 Hz and 880 Hz.

Solution: The procedure for generating a siren is as follows.

Step 1 Connect an 8-Ω speaker to the T2 pin.

Step 2 Configure Timer 2 to operate in toggle output mode.

Step 3 Load the value 37695 (= 65536 − (24,500,000 ÷ (440 ×2))) to RCAP2H:RCAP2L and TMR2H:TMR2L register pairs to generate the 440-Hz waveform.

Step 4 Enable Timer 2 and wait for half of a second.

Step 5 Stop Timer 2 and load the value 51616 (= 65536 − (24,500,000 ÷ (880 × 2))) to RCAP2H:RCAP2L and TMR2H:TMR2L register pairs to generate the 880-Hz waveform.

Step 6 Reenable Timer 2 and wait for half of a second.

Step 7 Stop Timer 2 and go to Step 3.

The assembly program for generating the two-tone siren is as follows.

```
        $nomod51
        $include (c8051F040.inc)
        org     00H
        ljmp    start
        org     0xAB
start:  acall   sysInit
        mov     TMR2CF,#0x0A        ; select SYSCLK as timer clock, enable T2 output
againx: mov     TMR2H,#high 37695   ; count up from 37695
        mov     TMR2L,#low 37695    ;   "
        mov     RCAP2H,# high 37695 ; set reload value to 37695
        mov     RCAP2L,# low 37695  ;   "
        mov     TMR2CN,#0x04        ; enable timer 2 in auto-reload timer mode
```

```
            mov     R0,#5
            lcall   delayby100ms
            clr     TR2                     ; stop timer 2
            mov     TMR2H,#high 51616       ; count up from 51616
            mov     TMR2L,#low 51616        ;      "
            mov     RCAP2H,# high 51616     ; set reload value to 51616
            mov     RCAP2L,# low 51616      ;      "
            mov     TMR2CN,#0x04            ; enable timer 2 in auto-reload timer mode
            mov     R0,#5
            lcall   delayby100ms
            clr     TR2                     ; stop timer 2
            ajmp    againx
; ************************************************************************************************
; The following subroutine enables all peripheral functions and assign port pins
; to peripheral functions. It configures SPI to 3-wire mode and switch to SFR page 0.
; ************************************************************************************************
sysInit:    mov     SFRPAGE,#0xF            ; switch to SFR page F
            mov     WDTCN,#0xDE             ; watch dog timer must be disabled
            mov     WDTCN,#0xAD             ;      "
            mov     CLKSEL,#0               ; use internal oscillator as SYSCLK
            mov     OSCICN,#0x83            ;      "
            mov     XBR0,#0xF7              ; assign all peripheral signal to I/O pin
            mov     XBR1,#0xFF              ; route T2 pin to port pin (P2.7)
            mov     XBR2,#0x5D              ; enable crossbar decoder
            mov     XBR3,#0x8F
            mov     P2MDOUT,#0x80           ; enable P2.7 for output
            mov     SFRPAGE,#0              ; switch SFR page 0
            mov     SPI0CN,#0x01            ; enable and configure SPI to 3-wire mode
            ret
; ************************************************************************************************
; The following function creates a delay that is a multiple of 100 ms. The multiple is
; passed in R0.
; ************************************************************************************************
delayby100ms:
            push    SFRPAGE
            mov     SFRPAGE,#0              ; switch to page 0
            mov     TMOD,#0x11              ; configure Timer 0 and 1 as mode 1 timer
            mov     CKCON,#02               ; Timer 0 use system clock divided by 48 to count
repw3:      clr     TR0
            mov     TH0,#high 14493         ; place 14493 in Timer 0 so that it overflows in
            mov     TL0,#low 14493          ; 51042 clock cycles (in 100 ms)
            clr     TF0
            setb    TR0                     ; enable Timer 0
loopw3:     jnb     TF0,loopw3              ; wait until TF0 is set again
            djnz    R0,repw3
            pop     SFRPAGE
            ret
            end
```

The C language version of the program is as follows.

```
#include <c8051F040.h>
void sysInit(void)
{
    SFRPAGE  = 0x0F;            // switch to SFR page F
    WDTCN    = 0xDE;            // disable watchdog timer
    WDTCN    = 0xAD;
    CLKSEL   = 0;               // select internal oscillator as SYSCLK
    OSCICN   = 0x83;            //
    XBR0     = 0xF7;            // assign all peripheral signals to port pins
    XBR1     = 0xFF;            //    "
    XBR2     = 0x5D;            //    "
    XBR3     = 0x8F;            //    "
    P2MDOUT  = 0x80;            // enable T2 pin output
    SFRPAGE  = 0;               // switch to SFR page 0
    SPI0CN   = 0x01;            // enable SPI in 3-wire mode
}
void delayby100ms(unsigned char k)
{
    unsigned char i,tempage;
    tempage  = SFRPAGE;
    SFRPAGE  = 0;
    TMOD     = 0x11;            // configure Timer 0 and 1 to mode 1
    CKCON    = 0x02;            // Timer 0 use system clock divided by 48 as clock source
    for (i = 0; i < k; i++){
        TH0  = 0x38;            // place 14494 in TH0:TL0 so it overflows in 100 ms
        TL0  = 0x9E;
        TF0  = 0;
        TR0  = 1;               // enable Timer 0
        while(!TF0);            // wait for 100 ms
        TR0  = 0;               // stop Timer 0
    }
    SFRPAGE  = tempage; // restore original SFRPAGE
}
void main (void)
{
    sysInit();                  // configure SYSCLK and all peripheral pin assignments
    TMR2CF   = 0x0A;            // select SYSCLK as timer 2 clock, enable T2 output
    TMR2CN   = 0;
    while (1) {
        TMR2H   = 0x93;         // generate 880 Hz tone
        TMR2L   = 0x3F;         //    "
        RCAP2H  = 0x93;         //    "
        RCAP2L  = 0x3F;         //    "
        TR2     = 1;            // enable Timer 2 in auto reload mode
        delayby100ms(5);        // wait for 0.5 seconds
        TR2     = 0;
        TMR2H   = 0xC9;         // generate 440 Hz tone
        TMR2L   = 0xA0;         //    "
```

```
        TMR2CN = 0x04;            //   "
        RCAP2H = 0xC9;            //   "
        RCAP2L = 0xA0;
        TR2    = 1;
        delayby100ms(5);
    TR2         = 0;
    }
}
```

Playing a Song

The program that plays a song can be created by modifying the siren program. To make the switching of notes easier, the whole score should be stored in a table. A note in the score has two components: **pitch** (frequency) and **tempo** (duration). The duration of a quarter note is 400 ms. The durations of other notes can be derived proportionally. The complete list of the music notes and their frequencies is in Appendix I.

To play a song from the speaker, one places the frequencies and durations of all the notes of a music score in a table (or two separate tables). For every note, the user program uses the toggle-output mode of Timer 2 to generate the digital waveform with the specified frequency and lets the waveform last for the corresponding duration. One problem with this method is that contiguous identical notes become one long note. This problem can be avoided by inserting a short period of silence between two identical notes. The following example illustrates this idea.

Example 8.7 Use the T2 output to drive a speaker and write a program to play "The Star-Spangled Banner." Use the internal oscillator of the C8051F040 as the clock source of Timer 2.

Solution: For each note, the value to be stored in TMR2H:TMR2L and RCAP2H:RCAP2L register pairs to generate the waveform with the note frequency can be calculated using Equation 8.3.

The assembly program that plays "The Star-Spangled Banner" is as follows.

```
            $nomod51
            $include (c8051F040.inc)
G3H         equ     0x0B            ; reload value for G3 note (at 24.5MHz)
G3L         equ     0xDC            ;    "
B3H         equ     0x3E            ; reload value for B3 note
B3L         equ     0x39            ;    "
C4H         equ     0x49            ; reload value for C4 note
C4L         equ     0x1A            ;    "
C4SH        equ     0x53            ; reload value for C4# note
C4SL        equ     0x5D            ;    "
D4H         equ     0x5D            ; reload value for D4 note
```

```
D4L      equ    0x0D          ;   "
E4H      equ    0x6E          ; reload value for E4 note
E4L      equ    0xD5          ;   "
F4H      equ    0x76          ; reload value for F4 note
F4L      equ    0xFB          ;   "
F4SH     equ    0x7E          ; reload value for F4# note
F4SL     equ    0xAB          ;   "
G4H      equ    0x85          ; reload value for G4 note
G4L      equ    0xEE          ;   "
A4H      equ    0x93          ; reload value for A4 note
A4L      equ    0x3F          ;   "
B4FH     equ    0x99          ; reload value for B4b note
B4FL     equ    0x59          ;   "
B4H      equ    0x9F          ; reload value for B4 note
B4L      equ    0x1C          ;   "
C5H      equ    0xA4          ; reload value for C5 note
C5L      equ    0x8D          ;   "
D5H      equ    0xAE          ; reload value for D5 note
D5L      equ    0x87          ;   "
E5H      equ    0xB7          ; reload value for E5 note
E5L      equ    0x6B          ;   "
F5H      equ    0xBB          ; reload value for F5 note
F5L      equ    0x7D          ;   "
ZH       equ    0xFF          ; reload value for stop sound (high frequency
ZL       equ    0xFA          ; and is inaudible)
notes    equ    118           ; total number of notes in the song
k        set    R1            ; loop index
         org    00H
         ljmp   start
         org    0xAB
         mov    SP,#0x7F      ; initialize stack pointer to 30H
start:   lcall  sysinit       ; power-on initialization
         mov    TMR2CF,#0xA   ; select SYSCLK as clock source, enable T2 output
         mov    k,#0          ; initialize k to 0
sloop:   mov    DPTR,#scoreHI ; get the high byte of the count up value for current note
         mov    A,k           ;   "
         movc   A,@A+DPTR     ;   "
         mov    TMR2H,A       ;   "
         mov    RCAP2H,A      ;   "
         mov    DPTR,#scoreLO ; get the low byte of the count up value for current note
         mov    A,k           ;   "
         movc   A,@A+DPTR     ;   "
         mov    TMR2L,A       ;   "
         mov    RCAP2L,A      ;   "
         mov    TMR2CN,#04    ; enable Timer 2 in auto-reload mode
         mov    DPTR,#dura    ; get the duration of the current note
         mov    A,k           ;   "
         movc   A,@A+DPTR     ;   "
         mov    R0,A          ;   "
         lcall  delayby10ms   ; play the note
```

```
            inc     k                   ; go to the next note
            cjne    k,#notes,sloop      ; reach the end of the score?
            mov     TMR2CN,#0           ; stop timer 2
forever:    nop
            jmp     forever

; ****************************************************************************************
; The following routine selects SYSCLK, assign I/O pin to peripheral functions
; and disable watchdog timer (must be done).
; ****************************************************************************************
sysinit:    mov     SFRPAGE,#0x0F
            mov     WDTCN,#0xDE         ; disable watchdog timer
            mov     WDTCN,#0xAD         ;    "
            mov     CLKSEL,#0           ; use internal oscillator as SYSCLK
            mov     OSCICN,#0x83        ;    "
            mov     XBR0,#0xF7          ; assign I/O pins to all peripheral functions
            mov     XBR1,#0xFF          ;    "
            mov     XBR2,#0x5D          ;    "
            mov     XBR3,#0x8F          ;    "
            mov     P2MDOUT,#0x80       ; enable T2 output (P2.7 pin)
            mov     SFRPAGE,#0          ; switch to SFR page 0
            mov     SPI0CN,#01          ; enable 3-wire SPI
            ret
; ****************************************************************************************
; The following routine create a time delay that is a multiple of 10 ms. The
; multiple is passed in R0.
; ****************************************************************************************
delayby10ms:
            push    SFRPAGE
            mov     SFRPAGE,#0          ; switch to SFR page 0
            mov     TMOD,#0x11          ; configure Timer 0 and 1 to mode 1
            mov     CKCON,#0            ; timer 0 uses system clock divided by 12 to count
repws:      clr     TR0                 ; stop the timer
            mov     TH0,#high 45119     ; count up from 45119 and overflows in 10 ms
            mov     TL0,#low 45119      ;    "
            clr     TF0
            setb    TR0                 ; start Timer 0
            jnb     TF0,$               ; wait until TF0 is set
            djnz    R0,repws
            pop     SFRPAGE
            ret
; ****************************************************************************************
; high bytes of score delay count.
; ****************************************************************************************
scoreHI:    db      D4H,B3H,G3H,B3H,D4H,G4H,B4H,A4H,G4H,B3H,C4SH
            db      D4H,ZH,D4H,ZH,D4H,B4H,A4H,G4H,F4SH,E4H,F4SH,G4H,ZH,G4H,D4H,B3H,G3H
            db      D4H,B3H,G3H,B3H,D4H,G4H,B4H,A4H,G4H,B3H,C4SH,D4H,ZH,D4H,ZH,D4H
            db      B4H,A4H,G4H,F4SH,E4H,F4SH,G4H,ZH,G4H,D4H,B3H,G3H,B4H,ZH,B4H
            db      B4H,C5H,D5H,ZH,D5H,C5H,B4H,A4H,B4H,C5H,ZH,C5H,ZH,C5H,B4H,A4H,G4H
            db      F4SH,E4H,F4SH,G4H,B3H,C4SH,D4H,ZH,D4H,G4H,ZH,G4H,ZH,G4H,F4SH
```

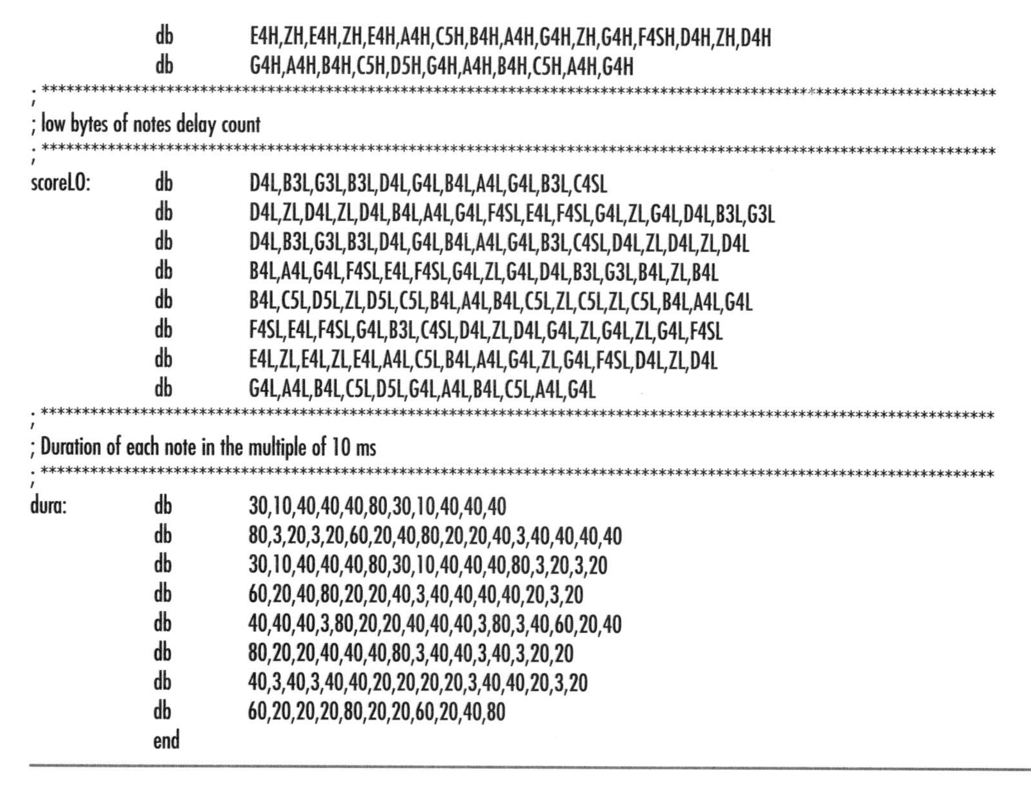

```
            db      E4H,ZH,E4H,ZH,E4H,A4H,C5H,B4H,A4H,G4H,ZH,G4H,F4SH,D4H,ZH,D4H
            db      G4H,A4H,B4H,C5H,D5H,G4H,A4H,B4H,C5H,A4H,G4H
; ****************************************************************************************************
; low bytes of notes delay count
; ****************************************************************************************************
scoreLO:    db      D4L,B3L,G3L,B3L,D4L,G4L,B4L,A4L,G4L,B3L,C4SL
            db      D4L,ZL,D4L,ZL,D4L,B4L,A4L,G4L,F4SL,E4L,F4SL,G4L,ZL,G4L,D4L,B3L,G3L
            db      D4L,B3L,G3L,B3L,D4L,G4L,B4L,A4L,G4L,B3L,C4SL,D4L,ZL,D4L,ZL,D4L
            db      B4L,A4L,G4L,F4SL,E4L,F4SL,G4L,ZL,G4L,D4L,B3L,G3L,B4L,ZL,B4L
            db      B4L,C5L,D5L,ZL,D5L,C5L,B4L,A4L,B4L,C5L,ZL,C5L,ZL,C5L,B4L,A4L,G4L
            db      F4SL,E4L,F4SL,G4L,B3L,C4SL,D4L,ZL,D4L,G4L,ZL,G4L,ZL,G4L,F4SL
            db      E4L,ZL,E4L,ZL,E4L,A4L,C5L,B4L,A4L,G4L,ZL,G4L,F4SL,D4L,ZL,D4L
            db      G4L,A4L,B4L,C5L,D5L,G4L,A4L,B4L,C5L,A4L,G4L
; ****************************************************************************************************
; Duration of each note in the multiple of 10 ms
; ****************************************************************************************************
dura:       db      30,10,40,40,40,80,30,10,40,40,40
            db      80,3,20,3,20,60,20,40,80,20,20,40,3,40,40,40,40
            db      30,10,40,40,40,80,30,10,40,40,40,80,3,20,3,20
            db      60,20,40,80,20,20,40,3,40,40,40,40,20,3,20
            db      40,40,40,3,80,20,20,40,40,40,3,80,3,40,60,20,40
            db      80,20,20,40,40,40,80,3,40,40,3,40,3,20,20
            db      40,3,40,3,40,40,20,20,20,20,3,40,40,20,3,20
            db      60,20,20,20,80,20,20,60,20,40,80
            end
```

The C language version of the program is as follows.

```
#include    <c8051F040.h>
#define     G3H     0x0B        // reload value of TMR2 for the G3 note
#define     G3L     0xDC        //    "
#define     B3H     0x3E        // reload value of TMR2 for the B3 note
#define     B3L     0x39        //    "
#define     C4H     0x49        // reload value of TMR2 for the C4 note
#define     C4L     0x1A        //    "
#define     C4SH    0x53        // reload value of TMR2 for the C4 sharp note
#define     C4SL    0x5D        //    "
#define     D4H     0x5D        // reload value of TMR2 for the D4 note
#define     D4L     0x0D        //    "
#define     E4H     0x6E        // reload value of TMR2 for the E4 note
#define     E4L     0xD5        //    "
#define     F4H     0x76        // reload value of TMR2 for the F4 note
#define     F4L     0xFB        //    "
#define     F4SH    0x7E        // reload value of TMR2 for the F4 sharp note
#define     F4SL    0xAB        //    "
#define     G4H     0x85        // reload value of TMR2 for the G4 note
#define     G4L     0xEE        //    "
#define     A4H     0x93        // reload value of TMR2 for the A4 note
#define     A4L     0x3F        //    "
#define     B4FH    0x99        // reload value of TMR2 for the B4 flat note
```

```
#define     B4FL    0x59    //    "
#define     B4H     0x9F    // reload value of TMR2 for the B4 note
#define     B4L     0x1C    //    "
#define     C5H     0xA4    // reload value of TMR2 for the C5 note
#define     C5L     0x8D    //    "
#define     D5H     0xAE    // reload value of TMR2 for the D5 note
#define     D5L     0x87    //    "
#define     E5H     0xB7    // reload value of TMR2 for the E5 note
#define     E5L     0x6B    //    "
#define     F5H     0xBB    // reload value of TMR2 for the F5 note
#define     F5L     0x7D    //    "
#define     ZH      0xFF    // stop sound
#define     ZL      0xFA    // (very high frequency and is inaudible)
#define     notes 118

void sysInit(void);
void delayby10ms(unsigned char kk);
code unsigned char scoreH[118] = {D4H,B3H,G3H,B3H,D4H,G4H,B4H,A4H,G4H,B3H,C4SH,
            D4H,ZH,D4H,ZH,D4H,B4H,A4H,G4H,F4SH,E4H,F4SH,G4H,ZH,G4H,D4H,B3H,G3H,
            D4H,B3H,G3H,B3H,D4H,G4H,B4H,A4H,G4H,B3H,C4SH,D4H,ZH,D4H,ZH,D4H,
            B4H,A4H,G4H,F4SH,E4H,F4SH,G4H,ZH,G4H,D4H,B3H,G3H,B4H,ZH,B4H,
            B4H,C5H,D5H,ZH,D5H,C5H,B4H,A4H,B4H,C5H,ZH,C5H,ZH,C5H,B4H,A4H,G4H,
            F4SH,E4H,F4SH,G4H,B3H,C4SH,D4H,ZH,D4H,G4H,ZH,G4H,ZH,G4H,F4SH,
            E4H,ZH,E4H,ZH,E4H,A4H,C5H,B4H,A4H,G4H,ZH,G4H,F4SH,D4H,ZH,D4H,
            G4H,A4H,B4H,C5H,D5H,G4H,A4H,B4H,C5H,A4H,G4H};
code unsigned char scoreL[118] = {D4L,B3L,G3L,B3L,D4L,G4L,B4L,A4L,G4L,B3L,C4SL,
            D4L,ZL,D4L,ZL,D4L,B4L,A4L,G4L,F4SL,E4L,F4SL,G4L,ZL,G4L,D4L,B3L,G3L,
            D4L,B3L,G3L,B3L,D4L,G4L,B4L,A4L,G4L,B3L,C4SL,D4L,ZL,D4L,ZL,D4L,
            B4L,A4L,G4L,F4SL,E4L,F4SL,G4L,ZL,G4L,D4L,B3L,G3L,B4L,ZL,B4L,
            B4L,C5L,D5L,ZL,D5L,C5L,B4L,A4L,B4L,C5L,ZL,C5L,ZL,C5L,B4L,A4L,G4L,
            F4SL,E4L,F4SL,G4L,B3L,C4SL,D4L,ZL,D4L,G4L,ZL,G4L,ZL,G4L,F4SL,
            E4L,ZL,E4L,ZL,E4L,A4L,C5L,B4L,A4L,G4L,ZL,G4L,F4SL,D4L,ZL,D4L,
            G4L,A4L,B4L,C5L,D5L,G4L,A4L,B4L,C5L,A4L,G4L};
code unsigned char dur[118] = {30,10,40,40,40,80,30,10,40,40,40,
            80,3, 20,3, 20,60,20,40,80,20,20,40,3, 40,40,40,40,
            30,10,40,40,40,80,30,10,40,40,40,80,3, 20,3, 20,
            60,20,40,80,20,20,40,3, 40,40,40,40,20,3, 20,
            40,40,40,3, 80,20,20,40,40,40,3, 80,3, 40,60,20,40,
            80,20,20,40,40,40,80,3, 40,40,3, 40,3, 20,20,
            40,3, 40,3, 40,40,20,20,20,20,3, 40,40,20,3, 20,
            60,20,20,20,80,20,20,60,20,40,80};
void main (void)
{
    int j;
    sysInit();                  // configure SYSCLK and assign peripheral functions to port pins
    TMR2CF  = 0x0A;             // select SYSCLK as timer 2 clock, enable T2 output
    TMR2CN  = 0;                // disable Timer 2
    j       = 0;
    while (j < notes) {
        TMR2H  = scoreH[j];     // play the jth note
```

```
            RCAP2H = scoreH[j];
            TMR2L  = scoreL[j];
            RCAP2L = scoreL[j];
            TR2    = 1;                  // enable Timer 2
            delayby10ms(dur[j]);
            TR2    = 0;                  // disable Timer 2 to change reload value
            j++;
        }
        while(1); // stay here
}
void sysInit(void)
{
        SFRPAGE = 0x0F;
        WDTCN   = 0xDE;                  // disable watchdog timer
        WDTCN   = 0xAD;
        CLKSEL  = 0;                     // select internal oscillator as SYSCLK
        OSCICN  = 0x83;                  //        "
        XBR0    = 0xF7;                  // assign port pins to all peripheral signals
        XBR1    = 0xFF;
        XBR2    = 0x5D;
        XBR3    = 0x8F;
        P2MDOUT |= 0x80;                 // enable T2 output (must be done)
        SFRPAGE = 0;                     // switch to SFR page 0
        SPI0CN  = 0x01;                  // enable 3-wire SPI
}
void delayby10ms(unsigned char k)
{
        unsigned char i, tempage;
        tempage = SFRPAGE;
        SFRPAGE = 0;
        TMOD    = 0x11;                  // configure Timer 0 and 1 to mode 1
        CKCON   = 0x00;                  // Timer 0 use system clock divided by 12 as clock source
        for (i = 0; i < k; i++){
            TR0    = 0;                  // stop Timer 0
            TH0    = 0xB0;               // place 45119 in TH0:TL0 so it overflows in 10 ms
            TL0    = 0x3F;
            TF0    = 0;
            TR0    = 1;                  // enable Timer 0
            while(!TF0);                 // wait for 10 ms
        }
        SFRPAGE = tempage;               // restore original SFRPAGE
}
```

These two programs can be tested by connecting a wire from the P2.7 pin to a buzzer or a speaker. Timers 2, 3, and 4 have many useful functions. However, they can neither capture the rising edge of a signal nor change the duty cycle of the toggle output. Some applications require these two features. The addition of the programmable counter array (PCA) is to satisfy these requirements.

8.5 Programmable Counter Array

The programmable counter array (PCA) provides enhanced timer functionality while requiring less CPU intervention than the standard 8051 counter/timers. The PCA0 of the C8051F040 consists of a dedicated 16-bit counter/timer and six 16-bit capture/compare modules. Each capture/compare module has its own associated I/O line (CEXn), which is routed through the crossbar to the I/O pin when enabled. The clock source can be selected from one of the six sources: system clock, system clock divided by 4, system clock divided by 12, the external oscillator clock divided by 8, Timer 0 overflow, or an external clock signal on the ECI pin. Each capture/compare module may be configured to operate independently in one of six modes:

- Edge-triggered capture
- Software timer
- High-speed output
- Frequency output
- 8-bit PWM
- 16-bit PWM

The block diagram of the C8051F040 PCA0 is shown in Figure 8.10.

8.5.1 PCA Timer/Counter

The PCA of the C8051F040 has a free-running 16-bit timer/counter consisting of registers PCA0H and PCA0L. PCA0H is the high byte and the PCA0L is the low byte of the 16-bit timer. This timer serves as the time base of all PCA operations.

The block diagram of the C8051F040 PCA timer is shown in Figure 8.11. Reading the PCA0L register automatically latches the value of PCA0H into a **snapshot** register, the following PCA0H read accesses this snapshot register. This guarantees an accurate reading of the entire 16-bit PCA0 timer. Reading PCA0L or PCA0H does not disturb the counter/timer operation. The rollover of the PCA0 counter/timer from 0xFFFF to 0x0000 sets the **CF** flag, which may interrupts the CPU if the **CF** interrupt is enabled. The operation of the PCA timer of the C8051F040 is controlled by a PCA control register (PCA0CN) and a PCA mode register (PCA0MD). The contents of these two registers are shown in Figures 8.12 and 8.13, respectively. The PCA0CN register enables the PCA timer to count and also keeps track of the capture/compare flags of all PCA channels, whereas the PCA0MD register selects the clock source, enables/disables PCA interrupts, and enables/disables the PCA module when the CPU is in the idle mode.

The original 8051 and many 8051 variants do not have the snapshot register of SiLabs devices. This would cause an error in reading the PCA timer value. The error could be as large as 256. One example is reading the PCA

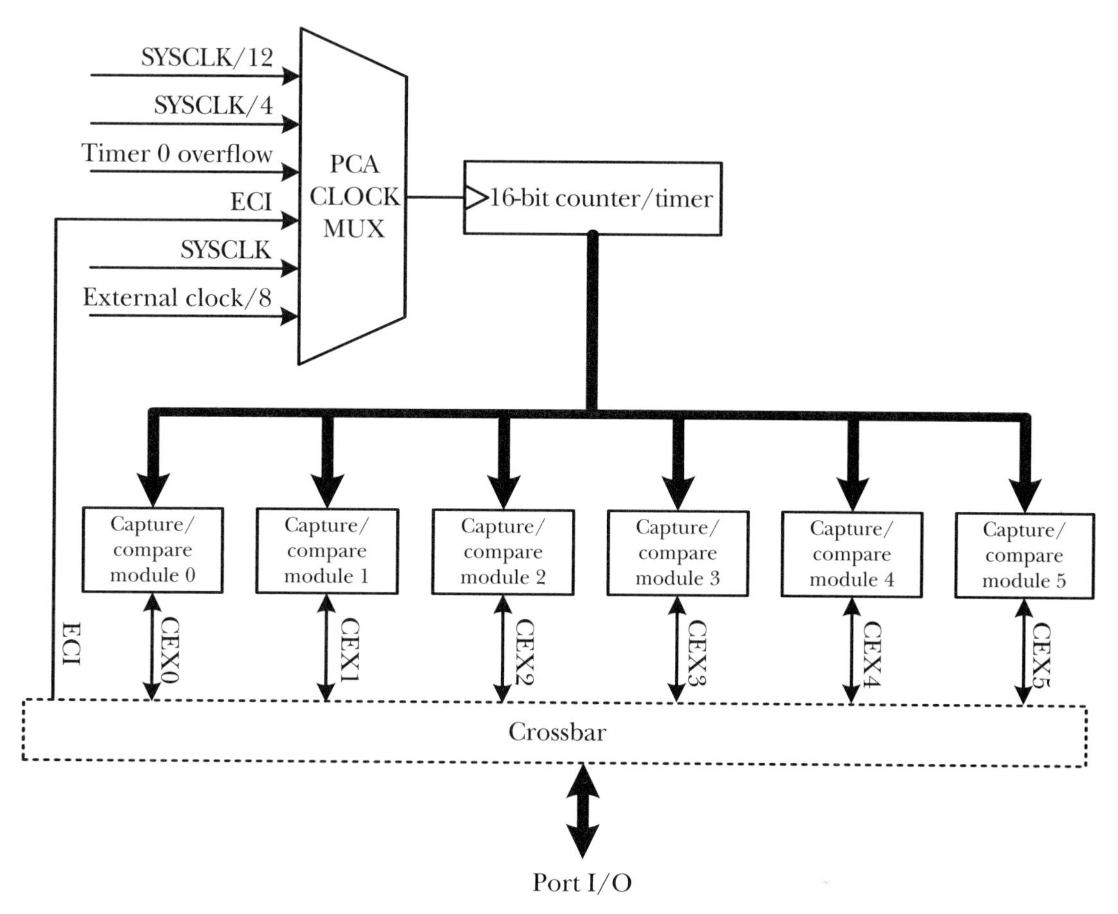

FIGURE 8.10
The C8051F040 PCA0 block diagram.

timer low byte when its value is 0xFF. Since the PCA timer is not stopped when it is being read, when the high byte is read, the low byte may have overflowed and caused the high byte to increment by one. This problem cannot be fixed by changing the read order. A simple solution is to stop the timer before reading it, and then reenable the timer after reading it.

8.5.2 Capture/Compare Modules

Each compare/capture module can perform the following functions:

- 16-bit capture, positive-edge (rising-edge) triggered
- 16-bit capture, negative-edge (falling-edge) triggered
- 16-bit capture, both positive-edge and negative-edge triggered.
- 16-bit software timer
- 16-bit high speed output

- Frequency output
- 8-bit pulse width modulation
- 16-bit pulse width modulation

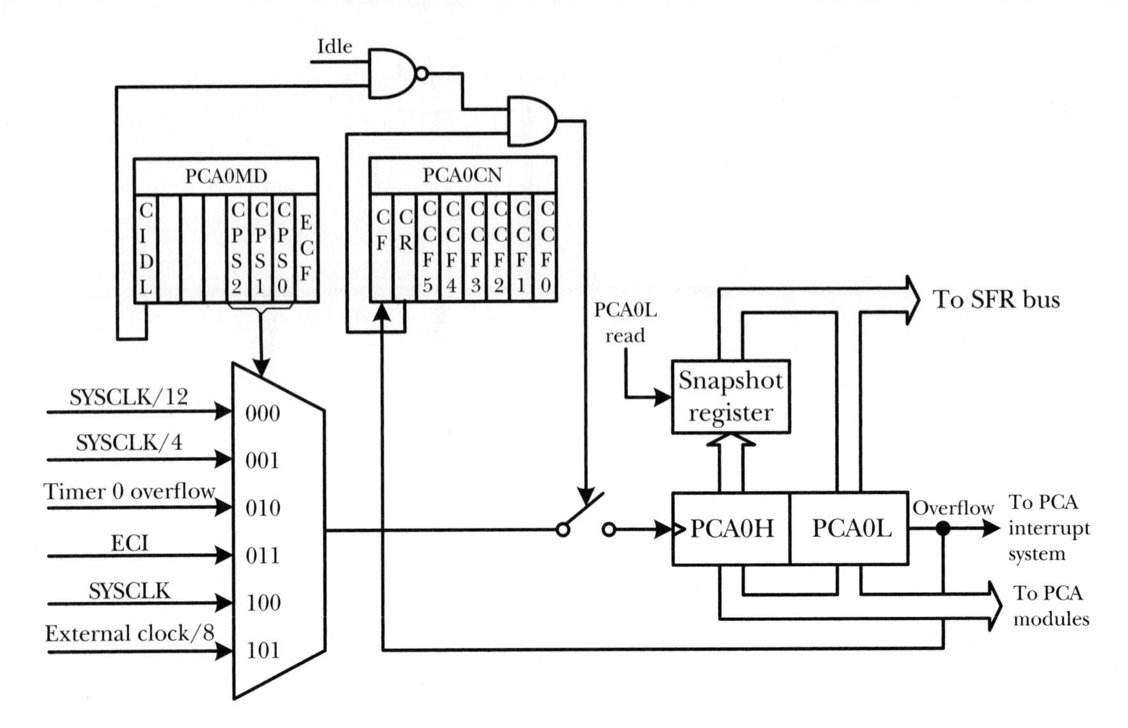

FIGURE 8.11
C8051F040 PCA timer block diagram.

FIGURE 8.12
PCA0 control register of C8051F040 (PCA0CN).

7	6	5	4	3	2	1	0	
CF	CR	CCF5	CCF4	CCF3	CCF2	CCF1	CCF0	Value after reset: 00h

CF: PCA timer overflow flag (needs to be cleared by software)
 0 = PCA0 or PCA timer did not overflow (rollover from 0xFFFF to 0x0000)
 1 = PCA0 or PCA timer has overflown.
CR: PCA timer run control
 0 = PCA0 timer disabled
 1 = PCA0 timer enabled
CCFn (n = 0...5): PCA0 module n capture/compare flag
 When set, this flag may cause an interrupt if it is enabled. This flag must be cleared by software.
 0 = a capture or a match did not occur
 1 = a capture or a match occurred

Each module has a **mode register** called **PCA0CPMn** to select which function it will perform. The contents of the mode register are shown in

7	6	5	4	3	2	1	0	
CIDL	--	--	--	CPS2	CPS1	CPS0	ECF	Value after reset: 00h

CIDL: PCA timer idle control
0 = PCA0 continues to function normally while the MCU is in idle mode
1 = PCA0 operation is suspended when the MCU is in idle mode
CPS2-CPS0: PCA timer clock source select (for PCA0MD)
000 = use SYSCLK / 12 as PCA timer clock source
001 = use SYSCLK / 4 as PCA timer clock source
010 = use Timer 0 overflow as PCA0's timer clock source
011 = use the high-to-low transition on ECI pin as PCA0 timer clock source
100 = SYSCLK is as selected as PCA0 timer clock source
101 = external clock (XTAL1 pin) is selected as PCA0 timer clock source
ECF: PCA timer overflow interrupt enable
0 = disable CF interrupt
1 = enable a PCA0 timer overflow interrupt when CF is set

FIGURE 8.13
PCA0 mode register of the C8051F040 (PCA0MD).

TABLE 8.3 *PCA0CPM Register Settings for PCA Capture/Compare Modules*

PWM16	ECOM	CAPP	CAPN	MAT	TOG	PWM	ECCF	Operation Mode
X	X	1	0	0	0	0	X	Rising-edge capture
X	X	0	1	0	0	0	X	Falling-edge capture
X	X	1	1	0	0	0	X	Capture triggered by CEXn pin transition
X	1	0	0	1	0	0	X	Software timer
X	1	0	0	1	1	0	X	High-speed output
X	1	0	0	0	1	1	X	Frequency output
0	1	0	0	0	0	1	X	8-bit PWM
1	1	0	0	0	0	1	X	16-bit PWM

Figure 8.14. The ECCFn bit enables the PCA interrupt when a module's event flag is set. The event flag **CCFn** is contained in the PCA0CN register and is set when a capture event, software timer, or high-speed output event occurs for a given module. A summary of the settings for entering each PCA mode is listed in Table 8.3.

Each module also has a pair of 8-bit **compare/capture registers (PCA0CPHn:PCA0CPLn)** associated with it. These registers store the timer value when a selected edge is detected or when a compare event should occur. The PCA0CPHn register controls the duty cycle of the 8-bit PWM waveform.

7	6	5	4	3	2	1	0	
PWM16n	ECOMn	CAPPn	CAPNn	MATn	TOGn	PWMn	ECCFn	PCA0CPMn reset: 00h

PWM16n: 16-bit pulse width modulation enable
 0 = 8-bit PWM selected
 1 = 16-bit PWM selected
ECOMn: Comparator function enable
 0 = disable compare function
 1 = enable compare function
CAPPn: Capture positive function enable
 0 = disable rising edge capture
 1 = enable rising edge capture
CAPNn: Capture negative edge enable
 0 = disable falling edge capture
 1 = enable falling edge capture
MATn: Match function enable
 Match function compares the PCA timer with a module's capture/compare register and causes CCFn (INTF) flag to set when match
 0 = disable match function
 1 = enable match function
TOGn: Toggle function enable
 When enabled, this function cause the logic level on the CEXn pin to toggle on a match between PCA timer and a module's capture/compare register
 0 = disable toggle function
 1 = enable toggle function
PWMn: Pulse width modulation mode enable
 If the TOGn bit is also set, the module operates in frequency output mode
 0 = PWM mode disabled
 1 = PWM mode is enabled
ECCFn : Capture/compare flag interrupt enable
 0 = disable capture/compare flag interrupt
 1 = enable capture/compare flag interrupt

FIGURE 8.14
PCA capture/compare mode register (PCA0CPMn, n = 0 . . . 5.

Example 8.8 Configure the PCA module 0 of the C8051F040 to capture the rising edge of the signal applied to the CEX0 pin. Disable the PCA overflow interrupt and use **SYSCLK ÷ 12** as the clock source of the timer. Allow PCA0 to function normally when idle.

Solution: The following registers need to be programmed properly.

PCA0MD Register: Set bit 7 to 0 so that PCA functions normally when the MCU is idle. Bits 3, 2, and 1 of the PCA0MD register should be set to 000 to select SYSCLK/12 as the PCA clock source. PCA overflow interrupt is not needed. Set bit 0 to 0. Write the value 0x00 into the PCA0MD register.

PCA0CN Register: The PCA counter must be enabled, and all flags should be cleared at the beginning. Write the value 0x40 into the PCA0CN register.

PCA0CPM Register: For module 0,

- Bit PWM160 (bit 7): set to 0 (this bit is don't care)
- Bit ECOM0 (bit 6): set to 0 to disable the comparator
- Bit CAPP0 (bit 5): set to 1 to capture the rising edge
- Bit CAPN0 (bit 4): set to 0 to disable the falling-edge capture
- Bit MAT0 (bit 3): set to 0 to disable the module 0 comparator
- Bit TOG0 (bit 2): set to 0 so that the CEX0 pin will not toggle when the PCA counter matches the compare/capture registers of module 0
- Bit PWM0 (bit 1): set to 0 to disable the pulse-modulation output
- Bit ECCF0 (bit 0): cleared to disable the CCF interrupt

Therefore, the value 0x20 should be written into the PCA0CPM0 register. The following instructions will configure the PCA0 to operate with the specified setting:

```
mov     PCA0MD,#0
mov     PCA0CN,#0x40
mov     PCA0CPM0,#0x20
```

8.5.3 Edge-triggered Capture Mode

In this mode, the PCA copies the 16-bit timer value into the capture register at the moment that the selected signal edge (rising and/or falling edge) on the CEXn pin is detected. This action allows the user to measure periods, pulse widths, duty cycles, and phase differences of signals of up to six separate inputs. The signal edge that triggers the capture action is selected by setting or clearing the CAPPn and CAPNn bits in the PCA mode register. The functioning of the capture mode is illustrated in Figure 8.15.

When a capture occurs, the module's event flag **CCFn** in the PCA0CN register is set to 1. An interrupt will be requested if the ECCFn bit in the PCA0CPMn register and the EA bit of the IE register are both set to 1. The **CCFn** flag must be cleared in the software.

In the interrupt-service routine, the 16-bit capture value must be saved in RAM before the next capture event occurs. A subsequent capture on the same

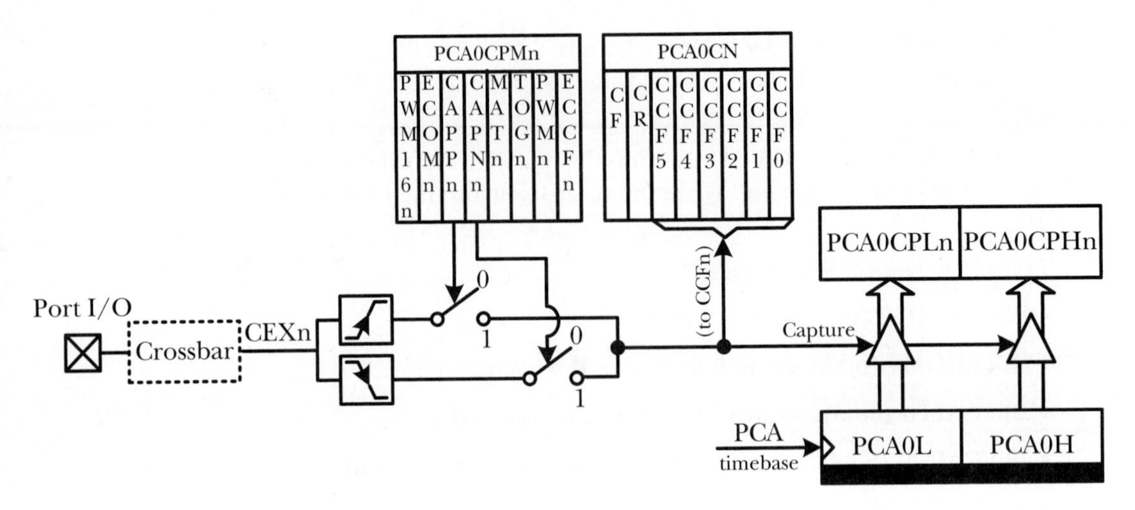

FIGURE 8.15
PCA capture mode diagram (C8051F040).

CEXn pin will write over the previous capture values in PCA0CPHn and PCA0CPLn.

The PCA 16-bit capture mode can be used to measure the period and also the pulse width of an unknown signal. The following example illustrates the measurement of pulse width.

Example 8.9 Write a program to measure the pulse width of a signal connected to the CEX0 pin, assuming that the system clock is 24 MHz.

Solution: The procedure for measuring the pulse width is as follows.

Step 1 Enable the PCA counter and set up module 0 to capture the signal rising edge.

Step 2 Wait for the arrival of the rising edge of the CEX0 signal.

Step 3 Reconfigure module 0 to capture the falling edge. Enable the PCA counter overflow interrupt and initialize the overflow count to 0.

Step 4 Write a service routine for the PCA timer-overflow interrupt, which increments the overflow count by 1.

Step 5 Wait until the falling edge occurs and compute the pulse width by taking the difference of the two captured edges.

The registers PCA0CN, PCA0MD, and PCA0CPM0 are configured as follows.

Configuration of PCA0CN: Enable the PCA timer and clear all PCA flags. Write the value **0x40** into the PCA0CN register.

Configuration of PCA0MD: Select **SYSCLK ÷ 12** as the clock source to the PCA timer, enable PCA timer-overflow interrupt, and turn off the PCA timer during idle mode. Load the value **0x81** into the PCA0MD register.

Configuration of PCA0CPM0: Disable the comparator function, enable rising-edge capture, disable match, disable toggling, disable pulse-width modulation, and disable capture interrupt. Write the value **0xA0** into the PCA0CPM0 register. Falling-edge capture must be enabled after the rising edge has been captured.

The program is as follows.

```
            $nomod51
            $include    (C8051F040.inc)
tovCnt      set         R5                  ; PCA counter overflow cnt
pwH         set         R6                  ; pulse width high byte
pwL         set         R7                  ; pulse width low byte
            org         0x00
            ljmp        start
            org         0x4B                ; PCA interrupt vector
            ljmp        pcaISR              ; jump to the service routine of PCA interrupt
            org         0xAB                ; starting address of the program
start:      mov         SP,#0x80            ; set up stack pointer
            lcall       sysInit             ; call for system initialization
            mov         SFRPAGE,#0          ; switch to SFR page 0 for PCA function
            mov         tovCnt,#0           ;          "
            mov         PCA0CN,#00          ; clear all PCA flags and disable PCA counter
            mov         PCA0MD,#0x81        ; initialize the PCA0MD register, enable PCA0 timer
                                            ; interrupt locally
            mov         PCA0CPM0,#0x20      ; select capture rising edge
            mov         IE,#00H             ; disable all interrupts at the beginning
            setb        CR                  ; enable the PCA timer to run
riseEdge:   jnb         CCF0,riseEdge       ; wait for the arrival of the rising edge
            mov         pwH,PCA0CPH0        ; save the rising edge
            mov         pwL,PCA0CPL0        ;     "
            mov         PCA0CPM0,#0x10      ; prepare to capture falling edge of the CEX0 signal
            clr         CF                  ; clear the PCA timer overflow flag
            clr         CCF0                ; also clear the module 0 capture flag
            mov         EIE1,#0x08          ; enable PCA0 interrupt
            setb        EA                  ; enable interrupt globally
fallEdge:   jnb         CCF0,fallEdge       ; wait for the arrival of the falling edge
            mov         EIE1,#0             ; disable PCA interrupt
            mov         A,PCA0CPL0          ; compare second edge with the first edge
            clr         C                   ;     "
            subb        A,pwL               ;     "
            mov         pwL,A               ;
            mov         A,PCA0CPH0          ;     "
            subb        A,pwH               ;     "
            mov         pwH,A               ;     "
```

```
            jnc         e2large
            dec         tovCnt              ; first edge is larger, so decrement tovCnt
e2large:    ajmp        $                   ; stay here forever
;*******************************************************************************************
; The following subroutine disables watchdog timer, selects external crystal oscillator to
; generate SYSCLK, assigns I/O pins to all peripheral signals, and selects 3-wire SPI.
;*******************************************************************************************
sysInit:    mov         SFRPAGE,#0x0F       ; switch to SFR page F
            mov         WDTCN,#0xDE         ; disable watchdog timer
            mov         WDTCN,#0xAD         ;     "
            mov         CLKSEL,#0           ; used internal oscillator to generate SYSCLK
            mov         OSCXCN,#0x67        ; configure external oscillator control
            mov         R7,#255
            djnz        R7,$                ; wait for about 1 ms
chkstable:  mov         A,OSCXCN            ; wait until external crystal oscillator is stable
            anl         A,#80H              ; before using it
            jz          chkstable           ;     "
            mov         CLKSEL,#1           ; use external crystal oscillator to generate SYSCLK
            mov         XBR0,#0xF7          ; assign all peripheral signals to port pins
            mov         XBR1,#0xFF          ;     "
            mov         XBR2,#0x5D          ;
            mov         XBR3,#0x8F          ;
            mov         SFRPAGE,#0          ; switch to SFR page 0
            mov         SPI0CN,#01          ; enable 3-wire SPI mode
            ret
;*******************************************************************************************
; The following is the PCA interrupt service routine.
;*******************************************************************************************
pcaISR:     anl         PCA0CN,#0x7F        ; clear CF flag
            inc         tovCnt              ; increment the PCA timer overflow count by 1
            reti
            end
```

The C language version of the program is as follows.

```
#include <c8051F040.h> // compiled using SDCC
void pcaISR (void) interrupt 9; // delete "interrupt 9" when compile using uVision
void sysInit (void);
unsigned char pwH, pwL,tovCnt;
unsigned long PulseWidth;
void main (void)
{
    unsigned int t1, t2;
    sysInit();
    SFRPAGE   = 0;              // switch to SFR page 0
    tovCnt    = 0;
    PCA0CN    = 0x00;           // clear all PCA flags
    PCA0MD    = 0x81;           // use SYSCLK/12 as the PCA timer clock input
    PCA0CPM0 = 0x20;            // prepare to capture the CEX0 rising edge
    IE        = 0;              // disable interrupt
```

```
        CR        = 1;                 // start the PCA timer
        while (!CCF0);                 // wait for the arrival of the rising edge
        pwH       = PCA0CPH0;          // save the captured rising edge value
        pwL       = PCA0CPL0;          //    "
        PCA0CPM0 = 0x10;               // prepare to capture CEX0's falling edge
        CCF0      = 0;
        CF        = 0;
        EIE1      = 0x08;              // enable PCA0 timer overflow interrupt
        EA        = 1;                 // enable interrupt globally
        while (!CCF0);                 // wait for the arrival of the falling edge
        EIE1      = 0;                 // disable PCA interrupts
        t1        = 256 * (unsigned int)pwH + (unsigned int)pwL;
        t2        = 256 * (unsigned int)PCA0CPH0 + (unsigned int)PCA0CPL0;
        if (t2 < t1)
            tovCnt--;                  // decrement PCA timer overflow count if edge 2 is smaller
        PulseWidth = (unsigned long) tovCnt * 65536 + (unsigned long)t2 – (unsigned long)t1;
}
void sysInit (void)
{
        unsigned char n;
        SFRPAGE = 0x0F;
        WDTCN   = 0xDE;                // disable watchdog timer
        WDTCN   = 0xAD;
        OSCXCN  = 0x67;                // start external oscillator; 24 MHz Crystal
                                       // system clock is 24 MHz
        for (n = 0; n < 255; n++);     // delay about 1 ms
        while ((OSCXCN & 0x80) == 0);  // wait for oscillator to stabilize
        CLKSEL  |= 0x01;               // switch to external oscillator
        XBR0    = 0xF7;                // assign all peripheral function pins to port pins
        XBR1    = 0xFF;                //    "
        XBR2    = 0x5D;                //    "
        XBR3    = 0x8F;                //    "
        SFRPAGE = 0;                   // switch to SFR page 0
        SPI0CN  = 0x01;                // enable 3-wire SPI
}
void pcaISR (void) interrupt 9
{
        CF      = 0;                   // clear the CF flag
        tovCnt++;                      // increment PCA timer overflow count
}
```

8.5.4 16-Bit Software Timer and High-Speed Output (Toggle)

The software timer mode also is referred to as the **compare** mode. When this mode is entered, the PCA circuit compares the PCA timer with the capture/compare register in every clock cycle. When these two 16-bit registers match, the **CCFn** flag of the PCA0CN register will be set. An interrupt will be requested if the associated interrupt-enable bit is set. This mode is entered by setting the ECOMn and MATn bits of the PCA0CPMn register.

The software timer is designed to create accurate time delays. There are three steps in using the software timer mode.

Step 1
Make a copy of the PCA timer (PCA0H, PCA0L) current value.

Step 2
Add a value that represents the desired time delay to this copy.

Step 3
Store the sum in the PCA compare registers (PCA0CPHn, PCA0CPLn).

When writing a 16-bit value into the PCA capture/compare registers, the low byte should be written first. Writing to PCA0CPLn clears the ECOMn bit to 0; writing to PCA0CPHn sets the ECOMn bit to 1 and starts the compare process.

The high-speed output (HSO) mode toggles a CEXn pin when a match occurs between the PCA timer and a module's compare register. For this mode, the TOGn bit in the PCA0CPMn register needs to be set in addition to the ECOMn and MATn bits, as seen in Figure 8.16. The user also has the option of flagging an interrupt when a match event occurs by setting the ECCFn bit.

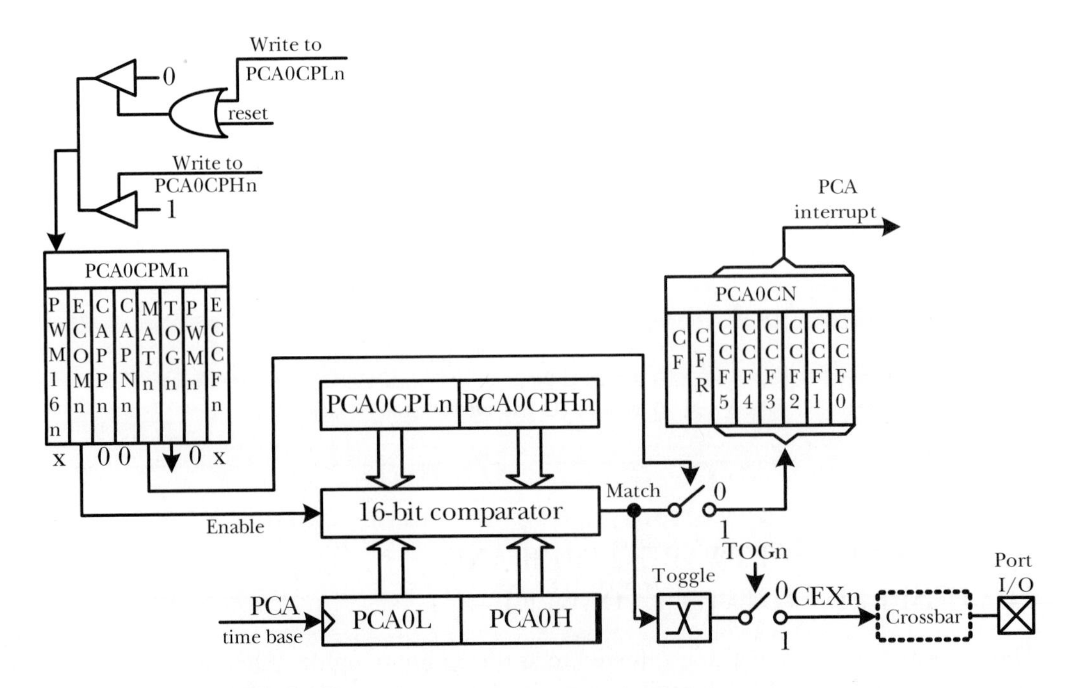

FIGURE 8.16
PCA software timer and HSO modes diagram.

The PCA software-timer mode simplifies time-delay creation. The following example illustrates this application.

Example 8.10 Write a subroutine that uses the PCA module 0 of the C8051F040 to create a time delay that is a multiple of 10 ms, assuming the system clock is 24 MHz.

Solution: Choose **SYSCLK ÷ 12** as the clock input to the PCA timer. The number of timer clocks required to create a 10-ms time delay is 20,000. This example will use register R0 to pass the multiple of the 10-ms delay to be created.

```
delayby10ms:
            push    SFRPAGE
            mov     SFRPAGE,#0
            mov     PCA0MD,#0x80        ; use SYSCLK/12 as PCA clock, suspend PCA when idle
            mov     PCA0CPM0,#0x48      ; select software timer mode
            mov     PCA0CN,#0           ; disable PCA timer
loopz:      mov     A,PCA0L             ; add 20,000 as the delay count
            add     A,#low 20000        ;     "
            mov     PCA0CPL0,A          ;     "
            mov     A,PCA0H             ;     "
            addc    A,#high 20000       ;     "
            mov     PCA0CPH0,A          ; start compare operation
            clr     CCF0
            setb    CR                  ; start PCA timer
            jnb     CCF0,$              ; wait until CCF0 is set to 1 again
            clr     CR                  ; disable PCA timer
            djnz    R0,loopz
            pop     SFRPAGE
            ret
```

The C language of the function is as follows.

```
void delayby10ms (unsigned char kk)
{
    unsigned char i, temp;
    temp      = SFRPAGE;
    SFRPAGE   = 0;
    PCA0MD    = 0x80;     // suspend PCA in idle mode and uses SYSCLK/12 as PCA clock
    PCA0CPM0 = 0x48;      // select software timer mode
    PCA0CN    = 0x00;     // disable PCA timer
    for (i = 0; i < kk; i++){
        PCA0CPL0 = PCA0L + 0x20;
        if (CY)           // check carry flag
            PCA0CPH0 = PCA0H + 0x4F;
        else
            PCA0CPH0 = PCA0H + 0x4E;
```

```
        CCF0 = 0;
        CR   = 1;
        while(!CCF0);
        CR   = 0;
    }
    SFRPAGE  = temp;
}
```

Example 8.11 Assume that there is a C8051F040 demo board running with a 24-MHz crystal oscillator. Write a program that enables the PCA module 0 to generate a 1-KHz digital waveform with 30 percent duty cycle.

Solution: A 1-KHz digital waveform with 30 percent duty cycle is shown in Figure 8.17.

We will use **OSC ÷ 12** as the clock input to the PCA timer. The period of the waveform to be generated is 1 ms and is equal to 2000 PCA clock cycles. With the 30 percent duty cycle, the high interval corresponds to 600 PCA counter counts, whereas the low interval corresponds to 1400 PCA counter counts.

There are several alternatives for generating this waveform. One of the alternatives is to pull the CEX0 pin to high at the beginning and then use compare toggle mode to generate the waveform. The main program starts the first compare operation and enables the CEX0 interrupt. The CEX0 interrupt-service routine continues to trigger the subsequent compare operations. This approach allows the CPU to perform other operations while generating the waveform.

The procedure for generating the specified waveform is as follows.

Step 1 Configure PCA module 0 to toggle mode and enable CCF0 interrupts. Pull the CEX0 pin to high.

Step 2 Use a register as a flag (called it **dlyflag**) to specify the delay to be used in the compare operation. When **dlyflag** is 1, use 600 as the delay count. When **dlyflag** is 0, use 1400 as the delay count.

Step 3 Make a copy of the PCA counter, add **600** to it, and store the sum in CAPCPH0:CAPCPL0 register pair. Set **dlyflag** to 0 and stay in a **wait** loop.

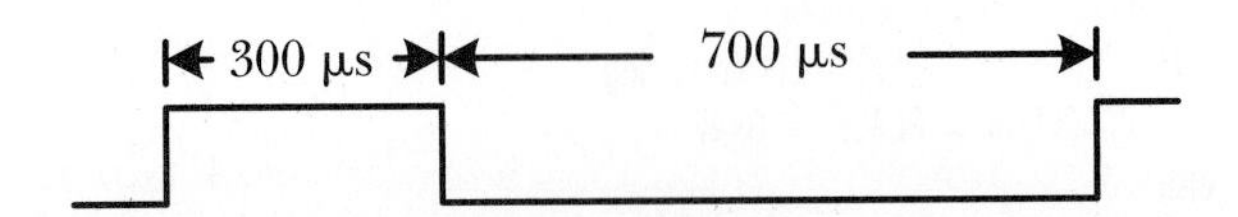

FIGURE 8.17
1-kHz 30percent duty-cycle periodic waveform.

Write an interrupt-service routine that performs the following operations.

Step 1 Clear the INTF0 flag and check the **dlyflag** value.

Step 2

```
If dlyflag = 1 then
    PCA0CPH0:PCA0CPL0 += 600;
    dlyflag = 0;
else
    PCA0CPH0:PCA0CPL0 += 1400;
    dlyflag = 1;
```

Step 3 Return from interrupt.

The assembly program that generates the specified waveform is as follows.

```
            $nomod51
            $include (C8051F040.inc)
dlyflag     set     R7                  ; use HT (LT) as delay if dlyflag == 1(0)
HTHB        equ     0x02                ; high byte of high interval (600)
HTLB        equ     0x58                ; low byte of high interval (600)
LTHB        equ     0x05                ; high byte of low interval (1400)
LTLB        equ     0x78                ; low byte of low interval (1400)

            org     0x00
            ljmp    start
            org     0x4B                ; PCA interrupt vector
            ljmp    pcaISR              ; jump to actual PCA interrupt service routine
            org     0xAB
start:      mov     SP,#0x80            ; set up the stack pointer
            lcall   sysInit             ; go to configure I/O port
            mov     SFRPAGE,#PCA0_PAGE  ; switch to PCA0 SFR page
            mov     PCA0CN,#0           ; disable PCA0, clear all PCA flags
            mov     PCA0MD,#0x80        ; use SYSCLK/12 as PCA clock, disable PCA
                                        ; overflow interrupt
            mov     PCA0CPM0,#0x4D      ; enable compare, match, toggle, and module 0 interrupt
            setb    P1.1                ; pull the CEX0 (P1.1) to high (important)
            mov     A,PCA0L             ; start the first compare operation
            add     A,#HTLB             ;     "
            mov     PCA0CPL0,A          ;     "
            mov     A,PCA0H             ;     "
            addc    A,#HTHB             ;     "
            mov     PCA0CPH0,A          ;     "
            setb    CR                  ; enable PCA0 counter
            mov     dlyflag,#0          ; next compare operation to produce the low time
            orl     EIE1,#0x08          ; enable PCA0 interrupt
            orl     EIP1,#0x08          ; place PCA0 in interrupt at high priority
```

```
            setb    EA                      ; enable interrupt globally
            ajmp    $
;*****************************************************************************************
; This subroutine disables watchdog timer, select external crystal oscillator to
; generate SYSCLK, assigns I/O pins to all peripheral signals, and selects 3-wire SPI.
;*****************************************************************************************
sysinit:    mov     SFRPAGE,#CONFIG_PAGE    ; switch to the SFR page that contains
                                            ; registers that configures the system
            mov     WDTCN,#0xDE             ; disable watchdog timer
            mov     WDTCN,#0xAD             ;    "
            mov     CLKSEL,#0               ; used internal oscillator to generate SYSCLK
            mov     OSCXCN,#0x67            ; configure external oscillator control
            mov     R7,#255
            djnz    R7,$                    ; wait for about 1 ms
chkstable:  mov     A,OSCXCN                ; wait until external crystal oscillator is stable
            anl     A,#80H                  ; before using it
            jz      chkstable               ;    "
            mov     CLKSEL,#1               ; use external crystal oscillator to generate SYSCLK
            mov     XBR0,#0xF7              ; assign I/O pins to all peripheral functions
            mov     XBR1,#0xFF              ; and enable crossbar
            mov     XBR2,#0x5D              ;    "
            mov     XBR3,#0x8F              ;    "
            mov     P1MDOUT,#0x07           ; enable P1.2, P1.1, & P1.0 output
            mov     SFRPAGE,#SPI0_PAGE      ;
            mov     SPI0CN,#0x01            ; select 3-wire SPI mode (to fix pin assignment)
            ret
;*****************************************************************************************
; PCA interrupt service routine.
;*****************************************************************************************
pcaISR:     anl     PCA0CN,#0x40            ; clear CF and all CCF flags
            mov     A,dlyflag               ; find out the delay value to use (HT or LT)
            jnz     addhi                   ; if dlyflg == 1, then use HT as delay.
            mov     A,PCA0CPL0              ; start the next compare operation with
            add     A,#LTLB                 ; LTHB:LTLB as delay
            mov     PCA0CPL0,A              ;    "
            mov     A,PCA0CPH0              ;    "
            addc    A,#LTHB                 ;    "
            mov     PCA0CPH0,A              ;    "
            mov     dlyflag,#1              ; change delay flag to 1
            reti
addhi:      mov     A,PCA0CPL0              ; start the next compare & toggle operation
            add     A,#HTLB                 ; using HTHB:HTLB as delay
            mov     PCA0CPL0,A              ;    "
            mov     A,PCA0CPH0              ;    "
            addc    A,#HTHB                 ;    "
            mov     PCA0CPH0,A              ;    "
            mov     dlyflag,#0              ; clear delay flag to 0
            reti
            end
```

The C language version of the program is as follows:

```
#include <C8051F040.h>
#define     HThi     0x02        // high interval delay count
#define     HTlo     0x58        //     "
#define     LThi     0x05        // low interval delay count
#define     LTlo     0x78        //     "
#define     HI       1
#define     LO       0

void PCA0_ISR (void);            // prototype of PCA0 interrupt service routine
void sysInit (void);             // system initialization function prototype
unsigned char hiORlo;
unsigned char temp;              // used to check if a carry to high byte is needed

void main (void)
{
    sysInit();
    SFRPAGE   = PCA0_PAGE;
    PCA0MD    = 0x80;            // stop PCA0 when idle, use SYSCLK/12 as PCA0 clock
    PCA0CN    = 0;               // clear all PCA0 related flag, disable PCA0 counter
    PCA0CPM0 = 0x4D;             // select high-speed toggle mode, enable compare interrupt
    P1       |= 0x02;            // pull CEX0 to high initially
    temp      = PCA0L;
    PCA0CPL0  = PCA0L + HTlo;
    PCA0CPH0  = PCA0H + HThi;
    if (PCA0CPL0 < temp)         // check carry out from low byte
        PCA0CPH0++;
    CR        = 1;               // enable PCA0 counter to count
    hiORlo    = LO;              // next time use LT as delay
    EIE1     |= 0x08;            // enable PCA0 interrupt as a group
    EIP1     |= 0x08;            // place PCA0 interrupt at high priority
    EA        = 1;               // enable interrupt globally
    while(1);                    // wait for PCA0 interrupt
}
void sysInit(void)
{
    unsigned char cx;
    SFRPAGE   = CONFIG_PAGE;
    WDTCN     = 0xDE;            // disable watchdog interrupt
    WDTCN     = 0xAD;            //     "
    OSCXCN    = 0x67;            // start external crystal oscillator
    for (cx = 0; cx < 255; cx++);
    while(!(OSCXCN & 0x80));     // wait for crystal oscillator to stabilize
    CLKSEL   |= 1;               // switch to external crystal oscillator as clock source
    XBR2      = 0x5D;            // enable crossbar decoder, and assign I/O pins to all
    XBR0      = 0xF7;            // peripheral signals
    XBR1      = 0xFF;            //     "
    XBR3      = 0x8F;            //     "
    P1MDOUT   = 0x07;            // enable CEX0 pin as output
```

```
    SFRPAGE   = SPI0_PAGE;        // switch to SFR page where the SPI is in
    SPI0CN    = 1;                // select 3-wire SPI (just to fixed I/O pin assignment)
}

void PCA0_ISR (void) interrupt 9
{
    PCA0CN   &= 0x40;             // clear all PCA0 flags
    temp      = PCA0CPL0;
    if (hiORlo == HI){
        PCA0CPL0 += HTlo;
        PCA0CPH0 += HThi;
        if (temp > PCA0CPL0)      // is there is a carry to the high byte?
            PCA0CPH0++;
        hiORlo = LO;
    } else {
        PCA0CPL0 += LTlo;
        PCA0CPH0 += LThi;
        if (temp > PCA0CPL0)      // is there is a carry to the high byte?
            PCA0CPH0++;
        hiORlo = HI;
    }
}
```

8.5.5 C8051F040 PCA Frequency-Output Mode

In this mode, the CEX pin is toggled whenever the values of the PCA0CPLn register and the PCA0L counter match. On a match, the value in the PCA0CPHn register is added to the PCA0CPLn register. The capture/compare module high byte holds the number of PCA clocks to count before the output is toggled. The frequency of the output waveform is given in the equation

$$f_{SQR} = f_{PCA} \div (2 \times \text{PCA0CPHn}) \tag{8.9}$$

A value of 0x00 in the PCA0CPHn register is equal to 256 for this equation.

This mode is entered by setting the ECOMn, TOGn, and PWMn bits in the PCA0CPMn register. The block diagram of the frequency output mode is shown in Figure 8.18.

Example 8.12 Write an assembly program that uses the frequency output mode of the PCA to generate a 10-kHz digital waveform from the CEX0 pin. This program is to be run on a C8051F040 demo board that uses the 24.5-MHz internal oscillator to generate the system clock.

Solution: We will use SYSCLK clock divided by 12 as the clock input to PCA. By using Equation 8.9, the value to be written into PCA0CPH0 is calculated to be 102. The following program will generate the 10-KHz waveform from the CEX0 pin.

```
        $nomod51
        $include (c8051F040.inc)
        org     00H
        ljmp    start
        org     0xAB
start:  acall   sysInit
        mov     SFRPAGE,#PCA0_PAGE
        mov     PCA0CPM0,#0x46      ; enable frequency output mode
        mov     PCA0MD,#0x80        ; use SYSCLK/12 as PCA clock
        mov     PCA0CPL0,#102       ; load 102 into PCA0CPL0
        mov     PCA0CPH0,#102       ;
        mov     PCA0L,#0            ; let PCA timer count up from 0
        mov     PCA0CN,#0x40        ; enable PCA counter and clear all flags
        ajmp    $                   ; stay here forever

sysInit: mov    SFRPAGE,#0x0F       ; switch to SFR page F
        mov     WDTCN,#0xDE         ; disable WDT timer
        mov     WDTCN,#0xAD         ;       "
        mov     CLKSEL,#00          ; use internal clock for SYSCLK
        mov     OSCICN,#0x83        ;       "
        mov     XBR0,#0xF7          ; assign all peripheral signals
        mov     XBR1,#0xFF          ; to port pins
        mov     XBR2,#0x5D          ; also enable crossbar
        mov     XBR3,#0x8F          ;       "
        mov     P1MDOUT,#02         ; enable CEX0 (P1.1) output
        mov     SFRPAGE,#0          ; switch to SFR page 0
        mov     SPI0CN,#01          ; enable 3-wire SPI
        ret
        end
```

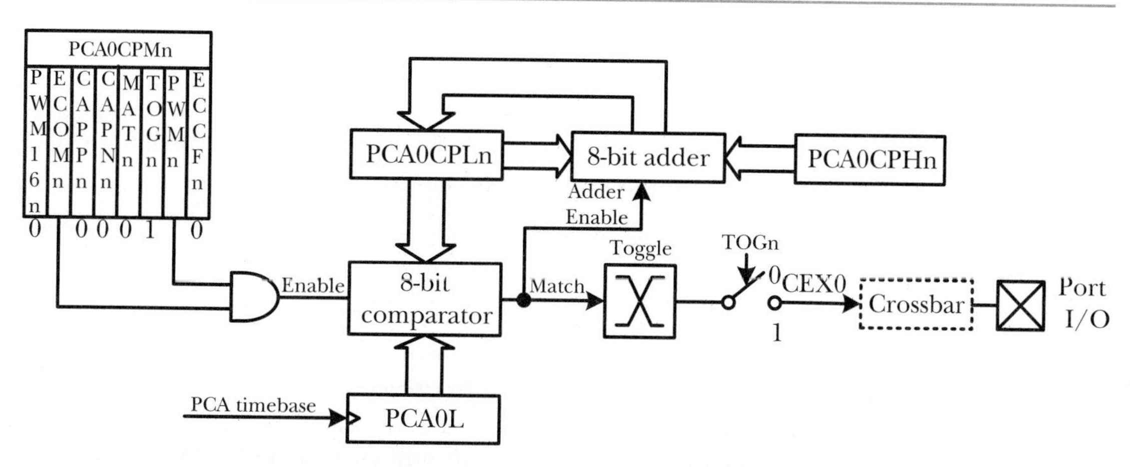

FIGURE 8.18
PCA frequency output mode diagram (C8051F040).

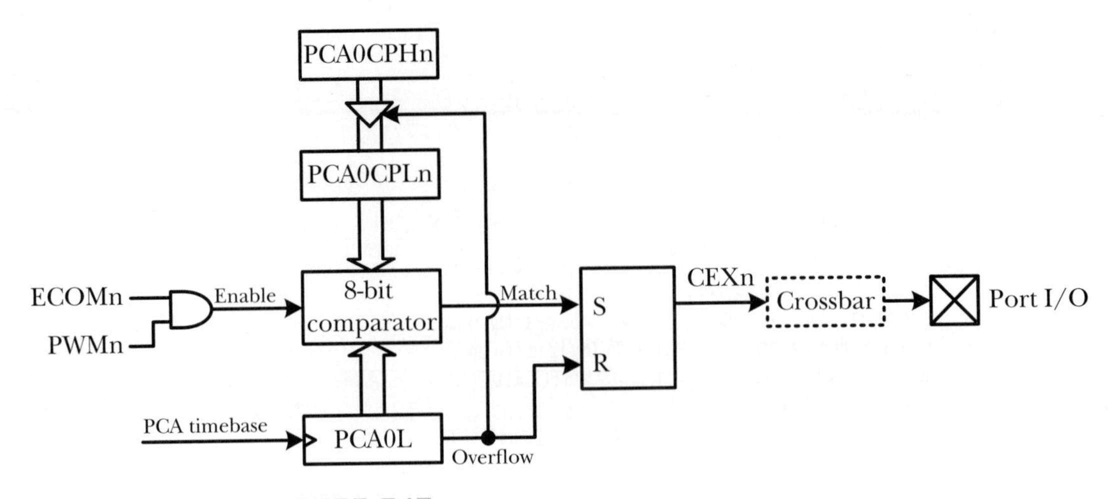

FIGURE 8.19
PCA 8-bit PWM mode diagram (C8051F040).

8.5.6 8-Bit Pulse-Width Modulation Mode (Fixed Frequency)

All of the PCA modules can be programmed to be a pulse-width modulator. The PWM output can be used to convert digital data to an analog signal by a simple external circuitry. The frequency of the PWM is dependent on the clock source for the PCA timer. The duty cycle of the PWM output signal is varied using the module's PCA0CPLn register. As shown in Figure 8.19, when the value in the low byte of the PCA0 timer is equal to the value in PCA0CPLn, the output of the CEXn pin will be pulled high. When the timer low-byte overflows from 0xFF to 0x00, the CEXn pin will be pulled low and the PCA0CPLn register is reloaded automatically with the value stored in the PCA0CPHn register. This mode is entered by setting the ECOMn and PWMn bits in the PCA0CPMn register.

By writing 0 to 255 into the PCA0CPHn register, the user can vary the duty cycle from 100 percent to 0.4 percent. The value to be written the PCA0CPHn value for a given duty cycle can be calculated by the equation

$$\text{PCA0CPHn} = 256\,(1 - \text{duty cycle}) \tag{8.10}$$

where the duty cycle is expressed as a fraction.

The frequency of the 8-bit PWM output is the chosen PCA timebase divided by 256.

Example 8.13 Write a program to configure the PCA channel 0 to generate an 8-bit PWM output with a 67 percent duty cycle. This program is to be run on a C8051F040 demo board running with a 24.5-MHz internal oscillator.

Solution: Load the value 85 into the PCA0CPH0, then the high time for the PWM output will be 171 clock cycles and the resultant PWM waveform will have about a 66.8 percent duty cycle. The following assembly program will generate the desired PWM waveform.

```
          $nomod51
          $include (c8051F040.inc)
          org       0x00
          ljmp      start
          org       0xAB
start:    mov       SP,#0x80           ; initialize the stack pointer
          lcall     sysinit
          mov       PCA0MD,#0x80       ; choose OSC/12 as clock input to PCA timer
          mov       PCA0L,#00          ; reset PCA timer to 0
          mov       PCA0H,#00          ;     "
          mov       PCA0CPM0,#0x42     ; configure module for PWM
          mov       PCA0CPL0,#85       ; set duty cycle to 66.8%
          mov       PCA0CPH0,#85       ;
          mov       PCA0CN,#0x40       ; enable the PCA timer and clear all PCA flags
forever:  nop
          ajmp      forever

sysInit:  mov       SFRPAGE,#0F        ; switch to SFR page F
          mov       WDTCN,#0xDE        ; disable WDT timer
          mov       WDTCN,#0xAD        ;     "
          mov       CLKSEL,#00         ; use internal clock for SYSCLK
          mov       OSCICN,#0x83       ;     "
          mov       XBR0,#0xF7         ; assign all peripheral signals
          mov       XBR1,#0xFF         ; to port pins
          mov       XBR2,#0x5D         ; also enable crossbar
          mov       XBR3,#0x8F         ;     "
          mov       P1MDOUT,#02        ; enable CEX0 (P1.1) output
          mov       SFRPAGE,#0         ; switch to SFR page 0
          mov       SPI0CN,#01         ; enable 3-wire SPI
          ret
          end
```

The C program that generates the specified waveform is as follows.

```
#include <C8051F040.h>
void sysInit(void);
void main (void)
```

```
{
    sysInit();
    PCA0MD    = 0x80;        // choose OSC/12 as clock input to PCA timer
    PCA0L     = 0;           // let PCA0 counter counts from 0
    PCA0H     = 0;           //     *
    PCA0CPM0 = 0x42;         // configure module for 8-bit PWM mode
    PCA0CPL0 = 85;           // set duty cycle to 66.8%
    PCA0CPH0 = 85;           //     "
    PCA0CN    = 0x40;        // enable the PCA timer and clear all PCA flags
    while(1);
}
void sysInit(void)
{
    ....
}
```

The PWM function has been used extensively in motor control. This application will be discussed in Section 8.6.

8.5.7 16-Bit Pulse-Width Modulator Mode (Fixed Frequency)

One limitation of the 8-bit PWM mode is that its output frequency is relatively high unless you use a low-frequency crystal oscillator. The C8051F040 PCA also provides a 16-bit width modulator mode which can divide down the PWM output frequency further. In this mode, the 16-bit capture/compare register defines the number of PCA clocks for the low time of the PWM signal. When the PCA timer matches the module capture/compare register contents, the output on the CEXn pin is asserted high; when the counter overflows, the CEXn pin is asserted low. To output a varying duty cycle, new value writes should be synchronous with PCA-CCFn match interrupts. The block diagram of the 16-bit mode PWM of the C8051F040 is shown in Figure 8.20.

The 16-bit PWM mode is entered by setting the ECOMn, PWMn, and PWM16n bits in the PCA0CPMn register. When writing a 16-bit value to the PCA0 capture/compare registers, the low byte should be written first. Writing to PCA0CPLn clears the ECOMn bit to 0; writing to PCA0CPHn sets ECOMn to 1 and starts the compare operation. The duty cycle of the 16-bit PWM mode is given in the equation

$$\text{DutyCycle}_{16} = (65536 - \text{PCA0CPn}) \div 65536 \quad (8.11)$$

Example 8.14 Write a program to configure the PCA module 0 so that a 16-bit PWM signal with a 60 percent duty cycle is generated. This

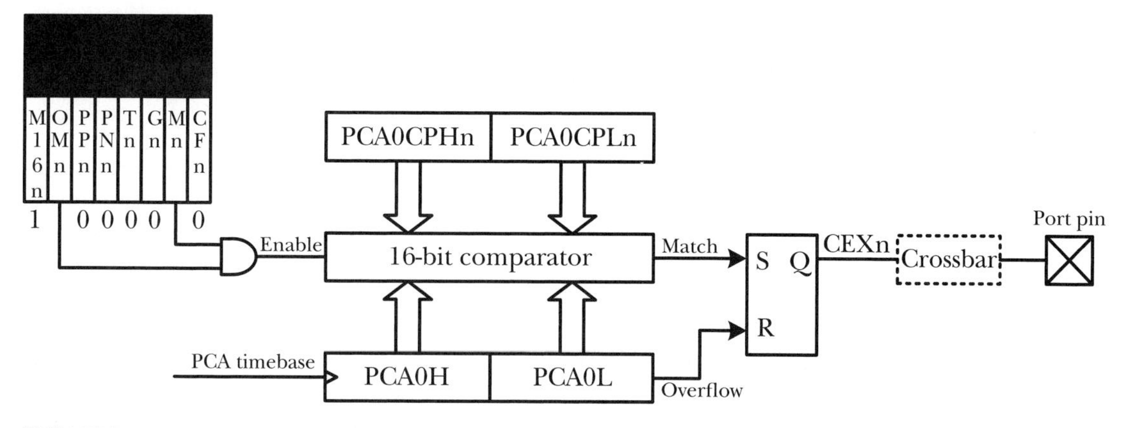

FIGURE 8.20
PCA 16-bit PWM mode (C8051F040).

program is to be run on a C8051F040 demo board running with a 24-MHz external oscillator.

Solution: Let's select the external oscillator as the system clock and use the system clock as the PCA timebase. By placing 26214 (6666H) in **PCA0CPH0:PCA0CPL0**, the duty cycle of the waveform will be set to 60 percent. The assembly program that uses the 16-bit PWM mode to generate the specified waveform is as follows.

```
$nomod51
$include    (c8051F040.inc)
            org     00H
            ljmp    start
            org     0xAB
start:      acall   sysInit
            mov     PCA0MD,#0x88        ; select SYSCLK as clock input to PCA timer
            mov     PCA0CPM0,#0xC2      ; select PCA 16-bit PWM mode
            mov     PCA0CPL0,#0x66      ; set duty cycle to about 60% (366.2 Hz)
            mov     PCA0CPH0,#0x66
            mov     PCA0L,#0            ; let PCA0 counter count up from 0
            mov     PCA0H,#0            ;     "
            mov     PCA0CN,#0x40        ; start PCA0 timer
forever:    nop
            ajmp    forever
sysInit:    ....                        ; same as Example 8.13
            end
```

8.6 DC Motor Control

A DC motor is an analog motor and is available in just about any size and is therefore common to many applications, especially those that require a large torque. Because the speed and torque of DC motors can be controlled precisely over a wide range, they are used extensively in control systems as positioning devices. The DC motor has a permanent magnetic field, and its armature is a coil. When a voltage is applied to the armature, the motor begins to spin. The applied voltage level determines the speed of rotation.

Among the several variations of DC motors, the brushless DC motor is by far the most popular one. Brushless DC motors are commonly used where precise speed control is necessary, for example, computer disk drives, spindles within CD, DVD drives, and mechanisms within office products (such as fans, laser printers, and photocopies). Modern brushless DC motors range in power from a fraction of a watt to many kilowatts. Larger brushless motors up to about 100-kW ratings are used in electric vehicles. They also find significant use in high-performance electric-model aircraft. To provide control to the speed, Hall-effect sensors often are used to provide speed information to the microcontroller that controls the DC motor.

The C8051F040 can interface with the DC motor through a driver, as shown in Figure 8.21. This configuration takes up only three I/O pins. The pin that controls the direction can be an ordinary I/O pin, but the pin that controls the speed must be a PWM pin. The pin that receives the feedback must be an input capture pin. It can be one of the unused PWM pin (n = 2, 3, 4, or 5) pin. Either pin must be configured in capture mode.

Although some DC motors operate at 5 V or less, the 8051 cannot supply the necessary current to drive a motor directly. The minimum current requirement of any practical motor is much higher than any microcontroller can supply. Depending on the size and rating of the motor, a suitable driver chip must be selected to take control signals from the 8051 and deliver the necessary voltage and current to the motor.

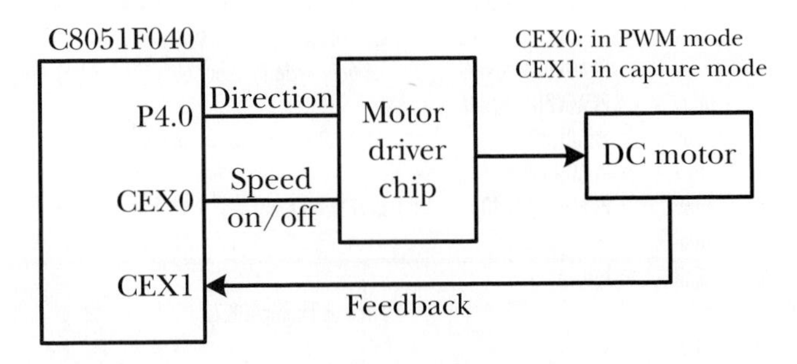

FIGURE 8.21
Simplified circuit for DC motor control.

8.6.1 DC Motor Driver ICs

Standard motor driver chips are available in many current and voltage ratings. One example is the SN754410 from TI. The SN754410 has four channels and can output up to 1 A of current per channel with a supply voltage from 4.5 to 36 V. Each chip has a separate logic supply and uses a logical input (0 or 1) to enable or disable each channel. Input pins are clamped with diodes to protect against the static and back electromotive force (EMF) generated during the reversing of a motor. The pin assignment of the SN754410 is shown in Figure 8.22. There are two supply voltages: V_{CC1} and V_{CC2}. V_{CC1} is the logic supply voltage, which is normally connected to 5.0 V. The V_{CC2} signal is the output-supply voltage and can be as high as 36 V. Drivers are enabled in pairs with drivers 1 and 2 enabled by the 1,2EN signal and drivers 3 and 4 enabled by the 3,4EN signal.

8.6.2 Driving a DC Motor Using the SN754410

A circuit that includes the C8051F040 and the SN754410 to drive the DC motor is shown in Figure 8.23. In the normal operation, the CEX0 pin outputs the PWM waveform, whereas the CEX2 pin outputs a 0 V. To rotate the DC motor in reverse direction, the CEX0 pin should be programmed to output a 0 V, whereas the CEX2 output should be programmed to output a PWM output. To brake the motor, the CEX0 should be programmed to output 0 V, whereas the CEX2 should be programmed to output 5 V for a duration that can stop the motor totally.

One way to provide feedback to the MCU is to add a Hall-effect sensor as in Figure 8.23. By attaching two small magnets on the two edges (180° apart) of the DC motor and configuring PCA module 1 in capture mode, the C8051 will be able to measure the motor rotation speed. Whenever a magnet passes by the Hall-effect sensor, it induces a pulse. This pulse can be captured by the PCA module. The difference of these two consecutive captured edges gives

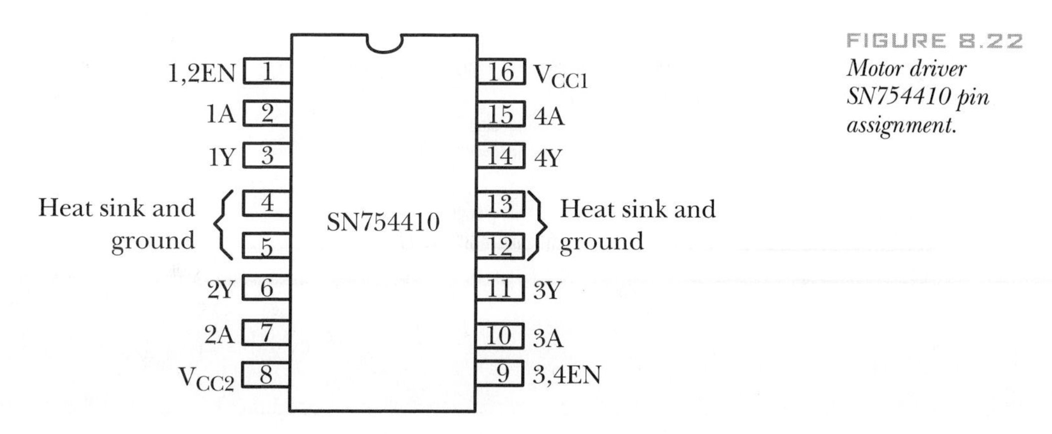

FIGURE 8.22 *Motor driver SN754410 pin assignment.*

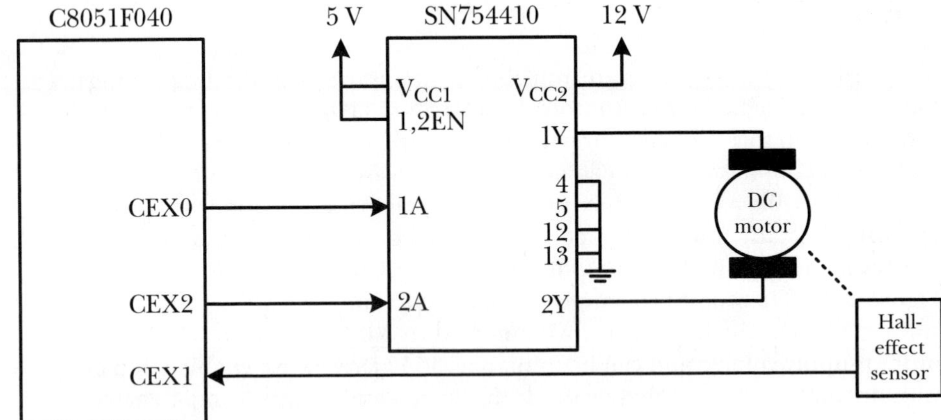

FIGURE 8.23 *A DC motor control circuit that uses Hall-effect transistor.*

the time for the motor to rotate half a cycle. Both of the programs in Examples 13 and 14 can be used to generate PWM waveforms to drive the motor in the circuit shown in Figure 8.23.

Example 8.15 For the circuit connection shown in Figure 8.25, write a C function to measure the motor speed (in rpm) assuming that the C8051F040 is running with a 24-MHz external oscillator.

Solution: To measure the motor speed, we need to capture two consecutive falling (or rising) edges. Let the difference of two consecutive edges be **diff** and the period of the timer be set to 0.5 μs. Then the motor speed (rpm) is

$$\text{Speed} = 60 \times 10^6 \div \text{diff}$$

The C function that measures the motor speed is as follows.

```
unsigned int motorRPM(void)
{
    unsigned int edge1,edge2,diff,rpm;
    long       int temp;

    PCA0CPM1 = 0x10;          /* configure PCA module 1 to capture falling edge */
    CCF1     = 0;             /* clear CCF1 flag */
    while (!CCF1);            /* wait for the first falling edge */
    edge1    = (unsigned int) PCA0CPH1 * 256 + (unsigned int)PCA0CPL1;
    CCF1     = 0;
    while (!CCF1);            /* wait for the second falling edge */
    edge2    = (unsigned int) PCA0CPH1 * 256 + (unsigned int)PCA0CPL1;
    diff     = edge2 - edge1;
    temp     = 60000000L     /(long) diff;
    rpm      = temp;
    return rpm;
}
```

8.7 Chapter Summary

Timer functions make many applications possible. With a timer, delay creation becomes much easier. Without a timer, signal parameter measurement and waveform generation become extremely difficult or even impossible. For many microcontrollers, capture, compare, and PWM are the three most popular timer functions that they provide. The programmable counter array (PCA) is the solution from SiLabs to support these three timer functions.

The original 8051 microcontroller provides Timers 0 and 1. Both timers have four operation modes. They can use a system clock or external pin divided by a prescale factor as their clock input. They often are used to create time delays and measure external events. They also can be used to measure signal frequency and pulse width.

Timer 2 is added to 8052 MCUs. Timer 2 provides capture, auto-reload, and baud-rate generation modes. Timer 2 can be programmed to count up or count down. The auto-reload mode of Timer 2 makes it easy to make sound or play songs. Timers 3 and 4 of the C8051F040 are identical to Timer 2.

The C8051F040 also adds a programmable counter array (PCA) that implements the following capabilities:

- Edge-triggered capture
- 16-bit software timer (called the compare function in some other microcontrollers)
- High-speed output (with programmable duty cycle)
- Frequency output (with a 50 percent duty cycle)
- 8-bit pulse-width modulation (PWM) with fixed frequency but programmable duty cycle
- 16-bit PWM with fixed frequency but programmable duty cycle

The PCA consists of a 16-bit timer/counter and six 16-bit compare/capture modules. Signal pins related to the operation of PCA include ECI, CEX0, CEX1, CEX2, CEX3, CEX4, and CEX5. The operation of the PCA main timer (PCA0H and PCA0L) is controlled by registers PCA0CN and PCA0MD. Each PCA module uses the PCA0CPMn ($n = 0, \ldots, 5$) register to control its operation and uses PCA0CPHn and PCA0CPLn to capture the arrival times of signal edges or hold the value for comparison during the compare operation. A PCA module may toggle its associated signal pin voltage on a match in a compare operation. This feature allows the PCA module to generate digital waveforms with a programmable frequency and duty cycle.

All PCA modules can be configured as pulse-width modulators (PWM). A pulse-width modulator can be used to approximate a D/A converter, which can be used to control a DC motor. The PWM mode of the C8051F040 does not provide much control on the frequency of the generated digital waveform. Fortunately, frequency does not matter for most applications that require PWM. One exception is the servo motor, which requires the driving signal to be exactly 50 Hz. When applications require the frequency to be in a certain range, the high-speed output mode can provide the desired frequency programmability.

DC motors are used wherever precise speed control is required. The speed of the DC motor is controlled by changing the voltage applied to the DC motor. The PWM mode of the PCA especially is suitable for this application. Almost every application that uses a DC motor requires it to reverse its direction of rotation or vary its speed. Reversing the direction is achieved by changing the polarity of the voltage applied to the motor. Sometimes there is a need to brake a DC motor. When a DC motor is turning, it picks up momentum. Turning off the voltage to the motor won't stop the motor immediately. To stop the DC motor abruptly, you need to reverse the voltage applied to the DC motor.

8.8 Exercise Problems

E8.1 Write a program to use Timer 2 of the C8051F040 to create a time delay of 20 ms, assuming that the C8051F040 is running with a 24-MHz external oscillator.

E8.2 Assume that you are using Timer 1 to measure the pulse width by connecting the signal to INT1 pin, and the C8051F040 is running with a 24-MHz external oscillator. What is the pulse width in μs if Timer 1 has the final count of 24000 (Timer 1 counts up from 0) and the clock input to Timer 1 is $f_{OSC}/12$?

E8.3 Write a program to measure the period of a signal connected to the CEX5 pin, assuming this program is to be run on a C8051F040 demo board running with a 24-MHz oscillator.

E8.4 Assume that there is a C8051F040 demo board running with a 24-MHz crystal oscillator. Write a program to measure the duty cycle of the digital signal connected to the CEX0 pin. Assume that the signal period is shorter than 30 ms, so that the PCA counter overflow need not be a concern during the measurement process.

E8.5 Write a program to generate a 10-KHz digital waveform with a 60 percent duty cycle from the CEX1 pin. Use the high-speed output mode to generate the waveform. This program is to be run on a C8051F040 demo board running with a 24-MHz external oscillator.

E8.6 Write a program to generate a 50-Hz digital waveform with a 5 percent duty cycle using the CEX5 pin, assuming that the C8051F040 is running with a 24-MHz crystal oscillator.

E8.7 Write a program to generate a 25-Hz digital waveform with a 50 percent duty cycle on the CEX0 pin as long as the voltage on the P4.0 pin is high. Your program should consist of two parts:

1. **Entry test.** As long as the P4.0 pin is low, it stays in this loop.
2. **Waveform generation body.** This part generates a pulse with 20-ms high time and 20-ms low time and, at the end of a period, tests the P4.0 signal. If the P4.0 signal is still high, it generates the next pulse. Otherwise, it jumps to entry test.

This program is to be run on a C8051F040 demo board running with a 24-MHz external oscillator. This program simulates the behavior of an **antilock** system.

E8.8 What value should be placed in PCA0CPHn in order to produce both 40 percent and 90 percent duty cycle PWM pulse outputs from the CEXn pin in the 8-bit PWM mode of the C8051F040?

E8.9 Write a program to generate ten pulses from the CEX4 pin. The pulse width is 160 μs, and two adjacent pulses are separated by 240 μs. This program is to be run on a C8051F040 demo board running with a 24-MHz oscillator.

E8.10 Write a program to generate a 500-kHz digital waveform with a 50 percent duty cycle from the T2 pin of a C8051F040 running with a 24-MHz crystal oscillator.

E8.11 Write a program to generate a 200-KHz digital waveform with a 50 percent duty cycle from the T2 pin of a C8051F040 running with a 24-MHz crystal oscillator.

E8.12 Compare the four waveform-generation methods mentioned in this chapter: (a) use software to set and reset a port pin, (b) use Timer 2 clock-out mode, (c) use the PCA high-speed output mode, and (d) use a pulse-width modulator.

E8.13 Write a program that uses the PCA module 1 (in high-speed output mode) to play the song "Home, Sweet Home," assuming that the C8051F040 demo board is running with a 24-MHz external oscillator.

E8.14 What are the highest and lowest frequencies of the PCA output in 8-bit PWM mode, 16-bit PWM mode, and frequency-output mode, assuming that the C8051F040 is running with a 24-MHz external crystal oscillator?

E8.15 Write a program to generate a 1-kHz waveform with a 30 percent duty cycle using the high-speed output mode of the PCA module of the C8051F040 running with a 24-MHz oscillator.

E8.16 Write a program to generate a 50-KHz digital waveform from the CEX5 pin. This program is to be run on a demo board that uses the 24-MHz external crystal oscillator to generate the system clock.

E8.17 Write a function to create a time delay of 100 μs, assuming the system clock is 24 MHz.

E8.18 Write a function to create a time delay of 1 ms, assuming the system clock is 24 MHz.

8.9 Laboratory Exercise Problems and Assignments

L8.1 **Three-Tone Siren Generation.** Using an output pin and a speaker to make sound, a vibration on something is needed. If we are to use a speaker to make sound, then we need to make the speaker to vibrate at certain frequency in the audible range. The audible-sound frequency range is from 20 Hz up to 20,000 Hz.

There is a permanent magnet and an electric magnet (a coil) in a speaker. When a high voltage is applied to one terminal of the speaker while at the same time the other terminal of the speaker is grounded, a current flows through the electric magnet and magnetizes it. Now the electric magnet and the permanent magnet attract each other and move close to each other. When both terminal of the speaker are at the ground level, there is no current flowing into the speaker, and the permanent magnet and the electric magnet are separated. By controlling the frequency of which these two magnets are attracting and separating each other, the air is vibrating, and hence, a sound is created.

An I/O pin can be used to make a sound. Before making the sound, connect the I/O pin to one terminal of the speaker (or a buzzer on the demo board) and the other terminal of the speaker to the ground. By writing a program to pull the pin to high and then pull it to low for the same duration and repeating it, a sound is created.

In this lab assignment, you are assigned to create a siren with three tones at **250 Hz**, **500 Hz**, and **1 kHz** using the P4.7 pin. Each tone lasts for half a second before changing to the next tone.

L8.2 **Period Measurement and Waveform Generation.** You are assigned to write a program to perform either **period measurement** using input capture or **waveform generation** using the high-speed mode of PCA. The procedure of this program is as follows.

Step 1
Configure the PCA timer to use the external crystal oscillator divided by 12 as the clock source so that the frequency of the clock signal to the timer is set to 2 MHz for a 24 MHz oscillator.

Step 2
Output a message on the LCD asking the user whether to perform period measurement or waveform generation.

Step 3
Wait for the user to enter **1** or **2** using the DIP switchs. Signal period measurement is performed if the user enters a **1**. Waveform generation is performed if the user enters a **2**. If the user enters any other choices, repeat Steps 2 and 3. This step must be performed using interrupt. Your program reserves a memory location called **avail** and clears it to 0 before waiting for you to set the DIP switches. You set a value in the DIP switch and then press a switchs connected to INT0 pin. The INT0 service routine simply sets the variable **avail** to 1 and return. The main program waits until **avail** is set to 1, then reads the DIP and continues.

Step 4
The signal of which the period is to be measured is taken from the function generator on the demo board or your laboratory bench. Make sure that the output of the function generator is between 0 and 3.3 V. You can achieve this by adjusting the magnitude and offset

of the function-generator output. If your demo board has a function generator, this requirement is satisfied already. You can use any input-capture channel (0 to 5) to perform period measurement. You will need to keep track of PCA timer overflow when measuring the period. Display the period on the LCD screen.

Step 5
Generate a 2-KHz waveform with a 70 percent duty cycle if you choose waveform generation. Use the oscilloscope to display the generated waveform.

L8.3 **Song Playing.** Configure module 1 of the PCA function to high-speed output mode to play a song of your choice. The lab procedure is as follows.

Step 1
Connect a speaker to the CEX1 pin.

Step 2
Convert the score of the song that you choose to two tables: one is the duration table; the other is the frequency table.

Step 3
Write a program to configure the PCA module 1 to high-speed output mode and looks up the two tables that you created in Step 2. Follow the algorithm similar to Example 8.9 to play the song.

One of the websites that you can find free song scores is www.cpdl.org/wiki/index.php/Category:Choral_music.

L8.4 Using the PWM and compare mode of the PCA to drive **DC** and **servo motors**, respectively. The 8051s PCA configured in PWM and high-speed output mode can be used to drive the DC motor, whereas the high-speed output mode can be used to drive the servo motor. The cooling fan with a DC motor (D24-B10A-04W4-000 from Globe motors) used in this lab is a single-phase motor and has only two terminals: one is to be connected to the motor-driver chip output (driven by PWM output), and the other pin is connected to ground. A picture of this fan is shown in Figure L8.4a.

A **servo** is a small device that incorporates a three-wire DC motor, a gear train, a potentiometer, an integrated circuit, and an output shaft bearing. The shaft of the servo motor can be positioned to specific angular positions by sending coded signals. As long as the coded signal exists on the input line, the servo motor will maintain the angular position of the shaft. If the coded signal changes, then the angular position of the shaft changes.

A common use of servo motors is in radio-controlled models (like cars, airplane, robots, and puppets). They also are used in powerful heavy-duty sail boats. Servos come in different sizes but use similar control schemes and are extremely useful in robotics. The motors are

FIGURE L8.4A
A cooling fan with DC motor for Lab Exercise 8.4.

small and extremely powerful for their size. They also draw power proportional to the mechanical load. A lightly loaded servo, therefore, doesn't consume much energy.

A typical servo looks like a rectangular box with a motor shaft coming out of one end and a connector with three wires out of the other end. The three wires are the power, control, and ground. Servos work with voltages between 4 and 6V. The control line is used to position the servo. Inexpensive servos have plastic gears, and more expensive servos have metal gears which are much more rugged but wear faster.

Servos are constructed from three basic pieces: a motor, a potentiometer that is connected to the output shaft, and a control board. The potentiometer allows the control circuitry to monitor the current angle of the servo motor. The motor, through a series of gears, turns the output shaft and the potentiometer simultaneously. The potentiometer is fed into the servo control circuit, and when the control circuit detects that the angle is not correct, it will turn the motor the correct direction until the angle is correct. Normally, a servo is used to control an angular motion between 0 and 180 degrees. It is not mechanically capable (unless modified) of turning any farther due to the mechanical stop built into the main output gear.

Servos are controlled by sending them a pulse of variable width. The control wire is used to send this pulse. As shown in Figure L8.4b, the pulse has a **minimum pulse**, **a maximum pulse**, and a **repetition**

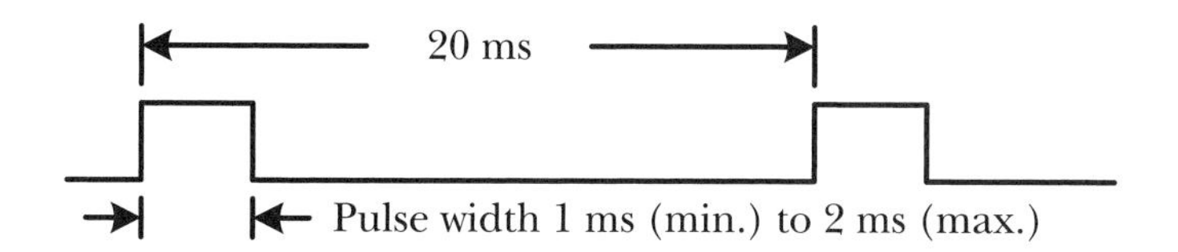

FIGURE L8.4B
Pulse patern of a servo motor.

rate. Given the rotation constraints of the servo, neutral is defined to be the position where the servo has exactly the same amount of potential rotation in the clockwise direction as it does in the counterclockwise direction. It is important to note that different servos will have different constraints on their rotation, but they all have a neutral position, and that position is always 1.5 ms.

The angle is determined by the duration of a pulse that is applied to the control line. The servo expects to see a pulse every 20 ms. The length of the pulse will determine how far the motor turns. For example, a 1.5-ms pulse will make the motor turn to the 90° position (neutral position).

When a servo is commanded to move, it will move to the position and hold that position. If an external force pushes against the servo while the servo is holding a position, the servo will resist from moving out of that position. The maximum amount of force the servo can exert is the torque rating of the servo. Servos will not hold their position forever though; the position pulse must be repeated to instruct the servo to stay in position.

As shown in Figure L8.4c, when a pulse is sent to the servo that is less than 1.5 ms, the servo rotates to a position and holds its output shaft some number of degrees counterclockwise from the neutral point. When the pulse is wider than 1.5 ms, the opposite occurs. The minimal width and the maximum width of pulse that will command the servo to turn to a valid position are functions of each servo. Different brands (and even different servos of the same brand) will have different maximums and minimums. Generally, the minimum pulse will be about 1-ms wide and the maximum pulse will be 2-ms wide.

Another parameter that varies from servo to servo is the turn rate. This is the time it takes from the servo to change from one position to another. The worst-case turning time is when the servo is holding at the minimum rotation and it is commanded to go to maximum rotation. This can take several seconds for very high-torque servos.

The diagram of the HS-311 servo motor to be used in this lab is shown in Figure L8.4d.

There are three terminals connected to the control circuit Figure L8.4a:

- Black wire: ground
- Red wire: connected to V_{CC}
- Yellow wire: connected to signal (PWM output)

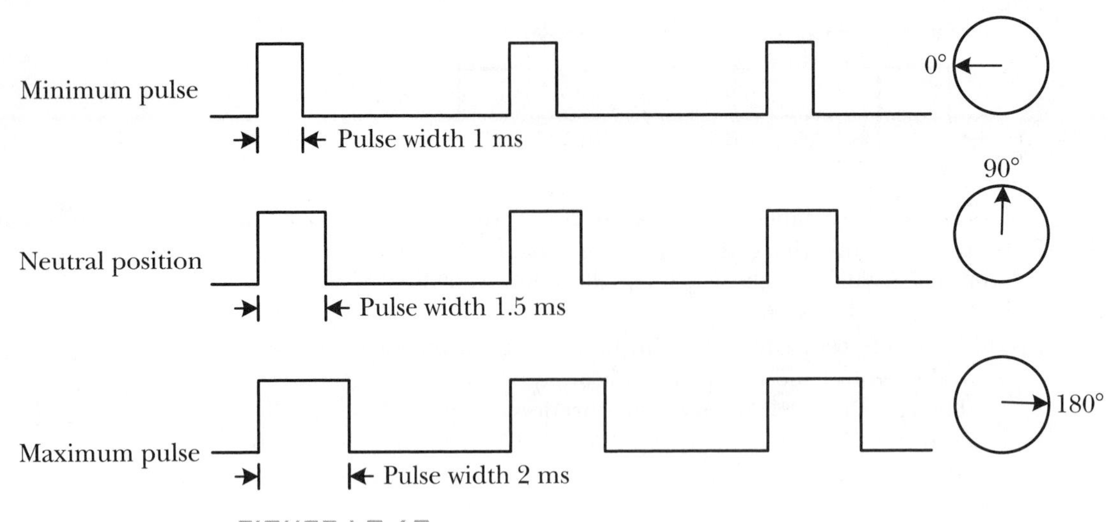

FIGURE L8.4C
Servo control pulse width and motor position.

FIGURE L8.4D
Hitec HS-311 hobby servo motor.

The 8051 PCA cannot supply enough current to drive the DC or servo motor directly. A motor-driver chip such as the **SN754410** from TI is needed to supply the current needed by the motor. The pin assignment of **SN754410** is shown in Figure 8.25. The lab procedure is as follows.

Step 1
Connect the circuit properly. Connect the CEX0 and CEX1 pins to the 1A and 2A pins of the SN754410, respectively. Connect the 1Y and 2Y pins to the DC motor and servo motor control input, respectively. Servo motors need the driving signal to be 50 Hz. This signal can be generated by using the PCA module 1 configured in high-speed output mode.

Step 2
Write a program that enables and configures the PCA function properly. The frequency of the PWM output for this lab is fixed (at what frequency?) but the duty cycle will be changed in response to the user input.

Step 3
Enter, assemble, and download the program onto the demo board for execution.

Step 4
Run the program. When the program runs, it outputs the message **duty cycle for DC:** to the UART port (displayed on the Hyper Terminal window). The user will enter from 00 to 99 (using the PC keyboard) to set the duty cycle for the DC motor and your program reads. After that, the MCU will update the duty cycle for the DC motor immediately. The C8051F040 next outputs the message **duty cycle for servo:** to the HyperTerminal window. The user will enter the new duty cycle for the digital waveform that drives the servo motor. Remember that the duty cycle for the servo motor is between 5 percent and 10 percent only. (See Figure L8.4c) **You can figure out a set of convenient values for specifying duty cycles for the servo motor**. The program reads in the **duty cycle** and then performs the required computation to derive the appropriate value that should be written into the duty cycle registers (PCA0CPH0:PCA0CPL). Your program will echo the new duty cycle on the LCD screen for the DC motor and servo motor, respectively.

After that, the speed of the DC motor and the position of the servo motor would be changed. Your program will wait for three seconds and then ask the user to enter the new duty cycles and repeat the same operation.

Universal Asynchronous Receiver Transceiver (UART)

9.1 Objectives

Upon successful completion of this chapter, you will be able to:

- Explain the pros and cons of parallel I/O and serial I/O methods
- Explain the electrical specification of the EIA-232 standard
- Explain the function specification of the EIA-232 standard
- Explain the procedural specification of the EIA-232 standard
- Explain the errors that can be detected by the UART module
- Explain how to use a null-modem connection to allow two processors to communicate with each other without using modems in the EIA-232 interface
- Explain the operation of the UART module of the C8051F040
- Configure the UART module to perform data transmission and reception
- Call UART functions to read data or transmit data via the UART module
- Use UART module to interface with shift registers, such as LV164, LV165, and 74HC595

9.2 Concepts of Serial I/O

The need to exchange data between the MCU and peripheral devices can be satisfied by using parallel data transfer (multiple bits in one transfer operation). However, there are a few drawbacks:

- Parallel data transfer requires many I/O pins. This requirement prevents the microcontroller from interfacing with as many devices as desired in the application.
- Many I/O devices do not have a high enough data rate to justify the use of parallel data transfer.

- Data synchronization for parallel transfer is difficult to achieve over a long distance. This requirement is one of the reasons that data communications always use serial transfer.
- Higher cost.

There are several widely used general-purpose serial interfaces today.

- ***Universal Asynchronous Receiver Transceiver* (UART):** This interface may also be referred to as Universal Synchronous Asynchronous Receiver Transceiver (USART) if it also supports synchronous mode of data transfer. A serial interface is called **synchronous** if the data transfer between the transmitter and the receiver is synchronized by a clock signal. Otherwise, it is referred to as **asynchronous**. This interface uses two wires in data transfer: TxD and RxD. The naming of these two signals may vary among different vendors.
- ***Serial Peripheral Interface* (SPI):** The SPI, introduced by Motorola, is a synchronous interface that uses three or four wires (MOSI, MISO, SCK, and SS) to carry out data transfer. It has been used to interface the microcontroller with peripheral chips, such as A/D converter, D/A converter, EEPROM, SRAM, seven-segment display driver, LCD driver, phase-lock loops, real-time clock chip, temperature sensor, and many other special-purpose chips. The **Microwire** interface from National Semiconductor is an SPI-compatible interface.
- ***Inter-Integrated Circuit* (I^2C):** This protocol was introduced by Philips in the 1980s. It uses two wires (SDA and SCL) to carry out all data-transfer activities. Similar to the SPI, the I^2C interface has been used to interface the microcontroller with a wide variety of peripheral chips.
- ***Control Area Network* (CAN):** CAN was introduced by Robert Bosch in 1980s. It initially was created for automotive applications as a method for enabling robust serial communication. Since its inception, the CAN protocol has gained widespread use in industrial automation and process control.

The original 8051 provides only the UART interface. The other three serial interfaces are implemented by 8051 variants optionally. Therefore, the implementations of them are not standardized and could be very different. These four serial interfaces will be discussed in this text. This chapter will be dedicated to the discussion of the UART interface. When operating in asynchronous mode, the UART often utilizes the industrial standard RS-232 protocol to communicate with a PC or other microcontrollers. The RS-232 protocol will be discussed in the next section.

9.3 The RS-232 Standard

The RS-232 standard was established in 1960 by the Electronic Industry Association (EIA) for interfacing between a computer and a modem. It has been revised several times since then. The latest revision, EIA-232E, was published in July 1991. In this revision, EIA replaced the prefix **RS** with the prefix **EIA**. This change represents no change in the standard, but allows users to identify the source of the standard. We refer to the standard as both RS-232 and EIA-232 throughout this text. In data communication terms, both computers and terminals are called *data terminal equipment* (DTE), whereas modems, bridges, and routers are referred to as *data communication equipment* (DCE).

There are four aspects to the EIA-232 standard:

1. Electrical specifications—specify the voltage level, rise time, and fall time of each signal, achievable data rate, and the distance of communication.
2. Functional specifications—specify the function of each signal.
3. Mechanical specifications—specify the number of pins and the shape and dimensions of the connectors.
4. Procedural specifications—specify the sequence of events for transmitting data based on the functional specifications of the interface.

9.3.1 EIA-232 Electrical Specification

The following electrical specifications of the EIA-232 are of interest to us.

1. ***Data Rates.*** The EIA-232 standard supports data rates of up to 20,000 bits per second (the usual upper limit is 19,200 baud). The EIA-232 standard did not set any fixed baud rate. The commonly used baud rate values are 9600, and 19,200.
2. ***Signal State Voltage Assignments.*** Voltages of -3 to -25 V with respect to signal ground are considered logic 1 (referred to as the *mark* condition), whereas voltages of $+3$ to $+25$ V are considered logic 0 (referred to as the *space* condition). The range of voltages between -3 and $+3$ V is considered a transition region for which a signal state is not assigned.
3. ***Signal Transfer Distance.*** The standard requires the EIA-232 implementation to transfer data correctly up to 15 m. Greater distance can be achieved with good design.

9.3.2 EIA-232 Functional Specification

The EIA-232 standard specifies 22 signals. These signals can be divided into six categories.

1. ***Signal Ground and Shield.***
2. ***Primary Communications Channel.*** This is used for data interchange and includes flow control signals.
3. ***Secondary Communications Channel.*** When implemented, this is used for control of the remote modem, requests for retransmission when errors occur, and governance over the setup of the primary channel.
4. ***Modem Status and Control Signals.*** These signals indicate modem status and provide intermediate checkpoints as the telephone voice channel is established.

5. ***Transmitter and Receiver Timing Signals.*** If a synchronous protocol is used, these signals provide timing information for the transmitter and receiver, which may operate at different baud rates.
6. ***Channel Test Signals.*** Before data is exchanged, the channel may be tested for its integrity and the baud rate automatically is adjusted to the maximum rate that the channel could support.

Signal Ground

Pins 7 and 1 and the shell are included in this category. Cables provide separate paths for each, but internal wiring often connects pin 1 and the cable shell/shield to the signal ground on pin 7. All signals are referenced to a common ground, as defined by the voltage on pin 7. This conductor may or may not be connected to the protective ground inside the DCE device.

Primary Communication Channel

Pin 2 carries the *transmit data* (TxD) signal, which is active when data is transmitted from the DTE device to the DCE device. When no data is transmitted, the signal is held in the mark condition (logic 1, negative voltage).

Pin 3 carries the *received data* (RxD), which is active when the DTE device receives data from the DCE device. When no data is received, the signal is held in the mark condition.

Pin 4 carries the *request to send* (RTS) signal, which will be asserted (logic 0, positive voltage) to prepare the DCE device for accepting transmitted data from the DTE device. Such preparation might include enabling the receive circuits or setting up the channel direction in half-duplex applications. When the DCE is ready, it acknowledges by asserting the CTS signal.

Pin 5 carries the *clear to send* (CTS) signal, which will be asserted (logic 0) by the DCE device to inform the DTE device that transmission may begin. RTS and CTS commonly are used as handshaking signals to moderate the flow of data into the DCE device.

Secondary Communication Channel

Pin 14 is the *secondary transmitted data* (STxD). Pin 16 is the *secondary received data* (SRxD). Pin 19 carries the *secondary request to send* (SRTS) signal. Pin 13 carries the *secondary clear to send* (SCTS) signal. These signals are equivalent to the corresponding signals in the primary communications channel. The baud rate, however, typically is much slower in the secondary channel, for increased reliability.

Modem Status and Control Signals

This group includes the following signals.

Pin 6—DCE Ready (DSR). When originating from a modem, this signal is asserted (logic 0) when all the following three conditions are satisfied.

1. The modem is connected to an active telephone line that is *off-hook.*
2. The modem is in data mode, not voice or dialing mode.
3. The modem has completed dialing or call setup functions and is generating an answer tone.

If the line goes off-hook, a fault condition is detected, or a voice connection is established, the DCE ready signal is de-asserted (logic 1).

Pin 20—DTE Ready (DTR). This signal is asserted (logic 0) by the DTE device when it wishes to open a communications channel. If the DCE device is a modem, the assertion of DTR prepares the modem to be connected to the telephone circuit and, once connected, maintains the connection. When DTR is de-asserted, the modem is switched to *on-hook* to terminate the connection (same as placing the phone back on the telephone socket).

Pin 8—Received Line Signal Detector (CD). Also called *carrier detect,* this signal is relevant when the DCE device is a modem. It is asserted (logic 0) by the modem when the telephone line is off-hook, a connection has been established, and an answer tone is being received from the remote modem. The signal is de-asserted when no answer tone is being received or when the answer tone is of inadequate quality to meet the local modem's requirements.

Pin 12—Secondary Received Line Signal Detector (SCD). This signal is equivalent to the CD (pin 8) signal but refers to the secondary channel.

Pin 22—Ring Indicator (RI). This signal is relevant when the DCE device is a modem and is asserted (logic 0) when a ringing signal is being received from the telephone line. The assertion time of this signal will equal approximately the duration of the ring signal, and it will be deasserted between rings or when no ringing is present.

Pin 23—Data Signal Rate Selector. This signal may originate in either the DTE or DCE devices (but not both) and is used to select one of two prearranged baud rates. The assertion condition (logic 0) selects the higher baud rate.

Transmitter and Receiver Timing Signals

This group consists of the following signals.

Pin 15—Transmitter Signal Element Timing (TC). Also called a *transmitter clock,* this signal is relevant only when the DCE device is a modem and is operating with a synchronous protocol. The modem generates this clock signal to control exactly the rate at which data is sent on TxD (pin 2) from the DTE device to the DCE device. The logic 1 to logic 0 (negative to positive) transition on this line causes a corresponding transition to the next data element on the TxD line. The modem generates this signal continuously, except when it is performing internal diagnostic functions.

Pin 17—Receiver Signal Element Timing (RC). Also called a *receiver clock,* this signal is similar to TC, except that it provides timing information for the DTE receiver.

Pin 24—Transmitter Signal Element Timing (ETC). Also called an *external transmitter clock* with timing signals provided by the DTE device for use by a modem, this signal is used only when TC and RC (pins 15 and 17) are not in use. The logic 1 to logic 0 transition indicates the time center of the data element. Timing signals will be provided whenever the DTE is turned on, regardless of other signal conditions.

Channel Test Signals

This group consists of the following signals.

Pin 18—Local Loopback (LL). This signal is generated by the DTE device and is used to place the modem into a test state. When LL is asserted (logic 0, positive voltage), the modem redirects its modulated output signal, which normally is fed into the telephone line and back into its receiving circuitry. This enables data generated by the DTE to be echoed back through the local modem to check the condition of the modem circuitry. The modem asserts its test mode signal on pin 25 to acknowledge that it has been placed in LL condition.

Pin 21—Remote Loopback (RL). This signal is generated by the DTE device and is used to place the remote modem into a test state. When RL is asserted (logic 0), the remote modem redirects its received data back to its transmitted data input, thereby remodulating the received data and returning it to its source. When the DTE initiates such a test, transmitted data is passed through the local modem, the telephone line, the remote modem, and back, to exercise the channel and confirm its integrity. The remote modem signals the local modem to assert test mode on pin 25 when the remote loopback test is underway.

Pin 25—Test Mode (TM). This signal is relevant only when the DCE device is a modem. When asserted (logic 0), it indicates that the modem is in a LL or RL condition. Other internal self-test conditions also may cause the TM signal to be asserted, depending on the modem and the network to which it is attached.

9.3.3 EIA-232 Mechanical Specification

The EIA-232 uses a 25-pin D-type connector (DB25). However, only a small subset of the 25 signals actually is used—a 9-pin connector (DB9) is used in most PCs. The signal assignment of the DB9 is shown in Figure 9.1. The DB9 is not part of the EIA-232 standard. Today, USB bus has become the most popular serial bus used in personal computers (PC). Most PCs no longer provide the DB9 connector. To use the UART port to communicate with the PC, the user will need to purchase a DB9-to-USB converter or an add-on serial card that plugs into the PCI bus on the PC.

9.3.4 EIA-232 Procedural Specification

The sequence of events that occurs during data transmission using the EIA-232 standard can be understood best by studying examples. Two examples are used to explain the procedure.

In the first example, two DTEs are connected with a point-to-point link using a modem. The modem requires only the following circuits to operate:

- Signal ground (GND)
- Transmitted data (TxD)
- Received data (RxD)
- Request to send (RTS)
- Clear to send (CTS)
- Data set ready (DSR)
- Carrier detect (CD)

Before the DTE can transmit data, the DSR circuit must be asserted to indicate that the modem is ready to operate. This signal should be asserted before the DTE attempts to make a request to send data. The DSR pin can be connected to the power supply of the DCE to indicate that it is switched on and ready to operate. When a DTE has data to send, it asserts the RTS signal. The modem responds, when ready, with the CTS signal asserted, indicating

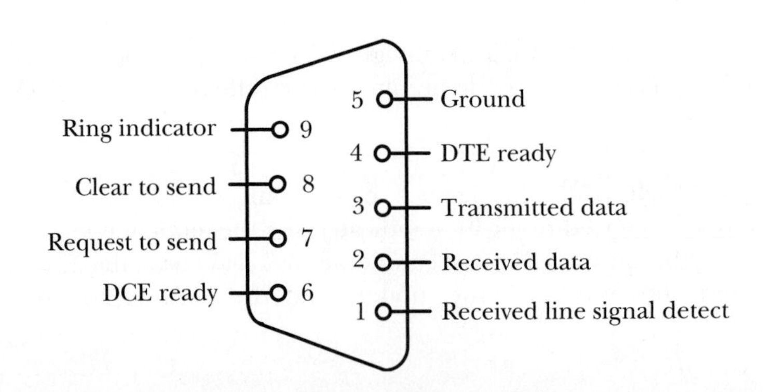

FIGURE 9.1 *EIA-232E DB9 connector and signal assignment.*

that data may be transmitted. If the arrangement is a half-duplex, then the assertion of the RTS signal also inhibits the receive mode. The DTE sends data to the local modem bit serially. The local modem modulates the data into the carrier signal and transmits the resultant signal over the dedicated communication lines. Before sending out modulated data, the local modem sends out a carrier signal to the remote modem so that it is ready to receive the data. The remote modem detects the carrier and asserts the CD signal. The assertion of the CD signal tells the remote DTE that the local modem is transmitting. The remote modem receives the modulated signal, demodulates it to recover the data, and sends it to the remote DTE over the RxD pin. The circuit connections are illustrated in Figure 9.2. After the local DTE finishes data transmission, it drops the RTS signal, and the local modem then drops CTS signal. The local modem also drops the carrier. The remote modem detects the loss of carrier; it then drops the CD signal, which then terminates the whole data transmission process.

The next example involves two computers exchanging data through a public telephone line. One of the computers (initiator) must dial the phone (automatically or manually) to establish the connection, just like people talking over the phone. Two additional leads are required for this application:

- Data terminal ready (DTR)
- Ring indicator (RI)

The data transmission in this setting is divided into three phases.

Phase 1

Connection Establishment. The following events occur in this phase.

1. The transmitting computer asserts the DTR signal to inform the local modem that it intends to make a call.
2. The local modem opens the phone line and dials the destination number, which is stored in the modem or transmitted to the modem by the computer via the TxD pin.
3. The remote modem detects a ring on the phone line and informs the remote computer that a call has arrived by asserting the RI signal.

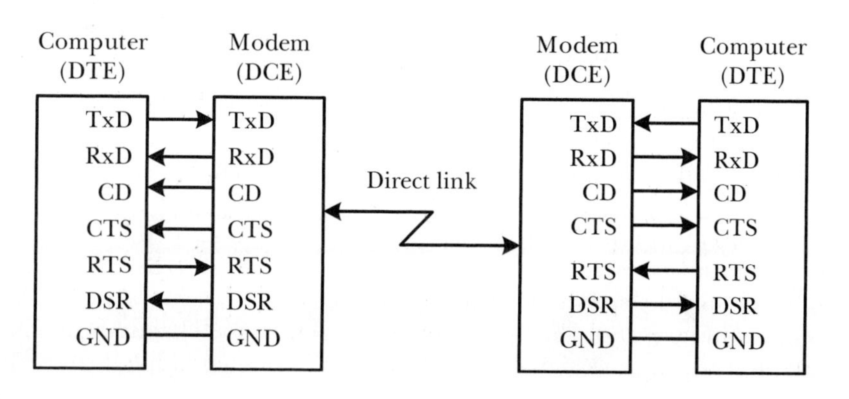

FIGURE 9.2 *Point-to-point asynchronous connection.*

4. The remote computer accepts the incoming call by asserting the DTR signal.
5. The remote modem answers the call by sending a carrier signal to the local modem via the phone line. It also informs the remote computer that it is ready for data transmission by asserting the DSR signal.
6. After detecting the carrier signal, the local modem asserts both the DSR and CD signals to inform the local computer that the connection is established and it is ready for data communication.
7. In a full-duplex data communication, the local modem also sends a carrier signal to the remote modem. The remote modem then asserts the CD signal to the remote computer.

Phase 2

Data transmission. The following events occur during this phase.

1. The local computer asserts the RTS signal when it has data to send.
2. The local modem answers the transmit request by asserting the CTS signal.
3. The local computer sends data bits serially to the local modem. The local modem then modulates the data over the carrier signal and sends the resultant signal to the remote modem.
4. The remote modem receives the modulated signal from the local modem, demodulates it to recover the data, and sends it to the remote computer over the RxD pin.

Phase 3

Disconnection. Disconnection requires only two steps.

1. After finishing data transmission, the local computer drops the RTS signal.
2. In response, the local modem de-asserts the CTS signal and drops the carrier (equivalent to hanging up the phone).

The circuit connection for this example is shown in Figure 9.3. A timing signal is not required in an asynchronous transmission.

9.3.5 Data Format

Using the EIA-232 protocol, data is transferred character by character. Each character is preceded by a start bit (a low), followed by eight or nine data bits, and terminated by a stop bit. The data format of a character is shown in Figure 9.4.

As shown in Figure 9.4, the least significant bit is transmitted first, and the most significant bit is transmitted last. The stop bit is high. The start bit and stop bit identify the start and end of a character.

Since there is no clock information in the asynchronous format, the receiver uses a clock signal with a frequency that is a multiple (can be 16,

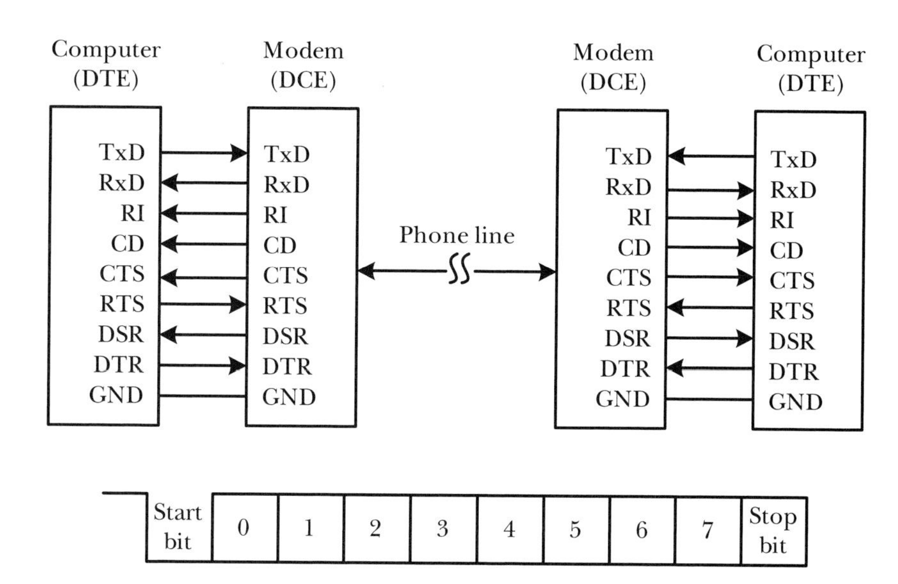

FIGURE 9.3 *Asynchronous connection over public phone line.*

FIGURE 9.4 *The format of a character.*

32, or 64) of the data rate to sample the incoming data in order to detect the arrival of the start bit and determine the logical value of each data bit. A clock with a frequency that is 16 times the data rate can tolerate a frequency difference in the clocks slightly over 3 percent at the transmitter and receiver.

The method for detecting the arrival of a start bit is similar among all microcontrollers:

> When the RxD pin is *idle* (*high*) for at least three sampling times and then is followed by a low voltage, the UART will look at the third, fifth, and seventh samples after the first low sample (these are called verification samples) to determine if a valid start bit has arrived. This process is illustrated in Figure 9.5. If the majority of these three samples are low, then a valid start bit is detected. Otherwise, the UART will restart the process. After detecting a valid start bit, the UART will start to shift in the data bits.

The method for determining the data bit value is also similar for most microcontrollers:

> The UART uses a clock with a frequency about 16 times (most often) that of the data rate to sample the RxD signal. If the majority of the seventh, eighth, and ninth samples are 1s, then the data bit is determined to be 1. Otherwise, the data bit is determined to be 0.

The stop bit is high. Using this format, it is possible to transfer data character by character without any gap.

The term *baud rate* is defined as the number of bit changes per second. Since the RS-232 standard uses a *non-return-to-zero* (NRZ) encoding method, baud rate is identical to bit rate. In a NRZ code, a logic 1 bit is sent as a high

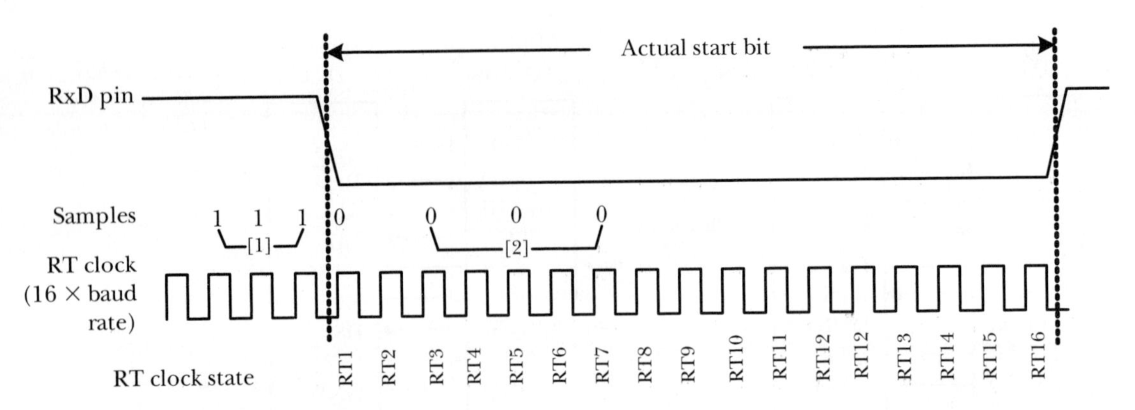

FIGURE 9.5
Detection of start bit (ideal case).

value and a logic 0 bit is sent as a low value. When a logic 1 follows another logic 1 during a data transfer, the voltage does not drop to zero before it goes high.

9.3.6 UART Detectable Errors

A UART can detect the following two types of communication errors.

1. ***Framing Error:*** The situation that a character is not terminated by a stop bit is referred to as a framing error.
2. ***Receiver Overrun Error:*** This error occurs when a new character arrives before the previously received character read by the CPU.

9.3.7 Null Modem Connection

When two DTE devices are located side by side and use the EIA-232 interface to exchange data, there really is no reason to use two modems to connect them. However, the EIA-232 standard does not allow the direct connection of two DTEs. In order to make this scheme work, a *null modem* is needed. The null modem interconnects leads in such a way as to fool both DTEs into thinking that they are connected to modems. The null modem connection is shown in Table 9.1.

In Table 9.1, the signals of DTE1 and DTE2 that are to be wired together are listed in the same row. The transmitter-timing and receiver-timing signals are not needed in asynchronous data transmission. Ring indicator is not needed either, because the transmission is not through a public phone line.

TABLE 9.1 *Null Modem Connection*

	DTE 1		DTE 2		
Signal Name	**DB25 pin**	**DB9 pin**	**DB9 pin**	**DB25 pin**	**Signal Name**
FG (frame ground)	1	–	–	1	FG
TD (transmit data)	2	3	2	3	RD
RD (receive data)	3	2	3	2	TD
RTS (request to send)	4	7	8	5	CTS
CTS (clear to send)	5	8	7	4	RTS
SG (signal ground)	7	5	5	7	SG
DSR (data set ready)	6	6	4	20	DTR
CD (carrier detect)	8	1	4	20	DTR
DTR (data terminal ready)	20	4	1	8	CD
DTR (data terminal ready)	20	4	6	6	DSR

9.4 The 8051 UART Module

The original 8051 has a full-duplex UART module. It has separate registers for the data to be transmitted and the data to be received. However, these two registers are accessed by referring to the same register *SBUF*. The original 8051 uses the serial-port control register (SCON) to control the overall operation of the UART port. The contents of the SCON register are shown in Figure 9.6.

The SiLabs C8051F040 has two UART modules: UART0 and UART1. The UART0 has all of the features of the original 8051 UART module and additional baud rate generation options. When in Modes 1 and 3, the UART0 module also can use Timers 3 and 4 overflows to generate the baud rate. The UART1 supports only Modes 1 and 3 and can use only Mode 2 (8-bit auto-reload mode) of Timer 1 to generate its baud rate.

9.4.1 UART Pins

There are two pins associated with each UART module. The pin association of UART modules as follows.

- ***The original* 8051:** RxD and TxD pins are multiplexed with Port P3 pins 0 and 1, respectively.
- **C8051F040:** TX0 and RX0 will be multiplexed with pins P0.0 and P0.1, respectively, whereas the assignment of the TX1 and RX1 pins will depend on whether I/O pins are assigned to SPI pins (SCK, MISO, MOSI, and NSS) and SMBus pins (SDA and SCL).

FIGURE 9.6
The SCON register (original 8051).

7	6	5	4	3	2	1	0	
SM0	SM1	SM2	REN	TB8	RB8	TI	RI	Value after reset = 00H
rw	rw	rw	rw	rw	rw	rw	rw	

SM0:SM1: Mode select bits
 00 = Mode 0, baud rate set to fosc/12
 01 = Mode 1, 8-bit UART Mode with variable baud rate generated by Timer 1 or 2 overflow
 10 = Mode 2, 9-bit UART Mode with baud rate = fosc/32 or fosc/64
 11 = Mode 3, 9-bit UART Mode with variable baud rate generated by Timer 1 or 2 overflow
SM2: Enables multiprocessor communication in Mode 2 and 3
 0 = disable multiprocessor communication in Mode 2 and 3
 1 = enable multiprocessor communication in Mode 2 and 3. The RI bit will not be set if the ninth data bit is 0. In Mode 1, if SM2 = 1 then RI will not be activated if a valid stop bit was not received
REN : Reception enable bit
 0 = reception disabled
 1 = reception enabled
TB8: The ninth bit that will be transmitted in Modes 2 and 3
RB8: The ninth bit that was received in Modes 2 and 3. If SM2 = 0, RB8 is the stop bit that was received. In Mode 0, RB8 is not used
TI: Transmit interrupt bit
 This bit is set by the hardware at the end of the eighth bit in Mode 0, or at the beginning of the stop bit in other Modes, in any serial transmission. This bit must be cleared in software
RI: Receive interrupt bit
 This bit is set at the end of eighth incoming bit time in Mode 0, or halfway through the stop bit time in other Modes, in any serial reception. This bit must be cleared in software

9.4.2 UART Operation Mode

The UART modules of the original 8051 and the C8051F040 have four operation modes.

- ***Mode* 0:** In this mode, data is shifted in or out via the RxD pin, whereas the TxD pin outputs the clock for shifting data. The least significant bit is shifted in or out first. Eight bits are shifted in or out each time. The shift clock is fixed at 1/12 of the oscillator frequency.
- ***Mode* 1:** In this mode, ten bits (including a start bit), eight data bits, and one stop bit are transmitted or received. On reception, the stop bit is stored in the RB8 bit of the SCON register. The baud rate is generated by Timer 1 or Timer 2 overflow (UART0 of the C8051F040 can also use Timer 3 or Timer 4 overflow).
- ***Mode* 2:** In this mode, eleven bits including a start bit, eight data bits, a programmable ninth data bit, and a stop bit are transmitted or received.

On reception, the ninth bit is stored in the RB8 bit of the SCON register, while the stop bit is ignored. The baud rate can be programmed to either 1/32 or 1/64 of the oscillator frequency.

- **Mode 3:** In this mode, eleven bits including a start bit, eight data bits, a programmable ninth data bit, and a stop bit are transmitted or received. The baud rate is generated by Timer 1 or Timer 2 overflow (UART0 of the C8051F040 can also use Timer 3 or Timer 4 overflow).

In all four operation modes, transmission is started by writing into the SBUF register. Reception is started in Mode 0 by setting the REN bit to 1 when the RI bit is 0. In other modes, reception is initiated by the incoming start bit when the REN bit is 1.

9.4.3 Registers Associated with the UART

The UART of the original 8051 uses the SCON register to configure data transmission and reception. Data to be transmitted is written into the SBUF register, whereas received data can be read from the same register.

The UART0 of the C8051F040 uses the following registers to support its operation.

- ***Serial Control Register* 0 (SCON0):** The contents of this register are shown in Figure 9.7.
- ***UART0 Status and Clock Selection Register* (SSTA0):** The contents of this register are shown in Figure 9.8.
- ***UART0 Data Buffer* (SBUF0):** This register serves the same function as that of SBUF in the original 8051.
- ***UART0 Slave Address* (SADDR0):** This register holds the slave address of UART0 when it is communicating in the multiprocessor environment.
- ***UART0 Slave Address Enable* (SADEN0):** This register enable or disable any address bit to be compared against a received address pattern. A received address bit will be compared if the corresponding bit in this register is set to 1.

The UART1 uses the SCON1 to configure its operation characteristics and uses SBUF1 to hold the data to be transmitted and the received data. The contents of SCON1 are shown in Figure 9.9.

9.4.4 Baud Rate Generation for the Original 8051

The baud rate in Mode 0 is set to $f_{OSC}/12$. The baud rate in Mode 2 depends on the value of the SMOD bit in the PCON register (shown in Figure 6.13) and is given by

$$\textbf{Baud rate} = 2^{\textbf{SMOD}} \times (\textbf{SYSCLK} \div 2^6) \qquad \textbf{(9.1)}$$

FIGURE 9.7
The SCON0 register of C8051F040.

7	6	5	4	3	2	1	0	
SM00	SM10	SM20	REN0	TB80	RB80	TI0	RI0	Value after reset = 00H
rw	rw	rw	rw	rw	rw	rw	rw	

SM00: SM10: Mode select bits
00 = Mode 0, baud rate set to fosc/12
01 = Mode 1, 8-bit UART Mode with variable baud rate generated by Timer 1 or 2 or 3 or 4 overflow
10 = Mode 2, 9-bit UART Mode with baud rate = $f_{osc}/32$ or $f_{osc}/64$
11 = Mode 3, 9-bit UART Mode with variable baud rate generated by Timer 1 or 2 overflow
SM20: Enables multiprocessor communication in Mode 2 and 3
0 = disable multiprocessor communication in Mode 2 and 3
1 = enable multiprocessor communication in Mode 2 and 3. The RI0 bit will not be set if the ninth data bit is 0. In Mode 1, if SM20 = 1 then RI0 will not be activated if a valid stop bit was not received
REN0: Reception enable bit
0 = UART0 reception disabled
1 = UART0 reception enabled
TB80: The ninth bit that will be transmitted in Modes 2 and 3
RB80: The ninth bit that was received in Mode 2 and 3. If SM2 = 0, RB8 is the stop bit that was received. In Mode 0, RB8 is not used.
TI0: Transmit interrupt bit
This bit is set by the hardware at the end of the eighth bit in Mode 0, or at the beginning of the stop bit in other Modes, in any serial transmission. This bit must be cleared in software
RI0: Receive interrupt bit
This bit is set at the end of eighth incoming bit time in Mode 0, or halfway through the stop bit time in other Modes, in any serial reception. This bit must be cleared in software

The baud rate generation methods in Modes 1 and 3 of C8051F040 are slightly different from those of the original 8051.

Baud Rate Generation of Modes 1 and 3

For the original 8051, the baud rate in Modes 1 and 3 are determined by the overflow rate of Timer 1, or by Timer 2, or by both (one for transmission and the other for reception).

Using Timer 1 to Generate Baud Rate. If the UART uses Timer 1 to generate baud rate in Modes 1 and 3, then the UART baud rate is determined by the Timer 1 overflow rate and the value of the SMOD bit as

$$\textbf{Modes 1, 3 baud rate} = (2^{SMOD} \div 2^5) \times (\textbf{Timer 1 overflow rate}) \quad \textbf{(9.2)}$$

The user should disable the Timer 1 interrupt for this application. The user can configure Timer 1 for "timer" or "counter" operation, and in any of its

7	6	5	4	3	2	1	0	
FE0	RXOV0	TXCOL0	SMOD0	SoTCLK1	S0TCLK0	S0RCLK1	S0RCLK0	Value after reset = 00H
rw	rw	rw	rw	rw	rw	rw	rw	

FE0: Frame error flag
0 = frame error has not been detected
1 = frame error has been detected
RXOV0: Receive overrun flag
0 = receive overrun has not been detected
1 = receive overrun has been detected
TXCOL0: Transmit collision flag
0 = transmission collision has not been detected
1 = transmission collision has been detected
SMOD0: UART0 baud rate doubler
0 = UART0 baud rate divide-by-two enabled
1 = UART0 baud rate divide-by-two disabled
S0TCLK1:S0TCLK0: Transmit baud rate clock selection bits
00 = Timer 1 overflow generates UART0 TX baud rate
01 = Timer 2 overflow generates UART0 TX baud rate
10 = Timer 3 overflow generates UART0 TX baud rate
11 = Timer 4 overflow generates UART0 TX baud rate
S0RCLK1:S0RCLK0: Receive baud rate clock selection bits
00 = Timer 1 overflow generates UART0 RX baud rate
01 = Timer 2 overflow generates UART0 RX baud rate
10 = Timer 3 overflow generates UART0 RX baud rate
11 = Timer 4 overflow generates UART0 RX baud rate

FIGURE 9.8 *The UART0 Status and Clock Selection Register (SSTA0) of C8051F040.*

three running modes. However, it is most common to configure Timer 1 for timer operation in 8-bit auto-reload mode. The UART baud rate is given by the following formula for this configuration.

$$\textbf{Modes 1, 3 baud rate} = (2^{SMOD} \div 2^5) \times f_{OSC} \div [12 \times (2^8 - TH1)] \quad (9.3)$$

A sample of commonly used baud rate generated using Timer 1 is shown in Table 9.2.

Using Timer 2 to Generate Baud Rate. By setting the bit 5 of the T2CON register, Timer 2 will be used to generate the baud rate for reception of the UART module. By setting the bit 4 of the same register, Timer 2 will be used to generate the baud rate for transmission of the UART module.

The baud rate generator mode of Timer 2 is similar to the auto-reload mode. Whenever the TH2:TL2 register pair overflows, they will be reloaded with the 16-bit value in registers RCAP2H and RCAP2L. The baud rates in Modes 1 and 3 are

$$\textbf{Modes 1, 3 baud rate} = \text{Timer 2 overflow rate} \div 2^4 \quad (9.4)$$

For baud rate generation, Timer 2 is normally configured for "timer" operation. The timer operation is a little different for Timer 2 when it is used

FIGURE 9.9 *The SCON1 register of C8051F040.*

7	6	5	4	3	2	1	0	
S1MODE	--	MCE1	REN1	TB81	RB81	TI1	RI1	Value after reset = 00H
rw	rw	rw	rw	rw	rw	rw	rw	

S1MODE: Serial port 1 operation mode
- 0 = Mode 0: 8-bit UART with variable baud rate
- 1 = Mode 1, 9-bit UART with variable baud rate

MCE1: Multiprocessor communication enable
- Mode0: checks for valid stop bit
  - 0 = logic level of stop is ignored
  - 1 = RI1 will only be activated if stop bit is logic level 1
- Mode 1: Multiprocessor communication enable
  - 0 = logic level of ninth bit is ignored
  - 1= RI1 is set and an interrupt is generated only when the ninth bit is logic 1

REN1: Reception enable bit
- 0 = UART1 reception disabled
- 1 = UART1 reception enabled

TB81: The 9th bit that will be transmitted in Mode 1 (not used in 8-bit mode)

RB81: The ninth bit that was received in Mode 1. This bit holds the stop bit in Mode 0

TI1: Transmit interrupt bit
This bit is set by the hardware at the end of the eighth bit in Mode 0, or at the beginning of the stop bit in other modes, in any serial transmission. This bit must be cleared in software.

RI1: Receive interrupt bit
This bit is set at the end of eighth incoming bit time in Mode 0, or halfway through the stop bit time in other modes, in any serial reception. This bit must be cleared in software

TABLE 9.2 *Commonly Used Baud Rates Generated From Timer 1*

						Timer 1		
UART mode	f_{OSC} (MHz)	Desired Baud Rate	Resultant Baud Rate	Baud Rate Deviation	SMOD Bit in PCON	C/T Bit in TMOD	Timer Mode in TMOD	TH1 Reload Value
Mode 0 max	40.0	3.33 MHz	3.33 MHz	0	X	X	X	X
Mode 2 max	40.0	1250 kHz	1250 kHz	0	1	X	X	X
Mode 2 max	40.0	625 kHz	625 kHz	0	0	X	X	X
Modes 1 or 3	40.0	19200 Hz	18939 Hz	−1.36%	1	0	2	F5
Modes 1 or 3	40.0	9600 Hz	9470 Hz	−1.36%	1	0	2	EA
Modes 1 or 3	24.0	9600 Hz	9615 Hz	0.16%	1	0	2	F3
Modes 1 or 3	11.0592	57600 Hz	57600 Hz	0	1	0	2	FF
Modes 1 or 3	11.0592	28800 Hz	28800 Hz	0	1	0	2	FE
Modes 1 or 3	11.0592	19200 Hz	19200 Hz	0	1	0	2	FD
Modes 1 or 3	11.0592	9600 Hz	9600 Hz	0	1	0	2	FA

as a baud rate generator. As a timer, it would increment every machine cycle (thus at 1/12 the oscillator frequency). When used as a baud rate generator, Timer 2 increments at 1/2 the machine cycle clock frequency, and its overflow won't generate an interrupt to the CPU. Thus, the baud rate is given by the formula

$$\textbf{Modes 1, 3 baud rate} = f_{OSC} \div \{2^5 \times [2^{16} - (\text{RCAP2H, RCAP2L})]\} \quad (9.5)$$

where (RCAP2H, RCAP2L) is the contents of the 16-bit capture register of Timer 2.

A sample of commonly used baud rates generated using Timer 2 is shown in Table 9.3.

9.4.5 Baud Rate Generation for the C8051F040 UART Modules

The UART0 of the C8051F040 offers the four UART operation modes of the standard 8051. However, the UART1 supports only Modes 1 and 3 of the standard 8051s UART.

The baud rate generation methods for Modes 0 and 2 of the UART0 are identical to those (Equation 9.1) of the standard 8051 UART. Modes 1 and 3 of UART0 may use Timers 1, 2, 3, or 4 to generate its baud rate.

TABLE 9.3 *Commonly Used Baud Rates Generated from Timer 2*

f_{OSC} (MHz)	Desired Baud Rate	Timer 2 SFRs RCAP2H	Timer 2 SFRs RCAP2L	Resultant Baud Rate	Baud Rate Deviation
40.0	115200	FF	F5	113636	−1.36%
40.0	57600	FF	EA	56818	−1.36%
40.0	28800	FF	D5	29070	0.94%
40.0	19200	FF	BF	19231	0.16%
40.0	9600	FF	7E	9615	0.16%
24.0	57600	FF	F3	57692	0.16%
24.0	28800	FF	E6	28846	0.16%
24.0	19200	FF	D9	19231	0.16%
24.0	9600	FF	B2	9615	0.16%
11.0592	115200	FF	FD	115200	0
11.0592	57600	FF	FA	57600	0
11.0592	28800	FF	F4	28800	0
11.0592	19200	FF	EE	19200	0
11.0592	9600	FF	DC	9600	0

Using Timer 1 to Generate Baud Rate. Timer 1 must be configured to 8-bit reload mode. The baud rate is given by

$$\textbf{Model\_BaudRate} = (2^{SMOD0}/32) \times \textbf{Timer1\_OverflowRate} \quad (9.6)$$

Where the SMOD0 bit is the bit 4 of the SSTA0 register whereas **Timer1_OverflowRate** is given by

$$\textbf{Timer1\_OverflowRate} = \textbf{T1CLK}/(2^8 - \textbf{TH1}) \quad (9.7)$$

T1CLK can be SYSCLK divided by 1, 4, 8, 12, or 48.

Using Timers 2, 3, or 4 to Generate the Baud Rate. The baud rate is given by

$$\textbf{Model\_BaudRate} = (1/2^4 \times \textbf{Timer234\_OverflowRate}) \quad (9.8)$$

where

$$\textbf{Timer234\_OverflowRate} = \textbf{TnCLK} / (2^{16} - \textbf{RCAPn}) \quad (9.9)$$

TABLE 9.4 *Oscillator Frequencies for Standard Baud Rates*

Oscillator Frequency (MHz)	Divide Factor	Timer 1 Reload Value[1]	Timers 2, 3, 4 Reload Value	Resulting Baud Rate (Hz)[2]
24.0	208	0xF3	0xFFF3	115200 (115384)
22.1184	192	0xF4	0xFFF4	115200
18.432	160	0xF6	0xFFF6	115200
11.0592	96	0xFA	0xFFFA	115200
3.6864	32	0xFE	0xFFFE	115200
1.8432	16	0xFF	0xFFFF	115200
24.0	832	0xCC	0xFFCC	28800 (28846)
22.1184	768	0xD0	0xFFD0	28800
18.432	640	0xD8	0xFFD8	28800
11.0592	348	0xE8	0xFFE8	28800
3.6864	128	0xF8	0xFFF8	28800
1.8432	64	0xFC	0xFFFC	28800
24.0	2496	0x64	0xFF64	9600 (9615)
22.1184	2304	0x70	0xFF70	9600
18.432	1920	0x88	0xFF88	9600
11.0592	1152	0xB8	0xFFB8	9600
3.6864	384	0xE8	0xFFE8	9600
1.8432	192	0xF4	0xFFF4	9600

Notes:
1. Assumes SMOD0 = 1 and T1M = 1.
2. Numbers in parenthesis show the actual baud rate.

TnCLK can be SYSCLK or SYSCLK ÷ 12, or external clock ÷ 2, or SYSCLK ÷ 8.

A sample of commonly used baud rates, oscillator frequencies, and timer reload values for C8051F040 is given in Table 9.4.

9.5 The Operation of UART Module

This section provides a more detailed description on the four UART operation modes.

9.5.1 UART Mode 0

Mode 0 is available in the UART of the original 8051 and the UART0 of the C8051F040. Mode 0 provides synchronous, half-duplex communication. Serial data is transmitted and received on the RX0 (or RX) pin. The TX0 (or TX) pin provides the shift clock to both transmit and receive. The MCU must be the master, since it generates the shift clock for transmission in both directions.

Data transmission begins when an instruction writes a data byte to the SBUF0 (or SBUF) register. Eight bits are shifted out with least significant bit (LSB) first, and the TI0 (or TI) flag is set at the end of the eighth bit time. Data reception begins when the REN0 (or REN) bit is 1 and the RI0 (or RI) flag is 0. When this condition is satisfied, the UART generates eight clock pulses to shift in data bits. One cycle after the eighth bit is shifted in, the RI0 flag is set and reception stops until the software clears the RI0 flag. Interrupt will be requested if enabled when either the TI0 or RI0 flag is set.

The Mode 0 baud rate is SYSCLK/12. The RX0 (RX) pin is forced to open-drain in Mode 0, and an external pullup resistor is needed.

9.5.2 UART Mode 1

Mode 1 provides standard asynchronous, full-duplex communication using a total of 10 bits per byte: one start bit, eight data bits, and one stop bit. Data is transmitted LSB first. Data are transmitted from the TX0 (or TX) pin and received at the RX0 (or RX). On reception, the eight data bits are stored in SBUF0 (SBUF, or SBUF1) and the stop bit goes into RB8 bit of the SCON0 (SCON, or SCON1) register.

Data transmission is started by writing a byte into the SBUF0 (SBUF or SBUF1) register, whereas data reception is initiated by a detected 1-to-0 transition at the RX0 (or RX1) pin with the REN bit set to 1. After the stop bit is received, the data byte will be loaded into the SBUF0 (or SBUF1) register and the stop bit is stored in the RB80 (RB8 or RB81) bit if the following conditions are met.

1. RI0 (RI, or RI1) flag is 0
2. Either the SM20 (SM2, or SM21) bit is 0 or the received stop bit is 1

If these two conditions are not met, the SBUF0 (SBUF or SBUF1) register and the RB80 (RB8 or RB81) bit will not be loaded, and the RI0 (RI or RI1) flag will not be set. An interrupt will be requested if enabled when either TI0 (TI or TI1) or RI0 (RI, or RI1) is set.

9.5.3 UART Mode 2

Mode 2 provides asynchronous, full-duplex communication using a total of 11 bits per data byte: a start bit, eight data bits, a programmable ninth data bit, and a stop bit. Mode 2 supports multiprocessor communications and hardware address recognition. On transmission, the ninth bit is determined by the value of the TB80 (TB8) bit in the SCON0 (SCON) register. On reception, the ninth data bit goes into the RB80 (RB8) bit of the SCON0 (SCON) register, and the stop bit is ignored.

Data transmission begins when an instruction writes a data byte to the SBUF0 (SBUF) register. The TI0 (TI) flag is set at the end of the transmission (i.e., the beginning of the stop bit time). Data reception can begin any time after the REN0 (REN) bit of the SCON0 (or SCON) register is set to 1 and is initiated by the detected 1-to-0 transition on the RX0 (or RX) pin. After the stop bit is received, the data byte will be loaded into the SBUF0 (or SBUF) register if the following two conditions are satisfied.

1. RI0 (RI) flag is 0
2. Either the SM2 bit (bit 5 of the SCON0 or SCON register) is 0 or the received ninth bit is 1

If these conditions are not met, the SBUF0 (SBUF) register and RB80 (RB8) bit will not be loaded, and the RI0 (RI) flag will not be set.

9.5.4 UART Mode 3

Mode 3 uses Mode 2 transmission protocol with the Mode 1 baud rate generation method. Mode 3 transmits 11 bits for each data byte: a start bit, eight data bits, a programmable ninth data bit, and a stop bit. The baud rate is derived from Timers 1, 2, 3, or 4 overflows, as defined in Equations 9.6 and 9.8. Multiprocessor communications and hardware address recognition are supported.

9.6 Applications of the UART Mode 0

One of the applications of the UART Mode 0 is to interface with a shift register and use the shift register to drive other I/O devices (such as DIP, LEDs, and seven-segment displays).

For the original 8051, the I/O pins assigned to UART are in quasi-bidirectional mode in which pullup resistors are not needed. However, the RX0 pin of the C8051F040 is forced to open-drain in Mode 0, and an external pull-up typically will be required.

Example 9.1 Write an instruction sequence to configure the transmitting MCU to operate in Mode 0.

Solution: The transmitting MCU must perform the following operations:

- Assign I/O pins to RX0 and TX0 and configure TX0 pin for output
- Configure the SCON0 register properly

The following instruction sequence will perform the desired configuration.

```
mov   SFRPAGE,#0FH    ; switch to SFR page F
mov   XBR0,#0x04      ; assign I/O pins to RX0 and TX0 and other peripherals
mov   XBR2,#0x40      ; enable crossbar
orl   P0MDOUT,#01H    ; enable TX0 pin output
mov   SFRPAGE,#0      ; switch to SFR page 0
mov   SCON0,#0        ; configure UART0 to mode 0 & disable reception
```

There are a few shift registers that can be interfaced to the 8051 by using Mode 0 of UART. Some of them can be used to expand the number of input ports to the MCU, whereas the others can be used to expand the number of output ports. For example, the 74LV165 and 74HC589 are parallel-in, serial-out shift registers that can be used to add input ports, whereas the 74LV164 and 74LV595 are serial-in, parallel-out shift registers that can be used to add parallel output ports to the microcontroller. These devices can operate from 1.0 V, up to 3.6 V, or even 5.5 V.

The pin assignment and truth table for the 74LV165 are shown in Figure 9.10 and Table 9.5, respectively. The 74LV165 will shift in data on the rising edge of the CP signal if the $\overline{PL}$ signal is high and the $\overline{CE}$ signal is low. The whole shift-register contents will be shifted from Q0 to Q7 during this shifting process. The maximum shift frequency is 20 MHz when the power

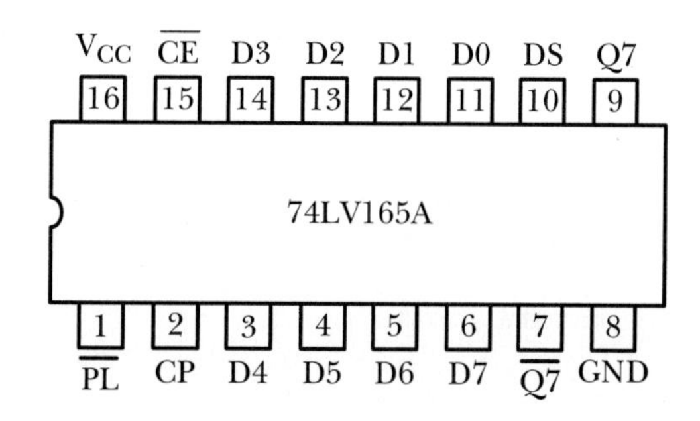

FIGURE 9.10 *Pin assignment of 74LV165.*

TABLE 9.5 *Function Table of 74LV165*

Operating Modes	Input					On Register		Output	
	PL	CE	CP	DS	D0 to D7	Q0	Q1 to Q6	Q7	Q7
Parallel load	L	X	X	X	L	L	L–L	L	H
	L	X	X	X	H	H	H–H	H	L
Serial shift	H	L	↑	l	X	L	q0–q5	q6	q6
	H	L	↑	h	X	H	q0–q5	q6	q6
Hold	H	H	X	X	X	q0	q1–q6	q7	q7

supply is from 3.0 to 3.6 V. When the $\overline{PL}$ signal is low, the data on D7 through D0 will be loaded into Q7 through Q0 of the shift register asynchronously.

Example 9.2 Show the circuit that uses one 74LV165 to add an input port to the C8051F040. The user will use a DIP switch to set up the number to be read by the MCU and use a debounced switch to interrupt the MCU, which will inform the MCU that there is data to be read. Write a program to configure the UART0 to Mode 0 and enable both reception and the reception interrupt.

Solution: A debounced switch should be connected to one of the INT inputs (for example, INT1) so that it will interrupt the MCU when being pressed. A circuit that allows the user to press a button to inform the C8051F040 to read in the data entered by the user is shown in Figure 9.11.

The circuit shown in Figure 9.11 allows the user to perform interactive input. Whenever the user program needs data from the user, it will output a message on the LCD to remind the user to enter a value. The user then will set a new value using the DIP switch and then will press the debounced switch that is connected to the INT1 pin. The pulse applied to the INT1 pin will interrupt the MCU, and the MCU will read the value in the interrupt service routine.

The program that performs the specified operations is as follows.

```
            $nomod51
            $include    (c8051F040.inc)
avail       set         R6              ; flag to indicate data is available
            org         0x00
            ljmp        start
            org         0x13            ; interrupt vector for INT1
            ljmp        INT1ISR
            org         0xAB
```

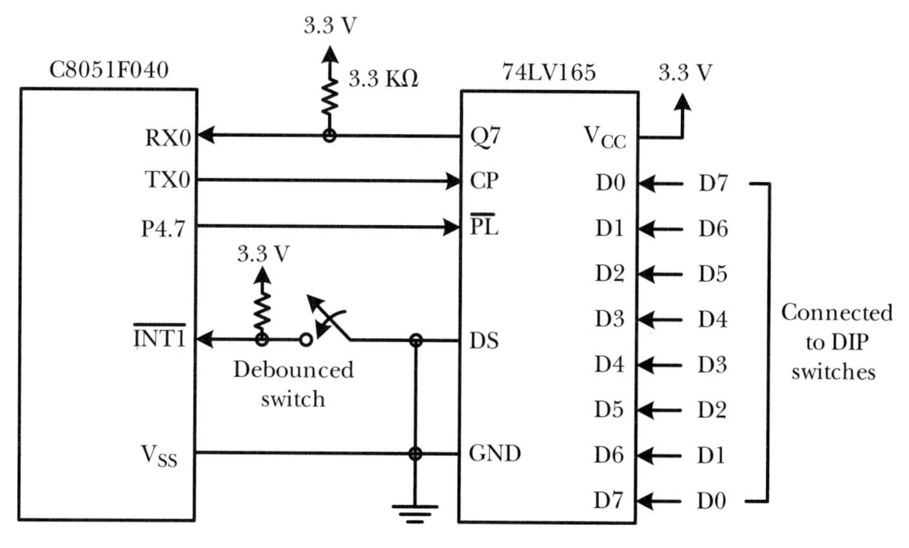

FIGURE 9.11
Circuit connection between the C8051F040 and 74LV165.

```
start:      mov     SP,#0x80            ; set up stack pointer
            lcall   sysinit             ; perform system initialization, switch to SFR page 0
            lcall   OpenLCD             ; configure LCD properly
            mov     avail,#0            ; indicate data not available yet
            setb    IT1                 ; configure INT1 to be edge-triggered
            setb    EX1                 ; enable INT1 interrupt
            setb    EA                  ;    "
            mov     DPTR,#msgx1         ; prompt user to enter data
            lcall   puts2lcd            ;    "
waitd:      mov     A,avail             ; check if data is available
            jz      waitd               ;    "
            mov     SFRPAGE,#0          ; switch SFRPAGE 0
            mov     SCON0,#10H          ; enable UART0 reception in mode 0
            jnb     RI0,$               ; wait until a byte has been shifted in
            clr     RI0                 ;
            mov     A,SBUF0             ; place the received byte in A
            mov     SFRPAGE,#0FH        ; switch to SFR page F
            mov     P5,A                ; output the received byte to LEDs
forever:    ajmp    $

msgx1:      db      "enter a number",0  ; message to prompt the user for input data

INT1ISR:    mov     SFRPAGE,#0FH        ; switch to SFR page F
            anl     P4,#7FH             ; pull PL pin to low to load data into shift register
            nop                         ; insert one cycle delay
            orl     P4,#80H             ; pull PL pin to high again
            mov     avail,#1            ; set avail flag to true
            reti
```

```
sysinit:    mov     SFRPAGE,#0FH        ; switch to SFR page F
            mov     WDTCN,#0DEH         ; disable watchdog timer
            mov     WDTCN,#0ADH         ;     "
            mov     CLKSEL,#0           ; select internal oscillator as SYSCLK
            mov     OSCICN,#83H         ;     "
            mov     XBR0,#04H           ; assign I/O pins to UART0
            mov     XBR2,#040H          ; also enable crossbar
            mov     P5MDOUT,#0FFH       ; enable Port P5 for output (LED port)
            mov     P7MDOUT,#0FFH       ; enable Port P7 for output (LCD port)
            mov     P4MDOUT,#0x80
            mov     P0MDOUT,#1
            ret
$include    (lcdUtil.a51)
$include    (delays.a51)
            end
```

The C language version of the program is as follows.

```
#include <C8051F040.h>   // include this file, delays.c and lcdUtil.c into the project
#include <lcdUtil.h>
void int1ISR (void);
void sysinit(void);
char avail;
data char ctmp;

void main (void)
{
    char *msg1 = "enter DIP data";
    sysinit();                      // select SYSCLK, assign I/O pins to peripheral signals
    openlcd();                      // initialize LCD
    avail = 0;
    puts2lcd (msg1);                // prompt the user to enter a value using the DIP switch
    SFRPAGE = LEGACY_PAGE;
    while (!avail);                 // wait for the avail flag to be set
    SCON0   = 0x10;                 // start to receive data
    while (!RI0);                   // wait until data has been shifted in
    ctmp    = SBUF0;                // read from the UART0
    SFRPAGE = 0x0F;                 // switch to LED page
    P5      = ctmp;                 // output the received data to LEDs
    while (1);
}
void sysinit(void)
{
    SFRPAGE  = 0x0F;
    WDTCN    = 0xDE;                // disable watchdog timer
    WDTCN    = 0xAD;                //     "
    CLKSEL   = 0;                   // select internal oscillator as SYSCLK
    OSCICN   = 0x83;                //     "
    XBR0     = 0x04;                // assign I/O pins to UART0 signals
    XBR2     = 0x40;                // also enable crossbar
    P5MDOUT = 0xFF;                 // enable LED port
```

```
        P4MDOUT = 0x80;              // enable PL pin
        P0MDOUT = 0x01;              // configure TX pin for output
        P7MDOUT = 0xFF;              // enable LCD port
}
void int1ISR (void) interrupt 2
{
        SFRPAGE  = 0x0F;
        P4       &= 0x7F;            // pull PL to low to load data into shift register
        P4       |= 0x80;            // pull PL to high to switch to serial shift mode
        avail    = 1;
}
```

Multiple 74LV165s can be cascaded to add multiple input ports to the microcontroller. In such a configuration, the DS input of a 74LV165 is connected to the Q7 output of the next 74LV165 in the chain. The 74LV165 that is connected directly to the MCU has its Q7 pin connected to the RX0 pin of the MCU.

The 74LV164 and 74LV595 are shift registers with serial input and parallel output. The user can use one or multiples of them to add one or multiple parallel output ports to the MCU. The diagram of pin assignment and the truth table for the operation of the 74LV164 are shown in Figure 9.12 and Table 9.6, respectively. Multiple 74LV164s can be cascaded. To cascade, connect the Q7

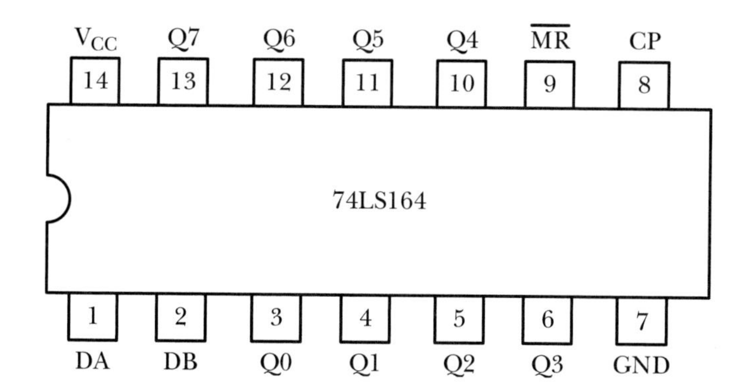

FIGURE 9.12 *74LV164 pin assignment.*

TABLE 9.6 *Function Table of 74LV164*

Operating Mode	Input				Output	
	MR	CP	DA	DB	Q0	Q1 to Q7
Reset	L	X	X	X	L	L to L
Shift	H	↑	l	l	L	q0 to q6
	H	↑	l	h	L	q0 to q6
	H	↑	h	l	L	q0 to q6
	H	↑	h	h	H	q0 to q6

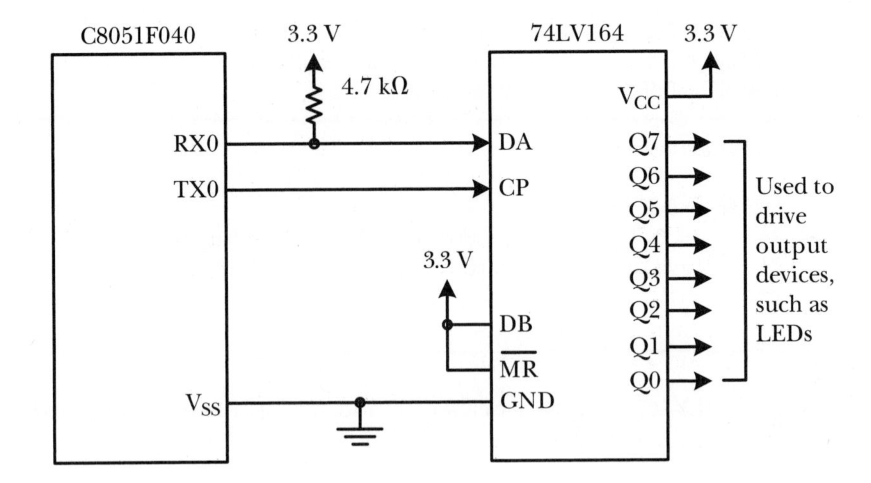

FIGURE 9.13 *Circuit connection between the C8051F040 and 74LV164.*

pin of the 74LV164 to the DA pin of the next 74LV164 in the chain and connect the 74LV164 closest to the MCU.

According to Table 9.6, the low level of the $\overline{MR}$ signal clears all Qs to 0. When the $\overline{MR}$ signal is high, Q0 receives the AND value of inputs DA and DB on the rising edge of the CP signal, whereas Q1 through Q7 receive the previous values of Q0 through Q6, respectively.

Example 9.3 Show the circuit connection for using one 74LV164 to add an output port to the C8051F040 via the UART0 module in Mode 0. Write a subroutine to output a byte in accumulator A to the 74LV164.

Solution: A typical circuit connection for the 74LV164 is shown in Figure 9.13. In the circuit, signal DB is tied to high, which will set the serial input to the value arrived at the DA pin.

The subroutine that outputs a byte to the 74LV164 is as follows.

```
putch:      jnb     TI0,$           ; wait until SBUF0 is empty
            mov     SBUF0,A         ; the value to be output is in A
            ret
```

9.7 Applications of UART Mode 1

Mode 1 of UART can be used to exchange data between MCUs. The connection can be direct or via an EIA232 interface.

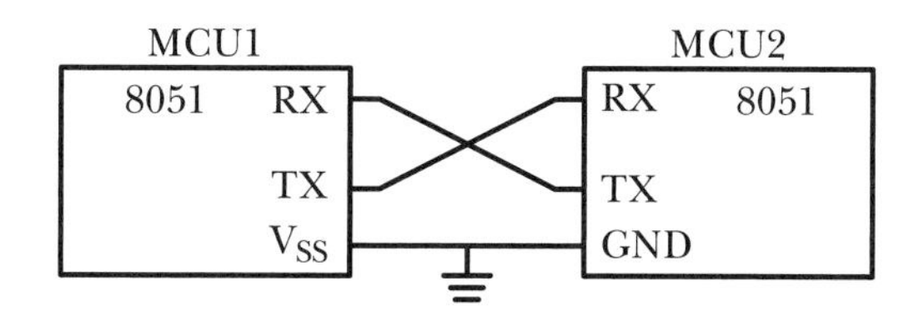

FIGURE 9.14 *UART Mode 1 interconnect diagram.*

9.7.1 Direct Connection of UART in Mode 1

The circuit connection for data exchange between two MCUs can be as simple as shown in Figure 9.14.

Example 9.4 For the circuit shown in Figure 9.14, write two programs to be run by MCU1 and MCU2. Assume both MCUs are the C8051F040 running with a 24-MHz external oscillator. The operations performed by Program 1 and Program 2 are as follows.

Program 1

Step 1 Disable the watchdog timer, select the external oscillator as SYSCLK, assign I/O pins to peripheral signals, and assign three I/O pins to the SPI.

Step 2 Initialize UART0 to operate in Mode 1.

Step 4 Initialize the variable **xcount** to 0.

Step 5 Call a delay function to wait for 200 ms, output **xcount** to UART0 so that it will be received by MCU2, and increment **xcount** by 1.

Step 6 Repeat Step 5 forever.

Program 2

Step 1 Disable the watchdog timer, select the external oscillator as SYSCLK, assign I/O pins to peripheral signals, and assign three I/O pins to the SPI. Enable Port P5 for output. Port P5 drives eight LEDs.

Step 2 Configure UART0 to operate in Mode 1.

Step 3 Enable the UART0 interrupt and stay in a **wait** loop to wait for receive interrupt.

Step 4 The UART0 interrupt service routine checks to see if it is a reception interrupt (RI0 = 1). If yes, reads the byte from the SBUF0 register and outputs it to LEDs. Otherwise, simply returns.

Solution: The assembly language version of **Program 1** is as follows.

```
            $nomod51
            $include    (c8051F040.inc)
xcount      set         R6
            org         0x00
```

```
            ljmp      start
            org       0xAB
start:      mov       SP,#80H
            lcall     sysinit
            mov       xcount,#0
            lcall     openUART0
againT:     mov       R0,#2                  ; wait for 200 ms
            lcall     delayby100ms           ;     "
            mov       SBUF0,xcount           ; write xcount to UART0
            jnb       TI0,$                  ; wait until the byte has been sent out
            inc       xcount
            clr       TI0
            ajmp      againT
; ***************************************************************************************************
; The following function configures UART0 to mode 1 and set its baud rate to 115384.
; ***************************************************************************************************
openUART0:
            mov       SFRPAGE,#TMR4_PAGE     ; switch to the page 2 of SFRs
            mov       RCAP4H,#0xFF           ; set TMR4 reload value to set baud rate
            mov       RCAP4L,#0xF3           ; to 115384 baud
            mov       TMR4H,#0xFF            ; also place reload value in TMR4
            mov       TMR4L,#0xF3            ;     "
            mov       TMR4CF,#0x08           ; select SYSCLK (= f_OSC) as TMR4's clock source
            mov       TMR4CN,#0x04           ; enable TMR4 in reload timer mode
            mov       SFRPAGE,#UART0_PAGE    ; switch back to the page 0 of SFRs
            mov       SCON0,#0x40            ; configure UART0 to mode 1
            mov       SSTA0,#0x0F            ; use timer 4 to generate UART0 baud rate
            setb      TI0                    ; set TI of SCON to Get Ready to Send
            clr       RI0                    ; clear RI of SCON to Get Ready to Receive
            ret
; ***************************************************************************************************
; The following function disable watchdog timer, select external CMOS clock as
; SYSCLK, assign I/O pins to UART0 signals.
; ***************************************************************************************************
sysinit:    mov       SFRPAGE,#0x0F          ; switch to SFR page F
            mov       WDTCN,#0xDE            ; disable watchdog timer
            mov       WDTCN,#0xAD            ;     "
            mov       CLKSEL,#0              ; used internal oscillator to generate SYSCLK
            mov       OSCXCN,#0x67           ; configure external oscillator control
            mov       R7,#255
            djnz      R7,$                   ; wait for about 1 ms
chkstable:  mov       A,OSCXCN               ; wait until external crystal oscillator is stable
            anl       A,#0x80                ; before using it
            jz        chkstable              ;     "
            mov       CLKSEL,#1              ; use external crystal oscillator to generate SYSCLK
            mov       XBR0,#0x04             ; assign I/O pins to UART0 signals
            mov       XBR2,#0x40             ; also enable crossbar
            mov       P0MDOUT,#1             ; configure TX pin for output
            ret
            $include  (delays.a51)           ; include delay functions here
            end
```

The C language version of **Program 1** is as follows (add both the **delays.c** and this file to the same project).

```
#include <c8051F040.h>
#include<delays.h>
void sysinit(void);
void openUART0(void);
void main (void)
{
    unsigned char xcount;
    sysinit();
    openUART0();
    xcount = 0;
    while (1) {
        delayby100ms(2);
        SBUF0 = xcount;             // output xcount via UART0
        while (!TI0);               // wait until the byte has been sent out
        TI0    = 0;                 // clear TI0 flag
        xcount1 1;
    }
}
void sysinit(void)
{
    unsigned char n;
    SFRPAGE  = 0x0F;
    WDTCN    = 0xDE;                // disable watchdog timer
    WDTCN    = 0xAE;                //    "
    OSCXCN   = 0x67;                // configure external oscillator to generate system clock
    for (n = 0; n < 255; n++);
    while(!(OSCXCN & 0x80));
    CLKSEL   = 0x01;                // use external oscillator generate system clock
    XBR0     = 0x04;                // assign I/0 pins to UART0
    XBR2     = 0x40;                // also enable crossbar
    P0MDOUT = 0x01;
}
void openUART0(void)
{
    SFRPAGE  = 0x02;
    RCAP4H   = 0xFF;                // set up TMR4 reload value for baud rate 115384
    RCAP4L   = 0xF3;                //    "
    TMR4H    = 0xFF;                // set TMR4 to reload value for baud rate 115384 initially
    TMR4L    = 0xF3;                //    "
    TMR4CF   = 0x08;                // select SYSCLK (= fOSC) as TMR4's clock source
    TMR4CN   = 0x04;                // enable TMR4 in reload mode
    SFRPAGE  = 0;                   // switch to SFR page 0
    SCON0    = 0x40;                // configure UART0 to mode 1, disable reception
    SSTA0    = 0x0F;                // use timer 4 to generate UART0 baud rate
    TI0      = 1;                   // get transmitter ready to send
    RI0      = 0;                   // get receiver ready to receive
}
```

The assembly language version of **Program 2** is as follows.

```
            $nomod51
            $include    (c8051F040.inc)
            org         0x00
            ljmp        start
            org         0x23                ; UART0 interrupt vector
            ljmp        Uart0ISR
            org         0xAB                ; start of main program
start:      mov         SP,#0x80
            lcall       sysinitb
            lcall       openUART0b          ; set up UART0 baud rate, enable receive in Mode 1
            setb        ES0                 ; enable UART0 interrupt
            setb        EA                  ; enable interrupt globally
againT:     ajmp        againT              ; stay in the wait loop to wait for receive interrupt
;*****************************************************************************************************
; The following function configures UART0 to mode 1 and set its baud rate to 115384.
;*****************************************************************************************************
openUART0b:
            mov         SFRPAGE,#2          ; switch to SFR page 2
            mov         RCAP4H,#0xFF        ; set TMR4 reload value to set baud rate
            mov         RCAP4L,#0xF3        ; to 115384
            mov         TMR4H,#0xFF         ; set TMR4 to be the same as reload value
            mov         TMR4L,#0xF3         ;       "
            mov         TMR4CF,#0x08        ; select SYSCLK (= fOSC) as TMR4's clock source
            mov         TMR4CN,#0x04        ; enable TMR4 to timer reload mode
            mov         SFRPAGE,#0          ; switch back to SFR page 0
            mov         SCON0,#0x50         ; configure UART0 to mode 1, enable reception
            mov         SSTA0,#0x0F         ; use timer 4 to generate UART0 baud rate
            setb        TI0                 ; set TI of SCON to Get Ready to Send
            clr         RI0                 ; clear RI of SCON to Get Ready to Receive
            ret
;*****************************************************************************************************
; The following function disables watchdog timer, selects external CMOS clock as
; SYSCLK, assigns I/O pins to peripheral signals, and enables SPI to 3-wire mode.
;*****************************************************************************************************
sysinitb:   mov         SFRPAGE,#0x0F       ; switch to SFR page F
            mov         WDTCN,#0xDE         ; disable watchdog timer
            mov         WDTCN,#0xAD         ;       "
            mov         CLKSEL,#0           ; used internal oscillator to generate SYSCLK
            mov         OSCXCN,#0x67        ; configure external oscillator control
            mov         R7,#255
againosc:   djnz        R7,againosc         ; wait for about 1 ms
chkstable:  mov         A,OSCXCN            ; wait until external crystal oscillator is stable
            anl         A,#0x80             ; before using it
            jz          chkstable           ;       "
            mov         CLKSEL,#1           ; use external crystal oscillator to generate SYSCLK
            mov         XBR0,#0x04          ; assign I/O pins to peripheral signals
            mov         XBR2,#0x40          ; also enable crossbar
            mov         P0MDOUT,#1          ; configure TX pin for output
            ret
```

```
; ************************************************************************************************
; The following routine is the actual interrupt service routine for UART0.
; ************************************************************************************************
Uart0ISR:   jnb     RI0,quitI           ; is it a reception interrupt?
            clr     RI0                 ; clear RI0 interrupt flag
            mov     A,SBUF0             ; get a copy of received byte
            mov     SFRPAGE,#0x0F       ; switch to SFRPAGE page F (to output to LEDs)
            mov     P5,A                ; output the received value to LEDs
quitI:      reti                        ; SFRPAGE will be restored after this instruction
            end
```

The C language version of **Program 2** is as follows.

```
#include <c8051F040.h>
void sysinitb (void);
void openUART0b (void);
void Uart0ISR (void);
data unsigned char ctmp;        // temporary storage to be used by Uart0ISR
void main (void)
{
    sysinitb();
    openUART0b();
    ES0 = 1;                    // enable UART0 interrupt
    EA  = 1;                    //      "
    while (1);                  // wait for reception interrupt
}
void sysinitb(void)
{
    unsigned char n;
    SFRPAGE  = 0x0F;
    WDTCN    = 0xDE;            // disable watchdog timer
    WDTCN    = 0xAE;            //      "
    CLKSEL   = 0;               // use internal oscillator as system clock temporarily
    OSCXCN   = 0x67;            // configure external oscillator to generate system clock
    for (n = 0; n < 255; n==);
    while (!(OSCXCN & 0x80));   // wait until external oscillator is stable
    CLKSEL   = 0x01;            // use external oscillator generate system clock
    XBR0     = 0x04;            // assign I/O pins to UART0 signals
    XBR2     = 0x40;            // also enable crossbar
    P0MDOUT = 0x01;
}
void openUART0b(void)
{
    SFRPAGE  = 0x02;
    RCAP4H   = 0xFF;            // set up TMR4 reload value
    RCAP4L   = 0xF3;            //      "
    TMR4H    = 0xFF;            // set TMR4 to reload value
    TMR4L    = 0xF3;            //      "
    TMR4CF   = 0x08;            // select SYSCLK (= fOSC) as TMR4's clock source
    TMR4CN   = 0x04;            // enable TMR4 in reload mode
```

```
        SFRPAGE = 0;            // switch to SFR page 0
        SCON0   = 0x50;         // configure UART0 to mode 1, enable reception
        SSTA0   = 0x0F;         // use timer 4 to generate UART0 baud rate
        TI0     = 1;            // get transmitter ready to send
        RI0     = 0;            // get receiver ready to receive
}
void Uart0ISR(void) interrupt 4
{
    if (RI0) {
        RI0     = 0;
        ctmp    = SBUF0;        // make a copy of received byte
        SFRPAGE = 0x0F;         // switch to SFR page to access LEDs
        P5      = ctmp;
    }
}
```

9.7.2 Data Exchange with UART Mode 1 via EIA232 Interface

Another way of exchange data between two MCUs or between an MCU and a PC is via the EIA232 interface. This method has been used widely in university laboratories for communicating a microcontroller demo board with a PC. Because the UART modules of the C8051F040 use 3.3 and 0 V to represent logic 1 and 0, respectively, they cannot be connected directly to the EIA232 interface. A voltage translation circuit, which is called the **EIA232 (or RS232) transceiver**, is needed to translate the voltage levels of the UART signals (RX and TX) to and from those of the corresponding EIA232 signals.

EIA232 transceivers are available from many vendors (including MAXIM, SIPEX, National Semiconductor, Philips, STMicroelectronics, Intersil, etc). These companies produce EIA232 transceiver chips that can operate with a single voltage supply over a range from 3.0 to 5.5 V.

The SP3223E series from Sipex are discussed in this section. The pin assignment and the use of each pin are illustrated in Figure 9.15. Adding an EIA232 transceiver chip will allow the 8051 variants to use the UART module to communicate with the EIA232 interface circuit. The circuit shown in Figure 9.16 is used in the SSE040 demo board and the SiLabs C8051F040 development kit. A null-modem connection is followed to allow the use of a straight-through cable to communicate with a PC.

Programming for data transmission and reception is the same with or without going through the EIA232 interface in UART Mode 1. There are a set of common routines that can be called by all programs to carry out data communications tasks. These routines include:

openUART0: This function configures UART0 to operate in Mode 1 with the baud rate of 19200.

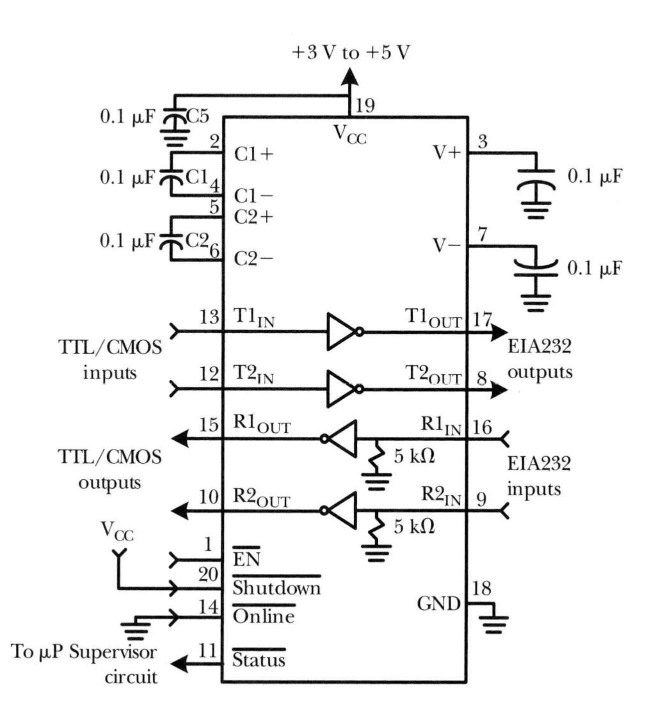

FIGURE 9.15 *SP3223E typical operating circuit.*

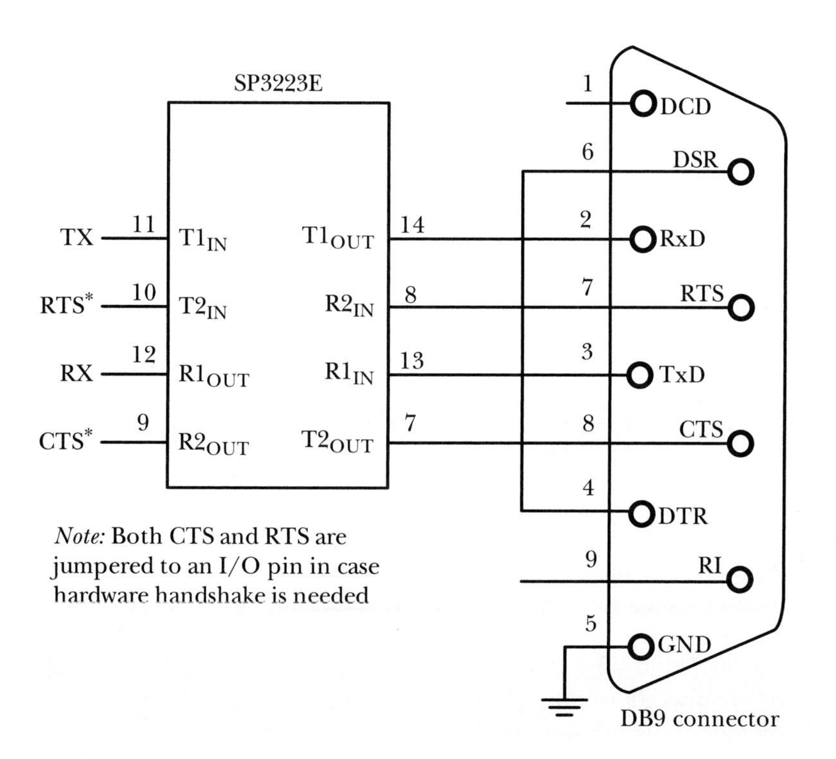

FIGURE 9.16 *Diagram of UART and EIA232 DB9 connector wiring in SSE040 demo board.*

putch: This routine outputs a character to the UART0 module.
putsc: This routine outputs a string in program memory to the UART0 module.
putsx: This routine outputs a string in XDATA memory to the UART module.
getch: This routine reads a character from the UART0 module.
gets: This routine reads a string from the UART0 module and stores the string in a buffer in XRAM.
newline: This routine outputs a carriage return followed by a new line to move the cursor to the leftmost position of the next row.

Example 9.5 Write the UART routines described previously in assembly language, assuming that the frequency of the system clock for the C8051F040 is 24 MHz.

Solution: Suppose we choose to use Timer 4 as the baud rate generator for UART0. Then the following subroutine will perform the desired configuration.

```
openUART0:
        mov     SFRPAGE,#TMR4_page      ; switch to SFR page 2
        mov     RCAP4H,#0xFF            ; set TMR4 reload value of the baud rate
        mov     RCAP4L,#0xB2            ; at 19200
        mov     TMR4H,#0xFF             ; set TMR4 to be the same as reload value
        mov     TMR4L,#0xB2             ;       "
        mov     TMR4CF,#0x08            ; select SYSCLK (= fOSC) as TMR4's clock source
        mov     TMR4CN,#0x04            ; enable TMR4 to timer reload mode
        mov     SFRPAGE,#UART0_PAGE     ; switch back to SFR page 0
        mov     SCON0,#0x50             ; configure UART0 to mode 1, enable reception
        mov     SSTA0,#0x0F             ; use timer 4 to generate UART0 baud rate
        setb    TI0                     ; Set TI of SCON to Get Ready to Send
        clr     RI0                     ; Clear RI of SCON to Get Ready to Receive
        ret
```

The character to be output by the **putch** routine is passed in accumulator A. The following subroutine uses the polling method to output the character in A.

```
putch:  jnb     TI0,$           ; wait until TI0 flag is set to 1
        clr     TI0
        mov     SBUF0,A         ; write the character to UART
        ret
```

The string to be output can be stored in program memory or data memory. Data memory is divided further into external memory (or XRAM) and data memory. Since data memory is very small, we will use XRAM to store the string to be output. There will be two versions of the **puts** subroutine. One version outputs the string in the program memory. The other version outputs the string in the XRAM.

The following subroutine outputs a string (pointed to by DPTR) in program memory.

```
putsc:    push   B              ; use B as the index to the string
          mov    B,#0
oloop:    mov    A,B            ; place index to string in A
          movc   A,@A+DPTR      ; get one character from string
          jz     odone          ; reach NULL character
          call   putch
          inc    B
          ajmp   oloop
odone:    pop    B
          ret
```

The following subroutine outputs a string (pointed to by DPTR) in XRAM.

```
putsx:    movx   A,@DPTR        ; get one character from string
          jz     xdone          ; reach NULL character
          call   putch
          inc    DPTR
          ajmp   putsx
xdone:    ret
```

The following subroutine reads a character from UART0 and returns it in A.

```
getch:    jnb    RI0,$          ; Is there a recently received character?
          clr    RI0
          mov    A,SBUF0        ; get the character
          ret
```

The following subroutine reads a string from the UART module and stores the string in a buffer pointed to by DPTR. The string is terminated by a **carriage return** character. The string entered by the user will be echoed by this subroutine for verification. This subroutine also supports a backspace feature.

```
gets:     call   getch
          cjne   A,#enter,notdone   ; received a carriage return?
          mov    A,#0
          movx   @DPTR,A            ; terminate the string with a NULL character
          ret
notdone:  movx   @DPTR,A            ; save the character in the buffer
          lcall  putch
          cjne   A,#BS,notBS
          mov    A,DPL
          clr    C
          subb   A,#1
          mov    DPL,A
```

```
          mov      A,DPH
          subb     A,#0
          mov      DPH,A
          mov      A,#space
          lcall    putch
          mov      A,#BS
          lcall    putch
          ajmp     gets
notBS:    inc      DPTR          ; move the buffer pointer
          ajmp     gets
```

The following subroutine moves the cursor to the leftmost position of the next row of the monitor screen.

```
newline:  mov      A,#enter      ; output carriage return character
          acall    putch
          mov      A,#LF         ; output line feed character
          acall    putch
          ret
```

The C language versions of the UART functions are similar and hence will be left to you as an exercise problem.

Example 9.6 Write a program to test the subroutines provided in Example 9.5.

Solution: The test program will perform the following operations.

1. Initialize the crossbar and configure system clock source.
2. Initialize UART0 to Mode 1, enable reception, and set baud rate to 19,200.
3. Call the **putsc** subroutine to output the message **Enter your name:**. (When seeing the message, you should enter your name.)
4. Call the **gets** function to read a string entered by the user from the keyboard.
5. Call the **putsx** function to echo the string entered by the user to the monitor screen.
6. Call the **newline** function to move cursor to the leftmost position of the next line.
7. Call the **putsc** subroutine to output the message **Enter your school:**. (When seeing the message, you should enter your school name.)
8. Call the **gets** subroutine to read a string entered by the user from the keyboard.

9. Call the **putsx** function to echo the string entered by the user to the monitor screen.
10. Call the **newline** function to move cursor to the leftmost position of the next line.

The test program in assembly language is as follows.

```
          $nomod51
          $include (c8051F040.inc)
LF        equ       0x0A              ; ASCII code of linefeed character
enter     equ       0x0D              ; ASCII code of enter key
BS        equ       0x08              ; ASCII code of backspace
space     equ       0x20              ; ASCII code of space character

          xseg at   0x00
buf:      ds        40                ; reserve 40 bytes as reception buffer
          cseg at   0x00
          ljmp      start
          org       0x23
          reti
          org       0xAB
start:    mov       SP,#0x80          ; set up stack pointer
          lcall     sysinit           ; select system clock source and enable crossbar
          lcall     openUART0         ; enable UART receive and transmit, set baud rate
          lcall     newline           ; move cursor to the leftmost position of the next line
          mov       DPTR,#msg1        ; output the first message
          lcall     putsc             ;     "
          mov       DPTR,#buf         ; read a string entered from the keyboard
          lcall     gets              ;     "
          lcall     newline
          mov       DPTR,#ans1        ; output "My name is "
          lcall     putsc             ;     "
          mov       DPTR,#buf         ; output my name entered from keyboard
          lcall     putsx             ;     "
          lcall     newline           ; move to next line
          mov       DPTR,#msg2        ; output the second message
          lcall     putsc             ;     "
          mov       DPTR,#buf         ; read the school name entered from the keyboard
          lcall     gets              ;     "
          lcall     newline
          mov       DPTR,#ans2
          lcall     putsc
          mov       DPTR,#buf         ; echo the school name on the monitor screen
          lcall     putsx             ;     "
          lcall     newline
againT:   ajmp      againT            ; infinite loop

msg1:     db        "Please enter your name: ",0
ans1:     db        "My name is ",0
msg2:     db        "Please enter your school name: ",0
ans2:     db        "My school is ",0
```

```
;***************************************************************************************************
; include openUARTOb subroutine here.
;***************************************************************************************************
;***************************************************************************************************
; include sysinitb subroutine here.
;***************************************************************************************************
;***************************************************************************************************
; include UART I/O routines here
;***************************************************************************************************
```

Before running this test program, you need to start the **HyperTerminal** program bundled with the windows and set the baud rate to 19,200. After running the program, you should see the screen similar to that in Figure 9.17.

9.8 Applications of UART Modes 2 and 3

Modes 2 and 3 are used mainly in multipoint data link (involves multiprocessor) communications between a master and one or more slave processors by special use of the ninth data bit and built-in UART address recognition hardware (the original 8051 does not have this address recognition hardware). Since the C8051F040 has special address hardware, it will be discussed separately. The diagram of a UART multiprocessor-mode interconnection is shown in Figure 9.18. A complete discussion of multipoint data link is beyond the scope of this text. We will only briefly explore how Modes 2 and 3 may be used in this environment.

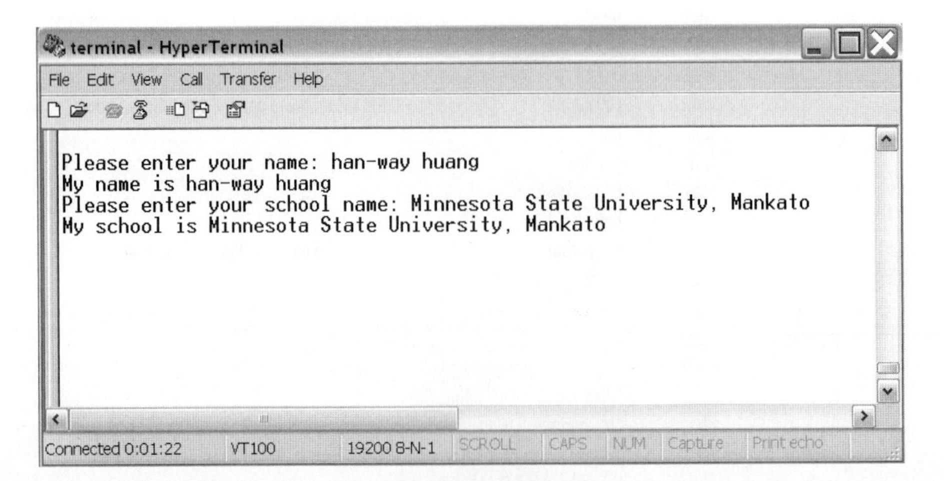

FIGURE 9.17 *Screen output after running the program in Example 9.6.*

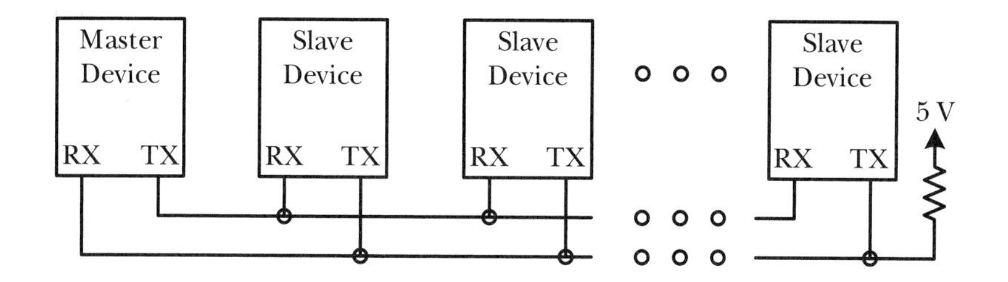

FIGURE 9.18 *UART multiprocessor mode interconnect diagram.*

9.8.1 Multiprocessor Application of UART of the Original 8051

In a multipoint data link, there is a controlling unit called a *master*. The remaining units are called *slaves*. The master directs the activities of the slaves. Each slave has its own unique address. The address is usually 8 bits. An address byte is sent on the same link. Slaves must be able to determine if a byte is an address byte or a data byte. Modes 2 and 3 provide easy ways to solve this problem. Depending on the value of the tenth bit, a character is treated as an address (1) or data (0) byte.

A communication protocol is needed to govern the communication activities between the master and slaves. The message sent to a slave should follow a predefined format so that no misunderstanding will occur. A message consists of several characters. The first character is always the address byte. Following the address byte is a command that tells the slave what to do. A command may consist of one or multiple bytes depending on the size of the command set. Following the command are the optional data bytes. A slave can send or receive data to or from the master in response to the request from the master.

After the power-on reset, all of the slaves may have their SM2 of the SCON registers set to 1 so that they are expecting to receive address bytes. If the tenth bit (11-bit character) of the incoming byte is a 1, all slaves will compare (using hardware or software) it with their addresses. The slave that was addressed will clear the SM2 bit of its SCON register and then wait for command or data bytes to come. Those slaves that were not selected will not change their SM2 bits and continue on their business. A special character such as **$** is used to mark the end of a message.

9.8.2 Multiprocessor Application of UART of C8051F040

The C8051F040 has additional address recognition support than the original 8051. The C8051F040 (and other SiLabs MCUs) uses the **SADDR0** register to hold the slave address and the **SADEN0** register to hold the address mask. The UART0 of C8051F040 recognizes two types of addresses:

- Masked address
- Broadcast address

Configuration of a Masked Address

When a bit in the SADEN0 is set to 1, the corresponding bit in SADDR0 will be checked against the received address byte. Bits set to 0 in SADEN0 correspond to "don't care" bits in SADDR0. Setting a few bits in SADEN0 to 0s reduces the number of address bits to be compared.

Setting the SM20 bit in the SCON0 register configures the UART0 such that, when a stop bit is received, the UART0 will generate an interrupt only if the ninth bit is 1 and the received data byte matches the UART0 slave address. Following the received address interrupt, the slave will clear its SM20 bit to enable interrupts on the reception of the following data byte(s). Once the entire message is received, the addressed slave resets its SM20 bit to 1 to ignore all transmissions until it receives the next address byte. While SM20 is 1, UART0 ignores all bytes that do not match the UART0 address and include a ninth bit that is logic 1.

Broadcast Addressing

Multiple addresses can be assigned to a single slave, and/or a single address can be assigned to multiple slaves, thereby enabling broadcast transmissions to more than one slave simultaneously. The broadcast address is the logical OR of registers SADDR0 and SADEN0. The resultant bits that are 0s are "don't cares." Typically, a broadcast address of 0xFF is acknowledged by all slaves, assuming don't care bits as 1s. The master processor can be configured to receive all transmissions, or a protocol can be implemented such that the master/slave role is temporarily reversed to enable half-duplex transmission between the original master and the slave(s).

Example 9.7 Suppose we have three slaves with their SADDR0 and SADEN0 registers set to values shown in Table 9.7. Find out their valid UART0 matching addresses and their broadcast addresses.

Solution: The resultant UART0 address and broadcast address are shown in Table 9.8.

TABLE 9.7 *Example Addresses and Mask Register Values*

	Slave 1	Slave 2	Slave 3
SADDR0	00110110	00110110	00110110
SADEN0	00001111	11110011	11000000

TABLE 9.8 *Example Addresses and Broadcast Addresses*

	Slave 1	Slave 2	Slave 3
SADDR0	00110110	00110110	00110110
SADEN0	00001111	11110011	11000000
UART0 address	xxxx0110	0011xx10	00xxxxxx
Broadcast address	00111111	11110111	11110110

9.9 Chapter Summary

When high-speed data transfer is not necessary, using serial data transfer enables the chip designer to make the most use of the limited number of I/O pins. Serial data transfer can be performed asynchronously or synchronously. Examples of popular serial interfaces include UART, SPI, I^2C (SMBus), and CAN.

The UART interface transfers data byte-by-byte and can be used to transfer data asynchronously or synchronously. This interface uses two signal pins (TX and RX) to perform data transfer. When data is transferred synchronously, the TX pin is used to carry the clock signal for synchronizing data transfer. Synchronous data transfer can be only half-duplex. Asynchronous data transfer can be full-duplex.

The UART module of the 8051 variants supports four operation modes: Mode 0, Mode 1, Mode 2, and Mode 3. Mode 0 is a synchronous mode in which the 8051 variant must be responsible for generating the clock signal to synchronize data transfer. Eight bits of data are transferred in Mode 0. This mode can be used to interface shift registers to the 8051. Adding shift registers is equivalent to expanding the number of parallel I/O ports. Most shift registers also can be cascaded.

Mode 1 is an asynchronous mode. Ten bits are transferred for each byte, including one start bit, eight data bits, and one stop bit. This mode is used mainly to exchange data between two processors. Two microcontrollers can exchange data using the UART interface directly or going through the EIA232 interface. When the EIA232 interface is used between two UART modules, a transceiver is needed to perform voltage-level translation, because UART and EIA232 operate at different voltage levels. EIA232 transceivers that operate with voltage supply from 3.0 to 5.0 V are available from many vendors.

Both Modes 2 and 3 are asynchronous and are used mainly in multiprocessor communication. For each data byte, eleven bits are transferred including one start bit, nine data bits, and one stop bit. One processor is the master, whereas the other processors are slaves. For each data-transfer transaction, the first byte sent out by the master is always the address byte. The address byte has its ninth data bit set to 1, whereas the data byte has its ninth bit set to 0. Only the slave with its address matching the received address will respond to the data bytes.

Baud rate is important to the successful data transfer of UART. The baud rate of Mode 0 is fixed at $f_{OSC}/12$. The baud rate of Mode 2 can be either SYSCLK divided by 64 or divided by 32. The baud rates of Modes 1 and 3 are generated by the overflows of Timers 1, 2, 3, or 4.

9.10 Exercise Problems

E9.1 Sketch the waveform of letter **h** when it is transmitted via the UART interface using the format of one start bit, eight data bits, and one stop bit.

E9.2 Delineate the waveform of letter **k** when it is transmitted via the UART interface using the format of one start bit, eight data bits, and one stop bit.

E9.3 Write an instruction sequence to configure the UART1 of the C8051F040 to operate in Mode 1 at 19,200 baud (use Timer 1 overflow to generate baud rate), assuming that the C8051F040 is operating with a 24-MHz oscillator.

E9.4 Write a group of C statements to configure the UART0 of the C8051F040 to operate in Mode 3 at 28,800 baud, assuming that the C8051F040 is operating with a 25-MHz crystal oscillator. Use Timer 4 overflow to generate the baud rate.

E9.5 Write a group of C statements to configure the UART0 of C8051 to operate in Mode 1 and use Timer 2 overflow to generate the baud rate of 28,800, assuming that the MCU is operating with a 24-MHz crystal oscillator.

E9.6 Write a sequence of instructions to configure the UART0 of C8051 to operate in Mode 1 and use Timer 1 overflow to generate the baud rate of 28,800, assuming that the MCU is operating with a 24-MHz crystal oscillator.

E9.7 For the circuit in Figure 9.13, write a program to configure UART0 to Mode 0, initialize a variable **xcnt** to 0, and then perform the following operations forever:

- Wait for 200 ms by calling a delay function
- Increment **xcnt** by 1
- Output **xcnt** to 74LV164

Assume that the 74LV164 is driving eight LEDs.

E9.8 Draw a circuit to illustrate how to add three input ports using three 74LV165s to the C8051F040 using the UART Mode 0. Write a program to read a 24-bit value from these three 74LV165s.

E9.9 Draw a circuit to illustrate how to cascade four 74LV164s to drive four seven-segment displays. Write a program to configure UART0 to Mode 0 and send 32-bit data to display 1234 on these four seven-segment displays.

E9.10 Write a program to output the value in B as two hex digits (ASCII code) to the UART0 module of the_C8051F040.

E9.11 Write a C version of the subroutines in Example 9.5 for the original 8051.

E9.12 Suppose we have four C8051F040 slaves with their SADDR0 and SADEN0 register set to values as shown in Table 9E.12. Find out their valid UART0 matching addresses and broadcast addresses.

TABLE 9.E12 *Example Addresses and Mask Register Values*

Register	Slave 1	Slave 2	Slave 3	Slave 4
SADDR0	00101011	00101110	11010011	11010011
SADEN0	11001100	00110111	10110110	01111001

E9.13 Modify the **getch** subroutine by adding the **echo** capability. Use an **echo** flag for this purpose. The **getch** subroutine will output the received character to the UART module if the **echo** flag is set to 1.

9.11 Laboratory Exercise Problems and Assignments

L9.1 Connect a circuit (as shown in Figure 9.13) and add eight LEDs to the Q0 to Q7 pins of the 74LV164. Write a program to perform the following operations.

Step 1 Configure UART0 to Mode 0.

Step 2 Initialize a variable **xcount** to 0.

Step 3 Call a delay function to wait for 200 ms.

Step 4 Output **xcount** to 74LV164.

Step 5 Increment the variable **xcount** by 1.

Repeat Steps 3, 4, and 5 forever.

L9.2 Connect two C8051F040 demo boards, as shown in Figure 9.14. Use the LEDs on the demo board to display a received value. If the demo board does not have LEDs but has LCD, then use LCD to display the received value. If the demo board has neither LEDs nor LCD, then add them. Write a program to be run on each demo board to perform the following operations.

Step 1 Configure UART0 to Mode 1 and set baud rate to 115,384.

Step 2 Initialize a variable *xyz* to 0.

Step 3 Call a delay function to wait for 200 ms.

Step 4 Output *xyz* to UART0 so that the value will be received by the other MCU.

Step 5 Enable UART0 receive interrupt.

Step 6 For every received value, display it on the LEDs (or LCD).

Step 7 Increment the variable *xyz* by 1.

Repeat Steps 3 through 7 forever.

L9.3 Real Time Clock Practice.

Step 1 Run a terminal program such as **HyperTerminal** on the PC. Make sure that the **HyperTerminal** operates at the same baud rate as does the UART0 module. Set baud rate to 19,200 for this lab practice.

Step 2 Use an EIA232 cable to connect the UART0 port of the demo board to the PC COM port.

Step 3 Write a program to configure UART0 to operate in Mode 1 and output this message on the **HyperTerminal**:

```
Enter the current time in hh:mm:ss format:
```

Step 4 Enter the current hours, minutes, and seconds in the format indicated in Step 3.

Step 5 Read in the current time, update it once every second, and display it on the monitor screen.

The SPI Function

10.1 Objectives

Upon successful completion of this chapter, you will be able to:

- Describe the SPI module of the C8051F040
- Configure the SPI operation parameters
- Read and transmit data via the SPI interface
- Interface with peripheral devices with SPI interface
- Use the SPI function to interface with the shift register 74LV595
- Use SPI to interface with the EEPROM 25AA080A
- Use SPI to interface with the 10-bit D/A converter LTC1661
- Use SPI to interface with the matrix LED display driver MAX6952

10.2 Introduction to the SPI Function

The serial peripheral interface (SPI) protocol initially was proposed by Freescale to simplify the interfacing of peripheral devices (ICs) to the microcontroller. The SPI has low software overhead, requires simple hardware connection, and provides good performance. Since its debut, the SPI has been adopted by almost all 8-bit, 16-bit, and 32-bit microcontrollers. The SPI allows microcontrollers (including the 8051 variants) to communicate synchronously with peripheral devices and other microcontrollers. The SPI module can operate as a **master** or as a **slave** in both three-wire or four-wire modes and supports multiple masters and slaves on a single SPI bus. When the SPI module is configured as a master, it is responsible for generating the clock signal (SCK) required for synchronizing data transfer. An SPI slave can respond only to the data-transfer request. The SPI data transfer is full-duplex, because the master and the slave exchange a byte with each other.

Although the SPI module allows the microcontroller to communicate with other microcontrollers, it is used mainly in interfacing with peripheral devices (such as shift registers, LED/LCD display drivers, real-time clock (RTC) chips, phase-locked loop (PLL) chips, serial memory components, digital temperature sensors, CAN controllers, Ethernet controllers, digital potentiometers, or A/D and D/A converter chips). As long as a peripheral device is compatible with the SPI protocol, it can communicate with the MCU via the SPI interface.

This chapter will discuss the signal pins and registers used in the SPI transfer, present the operation details of the SPI transfer operation, and then examine a few peripheral chips that incorporate the SPI interface.

10.3 SPI Signal Pins

The SPI function uses the following four signals to communicate.

1. **Serial Clock (SCK).** The SCK signal is an output from the master device and an input to the slave device. It is used to synchronize the transfer of data between the master and the slave on the MOSI and MISO pins. This signal is generated by the master and will be ignored by a slave device if it is not selected.
2. **Master In, Slave Out (MISO).** This signal is an output from the slave device and an input to the master device. It is used to transfer data serially from the slave to the master. Data is transferred most-significant-bit first. The MISO pin is placed in a high-impedance state when the SPI module is disabled and when the SPI is not selected when in slave mode.
3. **Master Out, Slave In (MOSI).** This signal is an output from the master device and an input to slave devices. It is used to transfer data serially from the master to the slave. Data is transferred most-significant-bit first. Some MCUs also may support the least significant bit to be transferred first. When configured as a master, the MOSI pin is driven by the most significant bit of the shift register in both three- and four-wire mode.
4. **Slave Select ($\overline{\text{NSS}}$ or $\overline{\text{SS}}$).** The C8051F040 used $\overline{\text{NSS}}$ to refer to this signal, whereas many other 8051 variants use $\overline{\text{SS}}$ to refer to this signal. This signal is used in four-wire mode but not in three-wire mode. In three-wire mode, this signal is disabled and is not routed to the port pin. In four-wire slave mode or four-wire multi-master mode, this signal is used as an input. In four-wire single-master mode, this signal is configured as an output and can be used to select one of the slave devices connected to the SPI bus.

For the C8051F040, SPI pins would be assigned to Port P0 pins. If UART0 is used, then Port P0 pins 2 through 5 would be assigned to signals SCK, MISO, MOSI, and NSS, respectively. If UART0 is disabled, then they will be assigned to Port P0 pins 0 to 3.

10.4 Registers Related to SPI

The C8051F040 uses the following four registers to support the operation of SPI data transfer.

- SPI0 configuration register (SPI0CFG)
- SPI0 control register (SPI0CN)
- SPI0 clock rate register (SPI0CKR)
- SPI0 data register (SPI0DAT)

The contents of SPI0CFG, SPI0CN, and SPI0CKR are shown in Figures 10.1, 10.2, and 10.3, respectively. The register SPI0CFG allows the user to choose between slave and master mode and select the clock phase and polarity for data shifting. The SPI0CN register is bit-addressable and allows the user to

7	6	5	4	3	2	1	0	
SPIBSY	MSTEN	CKPHA	CKPOL	SLVSEL	NSSIN	SRMT	RXBMT	Reset value 0x07
r	rw	rw	rw	r	r	r	r	

SPIBSY: SPI busy
This bit is set to 1 when an SPI transfer is in progress and will be cleared to 0 when data transfer is completed
MSTEN: Master mode enable
0 = operate in slave mode
1 = enable master mode
CKPHA: SPI0 clock phase
0 = data sampled on the first edge of SCK period
1 = data sampled on the second edge of SCK period
CKPOL: SPI0 clock polarity
0 = SCK is idle low (not in data transfer mode)
1 = SCK is idle high
SLVSEL: Slave selected flag
This flag is set to1 whenever the NSS pin is low indicating that the SPI0 is the selected slave It is cleared to 0 when the NSS input is high
NSSIN: NSS instantaneous pin input
This bit mimics the instantaneous value that is present on the NSS port pin at the time that the register is read
SRMT: Shift register empty(valid in slave mode, SRMT = 1 when in master mode)
This bit will be set to 1 when all data has been transferred in/out of the shift register, and there is no new information available to read from the transmit buffer or write to the receive buffer. It returns to 0 when a byte is transferred to the shift register
RXBMT: Receive buffer empty (valid in slave mode, RXBMT = 1 when in master mode)
0 = the data in receive buffer has been read
1 = the data in receive buffer has not been read yet

FIGURE 10.1
SPIO Configuration register (SPIOCFG of C8051F040).

7	6	5	4	3	2	1	0	
SPIF	WCOL	MODF	RXOVRN	NSSMD1	NSSMD0	TXBMT	SPIEN	Reset value 0x06
rw	rw	rw	rw	rw	rw	r	rw	

SPIF: SPI interrupt flag
This bit is set to 1 at the end of a data transfer. This bit is not automatically cleared by hardware and must be cleared by software
WCOL: Write collision flag
0 = no write collision
1 = a write to SPI0DAT register was attempted while a data transfer was in progress. This bit must be cleared by software
MODF: Mode fault flag
0 = no mode fault
1 = the NSS input is low while the SPI module is configured to be a master. This bit must be cleared by software
RXOVRN: Receiver overrun flag
0 = no receiver overrun
1 = a new byte is shifted in when the receive buffer still holds the previously received data byte. This bit must be cleared by software
NSSMD1:NSSMD0: Slave select mode
00 = 3-wire slave or 3-wire master mode (NSS is not routed to port pin)
01 = 4-wire slave or multi-master mode (default). NSS is always an input
1x = 4-wire single-master mode. NSS signal is mapped as an output from the device and will assume the value of NSSMD0
TXBMT: Transmit buffer empty
0 = transfer buffer contains data byte to be output
1 = the data in transmit buffer has been transferred to the SPI shift register
SPIEN: SPI0 enable
0 = SPI0 is disabled
1 = SPI0 is enabled

FIGURE 10.2
SPIO Control register (SPIOCN of C8051F040).

7	6	5	4	3	2	1	0	
SCR7	SCR6	SCR5	SCR4	SCR3	SCR2	SCR1	SCR0	Reset value 0x00
rw	rw	rw	rw	rw	rw	rw	rw	

SCR7-SCR0: SPI0 clock rate
These bits determine the frequency of the SCK output when the SPI0 is configured in master mode using the following expression:

$$f_{SCK} = \frac{f_{SYSCLK}}{2 \times (SPI0CKR + 1)} \tag{10.1}$$

FIGURE 10.3
SPIO clock rate register (SPIOCKR of C8051F040).

enable or disable the SPI and select between three- and four-wire modes. The SPI0CKR register sets the SPI data shift rate.

Example 10.1 Give a value to be programmed into the SPI0CKR register to set the SPI baud rate to 4 MHz, assuming that the C8051F040 is using a 24-MHz external oscillator as its system clock source.

Solution: The value to be written into the SPI0CKR register is given by the expression

$$SPI0CKR = f_{SYSCLK} \div (f_{SCK} \times 2) - 1 \qquad (10.2)$$

Since $f_{SYSCLK} = 24$ MHz and $f_{SCK} = 4$ MHz, we should write **0x02** into the SPI0CKR register.

Example 10.2 What is the highest possible baud rate for the SPI with a 24-MHz system clock?

Solution: The highest SPI baud rate occurs when 0 is written into the SPI0CKR register. Under this condition, the baud rate is 24 MHz / 2 = 12 MHz.

10.5 SPI Operation

Only a master SPI module can initiate data transmission. A transmission begins by writing to the master SPI data register. Data is transmitted and received simultaneously. The serial clock (SCK) synchronizes shifting and sampling of the information on the two serial data lines. The $\overline{NSS}$ pin allows selection of an individual slave SPI device; slave devices that are not selected do not interfere with SPI bus activities. Optionally, on a master SPI device, the $\overline{NSS}$ signal can be used to indicate multiple-master bus contention (more than one master starts SPI data transfer at about the same time). The $\overline{NSS}$ pin of the C8051F040 SPI can be used to enable a slave SPI device for data transmission in the four-wire, single-master mode. In four-wire, single-master mode, the MCU can enable or disable the slave device by setting the NSSMD0 bit properly. The C8051F040 supports both master and slave modes of operation.

10.5.1 Transmission Formats

The CKPHA and CKPOL bits in the SPI0CFG register of the C8051F040 allow the user to select one of the four combinations of serial clock phase and polarity. The clock phase control bit (CKPHA) selects one of two fundamentally different transmission formats. Clock phase and polarity should be identical for the master SPI device and the communicating slave device.

When the CKPHA bit is set to 0, the first edge on the SCK line is used to clock the first data bit of the slave into the master and the first data bit of the master into the slave. In some peripheral devices, the first bit of the slave's data is available at the slave D_{OUT} pin as soon as the slave is selected and, hence, can take advantage of this format. When the CKPHA bit is 1, data is sampled on the second edge of the clock period.

The CKPOL bit determines whether the SCK signal is set to idle low or high. When the SPI module is not transmitting data actively and the CKPOL bit is 0, then the SCK signal will be low. When the SPI module is not actively transmitting data and the CKPOL bit is 1, then the SCK signal will be high.

Data sampling occurs in the middle of a bit time. As shown in Figure 10.4, there are four possible data transmission formats for the SPI.

10.5.2 SPI Master Mode Operation

An SPI master device initiates all data transfers on an SPI bus. The SPI module is placed in master mode by setting the master enable bit (bit 6 of SPI0CFG). Writing a byte of data into the SPI data register when in master mode writes to the transmit buffer. If the SPI shift register is empty, the byte in the transmit buffer will be moved to the shift register, and a data transfer begins. The SPI master immediately shifts out data serially on the MOSI pin while providing the serial clock on the SCK pin. The **SPIF** flag is set to 1 at the end of the transfer. If interrupt is enabled, an interrupt request is generated when the **SPIF** flag is set to 1.

While the SPI master transfers data to a slave on the MOSI pin, the addressed (enabled) slave device simultaneously transfers the contents of its

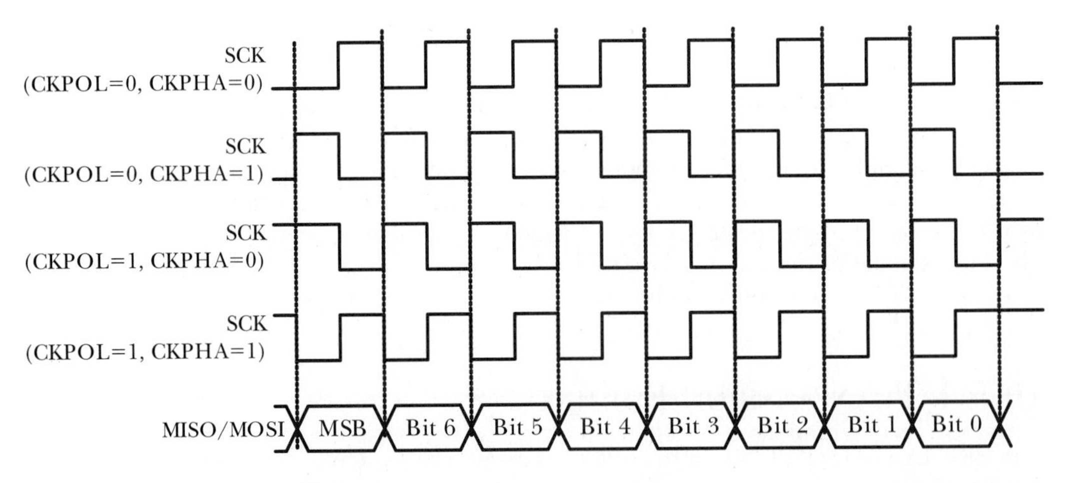

FIGURE 10.4
Data/clock timing diagram for SPI.

shift register to the SPI master on the MISO pin in a full-duplex operation. Therefore, the SPIF flag serves as both a transmit-complete and receive-data-ready flag. The data byte received from the slave is transferred most-significant-bit first into the master's shift register. When a byte is shifted fully into the register, it is moved to the receive buffer, from where it can be read by the processor by reading the SPI data register.

The C8051F040 can disable the slave select pin. When the $\overline{\text{NSS}}$ signal is disabled, it will not be routed through the crossbar and hence will not be available on the I/O port pin. If the $\overline{\text{NSS}}$ pin is enabled in the four-wire, single-master mode, it can be used as the slave-enable signal of a certain slave device. Since the signal level of the $\overline{\text{NSS}}$ pin is programmable, it can enable or disable the slave device that it is connected.

10.5.3 SPI Slave Mode Operation

When the SPI module is enabled and not configured as a master, it will operate as a SPI slave for the C8051F040. As a slave, bytes are shifted in through and out from the MISO pin by a master device controlling the SCK signal. When 8 bits have been shifted through the shift register, the SPIF flag is set to 1 and the byte is copied into the receive buffer.

As a slave, the SPI0 cannot initiate transfers. Data to be transferred to the master device is preloaded into the shift register by writing to the SPI0DAT register. When the master device sends in the SCK pulses, the data in SPI0DAT will then be shifted out. When configured as a slave, SPI0 can be configured for four- or three-wire operation. In the four-wire mode, the $\overline{\text{NSS}}$ signal is routed to a port pin and configured as a digital input. SPI0 is enabled when the $\overline{\text{NSS}}$ signal is low and disabled when the $\overline{\text{NSS}}$ signal is high. In the three-wire mode, the $\overline{\text{NSS}}$ pin does not get routed to an I/O pin. Since there is no way of uniquely addressing the device in the three-wire slave mode, the SPI0 must be the only slave present in the system in order to use this mode. It is important to note that (in the three-wire slave mode) there is no external means of resetting the bit counter that determines when a full byte has been received. The bit counter can be reset only by disabling and re-enabling the SPI0 with the SPIEN bit.

10.5.4 SPI Baud Rate

The C8051F040 uses a dedicated register (SPI0CKR) to derive the SPI clock rate. The equation for setting the SPI baud rate is given in Figure 10.3.

10.6 SPI Interrupt Sources of the C8051F040

When the SPI interrupts are enabled, the following four flags will generate an interrupt when any one of them is set to 1.

1. The SPI-interrupt flag **SPIF**. This flag is set to 1 at the end of a byte transfer. This flag can be set in any mode.
2. The write-collision flag **WCOL**. This flag is set to 1 if the MCU attempts to write to the SPI data register when the SPI transfer buffer has not been emptied to the SPI shift register. When this occurs, the write operation to the SPI0DAT register will be ignored. This fault can occur in all SPI modes.
3. The mode-fault flag **MODF**. This flag is set to 1 when the $\overline{\text{NSS}}$ pin is pulled low when the SPI is in master mode. When mode fault occurs, the SPI module will be changed to slave mode and the SPI module will be disabled to allow another master to control the SPI bus.
4. The receiver-overrun flag **RXOVRN**. When the C8051F040 is configured as a slave, this flag is set to 1 when a new byte is shifted in while a previous byte is still in the receive buffer register. When this occurs, the new byte will not be transferred to the receive buffer, allowing the previously received byte to be read.

Example 10.3 Write an instruction sequence to configure the 24-MHz C8051F040 SPI to operate with the characteristics:

- 4-wire, single-master mode
- 4-MHz data-shift rate
- Disable SPI interrupt
- SCK idle low
- Shift data on the falling edge of SCK and set most-significant-bit first

Solution: The following instruction sequence will achieve the specified setting for the C8051F040.

```
openSPI0:   push    SFRPAGE             ; save the current SFRPAGE
            mov     SFRPAGE,#SPI0_PAGE  ; switch to SPI SFR page
            mov     SPI0CFG,#0x67       ; enable master mode, select falling edge of SCK
            mov     SPI0CN,#0x0B        ; 4-wire single master mode, enable SPI
            mov     SPI0CKR,#0x02       ; set baud rate to 4 MHz
            anl     EIE1,#0xFE          ; disable SPI interrupt
            pop     SFRPAGE             ; restore SFRPAE
            ret
```

There is a set of operations common to all SPI applications. This set of operations should be written into subroutines so that they can be called by all SPI applications. The following example provides these subroutines.

Example 10.4 Write the following subroutines in both assembly and C languages for the C8051F040:

- **putc2SPI:** output the byte in accumulator A to SPI

- **puts2SPI:** output the string in XRAM pointed to by DPTR to SPI; the string is terminated by a NULL character
- **getcSPI:** read a byte from SPI and returns it in accumulator A
- **getsSPI:** read a string from SPI and store it in a buffer

Solution: The set of the SPI subroutines for the C8051F040 is as follows.

```
;*************************************************************************************************
; The following subroutine outputs the byte in A to SPI.
;*************************************************************************************************
putc2SPI:    push     SFRPAGE              ; save the current SFRPAGE value
             mov      SFRPAGE,#SPI0_PAGE
             clr      SPIF
             mov      SPI0DAT,A
             jnb      SPIF,$               ; wait until SPIF is set
             pop      SFRPAGE              ; restore the previous SFRPAGE value
             ret
;*************************************************************************************************
; The following subroutine outputs the string pointed to by DPTR in XRAM to the SPI.
;*************************************************************************************************
puts2SPI:    movx     A,@DPTR
             jz       donespi              ; is it a NULL character?
             acall    putc2SPI
             inc      DPTR
             ajmp     puts2SPI
donespi:     ret
;*************************************************************************************************
; The following subroutine reads a byte from SPI and returns it in A.
;*************************************************************************************************
getcSPI:     push     SFRPAGE
             mov      SFRPAGE,#SPI0_PAGE
             clr      SPIF                 ; make sure SPIF is cleared
             mov      SPI0DAT,#0           ; trigger eight SCK pulses
             jnb      SPIF,$               ; wait until a byte has been shifted in
             mov      A,SPI0DAT            ; place the byte in A
             pop      SFRPAGE
             ret
;*************************************************************************************************
; The following subroutine reads a string of k bytes from SPI and store the string
; in an XRAM buffer pointed to by DPTR. The number of bytes to read is passed in R0.
;*************************************************************************************************
getsSPI:     acall    getcSPI
             movx     @DPTR,A
             inc      DPTR
             djnz     R0,getsSPI           ; read k bytes yet?
             mov      A,#0
             movx     @DPTR,A              ; terminate the buffer with NULL
             ret
```

The C language versions of the subroutines for the C8051F040 are as follows.

```
void putc2SPI(unsigned char xc)
{
     unsigned char temp;
     temp    = SFRPAGE;
     SFRPAGE = SPI0_PAGE;              // switch to SFR page 0
     SPIF    = 0;
     SPI0DAT = xc;                     // output the char to SPI
     while(!(SPIF));                   // wait until the byte has been shifted out
     SFRPAGE = temp;                   // switch back to the original SFR page
}
void puts2SPI(unsigned char *ptr)
{
     unsigned char xc;
     while (xc = *ptr++)               // reach the end of the string?
          putc2SPI(xc);
}
unsigned char getcSPI(void)
{
     unsigned char temp, value;
     temp    = SFRPAGE;
     SFRPAGE = SPI0_PAGE;              // switch to the SFR page 0
     SPIF    = 0;                      // clear the SPIF flag
     SPI0DAT = 0;                      // start to shift in a byte
     while (!(SPIF));                  // wait until the byte has been shifted in
     value   = SPI0DAT;                // get the value shifted in from slave
     SFRPAGE = temp;                   // switch back to the original SFR page
     return value;
}
void getsSPI(unsigned char *ptr, unsigned char n)
{
     unsigned char i;
     for (i = 0; i < n; i++)
          *ptr++ = getcSPI();
     *ptr = 0;                         // terminate the buffer with a NULL
}
```

10.7 SPI Circuit Connection

The SPI protocol allows multiple masters to coexist in the system. However, when one MCU wants to send data to another MCU using the SPI bus, it needs to assert the SS pin of the other MCU, which will cause a mode fault and force the other MCU to become a slave with its SPI module disabled. The mode fault error can cause an interrupt to the MCU and allow the MCU to reenable the SPI to accept the data exchange. The SPI protocol does not provide a rule for the transfer of mastership or the arbitration of simultaneous data-transfer attempts. If the user decides to use the SPI in a multi-master environment, he/she will need to come up with a scheme to handle these issues.

It is more common for microcontrollers to use the SPI protocol to interface with peripheral devices. The three- and four-wire, single-master modes do not differ much in the sense of I/O pin usage. All of the examples before this chapter selected three-wire SPI mode to fix I/O pin assignment.

There could be several connection methods in a multi-slave SPI environment. One possibility is shown in Figure 10.5. In this connection method, the MCU can choose any peripheral device for data transfer. In Figure 10.5, Port 7 pins are used to drive the $\overline{SS}$ inputs of peripheral devices. Any other unused, general-purpose I/O pins can be used for this purpose.

If one doesn't need the freedom of selecting an arbitrary peripheral device for data transfer, then the connection method shown in Figure 10.6 can be used. This method can save a few I/O pins. Figure 10.6 differs from Figure 10.5 in that:

1. The MISO pin of each slave is wired to the MOSI pin of the slave device to its right. The MOSI pins of the master and slave 0 are still wired together.
2. The MISO pin of the master is wired to the same pin of the last slave device.
3. The $\overline{SS}$ inputs of all slaves are tied to ground to enable all slaves.

Thus, the shift registers of the SPI master and slaves become a ring. The data of slave ***k*** are shifted to the master SPI, the data of the master are shifted to slave 0, and the data of slave 0 are shifted to slave 1, and so on. In this configuration, a minimal number of pins control a large number of peripheral

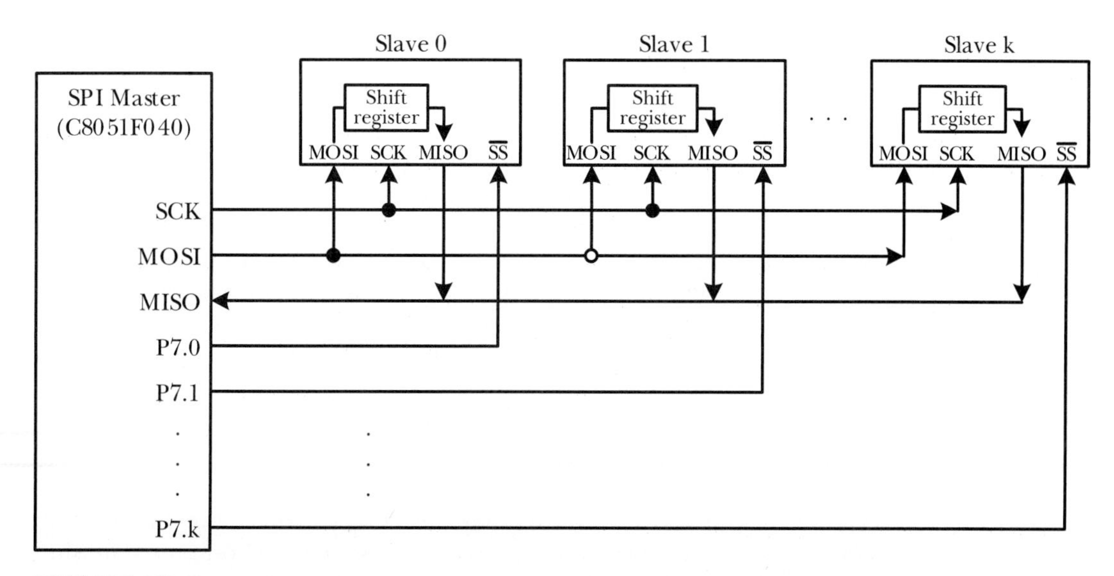

FIGURE 10.5
Single-master and multiple-slave device connection (Method 1).

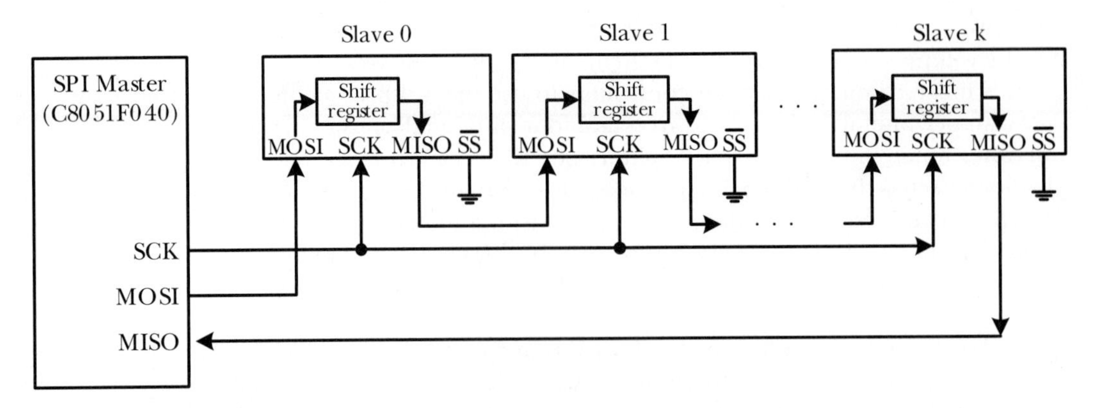

FIGURE 10.6
Single-master and multiple-slave device connection (Method 2).

devices. However, the master does not have the freedom to select an arbitrary slave device for data transfer without going through other slave devices.

This type of configuration often is used to extend the capability of the SPI slave. For example, suppose there is an SPI-compatible seven-segment display driver/decoder that can drive only four digits. By using this configuration, up to $4 \times k$ digits can be displayed when k driver/decoders are cascaded together.

Depending on the capability and the role of the slave device, either the MISO or MOSI pin may not be used in the data transfer. Many SPI-compatible peripheral chips do not have the MISO pin.

10.8 SPI-Compatible Chips

The SPI was proposed initially by Freescale to interface peripheral devices to a microcontroller. As long as a peripheral device supports the SPI interface protocol, it can be used with any microcontroller that implements the SPI subsystem. Many peripheral chips have been designed to provide the SPI interface and, hence, can be used with many different microcontrollers with the SPI interface. The Freescale SPI protocol is compatible with the National Semiconductor Microwire protocol. Therefore, any peripheral device that is compatible with the SPI also can be interfaced with the Microwire protocol.

10.9 The 74LV595 Shift Register

As shown in Figure 10.7, the 74LV595 from Philips consists of an 8-bit shift register and an 8-bit D-type latch with three-state parallel outputs. The shift register accepts serial data and provides a serial output. The shift register also

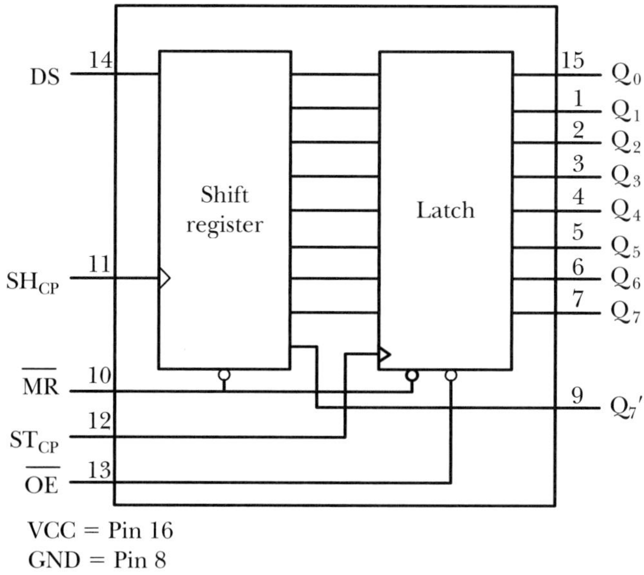

FIGURE 10.7
The 74LV595 block diagram and pin assignment.

provides parallel data to the 8-bit latch. The shift register and the latch have different clock sources. This device also has an asynchronous reset input. The frequency of the shift clock can be as high as 77 MHz.

The functions of the pins in Figure 10.7 are as follows.

DS: ***Serial Data Input.*** The data on this pin is shifted into the 8-bit shift register.
SH$_{CP}$: ***Shift Clock.*** The rising edge of this signal causes the data at the serial input pin to be shifted into the 8-bit shift register.
$\overline{MR}$. A low on this pin resets the shift register and the output latch of this device.
ST$_{CP}$: ***Store Clock.*** The rising edge of this signal loads the contents of the shift register into the output latch.
$\overline{OE}$: ***Output Enable.*** A low on this pin allows the data from the latch to appear at the output pins $Q_7, \ldots, Q_0$.
Q_7 to Q_0: Noninverted, tri-state latch outputs.
Q_7': ***Serial Data Output.*** This is the output of the eighth stage of the 8-bit shift register. This output does not have tri-state capability.

The 74LV595 is designed to shift in 8-bit data serially and then transfer it to the latch to be used as parallel data. The 74LV595 can be used to add parallel output ports to the MCU by using the connection methods shown in Figures 10.5 and 10.6.

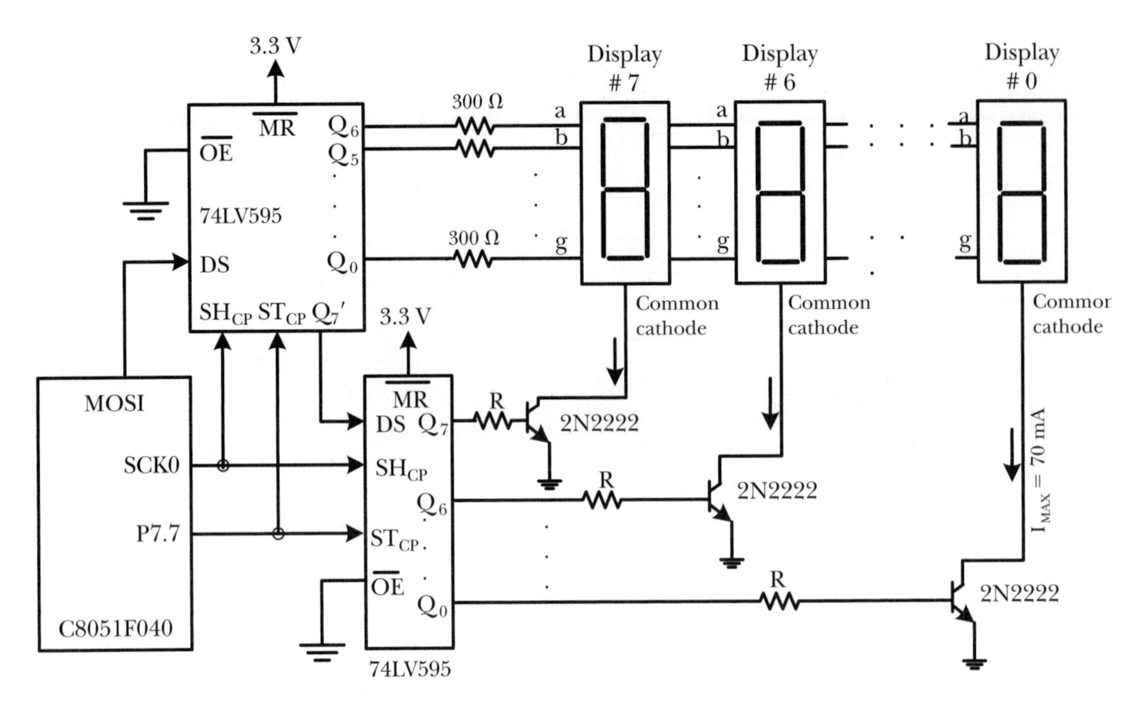

FIGURE 10.8
Two 74V595s together drive eight seven-segment displays.

Example 10.5 Describe how to use two 74LV595s to drive eight common-cathode, seven-segment displays, assuming that the bus clock frequency of the C8051F040 is 24 MHz.

Solution: Two 74LV595s can be cascaded using the method shown in Figure 10.6. One 74LV595 is used to hold the seven-segment pattern, whereas the other 74LV595 is used to carry digit-select signals. The circuit connection is shown in Figure 10.8.

Since there are only seven segments, the Q_7 bit of the segment-control 74LV595 is not needed. The P7.7 pin is used to control the ST_{CP} input of the 74LV595. The time-multiplexing technique illustrated in Example 3.8 will be used to display multiple digits in Figure 10.8. To light the digit on Display 7, the voltage at Q_7 of the digit-select 74LV595 must be driven to high. To light the digit on Display 6, the voltage at Q_6 of the digit-select 74LV595 must be driven to high, and so on.

There are four parts in the program to be written.

Part 1. Display Table Setup. The segment patterns and digit-select values are stored in a table so that the table lookup method can be used to display the desired pattern.

Part 2. SPI Module Initialization. Configure the SPI module to shift data at 12 MHz on the rising clock edge and the most-significant-data bit first.

Part 3. Timer Delay Function. Include the library function delayby1ms to multiplex the digits per millisecond.

Part 4. SPI Data Transfer. Include the library function putc2spi.a51 (or **putc2spi.c**) to output the display patterns and digit-select data.

The following program will display 87654321 on Displays 7 through 0.

```
            $nomod51
            $include    (c8051F040.inc)
icnt        set         R7
            org         0x00
            ljmp        start
            org         0xAB
start:      mov         SP,#0x80                ; set up stack pointer
            acall       sysInit
            acall       openSPI
forever:    mov         DPTR,#dispTab           ; use DPTR as a display pattern pointer
            mov         icnt,#0                 ; set loop count to 0
loopd:      mov         A,icnt                  ; send out digit select value
            movc        A,@A+DPTR               ;     "
            acall       putc2SPI                ;     "
            inc         icnt
            mov         A,icnt                  ; send out segment pattern
            movc        A,@A+DPTR               ;     "
            acall       putc2SPI                ;     "
            mov         SFRPAGE,#0x0F           ; switch to SFR page F
            clr         P7.7                    ; create a rising edge on STCP pin
            setb        P7.7                    ; to transfer new data to output latch
            inc         icnt
            mov         R0,#1                   ; wait for 1 ms
            acall       delayby1ms              ;     "
            cjne        icnt,#16,loopd          ; reach the end of the table?
            ajmp        forever

dispTab:    db          80H,7FH,40H,70H,20H,5FH,10H,5BH
            db          08H,33H,04H,79H,02H,6DH,01H,30H
;**************************************************************************************************
; The following function disable watchdog timer, select internal oscillator output as
; system clock, and assign all peripheral signals to port pins.
;**************************************************************************************************
sysInit:    mov         SFRPAGE,#CONFIG_PAGE    ; switch to SFR page F
            mov         WDTCN,#0DEH             ; disable watchdog timer
            mov         WDTCN,#0ADH             ;     "
            mov         CLKSEL,#0               ; select internal oscillator as SYSCLK
            mov         OSCICN,#83H             ;     "
            mov         XBR0,#0F7H              ; assign all peripheral signals to port pins
```

```
            mov      XBR1,#0FFH          ;    "
            mov      XBR2,#5CH           ;
            mov      XBR3,#8BH           ;
            mov      P7MDOUT,#0x80       ; configure P7.7 pin for output
            mov      P0MDOUT,#0xD5       ; configure SCK, MOSI, TX0, SCL, & TX1 for output
            ret
; ************************************************************************************************
; The following function enable SPI, select 3-wire single master mode, rising
; edge of SCK to shift data, set SPI baud rate to 12 MHz.
; ************************************************************************************************
openSPI:    push     SFRPAGE             ; save the current SFRPAGE
            mov      SFRPAGE,#SPI0_PAGE  ; switch to SFR page 0
            mov      SPI0CFG,#40H        ; master mode, rising edge of SCK to shift data
            mov      SPI0CN,#01H         ; 3-wire single master mode, enable SPI
            mov      SPI0CKR,#0          ; set baud rate to 12 MHz
            anl      EIE1,#0xFE          ; disable SPI interrupt
            pop      SFRPAGE             ; restore SFRPAE
            ret
            $include (delays.a51)
            $include (spiUtil.a51)
            end
```

The C language version of the program is as follows.

```
/*************************************************************************************************/
/* The project should include delays.c, spiUtil.c and this file.                                 */
/*************************************************************************************************/
#include <c8051F040.h>
#include "spiUtil.h"
#include "delays.h"
void openSPI (void);
void sysInit (void);
void main (void)
{
    unsigned char disp_tab[8][2] = {{0x80,0x7F},{0x40,0x70},{0x20,0x5F},{0x10,0x5B},
                                    {0x08,0x33},{0x04,0x79},{0x02,0x6D},{0x01,0x30}};
    char i;
    sysInit();
    openSPI(); // configure the SPI module
    while(1) {
        for (i = 0; i < 8; i++) {
            putc2SPI (disp_tab[i][0]);      // send out digit select value
            putc2SPI (disp_tab[i][1]);      // send out segment pattern
            SFRPAGE = 0x0F;
            P7 & = 0x7F;                    // transfer values to latches of 74LV595s
            P7 | = 0x80;                    //    "
            delayby1ms(1);                  // display a digit for 1 ms
        }
    }
}
```

```
/*****************************************************************************************************
; The following function disable watchdog timer, select internal oscillator clock as
; system clock, and assign all peripheral signals to port pins.
;*****************************************************************************************************/
void sysInit(void)
{
    SFRPAGE  = 0x0F;          // switch SFR page F
    WDTCN    = 0xDE;          // disable watchdog timer
    WDTCN    = 0xAD;          //     "
    CLKSEL   = 0x00;          // select internal oscillator as SYSCLK
    OSCICN   = 0x83;          //
    XBR0     = 0xF7;          // assign all peripheral signals to port pin
    XBR1     = 0xFF;          //     "
    XBR2     = 0x5C;          //     "
    XBR3     = 0x8B;          //     "
    P7MDOUT |= 0x80;          // configure P7.7 for output
    P0MDOUT  = 0xD5;          // configure TX0,SCK,MOSI, SCL, & TX1 for output
}
/*****************************************************************************************************
; The following function enable SPI, select 3-wire single master mode, rising
; edge of SCK to shift data, set SPI baud rate to 12 MHz.
;*****************************************************************************************************/
void openSPI (void)
{
    unsigned char temp;
    temp     = SFRPAGE;
    SFRPAGE  = 0;
    SPI0CFG  = 0x40;          // select master mode, rising edge of SCK to shift data
    SPI0CN   = 0x01;          // enable SPI to 3-wire single master mode
    SPI0CKR  = 0;             // set SPI baud rate to 12 MHz
    EIE1    &= 0xFE;          // disable SPI interrupt
    SFRPAGE  = temp;          // restore original SFR page
}
```

This example illustrates one of the applications of the shift register 74LV595. The main drawback of the example program is that the CPU spends all of its time on the task of multiplexing the displays. This program can be modified to use interrupt and free the CPU from idle waiting. Several manufacturers produce seven-segment display drivers to eliminate the display multiplexing task altogether. The MC14489 from Freescale and the MAX7221 from Maxim are two examples.

10.10 The LTC1661 D/A Converter

The LTC1661 from Linear Technology has two 10-bit voltage output digital-to-analog converters (DAC) that can interface with many microcontrollers using the SPI interface.

The LTC1661 has an output settling time of 30 μs. The D/A conversion is started by writing a 16-bit serial string that contains four control and ten data bits to the LTC1661. The LTC1661 can operate with a power supply ranging from 2.7 to 5.5 V.

10.10.1 Signal Pins

The pin assignment of the LTC1661 is shown in Figure 10.9. The functions of the LTC1661 signals are as follows.

GND: analog ground
$\overline{\text{CS}}$/LD: chip select or load—when this signal is low, the value on DIN can be shifted in to the shift register, and when this signal is high, the contents of the shift register will be loaded to the DAC register to be converted to voltage
DIN: serial digital data input
REF: reference analog input voltage; the reference voltage can be tied to V_{CC}
SCK: serial clock input; the highest frequency of SCK is 16.7 MHz
V_{CC}: positive power supply
V_{OUT}A and V_{OUT}B: channel A and channel B voltage output

10.10.2 Data Format

When shifting in data to be converted, the LTC1661 expects the most significant bit to be shifted in first. The shift-register inside the LTC1661 is 16-bit, which requires the MCU to perform two SPI transfers. When performing SPI

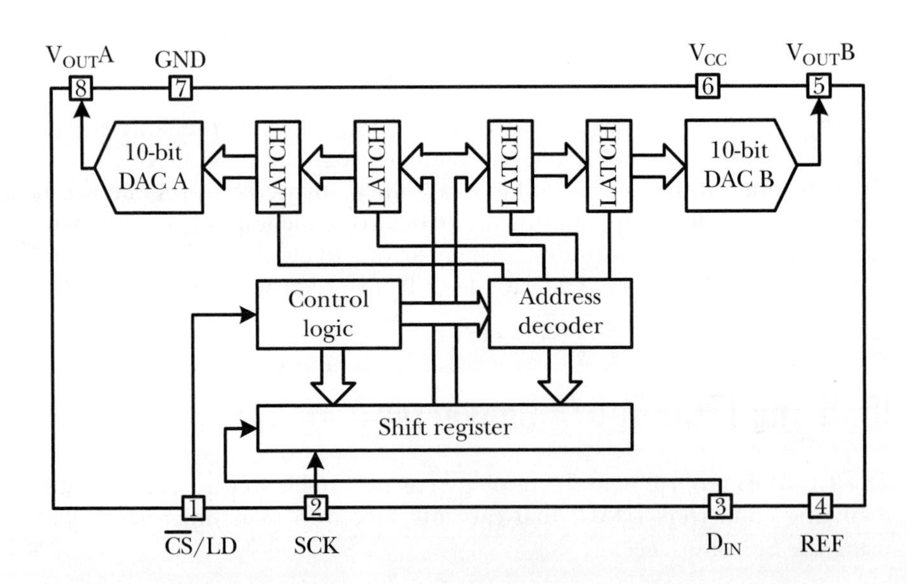

FIGURE 10.9 *LTC1661 DAC pin assignment and block diagram.*

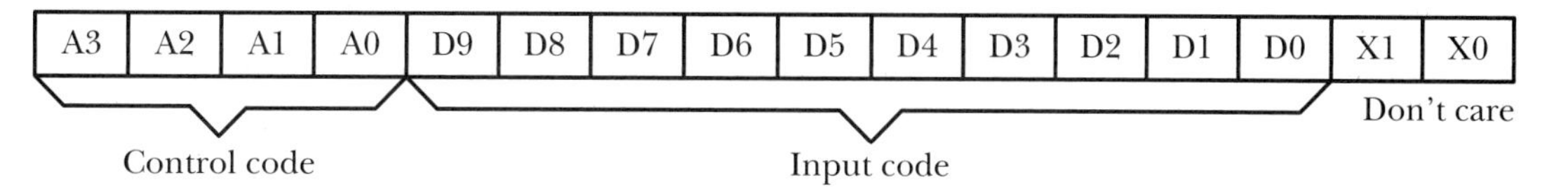

FIGURE 10.10
LTC1661 input data format.

transfer, the most significant four bits of the value to be converted must be sent as the lower four bits in the first SPI transfer. The lower six bits of the value to be converted should be sent in the second SPI transfer. The input data arrangement for the LTC1661 SPI transfer is illustrated in Figure 10.10.

10.10.3 DAC Control Functions

As shown in Figure 10.10, the first four bits of the data to be sent to the LTC1661 select the function performed by the LTC1661. The meanings of these four bits are illustrated in Table 10.1.

10.10.4 LTC1661 Output Voltage

The output voltage of the LTC1661 is given by

$$V_{OUT} = REF \times code \div (2^{10} - 1)$$

where, REF is the reference voltage. REF is often tied to V_{CC} to simplify the circuit connection unless there is a special voltage-level requirement for the output.

10.10.5 Interfacing the LTC1661 with the C8051F040

The LTC1661 is used mainly to generate waveforms. A typical circuit connection for interfacing the LTC1661 with the C8051F040 is shown in Figure 10.11.

Example 10.6 Write a program that uses the circuit in Figure 10.11 to generate a 1-kHz periodic square wave, assuming that the C8051F040 is running with a 24-MHz crystal oscillator.

Solution: The period of a 1-kHz periodic square waveform is 1 ms. By outputting a high at $V_{OUT}A$ for 0.5 ms, then outputting a low at $V_{OUT}A$ for 0.5 ms, and then repeating, a 1-kHz periodic square wave can be generated. The assembly program that generates this waveform is as follows.

TABLE 10.1 *LTC1661 DAC Control Functions*

Control A3	A2	A1	A0	Input Register Status	DAC Register Status	Power-Down Status	Comments
0	0	0	0	No change	No update	No change	No operation. Power-down status unchanged.
0	0	0	1	Load DAC A	No update	No change	Load input register A with data. DAC outputs and power-down status unchanged.
0	0	1	0	Load DAC B	No update	No change	Load input register B with data. DAC outputs and power-down status unchanged.
0	0	1	1		Reserved		
0	1	0	0		Reserved		
0	1	0	1		Reserved		
0	1	1	0		Reserved		
0	1	1	1		Reserved		
1	0	0	0	No change	Update outputs	Wake	Load both DAC regs with existing contents of input regs. Output update. Part wakes up.
1	0	0	1	Load DAC A	Update outputs	Wake	Load input reg A. Load DAC regs with new contents of input reg A and existing contents of reg B. Outputs update. Part wakes up.
1	0	1	0	Load DAC B	Update outputs	Wake	Load input reg B. Load DAC regs with existing contents of input reg A and new contents of reg B. Outputs update. Part wakes up.
1	0	1	1		Reserved		
1	1	0	0		Reserved		
1	1	0	1	No change	No update	Wake	Part wakes up. Input and and DAC regs unchanged. DAC outputs reflect existing contents of DAC regs.
1	1	1	0	No change	No update	Sleep	Part goes to sleep. Input and DAC regs unchanged. DAC outputs set to high impedance state.
1	1	1	1	Load DACs A, B with same 10-bit code	Update outputs	Wake	Load both input regs. Load both DAC regs with new contents of input regs. Output update. Part wakes up.

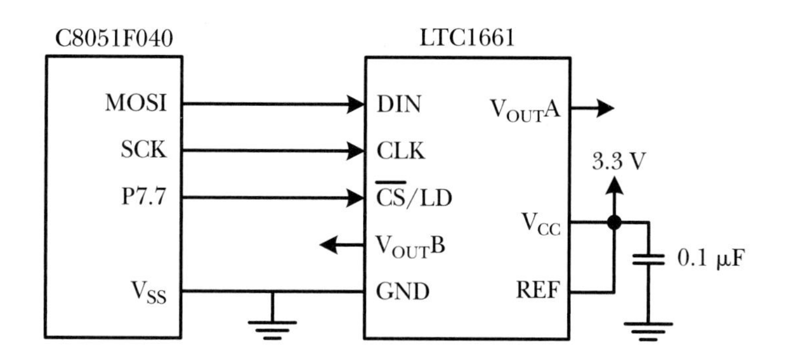

FIGURE 10.11 *LTC1661 to C8051F040 interface.*

```
            $nomod51
            $include    (c8051F040.inc)
icnt        set         R7
            org         0x00                ; reset vector
            ljmp        start
            org         0xAB
start:      mov         SP,#0x80            ; set up stack pointer
            acall       openSPI
            acall       sysInit
forever:    mov         SFRPAGE,#CONFIG_PAGE
            clr         P7.7
            mov         A,#0x9F             ; output a high
            lcall       putc2SPI            ;    "
            mov         A,#0xFC             ;    "
            lcall       putc2SPI            ;    "
            setb        P7.7                ;    "
            mov         R0,#10              ; stay for 0.5 ms
            acall       delayby50us         ;    "
            clr         P7.7                ; output a low
            mov         A,#0x90             ;    "
            lcall       putc2SPI            ;    "
            mov         A,#0x00             ;    "
            lcall       putc2SPI            ;    "
            setb        P7.7                ;    "
            mov         R0,#10              ; stay for 0.5 ms
            acall       delayby50us         ;    "
            ajmp        forever
;*************************************************************************************************************
; The following function disable watchdog timer, select internal oscillator clock as
; system clock, and assign all peripheral signals to port pins.
;*************************************************************************************************************
sysInit:    mov         SFRPAGE,#CONFIG_PAGE ; switch to SFR page F
            mov         WDTCN,#0xDE         ; disable watchdog timer
            mov         WDTCN,#0xAD         ;    "
            mov         CLKSEL,#0           ; select internal oscillator as SYSCLK
            mov         OSCICN,#0x83        ;    "
            mov         XBR0,#0xF7          ; assign all peripheral signals to port pins
```

```
            mov     XBR1,#0xFF          ;   "
            mov     XBR2,#0x5C          ;
            mov     XBR3,#0x8B          ;
            mov     P0MDOUT,#0xD5       ; configure SCK and MOSI for output
            mov     P7MDOUT,#0x80       ; configure P7.7 for output
            ret
;************************************************************************************************
; The following function enable SPI, select 3-wire single master mode, rising
; edge of SCK to shift data, set SPI baud rate to 12 MHz.
;************************************************************************************************
openSPI:    push    SFRPAGE             ; save the current SFRPAGE
            mov     SFRPAGE,#0          ; switch to SFR page 0
            mov     SPI0CFG,#0x40       ; master mode, rising edge of SCK to shift data
            mov     SPI0CN,#0x01        ; 3-wire single master mode, enable SPI
            mov     SPI0CKR,#0x01       ; set baud rate to 6 MHz
            anl     EIE1,#0xFE          ; disable SPI interrupt
            pop     SFRPAGE             ; restore SFRPAE
            ret
;************************************************************************************************
; include delayby50us here.
;************************************************************************************************
; include putc2SPI here
;************************************************************************************************
            end
```

The C language version of the program is straightforward and hence will be left to you as an exercise problem.

Example 10.7 Write a program to use the circuit in Figure 10.11 and a buzzer to generate a two-tone siren that oscillates between 500 Hz and 1 kHz. Each tone lasts for half a second.

Solution: The assembly language program that generates the specified two-tone siren is as given here.

```
            $nomod51
            $include  (c8051F040.inc)
outHi       macro                       ; macro that outputs a high voltage
            clr     P7.7
            mov     A,#0x9F
            lcall   putc2SPI
            mov     A,#0xFC
            lcall   putc2SPI
            setb    P7.7
            endm
outlo       macro                       ; macro that outputs a low voltage
            clr     P7.7
            mov     A,#0x90
            lcall   putc2SPI
```

```
            mov     A,#0x00
            lcall   putc2SPI
            setb    P7.7
            endm
cntLo       set     R1                      ; loop count for low frequency
cntHiH      set     R2                      ; outer loop count for high frequency
cntHiL      set     R3                      ; inner loop count for high frequency

            org     0x00
            ljmp    start
            org     0xAB
start:      mov     SP,#0x80                ; set up stack pointer
            acall   openSPI                 ; enable 3-wire SPI at 6 MHz shift rate
            acall   sysInit
forever:    mov     cntLo,#250              ; generate 500 Hz tone for half a second
loTone:     outHi                           ;     "
            mov     R0,#20                  ;     "
            acall   delayby50us             ;     "
            outLo                           ;     "
            mov     R0,#20                  ;     "
            acall   delayby50us             ;     "
            djnz    cntLo,loTone            ;     "
            mov     cntHiH,#2               ; generate 1000 Hz tone for half a second
hiTone1:    mov     cntHiL,#250             ;     "
hiTone2:    outHi                           ;     "
            mov     R0,#10                  ;     "
            acall   delayby50us             ;     "
            outLo                           ;     "
            mov     R0,#10                  ;     "
            acall   delayby50us             ;     "
            djnz    cntHiL,hiTone2          ;     "
            djnz    cntHiH,hiTone1          ;     "
            ajmp    forever
;************************************************************************************************
; The following function disable watchdog timer, select internal oscillator clock as
; system clock, and assign all peripheral signals to port pins.
;************************************************************************************************
sysInit:    mov     SFRPAGE,#CONFIG_PAGE    ; switch to SFR page F
            mov     WDTCN,#0xDE             ; disable watchdog timer
            mov     WDTCN,#0xAD             ;     "
            mov     CLKSEL,#0               ; select internal oscillator as SYSCLK
            mov     OSCICN,#0x83            ;     "
            mov     XBR0,#0xF7              ; assign all peripheral signals to port pins
            mov     XBR1,#0xFF              ;     "
            mov     XBR2,#0x5C              ;
            mov     XBR3,#0x8B              ;
            mov     P0MDOUT,#0xD5           ; configure SCK and MOSI for output
            mov     P7MDOUT,#0x80           ; configure P7.7 for output
            ret
;************************************************************************************************
```

```
; The following function enable SPI, select 3-wire single master mode, rising
; edge of SCK to shift data, set SPI baud rate to 6 MHz.
; ************************************************************************************
openSPI:    push    SFRPAGE         ; save the current SFRPAGE
            mov     SFRPAGE,#0      ; switch to SFR page 0
            mov     SPI0CFG,#0x40   ; master mode, rising edge of SCK to shift data
            mov     SPI0CN,#0x01    ; 3-wire single master mode, enable SPI
            mov     SPI0CKR,#0x01   ; set baud rate to 6 MHz
            anl     EIE1,#0xFE      ; disable SPI interrupt
            pop     SFRPAGE         ; restore SFRPAE
            ret
; ************************************************************************************
; include delayby50us here.
; ************************************************************************************
; ************************************************************************************
; include putc2SPI here.
; ************************************************************************************
            end
```

The C program that generates the same siren is as follows.

```
#include  <C8051F040.h>        // need to include delays.c and spiUtil.c in this project
#include  <delays.h>
#incldue  <spiUtil.h>
void sysinit(void);
void openSPI(void);

void main (void)
{
    int ik;
    sysinit();
    openSPI();

    while (1) {
        for (ik = 0; ik < 250; ik++) {
            P7 &= 0x7F;            // enable SPI transfer
            putc2SPI(0x9F);        // output 3.3 V
            putc2SPI(0xFC);
            P7 |= 0x80;            // start D/A conversion
            delayby100us(10);
            P7 &= 0x7F;            // enable SPI transfer
            putc2SPI(0x90);        // output 0 V
            putc2SPI(0x0);         //     "
            P7 | = 0x80;           // start D/A conversion
            delayby100us(10);
        }
        for (ik = 0; ik < 500; ik++) {
            P7 &= 0x7F;            // enable SPI transfer
            putc2SPI(0x9F);        // output 3.3 V
```

```
                putc2SPI(0xFC);
                P7 |= 0x80;                       // start D/A conversion
                delayby100us(5);
                P7 & = 0x7F;                      // enable SPI transfer
                putc2SPI(0x90);                   // output 0 V
                putc2SPI(0x0);                    //      "
                P7 | = 0x80;                      // start D/A conversion
                delayby100us(5);
            }
        }
}
void sysinit(void)
{
    int         n;
    SFRPAGE  = CONFIG_PAGE;
    WDTCN    = 0xDE;
    WDTCN    = 0xAD;
    OSCXCN   = 0x67;
    for (n = 0; n < 255; n++);
    while(!(OSCXCN & 0x80));
    CLKSEL   = 0x01;                              // use external crystal oscillator to generate SYSCLK
    XBR2     = 0x5D;                              // enable crossbar and assign all peripheral signals
    XBR0     = 0xF7;                              // to I/O pins
    XBR1     = 0xFF;                              //      "
    XBR3     = 0x8F;                              //      "
    P7MDOUT = 0x80;                               // configure P7.7 as output
    P0MDOUT = 0xD5;                               // configure SCK, MOSI for output
}
void openSPI(void)
{
    char temp;
    temp     = SFRPAGE;
    SFRPAGE  = SPI0_PAGE;
    SPI0CFG  = 0x40;                              // SCK idle low and shift data on the rising edge
    SPI0CN   = 0x01;                              // enable 3-wire SPI mode
    SPI0CKR  = 0x01;                              // set SPI baud rate to 6 MHz
    EIE1    & = 0xFE;                             // disable SPI interrupt
    SFRPAGE  = temp;                              // restore caller's SFRPAGE
}
```

10.11 EEPROM with SPI Interface

The serial EEPROMs with SPI interface are used widely in applications that need to store data for an extended period of time. The products of this nature are powered by batteries and require the EEPROM to consume little power.

10.11.1 The 25AA080A EEPROM

The 25AA080A made by Microchip is an EEPROM with an SPI interface and has an 8-k-bit capacity. The pin assignment is shown in Figure 10.12. The function of each pin is as follows.

$\overline{\text{CS}}$: ***Chip Select Input.*** The microcontroller must pull this signal to low in order to access the chip.
SO: ***Serial Data Output.*** The 25AA080A shifts out the data bit serially from this pin.
$\overline{\text{WP}}$: ***Write-Protection Input.*** This pin is used in conjunction with the WPEN bit of the status register to prohibit writes to the nonvolatile bits in the status register and the protected memory block.
SI: ***Serial Data Input.*** Incoming data is shifted into the EEPROM bit serially from this pin.
SCK: ***Clock Input.*** This signal synchronizes the data shifting. The highest-shift clock rate is 10 MHz.
$\overline{\text{HOLD}}$: The low voltage at this pin stops the communication to this device.

To allow data transfer to and from the 25AA080A, the $\overline{\text{CS}}$ pin must be low and the $\overline{\text{HOLD}}$ pin must be high. Data is shifted into the device on the rising edge of the SCK signal.

The 25AA080A has an instruction register and a status register to support its operation. The user sends instructions to inform the 25AA080A what operations are to be performed. The instructions that are implemented by the 25AA080A are listed in Table 10.2. When sending an instruction to

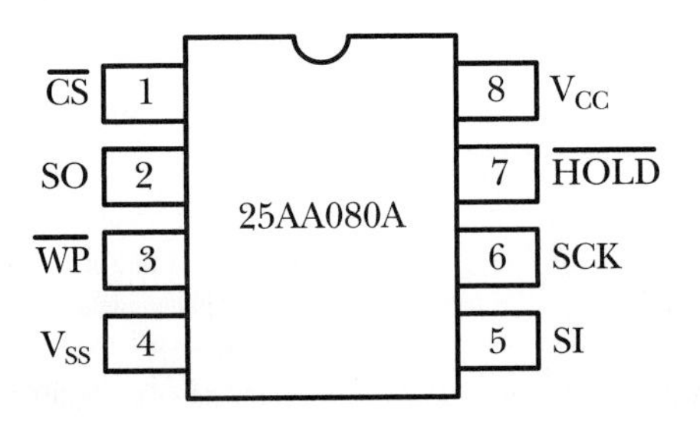

FIGURE 10.12 *The 25AA080A EEPROM.*

TABLE 10.2 *25AA080A Instruction Set*

Instruction	Instruction Code	Description
READ	0000 0011	Read data from memory array beginning at selected address
WRITE	0000 0010	Write data to memory array beginning at selected address
WRDI	0000 0100	Reset the write enable latch (disable write operation)
WREN	0000 0110	Set the write enable latch (enable write operations)
RDSR	0000 0101	Read status register
WRSR	0000 0001	Write status register

the 25AA080A, the $\overline{\text{CS}}$ pin must be pulled low. After the instruction has been shifted in, the $\overline{\text{CS}}$ pin must be pulled high in order for the instruction to take effect if the instruction is not followed by the 16-bit address.

10.11.2 Read Operation

To read data from the 25AA080A, the MCU sends in the 8-bit *read instruction* followed by the 16-bit address, with the six most significant bits (MSBs) of the address being don't care bits. After the correct read instruction and address are sent, the data stored in the memory at the selected address is shifted out on the SO pin. The data stored in the next address can be read sequentially by continuing to provide clock pulses. The internal address pointer is incremented automatically to the next higher address after each byte of data is shifted out. When the highest address (0x3FF) is reached, the address counter rolls over to 0x0000, allowing the read cycle to be continued indefinitely. The read operation is terminated by raising the $\overline{\text{CS}}$ pin.

10.11.3 Write Operation

Before writing data to the 25AA080A, the user must use the WREN instruction to set the write-enable latch inside the device. Once the write enable latch is set, the user may proceed by setting $\overline{\text{CS}}$ low, issuing a WRITE instruction followed by the 16-bit address, and then adding the data to be written. Up to 16 bytes of data can be sent to the device. The only restriction is that all of the bytes must reside in the same page. This write operation is referred to as a *page write.* The highest 12 address bits of the memory locations in the same page are identical. The $\overline{\text{CS}}$ signal must be pulled high after the least significant bit of the last byte has been shifted in for data to be written into the memory array. An internal write cycle is started after the $\overline{\text{CS}}$ signal goes to high to write data into the memory array. If the $\overline{\text{CS}}$ signal is pulled high at any other time, the write operation will not be completed. The internal write cycle takes about 5 ms to complete.

When the write operation is in progress, the status register can be read to check the WPEN, WIP, WEL, BP1, and BP0 bits. A read attempt of a memory-array location will not be possible during a write cycle. When the write cycle is completed, the write enable latch is reset.

10.11.4 Write Status Register

The contents of the status register are shown in Figure 10.13. The status register can be read in any time, even during a write cycle. The user may read this register by issuing the RDSR instruction and write to this register by issuing the WRSR instruction. The WPEN, BP1, and BP0 bits are nonvolatile, whereas the WEL and WIP bits are volatile. The WEL and WIP bits only can be

FIGURE 10.13
Status register of the 25AA080A.

7	6	5	4	3	2	1	0
WPEN	x	x	x	BP1	BP0	WEL	WIP
w/r				w/r	w/r	r	r

WPEN: Write protection enable (nonvolatile)
0 = disable write protection
1 = write protection is enabled if the WP pin is also low
BP1~BP0: Block protection bits
00 = none of the memory block is write protected
01 = the memory block from 0x300 to 0x3FF is write protected
10 = the memory block from 0x200 to 0x3FF is write protected
11 = the whole memory array is write protected
WEL: Write enable latch
0 = write to the memory array and status register is prohibited
1 = write to the memory array and status register is allowed
WIP: Write in progress
0 = no write operation is in progress
1 = a write operation is in progress

changed by sending appropriate instruction to the device. They cannot be changed by writing to the status register directly.

10.11.5 Data Protection

The 25AA080A is designed to prevent inadvertent writes to the memory array. After power-up, the write enable latch (the WEL bit is the status register) is reset to prevent any writes to the memory array. A write enable instruction must be issued to set the WEL bit. After a byte write, page write, or a status register write, the WEL bit is reset. The CS pin must be pulled high after a write operation to start an internal write cycle. Access to the memory array during an internal write cycle is ignored, and programming is continued.

When the WPEN bit is set and the WP pin is low, the write operation to the nonvolatile bits (WPEN, BP1, and BP0) of the status register and the protected memory block is disabled.

10.11.6 Interfacing the 25AA080A with the C8051F040

A circuit connection example between the C8051F040 and the 25AA080A is shown in Figure 10.14. The write protection feature is disabled in this circuit. The SPI should be enabled in four-wire mode, because the NSS pin is used to enable the 25AA080A.

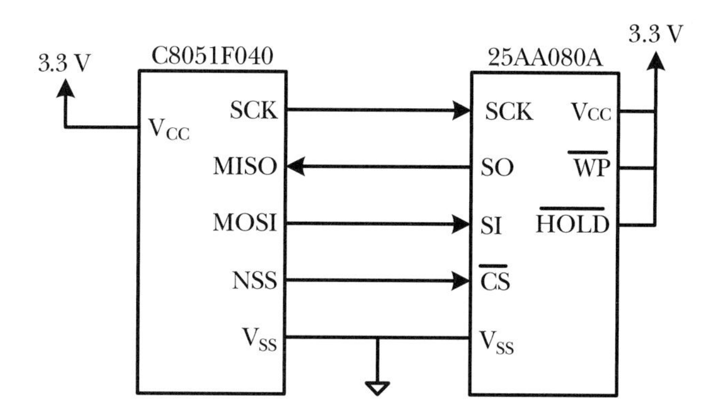

FIGURE 10.14 *Circuit connectoin between the 25AA080A and the C8051F040.*

Example 10.8 Write an assembly subroutine to write a byte into the EEPROM, an assembly subroutine to read a byte from the EEPROM, and an assembly program to test these two functions.

Solution: A macro for writing a byte through the SPI interface is created to be used in these two subroutines. The macro and the assembly subroutines that write a byte to the EEPROM and read a byte from the EEPROM at the specified address are as follows.

```
writeEE     macro     arg
            mov       SPI0DAT,arg          ; send out a byte via SPI interface
            jnb       SPIF,$               ; wait until the byte is shifted out
            clr       SPIF
            endm

;*************************************************************************************************
; The following subroutine writes a byte into the EEPROM at the specified address.
; The address high and low bytes are passed in R6 and R7 whereas the data byte is
; passed in R5. This function waits until write cycle is completed before returning
; to the caller.
;*************************************************************************************************
WrByteEE:   push      SFRPAGE
            mov       SFRPAGE,#SPI0_PAGE
            clr       NSSMD0               ; enable SPI transfer to 25AA080A
            writeEE   #WREN                ; set Write enable latch
            setb      NSSMD0               ; deactivate CE to 25AA080A
            mov       R0,#1
            acall     delayby1us           ; wait for T_NSS_DISABLE_MIN (> 500 ns)
            clr       NSSMD0               ; enable SPI transfer
            writeEE   #WREE                ; send the WREE command to EEPROM
            writeEE   R6                   ; send out address high byte
            writeEE   R7                   ; send out address low byte
            writeEE   R5                   ; send the value to be written
            setb      NSSMD0               ; start the internal EEPROM write cycle
```

```
            mov     R0,#1
            acall   delayby1us          ; wait for T_NSS_DISABLE_MIN (> 500 ns)
pollw:      clr     NSSMD0
            writeEE #RDSR               ; send out RDSR instruction
            writeEE #0                  ; shift in status register
            setb    NSSMD0
            mov     R0,#1
            acall   delayby1us
            mov     A,SPI0DAT
            anl     A,#0x01             ; check WIP bit (write in progress bit)
            jnz     pollw               ; continue to poll until internal write is complete
            pop     SFRPAGE
            ret
;*******************************************************************************************
; The following subroutine reads a byte from the 25AA080A. The high byte and low byte
; of the address of the location to be read are passed in R6 and R7. The data is to be
; returned in A.
;*******************************************************************************************
RdByteEE:   push    SFRPAGE
            mov     SFRPAGE,#SPI0_PAGE
            clr     NSSMD0              ; enable SPI transfer to 25AA080A
            writeEE #RDEE               ; send out read EE instruction
            writeEE R6                  ; send out address high byte
            writeEE R7                  ; send out address low byte
            writeEE #0                  ; shift in data
            setb    NSSMD0              ; deactivate slave select
            mov     R0,#1               ; wait for T_NSS_DISABLE_MIN (> 500 ns)
            lcall   delayby1us          ;     "
            mov     A,SPI0DAT           ; place data in A
            pop     SFRPAGE
            ret
```

The following test program performs the required initialization and then proceeds to write a byte into the EEPROM, read back the byte, and then display the value in LEDs (driven by P5). This operation starts from address 0 until 255 (a quarter of the EEPROM).

```
            $nomod51
            $include (C8051F040.inc)
writeEE     macro   arg
            mov     SPI0DAT,arg         ; send out a byte via SPI interface
            jnb     SPIF,$              ; wait until the byte is shifted out
            clr     SPIF
            endm
WREN        equ     0x06                // write enable command
WRDI        equ     0x04                // write disable command
WREE        equ     0x02                // write operation to EEPROM
RDEE        equ     0x03                // read operation to EEPROM
WRSR        equ     0x01                // write EEPROM status register
```

```
RDSR        equ       0x05               // read EEPROM status register
            cseg at   0x00
            ljmp      start
            org       0xAB
start:      mov       SP,#0x7F           ; establish stack pointer
            acall     openSPI
            acall     sysinit
            mov       R5,#0              ; send value starting from 0
            mov       R6,#0              ; address high byte
            mov       R7,#0              ; address low byte
tstloop:    acall     WrByteEE           ; write a byte into the EEPROM
            acall     RdByteEE           ; read back the byte written into EEPROM
            mov       P5,A               ; display that byte on LEDs
            mov       R0,#2              ; wait for 200 ms
            acall     delayby100ms       ;      "
            cjne      R5,#255,next       ; reach 255 yet?
            ajmp      stop
next:       inc       R7                 ; try next EEPROM location
            mov       A,R6               ;      "
            addc      A,#0               ;      "
            mov       R6,A               ;      "
            inc       R5                 ; increase the value to be tested
            ajmp      tstloop
stop:       ajmp      $

sysinit:    mov       SFRPAGE,#CONFIG_PAGE
            mov       WDTCN,#0xDE        ; disable watchdog timer
            mov       WDTCN,#0xAD        ;      "
            mov       CLKSEL,#0
            mov       OSCXCN,#0x67       ; configure external crystal oscillator
            mov       R7,#255            ; delay for about 1 ms
            djnz      R7,$               ;      "
waitOSC:    mov       A,OSCXCN           ; wait until external oscillator is stable
            anl       A,#0x80            ; before using it
            jz        waitOSC            ;      "
            mov       CLKSEL,#1          ; use external crystal oscillator for SYSCLK
            mov       XBR2,#0x40         ; enable crossbar decoder
            mov       XBR0,#0x07         ; enable UART0, SPI, SMBus
            mov       P5MDOUT,#0xFF      ; configure P5 for output
            mov       P0MDOUT,#0x35 ; configure TX0, SCK, MOSI, and NSS for output
            ret
;*****************************************************************************************************
; This function enables SPI to 4-wire single master mode, uses rising edge of SCK to
; shift data in and out, and sets baud rate to 1.5 MHz.
;*****************************************************************************************************
openSPI:    push      SFRPAGE
            mov       SFRPAGE,#SPI0_PAGE
            mov       SPI0CFG,#0x40      ; select master mode, rising edge to shift data
            mov       SPI0CN,#0x0D       ; enable 4-wire master mode, prepare SPI to transmit
            mov       SPI0CKR,#0x03      ; set SPI baud rate to 3 MHz
```

```
            pop         SFRPAGE
            ret
;**************************************************************************************************
; include WrByteEE subroutine here.
;**************************************************************************************************
;**************************************************************************************************
; include RdByteEE subroutine here.
;**************************************************************************************************
            $include    (spiUtil.a51)
            $include    (delays.a51)
            end
```

Example 10.9 Write a C function that writes a value into the specified location in the EEPROM, a C function that reads the value of the specified EEPROM location, and a program that tests these two functions.

Solution: The required constant definitions for these two functions are as follows.

```
#ifndef     BYTE
#define     BYTE unsigned char
#endif
#ifndef     UINT
#define     UINT unsigned int
#endif
#define     RDEE        0x03            // read command
#define     WREE        0x02            // write command
#define     WRDI        0x04            // reset write enable latch command
#define     WREN        0x06            // set write enable latch command
#define     RDSR        0x05            // read status register command
#define     WRSR        0x01            // write status register command
#define     EEPROM_SIZE 1024            // the capacity of EEPROM to be tested
```

The EEPROM write function follows a five-step procedure to write a byte into the specified memory location and waits until the write cycle is completed before it returns to the caller.

```
void WrByteEE(UINT address, BYTE value)
{
     BYTE saveSFRPAGE = SFRPAGE;
     SFRPAGE          = SPI0_PAGE;
// Step 1: set the Write Enable Latch to 1
     NSSMD0           = 0;            // enable SPI transfer to EEPROM
     SPI0DAT          = WREN;         // send the WREN command to enable write into EEPROM
     while (!SPIF);                   // wait until the WREN command is shifted out
     SPIF             = 0;
     NSSMD0           = 1;            // deactivate slave select
     delayby1us(1);                   // make sure T_NSS_DISABLE_MIN requirement is satisfied
```

```
// Step 2: send the WREE command
    NSSMD0 = 0;
    SPI0DAT = WREE;                                   // send out the WREE command
    while(!SPIF);
    SPIF    = 0;
// Step 3: Send the EEPROM destination address (MSB first)
    SPI0DAT = (BYTE)((address >> 8) & 0x00FF);        // send out address high byte
    while(!SPIF);
    SPIF    = 0;
    SPI0DAT = (BYTE)(address & 0x00FF);               // send out address low byte
    while(!SPIF);
    SPIF    = 0;
// Step 4: Send the value to be written
    SPI0DAT = value;                                  // send out the value to be written into EEPROM
    while(!SPIF);
    SPIF    = 0;
    NSSMD0 = 1;                                       // start internal write cycle
    delayby1us(1);
// Step 5: Wait until internal write cycle is completed before return
    do {
    NSSMD0 = 0;                                       // activate slave select
    SPI0DAT = RDSR;                                   // send read status register command
    while(!SPIF);
    SPIF    = 0;
    SPI0DAT = 0;                                      // trigger SCK clocks to shift in status register
    while(!SPIF);
    SPIF    = 0;
    NSSMD0 = 1;                                       // deactivate slave select after read
    delayby1us(1);
    } while (SPI0DAT & 0x01);
    SFRPAGE = saveSFRPAGE;
}
```

The EEPROM read function follows a three-step procedure to read the byte at the specified memory location.

```
BYTE RdByteEE(UINT address)
{
    BYTE value;
    BYTE saveSFRPAGE = SFRPAGE;
    SFRPAGE          = SPI0_PAGE;
// Step 1: send the read command
    NSSMD0           = 0;
    SPI0DAT          = RDEE;
    while(!SPIF);
    SPIF             = 0;
// Step 2: send the EEPROM source address (MSB first)
    SPI0DAT          = (BYTE)((address >> 8) & 0x00FF);
    while(!SPIF);
    SPIF             = 0;
```

```
    SPI0DAT = (BYTE)(address & 0x00FF);
    while(!SPIF);
    SPIF    = 0;
// Step 3: read the value and return
    SPI0DAT = 0;
    while (!SPIF);
    SPIF    = 0;
    NSSMD0 = 1;                          // deactivate slave select signal
    delayby1us(1);
    value   = SPI0DAT;                   // store the EEPROM value in local variable
    SFRPAGE = saveSFRPAGE;
    return value;
}
```

The test program tests the previous two functions by performing the following operations.

- Write the value 0x57 to every memory location and flash the LEDs driven by P5 in high speed.
- Read back the contents of every memory location and also flash the LEDs driven by P5 in high speed. The read back value is compared with the written value. Stop the program and output a message if there is any mismatch between the write and read values.
- Write the low byte of the memory address into every memory location and flash the LEDs driven by P5 in high speed.
- Read back the contents of every memory location and also flash the LEDs driven by P5 in high speed. The read back value is compared with the written value. Stop the program and output a message if there is any mismatch between the write and read values.
- When the verification process is complete, stay in an infinite loop and flash the LEDs driven by P5 at a slower speed.

The test program is as follows.

```
// include this file, uartUtil.c, and delays.c to the same project
#include <C8051F040.h>
#include <stdio.h>
#include <uartUtil.h>
#include <delays.h>
#ifndef     BYTE
#define     BYTE unsigned char
#endif
#ifndef     UINT
#define     UINT unsigned int
#endif
#define     RDEE      0x03              // read command
```

```
#define     WREE     0x02                          // write command
#define     WRDI     0x04                          // reset write enable latch command
#define     WREN     0x06                          // set write enable latch command
#define     RDSR     0x05                          // read status register command
#define     WRSR     0x01                          // write status register command
#define     EEPROM_SIZE 1024                       // the capacity of EEPROM to be tested

void WrByteEE(UINT address, BYTE value);
BYTE RdByteEE(UINT address);
void sysInit(void);
void openSPI (void);

void main (void)
{
    UINT address;
    BYTE test_byte, saveSFRPAGE;

    openSPI();
    sysInit();
    openUART0();                                   // configure UART0 to mode 1, baud rate set to 19200
                                                   // fill every EEPROM location with 0x57
    printf("Filling with 0x57's . . .\n");
    for (address = 0; address < EEPROM_SIZE; address++) {
        test_byte = 0x57;
        WrByteEE(address, test_byte);
        if((address % 16) == 0) { // each line print 16 locations
            printf("\nWriting 0x%4x: %2x ", address, (UINT)test_byte);
            saveSFRPAGE = SFRPAGE;
            SFRPAGE     = 0x0F;
            P5          = ~P5;
            SFRPAGE     = saveSFRPAGE;
        }
        else
            printf("%2x ", (UINT)test_byte);
    }
// Verify EEPROM with 0x57's
    printf("\n\nVerify 0x57's . . . \n");
    for (address = 0; address < EEPROM_SIZE; address++) {
        test_byte = RdByteEE(address);
        // print status to UART0
        if((address % 16) == 0){
            printf("\nVerifying 0x%4x: %2x ",address, (UINT)test_byte);
            saveSFRPAGE = SFRPAGE;
            SFRPAGE     = 0x0F;
            P5          = ~P5;
            SFRPAGE     = saveSFRPAGE;
        }
        else
            printf("%2x ", (UINT)test_byte);
        if(test_byte != 0x57){
```

```
                P5 = 0;
                printf("Error at %u\n",address);
                while(1);       //stop here on error
            }
        }
// Fill EEPROM with LSB of EEPROM addresses
        printf("\n\nFilling with LSB of EEPROM addresses . . . \n");
        for(address = 0; address < EEPROM_SIZE; address++){
            test_byte = address & 0xFF;
            WrByteEE(address,test_byte);
            //print status to UART0
            if((address % 16) == 0){
                printf("\nWriting 0x%4x: %2x ", address, (UINT)test_byte);
                saveSFRPAGE = SFRPAGE;
                SFRPAGE     = 0x0F;
                P5          = ~P5;
                SFRPAGE     = saveSFRPAGE;
        }
            else
                printf("%2x ", (UINT)test_byte);
        }
// Verify EEPROM with LSB of EEPROM addresses
        printf("\n\nVerifying LSB of EEPROM addresses . . . \n");
        for (address = 0; address < EEPROM_SIZE; address++){
            test_byte = RdByteEE(address);
            // print status to UART0
            if((address % 16) == 0){
                printf("\nVerifying 0x%4x: %2x ", address, (UINT)test_byte);
                saveSFRPAGE = SFRPAGE;
                SFRPAGE     = 0x0F;
                P5          = ~P5;
                SFRPAGE     = saveSFRPAGE;
            }
            else
                printf("%2x ", (UINT)test_byte);
            if (test_byte != (address & 0xFF)){
                SFRPAGE = 0x0F;
                printf("Error at %u\n", address);
                while(1);       // Stop here on error
            }
        }
        printf("\n\nVerification success!\n");
        while(1){
        //flash LEDs when done
            SFRPAGE = 0xF;
            P5      = ~P5;
            delayby100ms(2);
        }
    }
    void sysInit(void)
```

```
{
    UINT n;
    SFRPAGE  = CONFIG_PAGE;
    WDTCN    = 0xDE;              // disable watchdog timer
    WDTCN    = 0xAD;              //     "
    OSCXCN   = 0x67;              // configure external crystal oscillator
    for (n = 0; n < 255; n++);    // wait for about 1 ms
    while(!(OSCXCN & 0x80));      //wait until crystal oscillator is stable
    CLKSEL   = 0x01;              // use external crystal oscillator to generate SYSCLK
    XBR2     = 0x40;              // enable crossbar decoder
    XBR0     = 0x06;              // assign UART0 and SPI signals to I/O pins
    P0MDOUT = 0x35;               // configure TX0, SCK, MOSI, and NSS for output
    P5MDOUT = 0xFF;               // configure P5 for push-pull
}
void openSPI (void)
{
    SFRPAGE  = SPI0_PAGE;
    SPI0CFG  = 0x40;              // enable master mode
    SPI0CN   = 0x0D;              // enable SPI to 4-wire, single master mode
    SPI0CKR  = 3;                 // set SPI baud rate to 3 MHz
}
//------------------------------------------------------------------------
// include WrByteEE() here
//------------------------------------------------------------------------
//------------------------------------------------------------------------
// include RdByteEE() here
//------------------------------------------------------------------------
```

10.12 Matrix LED Displays

Many organizations have the need to display important information at the entrance or corners of their buildings. The information need not be displayed all at once but can be rotated. Temperature, date, humidity, and other data in turn are displayed at many crossroads. Schools display their upcoming games, the result of games yesterday, or other important events using large display panels. Many of these display panels use the matrix LED displays because of their brightness and versatility.

10.12.1 The Organization of Matrix LED Displays

Matrix LED displays are denoted by the number of their columns and rows. The most popular matrix LED display has five columns and seven rows (5×7). Other configurations (such as 5×8 and 8×8) also are available. It is obvious that the more rows and columns of LEDs used the better the resolution will be.

10.12.2 Colors of Matrix LED Displays

Several colors, including green, yellow, red, and bicolor (red/green), are available for matrix LED displays.

10.12.3 Connection Method

Matrix LED displays can be organized as **cathode-row** (anode column) or **anode-row** (cathode column). In a cathode-row organization, the LEDs in a row have a common cathode. In an anode-row organization, the LEDs in a row have a common anode. The cathode-row and anode-row organizations are shown in Figures 10.15 and 10.16, respectively.

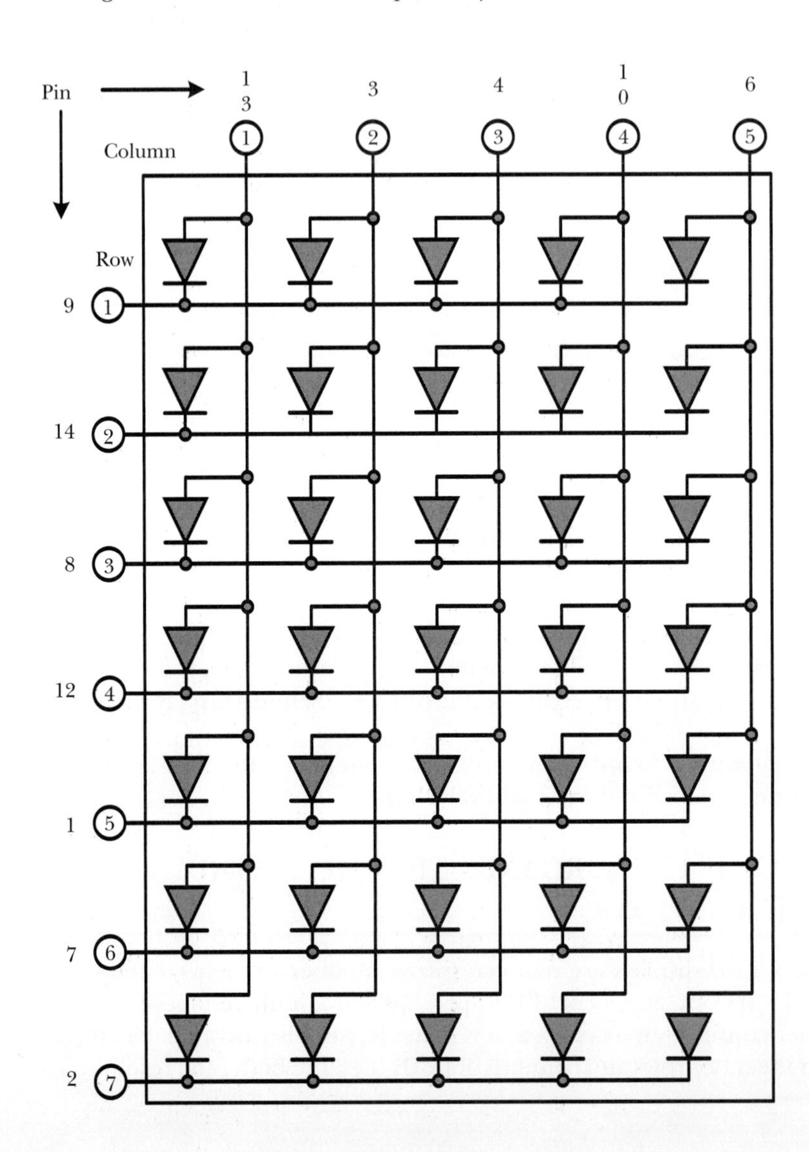

FIGURE 10.15 *Cathode-row matrix LEDs (Fairchild GMC8X75C). Based on data derived from http://www.fairchildsemi.com.*

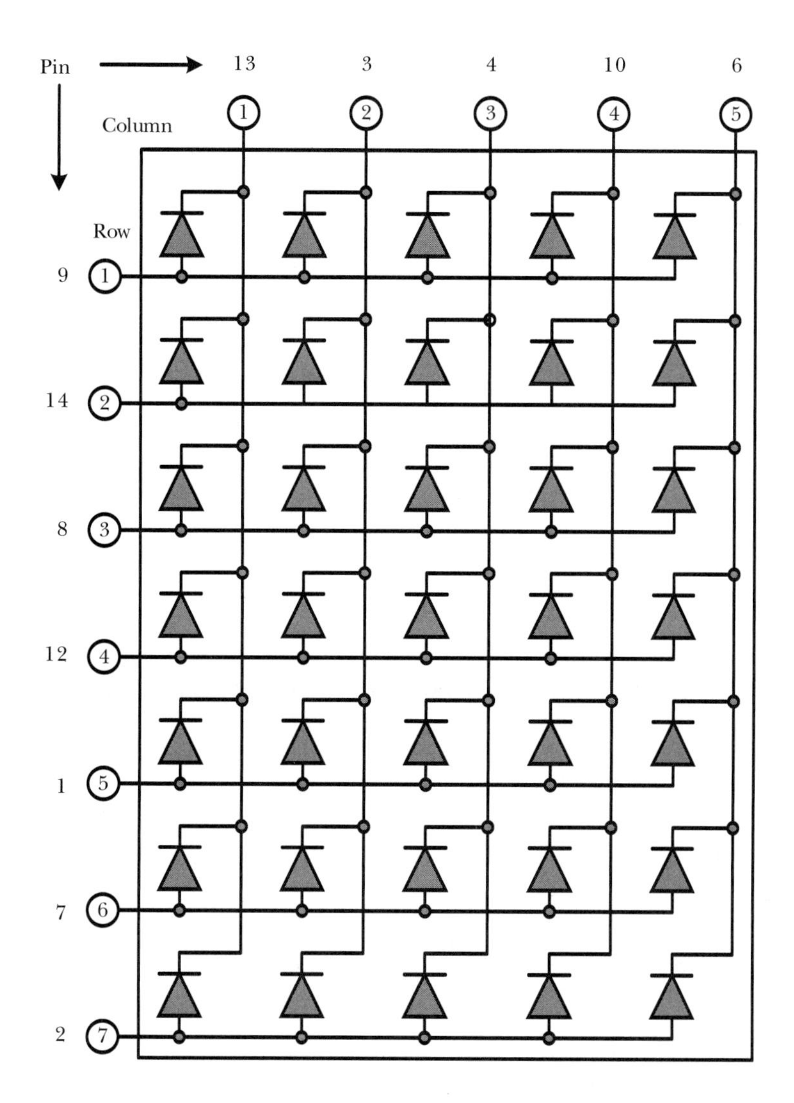

FIGURE 10.16 *Anode-row matrix LEDs (Fairchild GMA8X75C). Based on data derived from http://www.fairchildsemi.com.*

10.12.4 Dimension of Matrix LED Displays

The dimension of a matrix LED display often is indicated by its height. Common heights for matrix LED displays include 0.7″, 1.2″, 1.7″, and 2.3″. Larger matrix LED displays also are available.

10.12.5 Method of Driving Matrix LED Displays

Because of the internal connection of a matrix LED display, two parallel ports are needed to drive it. The matrix LED displays need to be scanned one row at a time, from top to bottom (or bottom to top). Usually, multiple matrix

LED displays are needed in the application, so a time-multiplexing technique needs to be used, which is quite demanding on CPU time.

Because of the popularity of matrix LED displays, a few companies have produced driver chips for them. Among them, Maxim provides driver chips with an SPI (MAX6952) or I^2C (MAX6953) interface for cathode-row matrix LEDs.

10.13 The MAX6952 Matrix LED Display Driver

The MAX6952 is designed to drive cathode-row matrix LED displays with a 5×7 organization. This chip can operate with a power supply ranging from 2.7 to 5.5 V and can drive four monocolor or two bicolor cathode-row matrix LED displays. It has a built-in ASCII 104-character font and 24 user-definable characters. The built-in characters follow Arial font, with the addition of the following common symbols: |, €, ¥, °, μ, ±, ↑, and ↓. The 24 user-definable characters are uploaded by the user into the on-chip RAM through the serial interface and are lost when the device is powered down. It allows automatic blinking control for each segment and provides 16-step digital brightness control. Both the 36-pin SSOP and 40-pin packages are available. The device includes a low-power shutdown mode, segment blinking (synchronized across multiple drivers, if desired), and a test mode that forces all LEDs on.

10.13.1 Pin Functions

The LED display pin functions are described in Table 10.3. The connection for four monocolor digits and two bicolor digits are shown in Tables 10.4 and 10.5, respectively. The typical circuit connection that drives four monocolor digits using the MAX6952 is shown in Figure 10.17.

The MAX6952 uses four pins to interface with the MCU: CLK, $\overline{CS}$, D_{IN}, and D_{OUT}. The $\overline{CS}$ signal must be low to clock data in and out of the device, and the D_{IN} signal is shifted in on the rising edge of the CLK pin. The maximum data-shift rate is 26 MHz. When the MAX6952 is not being accessed, D_{OUT} is not in high impedance, contrary to the SPI standard.

Multiple MAX6952s can be daisy-chained by connecting the D_{OUT} pin of one device to the D_{IN} pin of the next and driving the CLK and $\overline{CS}$ lines in parallel (Figure 10.18). Data at D_{IN} propagates through the internal shift registers and appears at D_{OUT} 15.5 clock cycles later, clocking out on the falling edge of CLK. When sending commands to daisy-chained MAX6952s, all devices are accessed at the same time. An access requires $16 \times n$ clock cycles, where n is the number of MAX6952s connected together. To update just one device in a daisy chain, the user can send the no-op command (0x00) to the others.

TABLE 10.3 *MAX6952 4-Digit Matrix LED Display Driver Pin Functions*

	Pin		
Name	**SSOP**	**PDIP**	**Function**
00 to 013	1, 2, 3, 6-14, 23, 24	1, 2, 3, 7-15, 26, 27	LED cathode drivers. 00 to 013 output sink current from the displays's cathode rows.
GND	4, 5, 6	4, 5, 6, 18	Ground.
I_{SET}	15	17	Segment current setting. Connect ISET to GND through series resistor R_{SET} to set the peak current.
BLINK	17	19	Blink clock output. Output is open-drain.
D_{IN}	18	20	Serial data input. Data is loaded into the internal 16-bit shift register on the rising edge of the CLK.
CLK	19	21	Serial-clock input. On the rising edge of CLK, data is shifted into the internal shift register. On the falling edge of CLK, data is clocked out of D_{OUT}. CLK input is active only when CS is low.
D_{OUT}	20	22	Serial data output. Data clocked into D_{IN} is output to D_{OUT} 15.5 clock cycles later. Data is clocked out on the falling edge of CLK. Output is push-pull.
CS	21	23	Chip-select input. Serial data is loaded into the shift register while CS is low. The last 16 bits of serial data are latched on CS's rising edge.
OSC	22	24	Multiplex clock input. To use the internal oscillator, connect capacitor C_{SET} from OSC to GND. To use the external clock, drive OSC with a 1-MHz to 8-MHz CMOS clock.
014 to 023	25-31, 34, 35, 36	28-34, 38, 39, 40	LED anode drivers. 014 to 023 output source current to the display's anode columns.
V+	32, 33	35, 36, 37	Positive supply voltage. Bypass V+ to GND with a 47−μF bulk capacitor and a 0.1−μF ceramic capacitor.

TABLE 10.4 *Connection Scheme for Four Monocolor Digits*

Digit	**00~06**	**07~013**	**014~018**	**019~023**
1	Digit 0 rows (cathodes) R1 to R7 Digit 1 rows (cathodes) R1 to R7		Digit 0 columns (anodes) C1 to C5	Digit 1 columns (anodes) C6 to C10
2		Digit 2 rows (cathodes) R1 to R7 Digit 3 rows (cathodes) R1 to R7	Digit 2 columns (anodes) C1 to C5	Digit 3 columns (anodes) C6 to C10

TABLE 10.5 *Connection Scheme for Two Bi-Color Digits*

Digit	00~06	07~013	014~018	019~023
1	Digit 0 rows (cathodes) R1 to R14		Digit 0 columns (anodes) C1 to C10	
			The 5 green anodes	The 5 red anodes
2		Digit 1 rows (cathodes) R1 to R14	Digit 1 columns (anodes) C1 to C10	
			The 5 green anodes	The 5 red anodes

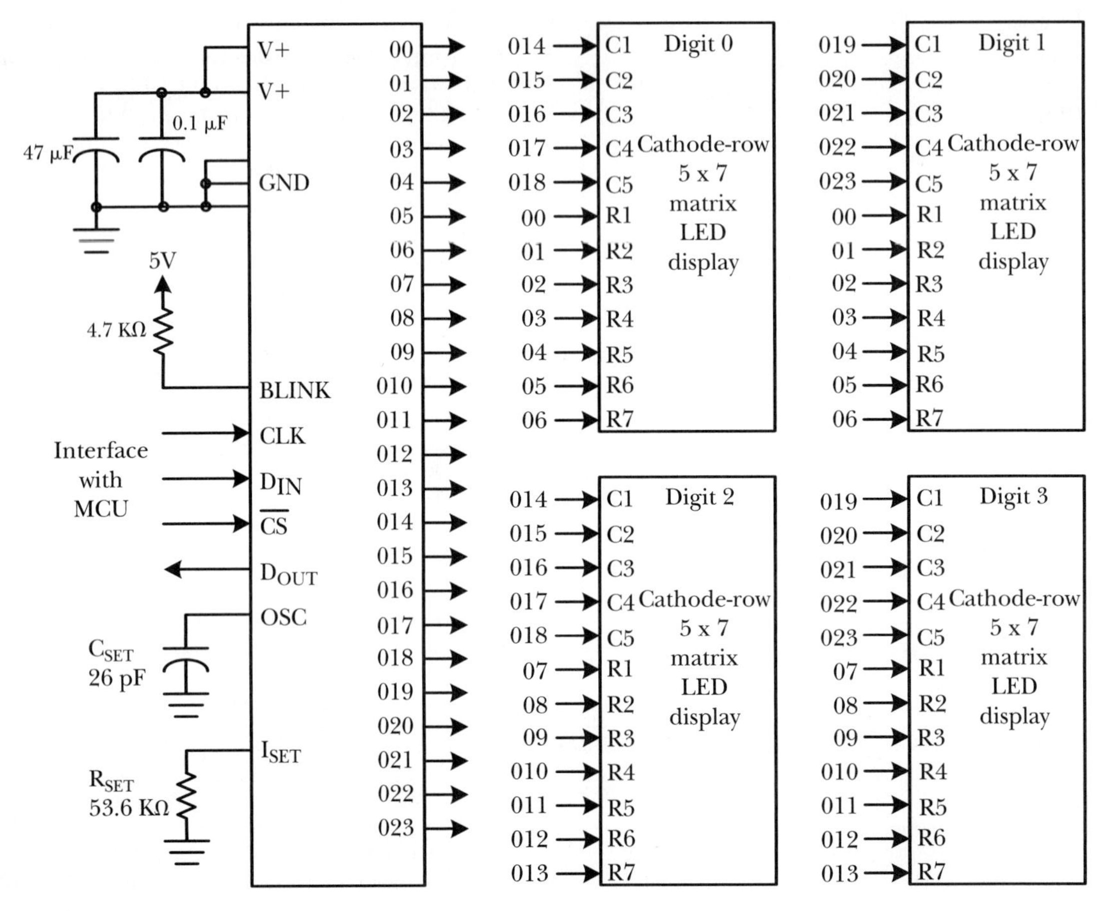

FIGURE 10.17
MAX6952 driving four matrix LED displays.

10.13.2 Internal Registers

The block diagram of the MAX6952 is shown in Figure 10.19. The MAX6952 contains a 16-bit shift register (in the serial interface block in Figure 10.19) into which DIN data are shifted on the rising edge of the CLK signal when the

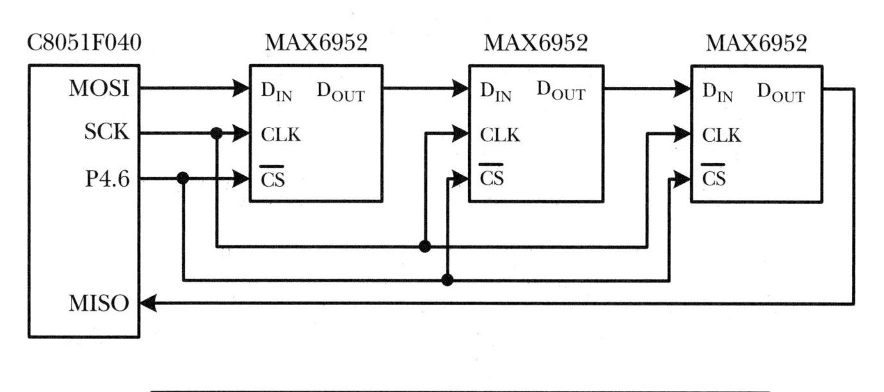

FIGURE 10.18 *MAX6952 daisy-chain connection.*

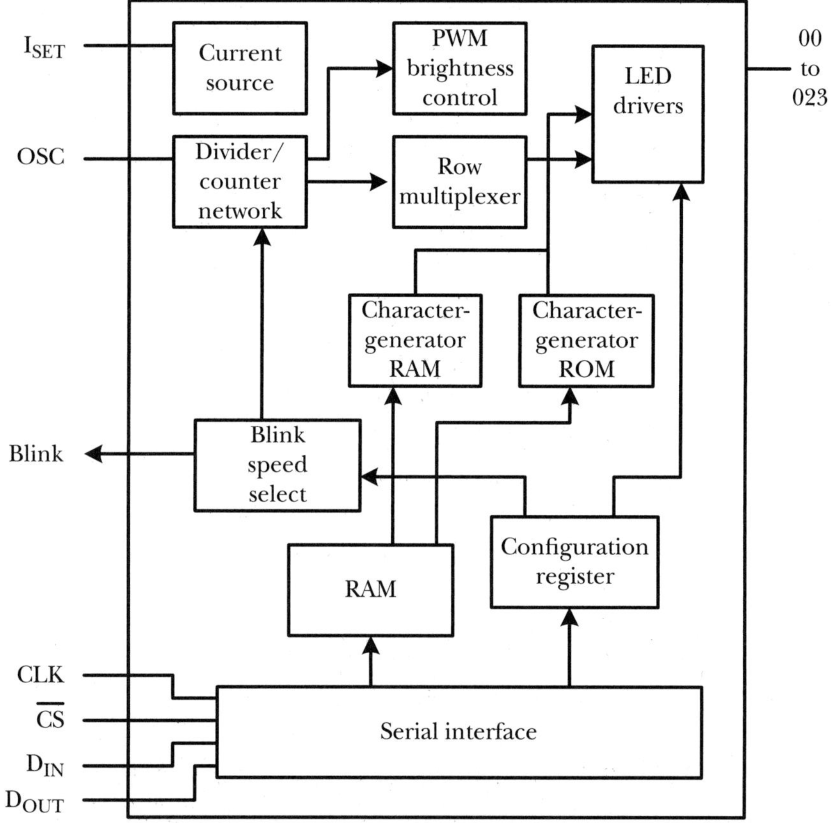

FIGURE 10.19 *MAX6952 functional diagram.*

CS signal is low. The 16 bits in the shift register are loaded into a latch on the rising edge of the CS signal. The 16 bits in the latch are then decoded and executed. The CLK input must be idled low. This signal must be taken to low before data transfer is started.

The upper 8 bits of the shift register select the destination register to which the lower 8 bits of the shift register are to be transferred. The address map of the MAX6952 is shown in Table 10.6.

TABLE 10.6 *MAX6952 Register Address Map*

	Address (Command Byte)								
Register	**D15**	**D14**	**D13**	**D12**	**D11**	**D10**	**D9**	**D8**	**Hex Code**
No op	$R/\overline{W}$	0	0	0	0	0	0	0	0x00
Intensity10	$R/\overline{W}$	0	0	0	0	0	0	1	0x01
Intensity32	$R/\overline{W}$	0	0	0	0	0	1	0	0x02
Scan limit	$R/\overline{W}$	0	0	0	0	0	1	1	0x03
Configuration	$R/\overline{W}$	0	0	0	0	1	0	0	0x04
User-defined fonts	$R/\overline{W}$	0	0	0	0	1	0	1	0x05
Factory reserved (do not write into)	$R/\overline{W}$	0	0	0	0	1	1	0	0x06
Display test	$R/\overline{W}$	0	0	0	0	1	1	1	0x07
Digit 0 plane P0	$R/\overline{W}$	0	1	0	0	0	0	0	0x20
Digit 1 plane P0	$R/\overline{W}$	0	1	0	0	0	0	1	0x21
Digit 2 plane P0	$R/\overline{W}$	0	1	0	0	0	1	0	0x22
Digit 3 plane P0	$R/\overline{W}$	0	1	0	0	0	1	1	0x23
Digit 0 plane P1	$R/\overline{W}$	1	0	0	0	0	0	0	0x40
Digit 1 plane P1	$R/\overline{W}$	1	0	0	0	0	0	1	0x41
Digit 2 plane P1	$R/\overline{W}$	1	0	0	0	0	1	0	0x42
Digit 3 plane P1	$R/\overline{W}$	1	0	0	0	0	1	1	0x43
Write digit 0 plane P0 and plane P1 with same data (reads as 0x00)	$R/\overline{W}$	1	0	0	0	0	0	0	0x60
Write digit 1 plane P0 and plane P1 with same data (reads as 0x00)	$R/\overline{W}$	1	1	0	0	0	0	1	0x61
Write digit 2 plane P0 and plane P1 with same data (reads as 0x00)	$R/\overline{W}$	1	1	0	0	0	1	0	0x62
Write digit 3 plane P0 and plane P1 with same data (reads as 0x00)	$R/\overline{W}$	1	1	0	0	0	1	1	0x63

The procedure for writing the MAX6952 is as follows.

Step 1
Pull the CLK signal to low.

Step 2
Pull the CS signal to low to enable the internal 16-bit shift register.

Step 3
Shift in 16 bits of data from the D_{IN} pin with the most-significant-bit first. The most significant bit (D15) must be low for a write operation.

Step 4
Pull the CS signal to high.

Step 5
Pull the CLK signal to low.

Any register data within the MAX6952 may be read by setting the D15 bit to 1. The procedure to read a register is as follows.

Step 1
Pull CLK to low.

Step 2
Pull the $\overline{CS}$ signal to low to enable the internal shift register.

Step 3
Clock 16 bits of data into the D_{IN} pin with bit 15 first. Bit 15 must be 1. Bits 14 through 8 contain the address of the register to be read. Bits 7 through 0 contain dummy data.

Step 4
Pull the $\overline{CS}$ signal to high. Bits 7 through 0 of the serial shift register will be loaded with the data in the register addressed by bits 15 through 8.

Step 5
Pull CLK to low.

Step 6
Issue another read command (which can be a no-op), and examine the bit stream at the D_{OUT} pin. The second 8 bits are the contents of the register addressed by bits 14 to 8 in Step 3.

Digit Registers

The MAX6952 uses eight digit registers to store the characters that the user wishes to display on the four 5 × 7 LED digits. These digit registers are implemented with two planes of 4 bytes, called P0 and P1. Each LED digit is represented by 2 bytes of memory, one byte in plane P0 and the other in plane P1. The digit registers are mapped so that a digit's data can be updated in plane P0, or plane P1, or both at the same time, as shown in Table 10.6.

If the blink function is disabled through the blink enable bit E in the configuration register, then the digit-register data in plane P0 is used to multiplex the display. The digit-register data in P1 is not used. If the blink function is enabled, then the digit-register data in both plane P0 and P1 are used alternately to multiplex the display. Blinking is achieved by multiplexing the LED display using data planes P0 and P1 on alternate phases of the blink clock (shown in Table 10.7).

The data in the digit registers does not control the digit segments directly. Instead, the register data is used to address a character generator, which stores the data of a 128-character font. The lower seven bits of the digit data (D6 to D0) select the character font. The most significant bit of the register data (D7) selects whether the font data is used directly (D7 = 0) or whether the font is inverted (D7 = 1). The inversion feature can be used to enhance the appearance of bicolor displays (by displaying, for example, a red character on a green background).

TABLE 10.7 *Digit Register Mapping with Blink Globally Enabled*

Segment's Bit Setting in Plane P1	Segment's Bit Setting in Plane P0	Segment Behavior
0	0	Segment off.
0	1	Segment on only during the first half of each blink period.
1	0	Segment on only during the second half of each blink period.
1	1	Segment on.

Configuration Register

The configuration register is used to enter and exit shutdown, select the blink rate, globally enable and disable the blink function, globally clear the digit data, and reset the blink timing. The contents of the configuration register are shown in Figure 10.20.

Intensity Registers

Display brightness is controlled by four pulse-width modulators, one for each display digit. Each digit is controlled by a nibble of one of the two intensity registers, *Intensity10* and *Intensity32*. The upper nibble of the Intensity10 register controls the intensity of the matrix display 1, whereas the lower nibble of the same register controls the intensity of the matrix display 0. Matrix displays 3 and 2 are controlled by the upper and lower nibbles of the Intensity32 register, respectively. The modulator scales the average segment current in 16 steps from a maximum of 15/16 down to 1/16 of the peak current. The minimum interdigit blanking time is therefore 1/16 of a cycle. The maximum duty cycle is 15/16.

Scan-Limit Register

The scan-limit register sets the number of monocolor digits to be displayed, either two or four. A bicolor digit is connected as two monocolor digits. The multiplexing scheme drives digits 0 and 1 at the same time, then digits 2 and 3 at the same time. To increase the effective brightness of the displays, the MAX6952 drives only two digits instead of four. By doing this, the average segment current doubles, but this also doubles the number of MAX6952s required for driving a given number of digits.

The contents of the scan-limit register are shown in Figure 10.21. This register has only one bit implemented. When this bit is 0, only digits 0 and 1 are displayed. Otherwise, all four digits are displayed.

Display-Test Register

The display-test register switches the drivers between one of two modes: normal and display test. Display-test mode turns on all LEDs by overriding (but not altering) all control and digit registers (including the shutdown register). In display-test mode, eight digits are scanned, and the duty cycle is 7/16 (half

7	6	5	4	3	2	1	0
P	x	R	T	E	B	x	S

P: Blink phase read back select
  0 = P1 blink phase
  1 = P0 blink phase
R: Global clear digit data
  0 = digit data on both plane P0 and P1 are not affected
  1 = clear digit data on both planes P0 and P1
T: Global blink timing synchronization
  0 = blink timing counter are unaffected
  1 = blink timing counters are reset on the rising edge of $\overline{CS}$
E: Global blink enable/disable
  0 = blink function is disabled
  1 = blink function is enabled
B: Blink rate selection
  0 = select slow blinking (refreshed for 1 s by plane P0, then 1 s by P1 at 4MHz)
  1 = select fast blinking
S: Shutdown mode
  0 = shutdown mode
  1 = normal operation

FIGURE 10.20 *The MAX6952 configuration register.*

7	6	5	4	3	2	1	0
X	X	X	X	X	X	X	2or4

2or4: Scan two digits (0 and 1) or all four digits
  0 = display digits 0 and 1 only
  1 = display digits 0, 1, 2, and 3

FIGURE 10.21 *The MAX6952 scan-limit register.*

power). The contents of the display test register are shown in Figure 10.22. Only bit 0 of this register is implemented.

Character-Generator Font Mapping

The character font is a 5 × 7 matrix. The character generator comprises of 104 characters in ROM and 24 user-definable characters. The selection from a total of 128 characters is represented by the lower seven bits of the 8-bit digit registers. The character map follows Arial font for 88 characters in the range from 0x28 to 0x7F. The first 32 characters map the 24 user-defined positions (RAM00 to RAM23), plus 8 extra common characters in ROM. When the most significant bit is 0, the device will display the font normally. Otherwise, the chip will display the font inversely.

User-Defined Font Register

The 24 user-definable characters are represented by 120 entries of 7-bit data (five entries per character) and are stored in the MAX6952's internal RAM. The 120 user-definable font data are written and read through a single

7	6	5	4	3	2	1	0
X	X	X	X	X	X	X	Test

Test: Test bit
 0 = normal operation
 1 = display test

FIGURE 10.22 *The MAX6952 display test register.*

register at the address 0x05. An auto-incrementing font address-pointer register in the MAX6952 indirectly accesses the font data. The font address pointer can be written, setting one of 120 addresses between 0x00 and 0xF7 but cannot be read back. The font data is written to and read from the MAX6952 indirectly, using this font address pointer. Unused font locations can be used as general-purpose scratch RAM. Font registers are only 7-bits wide.

To define new fonts, the user first needs to set the font address pointer. This is done by placing the address in the font address-pointer register and setting bit 7 to 1. After this, one can write the font data in the lower seven bits (to the font address pointer position) and clear bit 7.

The font address-pointer auto-increments after a valid access to the user-definable font data. Auto-incrementing allows the 120 font data entries to be written and read back very quickly, because the font pointer address needs to be set only once. When the last data location (0xF7) is written into, the font address-pointer increments to 0x80 automatically. If the font address pointer is set to an out-of-range address by writing data in the range from 0xF8 to 0xFF, then the address is set to 0x80 instead.

The memory mapping of user-defined font register 0x05 is detailed in Table 10.8. The behavior of the font pointer address is shown in Table 10.9. To display the font defined by the user, one must send in the RAM address from 0x00 through 0x17, corresponding to the font address-pointer value that is 5 × RAM address (one character needs five bytes).

10.13.3 Blinking Operation

The display blinking facility, when enabled, makes the LED drivers flip automatically between displaying the digit register data in planes P0 and P1. If the digit-register data for any digit is different in two planes, then that digit appears to flip between two characters. To make a character appear to blink on or off, write the character to one plane and use the blank character (0x20) for the other plane. Once blinking has been configured, it continues automatically without further intervention. Blinking is enabled by setting the E bit of the configuration register.

The blink speed can be programmed to be fast or slow and is determined by the frequency of the multiplex clock, OSC, and by setting the B bit of the configuration register. The blink rate selection bit B of the configuration register sets either a fast or slow blink speed for the whole display.

TABLE 10.8 *Memory Mapping of User-Defined Font Register 0x05*

Address Code (hex)	Register Data	SPI Read or Write	Function
0x85	0x00-0x7F	Read	Read 7-bit user-definable font data entry from current font address. MSB of the register data is clear. Font address pointer is incremented after the read.
0x05	0x00-0x7F	Write	Write 7-bit user-definable font data entry to current font address. Font address pointer is incremented after the write.
0x05	0x80-0xFF	Write	Write font address pointer with the register data.

TABLE 10.9 *Font Pointer Address Behavior*

Font Pointer Address	Action
0x80-0xF6	Valid range to set the font address pointer. Pointer autoincrements after a font data read or write, while pointer address remains in this range.
0xF7	Font address resets to 0x80 after a font data read or write to this pointer address.
0xF8 to 0xFF	Invalid range to set the font address pointer. Pointer is set to 0x80.

Blink Synchronization

When multiple digits are displayed, one can choose to synchronize the blinking operation of these digits. Internally, blink synchronization is achieved by resetting the display multiplexing sequence. As long as all MAX6952s are daisy-chained with one device's D_{OUT} connected to the D_{IN} of the next device, global synchronization is achieved by toggling the $\overline{CS}$ pin for each device, either together or in quick succession.

Blink Output

The blink output (the BLINK pin) indicates the blink phase and is high during the P0 period and low during the P1 period. Blink-phase status can be read back as the P bit in the configuration register. Typical uses for this output are

- To provide an interrupt to the processor so that segment data can be changed synchronously to the blinking. For example, a clock application may have colon segments blinking every second between hour and minute digits, and the minute display is best changed in step with the colon segments. Also, if the rising edge of the blink is detected, there is half a blink period to change the P1 data (P0 data drives the displays during this interval). Similarly, if the falling edge of the blink is detected, there is half a blink period to change the P0 digit data.
- If OSC is driven with an accurate frequency, the blink can be used as a seconds counter.

10.13.4 Choosing Values for R_{SET} and C_{SET}

The RC oscillator uses an external resistor (R_{SET}) and an external capacitor (C_{SET}) to set the oscillator frequency (f_{OSC}). The allowed range of f_{OSC} is 1 to 8 MHz. R_{SET} also sets the peak segment current. The recommended values for R_{SET} and C_{SET} set the oscillator to 4 MHz, which sets the slow and fast blink frequencies to 0.5 and 1 Hz. The recommended value of R_{SET} also sets the peak current to 40 mA, which makes the segment current adjustable from 2.5 to 37.5 mA in 2.5-mA steps.

$$I_{SEG} = K_I / R_{SET} \text{ mA}$$

$$f_{OSC} = K_F / (R_{SET} \times C_{SET} + C_{STRAY}) \text{ MHz}$$

where

K_I = 2144
K_F = 6000
R_{SET} = external resistor in kΩ
C_{SET} = external capacitor in pF
C_{STRAY} = stray capacitance from OSC pin to GND in pF, typically 2 pF

The recommended value for R_{SET} is 53.6 KΩ, and the recommended value for C_{SET} is 26 pF. The recommended value for R_{SET} is the minimum allowed value, since it sets the display driver to the maximum allowed segment current. R_{SET} can be set to a higher value and hence reduces the peak segment current to a lower value whenever it is desirable. The effective value of C_{SET} includes not only the actual external capacitor used but also the stray capacitance from OSC to GND.

Example 10.10 Daisy-chain two MAX6952s to drive eight cathode-row, monocolor, matrix LED displays GMC8975C made by Fairchild, assuming that the P4.6 pin is used to drive the $\overline{CS}$ input of two MAX6952s. The connection of two MAX6952s with the C8051F040 (running with a 24-MHz oscillator) is shown in Figure 10.23. Each MAX6952 is driving four matrix LED displays. Write a program to configure the SPI module to operate in master mode, transfer data at 6 MHz, and shift data in/out on the rising edge of the SCK clock. Display **SMARTBOY** on the matrix LED display without blinking.

Solution: The SPI0 module should be configured with the features

- 6-MHz baud rate
- Master mode with interrupts disabled
- Shift data on the rising edge with clock idle low
- Three-wire mode

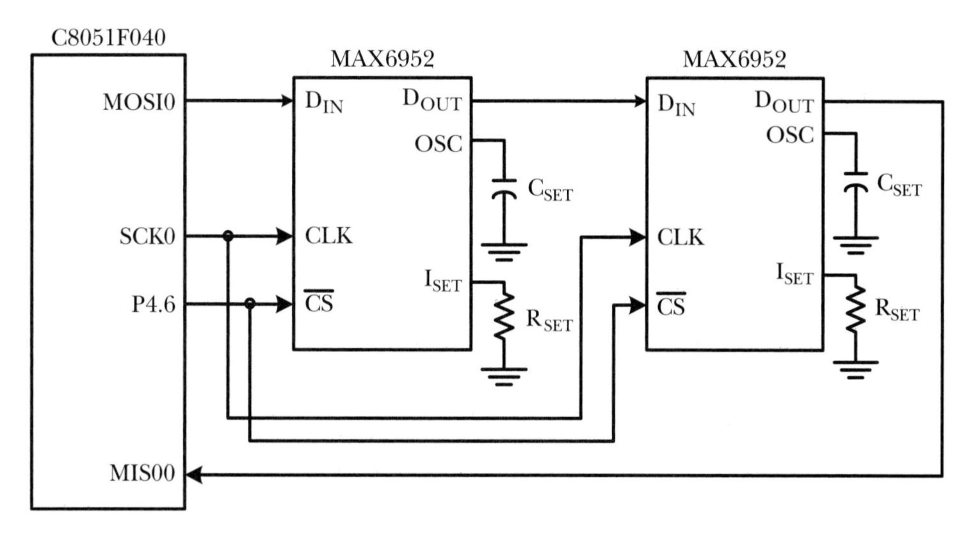

FIGURE 10.23
C8051F040 drives two MAX6952s.

Two MAX6952 chips should be configured as follows.

Intensity10 Register. We will set the intensity of both MAX6952s to maximum by sending out the value:

0x01, 0xFF, 0x01, 0xFF

Sixteen bits need to be written to each MAX6952, of which the upper eight bits are the address to select the Intensity10 register.

Intensity32 Register. Set the second and third displays to maximum intensity by sending the same data as to Intensity10:

0x02, 0xFF, 0x02, 0xFF

Scan Limit Register. Configure the MAX6952 to drive four monocolor displays by writing the following value to the MAX6952:

```
0x03, 0x01, 0x03, 0x01      ; drive four displays
```

Configuration Register. Configure this register to

- Select P1 blink phase
- Not to clear digit data on both plane P0 and P1
- Reset the blink timing counter on the rising edge of $\overline{CS}$
- Disable the blink function
- Select slow blinking (don't care)
- Select normal mode

Send the following values to the MAX6952s:

0x04, 0x11, 0x04, 0x11

Display Test Register. Disable display test as the start by writing the following values to the MAX6952:

0x07, 0x00, 0x07, 0x00

Digit 0 (rightmost digit) Plane P0. We will display letter **R** and letter **Y**, respectively, on display 0 of the first and second MAX6952 by sending the data:

0x20, 0x59, 0x20, 0x52

Digit 1 (Second Rightmost Digit) Plane P0. We will display letter **A** and letter **O** respectively, on display 1 of the first and second MAX6952 by sending the data:

0x21, 0x4F, 0x21, 0x41

Digit 2 (Second Leftmost Digit) Plane P0. We will display letter **M** and letter **B**, respectively, on display 2 of the first and second MAX6952 by sending the data:

0x22, 0x42, 0x22, 0x4D

Digit 3 (Leftmost Digit) Plane P0. We will display letter **S** and letter **T**, respectively, on the display 3 of the first and second MAX6952 by sending the data:

0x23, 0x54, 0x23, 0x53

The following C program will configure the SPI and MAX6952 properly.

```
#include <C8051F040.h>  // include this file and spiUtil.c to the same project
#include "spiUtil.h"
void sendtomax(char x1, char x2, char x3, char x4);
void sysInit();
void openSPI(void);
void main (void)
{
    sysInit();
    SFRPAGE = SPI0_PAGE;                  // switch to SFR page 0
    openSPI();
    sendtomax(0x01, 0xFF, 0x01, 0xFF);    // set intensity for digits 0 and 1
    sendtomax(0x02, 0xFF, 0x02, 0xFF);    // set intensity for digits 2 and 3
    sendtomax(0x03, 0x01, 0x03, 0x01);    // set scan limit to drive four digits
    sendtomax(0x04, 0x11, 0x04, 0x11);    // set configuration register
    sendtomax(0x07, 0x00, 0x07, 0x00);    // disable test
    sendtomax(0x20, 0x59, 0x20, 0x52);    // value for digit 0
```

```
    sendtomax(0x21, 0x4F, 0x21, 0x41);          // value for digit 1
    sendtomax(0x22, 0x42, 0x22, 0x4D);          // value for digit 2
    sendtomax(0x23, 0x54, 0x23, 0x53);          // value for digit 3
}
/**************************************************************************************************************/
/* place sysInit(void) here */
/**************************************************************************************************************/
void sendtomax (char c1, char c2, char c3, char c4)
{
    unsigned char temp;
    temp     = SFRPAGE;
    SFRPAGE = 0x0F;
    P4      &= 0xBF;                   // enable SPI transfer to MAX6952
    putc2SPI(c1);                      // send c1 to MAX6952
    putc2SPI(c2);                      // send c2 to MAX6952
    putc2SPI(c3);                      // send c3 to MAX6952
    putc2SPI(c4);                      // send c4 to MAX6952
    P4      |= 0x40;                   // load data from shift register to latch
    SFRPAGE = temp;
}
void openSPI(void)
{
    char saveSFRPAGE;
    saveSFRPAGE = SFRPAGE;
    SFRPAGE     = SPI0_PAGE;
    SPI0CKR     = 0x01;                // set baud rate to 6 MHz
    SPI0CFG     = 0x40;                // disable interrupt, set master mode, shift data on
                                       // rising edge, clock idle low
    SPI0CN      = 0x01;                // enable SPI in 3-wire master mode
    SFRPAGE     = saveSFRPAGE;
}
void sysInit(void)
{
    int  n;
    SFRPAGE  = CONFIG_PAGE;
    WDTCN    = 0xDE;
    WDTCN    = 0xAD;
    OSCXCN   = 0x67;
    for (n = 0; n < 255; n++);
    while(!(OSCXCN & 0x80));
    CLKSEL   = 0x01;                   // use external crystal oscillator to generate SYSCLK
    XBR2     = 0x5D;                   // enable crossbar and assign all peripheral signals
    XBR0     = 0xF7;                   // to I/O pins
    XBR1     = 0xFF;                   //      "
    XBR3     = 0x8F;                   //      "
    P4MDOUT = 0x40;                    // configure P4.6 as output
    P0MDOUT = 0xD5;                    // configure SCK, MOSI for output
}
```

Example 10.11 Modify the previous example to blink the display at a slow rate.

Solution: We need to change the setting of the configuration register and also send the space character (0x20) to the four digits in plane P1. New data to be sent to the configuration registers are

0x04, 0x19, 0x04, 0x19

The main program should be modified to as follows.

```
void main (void)
{
    sysInit();
    SFRPAGE = 0;                              /* switch to SFR page 0 */
    openspi();
    sendtomax(0x01, 0xFF, 0x01, 0xFF);        // set intensity for digits 0 and 1
    sendtomax(0x02, 0xFF, 0x02, 0xFF);        // set intensity for digits 2 and 3
    sendtomax(0x03, 0x01, 0x03, 0x01);        // set scan limit to drive four digits
    sendtomax(0x04, 0x19, 0x04, 0x19);        // set configuration register
    sendtomax(0x07, 0x00, 0x07, 0x00);        // disable test
    sendtomax(0x20, 0x59, 0x20, 0x52);        // value for digit 0 on plane P0
    sendtomax(0x21, 0x4F, 0x21, 0x41);        // value for digit 1
    sendtomax(0x22, 0x42, 0x22, 0x4D);        // value for digit 2
    sendtomax(0x23, 0x54, 0x23, 0x53);        // value for digit 3
    sendtomax(0x40, 0x20, 0x40, 0x20);        // value for digit 0 on plane P1 (space)
    sendtomax(0x41, 0x20, 0x41, 0x20);        // value for digit 1      "
    sendtomax(0x42, 0x20, 0x42, 0x20);        // value for digit 2      "
    sendtomax(0x43, 0x20, 0x43, 0x20);        // value for digit 3      "
}
```

Example 10.12 For the circuit shown in Figure 10.23, write a program to display the following message and shift the information from right-to-left every second and enable blinking.

08:30:40 Wednesday, 72°F, humidity: 60%

Solution: We will use plane P0 to shift the message once every half second. The message in plane P0 will be used to multiplex the display in half of a second and the message sent to plane P1 will be used to multiplex the display in the next half of a second. We will display the space character in each matrix display using plane P1 and display the normal characters using plane P0. A delay function will be invoked once every second to shift the message to plane P0. The main program is modified to as follows.

```
#include <C8051F040.h>                    // include this file, spiUtil.c, delays.c in the project
#include "spiUtil.h"
#include "delays.h"
void sendtomax(char x1, char x2, char x3, char x4);
void sysInit();
void openSPI(void);
char msgP0[41] = "08:30:40 Wednesday, 72°F, humidity: 60% ";   // 40 characters
void main (void)
{
    char k,i1,i2,i3,i4,j1,j2,j3,j4;
    sysInit();
    SFRPAGE = 0;                          // switch to SFR page 0
    openSPI();
    sendtomax(0x01, 0xFF, 0x01, 0xFF);    // set intensity for digits 0 and 1
    sendtomax(0x02, 0xFF, 0x02, 0xFF);    // set intensity for digits 2 and 3
    sendtomax(0x03, 0x01, 0x03, 0x01);    // set scan limit to drive four digits
    sendtomax(0x04, 0x1D, 0x04, 0x1D);    // configuration register
    sendtomax(0x40, 0x20, 0x40, 0x20);    // send space character to plane P1
    sendtomax(0x41, 0x20, 0x41, 0x20);
    sendtomax(0x42, 0x20, 0x42, 0x20);
    sendtomax(0x43, 0x20, 0x43, 0x20);
    k = 0;
    while (1) {
        i1 = k;
        i2 = (k+1)%40;
        i3 = (k+2)%40;
        i4 = (k+3)%40;
        j1 = (k+4)%40;
        j2 = (k+5)%40;
        j3 = (k+6)%40;
        j4 = (k+7)%40;
        sendtomax(0x20, msgP0[i1], 0x20, msgP0[j1]);
        sendtomax(0x21, msgP0[i2], 0x21, msgP0[j2]);
        sendtomax(0x22, msgP0[i3], 0x22, msgP0[j3]);
        sendtomax(0x23, msgP0[i4], 0x23, msgP0[j4]);
        delayby100ms(10);                 // wait for 1 s
        k = (k+1)%40;
        }
}
```

Example 10.13 Write a program to define fonts for three Chinese characters, as shown in Figure 10.24. Store the font of these three special characters at locations from 0x00 to 0x0E of the MAX6952.

Solution: The program that sets fonts is very straightforward. The **main** and **send_font** functions will set up the fonts as specified.

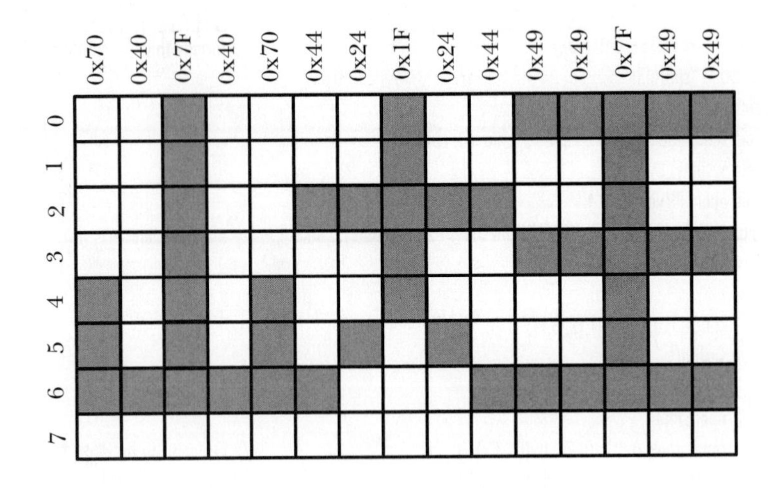

FIGURE 10.24 *User-definable font example.*

```
#include <C8051F040.h>          // include this file, spiUtil.c, and delays.c to the project
#include "spiUtil.h"
#include "delays.h"
void sendtomax(char x1, char x2, char x3, char x4);
void sysInit();
void openSPI(void);
void send_font (char xc);
char fonts [15] = {0x70,0x40,0x7F,0x40,0x70,0x44,0x24,0x1F,0x24,0x44,0x49,
                   0x49,0x7F,0x49,0x49};
void main (void)
{
    char i;
    sysInit();
    SFRPAGE = 0;                // switch to SFR page 0
    openSPI();
    send_font(0x80);            // set font address pointer address to 0x00
    for (i = 0; i < 15; i++)
        send_font(fonts[i]);
}
void send_font(char xx)
{
    unsigned char temp;
    temp    = SFRPAGE;
    SFRPAGE = 0x0F;
    P4     &= 0xBF;             // enable SPI transfer to MAX6952
    putc2SPI(0x05);             // specify font address pointer
    putc2SPI(xx);               // send a font value
    P4      |= 0x40;            // load data from shift register to latch
    SFRPAGE = temp;
}
/*************************************************************************************/
/* include sysInit() and openspi() functions here.
/*************************************************************************************/
```

To display user-defined fonts, one needs to send in the appropriate addresses in the range from 0x00 to 0x17. In this example, the addresses 0x00~0x02 should be used to access those three special characters. The following C statements will display those three Chinese characters followed by letters A, B, C, D, and E from left to right on the matrix displays shown in Figure 10.23:

```
sendtomax(0x20, 0x42, 0x20, 0x00); // 0x00 is the address of the first character font
sendtomax(0x21, 0x43, 0x21, 0x01); // 0x01 is the address of the second character font
sendtomax(0x22, 0x44, 0x22, 0x02); // 0x02 is the address of the third character font
sendtomax(0x23, 0x45, 0x23, 0x41);
```

10.14 Chapter Summary

When high-speed data transfer is not needed, using serial data transfer enables us to make the most of the limited number of I/O pins available on the MCU device. Serial data transfer can be performed asynchronously or synchronously. The SPI is a synchronous protocol introduced by Freescale for serial data exchange between peripheral chips and microcontrollers.

In the SPI format, a device (must be a microcontroller) is responsible for initiating the data transfer and generating the clock pulses for synchronizing data transfer. This device is referred to as the *SPI master*. All other devices in the same system are referred to as *SPI slaves*. The master device needs three signals to carry out the data transfer:

- **SCK:** a clock signal for synchronizing data transfer
- **MOSI:** serial data output from the master
- **MISO:** serial data input to the master

To transfer data to one or more SPI slaves, the MCU writes data into the SPI data register, and eight clock pulses will be generated to shift out the data in the SPI data register from the MOSI pin. If the MISO pin of the MCU also is connected (to the slave), then eight data bits also will be shifted into the SPI data register. To read data from the slave, the MCU also needs to write data into the SPI data register to trigger clock pulses to be sent out from the SCK pin. However, the value written into the SPI data register is unimportant in this case.

When configured as a slave device, the 8051 MCU also needs the fourth signal called Slave Select (SS). The SS signal enables the 8051 slave to respond to an SPI data transfer. Most slave peripheral devices have signals called CE (chip enable) or CS (chip select) to enable/disable the SPI data transfer.

Multiple peripheral devices with an SPI interface can be interfaced with a single MCU simultaneously. There are many different methods for interfacing multiple peripheral devices (with SPI interface) to the MCU. Two popular connection methods are given here.

1. ***Parallel Connection.*** In this method, the MISO, MOSI, and SCK signals of all the peripheral devices are connected to the same signals of the 8051 MCU. The 8051 MCU also needs to use certain unused I/O pins to control the CS (or CE) inputs of each individual peripheral device. Using this method, the 8051 MCU can exchange data with any selected peripheral device without affecting other peripheral devices.
2. ***Serial Connection.*** In this method, the MOSI input of a peripheral device is connected to the MISO pin of its predecessor, and the MISO output of a peripheral device is connected to the MOSI input of its successor. The MOSI input of the peripheral device that is closest (in terms of connection) to the MCU is connected to the MOSI output of the MCU. The MISO output of the last peripheral device (in the loop) is connected to the MISO input of the MCU. The SCK inputs of all peripheral devices are tied to the SCK pin of the MCU. Using this method, the data to be sent to the last device in the loop will need to go through all other peripheral devices. The CE (or CS) signals of all peripheral devices are controlled by the same signal.

In all of the peripheral devices with the SPI interface, shift registers such as 74LV595 and 74LV589 can be used to add parallel output and input ports to the MCU. The widely used digital temperature sensors are quite useful for displaying the ambient temperature. The TC72 and TC77 from Microchip and LM74 from National Semiconductor are digital temperature sensors with the SPI interface. The SPI interface also can be found in many LED and LCD display drivers. The MAX7221 from Dallas-Maxim can drive eight seven-segment displays. The MC14489 from Freescale can drive five seven-segment displays. Both driver chips can be cascaded to drive more displays. Many A/D and D/A converters also have an SPI interface (for example, the 10-bit D/A converter LTC1661). Matrix displays have been used widely in recent years. They use the time-multiplexing method to display multiple digits. Due to their high demand on CPU time in performing time multiplexing, driver chips (such as MAX6952 and MAX6953) have been designed to offload the CPU from time-multiplexing matrix displays.

10.15 Exercise Problems

E10.1 Configure the SPI module of the C8051F040 (running with a 24-MHz crystal oscillator) to operate with these setting:

- Master mode with all interrupt enabled
- SCK idle high and data shifted in falling edge
- Three-wire master mode
- Baud rate set to 6 MHz

E10.2 Configure the SPI module of the C8051F040 (running with a 24-MHz crystal oscillator) to operate with the setting:

- Master mode with all interrupt disabled
- SCK idle high and data shifted on the rising edge
- Four-wire mode

- Baud rate set to 3 MHz
- f_{OSC} is 24 MHz

E10.3 Assume that there is an SPI-compatible peripheral output device that has the characteristics:

- A CLK input pin that is used as the data-shifting clock signal. This signal is set to idle low and data is shifted in on the rising edge.
- A SI pin to shift in data on the rising edge of the CLK input.
- A $\overline{CE}$ pin, which enables the chip to shift in data when it is low.
- The highest data-shifting rate is 1 MHz.
- Most significant bit shifted in first.

Describe how to connect the SPI pins for the C8051F040 and this peripheral device and write an instruction sequence to configure the SPI subsystem properly for data transfer. Assume that the C8051F040 is running with a 24-MHz crystal oscillator.

E10.4 The 74LV165A is another SPI-compatible shift register. This chip has both serial and parallel inputs and often is used to expand the number of parallel input ports. The block diagram of the 74LV165A is shown in Figure E10.4. The operation of this chip is illustrated in the function table of Table E10.4.

The 74LV165A can be cascaded. To cascade, the Q7 output is connected to the DS input of its adjacent 74LV165A. All 74LV165As should share the same shift clock SC. Suppose we want to use two 74LV165As to interface with two DIP switches so that four hex digits can be read in by the C8051F040. Describe the circuit connection and write an instruction sequence to read the data from these two DIP switches.

E10.5 Suppose you are going to use four 74LV595s to drive four seven-segment displays using the circuit shown in Figure 10.6. Write an instruction sequence to display the number 1982 on these four seven-segment displays.

E10.6 Use the circuit shown in Figure 10.6 to connect eight 74LV595Vs with each 74LV595 driving one seven-segment display. Write a program to display the numbers 1, 2, . . . , 8 on those eight seven-segment displays in Figure 10.6. Display one digit at a time and turn off the other seven digits. Each digit is displayed for 1 second. Perform this operation continuously.

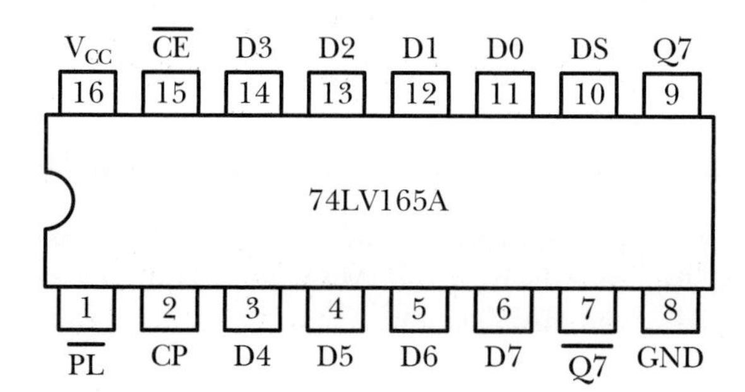

FIGURE E10.4 *The 74LV165A pin assignment.*

TABLE 10E.4 *Function Table of the 74LV165A*

Operating Mode	Inputs					Qn Flip-Flops		Outputs	
	PL	CE	CP	D_S	D0-D7	Q0	Q1-Q6	Q7	Q7
Parallel	L	X	X	X	L	L	L-L	L	H
load	L	X	X	X	H	H	H-H	H	L
Serial	H	L	↑	l	X	L	q0-q5	q6	$\overline{q6}$
shift	H	L	↑	h	X	H	q0-q5	q6	q6
Serial	H	↑	L	l	X	L	q0-q5	q6	$\overline{q6}$
shift	H	↑	L	h	X	H	q0-q5	q6	q6
Hold	H	H	X	X	X	q0	q1-q6	q7	q7
Hold	H	X	H	X	X	q0	q1-q6	q7	$\overline{q7}$

E10.7 Write a C program that use the LTC1661 in Figure 10.11 to generate a waveform, as shown in Figure E10.7.

E10.8 Write a C program to generate a sine waveform using the circuit shown in Figure 10.11. Divide one period of the sine wave into 120 points and use these 120 points to represent the waveform. Every two adjacent points are separated by 3°.

E10.9 Use the V_{OUT} pin of the LTC1661 shown in Figure 10.11 to drive a buzzer. Write a C program to generate a four-tone siren. The four frequencies of the siren are 200 Hz, 400 Hz, 800 Hz, and 1 kHz.

E10.10 The Maxim MAX7221 is a eight-digit, seven-segment display driver chip. The data sheet is on the CD for this text (also available on Maxim's website). Describe how to interface one MAX7221 with the C8051F040 using the SPI subsystem. Write a program to display the value 12345678 on the seven-segment displays driven by the MAX7221.

E10.11 The TC72 is a 10-bit resolution digital temperature sensor (in Celsius scale) with SPI interface. This sensor is available on the SSE040 demo board. Write a C program to measure temperature once every 200 ms and display the temperature on the LCD using three integral digits and one fractional digit.

E10.12 For the 25AA080A EEPROM shown in Figure 10.14, write a C function that writes data into a page of 16 bytes and a C function that reads a page of 16 bytes. The starting addresses of the EEPROM location to be written and the array of data to be written are passed to the page-write function. The starting addresses of the EEPROM location to be read and the data memory buffer to hold the read out data are passed to the page-read function.

E10.13 Write a test program to test the two functions written in Problem 10.12.

E10.14 Define a new font for each of the characters shown in Figure E10.14 and write a program to display these four characters on the four matrix displays driven by one MAX6952. The P3.7 signal drives the $\overline{CS}$ signal of the MAX6952.

E10.15 Digital potentiometers can be used to control gain of an amplifier, control the volume of a sound system, provide reference voltage to a D/A converter or an A/D converter, etc. The MCP41010I/P256 is a digital potentiometer with 256 tab positions and an SPI interface. The pin assignment of this chip is shown in Figure E10.15. Pins PA0 and PB0 are to external terminals of the potentiometer, whereas the PW0 is the wiper point position. Study the datasheet of the MCP41010 carefully. Describe how to interface this device to the C8051F040 and write a function to set the wiper-point voltage PW0 to any value between 0 V and V_{DD} (set V_{DD} to 3.3 V).

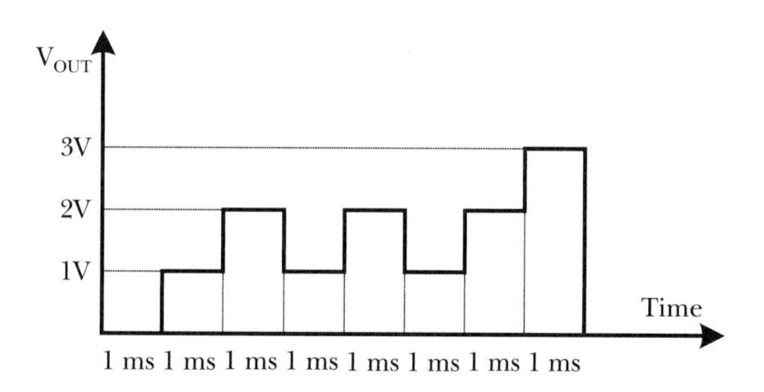

FIGURE E10.7 *Waveform to be generated.*

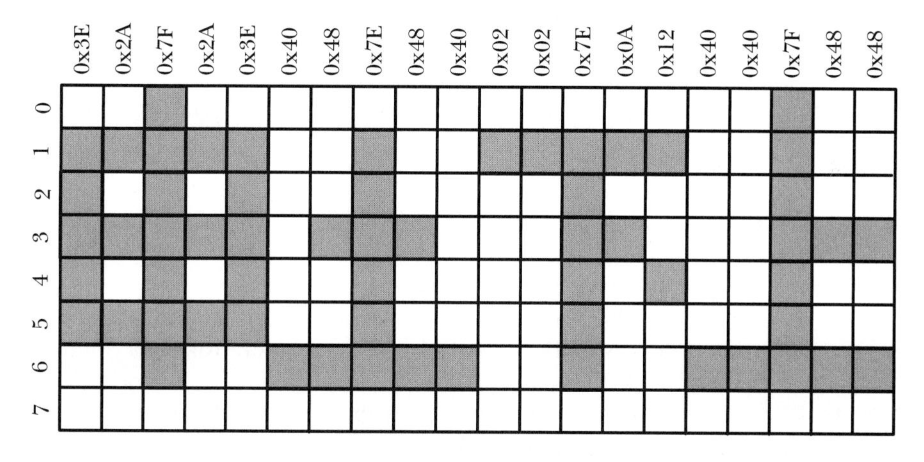

FIGURE E10.14 *User-definable font example.*

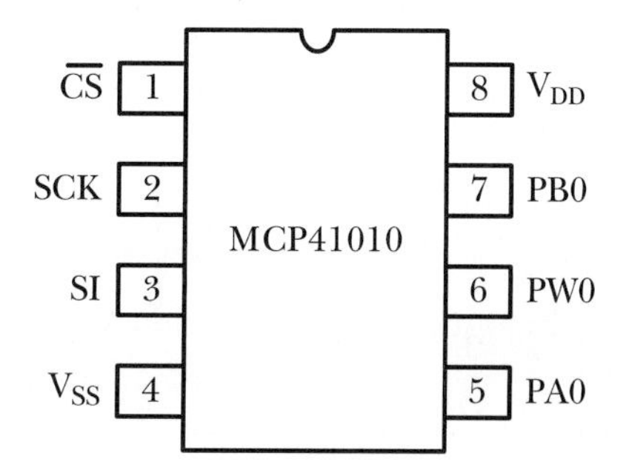

FIGURE E10.15 *Digital potentiometer MCP41010.*

10.16 Laboratory Exercise Problems and Assignments

L10.1 Use the Maxim seven-segment display driver MAX7221 (with SPI interface) to drive five seven-segment displays and write a program to display the following pattern forever.

```
0
10
210
3210
43210
54321
65432
76543
87654
98765
09876
10987
21098
32109
```

go to the 5th pattern (43210) and repeat.

L10.2 Use the TC72 to measure the ambient temperature and the LCD on the demo board to display the current temperature.

(a) Write a program to read the temperature once every 200 ms and display the new temperature reading on the LCD in two rows:

Temperature is
XXX.X°C

(b) Change the ambient temperature using your hand, hot water in a plastic bag, ice in a plastic bag, and so on to touch the TC72 and record the temperature reading.

L10.3 Use one MAX6952 and four matrix displays (e.g., common-cathode-row Fairchild GMC7175C) to perform the operation described in Exercise E10.14.

L10.4 Use a 10-bit resolution D/A converter LTC1661 from Linear Technology to perform the following operations:

(a) Generate a 3-tone siren with frequencies switching from 250 Hz to 500 Hz and then to 1000 Hz and repeat from the V_{OUTA} pin. You need a speaker (or a buzzer) for this (connect the speaker to V_{OUTA}).

(b) Generate 1-KHz and 2-KHz square waveforms simultaneously from V_{OUTA} and V_{OUTB}. Use two channels of the oscilloscope to display the waveform.

(c) Generate a triangular waveform from V_{OUTA} at the highest possible frequency.

I^2C Bus and SMBus

11.1 Objectives

Upon successful completion of this chapter, you will be able to:

- Describe the I^2C and SMBus protocols in general
- Explain the I^2C and SMBus signal components
- Configure the I^2C and SMBus operation parameters
- Use I^2C and SMBus to exchange data with peripheral devices
- Use I^2C and SMBus to interface with the digital thermostat DS1631A
- Use I^2C and SMBus to access the serial EEPROM 24LC08B
- Use I^2C and SMBus to interface with the real-time clock DS1337

11.2 Overview of Protocols

Inter-Integrated Circuit (I^2C) Bus and System Management Bus (SMBus) are two very similar serial buses that are widely implemented in 8051 variants to allow easy interfacing between microcontrollers and peripheral devices.

11.2.1 Overview of I^2C

The I^2C bus was developed by Philips in the 1980s with the objective of providing a simple way to communicate between integrated circuits by using a minimum number of pins. Originally, the I^2C bus was designed to link a small number of devices on a single card, such as to mange the tuning of a car radio or TV. The maximum allowable bus capacitance was set at 400 pF to allow proper rise and fall times for optimum clock and data signal integrity with a top speed of **100 kbps** (*standard speed mode*). In 1992 the standard bus speed was increased to **400 kbps** (*fast mode*), to keep up with the ever-increasing performance requirements of new ICs. The 1998 I^2C specification increased the top speed to **3.4 Mbps** (*high-speed mode*). All I^2C devices are designed to be able to communicate together on the same two-wire bus, and system functional architecture is limited only by the imagination of the designer.

Most applications set the I^2C bus length within the confines of consumer products (such as PCs, cellular phones, car radios, or TV sets). Only a few system integrators were using it to span a room or a building. The I^2C bus is now being used increasingly in multiple card systems (such as a blade servers) where the I^2C bus to each card needs to be isolatable to allow for card insertion and removal while the rest of the system is in operation or in systems where many more devices need to be located onto the same card where the total device and trace capacitance would have exceeded 400 pF.

New bus extensions and control devices help expand the I^2C bus beyond the 400-pF limit of about 20 devices and allow the control of more devices, even those with the same address. These new devices are popular with designers, as they continue to expand and increase the range of use of I^2C devices in maintenance and control applications.

I^2C Features

The I^2C bus consists of only two wires: a **serial data** line (SDA) and a **serial clock** line (SCL). Each device connected to the bus is software addressable by a unique address, and simple master/slave relationships exist at all times; masters can operate as master transmitters or as master receivers.

The I^2C bus allows multiple masters to coexist on the bus and includes collision detection and bus-use arbitration to prevent data corruption if two or more masters simultaneously initiate data transfer. Data transfer on the I^2C bus is bidirectional. The number of ICs that can be connected to the same bus segment is limited only by the maximum bus capacitive loading of 400 pF.

11.2.2 Overview of SMBus

The SMBus is a two-wire bus based on the principles of operations of I^2C and was designed to provide a control bus for system and power management-related tasks. The original goal of the SMBus is to define the communication link between:

- Intelligent battery
- Charger
- Microcontroller

The latest version of the SMBus is 2.0, which includes a low-power version and a normal-power version. The SMBus is compatible with the I^2C bus but has some minor differences in electrical characteristics, timing, and operating modes.

11.2.3 Differences between I^2C and SMBus Protocol

The differences between SMBus and I^2C bus are minor. A SMBus master will be able to control I^2C devices and vice versa. In the timing area, the SMBus clock is defined from 10 to 100 KHz, while I^2C can be a DC bus (0 to 100 kHz, 0 to 400 kHz, and 0 to 3.4 MHz). This means that an I^2C bus running at a frequency lower than 10 kHz will not be SMBus compliant, since the specification does not allow it.

Logic levels are slightly different in two buses. SMBus uses TTL levels, i.e., it recognizes a voltage level as 0 and 1 when its value is below 0.8 V and above 2.1 V, respectively. The I^2C bus uses CMOS levels, i.e., it uses 30 percent and 70 percent of V_{DD} as V_{IL} and V_{IH}, respectively. This causes no problem when $V_{DD} > 3.0$ V. If the V_{DD} of the I^2C device is below 3.0 V, then there is a problem, since the logic high/low levels may not be recognized. Table 11.1 summarizes the main differences among the DC parameters for I^2C and SMBus.

SMBus has a timeout feature, which will reset the devices if a communication takes too long. I^2C can be a "DC" bus, meaning that a slave device stretches the master clock when performing some routine while the master is accessing it. The SMBus protocol assumes that if something takes too long, then it means that there is a problem in the bus and that everybody must reset in order to clear this mode. Slave devices are not allowed to hold the clock low

TABLE 11.1 *DC Parameter Differences between SMBus and I2C*

		Std I^2C Mode Device		Fast I^2C Mode Device		SMBus Device		
Symbol	**Parameter**	**Min**	**Max**	**Min**	**Max**	**Min**	**Max**	**Unit**
V_{IL}	Fixed input level	−0.5	1.5	−0.5	1.5	—	0.8	V
	V_{DD} related input level	−0.5	0.3 V_{DD}	−0.5	0.3 V_{DD}	NA	NA	V
V_{IH}	Fixed input level	3.0	$V_{DD_{max}}$ +0.5	3.0	$V_{DD_{max}}$ +0.5	2.1	5.5	V
	V_{DD} related input level	0.7 V_{DD}	$V_{DD_{max}}$ +0.5	0.7 V_{DD}	$V_{DD_{max}}$ +0.5	NA	NA	V
V_{HYS}	$V_{IH} - V_{IL}$	NA	NA	0.05VDD	—	NA	NA	V
V_{OL}	V_{OL} @ 3 mA	0	0.4	0	0.4	NA	NA	V
	V_{OL} @ 6 mA	NA	NA	0	0.6	NA	NA	V
	V_{OL} @350μA	NA	NA	NA	NA	—	0.4	V
I_{PULLUP}		NA	NA	NA	NA	100	350	μA
I_{LEAK}		−10	10	−10	10	−5	5	μA

for too long. SMBus defines a clock low timeout interval to be 35 ms. I^2C does not specify any timeout limit. The SMbus timeout specifications do not preclude I^2C devices cooperating reliably on the SMbus. It is the responsibility of the designer to ensure that I^2C devices are not going to violate these bus timing parameters.

In I^2C bus, a slave does not need to acknowledge each time that its address matches the incoming address. However, the SMBus requires slaves to acknowledge their address all the time.

11.3 I^2C and SMB Protocols

A summary of the I^2C and SMBus protocols are given in this section.

11.3.1 Communication Procedure of I^2C and SMBus

An IC that wants to talk to another IC must follow the procedure prescribed by the following protocol.

Step 1
Wait until it sees no activity on the I^2C bus. When the SDA and SCL lines are both high, the I^2C bus is considered to be **free**.

Step 2
Assert the **start** signal to inform other devices that it wants to use the bus. All other devices must listen to the bus data to see whether they might be the one who will be called up (addressed).

Step 3
Provide a clock signal on the SCL line. It will be used by all the ICs as the reference time during which each bit of data on the SDA wire is correct. The data on the data wire (SDA) must be valid at the time the clock signal (SCL) switches from low to high.

Step 4
Send out (in serial form) the unique address of the IC that it wants to communicate with.

Step 5
Indicate on the bus whether it wants to send or receive data from the other chip. This action takes only one bit of time.

Step 6
Ask the other IC to acknowledge that it recognized its address and is ready to communicate.

Step 7
After the other IC acknowledges all is OK, data can be transferred.

Step 8
The first IC sends or receives as many 8-bit data bytes as it wants. After the transmission of each byte, the sending IC expects the receiving IC to acknowledge the transfer is going OK.

Step 9
When all the data is finished, the first chip must free up the bus. It does that by asserting a stop message.

11.3.2 Signal Levels

The I^2C and SMBus can have only two possible electrical states: *float high* and *driven low*. A master or a slave device drives the bus using an open-drain (or open-collector) driver. As shown in Figure 11.1, both the SDA and SCL lines are pulled up to V_{DD} via pull-up resistors. Because the driver circuit is open-drain, it can pull only the I^2C bus (SMBus) to low. When the clock or data output is low, the NMOS transistor is turned off. Therefore, no current flows from the bus through the *N* transistor and the bus will be pulled up to high by the pull-up resistor. Otherwise, the *N*-transistor is turned on and pulls the bus line to low.

The SMBus allows the V_{DD} to be in the range from 3.0 to 5.0 V, whereas the I^2C bus allows the ICs to operate with V_{DD} from 2.3 V up to 5.5 V. Both bus protocols allow ICs operating with different V_{DD} to be connected to the same bus.

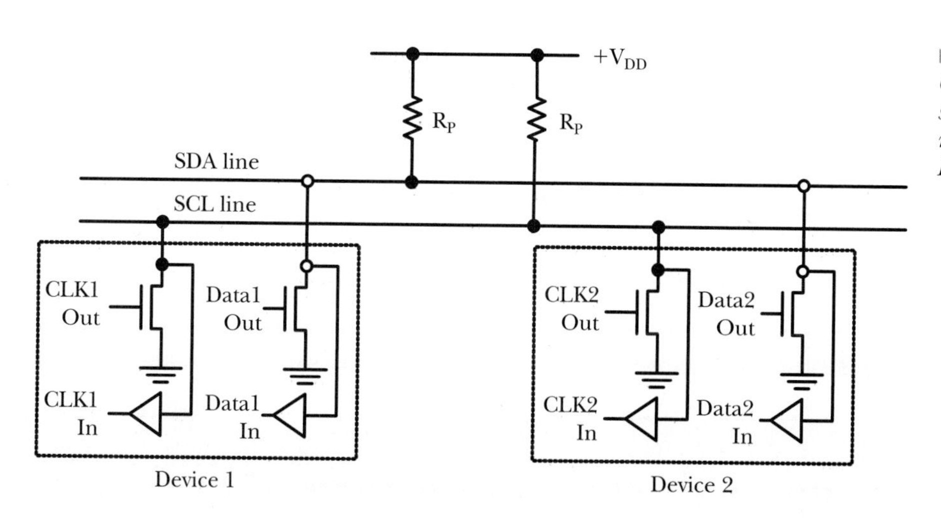

FIGURE 11.1 *Connecting standard- and fast-mode devices to the I^2C bus (SMBus).*

The designer is free to use appropriate resistor values for various speeds. But the calculation of what value to use will depend on the capacitance of the driven line, consideration of power-consumption, and the speed of the I²C (or SMBus) communication. In general, the recommended values for the pull-up resistors are 2.2 kΩ and 1 kΩ for standard mode and fast mode, respectively. When the data rate is equal to or below 100 kbps, the pull-up resistor should be set to 4.7 kΩ.

11.3.3 I²C AND SMBUS DATA-TRANSFER SIGNAL COMPONENTS

An I²C or SMBus data transfer consists of these fundamental signal components:

- Start (S)
- Stop (P)
- Repeated start (R)
- Data
- Acknowledge (A)

Start (S) Condition

A start condition indicates that a device would like to transfer data on the I²C bus (or SMBus). As shown in Figure 11.2, a start condition is represented by the SDA line going low when the clock (SCL) signal is high. The start condition will initialize the bus. The timing details will be taken care of by the MCU that implements the I²C bus (or SMBus). Whenever a data transfer using the I²C bus (or SMBus) is to be initiated, the application program instructs the MCU to trigger a start condition.

Stop Condition (P)

A stop condition indicates that a device wants to release the I²C bus (or SMBus). Once it is released (the driver is turned off), other devices may use the bus to transmit data. As shown in Figure 11.3, a stop condition is represented

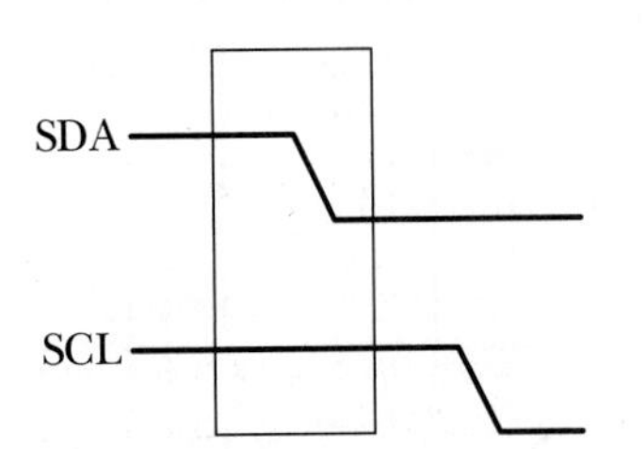

FIGURE 11.2
I²C bus (SMBus) start condition.

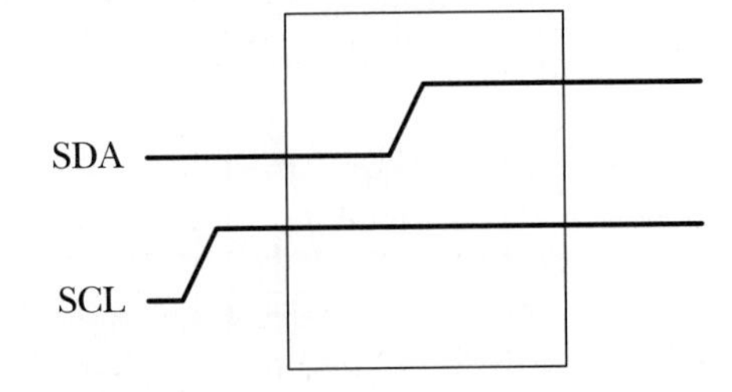

FIGURE 11.3
Stop (P) condition.

by the SDA signal going high when the clock (SCL) signal is high. Once the stop condition completes, both the SCL and the SDA signals will be high and the bus becomes idle. A new start condition can be used to send more data. Only a master device can assert the stop condition.

Repeated Start (R) Condition

A repeated start signal is a start signal generated without first generating a stop signal to terminate the current communication. This is used by the master to communicate with another slave or with the same slave in different mode (transmit/receive mode) without releasing the bus. A repeated start condition indicates that a device would like to send more data instead of releasing the bus lines. This is done when a start signal must be sent but a stop condition has not been generated yet. It prevents other device from grabbing the bus between transfers. The timing diagram of a repeated start condition is shown in Figure 11.4. The repeated start condition also is called the **restart** condition. In the figure, there is no stop condition occurring between the start condition and the restart condition.

Data

The data block represents the transfer of 8 bits of information with most significant bit transferred first. The data is sent on the SDA line, whereas clock pulses are carried on the SCL line. The clock can be aligned with the data to indicate whether each bit is a 1 or a 0.

Data on the SDA line is considered valid only when the SCL signal is high. When the SCL is not high, the data is permitted to change. This is how the timing of each bit works. Data bytes are used to transfer all kinds of information. When communicating with another I^{2}C device, the eight bits of data may be a control byte, an address, or data. An example of eight bits of data is shown in Figure 11.5.

Acknowledge (ACK) Condition

Data transfer in the I^{2}C (or SMBus) protocol needs to be acknowledged either positively (A) or negatively (NACK). As shown in Figure 11.6, a device can acknowledge (A) the transfer of each byte by bringing the SDA line low during the ninth clock pulse of the SCL.

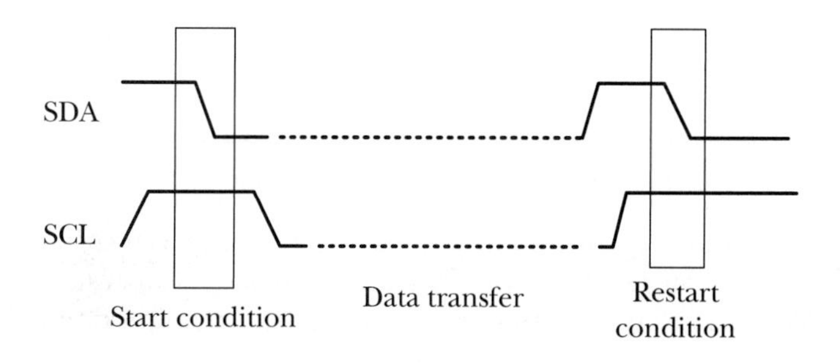

FIGURE 11.4
Restart condition.

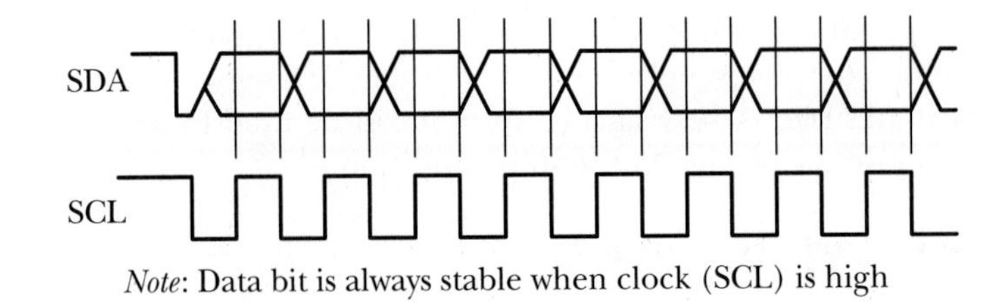

FIGURE 11.5
I^2C bus (or SMBus) data elements.

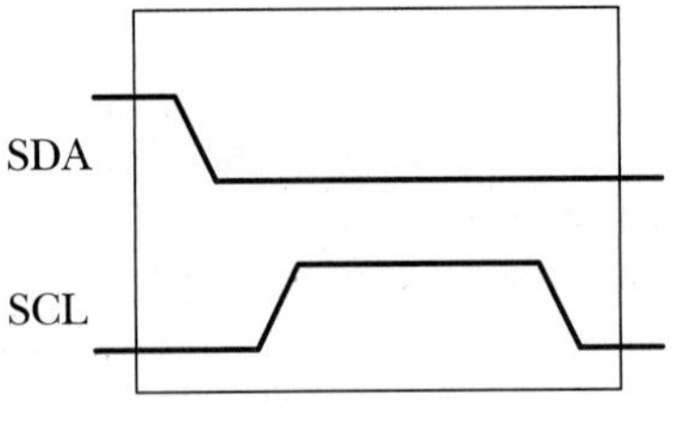

FIGURE 11.6
ACK condition.

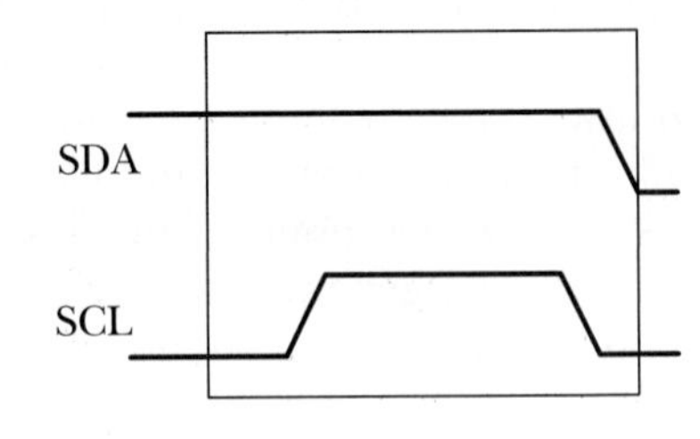

FIGURE 11.7
NACK condition.

11.3.4 Bus Arbitration

Both the I^2C and SMBus allow multiple master devices to coexist in the system. In the event that two or more master devices attempt to begin a transfer at the same time, an arbitration scheme is employed to force one or more masters to give up the bus. The master devices continue transmitting until one attempts a high while the other transmits a low. Since the bus driver has open-drain, the bus will be pulled low. The master attempting to transfer a high signal will detect a low on the SDA line and give up the bus by switching off its data-output stage. The winning master continues its transmission without interruption. This arbitration scheme is nondestructive: One device always wins, and no data is lost.

An example of the arbitration procedure is shown in Figure 11.8, where Data1 and Data2 are data driven by device 1 and 2 and the SDA is the resultant data on the SDA line. The moment there is a difference between the internal-data level of the master generating Data1 and the actual level on the SDA line, its output is switched off, which means that a high-output level is then connected to the bus. This will not affect the data transfer initiated by the winning master.

11.3.5 Synchronization

Whenever there are multiple I^2C or SMBus masters with start data transfer on the bus and before the arbitration process is completed, all masters generate their own clocks on the SCL line to transfer messages on the I^2C bus (or SMBus). Data is valid during the high period of the clock. A defined clock is

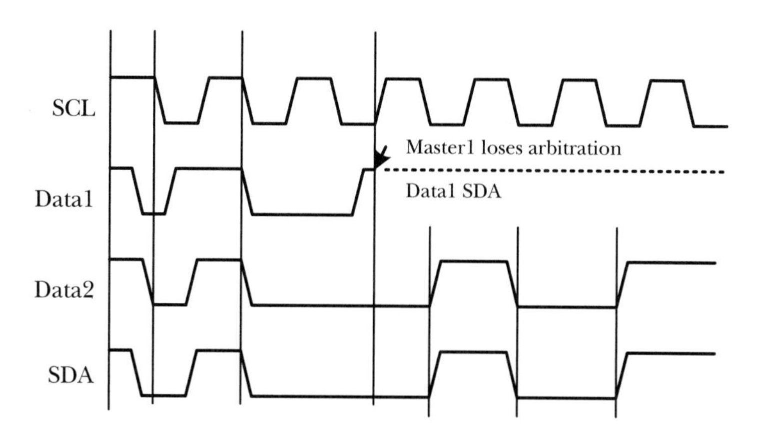

FIGURE 11.8 *Arbitration procedure of two masters.*

therefore needed for the bit-by-bit arbitration procedure to take place. For most microcontrollers, the SCL clock is generated by counting down a programmable reload value using the instruction clock signal.

Clock synchronization is performed using the wired-AND connection of I^{2}C interfaces to the SCL line. This means that a high-to-low transition on the SCL line will cause the devices concerned to start counting off their low period, and once a device clock has gone low, it will hold the SCL line in that state until the high state is reached (CLK1 in Figure 11.9). However, the transition from low to high of this clock may not change the state of the SCL line if another clock (CLK2) is still within its low period. The SCL line therefore will be held low by the device with the longest low period. Devices with shorter low periods enter a **high wait** state during this time.

When all devices concerned have counted off their low period, the clock line will be released and go high. There then will be no difference between the device clocks and the state of the SCL line, and all the devices will start counting their high periods. In this way, a synchronized SCL clock is generated with its low period determined by the device with the longest clock low period and its high period determined by the one with the shortest clock high period.

Handshaking

Clock synchronization mechanism can be used as a handshake in data transfer. Slave devices may hold the SCL low after completion of a 1-byte transfer. In such a case, it halts the bus clock and forces the master clock into wait states until the slave releases the SCL line. The master device also can hold the SCL low to indicate that it needs more time to handle a byte.

Clock Stretching

The clock synchronization mechanism can be used by slaves to slow down the bit rate of a transfer. After the master has driven the SCL low, the slave can drive the SCL low for the required period and then release it. If the slave SCL

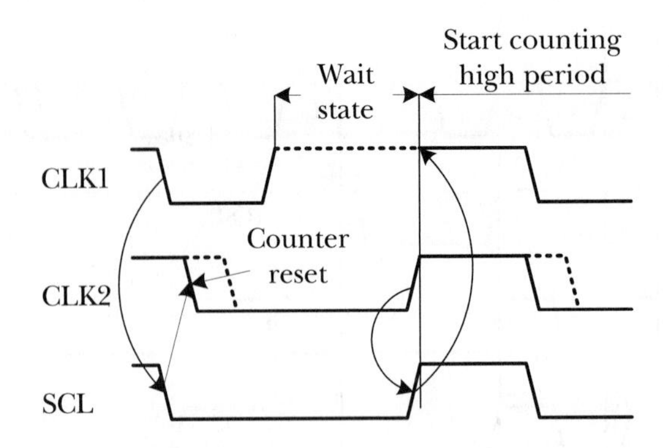

FIGURE 11.9 *Clock synchronization during the arbitration procedure.*

low period is greater than the master SCL low period, the resulting SCL bus signal low period is stretched.

11.3.6 Data Transfer Format

Data transfer format is related directly to the slave addressing method. The I^2C protocol supports both the 7-bit and the 10-bit addressing, whereas the SMBus protocol supports only the 7-bit addressing. Since most peripheral chips used with the 8-bit microcontrollers use only 7-bit address, we will only discuss 7-bit addressing in this book.

The first byte sent to the slave after the start condition is the 7-bit slave address and the R/ W bit. The format of this byte is shown in Figure 11.10. When the least significant bit is 1, the master device will read information from the selected slave. Otherwise, the master device will write information to a selected slave. When an address is sent, each device in the system compares the first seven bits after the start condition with its address. If they match, the device considers itself selected by the master as a slave receiver or transmitter, depending on the R/ W bit. The address of **seven 0s** is the **general call address** that is used to address every device connected to the I^2C bus.

The possible I^2C and SMBus data transfer formats are as follows.

- **Master transmitter to slave receiver.** The transfer direction is not changed. An example of this format using 7-bit addressing is shown in Figure 11.11.
- **Master reads slave immediately after the address byte.** At the moment of the first acknowledgement, the master transmitter becomes a master receiver, and the slave receiver becomes a slave transmitter. The first acknowledgement is still generated by the slave. The stop condition is generated by the master, which has previously sent a negative acknowledgement (A). An example of this format using the 7-bit addressing is shown in Figure 11.12.
- **Combined format.** During a change of direction within a transfer, both the start condition and the slave address are repeated, but with the R/ W

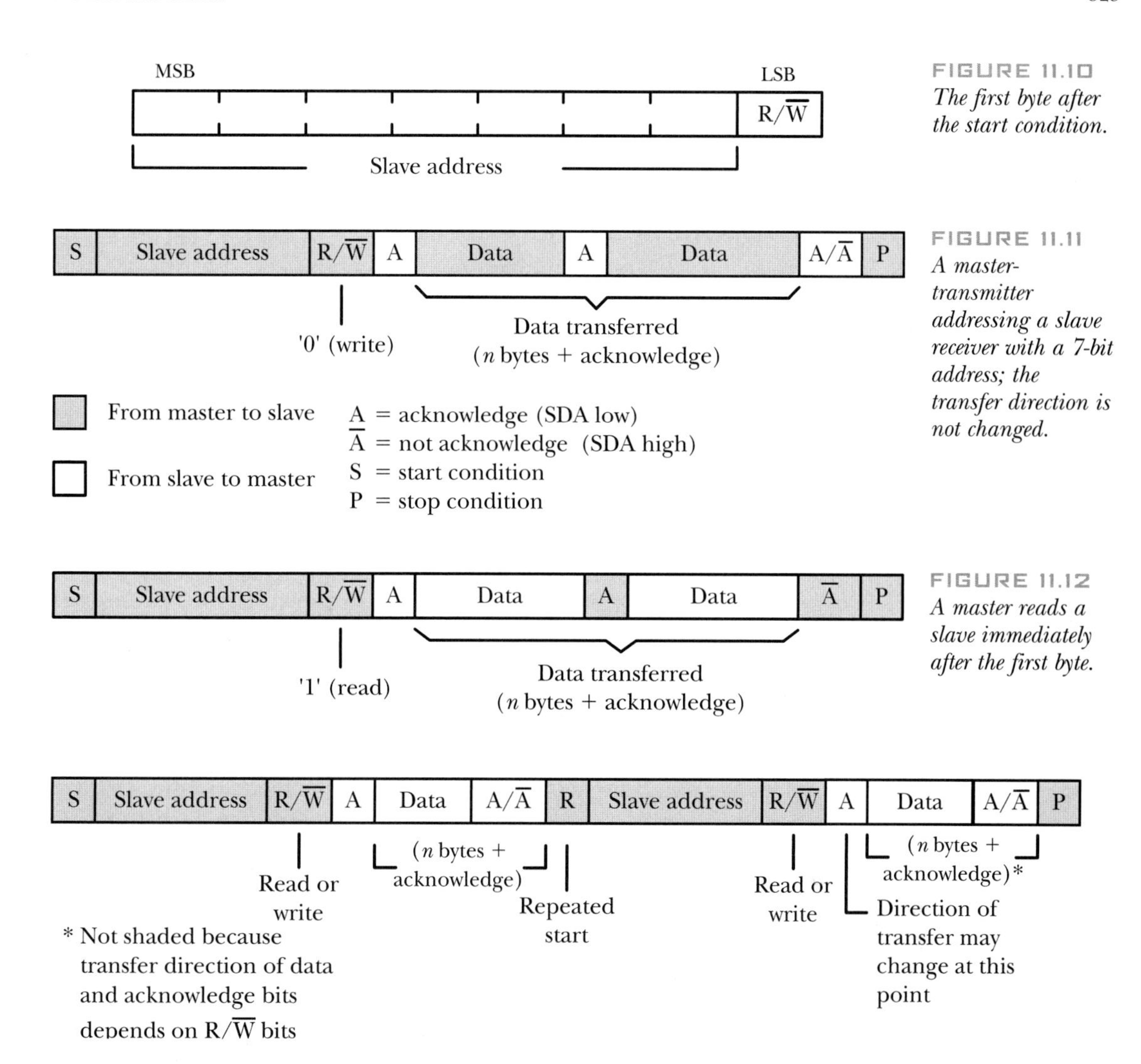

FIGURE 11.10
The first byte after the start condition.

FIGURE 11.11
A master-transmitter addressing a slave receiver with a 7-bit address; the transfer direction is not changed.

FIGURE 11.12
A master reads a slave immediately after the first byte.

FIGURE 11.13
Combined format.

bit reversed. If a master receiver sends a repeated start condition, it previously has sent a negative acknowledgement. An example of this format in the 7-bit addressing is shown in Figure 11.13. There are many possibilities for this format:

(a) The master reads data from slave 1 and then reads data from slave 2.
(b) The master writes data to slave 1 and then writes data to slave 2.
(c) The master reads data from a slave and then writes data to the same slave.
(d) The master writes data to a slave and then reads data from the same slave.

(e) The master reads data from slave 1 and then writes data to slave 2.
(f) The master writes data to slave 1 and then reads data from slave 2.

11.4 The SMBus of the C8051F040

The C8051F040 uses two pins (SDA and SCL) to support the data transfer of the SMBus. Since the routing priority of these two pins is very high, they always will be assigned to Port 0 pins (see Section 3.4.1). The SMBus serial interface is accessed and controlled through five registers: control register SMB0CN, clock rate register SMB0CR, address register SMB0ADR, data register SMB0DAT, and status register SMB0STA.

11.4.1 The SMB0CN Register

The contents of the SMB0CN register are shown in Figure 11.14. This register is used to configure and control the SMBus interface. All bits of this register except for bit 7 are readable/writable.

Setting the ENSMB bit to 1 enables the SMBus interface. Clearing the ENSMB bit disables the SMBus interface.

Setting the STA bit to 1 will place the SMBus in a master mode. If the bus is free, the SMBus will generate a start condition. If the bus is not free, SMBus waits for a stop condition to free the bus and then generates a start condition after a 5-μs delay per the SMB0CR value. If STA is set to logic 1 while SMBus is in master mode and one or more bytes have been transferred, a repeated start condition will be generated. After the start condition has been generated, the STA bit will be cleared to 0 automatically.

When the STO bit is set to 1 while the SMBus interface is in master mode, the interface generates a stop condition. In a slave mode, the STO flag may be used to recover from an error condition. In this case, a stop condition is not generated on the bus, but the SMBus hardware behaves as if a stop condition has been received and enters the **not addressed** slave receiver mode. Hardware automatically clears the STO bit when a stop condition is detected on the bus.

The **serial interrupt flag** (SI) is set to 1 by hardware when the SMBus interface enters any one of the 28 possible states, except the idle state. An interrupt may be generated if it is enabled. If the SI bit is set to 1 while the SCL line is low, the clock-low period of the serial clock will be stretched and the serial transfer is suspended until the SI bit is cleared to 0. The **SI** flag must be cleared by software. A high level on the SCL line is not affected by the setting of the **SI** flag.

The AA bit is used to set the level of the SDA line during the acknowledge clock cycle on the SCL line. Setting the AA bit to 1 will cause an ACK to be sent during the acknowledge cycle if the device is a receiver or has been addressed.

7	6	5	4	3	2	1	0	
BUSY	ENSMB	STA	STO	SI	AA	FTE	TOE	Reset value 00000000
R	R/W	R/W	R/W	R/W	R/W	R/W	R/W	

FIGURE 11.14 *The SMB0CN register of the C8051F040.*

BUSY: Busy status flag
 0 = SMBus is free
 1 = SMBus is busy
ENSMB: SMBus enable
 0 = SMBus disabled
 1 = SMBus enabled
STA: SMBus start flag
 0 = No START condition transmitted
 1 = Enable a master to send out a START condition when the bus is free. If STA is set after one or more multiple bytes have been transmitted or received and before a STOP is received, a repeated START condition is transmitted
STO: SMBus stop flag
 0 = No stop condition is transmitted
 1 = Cause a STOP condition to be transmitted. When a STOP condition is received, hardware clears this bit. If both STA and STO are set, a STOP condition is transmitted followed by a START condition. In slave mode, setting STO bit causes SMBus to behave as if a STOP condition was received
SI: SMBus interrupt flag
 0 = no interrupt condition occurred
 1 = One of the 27 possible SMBus states is entered
AA: SMBus assert acknowledge flag
 0 = A NACK (high on SDA line) is returned during the acknowledge cycle
 1 = An ACK (low on SDA line) is returned during the acknowledge cycle
FTE: SMBus free timer enabled
 0 = No timeout when SCL is high
 1 = Timeout when SCL high time exceeds limit specified by the SMBCR value
TOE: SMBus timeout enable bit
 0 = No timeout when SCL is low
 1 = Timeout when SCL low time exceeds limit specified by Timer 4, if enabled

Setting the AA bit to 0 will cause a NACK to be sent during the acknowledge cycle. After the transmission of a byte in slave mode, the slave can be removed temporarily from the bus by clearing the AA bit. The slave's own address and general call address will be ignored. To resume operation on the bus, the AA bit must be reset to 1 to allow the slave's address to be recognized.

Setting the FTE bit to 1 enables the timer in SMB0CR. When the SCL line goes high, the timer in SMB0CR counts up. A timer overflow indicates a free bus timeout: If SMBus is waiting to generate a start condition, it will do so after this timeout. The bus free period should be less than 50 μs.

When the TOE bit is set to 1, Timer 4 is used to detect SCL low timeout. If Timer 4 is enabled, Timer 4 is forced to reload when SCL is high and forced to count when SCL is low. With Timer 4 enabled and configured to overflow after 25 ms (and TOE set), a Timer 4 overflow indicates a SCL low timeout; the Timer 4 interrupt service routine then can be used to reset SMBus communication in the event of an SCL low timeout.

11.4.2 The SMB0CR register

This register sets the frequency of the serial clock SCL in master mode. The 8-bit value stored in the SMB0CR register is used as the preload value of a dedicated 8-bit timer. The timer counts up, and when it rolls over to 0x00, the SCL logic state toggles.

The value written into the SMB0CR register is bounded by the equation

$$\mathbf{SMB0CR < ((288 - 0.85 \times SYSCLK) \div 1.125)} \qquad (11.1)$$

where SYSCLK is the system clock frequency in MHz. For example, for a 24-MHz system clock, SYSCLK is 24.

The resultant high interval and low interval of the SCL signal are given by the equations:

$$T_{LOW} = (256 - SMB0CR) \div f_{SYS} \qquad (11.2)$$

$$T_{HIGH} \approx (258 - SMB0CR) \div f_{SYS} + 625\text{ ns} \qquad (11.3)$$

Using the previous value of SMB0CR, the bus-free timeout period is given in the equation:

$$T_{BFT} \approx 10 \times (256 - SMB0CR + 1) / f_{SYS} \qquad (11.4)$$

Let f_{SMBus} be the frequency of the SMBus, then f_{SMBus} and the corresponding value to be written into the SMB0CR register can be computed by using Equations 11.5 and 11.6, respectively.

$$f_{SMBus} = \frac{1}{T_{HIGH} + T_{LOW}} = \frac{1}{\frac{1}{f_{SYS}} \times [514 - 2 \times SMB0CR] + 625\text{ ns}} \qquad (11.5)$$

$$SMB0CR = \frac{1}{2} \times \left(514 - \frac{f_{SYS}}{f_{SMBus}} + 625\text{ ns} \times f_{SYS}\right) \qquad (11.6)$$

Example 11.1 Give a value to configure the SMBus to transfer data at the highest baud rate of 100 KHz, assuming that f_{SYS} = 24 MHz.

Solution: The value to be written into the SMB0CR register can be computed by using Equation 11.6 as

$$SMB0CR = (514 - (24 \times 10^6)/10^5 + 625 \times 10^{-9} \times 24 \times 10^6)/2$$
$$= 144.5 \approx 145$$

Obviously, this value satisfies the requirement set out by Equation 11.1.

11.4.3 SMBus Data Register (SMB0DAT)

The SMB0DAT register holds a byte of serial data to be transmitted or one that has just been received. Software can read or write to this register while the **SI** flag is set to 1. Software should not attempt to access the SMB0DAT register when the SMBus is enabled and the **SI** flag is 0, because hardware may be in the process of shifting a byte of data in or out of the register. Data in this register always is shifted out most-significant-bit first.

11.4.4 Address Register

This register holds the slave address for the SMBus interface, as shown in Figure 11.15. In slave mode, the seven most significant bits hold the 7-bit slave address. The least significant bit is used to enable the recognition of the general call address (0x00). If bit 0 is set to logic 1, the general call address will be recognized. Otherwise, the general call address is ignored. The contents of this register are ignored when SMBus is operating in master mode.

11.4.5 Status Register

The SMB0STA status register holds an 8-bit status code indicating the current state of the SMBus interface. There are 28 possible SMBus states, each with a corresponding unique status code. The upper five bits of the status codes vary, while the lowest three bits of a valid status code are fixed at zero when the SI bit is 0. Therefore, all possible status codes are multiples of eight.

For the purpose of application software, the contents of the SMB0STA register is only defined when the **SI** flag is logic 1. Software should never write to the SMB0STA register; doing so will yield indeterminate results. The 28 SMBus states along with their corresponding status codes are given in Table 11.2.

7	6	5	4	3	2	1	0	
SLV7	SLV6	SLV5	SLV4	SLV3	SLV2	SLV1	GC	Reset value 00000000
R/W	R/W	R/W	R/W	R/W	R/W	R/W	R/W	

SLV7-SLV0: SMBus slave address
GC: General call address enable
0 = general call address ignored
1 = general call address is recognized

FIGURE 11.15 *SMBus address.*

TABLE 11.2 *SMB0STA Status Codes and States*

Mode	Status Code	SMBus State	Typical Action
MT/MR	0x08	START condition transmitted.	Load SMB0DAT with slave address + R/W, clear STA.
	0x10	Repeated START condition transmitted.	Load SMB0DAT with slave address + R/W, clear STA.
Master Transmitter	0x18	Slave address + W transmitted. ACK received.	Load SMB0DAT with data to be transmitted.
	0x20	Slave address + W transmitted. NACK received.	Acknowledge poll to retry, set STO + STA.
	0x28	Data byte transmitted. ACK received.	1) Load SMB0DAT with next byte, OR 2) Set STO, OR 3) Clear STO then set STA for repeated START.
	0x30	Data byte transmitted. NACK received.	1) Retry transfer OR 2) Set STO.
	0x38	Arbitration lost.	Save current data.
Master Receiver	0x40	Slave address + R transmitted. ACK received.	If only receiving one byte, clear AA, wait for received data.
	0x48	Slave address + R transmitted. NACK received.	Acknowledge poll to retry, set STO + STA.
	0x50	Data byte received. ACK transmitted.	Read SMB0DAT. Wait for next byte, if next byte is last byte, clear AA.
	0x58	Data byte received. NACK transmitted.	Set STO.
Slave Receiver	0x60	Own slave address + W received. ACK sent.	Wait for data.
	0x68	Arbitration lost in sending SLA + R/W as master. Own address + W received. ACK transmitted.	Save current data for retry when bus is free, wait for data.
	0x70	General call address received. ACK sent.	Wait for data.
	0x78	Arbitration lost in sending SLA + R/W as master. General call address received. ACK transmitted.	Save current data for retry when bus is free.
	0x80	Data byte received. ACK transmitted.	Read SMB0DAT. Wait for next byte or STOP.
	0x88	Data byte received. NACK transmitted.	Set STO to reset SMBus.
	0x90	Data byte received after general call address. ACK transmitted.	Read SMB0DAT. Wait for next byte or STOP.
	0x98	Data byte received after general call address. NACK transmitted.	Set STO to reset SMBus.
	0xA0	STOP or repeated START received.	No action necessary.

TABLE 11.2 *SMB0STA Status Codes and States (continued)*

Mode	Status Code	SMBus State	Typical Action
Slave Transmitter	0xA8	Own address + R transmitted. ACK transmitted.	Load SMB0DAT with data to transmit.
	0xB0	Arbitration lost in transmitting SLA+R/W as master. Own address + R received. ACK sent.	Save current data for retry when bus is free. Load SMB0DAT with data to transmit.
	0xB8	Data byte transmitted. ACK received.	Load SMB0DAT with data to transmit.
	0xC0	Data byte received. NACK transmitted.	Wait for STOP.
	0xC8	Last data byte transmitted (AA=0). ACK received.	Set STO to reset SMBus.
Slave	0xD0	SCL clock high timer per SMB0CR timed out.	Set STO to reset SMBus.
All	0x00	Bus error (illegal START or STOP).	Set STO to reset SMBus.
	0xF8	Idle.	State does not set SI bit.

11.5 Using the C8051F040 SMBus

The SMBus can operate in both the master and slave modes. The hardware provides timing and shifting control for the serial transfers; byte-wise control is user-defined. The SMBus hardware performs the following application-independent tasks.

- ***Timing Control:*** In master mode, the hardware generates the clock signal on the SCL pin and synchronizes the data on the SDA pin. Hardware also recognizes time-outs and bus errors.
- ***Serial Data Transfers:*** The hardware controls all shifting of data to and from the SDA bus, including the acknowledgement level. The acknowledgement level is user-defined.
- ***Slave Address Recognition:*** The hardware recognizes a start signal from another device and reads the following slave address. If the slave address matches the contents of the SMBus address register, then the hardware acknowledges the address. This feature is enabled when the AA bit of the SMB0CN register is set to 1.

11.5.1 Implementation Choices

User software controls the SMBus on a state-by-state basis. Upon each state change, the SI bit is set by hardware, and an interrupt is generated if it is enabled. The SMBus then is halted until user software services the state change and clears the SI bit. The SMBus operation is defined most easily in a state table; however, it is not necessary to define all 28 states. For example, if the SMBus is the only master in the system, the slave and arbitration states may be left undefined. If the SMBus will never operate as a master, the master states may be left undefined. If states are left undefined, a default response should be programmed to account for unexpected or error situations.

The SMBus state table lends itself to a **case-switch** statement definition in C language. However, for simple or time-restricted systems, an assembly state-decoding can be more efficient. Note that the status codes held in SMB0STA are multiples of 8. If the SMBus states are programmed in 8-byte segments, SMB0STA may be used as a software index. In this case, a status code is decoded in three assembly commands. However, only 8 bytes of code space are available for each state definition. For states that require more than 8 bytes, the program must branch out of the state table so that subsequent states are not disturbed.

The easiest way to implement the state approach is to start the data transfer by generating the start condition and then let interrupts take care of the remaining transactions. The next section discusses the features of an EEPROM chip with I^2C interface and illustrates how it is interfaced with the SMBus of the C8051F040.

11.5.2 Interfacing the Serial EEPROM 24LC08 with SMBus

Some applications require the use of a large amount of nonvolatile memory, because these applications are powered by batteries and may be used in the field for an extended period of time. Many semiconductor manufacturers produce serial EEPROMs with a serial interface. Both serial EEPROMs with the SPI and I^2C interfaces are available.

The 24LC08B from Microchip is a serial EEPROM with the I^2C interface. This device is an 8-Kbit EEPROM organized as four blocks of 256×8-bit memory. A low-voltage design permits operation down to 2.5 V with standby and active currents of only 1 μA and 1 mA, respectively. The 24LC08B also has a page-write capability for up to 16 bytes of data.

Pin Assignment and Block Diagram

The pin assignment and the block diagram of the 24LC08B are shown in Figures 11.16 and 11.17, respectively.

The pins SCL and SDA are for I^2C bus or SMBus communications. The SCL line is a clock input that can be as high as 400 kHz. The WP pin is used as the write-protection input. When this pin is high, the 24LC08B cannot be written into. Pins 2, . . . ,0 are not used and can be left floating, grounded, or pulled to high.

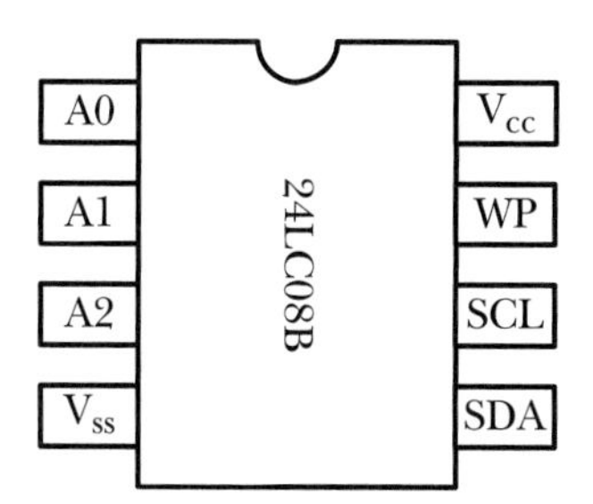

FIGURE 11.16 *24LC08B PDIP package pin assignment.*

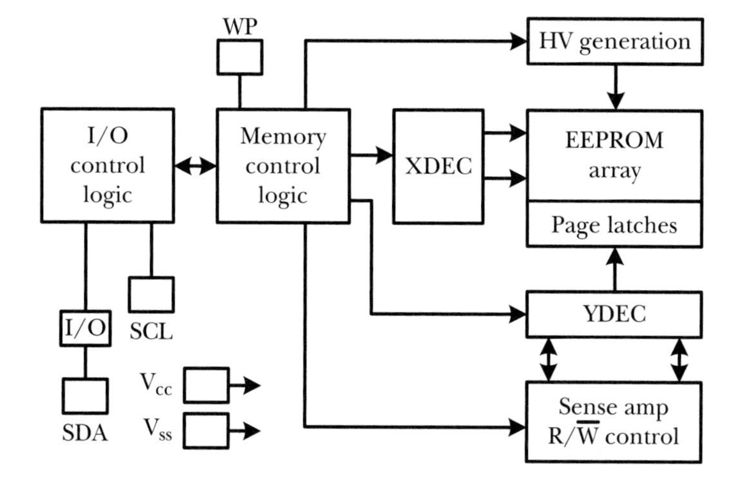

FIGURE 11.17 *Block diagram of 24LC08B.*

Device Addressing

Like any other I²C slave, the first byte sent to the 24LC08B after the start condition is the control byte. The contents of the control byte for the 24LC08B are shown in Figure 11.18. The upper 4 bits are the device address of the 24LC08B, whereas B1B0 are the block addresses of the memory location to be accessed. For any access to the 24LC08B, the master must also send an 8-bit byte address after the control byte. There is an address pointer inside the 24LC08B. After the access of each byte, the address pointer is incremented by 1.

Write Operation

The 24LC08B supports byte-write and page-write operations. In a byte-write operation, the following events will occur.

- The master asserts the START condition
- The master sends the control byte to the 24LC08B
- The 24LC08B acknowledges the data transmission
- The master sends the byte address to the 24LC08B

FIGURE 11.18
24LC08B Control byte contents.

7	6	5	4	3	2	1	0
1	0	1	0	X	B1	B0	$R/\overline{W}$

- The 24LC08B acknowledges the data transmission
- The master (MCU) sends the data byte to the 24LC08B
- The 24LC08B acknowledges the data transmission
- The master asserts the STOP condition

In a page-write operation, the master can send up to 16 bytes of data to the 24LC08B. The write-control byte, byte address, and the first data byte are transmitted to the 24LC08B in the same way as in a byte write. But instead of asserting a stop condition, the master transmits up to 16 data bytes to the 24LC08B that are temporarily stored in the on-chip page buffer and will be written into the memory after the master has asserted a stop condition. After the receipt of each byte, the four lower address-pointer bits are incremented internally by 1. The higher-order 6 bits of the byte address remain constant. If the master should transmit more than 16 bytes prior to generating the stop condition, the address counter will roll over, and the previously received data will be overwritten.

Acknowledge Polling

When the 24LC08B is writing the data held in the write buffer into the EEPROM array, it will not acknowledge any further write operation. This fact can be used to determine when the cycle is complete.

Once the stop condition for a write command has been issued from the master, the device initiates the internal write cycle. The ACK polling can be initiated immediately. This involves the master sending a start condition followed by the control byte for a write command ($R/\overline{W} = 0$). If the 24LC08B is still busy, then no ACK will be returned. If the cycle is complete, then the device will return the ACK, and the master then can proceed with the next read or write command. The polling process is illustrated in Figure 11.19.

Read Operation

The 24LC08B supports three types of read operations: *current-address read*, *random read*, and *sequential read.*

Current-Address Read

As was explained earlier, the internal-address counter is incremented by 1 after each access (read or write). The current-address read allows the master to read the byte immediately following the location accessed by the previous read or write operation. On receipt of the slave address with the $R/\overline{W}$ bit set to 1, the 24LC08B issues an acknowledgement and sends out a data byte. The

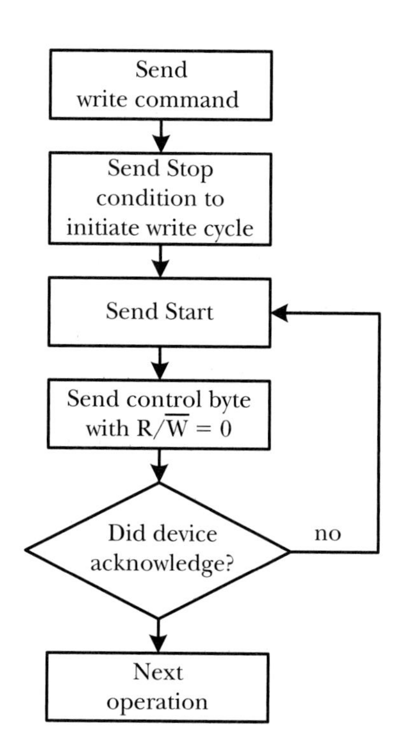

FIGURE 11.19
Acknowledge polling flow.

master will not acknowledge the transfer, asserts a stop condition, and the 24LC08B discontinues transmission.

Random Read

Random read operations allow the master to access any memory location in a random manner. To perform this type of read operation, the user must send the address of the memory location to be read. The procedure for performing a random read is as follows.

Step 1
The master asserts a start condition.

Step 2
The master sends the control byte to the 24LC08B.

Step 3
The 24LC08B acknowledges the control byte.

Step 4
The master sends the address of the byte to be read to the 24LC08B.

Step 5
The 24LC08B acknowledges the byte address.

Step 6
The master asserts a repeated start condition.

Step 7
The master sends the control byte with $R/\overline{W} = 1$.

Step 8
The 24LC08B acknowledges the control byte and sends the data to the master.

Step 9
The master asserts NACK to the 24LC08B.

Step 10
The master asserts the stop condition.

Sequential Read

If the master acknowledges the data byte returned by the random read operation, the 24LC08B will transmit the next sequentially addressed byte as long as the master provides the clock signal on the SCL line. The master can read the whole chip using sequential read. Using sequential read, the MCU can read multiple bytes by sending a single address to the 24LC08B.

Example 11.2 Describe the circuit connection between the C8051F040 and the 24LC08B. Write a function that writes a byte into the specified location of the 24LC08B, a function that reads a byte from the specified location of the 24LC08B using the interrupt-driven approach, and also a test program to test these two functions. The test program calls the write function to write a byte into the EEPROM, waits for 200 ms, then reads it out, and outputs it to LEDs driven by P5 for verification.

Solution: The circuit connection is as shown in Figure 11.20.

The function that writes a byte into and the function that reads a byte from the 24LC08 are written by using the interrupt-driven approach. Since SMBus interrupts can be generated by many different causes, it will be easier to use the state table approach to write the SMBus interrupt service routine. Both the **sndByte** and **getByte** subroutine starts

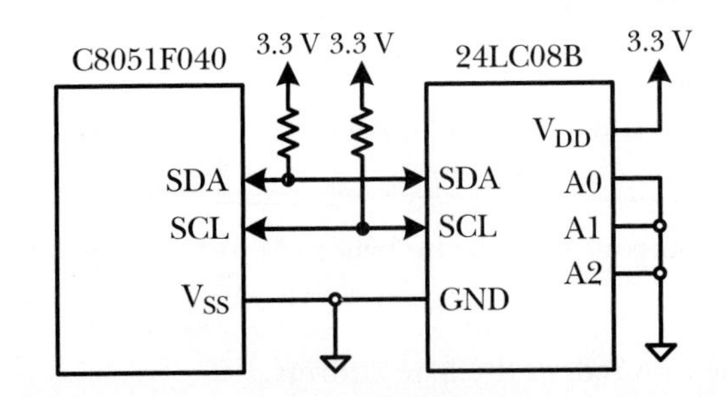

FIGURE 11.20 *Circuit connection between the 24LC08B and C8051F040.*

the operation by generating the start condition and then letting the SMBus interrupts take care of the data-transfer transactions. The **sndByte** and **getByte** functions and their test program are as follows.

```
                $nomod51
                $include (c8051F040.inc)
; MT stands for master transmitter; MR stands for master receiver
WRITE           equ         0x00
READ            equ         0x01
chipEE          equ         0xA0        ; EEPROM device ID
SMBusErr        equ         0x00        ; (all modes) BUS ERROR
SMBstart        equ         0x08        ; (MT & MR) START transmitted
SMBrestart      equ         0x10        ; (MT & MR) repeated START
SMBMtAddAck     equ         0x18        ; (MT) Slave address + W transmitted;
                                        ; ACK received
SMBMtAddNack    equ         0x20        ; (MT) Slave address + W transmitted;
                                        ; NACK received
SMBMtDbAck      equ         0x28        ; (MT) data byte transmitted; ACK rec'vd
SMBMtDbNack     equ         0x30        ; (MT) data byte transmitted; NACK rec'vd
SMBMtArbLost    equ         0x38        ; (MT) arbitration lost
SMBMrAddAck     equ         0x40        ; (MR) Slave address + R transmitted;
                                        ; ACK received
SMBMrAddNack    equ         0x48        ; (MR) Slave address + R transmitted;
                                        ; NACK received
SMBMrDbAck      equ         0x50        ; (MR) data byte rec'vd; ACK transmitted
SMBMrDbNack     equ         0x58        ; (MR) data byte rec'vd; NACK transmitted

dataZone        segment     data        ; declare DATA segment
                rseg        dataZone    ; select DATA segment
txByte:         ds          1           ; Holds a byte to be transmitted by the SMBus
rcvByte:        ds          1           ; Holds a byte just received by the SMBus
wrAdr:          ds          1           ; Holds the slave address + WRITE
RdAdr:          ds          1           ; Holds the slave address + READ
memLoc:         ds          1           ; EEPROM memory location to be accessed
                                        ; Variables used for testing.
tstCnt:         ds          1           ; Test counter variable
tstByte:        ds          1           ; Test data
tstAdr:         ds          1           ; Test memory location

bitZone         segment     bit
                rseg        bitZone
RW:             dbit        1           ; R/W command bit. 1=READ, 0=WRITE
SMbusy:         dbit        1           ; SMBus Busy flag (kept in software)
byteSent:       dbit        1           ; Used to indicate what byte was just sent:
                                        ; 1: EEPROM memory address sent
                                        ; 0: Data byte sent

                cseg
                org         00h
                ljmp        start
```

```
                org         03Bh                ; SMBus Interrupt Vector
                ljmp        SMBusISR
SMBcode         segment     code
                rseg        SMBcode
                using       0
start:          mov         SP,#0x7F            ; establish stack pointer
                acall       sysinit
                mov         P5MDOUT,#0xFF       ; enable port P5 for output
                mov         P5,#0               ; turn off LEDs when started
                mov         SFRPAGE,#SPI0_PAGE  ; switch to SFRPAGE 0
                mov         SPI0CN,#0x01        ; enable SPI to 3-wire mode
                acall       openSMBus           ; Initialize SMBus
                setb        EA                  ; Enable global interrupts
                mov         tstByte, #0xFF      ; Load initial test values
                mov         tstAdr, #0          ;     "
                mov         tstCnt, #0xFE       ;     "

; TEST CODE——————————————————————————————————————————————————————————————
TLoop:
; Send tstByte to memory location tstAdr
                mov         txByte, tstByte     ; Load transmit data into txByte
                mov         memLoc, tstAdr      ; Load memory address into memLoc
                acall       SndByte             ; Call send routine
                mov         SFRPAGE,#0x0F       ; output the complement of tstCnt
                mov         A,tstCnt            ; to LEDs
                cpl         A                   ;     "
                mov         P5,A                ;     "
                mov         SFRPAGE,#0          ;     "
; Read memory location at tstAdr into rcvByte
                mov         memLoc, tstAdr      ; Load memory address into memLoc
                acall       getByte             ; Call receive routine
                mov         R0,#2               ; wait for 200 ms before changing to next value
                lcall       delayby100ms        ;     "
; Compare byte received to byte sent
                mov         A, rcvByte          ; Load received byte into accumulator
                cjne        A, tstByte,done     ; Compare sent byte to received byte
                                                ; Jump to done if not equal
; Change test variables
                dec         tstByte             ; If sent=received, change test variables
                inc         tstAdr              ; and cycle through again.

; Cycle through again if tstCnt is not zero
                djnz        tstCnt,TLoop        ; Decrement counter, loop back to beginning
                mov         A,#99h              ; Load accumulator with 99h if test successful.
done:           jmp         $                   ; Spin
;*****************************************************************************************************
sysinit:        mov         SFRPAGE,#0x0F       ; switch SFR page F
                mov         WDTCN,#0xDE         ; disable watchdog timer
                mov         WDTCN,#0xAD         ;     "
                mov         CLKSEL,#0           ; use internal oscillator as SYSCLK
```

```
                mov     OSCICN,#x83             ;    "
                mov     XBR0,#0xF7              ; assign I/O pins to all peripheral
                mov     XBR1,#0xFF              ;    "
                mov     XBR2,#0x5D              ;    "
                mov     XBR3,#0x8F              ;    "
                ret
;********************************************************************************
; This function writes a byte into the EEPROM location at memLoc. The data to be
; written is stored at global memory location txByte.
;********************************************************************************
SndByte:        jb      SMbusy, $               ; Wait for SMBus to be free
                clr     RW                      ; RW = 0 (WRITE)
                mov     wrAdr,#chipEE+WRITE     ; store slave address + WRITE
                setb    SMbusy                  ; Occupy SMBus
                setb    STA                     ; Initiate Transfer
                ret
;********************************************************************************
; This function reads a byte from EEPROM location at memLoc. This function relies
; on SMB interrupt service routine to complete the read operation.
;********************************************************************************
getByte:        jb      SMbusy,$                ; Wait for SMBus to be free
                setb    RW                      ; RW = 1 (READ)
                mov     wrAdr,#chipEE+WRITE     ; store slave address + WRITE
                mov     RdAdr,#chipEE+READ      ; store slave address + READ
                setb    SMbusy                  ; Occupy SMBus
                setb    STA                     ; Initiate Transfer
                jb      SMbusy,$                ; Wait for receive to finish
                ret
;********************************************************************************
; This function enables SMBus, sets up its baud rate and also enables SMBus interrupt.
;********************************************************************************
openSMBus:
                mov     SMB0CN, #0x04           ; SMBus to send ACKs on acknowledge cycle
                mov     SMB0CR, #145            ; SMBus clock rate = 100KHz
                orl     SMB0CN, #0x40           ; enable SMBus
                orl     EIE1, #0x02             ; enable SMBus interrupts
                clr     SMbusy
                ret
;********************************************************************************
; The SMBus Interrupt service is constructed using a state table as shown in Table 11.2.
; Every state is allocated 8 bytes to handle the interrupt. For those interrupt causes
; that require more than eight bytes to handle, a jmp instruction brings the program
; control to the appropriate location.
;********************************************************************************
SMBusISR:       push    PSW
                push    DPH
                push    DPL
                push    ACC
                mov     A,SMB0STA               ; Load accumulator with current SMBus state.
                                                ; State corresponds to the address offset
                                                ; for each state execution
```

```
            anl     A,#0x7F                 ; Mask out upper bit, since any states that
                                            ; set this bit are not defined in this code.
            mov     DPTR,#SMBStateTable     ; Point DPTR to the beginning of the state table
            jmp     @A+DPTR                 ; Jump to the current state

SMBStateTable:
; SMBusErr
; All Modes: Bus Error
; Reset hardware by setting STOP bit
            org     SMBStateTable + SMBusErr
            setb    STO
            jmp     SMB_ISR_END             ; Jump to exit ISR

; SMBstart
; Master Transmitter/Receiver: START transmitted. The R/W bit will always be a zero (W)
; in this state because for both write and read, the memory address must first be written.
            org     SMBStateTable + SMBstart
            mov     SMB0DAT,wrAdr           ; Load slave address + W
            clr     STA                     ; Manually clear START bit
            jmp     SMB_ISR_END             ; Jump to exit ISR

; SMBrestart
; Master Transmitter/Receiver: Repeated START transmitted. This occurred during a read only.
            org     SMBStateTable + SMBrestart
            mov     SMB0DAT,RdAdr           ; Load slave address + R
            clr     STA                     ; Manually clear START bit
            jmp     SMB_ISR_END

; SMBMtAddAck
; Master Transmitter: Slave address + WRITE transmitted, ACK received
            org     SMBStateTable + SMBMtAddAck
            mov     SMB0DAT,memLoc          ; Load memory address
            setb    byteSent                ; byteSent=1: In the next ISR call,
                                            ; the memory address will have just been sent.
            jmp     SMB_ISR_END

; SMBMtAddNack
; Master Transmitter: Slave address + WRITE transmitted, received NACK. Should try again.
            org     SMBStateTable + SMBMtAddNack
            setb    STO
            setb    STA
            jmp     SMB_ISR_END

; SMBMtDbAck
; Master Transmitter: Data byte transmitted. ACK received. This state is used in both read
; and write operations. Check byteSent; if 1, memory address has just been sent.
; Else, data has been sent.
            org     SMBStateTable + SMBMtDbAck
            jbc     byteSent,AddrSent       ; If byteSent=1, clear bit and
```

```
                                                ; jump to AddrSent to process
                                                ; outside of state table.
            jmp         DATA_Sent               ; If byteSent=0, data has just been sent,
                                                ; transfer is finished.
                                                ; jump to end transfer

; SMBMtDbNack
; Master Transmitter: Data byte transmitted. NACK received.
; Slave not responding. Send STOP followed by START to try again.
            org         SMBStateTable + SMBMtDbNack
            setb        STO
            setb        STA
            jmp         SMB_ISR_END

; SMBMtArbLost
; Master Transmitter: Arbitration Lost. This condition should not occur. If so, restart transfer.
            org         SMBStateTable + SMBMtArbLost
            setb        STO
            setb        STA
            jmp         SMB_ISR_END

; SMBMrAddAck
; Master Receiver: Slave address + READ transmitted. ACK received.
; Set to transmit NACK after next transfer since it will be the last (only) byte.
            org         SMBStateTable + SMBMrAddAck
            clr         AA                      ; NACK sent on acknowledge cycle
            jmp         SMB_ISR_END

; SMBMrAddNack
; Master Receiver: Slave address + READ transmitted. NACK received.
; Slave not responding. Send repeated START to try again.
            org         SMBStateTable + SMBMrAddNack
            clr         STO
            setb        STA
            jmp         SMB_ISR_END

; SMBMrDbAck
; Master Receiver: Data byte received. ACK transmitted. This condition should not occur because
; AA is cleared in previous state. Send STOP if state does occur.
            org         SMBStateTable + SMBMrDbAck
            setb        STO
            jmp         SMB_ISR_END

; SMBMrDbNack
; Master Receiver: Data byte received. NACK transmitted. Read operation completed.
; Read data register and send STOP
            org         SMBStateTable + SMBMrDbNack
            mov         rcvByte, SMB0DAT
            setb        STO
            setb        AA                      ; Set AA for next transfer
```

```
                clr         SMbusy
                jmp         SMB_ISR_END

; End of State Table
;-----------------------------------------------------------------------------
; Program segment to handle SMBus states that require more than 8 bytes of program
; space. Address byte has just been sent. Check RW. If R (1), jump to RW_READ.
; If W, load data to transmit into SMB0DAT.
;-----------------------------------------------------------------------------
AddrSent:       jb          RW, RW_READ
                mov         SMB0DAT, txByte         ; Load data
                jmp         SMB_ISR_END             ; Jump to exit ISR

; Operation is a READ, and the address byte has just been sent. Send
; repeated START to initiate memory read.
RW_READ:        clr         STO
                setb        STA                     ; Send repeated START
                jmp         SMB_ISR_END             ; Jump to exit ISR

; Operation is a WRITE, and the data byte has just been sent. Transfer
; is finished. Send STOP, free the bus, and exit the ISR.
DATA_Sent:
                setb        STO                     ; Send STOP and exit ISR.
                clr         SMbusy                  ; Free SMBus
                jmp         SMB_ISR_END             ; Jump to exit ISR
;-----------------------------------------------------------------------------
; SMBus ISR exit. Restore registers, clear SI bit, and return from interrupt.
;-----------------------------------------------------------------------------
SMB_ISR_END:
                clr         SI
                pop         ACC
                pop         DPL
                pop         DPH
                pop         PSW
                reti
;******************************************************************************
; add delayby100 ms subroutine here.
;******************************************************************************
                end
```

The C language version of the program is as follows:

```
#include    <c8051F040.h>               // SFR declarations
#include    "delays.h"                  // add delays.c to the project
#define     WRITE       0x00            // SMBus WRITE command
#define     READ        0x01            // SMBus READ command
#define     chipEE      0xA0            // Device address for EEPROM
// SMBus states:
// MT = Master Transmitter
// MR = Master Receiver
```

```
#define     SMBusErr       0x00     // (all modes) BUS ERROR
#define     SMBstart       0x08     // (MT & MR) START transmitted
#define     SMBrestart     0x10     // (MT & MR) repeated START
#define     SMBMtAddAck    0x18     // (MT) Slave address + W transmitted;
                                    // ACK received
#define     SMBMtAddNack   0x20     // (MT) Slave address + W transmitted;
                                    // NACK received
#define     SMBMtDbAck     0x28     // (MT) data byte transmitted; ACK rec'vd
#define     SMBMtDbNack    0x30     // (MT) data byte transmitted; NACK rec'vd
#define     SMBMtArbLost   0x38     // (MT) arbitration lost
#define     SMBMrAddAck    0x40     // (MR) Slave address + R transmitted;
                                    // ACK received
#define     SMBMrAddNack   0x48     // (MR) Slave address + R transmitted;
                                    // NACK received
#define     SMBMrDbAck     0x50     // (MR) data byte rec'vd; ACK transmitted
#define     SMBMrDbNack    0x58     // (MR) data byte rec'vd; NACK transmitted

char        txByte;             // Byte to be transmitted
char        rcvByte;            // Byte to be returned
char        slaAdr;             // slave device ID
char        wrAdr, RdAdr;       // slave address + Write (or Read)
char        memLoc;             // EEPROM location to be accessed
char        tstCnt;             // test counter value
char        tstByte;            // test data
char        tstAdr;             // test memory location
bit         SMbusy;             // This bit is set when a send or receive
                                // is started. It is cleared by the
                                // ISR when the operation is finished
bit         RW;                 // R/W command bit. 1=read, 0=write
bit         byteSent;           // Used to indicate what byte was just sent:
                                // 0 = data byte sent
                                // 1 = EEPROM memory address sent
//-----------------------------------------------------------------------------
// Function PROTOTYPES
//-----------------------------------------------------------------------------
void sysInit(void);
void openSMBus(void);
void SMBus_ISR (void);
void SndByte (char chipID, unsigned char ByteAdr, char OutByte);
char GetByte (char chipID, unsigned int ByteAdr);

void main (void)
{
    char temp;
    sysInit();
    openSMBus();                        // initialize SMBus
    EA     = 1;                         // enable global interrupts
    tstByte = 0xFF;                     // load initial test values
    tstAdr  = 0;                        //     "
    tstCnt  = 0xFE;                     //     "
```

```
    while (tstCnt){
        slaAdr   = chipEE;                     // load slave address
        txByte   = tstByte;                    // load transmit data to txByte;
        memLoc = tstAdr;                       // load memory address into memLoc
        SndByte(slaAdr, memLoc, txByte);
        slaAdr   = chipEE;                     // load slave address
        memLoc = tstAdr;                       // load memory address into memLoc
        temp     = GetByte(slaAdr, memLoc);    // read back the same location
        SFRPAGE = 0x0F;                        // switch to SFR page F
        P5        = temp;                      // output the value written into EEPROM to LEDs
        SFRPAGE = 0;                           // switch back to SFR page 0
        delayby100ms(2);                       // wait for 200 ms
        tstAdr++;                              // move to next location
        tstByte++;                             // change the data to be written into
        tstCnt--;                              // decrement the byte count
    }
    while(1);                                  // infinite loop
}
void sysInit(void)
{
    SFRPAGE  = 0x0F;                           // switch to SFR page F
    WDTCN    = 0xDE;                           // disable watchdog timer
    WDTCN    = 0xAD;                           //      "
    CLKSEL   = 0;                              // use internal oscillator as SYSCLK
    OSCICN   = 0x83;                           //      "
    XBR0     = 0xF7;                           // assign I/O pins to peripheral function
    XBR1     = 0xFF;                           //      "
    XBR2     = 0x5D;                           //      "
    XBR3     = 0x8F;                           //      "
    P5MDOUT = 0xFF;                            // enable Port P5 for output
    P5       = 0;                              // turn off LEDs
    SFRPAGE  = 0;                              // switch to SFR page 0
    SPI0CN   = 1;                              // enable SPI to 3-wire mode
}
void openSMBus(void)
{
    SMB0CN  = 0x04;                            // SMBus to send ACK on acknowledge cycle
    SMB0CR  = 0x7F;                            // SMBus clock rate = 100 KHz
    SMB0CN |= 0x40;                            // enable SMBus
    EIE1     = 0x02;                           // enable SMBus interrupt
    SMbusy  = 0;                               // SMBus ready to transmit data
}
// SMBus byte write function————————————————————————————
// Writes a single byte at the specified memory location.
// OutByte = data byte to be written
// ByteAdr = memory location to be written into (2 bytes)
// chipID   = device address of EEPROM chip to be written to
void Sndbyte (char chipID, unsigned char ByteAdr, char OutByte)
{
    while (SMbusy);                            // Wait for SMBus to be free.
```

```
        SMbusy  = 1;                    // Occupy SMBus (set to busy)
        RW      = 0;
        SMB0CN  = 0x44;                 // SMBus enabled, ACK on acknowledge cycle
        wrAdr   = chipID & 0xFE;// store slaAdr+W
        txByte  = OutByte; // Data to be written
        memLoc  = ByteAdr;              // specify memory address to write
        STO     = 0;
        STA     = 1;                    // Start transfer
}
// SMBus random read function-----------------------------------------------
// Reads 1 byte from the specified memory location.
// ByteAdr   = memory address of byte to read
// chipID    = device address of EEPROM to be read from
char GetByte (char chipID, unsigned int ByteAdr)
{
        while (SMbusy);                 // Wait for bus to be free.
        SMbusy  = 1;                    // Occupy SMBus (set to busy)
        RW      = 1;                    // indicate read
        SMB0CN  = 0x44; // SMBus enabled, ACK on acknowledge cycle
        wrAdr   = chipID;               // Chip select + READ
        RdAdr   = chipID+1;             // set up read address
        memLoc  = ByteAdr;              // specify memory address to read
        STO     = 0;
        STA     = 1;                    // Start transfer
        while (SMbusy); // Wait for transfer to finish
        return rcvByte;
}
//------------------------------------------------------------------------
// Interrupt Service Routine
//------------------------------------------------------------------------
void SMBUS_ISR (void) interrupt 7
{
        switch (SMB0STA){               // Status code for the SMBus (SMB0STA register)
        case SMBusErr:
                STO = 1;                // reset SMBus
                break;
        case SMBstart:
                SMB0DAT = wrAdr;        // Load address of the slave to be accessed.
                STA     = 0;            // Manually clear START bit
                break;
        // Master Transmitter/Receiver: Repeated START condition transmitted.
        // This state should only occur during a read, after the memory address
        // has been sent and acknowledged.
        case SMBrestart:
                SMB0DAT = RdAdr;        // COMMAND should hold slave address + R.
                STA     = 0;            // manually clear START bit
                break;
        // Master Transmitter: Slave address + W transmitted. ACK received.
        case SMBMtAddAck:
                SMB0DAT = memLoc;       // Load high byte of memory address to be written.
```

```
        byteSent  = 1;                    // indicate sending out memory location address
        break;
    // Master Transmitter: Slave address + WRITE transmitted. NACK received.
    // The slave is not responding. Send a STOP followed by a START to try again.
    case SMBMtAddNack:
        STO = 1;
        STA = 1;
        break;
    // Master Transmitter: Data byte transmitted. ACK received.
    // This state is used in both READ and WRITE operations. Check RW to find
    // out data transfer type. RW = 1 is read byte; RW = 0 is write data
    case SMBMtDbAck:
        if (byteSent){
            if (RW) { // If R/W = READ, sent repeated START.
                STO = 0;
                STA = 1;
            } else {
                SMB0DAT = txByte;         // If a write, load byte to write.
                byteSent  = 0;            // indicate the byte sent is data
            }
        }
        else { // data byte has just been sent, transfer is completed
            STO    = 1;                   // send STOP and exit ISR
            SMbusy = 0;                   // Free SMBus
        }
        break;
    // Master Transmitter: Data byte transmitted. NACK received.
    // Slave not responding. Send STOP followed by START to try again.
    case SMBMtDbNack:
        STO = 1;
        STA = 1;
        break;
    // Master Transmitter: Arbitration lost.
    // Should not occur. If so, restart transfer.
    case SMBMtArbLost:
        STO = 1;
        STA = 1;
        break;
    // Master Receiver: Slave address + READ transmitted. ACK received.
    // Set to transmit NACK after next transfer since it will be the last
    // (only) byte.
    case SMBMrAddAck:
        AA = 0;                           // NACK sent on acknowledge cycle.
        break;
    // Master Receiver: Slave address + READ transmitted. NACK received.
    // Slave not responding. Send repeated start to try again.
    case SMBMrAddNack:
        STO = 0;
        STA = 1;
        break;
```

```
        // Data byte received. ACK transmitted.
    // State should not occur because AA is set to zero in previous state.
    // Send STOP if state does occur.
    case SMBMrDbAck:
        STO    = 1;
        SMbusy = 0;
        break;
    // Data byte received. NACK transmitted.
    // Read operation has completed. Read data register and send STOP.
    case SMBMrDbNack:
        rcvByte = SMB0DAT;
        STO    = 1;
        AA     = 1;                         // set AA for next transfer
        SMbusy = 0;                         // Free SMBus
        break;
    // All other status codes meaningless in this application. Reset communication.
    default:
        STO    = 1;                         // Reset communication.
        SMbusy = 0;
        break;
    }
    SI = 0;                                 // clear interrupt flag
}
```

The programs in this example can be modified to support other transfer modes (for example, page-write and current-address read mode). The modification will be left for you as an exercise problem.

11.6 Using the Digital Thermostat DS1631A

Many embedded products (such as network routers and switches) are used in a larger system, and their failures due to overheating can severely damage the functioning or even cause the total failure of the larger system. Using a thermostat to warn of potential system overheating is indispensable for the proper functioning of many embedded systems.

The digital thermostat device DS1631A from Dallas Semiconductor is one such product. The DS1631A will assert a signal (T_{OUT}) whenever the ambient temperature exceeds the *trip point* established by the user.

11.6.1 Pin Assignment

The pin assignment of the DS1631A is shown in Figure 11.21. The SDA and SCL pins are used as data and clock lines so that the DS1631A can be connected to an I^2C bus or SMBus. Pins A2, . . . , A0 are address inputs to the

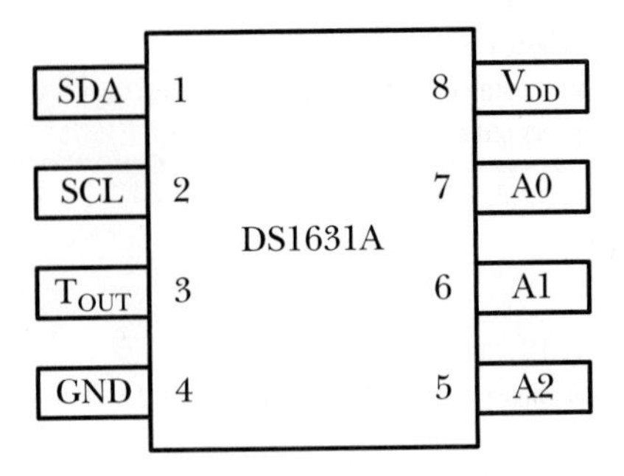

FIGURE 11.21
Pin assignment of DS1631A.

DS1631A. The T_{OUT} pin is the thermostat output, which is asserted whenever the ambient temperature is above the trip point set by the user.

11.6.2 Functional Description

The block diagram of the DS1631A is shown in Figure 11.22. The DS1631A converts the ambient temperature into 9-, 10-, 11-, or 12-bit readings over a range of -55 to 125°C. The accuracy of the thermometer is ± 0.5°C from 0 to $+70$°C with $3.0\ V \leq V_{DD} \leq 5.5\ V$.

The thermostat output T_{OUT} is asserted whenever the converted ambient temperature is equal to or higher than the value stored in the TH register. After being asserted, T_{OUT} will be deasserted only when the converted temperature reading drops below the value stored in the TL register. The DS1631A automatically begins taking temperature measurements at power-up, which allows it to function as a stand-alone thermostat. The DS1631A conforms to the I^2C bus specification.

11.6.3 DS1631A Registers

The operation of the DS1631A is supported by four registers: Config, TH, TL, and Temperature. The DS1631A represents a negative temperature in two's complement format. The Temperature register holds the converted temperature. The lowest four bits of this register are always zeros. When the most significant bit of this register is 1, the temperature is negative.

Both the TH and TL are EEPROM-based 2-byte registers. TH holds the upper-alarm temperature value. Whenever the converted temperature is equal to or higher than the value in TH, the T_{OUT} signal is asserted. Once being asserted, T_{OUT} can be deasserted only when the converted temperature drops below the value in the TL register.

The contents of the Config register are shown in Figure 11.23. The lower 2 bits of the Config register are EEPROM based, whereas the upper 6 bits are SRAM based. The Config register allows the user to program various DS1631A options, such as conversion resolution, T_{OUT} polarity, and operation mode. This register also provides information about conversion status, EEPROM

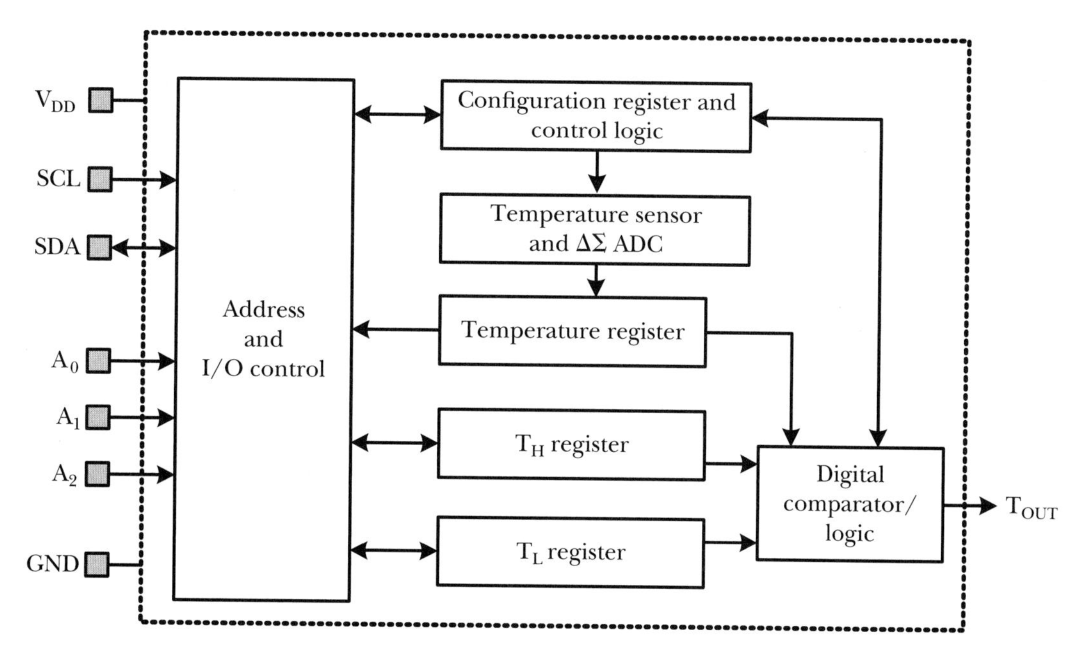

FIGURE 11.22
DS1631A functional diagram.

activity, and thermostat activity. This register can be read from and written into using the **Access Config** command. When writing into the Config register, conversions should first be stopped by using the **Stop Convert T** command if the device is in continuous conversion mode. Since the POL and 1SHOT bits are stored in EEPROM, they can be programmed prior to installation, if desired. All other bits are in SRAM and are powered up in state shown in Figure 11.23.

11.6.4 The DS1631A Operation

The DS1631A begins conversions automatically at power-up. It can be configured to operate in continuous conversion or single conversion mode. The default resolution of the DS1631A is 12 bits. However, the user can choose to use 9-, 10-, or 11-bit resolution when desired. The conversion resolution is set via the R1:R0 bits of the Config register. A few samples of temperatures and their converted values are shown in Table 11.3. The lowest 4 bits are always 0s in the table, because the resolution is 12 bits.

Both the TH and TL registers are in EEPROM. Their resolutions match the output temperature resolution and are determined by the R1:R0 bits. Writing to and reading from these two registers are achieved by using the **Access TH** and **Access TL** commands. When making changes to the TH and

7	6	5	4	3	2	1	0	
DONE	THF	TLF	NVB	R1	R0	POL*	1SHOT*	Reset value = 100011xxb

*NV (EEPROM)

Done: Temperature conversion done (read-only)
- 0 = Temperature conversion is in progress
- 1 = Temperature conversion is complete. Will be cleared when the Temperature register is read

THF: Temperature high flag (read/write)
- 0 = The measured temperature has not exceeded the value in T_H register
- 1 = The measured temperature has exceeded the value in T_H register. THF remains at 1 until it is overwritten with a 0 by the user, the power is recycled, or a software POR command is issued

TLF: Temperature low flag (read/write)
- 0 = The measured temperature has not been lower than the value in T_L register
- 1 = At some point after power up, the measured temperature is lower than the value stored in the T_L register. TLF remains at 1 until it is overwritten with a 0 by the user, the power is recycled, or a software POR command is issued

NVB: Nonvolatile memory busy (read only)
- 0 = NV memory is not busy
- 1 = A write to EEPROM memory is in progress

R1:R0 : Resolution bits (read/write)
- 00 = 9-bit resolution (conversion time is 93.75 ms)
- 01 = 10-bit resolution (conversion time is 187.5 ms)
- 10 = 11-bit resolution (conversion time is 375 ms)
- 11 = 12-bit resolution (conversion time is 750 ms)

POL: T_{OUT} polarity (read/write)
- 0 = T_{OUT} active low
- 1 = T_{OUT} active high

1SHOT: Conversion mode (read/write)
- 0 = Continuous conversion mode. The Start Convert T command initiates continuous temperature conversions.
- 1 = One-shot mode. The Start Convert T command initiates a single temperature conversion and then the device enters a low-power standby mode

FIGURE 11.23
DS1631A Configuration register.

TL registers, conversions first should be stopped using the **Stop Convert T** command if the device is in continuous mode.

Since the DS1631A automatically begins taking temperature measurements at power-up, it can function as a standalone thermostat. For standalone operation, the nonvolatile TH and TL registers and the POL and 1SHOT bits in the Config register should be programmed to the desired values prior to installation.

Depending on the sign of the value, the value in the Temperature register can be converted to temperature as follows.

TABLE 11.3 *12-bit Resolution Temperature/Data Relationship*

Temperture (°C)	Digital Output (Binary)	Digital Output (Hex)
+125	0111 1101 0000 0000	0x7D00
+25.0625	0001 1001 0001 0000	0x1910
+10.125	0000 1010 0010 0000	0x0A20
+0.5	0000 0000 1000 0000	0x0080
0	0000 0000 0000 0000	0x0000
−0.5	1111 1111 1000 0000	0xFF80
−10.125	1111 0101 1110 0000	0xF5E0
−25.0625	1110 0110 1111 0000	0xE6F0
−55	1100 1001 0000 0000	0xC900

Positive Conversion Result

Step 1
Truncate the lowest four bits.

Step 2
Divide the upper 12 bits by 16.

For example, the conversion result 0x7D00 corresponds to 0x7D0/16 = 125°C. The conversion result 0x6040 corresponds to 0x604/16 = 96.25°C.

Negative Conversion Result

Step 1
Compute the two's complement of the conversion result.

Step 2
Truncate the lowest 4 bits.

Step 3
Divide the upper 12 bits of the two's complement of the conversion result. For example, the conversion result 0xE280 corresponds to −0x1D8/16 = −29.5°C.

11.6.5 DS1631A Command Set

The DS1631A supports the following commands.

- **Start Convert T [0x51].** This command initiates temperature conversions. If the DS1631A is in one-shot mode, only one conversion is performed. In continuous mode, continuous temperature conversions are performed until a **Stop Convert T** command is issued.

- **Stop Convert T [0x22].** This command stops temperature conversions when the device is in continuous-conversion mode.
- **Read Temperature [0xAA].** This command reads the last converted temperature value from the 2-byte Temperature register.
- **Access TH [0xA1].** This command reads or writes the TH register.
- **Access TL [0xA2].** This command reads or writes the TL register.
- **Access Config [0xAC].** This command reads or writes the Config register.
- **Software POR [0x54].** This register initiates a software power-on-reset operation, which stops temperature conversions and resets all registers and logic to their power-on states. The software POR allows the user to simulate cycling the power without actually powering down the device.

11.6.6 Interfacing the DS1631A with the C8051F040

A typical circuit connection between the C8051F040 and a DS1631A is shown in Figure 11.24. The address input to the DS1631A is arbitrarily set to 001 in Figure 11.24.

To initiate SMBus communication, the C8051F040 MCU generates a start condition followed by a control byte that contains the DS1631A slave address. The R/$\overline{W}$ bit of the control byte must be a 0, because the C8051F040 MCU next will write a command byte to the DS1631A. The format for the control byte is shown in Figure 11.25. The DS1631A responds with an ACK after

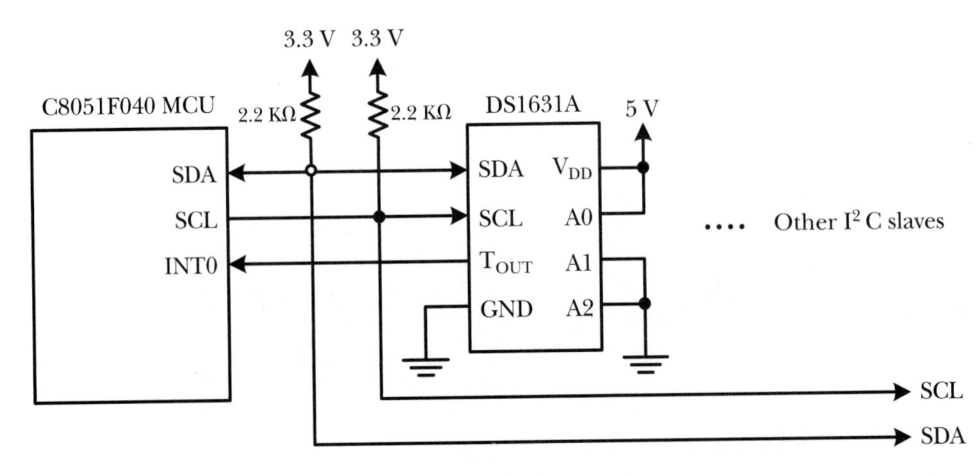

FIGURE 11.24
Typical circuit connection between the C8051F040 and DS1631A.

7	6	5	4	3	2	1	0
1	0	0	1	A_2	A_1	A_0	R/$\overline{W}$

FIGURE 11.25
Control byte for DS1631A.

receiving the control byte. This must be followed by a command byte from the master, which indicates what type of operation is to be performed. The DS1631A again responds with an ACK after receiving the command byte. If the command byte is a **Start Convert T** or a **Stop Convert T** command, the transaction is finished, and the master must issue a stop condition to signal the end of communication sequence. If the command byte indicates a write or read operation, additional actions must occur.

Write Data to DS1631A

The master can write data to the DS1631A by issuing an **Access Config**, **Access TH**, or **Access TL** command following the control byte. Since the R/ W bit in the control byte was a 0, the DS1631A already is prepared to receive data. Therefore, after receiving an ACK in response to the command byte, the C8051F040 can begin transmitting data immediately. When writing to the Config register, the MCU must send one byte of data, and when writing to the TH or TL register, the MCU must send two bytes of data. The most significant byte is sent first. After receiving each data byte, the DS1631A responds with an ACK, and the transaction is finished with a stop condition from the MCU.

Read Data from DS1631A

The C8051F040 can read data from the DS1631A by issuing an **Access Config**, **Access TH**, **Access TL**, or **Read Temperature** command following the control byte. After receiving an ACK in response to the command, the MCU must generate a repeated start condition followed by a control byte with the same slave address as the first byte but with the R/ W bit set to 1. After the DS1631A sends an ACK in response to this control byte, it begins transmitting the requested data on the next clock cycle. One byte of data will be transmitted when reading from the Config register, after which the MCU must respond with a NACK followed by a stop condition. For 2-byte reads, the MCU must respond to the first data byte with an ACK and to the second data byte with a NACK followed by a stop condition. The DS1631A sends out the most-significant byte first and then the least-significant byte. If only the most significant byte of data is needed, the master can issue a NACK followed by a stop condition after reading the first data byte.

Example 11.3 Write a function to configure the DS1631A to perform a one-shot conversion with 12-bit resolution, a function to start the temperature conversion, a function to read back the temperature, a function to format the temperature, and a function to display the temperature on the HyperTerminal. Write a main program to test these functions.

Solution: We will use the interrupt-driven approach to handle data transfers to and from the DS1631A. The write-byte count and read-byte count are used to control the state transition. For each command byte or data byte transmitted, the write-byte count is decremented by 1.

When the write-byte count is decremented to 0 and read-byte count is also 0, the interrupt service routine sets the STO bit and completes the data-transfer operation. When the write-byte count is decremented to 0 but the read-byte count is not 0, the interrupt service routine generates the restart condition to prepare the read operation. After performing the required initialization, the main function stays in an infinite loop, and performs temperature conversion once every second, and displays it on the **HyperTerminal** window.

The following variables are used to implement the desired functions.

```
char    WrAdr, RdAdr;           // slave address + Write (or Read)
bit     SMbusy;                 // This bit is set when a send or receive is started. It is cleared
                                // by the ISR when the operation is finished
bit     RW;                     // R/W command bit. 1=read, 0=write
char    txbuf[2];               // buffer of data to be transferred to DS1631A
char    rxbuf[2];               // buffer to hold data read from DS1631A
char    xdata dispBuf[6];       // buffer for formatted temperature to be displayed
char    command;                // command to be sent to DS1631A
char    Byte2Send;              // number of bytes to send to DS1631A
char    Byte2Read;              // number of bytes to read from DS1631A
char    wIndex;                 // index to txbuf
char    rIndex;                 // index to rxbuf
```

The specified C functions and the test program are as follows.

```
#include  <c8051F040.h>                    // SFR declarations
#include  <stdio.h>
#include  "uartUtil.h"                     // add uartUtil.c into the project
#include  "delays.h"                       // (need to add delays.c into the project)
#define   WRITE             0x00           // SMBus WRITE command
#define   READ              0x01           // SMBus READ command
#define   DS1631A           0x92           // Device address for DS1631A
// SMBus states:
// MT     = Master Transmitter
// MR     = Master Receiver
#define   SMBusErr          0x00           // (all modes) BUS ERROR
#define   SMBstart          0x08           // (MT & MR) START transmitted
#define   SMBrestart        0x10           // (MT & MR) repeated START
#define   SMBMtAddAck       0x18           // (MT) Slave address + W transmitted;
                                           // ACK received
#define   SMBMtAddNack      0x20           // (MT) Slave address + W transmitted;
                                           // NACK received
#define   SMBMtDbAck        0x28           // (MT) data byte transmitted; ACK rec'vd
#define   SMBMtDbNack       0x30           // (MT) data byte transmitted; NACK rec'vd
#define   SMBMtArbLost      0x38           // (MT) arbitration lost
#define   SMBMrAddAck       0x40           // (MR) Slave address + R transmitted;
                                           // ACK received
#define   SMBMrAddNack      0x48           // (MR) Slave address + R transmitted;
                                           // NACK received
```

```
#define   SMBMrDbAck       0x50    // (MR) data byte rec'vd; ACK transmitted
#define   SMBMrDbNack      0x58    // (MR) data byte rec'vd; NACK transmitted
// DS1631A commands
#define   StartConvertT    0x51    // command to start temperature conversion
#define   StopConvertT     0x22    // command to stop temperature conversion
#define   ReadTemp         0xAA    // command to read temperature
#define   AccessTH         0xA1    // command to access TH from DS1631A
#define   AccessTL         0xA2    // command to access TL from DS1631A
#define   AccessConfig     0xAC    // command to access Config register
#define   SoftwarePOR      0x54    // command to reset DS1631A

char      WrAdr, RdAdr;            // slave address + Write (or Read)
bit       SMbusy;                  // This bit is set when a send or receive is started. It is cleared
                                   // by the ISR when the operation is finished
bit       RW;                      // R/W command bit. 1=read, 0=write

char      txbuf[2];                // buffer of data to be transferred to DS1631A
char      rxbuf[2];                // buffer of to hold data read from DS1631A
char      xdata dispBuf[6];        // buffer for temperature to be displayed
char      command;                 // command to be sent to DS1631A
char      Byte2Send;               // number of bytes to send to DS1631A
char      Byte2Read;               // number of bytes to read from DS1631A
char      wIndex;                  // index to txbuf
char      rIndex;                  // index to rxbuf
// Function PROTOTYPES
void sysInit(void);
void openSMBus(void);
void SMBus_ISR (void);
void openDS1631(char chipID,char config);   // initialize DS1631A
void ConvertT(char chipID);                 // start temperature conversion
void getTemp(char chipID);                  // read temperature from DS1631A
void formatTemp(void);                      // format temperature to 3 integral, one fractional digits
void dispTemp(void);                        // display temperature on monitor screen

void main (void)
{
    sysInit();
    openUART0();
    openSMBus();                            // initialize SMBus

    EA = 1;                                 // enable global interrupts
    printf("\nConfigure Thermostat DS1631A . . . ");
    openDS1631(DS1631A, 0x0F);              // select 12-bit resolution, one shot mode
    while (1) {
        ConvertT(DS1631A);                  // start temperature conversion
        delayby1s(1);                       // wait for 1 second
        getTemp(DS1631A);                   // read back temperature
        formatTemp();
        dispTemp();                         // display temperature on HyperTerminal
    }
}
```

```
void sysInit(void)
{
    SFRPAGE   = 0;                  // switch to SFR page 0
    SPI0CN    = 1;                  // enable SPI to 3-wire mode
    SFRPAGE   = 0x0F;               // switch to SFR page F
    WDTCN     = 0xDE;               // disable watchdog timer
    WDTCN     = 0xAD;               //     "
    CLKSEL    = 0;                  // use internal oscillator as SYSCLK
    OSCICN    = 0x83;               //     "
    XBR0      = 0xF7;               // assign I/O pins to peripheral function
    XBR1      = 0xFF;               //     "
    XBR2      = 0x5D;               //     "
    XBR3      = 0x8F;               //     "
    P0MDOUT = 0x15;                 // configure SMBus to be open drain
}
void openSMBus(void)
{
    SMB0CN  = 0x04;                 // SMBus to send ACK on acknowledge cycle
    SMB0CR  = 145;                  // SMBus clock rate = 100 KHz
    SMB0CN |= 0x40;                 // enable SMBus
    EIE1    = 0x02;                 // enable SMBus interrupt
    SMbusy  = 0;                    // SMBus ready to transmit data
}
// ****************************************************************************************************
// The following function configure the DS1631A.
// ****************************************************************************************************
void openDS1631(char chipID, char config)
{
    while(SMbusy);                  // wait until SMBus is idle
    SMbusy    = 1;                  // occupies SMBus
    RW        = 0;                  // perform write operation
    SMB0CN    = 0x44;               // enable SMBus in master mode
    WrAdr     = chipID;             // set control byte to (DS1631A ID + W) chip
    command   = AccessConfig;
    txbuf[0]  = config;             // configuration value to be sent to DS1631A
    Byte2Send = 2;                  // number of bytes to be transferred
    Byte2Read = 0;                  // number of bytes to be read
    STO       = 0;                  // clear STO bit
    STA       = 1;                  // start transfer
}
// ****************************************************************************************************
// The following function sends the command to DS1631A to start temperature
// conversion.
// ****************************************************************************************************
void ConvertT(char chipID)
{
    while(SMbusy);                  // wait until SMBus is idle
    SMbusy    = 1;                  // occupies SMBus
    RW        = 0;                  // perform write operation
    SMB0CN    = 0x44;               // enable SMBus in master mode
```

```
    WrAdr      = chipID;              // set control byte to (DS1631A ID + W) chip
    command    = StartConvertT;
    Byte2Send  = 1;                   // number of bytes to be transferred
    Byte2Read  = 0;                   // number of bytes to be read
    STO        = 0;                   // clear STO bit
    STA        = 1;                   // start transfer
}

// ************************************************************************************
// The following function reads the current time from the RTC chip.
// ************************************************************************************
void getTemp(char chipID)
{
    while(SMbusy);
    SMbusy     = 1;                   // occupy SMbus (set to busy)
    RW         = 1;                   // indicate read
    SMB0CN     = 0x44;                // enable SMBus module to master mode
    WrAdr      = chipID;              // set control byte to (DS1631A ID + W)
    RdAdr      = chipID+1;            // set read control byte to (DS1631A ID + R)
    command    = ReadTemp;
    Byte2Send  = 1;                   // send one byte (command)
    Byte2Read  = 2;                   // read 2 bytes (temperature)
    STO        = 0;
    STA        = 1;                   // start transfer
    while(SMbusy);
}
// ************************************************************************************
// The following function format temperature to 3 integral digits and one
// fractional digit.
// ************************************************************************************
void formatTemp(void)
{
    unsigned int temp;
    unsigned char temp2;
    dispBuf[0] = 0x20;                // fill in space character
    dispBuf[1] = 0x20;                //      "
    dispBuf[2] = 0x30;                // decimal 0
    dispBuf[3] = '.';                 // decimal point
    dispBuf[4] = 0x30;                // decimal 0
    dispBuf[5] = 0;                   // NULL character
    if(rxbuf[0] & 0x80){              // is the number negative?
        dispBuf[0] = '-';
        temp       = (unsigned int)rxbuf[0]*256 + (unsigned int)rxbuf[1];
        temp       = -temp;           // compute the 2's complement of the temperature
        rxbuf[0]   = temp/256;
        rxbuf[1]   = temp%256;
    }
    dispBuf[2] = rxbuf[0] % 10 + 0x30;        // ASCII of the ones digit of temperature
    temp2      = rxbuf[0]/10;
    if (temp2) {
```

```
            dispBuf[1] = temp2 % 10 + 0x30;         // ASCII of the tens digit of temperature
            temp2      = temp2 / 10;
      }
      if(temp2)
            dispBuf[0] = 0x31;             // hundred's digit is 1
      switch (rxbuf[1] & 0xC0){            // check the highest two bits
            case 0x00:
                  dispBuf[4] = 0x30;       // decimal 0
                  break;
            case 0x40:
                  dispBuf[4] = 0x33;       // decimal 3
                  break;
            case 0x80:
                  dispBuf[4] = 0x35;       // decimal 5
                  break;
            case 0xC0:
                  dispBuf[4] = 0x38;       // decimal 8
                  break;
            default:
                  dispBuf[4] = 0x30;
                  break;
      }
}
// ******************************************************************************************************
// The following function displays the temperature reading on the HyperTerminal.
// ******************************************************************************************************
void dispTemp(void)
{
      SFRPAGE = UART0_PAGE;
      printf("\nTemperature is ");
      putsx(dispBuf);
}
// ******************************************************************************************************
// SMBus Interrupt Service Routine
// ******************************************************************************************************
void SMBUS_ISR (void) interrupt 7
{
      switch (SMB0STA){                    // Status code for the SMBus (SMB0STA register)
      case SMBusErr:
            STO = 1;                       // reset SMBus
            break;
      case SMBstart:
            SMB0DAT = WrAdr;               // send slave ID + W
            STA = 0;                       // Manually clear START bit
            break;
      // Master Transmitter/Receiver: Repeated START condition transmitted.
      // This state should only occur during a read, after the memory address
      // has been sent and acknowledged.
      case SMBrestart:
            SMB0DAT = RdAdr;               // send slave ID + R.
```

```
        STA = 0;                    // manually clear START bit
        break;
    // Master Transmitter: Slave address + W transmitted. ACK received.
    case SMBMtAddAck:
        SMB0DAT = command;          // send a command to DS1631A.
        wIndex   = 0;               // write buffer starting index
        rIndex   = 0;               // read buffer starting index
        Byte2Send--;
        break;
    // Master Transmitter: Slave address + WRITE transmitted. NACK received.
    // The slave is not responding. Send a STOP followed by a START to try again.
    case SMBMtAddNack:
        STO = 1;
        STA = 1;
        break;
    // Master Transmitter: Data byte transmitted. ACK received.
    // This state is used in both READ and WRITE operations. Check RW to find
    // out data transfer type. RW = 1 is read byte; RW = 0 is write data
    case SMBMtDbAck:
        if (Byte2Send==0){          // all bytes to be transmitted have been sent
            if (RW) { // If R/W = READ, sent repeated START.
                STO    = 0;
                STA    = 1;
            } else {
                STO    = 1;         // If a write, generate a STOP condition.
                SMbusy = 0;         // release SMBus
            }
        }
        else { // send next data byte
            SMB0DAT = txbuf[wIndex++];
            Byte2Send--;
        }
        break;
    // Master Transmitter: Data byte transmitted. NACK received.
    // Slave not responding. Send STOP followed by START to try again.
    case SMBMtDbNack:
        STO = 1;
        STA = 1;
        break;
    // Master Transmitter: Arbitration lost.
    // Should not occur. If so, restart transfer.
    case SMBMtArbLost:
        STO = 1;
        STA = 1;
        break;
    // Master Receiver: Slave address + READ transmitted. ACK received.
    // Set to transmit NACK after next transfer since it will be the last only byte.
    case SMBMrAddAck:
        if(Byte2Read > 1)
            AA  = 1;                // ACK sent on acknowledge cycle.
```

```
        else
            AA = 0;                 // read one byte only, assert NACK
        break;
    // Master Receiver: Slave address + READ transmitted. NACK received.
    // Slave not responding. Send repeated start to try again.
    case SMBMrAddNack:
        STO = 0;
        STA = 1;
        break;
    // Data byte received. ACK transmitted.
    case SMBMrDbAck:
        Byte2Read--;
        if(Byte2Read == 1)
            AA = 0;                 // NACK the last byte
        else
            AA = 1;
        rxbuf[rIndex++] = SMB0DAT;
        break;
    // Data byte received. NACK transmitted.
    // Read operation has completed. Read data register and send STOP.
    case SMBMrDbNack:
        rxbuf[rIndex] = SMB0DAT;
        STO    = 1;                 // assert STOP condition
        AA     = 1;                 // set AA for next transfer
        SMbusy = 0;                 // Free SMBus
        break;
    // All other status codes meaningless in this application. Reset communication.
    default:
        STO    = 1;                 // Reset communication.
        SMbusy = 0;
        break;
    }
    SI = 0;                         // clear interrupt flag
}
```

This example does not implement all of the functions required in the applications of the DS1631A. However, it is straightforward to add those functions, and hence, they will be left for you as exercise problems.

11.7 Interfacing with I^2C Serial Real-Time Clock DS1337

Keeping track of time-of-day is one of the important applications in many embedded systems. Without a dedicated chip, it can take up a significant amount of CPU time. The DS1337 from Maxim-IC is designed to ease the time-keeping task. The DS1337 provides the clock/calendar function that keeps track of seconds, minutes, day, date, month, and year information in BCD format. The date at the end of the month automatically is adjusted for months with fewer than 31 days, including corrections for leap year. The clock operates in either the 24-hour or 12-hour format with an AM/PM indicator. An I^2C interface is provided for MCU to communicate with the DS1337. The pin assignment and block diagram of the DS1337 are shown in Figure 11.26.

11.7.1 Signal Functions

- **V_{CC}, GND:** *DC Power Supply Inputs.* V_{CC} is the DC supply and can be from 1.8 to 5.5 V in normal operation. When V_{CC} is lower than 1.8 V but above 1.3 V, the DS1337 is in time-keeping mode only.
- **$\overline{\text{INTA}}$:** *Interrupt Output A.* When enabled, $\overline{\text{INTA}}$ is asserted low when the time/day/data matches the values set in the alarm registers. This pin is an open-drain output and requires an external pullup resistor.
- **SDA:** *Serial Data Input/Output.* This pin is the serial data pin in I^2C (or SMBus) bus.
- **SCL:** *Serial Clock Input.* This pin is used to synchronize data movement on the I^2C (or SMBus) bus.
- **SQW/$\overline{\text{INTB}}$:** *Square-Wave/Interrupt Output B.* This pin is used as a programmable square-wave or interrupt-output signal. It is an open-drain output and requires an external pullup resistor.
- **X1, X2:** *Connections for Crystal Oscillator.* These two pins are used to connect to a standard 32.768-kHz quartz crystal. The internal oscillator circuitry is designed for operation with a crystal having a specified load capacitance (C_L) of 6 pF. The user also can use an external 32.768-kHz oscillator

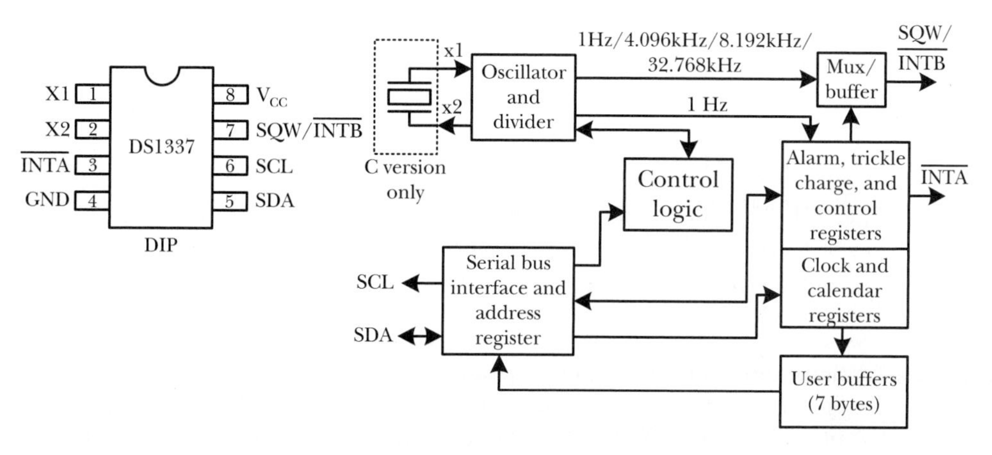

FIGURE 11.26
DS1337 pin assignment and block diagram.

instead of the quartz crystal. In this configuration, the external oscillator signal is connected to the X1 pin, and the X2 pin is floated. The DS1337C package already has integrated a standard 32.768-kHz crystal.

11.7.2 Address Map

The DS1337 has many registers to support its operation. The address map of the DS1337 is shown in Figure 11.27. The DS1337 has an address pointer that points to the register being accessed. After accessing a DS1337 register, the address pointer is incremented to point to the next register. During a multi-byte access, when the address pointer reaches the end of the register space (0x0F), it wraps to location 0x00. On an I^2C start, stop, or address pointer incrementing to location 0, the current time is transferred to a second set of registers. The time information is read from these secondary registers, while the clock may continue to run.

11.7.3 Clock and Calendar

The contents of the time and calendar registers are in the BCD format in Figure 11.27 and must be initialized by the software. The day-of-week register increments at midnight. Values that correspond to the day-of-week are user-defined but must be sequential (i.e., if 1 equals Sunday, then 2 equals Monday, and so on).

The clock and calendar registers are updated once per second. The DS1337 can operate in either 12-hour or 24-hour mode. Bit 6 of the hours register is defined as the 12- or 24-hour mode-select bit. When bit 6 is high, the 12-hour mode is selected. In the 12-hour mode, bit 5 is the $\overline{\text{AM}}$/PM bit with logic-high being PM. In the 24-hour mode, bit 5 is the second 10-hour bit (20 to 23 hours). All hour values, including the alarms, must be reinitialized whenever the 12/$\overline{24}$-hour mode bit is changed. The century bit (bit 7 of the month register) is toggled when the year register overflows from 99 to 00.

11.7.4 Special Registers

The DS1337 has two registers (control and status) that control the functioning of alarms and square-wave output. The contents of the control register and status register are shown in Figures 11.28 and 11.29, respectively.

The control register allows the user to perform the following functions.

- Start and stop the on-chip oscillator
- Select the frequency (there are four possibilities) of the SQW/ INTB output. The SQW/ INTB pin produces square wave output with the frequency specified by the RS2 and RS1 bits whenever the INTCN bit is 0. This signal can be used in many applications.

Address	Bit 7	Bit 6	Bit 5	Bit 4	Bit 3	Bit 2	Bit 1	Bit 0	Range
0x00	0	10 Second			Second				00–59
0x01	0	10 Minute			Minute				00–59
0x02	0	$12/\overline{24}$	$\overline{AM}$/PM 10-HR	10-HR	Hours				01–12 + $\overline{AM}$/PM 00-23
0x03	0	0	0	0	0	Day			1–7
0x04	0	0	10-Date		Date				1–31
0x05	century	0	0	10 month	Month				Century + 01–12
0x06	10-Year				Year				00–99
0x07	A1M1	10-Second alarm 1			Second alarm 1				00–59
0x08	A1M2	10-Minute alarm 1			Minute alarm 1				00–59
0x09	A1M3	$12/\overline{24}$	$\overline{AM}$/PM 10-HR	10-HR	Hour alarm 1				01–12 + $\overline{AM}$/PM 00–23
0x0A	A1M4	DY/$\overline{DT}$	10Date		Day alarm 1 Date alarm 1				1–7 1–31
0x0B	A2M2	10-Minute alarm 1			Minute alarm 2				00–59
0x0C	A2M3	$12/\overline{24}$	$\overline{AM}$/PM 10-HR	10-HR	Hour alarm 2				01–12 + $\overline{AM}$/PM 00–23
0x0D	A2M4	DY/$\overline{DT}$	10 Date		Day alarm 2 Date alarm 2				1–7 1–31
0x0E	EOSC	0	0	RS2	RS1	INTCN	A2IE	A1IE	Control register
0x0F	OSF	0	0	0	0	0	A2F	A1F	Status register

FIGURE 11.27
Timer keep registers of the DS1337.

- Select the interrupt output on alarm 1 and alarm 2 matches
- Enable and disable alarm 1 and alarm 2 interrupts

11.7.5 Alarms

The DS1337 contains two time-of-day/date alarms. Alarm 1 can be set by writing to registers 0x07 through 0x0A. Alarm 2 can be set by writing to registers 0x0B through 0x0D. The alarms can be programmed to operate in two different modes—each alarm can drive its own separate interrupt output (INTA or INTB) or both alarms can drive a common interrupt output (INTA). Bit 7 of each of the time-of-day/data alarm registers is a mask bit. When all of the mask bits of each alarm are set to 0, an alarm only occurs when the values

FIGURE 11.28
The RTC Control register.

7	6	5	4	3	2	1	0	
$\overline{\text{EOSC}}$	0	0	RS2	RS1	INTCN	A2IE	A1IE	Reset value = 0x18

$\overline{\text{EOSC}}$: Enable oscillator
  0 = oscillator is started
  1 = oscillator is stopped
RS2-RS1: Square wave output frequency
  00 = 1 Hz
  01 = 4.096 kHz
  10 = 8.192 kHz
  11 = 32.768 kHz
INTCN: Interrupt control
  0 = A match between the time-keeping registers and either alarm 1 or alarm 2 registers activates the $\overline{\text{INTA}}$ pin
  1 = A match between the time-keeping registers and the alarm 1 registers activates the $\overline{\text{INTA}}$ pin. A match of time-keeping registers and alarm 2 registers activates the SQW/$\overline{\text{INTB}}$ pin
A2IE: Alarm 2 interrupt enable (A2IE)
  0 = The setting of the A2F flag does not initiate an interrupt signal
  1 = The setting of the A2F flag asserts $\overline{\text{INTA}}$ (when INTCON = 0) or the SQW/$\overline{\text{INTB}}$ signal (when INTCN = 1)
A1IE: Alarm 1 interrupt enable (A1IE)
  0 = The setting of the A1F flag does not initiate an interrupt signal
  1 = The setting of the A1F flag asserts $\overline{\text{INTA}}$ signal when INTCN = 1

FIGURE 11.29
The RTC Status register.

7	6	5	4	3	2	1	0	
OSF	0	0	0	0	0	A2F	A1F	Reset value = 0x18

OSF: Oscillator stop flag
  0 = oscillator is running
  1 = oscillator is stopped
A2F: Alarm 2 flag
  0 = no alarm match between alarm 2 registers and timer registers
  1 = the time registers match alarm 2 registers. If the INTCN bit is set to 0 and the A2IE bit is 1, the $\overline{\text{INTA}}$ pin goes low. If the INTCN bit is 1 and the A2IE bit is 1, the SQW/$\overline{\text{INTB}}$ signal goes low. This bit can only be written to 0
A1F: Alarm 1 flag
  0 = no alarm match between alarm 1 registers and timer registers
  1 = the time registers match alarm 1 registers. If the INTCN bit is also set to 1, the $\overline{\text{INTA}}$ pin goes low. This bit can only be written to 0

in the time-keeping registers 0x00 through 0x06 match the values stored in the time-of-day/date alarm registers. The alarm also can be programmed to repeat every second, minute, hour, day, or date. Table 11.4 shows the possible settings. Configurations not listed in the table result in an illegal operation.

The DY/$\overline{\text{DT}}$ bits (bit 6 of the alarm day/date registers) control whether the alarm value stored in bits 0 through 5 of that register reflects the day of the week

TABLE 11.4 *Alarm Mask Bits*

DY/DT	Alarm 1 Register Mask Bits A1M4	A1M3	A1M2	A1M0	Alarm Rate
X	1	1	1	1	Once per second
X	1	1	1	0	When seconds match
X	1	1	0	0	When minutes and seconds match
X	1	0	0	0	When hours, minutes, and seconds match
0	0	0	0	0	When date, hours, minutes, and seconds match
1	0	0	0	0	When day, hours, minutes, and seconds match

DY/DT	Alarm 2 Register Mask Bits A2M4	A2M3	A2M2	Alarm Rate
X	1	1	1	Once per minute (00 seconds of every minute)
X	1	1	0	When minutes match
X	1	0	0	When hours and minutes match
0	0	0	0	When date, hours, and minutes match
1	0	0	0	When day, hours, and minutes match

or the date of the month. If DY/$\overline{\text{DT}}$ is written to logic 0, the alarm is the result of a match with the date of the month. Otherwise, the alarm is the result of a match with day of the week.

When the RTC register values match alarm register settings, the corresponding alarm flag (**A1F** or **A2F**) bit is set to logic 1. If the corresponding alarm interrupt enable (A1IE or A2IE) is also set to logic 1, the alarm condition activates one of the interrupt output ($\overline{\text{INTA}}$ or SQW/$\overline{\text{INTB}}$) signals. The match is tested on the once-per-second update of the time and date registers.

11.7.6 Interfacing the DS1337 with C8051F040

The DS1337 works in both the I²C standard and fast modes. Since the C8051F040 supports the SMBus, the DS1337 can operate only in standard mode when interfacing with the C8051F040. The 7-bit slave address of the DS1337 is 1101000_2. A typical circuit connection between the C8051F040 and the DS1337 is shown in Figure 11.30.

Write Data to DS1337

The C8051F040 needs to set up the time-of-day after power up. The MCU can obtain the time-of-day, alarm times, and control byte from the user by interactive I/O via a keypad, a DIP, or other methods and then send them to the DS1337 in a single block transfer.

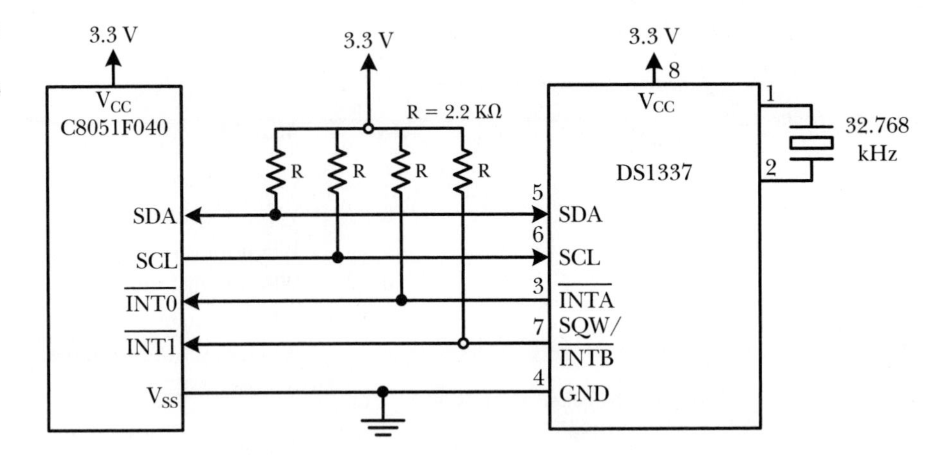

FIGURE 11.30 *Typical circuit connection between the C8051F040 and the DS1337.*

To send data to DS1337, the MCU generates a start condition and then sends out the DS1337's slave ID. The DS1337 will acknowledge and then the MCU will send out start register address, the time-of-day, alarms, and control byte to the DS1337 in one block transfer. After receiving time-of-day, alarm, and control byte from the MCU, the DS1337 starts to update the time once every second.

Reading Data from DS1337

The MCU needs to update and display time-of-day once every second. The procedure for reading data from the DS1337 is defined by the I^2C protocol.

Step 1
The MCU generates a start condition.

Step 2
The MCU sends out the slave ID to the DS1337 with the $R/\overline{W}$ bit set to 0.

Step 3
The MCU sends the start register address to the DS1337.

Step 4
After receiving acknowledgement from the DS1337, the MCU generates a restart condition.

Step 5
The MCU sends the slave ID to the DS1337 with the $R/\overline{W}$ bit set to 1.

Step 6
The MCU reads a block of data from the DS1337 and acknowledge every byte except the last byte.

Step 7

The MCU generates a stop condition to terminate the block read.

Example 11.4 Write a function to write time-of-day into the RTC, a function to read time-of-day from the RTC, and a test program to test these two functions.

Solution: We will use **byte count** to control the state transition in the state machine of the SMBus interrupt service routine. The number of bytes to be written into the RTC chip for the **sendTime** function is eight, which includes seven bytes of time values and one byte of register address. The numbers of bytes to be written and read for the **readTime** function are one and seven, respectively.

The following data structures are required for supporting these two functions and the test main program.

```
char    WrAdr, RdAdr;        // slave address + Write (or Read)
bit     SMbusy;              // This bit is set when a send or receive
                             // is started. It is cleared by the ISR when the operation is finished
bit     RW;                  // R/W command bit. 1=read, 0=write
char    xdata rdbuf[15];     // buffer to hold time read from keyboard
char    xdata txbuf[7];      // buffer of time to be transferred to RTC
char    xdata rxbuf[7];      // buffer of time to be read from RTC
char    timeLoc;             // RAM address in RTC for read and write
char    Byte2Send;           // number of bytes to send to DS1337
char    Byte2Read;           // number of bytes to read from DS1337
char    wIndex;              // index to txbuf
char    rIndex;              // index to rxbuf
```

The main function performs these functions.

- Selects internal oscillator as SYSCLK
- Disables watchdog timer
- Initializes UART and SMBus modules
- Reminds user to enter time-of-day (7 bytes) from keyboard
- Reads in time-of-day from keyboard
- Converts the time-of-day from ASCII string into BCD and stores in **txbuf**
- Calls the **sendTime** function to send the time to the RTC chip
- Calls the **readTime** function to read out the time from the RTC chip
- Outputs each time component on the LCD

The **sendTime**, **readTime**, and the **main()** function are as follows.

```
#include <c8051F040.h>                              // SFR declarations
#include <stdio.h>
#include "uartUtil.h"
#include "delays.h"
#include "lcd_util_040.h"

#define WRITE              0x00         // SMBus WRITE command
#define READ               0x01         // SMBus READ command
#define chipRTC            0xD0         // Device address for RTC (MAX1337)
#define SMBusErr           0x00         // (all modes) BUS ERROR
#define SMBstart           0x08         // (MT & MR) START transmitted
#define SMBrestart         0x10         // (MT & MR) repeated START
#define SMBMtAddAck        0x18         // (MT) Slave address + W transmitted;
                                        // ACK received
#define SMBMtAddNack       0x20         // (MT) Slave address + W transmitted;
                                        // NACK received
#define SMBMtDbAck         0x28         // (MT) data byte transmitted; ACK rec'vd
#define SMBMtDbNack        0x30         // (MT) data byte transmitted; NACK rec'vd
#define SMBMtArbLost       0x38         // (MT) arbitration lost
#define SMBMrAddAck        0x40         // (MR) Slave address + R transmitted;
                                        // ACK received
#define SMBMrAddNack       0x48         // (MR) Slave address + R transmitted;
                                        // NACK received
#define SMBMrDbAck         0x50         // (MR) data byte rec'vd; ACK transmitted
#define SMBMrDbNack        0x58         // (MR) data byte rec'vd; NACK transmitted

char                       WrAdr, RdAdr;  // slave address + Write (or Read)
bit                        SMbusy;        // This bit is set when a send or receive
                                          // is started. It is cleared by the
                                          // ISR when the operation is finished
bit                        RW;            // R/W command bit. 1=read, 0=write

char   xdata rdbuf[15];                   // buffer to hold time read from keyboard
char   *ptr;                              // Time of Day buffer.
const char   *dayOfWeek[] = {0,           // Day of Week buffer.
     "Sunday ",
     "Monday ",
     "Tuesday ",
     "Wednesday ",
     "Thursday ",
     "Friday ",
     "Saturday "};
char   xdata txbuf[7];                    // buffer of time for transfer to RTC
char   xdata rxbuf[7];                    // buffer of time for read from RTC
char   timeLoc;                           // RAM address in RTC for read and write
char   Byte2Send;                         // number of bytes to send to DS1337
char   Byte2Read;                         // number of bytes to read from DS1337
char   wIndex;                            // index to txbuf
char   rIndex;                            // index to rxbuf
```

```
//————————————————————————————————————————————
// Function PROTOTYPES
//————————————————————————————————————————————
void  sysInit     ( void );
void  openSMBus ( void );
void  SMBus_ISR ( void );
void  sendTime ( char chipID );              // send time to RTC chip
void  readTime ( char chipID );              // read time from RTC chip
void  convertTime ( void );                  // convert time read from keyboard
void  Init_TIMER0 ( void );

void main (void) {
    unsigned char seconds, day;

    sysInit();
    openUART0();
    openlcd();
    openSMBus();                             // initialize SMBus

    // request time data through UART
    SFRPAGE = UART0_PAGE;
    printf("\n\nEnter year:month:day (yymmdd) =>");
    getsUART(&rdbuf[0]);
    printf("\nEnter day-of-week (DD) =>");
    getsUART(&rdbuf[6]);
    printf("\nEnter hours:minute:second (hhmmss) =>");
    getsUART(&rdbuf[8]);
    newline();
    // Output the Day and Time on the LCD
    convertTime();
    EA      = 1;                             // enable global interrupts
    timeLoc = 0;                             // write to RTC at time register 0
    sendTime (chipRTC);
    timeLoc = 0;                             // read from RTC at time register 0
    readTime(chipRTC);

    while (1) {
        timeLoc = 0;
        SFRPAGE = SMB0_PAGE;
        readTime(chipRTC);

        if (seconds != rxbuf[0]) {// update the display when the seconds changes
            seconds = rxbuf[0];
            SFRPAGE = 0x0F;                  // change to the lcd page
            // display the time to the second row of LCD
            cmd2lcd(0xC0);                   // move the LCD index to home of the second row
            putc2lcd((rxbuf[2] >> 4) + 0x30);
            putc2lcd((rxbuf[2] & 0x0F) + 0x30);
            putc2lcd(':');
            putc2lcd((rxbuf[1] >> 4) + 0x30);
```

```
                putc2lcd((rxbuf[1] & 0x0F) + 0x30);
                putc2lcd(':');
                putc2lcd((rxbuf[0] >> 4) + 0x30);
                putc2lcd((rxbuf[0] & 0x0F) + 0x30);

                if (day != rxbuf[3]) { // update day of week when it changes
                    day = rxbuf[3];

                    // display the day on the first row of LCD
                    ptr = dayOfWeek[rxbuf[3]];
                    cmd2lcd(0x80);                 // move the LCD index to home of the first row
                    while (*ptr) {
                        putc2lcd(*ptr);
                        ptr++;
                    }
                }
            }
        }
}
void sysInit(void) {
    SFRPAGE  = 0;                          // switch to SFR page 0
    SPI0CN   = 1;                          // enable SPI to 3-wire mode
    SFRPAGE  = 0x0F;                       // switch to SFR page F
    WDTCN    = 0xDE;                       // disable watchdog timer
    WDTCN    = 0xAD;                       //     "
    CLKSEL   = 0;                          // use internal oscillator as SYSCLK
    OSCICN   = 0x83;                       //     "
    XBR0     = 0xF7;                       // assign I/O pins to peripheral function
    XBR1     = 0xFF;                       //     "
    XBR2     = 0x5D;                       //     "
    XBR3     = 0x8F;                       //     "
    P0MDOUT = 0x15;                        // configure SMBus to be open drain
    P7MDOUT = 0xFF ;                       // enable LCD port
}

void openSMBus(void) {
    SMB0CN   = 0x04;                       // SMBus to send ACK on acknowledge cycle
    SMB0CR   = 145;                        // SMBus clock rate = 100 KHz
    SMB0CN  |= 0x40;                       // enable SMBus
    EIE1     = 0x02;                       // enable SMBus interrupt
    SMbusy   = 0;                          // SMBus ready to transmit data
}

// ****************************************************************************************
// The following function converts the input string into BCD values. It combines
// two bytes (ASCII) into one byte (BCD).
// ****************************************************************************************
void convertTime(void) {
    char ix,jk;
    jk = 0;
```

```
        ix = 0;
        for (jk = 6; jk >= 0; jk--) {
            txbuf[jk] = (char)(rdbuf[ix] - 0x30) * 16 + (rdbuf[ix+1] & 0x0F);
            ix += 2;
        }
}

// ************************************************************************************
// The following function sends the current time to the RTC. The current time is
// stored in txbuf[7].
// ************************************************************************************
void sendTime(char chipID) {
        while(SMbusy);                          // wait until SMBus is idle
        SMbusy     = 1;                         // occupies SMBus
        RW         = 0;                         // perform write operation
        SMB0CN     = 0x44;                      // enable SMBus in master mode
        WrAdr      = chipID;                    // set control byte to (RTC ID + W) chip
        Byte2Send  = 8;                         // number of bytes to be transferred
        Byte2Read  = 0;                         // number of bytes to be read
        STO        = 0;                         // clear STO bit
        STA        = 1;                         // start transfer
}

// ************************************************************************************
// The following function reads the current time from the RTC chip.
// ************************************************************************************
void readTime(char chipID) {
        while(SMbusy);
        SMbusy     = 1;                         // occupy SMbus (set to busy)
        RW         = 1;                         // indicate read
        SMB0CN     = 0x44;                      // enable SMBus module to master mode
        WrAdr      = chipID;                    // set control byte to (RTC ID + W)
        RdAdr      = chipID+1;                  // set read control byte to (RTC ID + R)
        Byte2Send  = 1;                         // send one byte (register address)
        Byte2Read  = 7;                         // read 7 bytes
        STO        = 0;
        STA        = 1;                         // start transfer
        while(SMbusy);
}

// ************************************************************************************
// Interrupt Service Routine
// ************************************************************************************
void SMBUS_ISR (void) interrupt 7 {
        switch (SMB0STA) {// Status code for the SMBus (SMB0STA register)
            case SMBusErr:
                STO = 1; // reset SMBus
                break;
            case SMBstart :
                SMB0DAT = WrAdr;                // send slave ID + W
```

```
        STA      = 0;                       // manually clear START bit
        break;
// Master Transmitter/Receiver: Repeated START condition transmitted.
// This state should only occur during a read, after the memory address
// has been sent and acknowledged.
case SMBrestart :
        SMB0DAT = RdAdr;                    // send slave ID + R.
        STA      = 0;                       // manually clear START bit
        break;
// Master Transmitter: Slave address + W transmitted. ACK received.
case SMBMtAddAck :
        SMB0DAT = timeLoc;                  // Load address of RTC register to be accessed.
        wIndex   = 0;                       // write buffer starting index
        rIndex   = 0;                       // read buffer starting index
        Byte2Send--;
        break;
// Master Transmitter: Slave address + WRITE transmitted. NACK received.
// The slave is not responding. Send a STOP followed by a START to try again.
case SMBMtAddNack :
        STO      = 1;
        STA      = 1;
        break;
// Master Transmitter: Data byte transmitted. ACK received.
// This state is used in both READ and WRITE operations. Check RW to find
// out data transfer type. RW = 1 is read byte; RW = 0 is write data
case SMBMtDbAck :
        if (Byte2Send == 0) {               // all bytes to be transmitted have been sent
            if (RW) { // If R/W=READ, sent repeated START.
                STO    = 0;
                STA    = 1;
            }
            else {
                STO    = 1;                 // If a write, generate a STOP condition.
                SMbusy = 0;                 // release SMBus
            }
        }
        else { // send next data byte
            SMB0DAT    = txbuf[wIndex++];
            Byte2Send--;
        }
        break;
// Master Transmitter: Data byte transmitted. NACK received.
// Slave not responding. Send STOP followed by START to try again.
case SMBMtDbNack :
        STO = 1;
        STA = 1;
        break;
// Master Transmitter: Arbitration lost.
// Should not occur. If so, restart transfer.
case SMBMtArbLost :
```

```
            STO     = 1;
            STA     = 1;
            break;
        // Master Receiver: Slave address + READ transmitted. ACK received.
        // Set to transmit NACK after next transfer since it will be the last
        // (only) byte.
        case SMBMrAddAck :
            if(Byte2Read > 1)
                AA = 1;                        // ACK sent on acknowledge cycle.
            else
                AA = 0;                        // read one byte only, assert NACK
            break;
            // Master Receiver: Slave address + READ transmitted. NACK received.
            // Slave not responding. Send repeated start to try again.
        case SMBMrAddNack :
            STO     = 0;
            STA     = 1;
            break;
        // Data byte received. ACK transmitted.
        // State should not occur because AA is set to zero in previous state.
        // Send STOP if state does occur.
        case SMBMrDbAck :
            Byte2Read--;
            if(Byte2Read == 1)
                AA = 0;      // NACK the last byte
            else
                AA = 1;
            rxbuf[rIndex++] = SMB0DAT;
            break;
        // Data byte received. NACK transmitted.
        // Read operation has completed. Read data register and send STOP.
        case SMBMrDbNack :
            rxbuf[rIndex] = SMB0DAT;
            STO     = 1;                       // assert STOP condition
            AA      = 1;                       // set AA for next transfer
            SMbusy = 0;                        // Free SMBus
            break;
        // All other status codes meaningless in this application. Reset communication.
        default :
            STO     = 1;                       // Reset communication.
            SMbusy = 0;
            break;
        }

    SI =0;                                     // clear interrupt flag
}
```

The **readTime** and **sendTime** functions can be modified so that they can read and write from 1 to 16 bytes and start from different register addresses. The modification is straightforward and hence will be left for you as an exercise problem.

The frequency of the SQW output of the DS1337 can be set to 1 Hz and can interrupt the MCU once every second if the $\overline{INT1}$ interrupt is enabled. By doing so, the MCU can be reminded to update the time-of-day every second. The $\overline{INTA}$ output can be used to remind the MCU to turn on the alarm when the alarm time is matched.

11.8 Chapter Summary

The I^2C and SMBus are two compatible serial protocols that are alternatives to the SPI serial interface protocol. Compared with the SPI protocol, the I^2C bus (and SMBus) offers these advantages:

- No chip-enable or chip-select signal for selecting slave devices
- Allows multiple master devices to coexist in a system because it provides easy bus arbitration
- Allows many more devices in the same I^2C bus
- Allows resources to be shared by multiple master (microcontrollers) devices

However, the SPI has these advantages over these I^2C interface:

- Higher data rates (no longer true for I^2C high-speed mode)
- Much lower software overhead to carry out data transmission

Data transfer over the I^2C bus (and SMBus) requires the user to generate the signal components:

1. Start condition
2. Stop condition
3. ACK
4. Restart condition
5. Data

Whenever there are multiple master devices attempting to send data over the I^2C bus (or SMBus), bus arbitration is carried out automatically. The loser is decided whenever it attempts to drive the data line to high, whereas another master device drives the same data line to low.

To select the slave device without using the chip-select (or chip enable) signal, address information is used. Both 7-bit and 10-bit addresses are supported in the same I^2C bus. Ten-bit addressing will be used in a system that consists of many slave devices. The I^2C bus supports three speed rates:

- 100 KHz
- 400 KHz
- 3.4 MHz

The SMBus of the C8051F040 supports only 7-bit addressing and a 100-KHz clock rate.

Each data transfer starts with a start condition and ends with the stop condition. One or two control bytes will follow the start condition, which specifies the slave device to receive or send data. For each data byte, the receiver must assert either the ACK or NACK condition to acknowledge or unacknowledge, respectively, the data transfer.

The 24LC08B is an EEPROM with an I²C interface. The capacity of the 24LC08B is 8-Kbits. This chip has an internal-address pointer that will increment automatically after each access. This feature can increase the access efficiency when the access patterns are sequential.

The DS1631A is a thermostat and digital temperature sensor with the I²C interface. The user can use this device to monitor the ambient temperature and set the high-temperature threshold. When the ambient temperature exceeds the preset high-temperature threshold, the DS1631A may generate an interrupt to remind the MCU to take appropriate actions. The DS1631 can operate at 12-bit, 11-bit, 10-bit, or 9-bit resolutions and operate in one-shot mode or continuous-conversion mode.

The DS1337 is a real-time clock that can keep track of the current time and calendar information. After the current time and calendar information have been set up, the DS1337 will update it once per clock period. As long as the clock frequency input is accurate, the DS1337 can keep track of the time very accurately. The DS1337 also can interrupt the MCU optionally once per second so that the MCU can update the time display.

Data transfer in the SMBus is controlled by a finite-state machine in which the function (or subroutine) starts the data transfer, whereas the SMBus interrupt service routine takes care of the step-by-step transaction. Three examples have been given to illustrate how the SMBus interfaces with the slave devices in this chapter. Designers can modify these functions to deal with other I²C peripheral chips and also provide further processing or data formatting to make the data more user-friendly.

The approach that uses read- and write-byte counts to control the state transition of SMBus interrupt service can be modified easily to accommodate many applications that use SMBus peripheral devices.

11.9 Exercise Problems

E11.1 Does the I²C clock frequency need to be exactly equal to 100 KHz or 400 KHz?
Why?

E11.2 Suppose that the 7-bit address of an I²C slave is **B'10101 A1 A0'** with A1 tied to high and A0 pulled to low. What is the 8-bit hex write address for this device? What is the 8-bit hex read address?

E11.3 Assuming that the C8051F040 is running with a 16-MHz bus clock, compute the values to be written into the SMB0CR register to set the baud rate to 60 KHz and 100 KHz, respectively.

E11.4 Assuming that the C8051F040 is running with a 25-MHz bus clock, compute the values to be written into the SMB0CR register for setting the baud rate to 100 KHz and 40 KHz.

E11.5 Write a C statement to call the **openDS1631** function to configure the DS1631A to operate in one-shot mode, 10-bit resolution, and active-low polarity for T_{OUT} output.

E11.6 Assuming that the DS1631A has been configured to operate in one-shot mode with 12-bit resolution, modify the program in Example 11.3 to display the converted temperature in the format of three integer digits and one fractional digit in LCD.

E11.7 For the circuit shown in Figure 11.24, add an alarm speaker to the Timer 3 output and set the high-temperature trip point to 50°C. Use Timer 3 to generate a two-tone alarm with frequencies set to about 1 kHz and 2 kHz. Whenever the temperature exceeds 50°C, turn on the alarm until the temperature drops down to 23°C.

E11.8 The MAX5812 is a DAC with 12-bit resolution and I^2C interface. The MAX5821 datasheet can be downloaded from the website http://www.maxim-ic.com. What is the 7-bit address of this device? What is the highest operating frequency of this chip? How many commands are available to this chip? How is this device connected to the I^2C bus?

E11.9 For the MAX5812 mentioned in E11.8, write a program to generate a sine wave from the **OUT** pin.

E11.10 The MCP23016 is an I/O expander from Microchip. This chip has an I^2C interface and can add 16 I/O pins to the microcontroller. Download the data sheet of MCP23016 from Microchip's website. Show the circuit connection of this chip to the microcontroller, and write program to configure the GP1.0, . . . , GP1.7 pins for input and configure the GP0.0, . . . , GP0.7 pins for output. Configure input polarity to active-low.

E11.11 What are the corresponding temperatures for the conversion results 0x6800, 0x7200, 0x4800, 0xEE60, and 0xF280 output by the DS1631A? Assume that a 12-bit resolution is used.

E11.12 What will be the conversion results sent out by the DS1631A for the temperature values 40°C, 50°C, 80.5°C, −10.25°C, and −20.5°C? Assume that a 12-bit resolution is used.

E11.13 Write a function that performs a block read from the 24LC08B. The control byte, the starting address to be read, and the number of bytes to be read are passed in R2:R3, R4:R5, and R6, respectively. Return the error code in R7 and store the returned data in a buffer pointed to by DPTR. This function should check the error of no acknowledgement.

E11.14 Write a function that performs a current-address read to the 24LC08B. Pass the control byte in A to this subroutine. Return the data byte and error code in A and B, respectively. This function should check the error that the 24LC08B did not acknowledge.

E11.15 Write a function that performs a page write to the 24LC08B. Read back the data byte-by-byte and display them in the **HyperTerminal** window and LEDs for verification with each byte displayed for 200 ms.

E11.16 Modify the **sendTime** and **readTime** functions in Example 11.4 so that they can be called to read any number (1 to 16) and write any number of bytes starting from any available address.

E11.17 Invoke the two functions that you created in problem E11.16 to set the alarm times to 6:00 AM and 5:00 PM using alarm 1 and alarm 2, respectively. Write appropriate C statements to achieve this goal, assuming the time-of-day has been set to 12-hour mode.

E11.18 For the circuit in Figure 11.30, write a sequence of C statements to configure the SQW output frequency to 1 Hz and enable the INT1 interrupt. Write the INT1 interrupt service routine to read back the current time-of-day and display it on the LCD.

11.11 Laboratory Exercise Problems and Assignments

L11.1 Write a program to store 0 to 255 in the first block of the 24LC08B of your demo board, then stay in an infinite loop to read out one value from 24LC08B sequentially every half-second, and display the value on eight LEDs of the demo board.

L11.2 Connect the DS1631A to the SMBus of your C8051F040 demo board. Write a program to set up the high- and low-temperature trip points to 40°C and 20°C, respectively. Whenever the temperature goes above 40°C, turn on the alarm (speaker). Turn off the alarm when the temperature drops below 20°C. Display the temperature on the LCD display and update the display once every second.

L11.3 Use the DS1337 (if any) on your demo board to set up a time-of-day display using the **HyperTerminal** window. Enter the current time and calendar information using the keyboard. Update the current time-of-day display once a second. In addition, add an alarm time to your program. Whenever the current time matches the alarm, generate the alarm for 1 minute using the buzzer on the demo board.

CHAPTER 12

Analog-to-Digital and Digital-to-Analog Converters

12.1 Objectives

Upon successful completion of this chapter, you will be able to:

- Explain the A/D conversion methods
- Describe the resolution, the various channels, and the operation modes of the C8051F040 A/D converters
- Interpret the A/D conversion results
- Describe the procedure for using the A/D converters of the C8051F040
- Configure the A/D converter for the application
- Use the temperature sensor TC1047A from Microchip
- Use the humidity sensor HIH-4000 from Honeywell
- Use the barometric-pressure sensor MP3H6115A from Freescale
- Understand the operation of the D/A converters of the C8051F040
- Use the D/A converter of the C8051F040 to generate waveforms

12.2 Basics of A/D Conversion

Many embedded applications deal with nonelectric quantities, such as weight, humidity, pressure, weight, massflow, airflow, temperature, light intensity, and speed. These quantities are *analog* in nature because they have a continuous set of values over a given-range—in contrast to the discrete values of digital signals. To enable the microcontroller to process these quantities, they need to be represented in digital form. Thus, a special device that can covert an analog quantity into a digital value is required. This device is referred to as an *analog-to-digital* converter.

12.2.1 A Data Acquisition System

An ADC can only convert an electric voltage. Thus, a nonelectric quantity must be converted into an electric voltage before it can be processed by the ADC. The device that can convert a nonelectric quantity into an electric voltage is referred to as a *transducer*. In general, a transducer is a device that converts the quantity from one form to another. For example, a temperature sensor is a transducer that can convert the temperature into a voltage. A *load cell* is the transducer that can convert a weight into a voltage.

A transducer may not generate an output voltage in the range suitable for A/D conversion. This may happen when the range of a nonelectric quantity of interest is converted into a very narrow range of voltage output. Since the range of voltage output is very narrow, the nonelectric quantity will be converted into a very small range of digital values. Thus, the microcontroller cannot translate the digital value accurately back to the original nonelectric quantity.

If the transducer output is from 0 V to a small voltage (for example, 100 mV), then we can use an amplifier to amplify the transducer output voltage to an appropriate magnitude. If the transducer output is from a voltage to another slightly larger voltage (for example, from 2 to 2.05 V), then we will need a circuit that can shift and scale the transducer output (for example, shift and scale to the range of 0 V to V_{CC}) so that it can be processed by the A/D converter properly. The circuit that performs the scaling and shifting of the transducer output is called a *signal-conditioning circuit*. The overall A/D conversion process is illustrated in Figure 12.1. The A/D converter may be part of the computer.

12.2.2 Analog Voltage and Digital Code Characteristic

An ideal A/D converter should demonstrate the linear input/output relationship shown in Figure 12.2. However, the output characteristic shown in Figure 12.2 is unrealistic, because it requires the A/D converter to use an infinite number of bits to represent the conversion result. The output characteristic of an ideal A/D converter using n bits to represent the conversion result is shown in Figure 12.3. An n-bit A/D converter has 2^n possible output code values. The area between the dotted line and the staircase is called the *quantization error*. The value of $V_{DD}/2^n$ is the resolution of this A/D converter. Using n bits to represent the conversion result, the average *conversion error* is $V_{DD}/2^{n+1}$ if the converter is perfectly linear. For a real A/D converter, the output characteristic may have *nonlinearity* (the staircase may have unequal steps in some values) and *nonmonotonicity* (higher voltage may have smaller code).

Obviously, the more bits used in representing the A/D conversion result, the smaller the conversion error will be. Most microcontrollers use 8 bits, 10 bits, or 12 bits to represent the conversion result. Some microcontrollers (mainly 8051 variants from Silicon Laboratory, TI, and Analog Devices) use

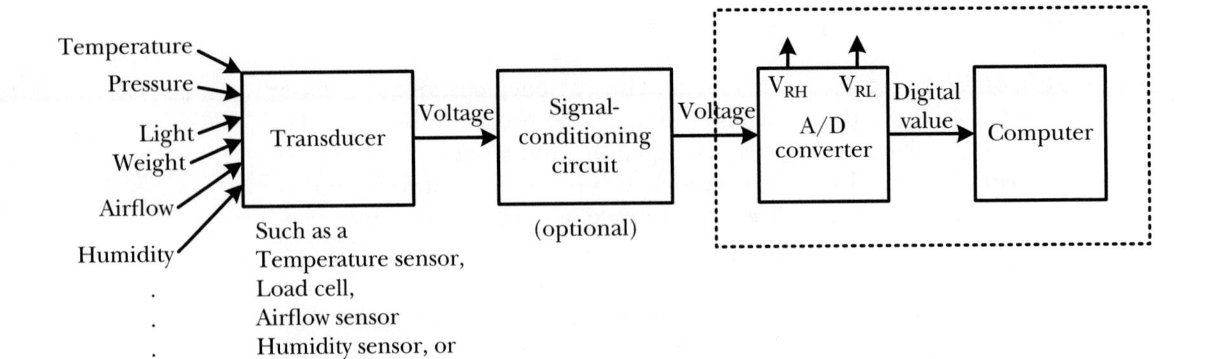

FIGURE 12.1
The A/D conversion process.

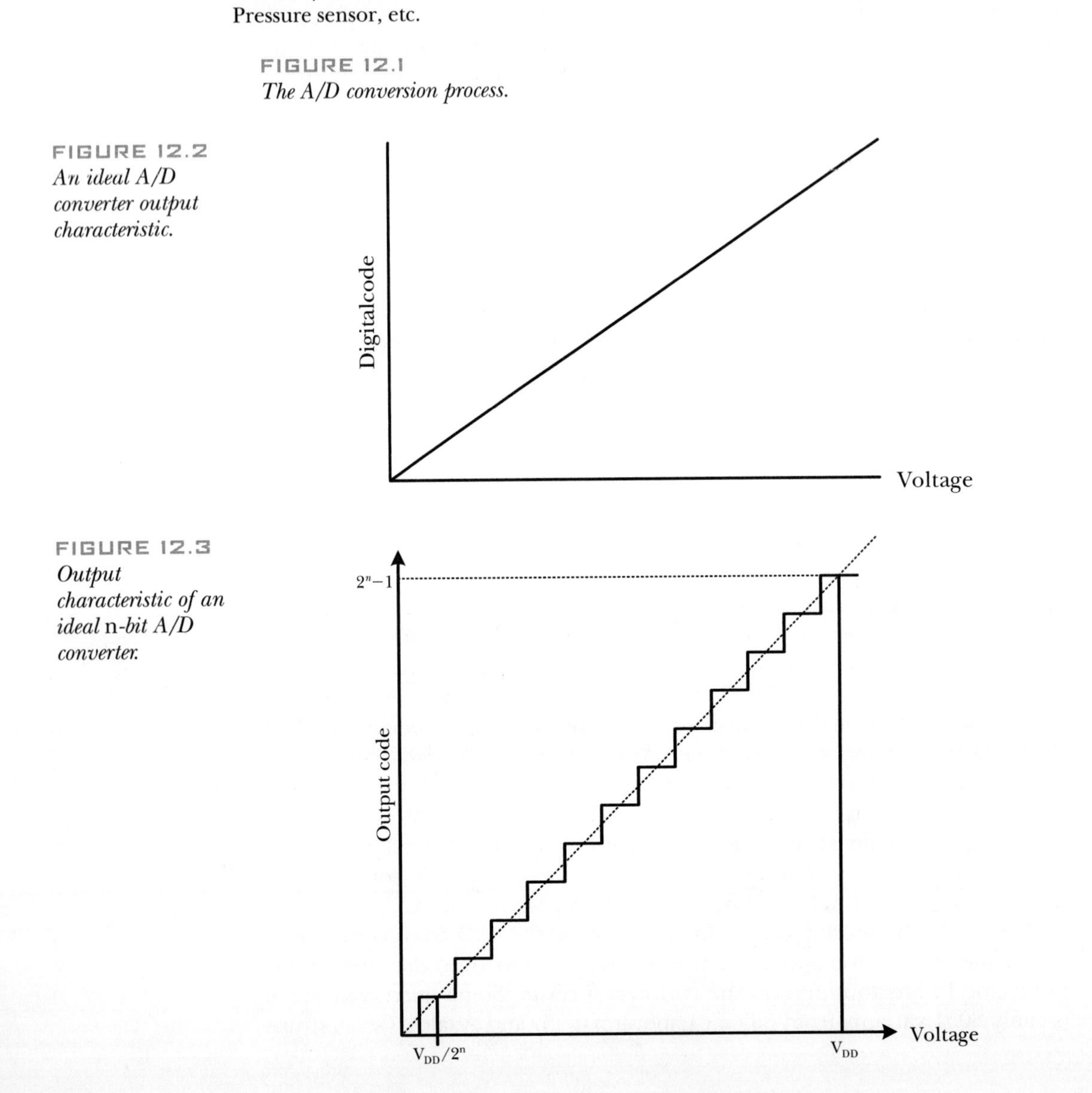

FIGURE 12.2
An ideal A/D converter output characteristic.

FIGURE 12.3
Output characteristic of an ideal n*-bit A/D converter.*

16 bits or even 24 bits to represent conversion results. Whenever the on-chip A/D converter cannot provide the required accuracy, an external A/D converter should be considered.

12.2.3 A/D Conversion Algorithms

Many A/D conversion algorithms have been introduced in the past. These algorithms can be divided into four categories:

1. Parallel (flash) A/D conversion method
2. Slope and double-slope A/D conversion method
3. Sigma-delta A/D conversion method
4. Successive-approximation A/D conversion method

Parallel (Flash) A/D Converters

In this type of A/D converter, 2^n comparators are used. One of the inputs to each comparator is the input voltage to be converted; the other input corresponds to the voltage that represents one of the 2^n combinations of n-bit values. The comparator output will be high whenever the analog input (to be converted) is higher than the voltage that represents one of the 2^n combinations of the n-bit value. The largest n-bit value that causes the comparator output to become true is selected as the A/D conversion value through a priority encoder. It is obvious that this type of A/D converter will be very fast. However, they require a lot of hardware resources to implement and therefore are not suitable for implementing high-resolution A/D converters. This type of A/D converter often is used in applications that require high speed but low resolution, such as a video signal. Over the years, many variations to this approach have been proposed to produce high-speed A/D converters. The most commonly used technique is to pipeline a flash A/D converter, which will reduce the amount of hardware required while still achieving high conversion speed.

Slope and Double-Slope A/D Converters

This type of A/D converter is used in Microchip PIC14000 microcontrollers in which the charging and discharging of a capacitor is used to perform A/D conversion. It requires relatively simple hardware and is popular in low-speed applications, such as digital multimeters. In addition, high resolution (10-bit to 16-bit) can be achieved.

Sigma-Delta A/D Converters

This type of A/D converter uses the *oversampling* technique to perform A/D conversion. It has good noise immunity and can achieve high resolution. Sigma-delta A/D converters are becoming more and more popular in implementing high-resolution A/D converters. The only disadvantage is its slow conversion speed. However, this weakness is improving because of advancements in CMOS technology.

Successive-Approximation A/D Converters

The successive-approximation method approximates the analog signal to *n*-bit code in *n* steps. It may be used for low-frequency applications with large DC noise, such as an electrocardiograph. Let Vx be the voltage to be converted. The procedure of this method is as follows.

Step 1

```
SAR ← 0, i ← n − 1
```

Step 2

```
SAR[i] ← 1
```

Step 3

Convert (using a DAC) the resultant value of SAR into a voltage and compare with Vx.

Step 4

if Vx is smaller, **SAR[i] ← 0.**

Step 5

If ($i == 0$) then stop. Otherwise, decrement i by 1 and go to Step 2.

Because of its balanced speed and precision, this method has become one of the most popular A/D conversion methods. Most microcontrollers use this method to implement the A/D converter. The C8051F040 and many 8051 variants also use this technique to implement A/D converters.

12.2.4 Interpreting A/D Conversion Result

An A/D converter needs a *low reference voltage* (V_{RL}) and a *high reference voltage* (V_{RH}) to perform the conversion. The V_{RL} voltage is often set to ground, whereas the V_{RH} voltage may be set to V_{DD}. Some microcontrollers simply tie V_{RL} to the ground voltage and leave only the V_{RH} voltage programmable. Most A/D converters are *ratiometric*, for the following reasons.

- A 0-V (or V_{RL}) analog input is converted to the digital code of n 0s
- A V_{DD} (or V_{RH}) analog input is converted to the digital code of $\mathbf{2^n - 1}$
- A k-V (or $V_{RL} + k$) input will be converted to the digital code of $\boldsymbol{k \times (2^n - 1) \div V_{DD}}$

Here, $\boldsymbol{n}$ is the number of bits used to represent the A/D conversion result.

The A/D conversion result would be most accurate if the value of the analog signal covers the whole voltage range from V_{RL} to V_{RH}. The A/D conversion result k corresponds to an analog voltage V_K given by the equation

$$V_K = V_{RL} + (\text{range} \times k) \div (2^n - 1) \tag{12.1}$$

where

$$\text{Range} = V_{RH} - V_{RL}$$

Example 12.1 Suppose that there is a 10-bit A/D converter with $V_{RL} = 1$ V and $V_{RH} = 4$ V. Find the corresponding voltage values for the A/D conversion results of 25, 80, 240, 500, 720, 800, and 900.

Solution:

$\text{Range} = V_{RH} - V_{RL} = 4\text{ V} - 1\text{ V} = 3\text{ V}$

The voltages corresponding to the A/D conversion results of 25, 80, 240, 500, 720, 800, and 900 are

$1\text{ V} + (3 \times 25) \div (2^{10} - 1) = 1.07\text{ V}$

$1\text{ V} + (3 \times 80) \div (2^{10} - 1) = 1.23\text{ V}$

$1\text{ V} + (3 \times 240) \div (2^{10} - 1) = 1.70\text{ V}$

$1\text{ V} + (3 \times 500) \div (2^{10} - 1) = 2.47\text{ V}$

$1\text{ V} + (3 \times 720) \div (2^{10} - 1) = 3.11\text{ V}$

$1\text{ V} + (3 \times 800) \div (2^{10} - 1) = 3.35\text{ V}$

$1\text{ V} + (3 \times 900) \div (2^{10} - 1) = 3.64\text{ V}$

12.2.5 Voltage Amplifying Circuit

As described in Section 12.2.2, some transducers have a voltage output ranging from 0 V to a value much smaller than V_{CC}, which may cause the ADC conversion result to be inaccurate. The voltage amplifying (scaling) circuit can be used to improve the accuracy, because it allows the A/D converter to utilize its full dynamic range. The diagram of a possible voltage amplifying circuit is shown in Figure 12.4. Because the OP AMP has infinite input impedance, the current that flows through the resistor R_2 will be the same as the current that flows through R_1. In addition, the voltage at the inverting input terminal (same as the voltage drop across R_1) would be the same as that at the noninverting terminal (V_{IN}). Therefore, the voltage gain of this circuit is given by the equation:

$$A_V = V_{OUT} \div V_{IN} = (R_1 + R_2) \div R_1 = 1 + R_2/R_1 \qquad (12.2)$$

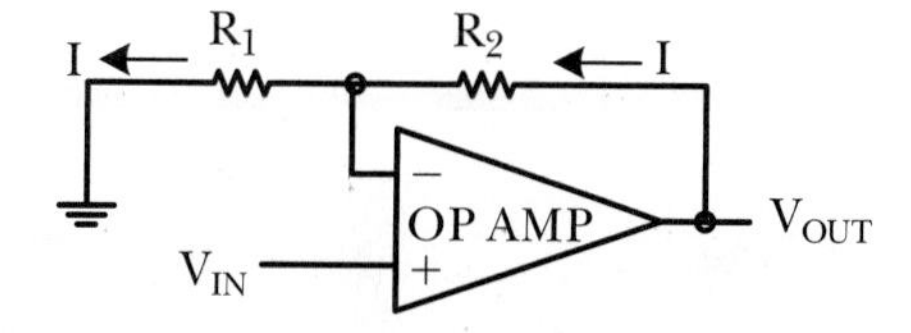

FIGURE 12.4 *A voltage amplifying circuit.*

Example 12.2 Suppose the transducer output voltage ranges from 0 V to 200 mV. Choose the appropriate values for R_1 and R_2 to scale this range to 0 ~ 5 V.

Solution:

$5 \text{ V} \div 200 \text{ mV} = 25 \therefore R_2/R_1 = 24$

By choosing 240 KΩ for R_2 and 10 KΩ for R_1, we obtain a R_2/R_1 ratio of 24 and achieve the desired scaling goal.

12.2.6 Voltage Translation Circuit

Some transducers have output voltage in the range of $V_1 \sim V_2$ (V_1 may be negative and V_2 may not be equal to V_{DD}) instead of 0 V ~ V_{DD}. The accuracy of A/D conversion can be improved by using a circuit that shifts and scales the transducer output so that it covers the whole range of 0 V ~ V_{DD}.

An OP AMP circuit that can shift and scale the transducer output is shown in Figure 12.5. This circuit consists of a summing circuit and an inverting circuit (also called a voltage follower). The voltage V_{IN} comes from the transducer output; V_1 is an **offsetting** voltage. By choosing appropriate values for V_1 and resistors R_0, R_1, R_2, and R_f, the desired voltage shifting and scaling can be achieved. Equation 12.5 shows that the resistance R_0 is an independent variable and can be set to a convenient value.

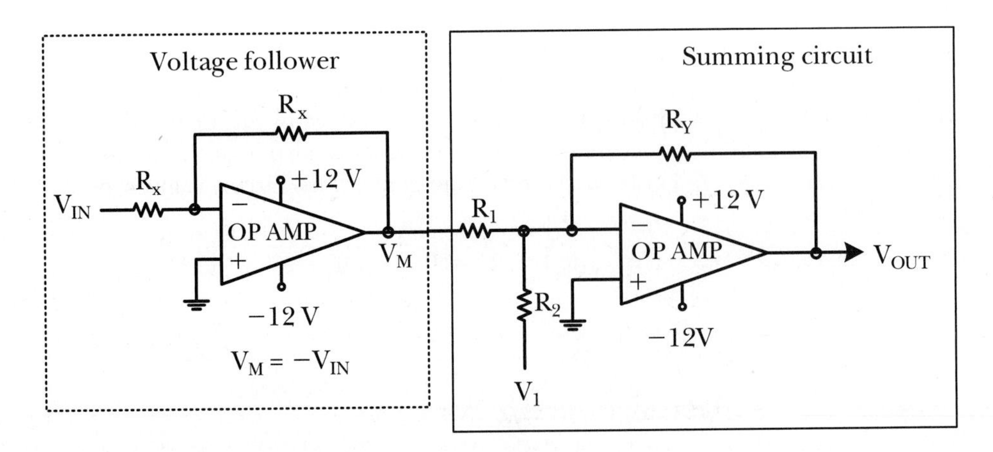

FIGURE 12.5
Level shifting and scaling circuit.

Example 12.3 Choose appropriate resistor values and the offsetting voltage so that the circuit shown in Figure 12.7c can shift the voltage from the range of $-1.2 \sim 3.0$ V to the range of $0 \sim 5$ V.

Solution: Applying Equation 12.5

$$0 = -1.2 \times (R_Y/R_1) - (R_Y/R_2) \times V_1$$

$$5 = 3.0 \times (R_Y/R_1) - (R_Y/R_2) \times V_1$$

By choosing $R_X = R_1 = 10$ KΩ, $R_2 = 50$ KΩ, $R_Y = 12$ KΩ, and $V_1 = -5$ V, one can translate and scale the voltage to the specified range. This example tells us that the selection of resistors and the voltage V_1 is a trial-and-error process.

12.3 The C8051F040 A/D Converters

The C8051F040 has a 12-bit A/D converter (ADC0) and an 8-bit A/D converter (ADC2). Each of these two A/D converters will be discussed in separate sections.

12.4 The 12-bit ADC0 A/D Converter

The ADC0 supports nine channels of input, provides 100-ksps (thousand samples per second) throughput, has a 12-bit resolution, and has a programmable window detector. The functional block diagram of ADC0 is shown in Figure 12.6. The detail of each subsection will be discussed in the following subsections.

12.4.1 Analog Multiplexer and Programmable Gain Amplifier

The 12-bit ADC0 converter uses a 9-to-1 analog multiplexer (AMUX) to handle analog inputs from the sources:

- Four external analog inputs (AIN0.0 through AIN0.3)
- Port 3 port pins
- High-voltage difference amplifier
- An internally connected on-chip temperature sensor

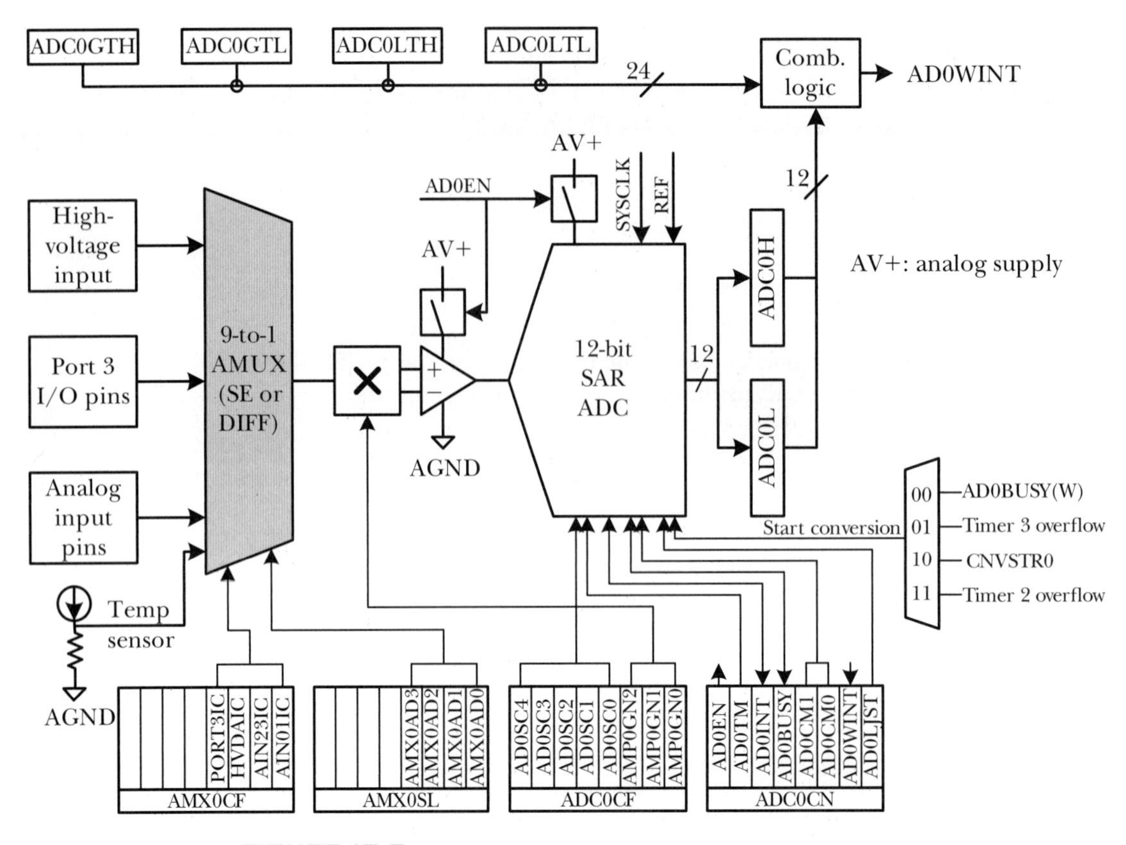

FIGURE 12.6
12-bit ADC0 functional block diagram.

The AMUX input pairs can be programmed to operate in either differential (**DIFF**) or single-ended (**SE**) mode. This allows the user to select the best measurement technique for each input channel and even accommodate mode changes on-the-fly. The AMUX defaults to all single-ended inputs upon reset. There are three registers associated with the functioning of the AMUX: the channel-selection register (AMX0SL), the AMUX configuration register (AMX0CF), and the port pin selection register (AMX0PRT). The analog input diagram is shown in Figure 12.7.

12.4.2 Dedicated External Analog Inputs

AIN0.0 through AIN0.3 are dedicated external analog input pins. These four inputs can be configured to be differential or single-ended analog inputs. The **differential input** configuration has the advantage that it can cancel out common mode signals (could be noise) and hence is suitable for noisy environment applications. In some situations, it can even eliminate the need for

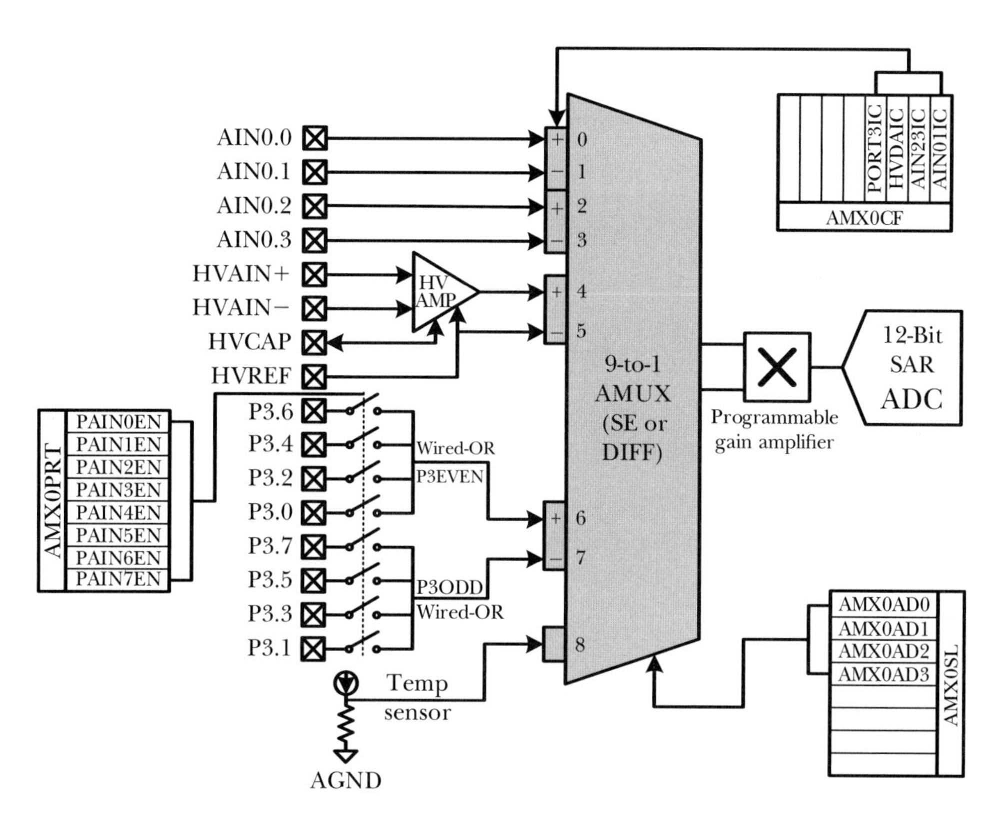

FIGURE 12.7
Analog input diagram.

signal conditioning circuit. The selection of input method for AIN0.0 through AIN0.3 is programmed via the AMX0CF register. The contents of AMX0CF are shown in Figure 12.8. This register also selects input method for analog inputs connected to Port 3 pins and high-voltage analog pins (HVAIN+ and HVAIN−).

12.4.3 High-Voltage Difference Amplifier (HVDA)

The **HVDA** can be used to measure high-differential voltage up to 60 V peak-to-peak, rejecting high common-mode voltages up to ±60 V, and condition the signal voltage range to be suitable for input to ADC0. The input signal may be below AGND to −60 V and as high as +60 V, making the device suitable for both single and dual supply applications. The HVDA provides a common-mode signal for the ADC input range using the on-chip circuitry. The HVDA has two stages. The first stage has a voltage gain equal to 0.05. The gain of the second stage is programmable and ranges from 1 to 280 and hence

7	6	5	4	3	2	1	0	
--	--	--	--	PORT3IC	HVDA2C	AIN23IC	AIN01IC	Value after reset = 0x00
				R/W	R/W	R/W	R/W	

PORT3IC: Port 3 even/odd pin input pair configuration bit
0 = Port 3 even and odd input channels are independent single-ended inputs.
1 = Port 3 even and odd input channels are (respectively) +, − difference input pair
HVDA2C: HVDA 2's complement bit
0 = HVDA output measured as an independent single-ended input
1 = HVDA result for 2's complement value
AIN23IC: AIN0.2 and AIN0.3 input pair configuration bit
0 = AIN0.2 and AIN0.3 are independent single-ended inputs
1 = AIN0.2 and AIN0.3 are (respectively) +, − difference input pair
AIN01IC: AIN0.0 and AIN0.1 input pair configuration bit
0 = AIN0.0 and AIN0.1 are independent single-ended inputs
1 = AIN0.0 and AIN0.1 are (respectively) +, − difference input pair

Note: The ADC0 data word is in 2's complement format for channels configured as difference

FIGURE 12.8
The analog multiplexer configuration register (AMX0CF).

results in an overall HVDA gain ranging from 0.05 to 14. The gain of the high-voltage difference is programmed via the HVA0CN register. The contents of HVA0CN are shown in Figure 12.9. By programming the high-voltage gain control bits properly, the high-voltage input can be scaled to a level that can be handled by the A/D converter.

The HVDA uses four external pins: +HVAIN, −HVAIN, HVCAP, and HVREF. The +HVAIN and −HVAIN inputs serve as the differential inputs to the HVDA. The HVREF input should be used to provide a common-mode reference for input to ADC0, and to prevent the output of the HVDA circuit from saturating. The output of the HVDA circuit, which is calculated by Equation 12.6, must remain within the output voltage range specification (0.1 V to V_{CC} − 0.1 V). As shown in Figure 12.9, the HVREF input serves as the inverting input when the differential input mode is selected.

$$\mathbf{V_{OUT} = [(HVAIN+) - (HVAIN-)] \times Gain + HVREF} \qquad (12.6)$$

12.4.4 Additional Analog Input from Port 3

Up to eight additional external analog signals can be connected to Port 3 pins. The even-numbered inputs are wired-OR together whereas the odd-numbered inputs are also wired-OR together. These analog inputs must be configured to be either single-ended inputs or differential input pairs all together. It is impossible to configure some of these inputs to be single-ended and others to be differential pairs. The Port 3 pin selection is controlled by the AMX0PRT register. When the corresponding bit in AMX0PRT of a Port 3 pin is set to 1, it is allowed to enter the AMUX and be converted. The contents of the AMX0PRT register are shown in Figure 12.10.

7	6	5	4	3	2	1	0	
HVDAEN	--	--	--	HVGAIN3	HVGAIN2	HVGAIN1	HVGAIN0	Value after reset = 0x00
R/W				R/W	R/W	R/W	R/W	

HVDAEN: High-voltage difference amplifier (HVDA) enable bit
0 = the HVDA is disabled
1 = the HVDA is enabled
HVGAIN3-HVGAIN0: HVDA gain control bits

0000 : gain = 0.05	1000 : gain = 1.0
0001 : gain = 0.1	1001 : gain = 1.6
0010 : gain = 0.125	1010 : gain = 2.0
0011 : gain = 0.2	1011 : gain = 3.2
0100 : gain = 0.25	1100 : gain = 4.0
0101 : gain = 0.4	1101 : gain = 6.2
0110 : gain = 0.5	1110 : gain = 7.6
0111 : gain = 0.8	1111 : gain = 14

FIGURE 12.9
High-voltage difference amplifier control (HVA0CN).

7	6	5	4	3	2	1	0	
PAIN7EN	PAIN6EN	PAIN5EN	PAIN4EN	PAIN3EN	PAIN2EN	PAIN1EN	PAIN0EN	Value after reset = 0x00
R/W				R/W	R/W	R/W	R/W	

PAINxEN: Pin x analog input enable bit (x = 0,...,7)
0 = P3.x is not selected as an analog input to the AMUX
1 = P3.x is selected as an analog input to the AMUX
Note: Any number of Port 3 pins may be selected simultaneously inputs to the AMUX. Odd numbered and even numbered pins that are selected simultaneously are shorted together as "wired-OR"

FIGURE 12.10
Port 3 pin selection register (AMX0PRT).

7	6	5	4	3	2	1	0	
--	--	--	--	AMX0AD3	AMX0AD2	AMX0AD1	AMX0AD0	Value after reset = 0x00
				R/W	R/W	R/W	R/W	

AMX0AD3-0: AMX0 address bits
0000−1111: ADC inputs selected are shown in Table 12.1

FIGURE 12.11
AMUX0 channel-selection register (AMX0SL).

12.4.5 Analog Multiplexer Channel Selection

The analog source selected for conversion is controlled by the AMX0CF register and the AMX0SL register. The contents of the AMX0SL register are shown in Figure 12.11. The resultant analog selection chart is shown in Table 12.1.

TABLE 12.1 *AMUX Selection Chart (AMX0AD3-0 and AMX0CF3-0 Bits)*

		AMX0SL Bits 3-0								
		0000	**0001**	**0010**	**0011**	**0100**	**0101**	**0110**	**0111**	**1xxx**
AMX0CF Bits 3-0	0000	AIN0.0	AIN0.1	AIN0.2	AIN0.3	HVDA	AGND	P3EVEN	P30DD	TEMP SENSOR
	0001	+(AIN0.0) −(AIN0.1)		AIN0.2	AIN0.3	HVDA	AGND	P3EVEN	P30DD	TEMP SENSOR
	0010	AIN0.0	AIN0.1	+(AIN0.2) −(AIN0.3)		HVDA	AGND	P3EVEN	P30DD	TEMP SENSOR
	0011	+(AIN0.0) −(AIN0.1)		+(AIN0.2) −(AIN0.3)		HVDA	AGND	P3EVEN	P30DD	TEMP SENSOR
	0100	AIN0.0	AIN0.1	AIN0.2	AIN0.3	+(HVDA) −(HVREF)		P3EVEN	P30DD	TEMP SENSOR
	0101	+(AIN0.0) −(AIN0.1)		AIN0.2	AIN0.3	+(HVDA) −(HVREF)		P3EVEN	P30DD	TEMP SENSOR
	0110	AIN0.0	AIN0.1	+(AIN0.2) −(AIN0.3)		+(HVDA) −(HVREF)		P3EVEN	P30DD	TEMP SENSOR
	0111	+(AIN0.0) −(AIN0.1)		+(AIN0.2) −(AIN0.3)		+(HVDA) −(HVREF)		P3EVEN	P30DD	TEMP SENSOR
	1000	AIN0.0	AIN0.1	AIN0.2	AIN0.3	HVDA	AGND	+P3EVEN −P3ODD		TEMP SENSOR
	1001	+(AIN0.0) −(AIN0.1)		AIN0.2	AIN0.3	HVDA	AGND	+P3EVEN −P3ODD		TEMP SENSOR
	1010	AIN0.0	AIN0.1	+(AIN0.2) −(AIN0.3)		HVDA	AGND	+P3EVEN −P3ODD		TEMP SENSOR
	1011	+(AIN0.0) −(AIN0.1)		+(AIN0.2) −(AIN0.3)		HVDA	AGND	+P3EVEN −P3ODD		TEMP SENSOR
	1100	AIN0.0	AIN0.1	AIN0.2	AIN0.3	+(HVDA) −(HVREF)		+P3EVEN −P3ODD		TEMP SENSOR
	1101	+(AIN0.0) −(AIN0.1)		AIN0.2	AIN0.3	+(HVDA) −(HVREF)		+P3EVEN −P3ODD		TEMP SENSOR
	1110	AIN0.0	AIN0.1	+(AIN0.2) −(AIN0.3)		+(HVDA) −(HVREF)		+P3EVEN −P3ODD		TEMP SENSOR
	1111	+(AIN0.0) −(AIN0.1)		+(AIN0.2) −(AIN0.3)		+(HVDA) −(HVREF)		+P3EVEN −P3ODD		TEMP SENSOR

Note. "P3EVEN" denotes even numbered and "P30DD" odd numbered port 3 pins selected in the AMX0PRT register.

12.4.6 Programmable Gain Amplifier (PGA)

The output of AMUX is further amplified by the PGA. The gain of the PGA is from 0.5 to 8 and is set by programming the ADC0CF register. The contents of ADC0CF are shown in Figure 12.12. By choosing the gain properly, the signal-conditioning circuit between the transducer and the A/D converter (MCU) sometimes may be eliminated.

12.4.7 A/D Conversion Clock

The ADC0 requires a clock signal (referred to as a **SAR clock**) to perform the A/D conversion. The highest clock frequency for the C8051F040 A/D converter is 2.5 MHz.

The frequency of the A/D conversion clock signal is derived by dividing the system clock, as shown in Figure 12.12. Excluding signal tracking time, a 12-bit A/D conversion takes 15 SAR clock cycles to complete.

12.4.8 A/D Conversion Start Methods

A/D conversion can be started in one of four ways, depending on the programmed states of the ADC0CM1 and ADC0CM0 bits of the ADC0CN register. These methods are

1. Writing a 1 to the AD0BUSY bit of the ADC0CN register
2. A Timer 3 overflow (i.e., timed continuous conversions)
3. A rising edge detected on the external ADC convert start signal, CNVSTR0
4. A Timer 2 overflow (i.e., timed continuous conversions)

7	6	5	4	3	2	1	0	
AD0SC4	AD0SC3	AD0SC2	AD0SC1	AD0SC0	AMP0GN2	AMP0GN1	AMP0GN0	Value after reset = 0xF8
R/W	R/W	R/W	R/W	R/W	R/W	R/W	R/W	

AD0SC4-0: ADC0 SAR conversion clock period bits

The selection of these four bits must satisfy the following expression:

$$\text{AD0SC} \geq (\text{SYSCLK} \div \text{CLK}_{\text{SAR0}}) - 1^* \text{ or } \text{CLK}_{\text{SAR0}} = \text{SYSCLK} \div (\text{AD0SC} + 1)$$

AMP0GN2-0: ADC0 internal amplifier gain (PGA)

000: gain = 1	011: gain = 8
001: gain = 2	10x: gain = 16
010: gain = 4	11x: gain = 0.5

Note: AD0SC is the rounded up result

FIGURE 12.12
ADC0 configuration register (ADC0CF).

The AD0BUSY bit is set to 1 during the conversion process and restored to 0 when conversion is complete. The falling edge of AD0BUSY sets the **AD0INT** flag (bit 5 of ADC0CN) and may trigger an interrupt (when enabled). Converted data is available in the ADC0 result registers ADC0H and ADC0L. Converted data may be left or right justified in the ADC0H:ADC0L register pair depending on the value of the AD0LJST bit in the ADC0CN register. The contents of the ADC0CN register are shown in Figure 12.13.

12.4.9 Analog Input Signal Tracking (Sampling)

Each ADC0 conversion must be preceded by a minimum tracking time for the converter result to be accurate. The AD0TM bit of the ADC0CN register controls the ADC0 track-and-hold mode. In its default state, the ADC0 input is continuously tracked when a conversion is not in progress. When the AD0TM bit is 1, ADC0 operates in low-power tracking mode. In this mode, each conversion is preceded by a tracking period of 3 SAR clock cycles (CLK_{SAR0}) after the start-of-conversion signal (CNVSTR0) is asserted. When the CNVSTR0 signal is used to initiate conversions in low-power tracking mode, ADC0 tracks

7	6	5	4	3	2	1	0	
AD0EN	AD0TM	AD0INT	AD0BUSY	AD0CM1	AD0CM0	AD0WINT	AD0LJST	Value after reset = 0x00
R/W	R/W	R/W	R/W	R/W	R/W	R/W	R/W	

AD0EN: ADC0 enable bit
  0 = ADC0 disabled. ADC0 is in low-power shutdown
  1 = ADC0 enabled. ADC0 is active and ready for data conversion
AD0TM: ADC track mode
  0 = Tracking ADC is enabled when ADC is enabled unless a conversion is in progress
  1 = Tracking is defined by AD0CM1-0 bits
AD0INT: ADC0 conversion complete interrupt flag
  0 = ADC0 has not completed a data conversion since the last time this flag is cleared
  1 = ADC0 has completed a data conversion
AD0BUSY: ADC0 busy bit
  0 = No ADC conversion is in progress
  1 = ADC0 conversion is in progress (MCU writes 1 to this bit to start a new conversion)
AD0CM1-AD0CM0: ADC0 start of conversion mode select
  See Table 12.2.
AD0WINT: ADC0 window compare interrupt flag
  0 = ADC0 window comparison data match has not occurred since this flag was last cleared
  1 = ADC0 window comparison data match has occurred
AD0LJST: ADC0 left justified select
  0 = Data in ADC0H:ADC0L registers are right-justified
  1 = Data in ADC0H:ADC0L registers are left-justified

FIGURE 12.13
ADC0 control register (ADC0CN).

TABLE 12.2 *ADC0 Tracking Mode Select*

AD0CM1-0	AD0TM = 0	AD0TM = 1
00	ADC0 conversion initiated on every write of 1 to the AD0BUSY bit	Tracking starts with the write of 1 to the AD0BUSY bit and lasts for 3 CLK_{SAR0}, followed by conversion
01	ADC0 conversion initiated on overflow of Timer 3	Tracking started by the overflow of Timer 3 and lasts for 3 CLK_{SAR0}, followed by conversion
10	ADC0 conversion initiated on the rising edge of the CNVSTR0 signal	ADC0 tracks only when CNVSTR0 input is low; conversion starts on rising edge of the CNVSTR0 signal
11	ADC0 conversion initiated on overflow of Timer 2	Tracking started by the overflow of Timer 2 and lasts for 3 CLK_{SAR0}, followed by conversion

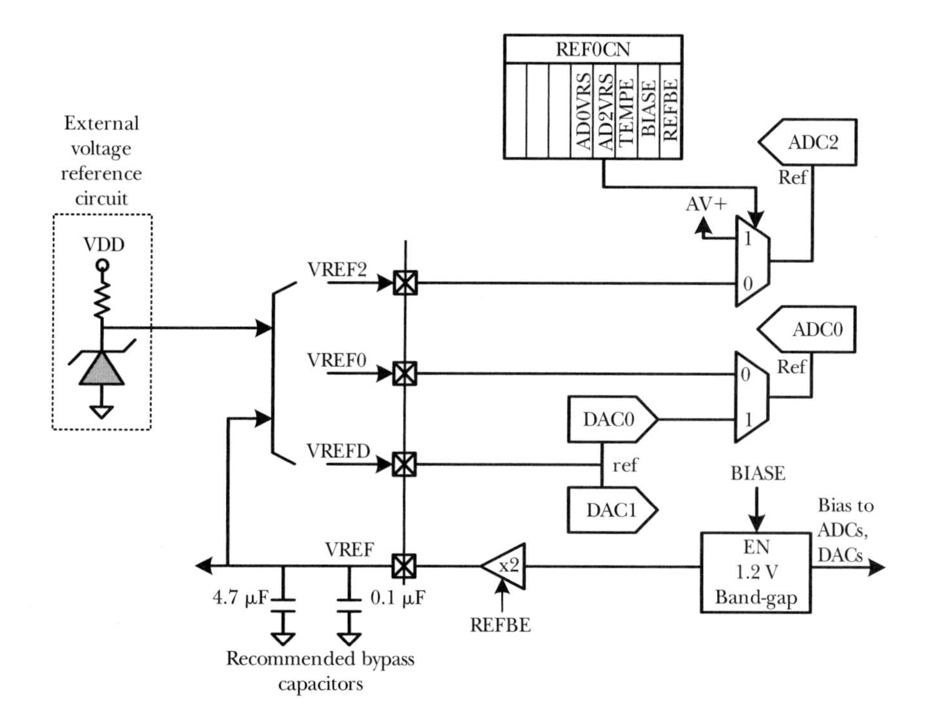

FIGURE 12.14 *C8051F040 ADC and DAC voltage reference functional block diagram.*

only when CNVSTR0 is low; conversion begins on the rising edge of the CNVSTR0 signal. Tracking also can be disabled when the entire chip is in low power, standby, or sleep mode.

12.4.10 A/D and D/A Reference Voltages

Both the analog-to-digital and digital-to-analog converters need reference voltages to perform their conversion operation. The C8051F040 has a voltage reference circuit to support the ADC and DAC modules. As shown in Figure 12.14, the internal reference voltage is a bandgap generator with a 1.2-V output, which can be further amplified by 2 before it is driven to the VREF pin. The internal-reference voltage generator must be enabled before it can be used as a reference voltage. Bypass capacitors are recommended from the VREF pin to AGND.

The ADC0 may use the DAC0 output or the VREF0 input as its reference voltage. The user may connect an external voltage or the VREF pin to the VREF0 pin. The ADC2 may use AV+ or VREF2 input as its reference voltage for conversion. Like VREF0, the VREF2 pin can be connected to an external voltage reference or the VREF pin (internal-reference voltage generator).

The reference control register REF0CN enables/disables the internal-reference generator and selects the reference inputs for ADC0 and ADC2. The BIASE bit in REF0CN enables the bandgap-reference generator, whereas the REFBE bit enables the gain-of-two buffer amplifier which drives the VREF pin. The BIASE bit must be set to 1 if either DAC or ADC is used, regardless of the voltage reference used. The contents of the REF0CN are shown in Figure 12.15.

The typical value of the VREF is 2.43 V when the REFBE bit is set to 1. The turn-on time of the VREF voltage is ranging from 10 μs to 2 ms, depending on whether none or one or two bypass capacitors are used where

- Both the 4.7 μF and 0.1 μF bypass capacitors are used: turn on time is 2 ms
- Only the 0.1 μF bypass capacitor is used: turn on time is 20 μs
- No bypass capacitor is used: turn on time is 10 μs

7	6	5	4	3	2	1	0	
--	--	--	AD0VRS	AD2VRS	TEMPE	BIASE	REFBE	Value after reset = 0x00
R/W	R/W	R/W	R/W	R/W	R/W	R/W	R/W	

AD0VRS: ADC0 voltage reference select
  0: ADC0 voltage reference from VREF0 pin
  1: ADC0 voltage reference from DAC0 output
AD2VRS: ADC2 voltage reference select
  0: ADC2 voltage reference from VREF2 pin
  1: ADC2 voltage reference from AV+
TEMPE: Temperature sensor off
  0: Internal temperature sensor off
  1: Internal temperature sensor on
BIASE: ADC/DAC bias generator enable bit (must be 1 if using ADC or DAC)
  0: Internal bias generator off
  1: Internal bias generator on
REFBE: internal reference buffer enable bit
  0: Internal reference buffer off
  1: Internal reference buffer on. Internal voltage reference is driven on the VREF pin

FIGURE 12.15
Reference voltage control register (REF0CN).

When the internal reference voltage is used, the application must provide enough time for the reference voltage to become stable.

When external-voltage reference is chosen, its value must be at least 1 V but no larger than the analog power supply minus 0.3 V (**AV+ −0.3 V**). AV+ is set to V_{DD} most of the time.

12.4.11 Interpreting the Conversion Result

The input voltage to the A/D converter (12-bit SAR ADC in Figure 12.7) is from **0** volts to **VREF** volts for the single-ended input mode and is from **−VREF/2** to **VREF/2** for the differential mode. The conversion result for the single-ended mode is from 0x0000 to 0x0FFF for a right-justified mode and is from 0x0000 to 0xFFF0 for the left-justified mode. The conversion result is represented in two's complement format when $+VIN_K - VIN_{K+1}$ is negative (*k* is 0 or 2). The range for the differential input is from 0xF800 to 0x07FF for the right-justified mode and is 0x8000 to 0x7FF0 in the left-justified mode. A sample of conversion results is shown in Table 12.3.

TABLE 12.3 *ADC0 Conversion Result Examples*

(a) Single-Ended Mode (AMX0CF = 0x00, AMX0SL = 0x00)

AIN0 (volts)	ADC0H:ADC0L (AD0LJST = 0)	ADC0H:ADC0L (AD0LJST = 1)
VREF * (4095/4096)	0x0FFF	0xFFF0
VREF/2	0x0800	0x8000
VREF * (2047/4096)	0x07FF	0x7FF0
0	0x0000	0x0000

(b) Differential Mode (AMX0CF = 0x01, AMX0SL = 0x00)

AIN0 – AIN1 (volts)	ADC0H:ADC0L (AD0LJST = 0)	ADC0H:ADC0L (AD0LJST = 1)
VREF * (2047/2048)	0x07FF	0x7FF0
VREF/2	0x0400	0x4000
VREF * (1/2048)	0x0001	0x0010
0	0x0000	0x0000
−VREF * (1/2048)	0xFFFF (−1)	0xFFF0
−VREF/2	0xFC00 (−1024)	0xC000
−VREF/1	0xF800 (−2048)	0x8000

12.4.12 ADC0 Programmable Window Detector

The ADC0 of the C8051F040 has the capability to find out if the input analog signal is within a preprogrammed limit and notifies the system. This feature is especially desirable when it is interrupt-driven by setting the AD0WINT bit of the ADC0CN register. The high and low bytes of the reference words are loaded into the **ADC0 Greater-Than** and **ADC0 Less-Than** registers (ADC0GTH, ADC0GTL, ADC0LTH, and ADC0LTL).

Let

x = the value loaded into the register pair ADC0GTH:ADC0GTL
y = the value loaded into the register pair ADC0LTH:ADC0LTL
k = the A/D conversion result

Then the **AD0WINT** flag will be set as follows.

Case 1 ($y > x$):

AD2WINT set to 1 if $x < k < y$
AD2WINT cleared to 0 if $k > y$ or $k < x$

Case 2 ($y < x$):

AD2WINT set to 1 if $k > x$ or $k < y$
AD2WINT cleared to 0 if $y < k < x$

12.4.13 On-Chip Temperature Sensor

The C8051F040 has an on-chip temperature sensor that converts the MCU temperature to voltage. The transfer function of the temperature sensor for PGA = 1 is shown in Figure 12.16. As shown in the figure, the temperature-sensor voltage output span is 0.32 (0.66 to 0.98 V) for the temperature range of −40 to 85°C. Assume that we set VREF to 1.8 V (the lowest breakdown voltage of the commercially available Zener diodes), then it only covers 18.9 percent (= 0.34 ÷ 1.8) of the dynamic range (0 to 1.83 V) of the ADC, and the accuracy may not be satisfactory for your applications. The temperature transfer functions for other PGA values are not available at the time of this writing. When running the MCU at high frequencies, the chip temperature may be much higher than the ambient temperature. Therefore, the temperature reading obtained from A/D conversion could be very misleading.

Example 12.4 A signal with a magnitude from 0.1 to 2.9 V is connected to the AIN0.0 pin. Make appropriate suggestions on how to configure the ADC0 so that this voltage can be measured correctly. Write an instruction sequence to carry out the suggested configuration. Assume that the C8051F040 is operating with a 24-MHz crystal oscillator.

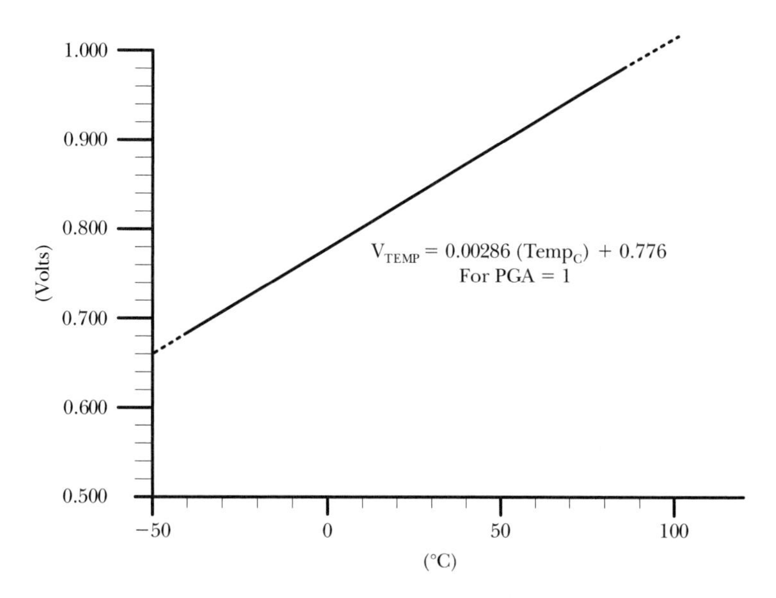

FIGURE 12.16
Temperature-sensor transfer function.

Solution: The considerations for ADC0 configuration are as follows.

- **Reference Voltage** VREF0: Because the unknown voltage ranges from 0.1 to 2.9 V, setting VREF0 to 3.0 V using a Zener diode is easy.
- **Input Mode:** This measurement can be handled by using the single-ended mode.
- **SAR Clock:** Set the SAR clock frequency to 2.4 MHz by setting the upper 5 bits of the ADC0CF register to 01001.
- **PGA Gain:** The PGA gain should be set to 1 to maximize the range of input voltage, while at the same time is not to exceed the VREF0 value. Set the lowest 3 bits of the ADC0CF to 000.
- **A/D Conversion Start Method:** This is not critical. We would use the method that writes a 1 to the AD0BUSY bit.
- **Analog Signal Tracking Method:** This is not critical. Let's use a low-power tracking method to save power.

The A/D circuit connection is shown in Figure 12.17.

The instruction sequence that carries out the specified configuration is as follows.

```
mov     AMX0CF,#0        ; configure all input to single-ended mode
mov     AMX0SL,#0        ; select channel AIN0.0
mov     ADC0CF,#0x48     ; set SAR clock to 2.4 MHz and PGA gain to 1
mov     ADC0CN,#0xC0     ; enable ADC0, select low-power tracking, start
                         ; A/D conversion by writing a 1 to AD0BUSY bit
                         ; A/D result right-justified
mov     REF0CN,#02       ; use VREF0 as voltage reference for ADC0
                         ; enable internal bias
```

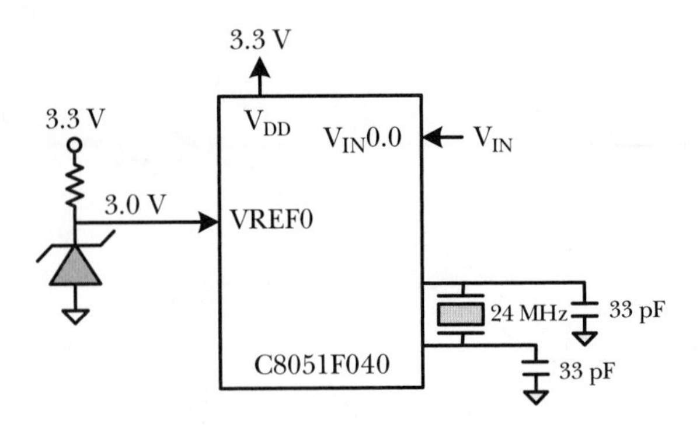

FIGURE 12.17
Circuit connection for Examples 12.4, 12.5, and 12.6.

Example 12.5 Write an instruction sequence to set up greater-than and less-than registers to remind the C8051F040 whenever the input is between 1.0 and 2.0 V for the circuit in Figure 12.17.

Solution: The A/D conversion results for 1.0 and 2.0 V are 1365 (0x555) and 2730 (0xAAA), respectively. The following instruction sequence will interrupt the MCU whenever the input voltage is between 1.0 and 2.0 V.

```
mov     ADC0GTH,#0x05       ; value for right-justified result
mov     ADC0GTL,#0x55       ;
mov     ADC0LTH,#0x0A       ;
mov     ADC0LTL,#0xAA       ;
orl     EIE1,#04            ; enable ADC0 window comparison interrupt
setb    EA                  ; enable interrupt globally
```

Example 12.6 Write a program to measure the voltage connected to the AIN0.0 pin (as shown in Figure 12.17) once every 100 ms and display the result on the LCD using the SSE040 demo board. Display the voltage in one integer digit and one fractional digit.

Solution: The procedure of the program is as follows.

Step 1 Configure the system clock source; assign I/O pins to peripheral functions.

Step 2 Configure the ADC0 to operate with the same setting as that in Example 12.4.

Step 3 Configure the LCD display.

Step 4 Start an A/D conversion and wait until the A/D conversion is complete.

Step 5 Convert the A/D conversion result back to voltage in one integer and one fractional digit format. Display the voltage on the LCD.

Step 6 Wait for 100 ms.

Step 7 Go to Step 4.

Because the conversion result of 3 V is 4095, the A/D conversion result can be translated back to voltage by dividing the conversion result by 1365. The quotient of this division is the integer digit. The fractional digit can be computed by multiplying the remainder by 10 and then dividing the product by 1365.

The assembly program that performs the specified operation is as follows.

```
        $nomod51
        $include (c8051F040.inc)
ii      set     R0              ; divide loop index
tmp     set     R1              ; temporary storage
Rhi     set     R2              ; high byte of register R
Rlo     set     R3              ; low byte of register R
Qhi     set     R4              ; high byte of register Q
Qlo     set     R5              ; low byte of register Q
Nhi     set     R6              ; high byte of register N
Nlo     set     R7              ; high byte of register N

rotate  macro   arg             ; a macro that rotate arg to the left through carry
        mov     A,arg           ;   "
        rlc     A               ;   "
        mov     arg,A           ;   "
        endm                    ;   "

        xseg    at 0x00
volt:   ds      6               ; buffer to hold voltage value
        cseg    at 0x00
        ljmp    start
        org     0xAB
start:  mov     SP,#0x7F        ; set up stack pointer
        lcall   sysinit
        lcall   openLCD         ; initialize LCD
        mov     A,#0x80         ; set cursor to row 0, column 0
        lcall   cmd2lcd         ;   "
        mov     DPTR,#voltage   ; output "voltage = " on LCD
        lcall   puts2lcd        ;   "
        lcall   openADC0        ; initialize ADC0 module
        mov     A,#'.'          ; place the ASCII code of '.' in buffer
        mov     DPTR,#volt+1    ;   "
        movx    @DPTR,A         ;   "
        mov     A,#'V'          ; store ASCII code of V
        mov     DPTR,#volt+3    ;   "
        movx    @DPTR,A         ;   "
        mov     A,#0            ; terminate the buffer with a NULL
        mov     DPTR,#volt+4    ;   "
        movx    @DPTR,A         ;   "
```

```
forever:    orl     ADC0CN,#0x10        ; start an A/D conversion on AIN0.0 pin
testa2d:    mov     A,ADC0CN            ; wait until A/D conversion is completed
            anl     A,#0x10             ;    "
            jnz     testa2d             ;    "
            mov     Qhi,ADC0H           ; pass the dividend
            mov     Qlo,ADC0L           ;    "
            mov     Nhi,#0x05           ; pass the divisor (1365)
            mov     Nlo,#0x55           ;    "
            lcall   div16u              ; call the subroutine to divide
            mov     DPTR,#volt          ; convert the integer digit to ASCII and
            mov     A,Qlo               ; save it in buffer
            add     A,#30H              ;    "
            movx    @DPTR,A             ;    "
            lcall   mulrem              ; multiply remainder by 10
            mov     Nhi,#0x05           ; pass the divisor
            mov     Nlo,#0x55           ;    "
            lcall   div16u              ; compute the fractional digit
            mov     DPTR,#volt+2        ;    "
            mov     A,Qlo               ; convert the fractional digit to ASCII
            add     A,#0x30             ; and save it in buffer
            movx    @DPTR,A             ;    "
            mov     A,#0x8A             ; set cursor position to the 11th byte from left
            lcall   cmd2lcd             ;    "
            mov     DPTR,#volt          ; output the voltage value
            lcall   putsr2lcd           ;    "
            mov     R0,#1               ; wait for 100 ms
            lcall   delayby100ms        ;    "
            ljmp    forever             ; repeat forever

voltage:    db      "voltage = ",0
;************************************************************************************************
; This routine multiplies the 16-bit remainder by 10 and returns the product in Qhi & Qlo
;************************************************************************************************
mulrem:     mov     A,#10
            mov     B,Rlo
            mul     AB
            mov     Qlo,A
            mov     Qhi,B
            mov     A,#10
            mov     B,Rhi               ; Rhi is no larger than 5
            mul     AB                  ;
            add     A,Qhi               ; B should contain 0
            mov     Qhi,A               ; save the high byte of remainder × 10
            ret
;************************************************************************************************
; The following routine performs a 16-bit by 16-bit division and returns the
; quotient and remainder in R4:R5 and R2:R3, respectively.
;************************************************************************************************
div16u:     mov     ii,#0               ; initialize the loop count to 0
            mov     Rhi,#0
```

```
            mov       Rlo,#0
repeat:     inc       ii
; The following instructions shift registers R and Q to the left one place
            clr       C
            rotate    Qlo
            rotate    Qhi
            rotate    Rlo
            rotate    Rhi

; The following instructions perform a division step by performing subtract operations
            clr       C
            mov       A,Rlo
            subb      A,Nlo            ; subtract the lower bytes of two numbers
            mov       tmp,A            ; save the difference in the temporary register
            mov       A,Rhi
            subb      A,Nhi            ; subtract the upper bytes of the 16-bit numbers
            jc        less             ; the minuend (R) is smaller
            mov       Rhi,A            ; store the difference back to the minuend
            mov       A,Qlo
            orl       A,#0x01          ; set the lsb of Q to 1
            mov       Qlo,A            ;     "
            mov       A,tmp            ; also store the lower byte of the difference back
            mov       Rlo,A            ; to the minuend
            ajmp      checkend
less:       mov       A,Qlo
            anl       A,#0xFE          ; set the lsb of Q to 0
            mov       Qlo,A            ;     "
checkend:   cjne      ii,#16,repeat
exit:       ret
; ***************************************************************************************************
; The following routine configures ADC0 to use single-ended input mode on AIN0.0,
; set SAR clock to 2.4 MHz, start A/D conversion by setting the AD0BUSY bit,
; set result to right-justified.
; ***************************************************************************************************
openADC0:   mov       AMX0CF,#0        ; configure all input to single-ended mode
            mov       AMX0SL,#0        ; select channel AIN0.0
            mov       ADC0CF,#0x48     ; set SAR clock to 2.4 MHz and PGA gain to 1
            mov       ADC0CN,#0xC0     ; enable ADC0, select low-power tracking, start
                                       ; A/D conversion by writing a 1 to AD0BUSY bit
                                       ; A/D result right-justified
            mov       REF0CN,#02       ; use VREF0 as voltage reference, enable internal bias
            ret
; ***************************************************************************************************
; The following function disable watchdog timer, select external CMOS clock as
; SYSCLK, assign I/O pins to peripheral signals, enable SPI to 3-wire mode.
; ***************************************************************************************************
sysinit:    mov       SFRPAGE,#0x0F
            mov       WDTCN,#0xDE      ; disable watchdog timer
            mov       WDTCN,#0xAD      ;     "
            mov       CLKSEL,#0        ; use external oscillator to generate SYSCLK
```

```
            mov     OSCXCN,#0x67        ; configure external oscillator control
            mov     R7,#255
againosc:   djnz    R7,againosc         ; wait for about 1 ms
chkstable:  mov     A,OSCXCN            ; wait until external crystal oscillator is stable
            anl     A,#0x80             ; before using it
            jz      chkstable           ;    "
            mov     CLKSEL,#1           ; use external crystal oscillator to generate SYSCLK
            mov     XBR0,#0xF7          ; assign I/O pins to all peripheral
            mov     XBR0,#0xF7          ; assign I/O pins to peripheral signals
            mov     XBR1,#0xFF          ;    "
            mov     XBR2,#0x5D          ; also enable crossbar
            mov     XBR3,#0x8F          ;    "
            mov     SFRPAGE,#0
            mov     SPI0CN,#01          ; enable SPI to 3-wire mode
            ret
            $include (delays.a51)
            $include (lcdUtil.a51)
            end
```

The C language version of the program is as follows.

```
#include <c8051F040.h>
#include <delays.h>                     // add delays.c to the project
#include <lcdUtil.h>                    // add lcdUtil.c and lcdUtil.SRC to the project
void sysinit(void);
void openADC0(void);
void main (void)
{
    char *msg = "voltage = ";
    char  volt[5];
    unsigned int temp;
    sysinit();
    openlcd();
    cmd2lcd(0x80);                      // set cursor to first row, first column
    puts2lcd(msg);                      // output the string "voltage = " on LCD
    openADC0();                         // configure ADC0
    volt[1] = '.';                      // period character
    volt[3] = 'V';                      // letter 'V'
    volt[4] = 0;                        // NULL character
    while(1) {
        ADC0CN |= 0x10;                 // start A/D conversion
        while(ADC0CN & 0x10);           // wait until A/D conversion is done
        temp    = ADC0H*256 + ADC0L;    // combine the low and high bytes
        volt[0] = temp / 1365 + 0x30;   // save integer digit in ASCII code
        volt[2] = ((temp % 1365) * 10)/1365 + 0x30; // store fractional digit in ASCII
        cmd2lcd(0x8A);                  // set LCD cursor to 11th byte from left
        puts2lcd(&volt[0]);             // output the voltage value on LCD
        delayby100ms(1);                // wait for 100 ms
    }
}
```

```
void sysinit(void)
{
    unsigned char cx;
    SFRPAGE = 0x0F;
    WDTCN   = 0xDE;                    // disable watchdog timer
    WDTCN   = 0xAE;                    //    "
    OSCXCN  = 0x67;                    // start external crystal oscillator
    for (cx = 0; cx < 255; cx++);      // wait for about 1 ms
    while(!(OSCXCN & 0x80));           // wait for crystal oscillator to stabilize
    CLKSEL |= 1;                       // switch to external crystal oscillator as clock source
    XBR0    = 0xF7;                    // assign I/O pins to peripheral signals
    XBR1    = 0xFF;                    //    "
    XBR2    = 0x5D;                    // also enable crossbar
    XBR3    = 0x8F;                    //    "
    SFRPAGE = 0;                       // switch to SFR page 0
    SPI0CN  = 0x01;                    // enable SPI to 3-wire mode
}
void openADC0(void)
{
    AMX0CF  = 0;                       // configure all inputs to single-ended mode
    AMX0SL  = 0;                       // select channel AIN0.0
    ADC0CF  = 0x48;                    // set SAR clock to 2.4 MHz and PGA = 1
    ADC0CN  = 0xC0;                    // enable ADC0, low-power tracking, result right-justified
                                       // A/D conversion started by setting AD0BUSY to 1
    REF0CN  = 2;                       // use VREF0 as voltage reference (set to 3 V)
}
```

12.5 The C8051F040 ADC2 A/D Converter

The ADC2 subsystem of the C8051F040 consists of an 8-channel, 8-bit resolution, successive-approximation A/D converter. The ADC2 can complete 500,000 A/D conversions per second and is suitable for those applications that require high throughput but not high resolution. The functional diagram of ADC2 is shown in Figure 12.18.

12.5.1 Analog Multiplexer and PGA

The main function of the analog multiplexer is to select an analog input or the difference of an analog input pair to be converted into a digital value. Each even input and its next adjacent odd input can form a differential pair and converted by the A/D converter. The decision regarding whether to treat two adjacent even and odd inputs as a differential pair is made by programming a bit in the AMX2CF register. The contents of the AMX2CF register are

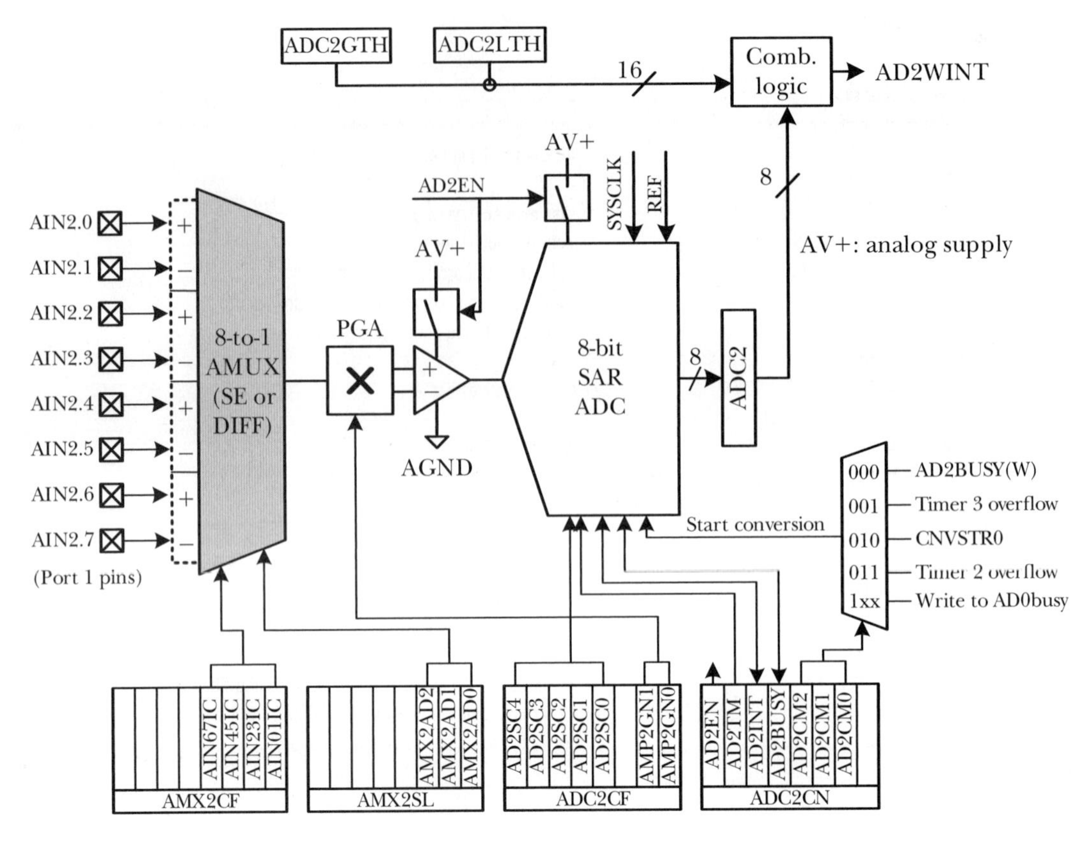

FIGURE 12.18
8-bit ADC2 functional block diagram.

shown in Figure 12.19. The selection of an analog input or a differential input pair to convert is controlled by the AMX2SL register. The contents of AMX2SL are shown in Figure 12.20. The resultant analog input selection for ADC2 is shown in Table 12.4.

The gain of PGA can be 0.5, 1, 2, or 4. The choice of gain is controlled by the ADC2 configuration register (ADC2CF). The contents of ADC2CF are shown in Figure 12.21. The selection of ADC2 clock frequency also is controlled by the ADC2CF register.

12.5.2 ADC2 Operation

ADC2 has a maximum conversion speed of 500 ksps. Like ADC0, the ADC2 conversion clock is a scaled version of the system clock, determined by the ADC2SC bits in the ADC2CF register. The resultant ADC2 conversion clock is the system clock divided by (AD2SC + 1). The maximum ADC2 conversion clock is 6 MHz.

7	6	5	4	3	2	1	0	
--	--	--	--	PIN67IC	PIN45IC	PIN23IC	PIN01IC	Value after reset: 0x00

PIN67IC: P1.6, P1.7 input pair configuration
0 : P1.6 and P1.7 are independent single-ended inputs
1: P1.6, P1.7 are non-inverting and inverting input pair
PIN45IC: P1.4, P1.5 input pair configuration
0 : P1.4 and P1.5 are independent single-ended inputs
1: P1.4, P1.5 are non-inverting and inverting input pair
PIN23IC: P1.2, P1.3 input pair configuration
0: P1.2 and P1.3 are independent single-ended inputs
1: P1.2, P1.3 are non-inverting and inverting input pair
PIN01IC: P1.0, P1.1 input pair configuration
0: P1.0 and P1.1 are independent single-ended inputs
1: P1.0, P1.1 are non-inverting and inverting input pair

FIGURE 12.19
The AMUX2 configuration register (AMX2CF).

7	6	5	4	3	2	1	0	
--	--	--	--	--	AMX2AD2	AMX2AD1	AMX2AD0	Value after reset: 0x00

AMX2AD2-0: AMX2 address bits
000 -111b: ADC inputs selected per Table 12.4.

FIGURE 12.20
The AMUX2 channel select register (AMX2SL).

A conversion can be initiated in one of five ways, depending on the programmed states of the ADC2 **start-of-conversion-mode** bits in the ADC2CN register. The contents of the ADC2CN register are shown in Figure 12.22. The five starting methods of ADC2 are

1. Writing a 1 to the AD2BUSY bit of the ADC2CN register
2. A Timer 3 overflow (i.e., timed continuous conversions)
3. A rising edge detected on the external ADC conversion start signal (CNVSTR2 or CNVSTR0)
4. A Timer 2 overflow (i.e., timed continuous conversions)
5. Writing a 1 to the AD0BUSY of the register ADC0CN

If the signal CNVSTR2 is enabled in the digital crossbar, CNVSTR2 will be the external conversion start signal for ADC2. However, if only CNVSTR0 is enabled in the digital crossbar and CNVSTR2 is not enabled, then CNVSTR0 may serve as the conversion start signal for both the ADC0 and ADC2. This makes synchronous sampling of both ADC0 and ADC2 possible.

During conversion, the AD2BUSY bit is set to logic 1 and restored to 0 when conversion is complete. The falling edge of AD2BUSY triggers an

TABLE 12.4 *AMUX Selection Chart (AMX2AD2-0 and AMX2CF3-0 bits)*

		AMX2AD2-0							
		000	001	010	011	100	101	110	111
AMX2CF Bits 3-0	0000	P1.0	P1.1	P1.2	P1.3	P1.4	P1.5	P1.6	P1.7
	0001	+(P1.0) −(P1.1)	−(P1.0) +(P1.1)	P1.2	P1.3	P1.4	P1.5	P1.6	P1.7
	0010	P1.0	P1.1	+(P1.2) −(P1.3)	−(P1.2) +(P1.3)	P1.4	P1.5	P1.6	P1.7
	0011	+(P1.0) −(P1.1)	−(P1.0) +(P1.1)	+(P1.2) −(P1.3)	−(P1.2) +(P1.3)	P1.4	P1.5	P1.6	P1.7
	0100	P1.0	P1.1	P1.2	P1.3	+(P1.4) −(P1.5)	−(P1.4) +(P1.5)	P1.6	P1.7
	0101	+(P1.0) −(P1.1)	−(P1.0) +(P1.1)	P1.2	P1.3	+(P1.4) −(P1.5)	−(P1.4) +(P1.5)	P1.6	P1.7
	0110	P1.0	P1.1	+(P1.2) −(P1.3)	−(P1.2) +(P1.3)	+(P1.4) −(P1.5)	−(P1.4) +(P1.5)	P1.6	P1.7
	0111	+(P1.0) −(P1.1)	−(P1.0) +(P1.1)	+(P1.2) −(P1.3)	−(P1.2) +(P1.3)	+(P1.4) −(P1.5)	−(P1.4) +(P1.5)	P1.6	P1.7
	1000	P1.0	P1.1	P1.2	P1.3	P1.4	P1.5	+(P1.6) −(P1.7)	−(P1.6) +(P1.7)
	1001	+(P1.0) −(P1.1)	−(P1.0) +(P1.1)	P1.2	P1.3	P1.4	P1.5	+(P1.6) −(P1.7)	−(P1.6) +(P1.7)
	1010	P1.0	P1.1	+(P1.2) −(P1.3)	−(P1.2) +(P1.3)	P1.4	P1.5	+(P1.6) −(P1.7)	−(P1.6) +(P1.7)
	1011	+(P1.0) −(P1.1)	−(P1.0) +(P1.1)	+(P1.2) −(P1.3)	−(P1.2) +(P1.3)	P1.4	P1.5	+(P1.6) −(P1.7)	−(P1.6) +(P1.7)
	1100	P1.0	P1.1	P1.2	P1.3	+(P1.4) −(P1.5)	−(P1.4) +(P1.5)	+(P1.6) −(P1.7)	−(P1.6) +(P1.7)
	1101	+(P1.0) −(P1.1)	−(P1.0) +(P1.1)	P1.2	P1.3	+(P1.4) −(P1.5)	−(P1.4) +(P1.5)	+(P1.6) −(P1.7)	−(P1.6) +(P1.7)
	1110	P1.0	P1.1	+(P1.2) −(P1.3)	−(P1.2) +(P1.3)	+(P1.4) −(P1.5)	−(P1.4) +(P1.5)	+(P1.6) −(P1.7)	−(P1.6) +(P1.7)
	1111	+(P1.0) −(P1.1)	−(P1.0) +(P1.1)	+(P1.2) −(P1.3)	−(P1.2) +(P1.3)	+(P1.4) −(P1.5)	−(P1.4) +(P1.5)	+(P1.6) −(P1.7)	−(P1.6) +(P1.7)

7	6	5	4	3	2	1	0	
AD2SC4	AD2SC3	AD2SC2	AD2SC1	AD2SC0	--	AMP2GN1	AMP2GN0	Value after reset: 0xF8

AD2SC4-0: ADC2 SAR conversion clock period bits
SAR conversion clock is derived by the following equation, where AD2SC refers to the 5-bit value held in AD2SC4-0. The highest SAR clock rate is 6 MHz

$$AD2SC \geq (SYSCLK \div CLK_{SAR2}) - 1 \text{ or } CLK_{SAR2} = SYSCLK \div (AD2SC + 1)$$

AMP2GN1-AMP2GN0: ADC2 internal amplifier gain
00: gain = 0.5
01: gain = 1
10: gain = 2
11: gain = 4

FIGURE 12.21
The ADC2 configuration register (ADC2CF).

7	6	5	4	3	2	1	0	
AD2EN	AD2TM	AD2INT	AD2BUSY	AD2CM2	AD2CM1	AD2CM0	AD2WINT	Value after reset: 0x00

AD2EN: ADC2 enable bit
0 = ADC2 disabled. ADC2 is in low-power shutdown mode
1 = ADC2 enabled. ADC2 is active and ready for data conversion
AD2TM: ADC2 track mode bit
0 = normal track mode. When ADC2 is enabled, tracking is continuous unless a conversion is in progress
1 = low-power track mode. Tracking is defined by AD2CM2-0 bits
AD2INT: ADC2 conversion complete interrupt flag
0 = ADC2 has not completed a data conversion since the last time this flag was cleared
1 = ADC2 has completed a conversion
AD2BUSY: ADC2 busy bit (write 1 to this bit starts the ADC2 conversion)
0 = ADC2 conversion is complete or not busy (write 0 this bit has not effect)
1 = ADC2 conversion is in progress
ADC2CM2-ADC2CM0: ADC2 start of conversion mode select
See Table 12.5.
AD2WINT: ADC2 window compare interrupt flag
0 = ADC2 window comparison data match has not occurred since this flag was last cleared
1 = ADC2 window comparison data match has matched. This flag must be cleared in software

FIGURE 12.22
The ADC2 control register (ADC2CN).

TABLE 12.5 *ADC2 Conversion Start Method Selection*

ADC2CM2-ADC2CM0	AD2TM = 0	AD2TM = 1
000	On every write of 1 to AD2BUSY	Tracking initiated on write of 1 to AD2BUSY and lasts for 3 SAR2 clocks, followed by conversion
001	On every overflow of Timer 3	Tracking initiated on overflow of Timer 3 and lasts for 3 SAR2 clocks, followed by conversion
010	On rising edge of external CNVSTR2 or CNVSTR0	ADC2 tracks only when CNVSTR2 (or CNVSTR0) input is logic low; conversion starts on rising edge of CNVSTR2.
011	On overflow of Timer 2	Tracking initiated on overflow of Timer 2 and lasts for 3 SAR2 clocks, followed by conversion
1xx	On write of 1 to AD0BUSY	Tracking initiated on write 1 to AD0BUSY and lasts for 3 SAR2 clocks, followed by conversion

interrupt and sets the interrupt flag in ADC2CN. Converted data is available in the ADC2 data word, ADC2.

12.5.3 Tracking Modes

Each ADC2 conversion must be preceded by a minimum tracking time for the converted result to be accurate. The AD2TM bit in the ADC2CN register controls the ADC2 track-and-hold mode. In its default state, the ADC2 input is tracked continuously, except when a conversion is in progress. When the AD2TM bit is 1, ADC2 operates in low-power mode. In this mode, each conversion is preceded by a tracking period of 3 SAR2 clocks. When the CNVSTR2 (or CNVSTR0) signal is used to initiate conversions in low-power tracking mode, ADC2 tracks only when CNVSTR2 is low, and conversion begins on the rising edge of CNVSTR2. Tracking also can be disabled when the entire chip is in low-power standby or sleep modes. Low-power track-and-hold mode also is useful when AMUX or PGA settings are changed frequently.

12.5.4 ADC2 Programmable Window Detector

Like the ADC0, the ADC2 also has a programmable window detector, which continuously compares the ADC2 output to user-programmed limits and notifies the system when an out-of-bound condition is detected. This is effective especially in an interrupt-driven system, saving code space and CPU bandwidth while delivering faster system-response times. The window-detector interrupt flag (**AD2WINT** in **ADC2CN**) can also be used in polled mode. The reference words are loaded into the ADC2 greater-than and ADC2 less-than registers (ADC2GT and ADC2LT). The window-detector flag can be asserted when the measured data is inside or outside the user-programmed limits, depending on the programming of the ADC2GT and ADC2LT registers.

Let

x = the value loaded into register pair ADC2GT
y = the value loaded into register pair ADC2LT
k = the A/D conversion result

Then the AD2WINT flag will be set as follows.

Case 1 (y > x).

AD2WINT set to 1 if $x < k < y$
AD2WINT cleared to 0 if $k > y$ or $k < x$

Case 2 (y < x).

AD2WINT set to 1 if $k > x$ or $k < y$
AD2WINT cleared to 0 if $y < k < x$

Example 12.7 Write an instruction sequence to configure ADC2 to operate with the following settings by assuming that the C8051F040 is operating with a 24-MHz external oscillator.

- Enable ADC2
- Convert differential pair P1.6 and P1.7 (P1.6 – P1.7)
- Set SAR2 clock rate to 4 MHz
- PGA gain set to 2
- Use normal track mode and start ADC2 conversion on every write of 1 to AD2BUSY bit
- Use VREF2 as voltage reference
- Set the **AD2WINT** flag whenever the conversion result is between 70H and 90H

Solution: The following instruction sequence will configure ADC2 accordingly.

```
mov     AMX2CF,#080H
mov     AMX2SL,#06H
mov     ADC2CF,#2AH        ; set CLKSAR2 to 4 MHz, set gain to 2
mov     ADC2CN,#80H        ; enable ADC2, start ADC by setting AD2BUSY bit
mov     ADC2GT,#70H        ; set up window values
mov     ADC2LT,#90H        ;    "
mov     REF0CN,#02         ; enable internal bias generator, use VREF2 as
                           ; voltage reference
```

12.6 Using the Temperature Sensor TC1047A

The TC1047A from Microchip is a three-pin temperature sensor whose voltage output is proportional directly to the measured temperature. The TC1047A can accurately measure temperature from −40 to 125°C with a power supply from 2.7 to 5.5 V.

The output voltage range of the TC1047A is 0.1 to 1.75 V for temperature from −40°C to 125°C. The voltage versus temperature relationship is shown in Figure 12.23. The TC1047A has a 10 mV/°C voltage slope output response.

Example 12.8 Describe a circuit connection and write the required program to build a digital temperature thermometer. Display the temperature in three integral and one fractional digits using the LCD. Measure and display the temperature over the whole range of the TC1047A, that is, −40 to +125°C. Update the display data 10 times per second and assume that the C8051F040 operates with a 24-MHz crystal oscillator.

Solution: The voltage-output span of the TC1047A is 1.75 V (1.75 V − 0 V) if we extrapolate the temperature down to −50°C. If we use the ADC0 to perform the conversion and set **VREF0** to 3 V, then the span of the conversion result would be 2389. By dividing the conversion result by 13.65 (or 273/20), the conversion result can be translated back to temperature. Therefore, the inaccuracy would be less than 0.1°C. There is no need to add a signal-conditioning circuit, because the temperature is to be displayed in three integral and one fractional digits. The circuit connection of the TC1047A and C8051F040 is shown in Figure 12.24.

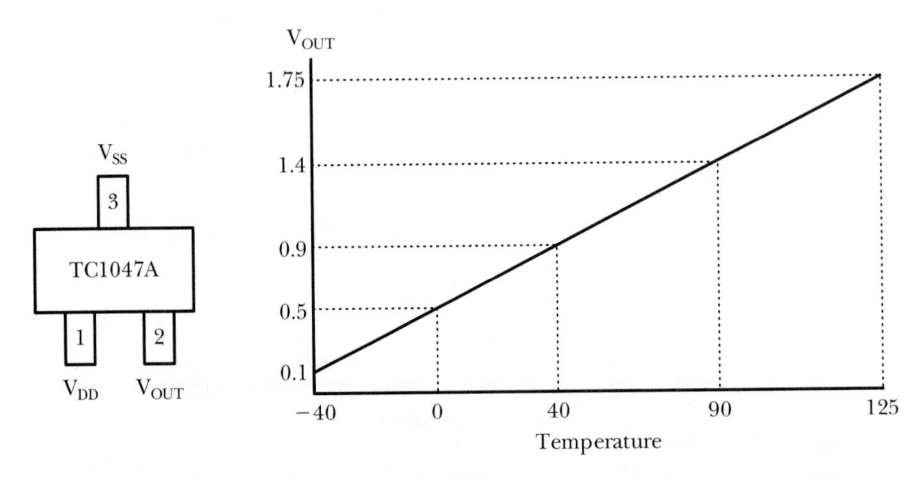

FIGURE 12.23
TC1047A V_{OUT} versus temperature characteristic.

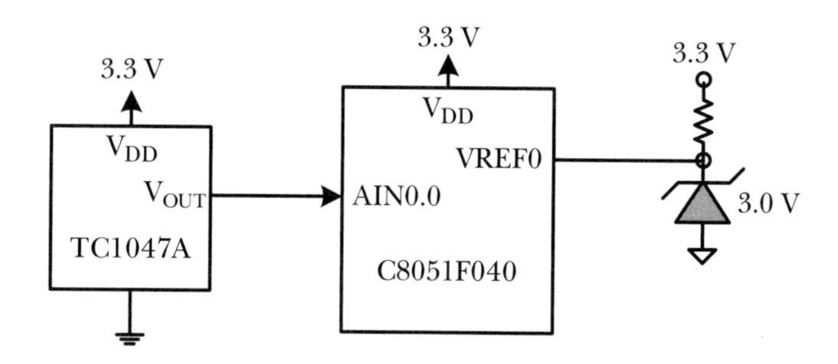

FIGURE 12.24 *Circuit connection between the TC1047A and the C8051F040.*

Since a change of 1°C causes the temperature-sensor outputs to change 10 mV/°C, the voltage output of 0 V will correspond to −50°C by extrapolation. A temperature of 250°C would (corresponding to sensor output 3.0 V) correspond to the A/D conversion result of 4095. After translating the A/D conversion result to temperature, we need to subtract 50 from it to obtain the correct temperature. In this step, the temperature must be complemented in order to compute its magnitude.

The C program that performs the desired operation is as follows.

```
#include <c8051F040.h>
#include <delays.h>                          // include delays.c to the project
#include <lcdUtil.h>                         // include lcdUtil.c & lcdUtil.SRC to the project
void sysinit(void);
void openADC0(void);

void main (void)
{
    char *msg = "temperature = ";
    char temp[8],tmp1;
    unsigned int tmp,tmp2;
    sysinit();
    openlcd();
    cmd2lcd(0x80);                           // set cursor to first row, first column
    puts2lcd(msg);                           // output the string "temperature = " on LCD
    openADC0();                              // initialize ADC0 module
    temp[3] = '.';                           // period character
    temp[5] = 223;                           // ASCII code of degree character
    temp[6] = 'C';                           // letter C
    temp[7] = 0;                             // NULL character
    while(1) {
        ADC0CN |= 0x10;                      // start A/D conversion
        while(ADC0CN & 0x10);                // wait until A/D conversion is done
        tmp = (unsigned int)ADC0H*256 + (unsigned int) ADC0L; // combine low & high bytes
        tmp1 = (tmp * 20)/273;               // integer part of temperature
        tmp2 = (tmp * 20)%273;               // remainder of translation to temperature
        temp[4] = (tmp2 * 10)/273 + 0x30;    // ASCII code of the fractional digit
        tmp1 -= 50;                          // translate to correct temperature scale
```

```
            if (tmp1 < 0){
                tmp1 = −tmp1;                        // takes two's complement of tmp1 to get magnitude
                if (temp[4] != 0x30){
                    temp[4] = 0x3A − temp[4] + 0x30; // ten's complement fractional digit
                    tmp1−−;
                }
                temp[0] = 0x2D;                      // set sign to minus
                temp[2] = tmp1 % 10 + 0x30;          // one's digit ASCII code
                tmp1 /= 10;
                if (tmp1==0)
                    temp[1] = 0x20;                  // place space in case that ten's digit is 0
                else
                    temp[1] = tmp1 + 0x30;           // ten's digit ASCII code
            }
            else {
                temp[2] = tmp1 % 10 + 0x30;          // separate one's digit
                tmp1 /= 10;
                if (tmp1 == 0){
                    temp[0] = 0x20;                  // hundred's digit is 0, place space in it
                    temp[1] = 0x20;                  // ten's digit is 0, place space in it
                }
                else if (tmp1 < 10) {
                    temp[1] = 0x30 + tmp1;
                    temp[0] = 0x20;                  // hundred's digit is 0, place space
                }
                else {
                    temp[1] = tmp1 % 10 + 0x30;
                    temp[0] = 0x31;                  // hundred's digit is 1
                }
            }
            cmd2lcd(0x88);                           // set LCD cursor to 8th byte from left
            puts2lcd(&temp[0]);                      // output the temperature value on LCD
            delayby100ms(1);                         // wait for 100 ms and then repeat
        }
}
void sysinit(void)
{
    unsigned char cx;
    SFRPAGE  = 0x0F;
    WDTCN    = 0xDE;                             // disable watchdog timer
    WDTCN    = 0xAE;                             //      "
    OSCXCN   = 0x67;                             // start external crystal oscillator
    for (cx = 0; cx < 255; cx++);                // wait for about 1 ms
    while(!(OSCXCN & 0x80));                     // wait for crystal oscillator to stabilize
    CLKSEL   |= 1;                               // switch to external crystal oscillator as clock source
    XBR0     = 0xF7;                             // assign I/O pins to peripheral signals
    XBR1     = 0xFF;                             //      "
    XBR2     = 0x5D;                             // also enable crossbar
    XBR3     = 0x8F;                             //      "
    P7MDOUT = 0xFF;                              // enable Port 7 push-pull
```

```
        SFRPAGE  = 0;                 // switch to SFR page 0
        SPIOCN   = 0x01;              // enable SPI to 3-wire mode
}
void openADC0(void)
{
        AMX0CF = 0;                   // configure all inputs to single-ended mode
        AMX0SL = 0;                   // select channel AIN0.0
        ADC0CF = 0x48;                // set SAR clock to 2.4 MHz and PGA = 1
        ADC0CN = 0xC0;                // enable ADC0, low-power tracking, result right-justified
                                      // A/D conversion started by setting AD0BUSY to 1
        REF0CN = 2;                   // use VREF0 as voltage reference (set to 3 V)
}
```

12.7 Measuring Barometric Pressure

Barometric pressure refers to the pressure existing at any point within the earth's atmosphere. This pressure can be measured as an absolute pressure (with reference to absolute vacuum) or can be referenced to some other value or scale. The meteorology and avionics industries traditionally measure the absolute pressure and then reference it to a sea-level pressure value. This complicated process is used in generating maps of weather systems.

Mathematically, atmospheric pressure is related exponentially to altitude. Once the pressure at a particular location and altitude is measured, the pressure at any other altitude can be calculated.

Several units have been used to measure the barometric pressure: in-Hg, kPa, mbar, or psi. A comparison of barometric pressure using four different units at sea level and up to 15,000 feet above sea level is shown in Table 12.6.

There are three forms of pressure transducers: gauge, differential, and absolute. Both the gauge pressure (psig) and differential pressure (psid) transducers measure pressure differentially. The acronym "psi" stands for "pounds per square inch", whereas the letters "g" and "d" stand for "gauge" and "differential," respectively. A gauge pressure transducer measures pressure against

TABLE 12.6 *Altitude versus Pressure Data*

Altitude (ft)	Pressure (in-Hg)	Pressure (mbar)	Pressure (kPa)	Pressure (psi)
0	29.92	1013.4	101.4	14.70
500	29.38	995.1	99.5	14.43
1000	28.85	977.2	97.7	14.17
6000	23.97	811.9	81.2	11.78
10000	20.57	696.7	69.7	10.11
15000	16.86	571.1	57.1	8.28

ambient air, whereas the differential pressure transducer measures against a reference pressure. An absolute pressure transducer measures the pressure against a vacuum (0 psia), and hence, it measures the barometric pressure.

The Freescale MP3H6115A series is an integrated silicon pressure sensor that measures absolute pressure. The MP3H6115A is signal-conditioned, temperature compensated, and calibrated and is suitable for the applications:

- Aviation altimeters
- Industrial controls
- Engine control/manifold absolute pressure (MAP)
- Weather station and weather reporting device barometers

The MP3H6115A is housed in a small package and must be surface-mounted to the printed circuit board. The pressure range that can be measured by the MP3H6115A is from 15 to 115 kPa. The pin assignment of the MP3H6115A is given in Figure 12.25, whereas its operation characteristics are given in Table 12.7. The transfer function of the voltage output of the MP3H6115A is given by the equation:

$$V_{OUT} = V_S \times (0.009 \times P - 0.095) \pm (\text{pressure error} \times \text{temp. factor} \times 0.009 \times V_S) \quad (12.9)$$

where V_S is the power supply to the MP3H6115A and is in the range from 2.7 to 3.3 V. The temperature error factor and pressure error band are shown in Figures 12.26 and 12.27, respectively.

Example 12.9 Compute V_{OUT} of MP3H6115A for P = 115 kPa with V_S = 3.3 V at a room temperature of 25°C.

Solution: Figure 12.27 shows that the pressure error band is 1.5 kPa, whereas Figure 12.26 indicates that temperature error factor at 25°C is 1. Substituting these values along with the V_S value (3.3 V) into Equation 12.9, we obtain

$$\begin{aligned} V_{OUT} &= 3.3 \times (0.009 \times P - 0.095) \pm (1.5 \times 1 \times 0.009 \times 3.3) \\ &= 3.10 \pm 0.04 \text{ V} \end{aligned}$$

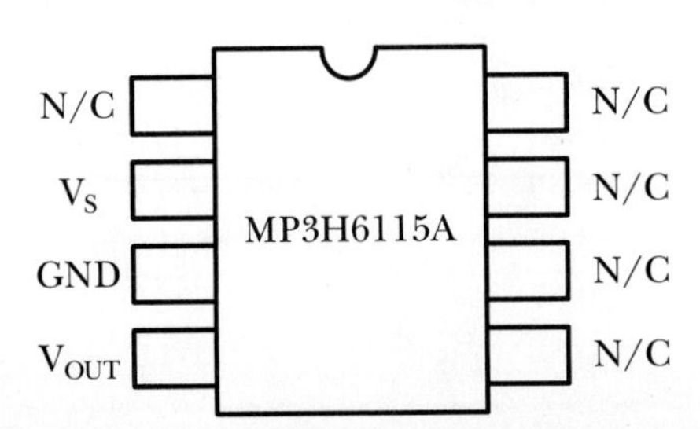

FIGURE 12.25 *The MP3H6115A pressure sensor.*

TABLE 12.7 *Operating Characteristics of the MP3H6115A (V_S = 3.0 V, T_A = 25°C)*

Characteristics	Symbol	Min	Typ	Max	Unit
Pressure range	P_{OP}	15	—	115	kPa
Supply voltage[1]	V_S	2.7	3.0	3.3	V
Supply current	I_O	—	4.0	8.0	mA
Minimum pressure offset[2] @V_S = 3.0 V (0 to 85°C)	V_{OFF}	0.079	0.12	0.161	V
Full scale output[3] @V_S = 3.0 V (0 to 85°C)	V_{FSO}	2.780	2.82	2.861	V
Full scale span[4] @V_S = 3.0 V (0 to 85°C)	V_{FSS}	2.660	2.70	2.741	V
Accuracy[5] (0 to 85°C)	—	—	—	±1.5	$\%V_{FSS}$
Sensitivity	V/P	—	27	—	mV/kPa
Response time[6]	t_R	—	1.0	—	ms
Warm-up time[7]	—	—	20	—	ms
Offset stability[8]	—	—	±0.25	—	$\%V_{FSS}$

Notes:

1. Device is ratiometric within this specified excitation range
2. Offset (V_{OFF}) is defined as the output voltage at the minimum rated pressure
3. Full scale output (V_{FSO}) is defined as the output voltage at the maximum or full rated pressure.
4. Full scale span (V_{FSS}) is defined as the algebraic difference between the output voltage at full rated pressure and the output at the minimum rated pressure.
5. Accuracy is the deviation in actual output from nominal output over the entire pressure range and temperature range as a percent of span at 25°C due to errors in linearity, temperature hysteresis, pressure hysteresis, TcSpan, and TcOffset
6. Response time is defined as the time for the incremental change in the output to go from 10% to 90% of its final value when subjected to a specified step change in pressure.
7. Warm-up time defined as the time required for the product to meet the specified output voltage after the pressure has been stabilized.
8. Offset stability is the product's output deviation when subjected to 1000 cycles of pulsed pressure, temperature cycling with bias test.

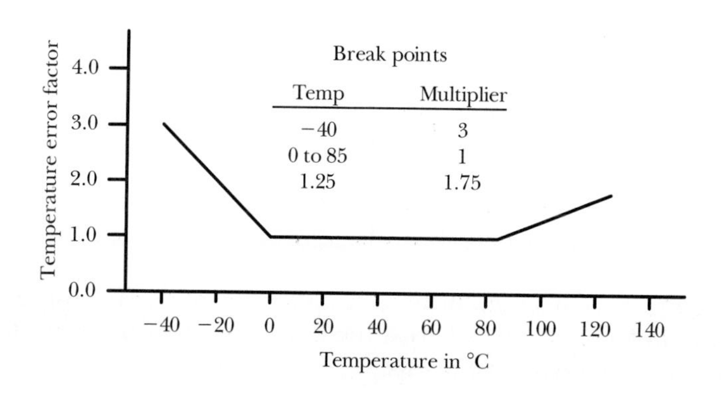

FIGURE 12.26 *Temperature error band of MP3H6115A.*

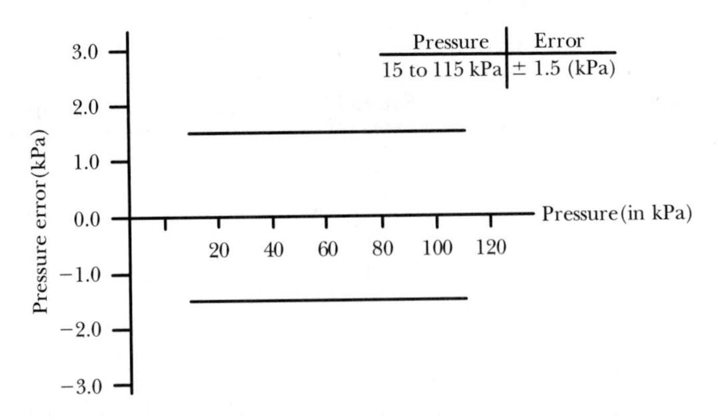

FIGURE 12.27
Pressure error band of MP3H6115A.

Example 12.10 Describe a circuit connection and write the required program to build a digital barometric-pressure meter. Display the pressure in three integral and one fractional digit using the LCD. Measure and display the pressure over the whole range of the MP3H6115A, that is, 15 to 115 kPa. Update the display data 10 times per second and assume that the C8051F040 operates with a 24-MHz crystal oscillator at room temperature.

Solution: Refer to Table 12.7. The full scale span of the MP3H6115A output is 2.7 V at a 3.0-V power supply. There is no need to add a signal-conditioning circuit by following the similar reasoning in Example 12.9.

The pressure span is 100 kPa (from 15 to 115 kPa), whereas the voltage output span is 2.7 V (from 0.12 to 2.82 V). Suppose we set the voltage reference (V_{REF0}) to ADC0 3.0 V, then the ADC conversion result span for the pressure will be 3686 (164 to 3850). To translate the conversion result back to pressure (in the unit of kPa), this equation should be used:

Pressure = (A/D conversion result at P − A/D conversion result at 15 kPa) / 36.86 + 15 (12.10)

≈ (A/D conversion result at P −A/D conversion result at 15 kPa) × 10 / 369 + 15

The circuit connection for measuring the barometric pressure is shown in Figure 12.28, whereas the C program that performs the pressure measurement is as follows.

```
#include <c8051F040.h>
#include < delays.h>        // include delays.c to the project
#include <lcdUtil.h>        // include lcdUtil.c & lcdUtil.SRC to the project
void sysinit(void);
void openADC0(void);

void main (void)
{
    char *msg = "Pressure = ";
    char press[9], tmp1;
    unsigned int tmp, tmp2, tmp3;
    sysinit();
    openlcd();
    cmd2lcd(0x80);                      // set cursor to first row, first column
    puts2lcd(msg);                      // output the string "Pressure = " to the LCD
    openADC0();                         // initialize ADC0 module
    press[3] = '.';                     // period character
    press[5] = 'k';                     // letter k
    press[6] = 'P';                     // letter P
    press[7] = 'a';                     // letter a
    press[8] = 0;                       // NULL character
    delayby10ms(2);                     // wait for pressure sensor to warm up (need 20 ms according to table 12.7)
    while(1) {
        ADC0CN |= 0x10;                 // start A/D conversion
        while(ADC0CN & 0x10);           // wait until A/D conversion is done
        tmp = ((unsigned int)ADC0H*256 + (unsigned int)ADC0L - 164) * 10; // combine low
                                        // and high bytes
        tmp1 = tmp / 369 + 15;          // integer part of the pressure
        tmp2 = tmp % 369;               // remainder of the pressure (after translation)
        press[4] = (tmp2 * 10) / 369;   // compute fractional digit
        tmp3 = (tmp2 * 10) % 369;       // compute the remainder for the fractional digit
        if (tmp3 > 185 || tmp3 == 185){// need to round up?
            press[4]++;
            if (press[4] == 10){ // fractional digit becomes 10 after round up?
                tmp1++;
                press[4] = 0;
            }
        }
        press[4] += 0x30;               // convert fractional digit to ASCII code
        press[2] = tmp1 % 10 + 0x30;    // separate one's digit
        tmp1 /= 10;                     // divide integer part by 10
        if (tmp1 == 0) {
            press[0] = 0x20;            // leading two digits are zero, fill with spaces
            press[1] = 0x20;            //    "
        }
        else {
            press[1] = tmp1 % 10 + 0x30;   // separate ten's digit
            tmp1 /= 10;
            if (tmp1 == 0)
                press[0] = 0x20;        // hundred's digit is zero
```

```
                else
                        press[0] = 0x31;            // otherwise, hundred's digit is 1
            }
            cmd2lcd(0xC8);                          // set LCD cursor to column 8, row 2
            puts2lcd(&press[0]);                    // output the pressure value on LCD
            delayby100ms(1);                        // wait for 100 ms
        }
}
void sysinit(void)
{
        unsigned char cx;
        SFRPAGE  = SPI0_PAGE;
        SPI0CN   = 0x01;                            // enable 3-wire SPI
        SFRPAGE  = 0x0F;
        WDTCN    = 0xDE;                            // disable watchdog timer
        WDTCN    = 0xAE;                            //      "
        OSCXCN   = 0x67;                            // start external crystal oscillator
        for (cx = 0; cx < 255; cx++);               // wait for about 1 ms
        while(!(OSCXCN & 0x80));                    // wait for crystal oscillator to stabilize
        CLKSEL  |= 1;                               // switch to external crystal oscillator as clock source
        XBR0     = 0xF7;                            // assign I/O pins to peripheral signals
        XBR1     = 0xFF;                            //      "
        XBR2     = 0x5D;                            // also enable crossbar
        XBR3     = 0x8F;                            //      "
        P7MDOUT  = 0xFF;                            // enable Port 7 push-pull
        SFRPAGE  = 0;                               // switch to SFR page 0
}
void openADC0(void)
{
        AMX0CF   = 0;                               // configure all inputs to single-ended mode
        AMX0SL   = 0;                               // select channel AIN0.0
        ADC0CF   = 0x48;                            // set SAR clock to 2.4 MHz and PGA = 1
        ADC0CN   = 0xC0;                            // enable ADC0, low-power tracking, result right-justified
                                                    // A/D conversion started by setting AD0BUSY to 1
        REF0CN   = 2;                               // use VREF0 as voltage reference (set to 3 V)
}
```

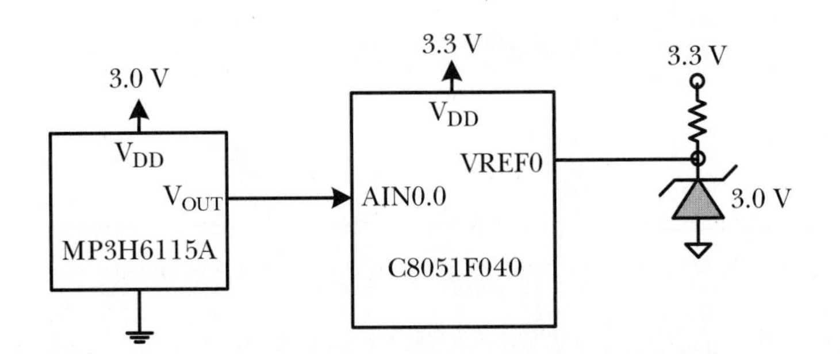

FIGURE 12.28 *Circuit connection between the MP3H6115A and the C8051F040.*

12.8 Measuring Humidity

Honeywell designs and manufactures a family of humidity sensors for a wide variety of applications. The HIH-4000 series humidity sensors are designed to measure relative humidity (RH) for large volume users. This sensor has linear voltage output and hence can be connected directly to a microcontroller. With a typical current draw of 200 μA, the HIH-4000 series is suited for low-drain, battery-powered systems. The HIH-4000 provides good resistance to most application hazards, such as wetting, dust, dirt, oils, and common environment chemicals.

The power supply to the HIH-4000 sensor can be from 4 to 5.8 V but with 5.0 V as its typical value. The voltage output of the HIH-4000 with power supply V_S at 25°C is expressed as

$$V_{OUT} = V_S \times (0.0062 \times (RH) + 0.16) \quad (12.10)$$

The HIH-4000 has three signal pins, as shown in Figure 12.29. At room temperature and a 5-V power supply, the voltage output corresponding to 0 to 100 percent relative humidity is from 0.8 to 3.9 V. The high end of this output range is higher than the C8051F040 power supply (must be lower than 3.6 V). In order to perform A/D conversion, the HIH-4000 voltage output must be scaled down. There are two ways to achieve this goal for the C8051F040:

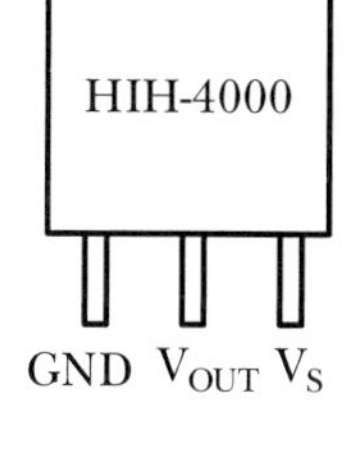

FIGURE 12.29 *Honeywell HIH-4000 humidity sensor.*

- Use voltage divider (passive or active) to scale down the voltage
- Apply the HIH-4000 input to HVAIN+ pin, set HVAIN− and HVREF to 0 V, set VREF0 to 3.0 V, and set the HVGAIN to 0.5.

The second method is more attractive because it does not need extra components.

Example 12.11 Describe a circuit connection and the required program to build a digital humidity meter. Display the pressure in three integral and one fractional digits using the LCD. Measure and display the humidity over the whole range of the HIH-4000, that is, 0 to 100 percent RH. Update the display data 10 times per second and assume that the C8051F040 operates with a 24-MHz crystal oscillator.

Solution: The circuit connection between the HIH-4000 and C8051F040 is shown in Figure 12.30. In the figure, the voltage output of HIH-4000 is treated as a high-voltage input to ADC0.

By setting the high-voltage gain to 0.5, the actual voltage to be converted is from 0.4 to 1.95 V which will be converted to 819 (0% RH) to 3993 (100% RH) by ADC0. The A/D conversion result span is 3174.

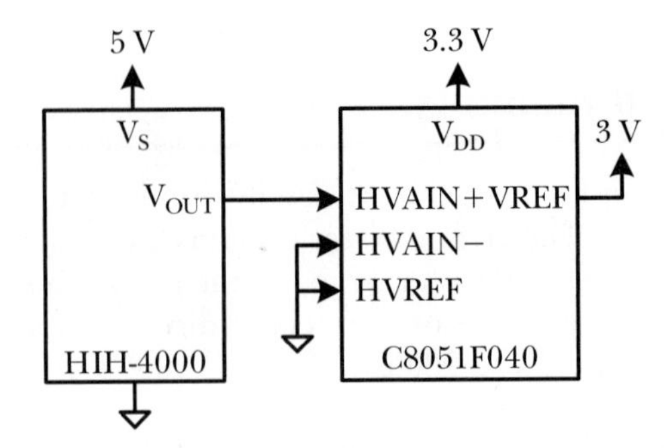

FIGURE 12.30
Circuit connection between HIH-4000 and C8051F040.

To translate back to relative humidity, this equation should be used:

$$\begin{aligned} RH &= (\text{conversion result} - 819) \div 31.74 \\ &\approx (\text{conversion result} - 819) \times 10 \div 317 \end{aligned} \quad (12.11)$$

The C program that performs the ADC0 configuration, A/D conversion, conversion-result formatting, and LCD display updating is as follows.

```
#include <c8051F040.h>
#include <delays.h>                  // include delays.c to the project
#include <lcdUtil.h>                 // include lcdUtil.c & lcdUtil.SRC to the project
void sysinit (void);
void openADC0 (void);

void main (void)
{
    char *msg = "Humidity = ";
    char  humid[7], tmp1;
    unsigned int tmp, tmp2, tmp3;
    sysinit();
    openlcd();
    cmd2lcd(0x80);                   // set cursor to first row, first column
    puts2lcd(msg);                   // output the string "Humidity = " on LCD
    openADC0();                      // initialize ADC0 module
    humid[3] = '.';                  // period character
    humid[5] = '%';                  // ASCII code of %
    humid[6] = 0;                    // NULL character

    while(1) {
        ADC0CN |= 0x10;              // start A/D conversion
        while(ADC0CN & 0x10);        // wait until A/D conversion is done
        tmp = ((unsigned int)ADC0H*256 + (unsigned int)ADC0L - 819) * 10; // combine low &
                                     // high bytes
        tmp1 = tmp / 317;            // integer part of the humidity
        tmp2 = tmp % 317;            // remainder of the humidity (after translation)
        humid[4] = (tmp2 * 10)/317;  // compute fractional digit
```

```
            tmp3 = (tmp2 * 10) % 317;             // compute the remainder for the fractional digit
            if (tmp3 > 158 || tmp3 == 158){// need to round up?
                humid[4]++;
                if (humid[4] == 10) { // fractional digit becomes 10 after round up?
                    tmp1++;
                    humid[4] = 0;
                }
            }
            humid[4] += 0x30;                     // convert fractional digit to ASCII code
            humid[2] = tmp1 % 10 + 0x30;          // separate one's digit
            tmp1 /= 10;                           // divide integer part by 10
            if (tmp1 == 0) {
                humid[0] = 0x20;                  // leading two digits are zero, fill with spaces
                humid[1] = 0x20;                  //     "
            }
            else {
                humid[1] = tmp1 % 10 + 0x30;      // separate ten's digit
                tmp1 /= 10;
                if (tmp1 == 0)
                    humid[0] = 0x20;              // hundred's digit is zero
                else
                    humid[0] = 0x31;              // otherwise, hundred's digit is 1
            }
            cmd2lcd(0x8B);                        // set LCD cursor to column 11, row 1
            puts2lcd(&humid[0]);                  // output the humidity value to LCD
            delayby100ms(1);                      // wait for 100 ms
        }
}
void sysinit(void)
{
    unsigned char cx;
    SFRPAGE   = 0x0F;
    WDTCN     = 0xDE;                         // disable watchdog timer
    WDTCN     = 0xAE;                         //     "
    OSCXCN    = 0x67;                         // start external crystal oscillator
    for (cx = 0; cx < 255; cx++);             // wait for about 1 ms
    while(!(OSCXCN & 0x80));                  // wait for crystal oscillator to stabilize
    CLKSEL    |= 1;                           // switch to external crystal oscillator as clock source
    XBR0      = 0xF7;                         // assign I/O pins to peripheral signals
    XBR1      = 0xFF;                         //     "
    XBR2      = 0x5D;                         // also enable crossbar
    XBR3      = 0x8F;                         //     "
    P7MDOUT = 0xFF;                           // enable Port 7 push-pull
    SFRPAGE   = 0;                            // switch to SFR page 0
    SPI0CN    = 0x01;                         // enable SPI to 3-wire mode
}
void openADC0(void)
{
    AMX0CF = 0;                               // configure all inputs to single-ended mode
    AMX0SL = 0x04;                            // select channel HVDA
```

```
    HVA0CN = 0x86;              // enable HVDA and set HVGA to 0.5
    ADC0CF = 0x48;              // set SAR clock to 2.4 MHz and PGA = 1
    ADC0CN = 0xC0;              // enable ADC0, low-power tracking, result right-justified
                                // A/D conversion started by setting AD0BUSY to 1
    REF0CN = 2;                 // use VREF0 as voltage reference (set to 2 V)
}
```

12.9 Digital-to-Analog Converter of the C8051F040

The C8051F040 provides two on-chip 12-bit digital-to-analog converters. Each DAC has an output swing of **0 V** to **(1 − 1/4096) VREFD** for a corresponding input code range of 0x000 to 0xFFF. The DACs may be enabled/disabled via their corresponding control registers, DAC0CN and DAC1CN. The DAC output will be in a high-impedance state if it is disabled. A DAC needs a voltage reference to perform D/A conversion. The voltage reference for each DAC is supplied at the VREFD pin. The block diagram of DAC0 is shown in Figure 12.31. The functional block diagram of DAC1 is identical to that of DAC0. Each value takes 10 μs to convert to voltage.

12.9.1 Signal Pins Related to DACs

The AV+ signal supplies power to DAC0 and DAC1 to perform digital-to-analog conversion. The DAC0 and DAC1 pins are the voltage outputs for DAC0 and DAC1 modules, respectively. The VREFD pin is used as the voltage reference for digital-to-voltage conversion.

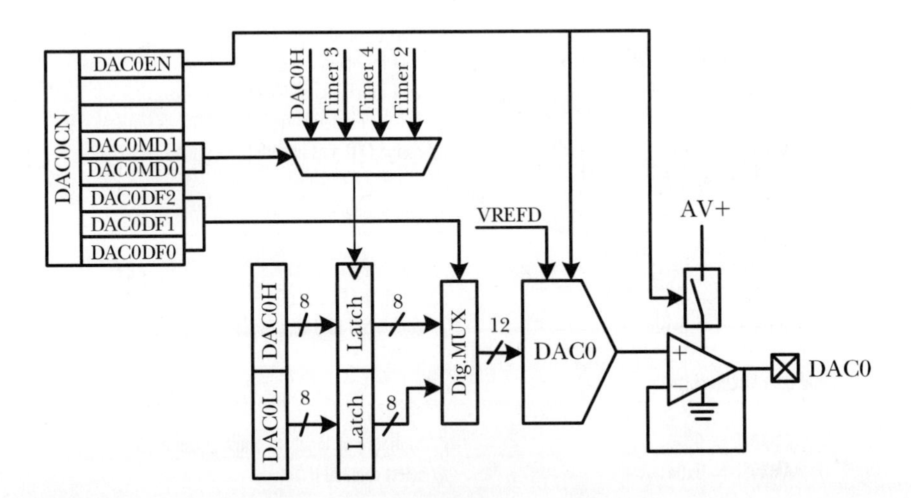

FIGURE 12.31 *DAC0 functional block diagram.*

12.9.2 Registers Related to DACs

The DAC0CN and DAC1CN control the operations of DAC0 and DAC1, respectively. The contents of DAC0CN and DAC1CN are shown in Figures 12.32 and 12.33, respectively. The value (12-bit) to be converted by DAC0 is written into DAC0H and DAC0L, whereas the 12-bit value to be converted by DAC1 is written into DAC1H and DAC1L. The lower 8 bits of the value to be converted is written into DAC0L (or DAC1L) first. Registers related to DAC0 reside in SFR page 0, whereas the registers related to DAC1 reside in SFR page 1.

7	6	5	4	3	2	1	0	
DAC0EN	--	--	DAC0MD1	DAC0MD0	DAC0DF2	DAC0DF1	DAC0DF0	Value after reset: 0x00

DAC0EN: DAC0 enable bit
0 = DAC0 disabled; DAC0 is in low-power shutdown mode
1 = DAC0 enabled. DAC0 output pin is active; DAC0 is operational
DAC0MD1~DAC0MD0: DAC0 mode bit
00 = DAC output updates occur on a write to DAC0H
01 = DAC output updates occur on Timer 3 overflow
10 = DAC output updates occur on Timer 4 overflow
11 = DAC output updates occur on Timer 2 overflow
DAC0DF2~DAC0DF0: DAC0 data format bits
000: The most significant nibble of data word are in DAC0H[3:0]

DAC0H								DAC0L							
				msb											lsb

001: The most significant 5 bits of DAC0 data word are in DAC0H[4:0]

DAC0H								DAC0L							
			msb											lsb	

010: The most significant 6 bits of DAC0 data word are in DAC0H[5:0]

DAC0H								DAC0L							
		msb											lsb		

011: The most significant 6 bits of DAC0 data word are in DAC0H[6:0]

DAC0H								DAC0L							
	msb											lsb			

1xx: The most significant 8 bits of DAC0 data word are in DAC0H[7:0]

DAC0H								DAC0L							
msb											lsb				

FIGURE 12.32
The DAC0 control register (DAC0CN).

7	6	5	4	3	2	1	0	
DAC1EN	--	--	DAC1MD1	DAC1MD0	DAC1DF2	DAC1DF1	DAC1DF0	Value after reset: 0x00

DAC1EN: DAC1 enable bit
  0 = DAC1 disabled; DAC1 is in low-power shutdown mode
  1 = DAC1 enabled. DAC1 output pin is active; DAC1 is operational
DAC1MD1~DAC1MD0: DAC1 mode bit
  00 = DAC output updates occur on a write to DAC1H
  01 = DAC output updates occur on Timer 3 overflow
  10 = DAC output updates occur on Timer 4 overflow
  11 = DAC output updates occur on Timer 2 overflow
DAC1DF2~DAC1DF0: DAC1 data format bits
  000: The most significant nibble of data word are in DAC1H[3:0]

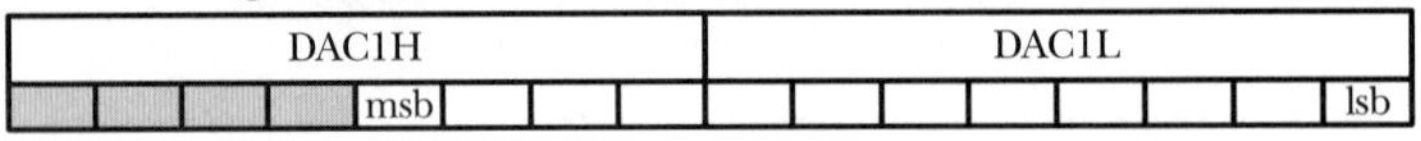

DAC1H								DAC1L							
				msb											lsb

  001: The most significant 5 bits of DAC1 data word are in DAC1H[4:0]

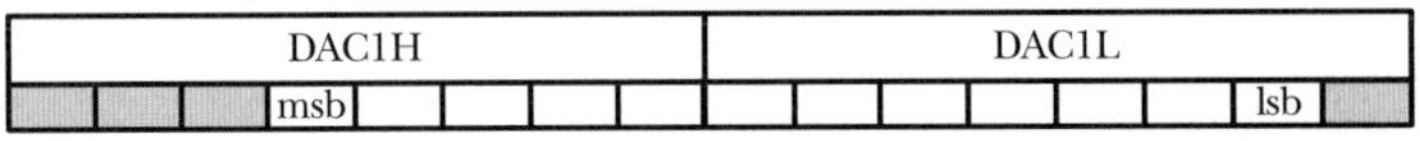

DAC1H								DAC1L							
			msb											lsb	

  010: The most significant 6 bits of DAC1 data word are in DAC1H[5:0]

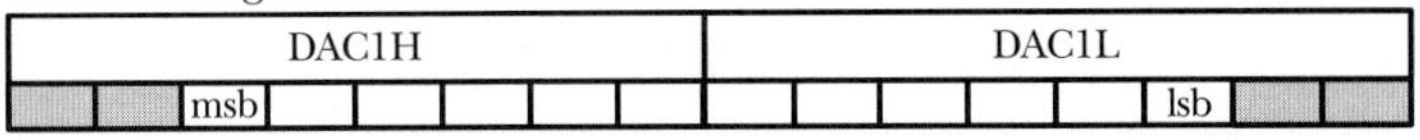

DAC1H								DAC1L							
		msb											lsb		

  011: The most significant 7 bits of DAC1 data word are in DAC1H[6:0]

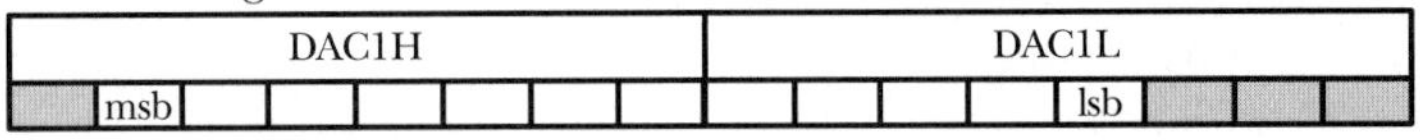

DAC1H								DAC1L							
	msb											lsb			

  1xx: The most significant 8 bits of DAC1 data word are in DAC1H[7:0]

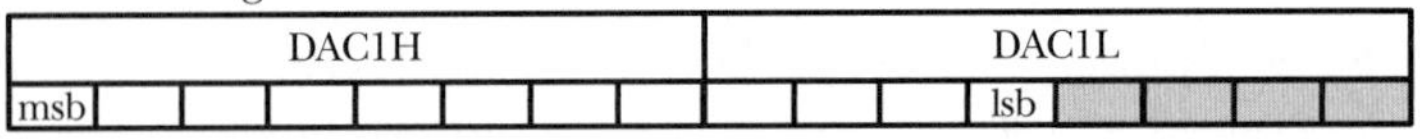

DAC1H								DAC1L							
msb											lsb				

FIGURE 12.33
The DAC1 control register (DAC1CN).

12.9.3 DAC Operation

In order to convert a digital value to a voltage, the user needs to enable the DAC and choose one of the four methods to start D/A conversion.

1. Write the high byte to the DACxH (x = 0 or 1) register: For this start method, the user should write the low byte into DACxL and then write the high byte into the DACxH register.
2. Timer 3 overflow
3. Timer 4 overflow
4. Timer 2 overflow

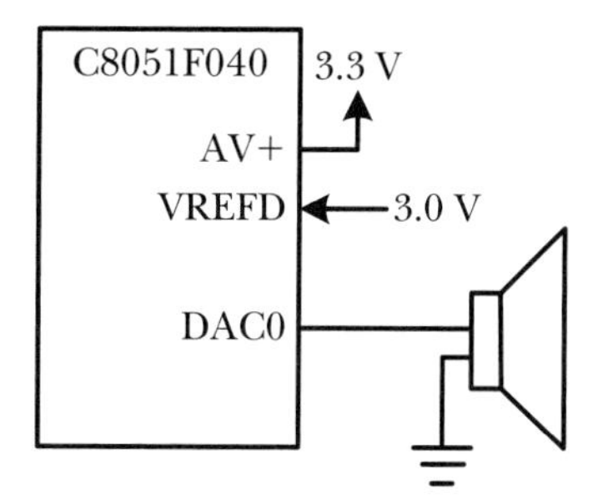

FIGURE 12.34
Circuit for generating a siren.

When one of these four events occurs, the value in DACxH:DACxL will be copied into the DAC latch and D/A conversion starts. As shown in Figures 12.34, the value to be converted can be right-justified, left-justified, or somewhere in between.

Example 12.12 Use DAC0 to generate a 1-kHz digital waveform. Use Timer 2 overflow to start the D/A conversion. Assume that the C8051F040 is running with a 24-MHz external crystal oscillator and VREFD is connected to a 3.0-V power supply.

Solution:

```
            $nomod51
            $include (C8051F040.inc)
HiLo        bit         0
            org         0x00
            ljmp        start
            org         0x2B            ; TMR2 interrupt vector
            ljmp        T2ISR           ; jump to the actual TMR2 interrupt service routine
            org         0xAB
start:      mov         SP,#0x7F        ; setup stack pointer
            acall       sysInit         ; set up SYSCLK, assign I/O pins to peripheral functions
            acall       DAC0init
            acall       openTimer2      ; initialize Timer 2
            setb        HiLo            ; next time to pull DAC0 pin to high
            mov         DAC0L,#0x00     ; start with the DAC0 pin low
            mov         DAC0H,#0x00     ;    "
            setb        TR2             ; start TMR2
            mov         IE,#0xA0        ; enable Timer 2 interrupt, disable other interrupts
            setb        PT2             ; set Timer 2 interrupt at high priority
forever:    nop                         ; stay here to perform operations, the waveform will
            ajmp        forever         ; repeat forever

;**************************************************************************************************
; The following subroutine enables all peripheral functions and assign port pins
; to peripheral functions. It configures SPI to 3-wire mode and switch to SFR page 0.
;**************************************************************************************************
```

```
sysInit:        mov     SFRPAGE,#0x0F
                mov     WDTCN,#0xDE         ; disable watchdog timer (must be done)
                mov     WDTCN,#0xAD         ;       "
                mov     CLKSEL,#0           ; use external oscillator to generate SYSCLK
                mov     OSCXCN,#0x67        ; configure external oscillator control
                mov     R7,#255
againosc:       djnz    R7,againosc         ; wait for about 1 ms
chkstable:      mov     A,OSCXCN            ; wait until external crystal oscillator is stable
                anl     A,#0x80             ; before using it
                jz      chkstable           ;       "
                mov     CLKSEL,#1           ; use external crystal oscillator to generate SYSCLK
                mov     XBR0,#0xF7          ; route all peripheral function signals
                mov     XBR1,#0xFF          ; to I/O pins
                mov     XBR2,#0x5D          ; enable digital crossbar
                mov     XBR3,#0x8F          ;       "
                mov     SFRPAGE,#0          ; switch SFR page 0
                mov     SPI0CN,#01          ; enable and configure SPI to 3-wire mode
                ret
;************************************************************************************************
; The following function enables the DAC0 to start conversion with Timer 2
; overflow, DAC0H:DAC0L right-justified.
;************************************************************************************************
DAC0init:       mov     DAC0CN,#0x80        ; enable DAC0, conversion is starting by writing into
                                            ; DAC0H
                mov     REF0CN,#02          ; enable internal bias generator
                ret
;************************************************************************************************
; The following function initializes Timer 2 in timer mode, auto-reload on overflow, and
; count up from 64536.
;************************************************************************************************
openTimer2:
                mov     TMR2CN,#0           ; TMR2 operates in timer mode, reload
                mov     TMR2CF,#0           ; TMR2 counts up and uses SYSCLK/12 as clock
                mov     RCAP2L,#0x18        ; TMR2 auto reload 64536 every 1000 clock cycles
                mov     RCAP2H,#0xFC        ;       "
                mov     TMR2L,#0x18         ; TMR2 counts up from 64536 and overflow in
                mov     TMR2H,#0xFC         ; 0.5 ms
                clr     TF2
                ret
;************************************************************************************************
; The following is the actual TMR2 interrupt service routine.
;************************************************************************************************
T2ISR:          clr     TF2                 ; clear TMR2 interrupt flag
                jnb     HiLo,islow          ; is HiLo flag equal to 0?
                mov     DAC0L,#0xFF         ; next time DAC0 outputs high
                mov     DAC0H,#0x0F         ;       "
                clr     HiLo                ; next time load 0 to DAC0H:DAC0L
                reti
```

```
islow:      mov     DAC0L,#0        ; next time DAC0 outputs low
            mov     DAC0H,#0        ;    "
            setb    HiLo            ; next time load 0xFFF to DAC0H:DAC0L
            reti
            end
```

The C language version of the program is as follows.

```
#include <c8051F040.h>
void TMR2ISR(void);
void sysInit(void);
void openDAC0 (void);
void openTimer2 (void);
char HiLo;
void main (void)
{
    HiLo = 1;               // specify to load 0xFFF to DAC0 for next TMR2 overflow
    sysInit();              // select SYSCLK source, assign I/O pins to peripheral signals
    openDAC0();             // configure DAC0
    openTimer2();           // configure TMR2 in timer mode, auto-reload
    DAC0L = 0;              // DAC0 output 0 at the beginning
    DAC0H = 0;              //    "
    IE    = 0xA0;           // enable TMR2 interrupt
    PT2   = 1;              // place TMR2 interrupt in high priority
    TR2   = 1;              // start TMR2
    while(1);               // infinite loop, other operations can be performed here
}
void sysInit(void)
{
    SFRPAGE = 0x0F;                 // switch to SFR page F
    WDTCN  = 0xDE;                  // disable watchdog timer
    WDTCN  = 0xAD;                  //    "
    OSCXCN = 0x67;                  // start external crystal oscillator
    for (cx = 0; cx < 255; cx++);   // wait for about 1 ms
    while(!(OSCXCN & 0x80));        // wait for crystal oscillator to stabilize
    CLKSEL |= 1;                    // switch to external crystal oscillator as clock source
    XBR0   = 0xF7;                  // assign I/O pins to peripheral functions
    XBR1   = 0xFF;                  //    "
    XBR2   = 0x5D;                  //    "
    XBR3   = 0x8F;                  //    '
    SFRPAGE = 0;                    // switch to SFR page 0
    SPI0CN  = 1;                    // enable SPI in 3-wire mode
}
void openDAC0 (void)
{
    DAC0CN = 0x80;                  // enable DAC0, start DAC0 by writing into DAC0H
    REF0CN = 0x02;                  // enable internal bias generator
}
```

```
void TMR2ISR(void) interrupt 5
{
    TF2 = 0;
    if (HiLo){
        DACOL = 0xFF;           // next TMR2 overflow to write high
        DACOH = 0x0F;           // value into DACO
        HiLo  = 0;
    }
    else {
        DACOL = 0;              // next TMR2 overflow to write low value
        DACOH = 0;              // into DACO
        HiLo  = 1;
    }
}
void openTimer2 (void)
{
    TMR2CN = 0;                 // Timer 2 in timer mode, auto reload
    TMR2CF = 0;                 // TMR2 counts up and use SYSCLK/12 as clock source
    RCAP2L = 0x18;              // TMR2 reload value 64536 (0xFC18)
    RCAP2H = 0xFC;              //     "
    TMR2L  = 0x18;              // start value
    TMR2H  = 0xFC;
    TF2    = 0;                 // clear the interrupt flag
}
```

Example 12.13 Describe how to use the DAC0 to generate a two-tone siren. The frequency of the low tone is 440 Hz, whereas the frequency of the high tone is 880 Hz. Each frequency lasts for half a second and then switches to the other frequency. The C8051F040 operates with a 24-MHz crystal oscillator.

Solution: The user must connect the circuit as shown in Figure 12.34.

Suppose we choose to start the DAC0 conversion by writing into the DAC0H register and setting the clock source to Timer 2 to SYSCLK. In each siren period, the DAC0 output toggles twice (from 0 to 3 V and then back to 0 V). The count value for DAC0 to toggle its output can be computed as

at 440 Hz count value: is $(24 \times 10^6) \div 440 \div 2 = 27273$

at 880 Hz count value: is $(24 \times 10^6) \div 880 \div 2 = 13636$

The 440-Hz siren tone will toggle the DAC0 output 440 times in half a second, whereas the 880-Hz siren tone will toggle the DAC0 output 880 times in the same period of time.

The procedure for generating a two-tone siren with the specified frequencies using DAC0 is as follows.

Step 1 Use the variable count to hold the number of times that a siren tone needs to be repeated in 0.5 s.

Step 2 Use the variable HiLo to indicate whether to write the value 0xFFF or 0x000 into DAC0H:DAC0L in the next Timer 2 overflow interrupt.

Step 3 Write a Timer 2 interrupt service routine that performs the operations:

HiLo = 0:
Write 0 into DAC0H and DAC0L, set HiLo to 1, and decrement count by 1.

HiLo = 1:
Write 0x0F and 0xFF into DAC0H and DAC0L, respectively, clear HiLo to 0, and decrement count by 1.

Step 4 Configure DAC0 to start conversion on Timer 2 interrupt.

Step 5 Configure the MCU to use external oscillator as SYSCLK. Enable Timer 2 interrupt.

Step 6 Initialize the variable count to 440 and load 0x9577 (= 65536 − 27273) into the RCAP2H:RCAP2L register pair.

Step 7 Wait until count is decremented to zero. Reinitialize **count** to 880 and load 0xCABC (65536 − 13636) into the RCAP2H:RCAP2L register pair.

Step 8 Wait until **count** is decremented to 0 and go to Step 6.

The assembly program that implements this algorithm is as follows.

```
        $nomod51
        $include (C8051F040.inc)
#define lohi    0x95    ; high byte of the reload value for 440 Hz
#define lolo    0x77    ; low byte of the reload value for 440 Hz
#define hihi    0xCA    ; high byte of the reload value for 880 Hz
#define hilo    0xBC    ; low byte of the reload value for 880 Hz
#define F1hi    0x01    ; repeat count for 440 Hz
#define F1lo    0xB8    ;    "
#define F2hi    0x03    ; repeat count for 880 Hz
#define F2lo    0x70    ;    "
loophi  set     R0      ; loop count for toggling (decrement in TMR2ISR)
looplo  set     R1      ;    "
HiLo    bit     0       ; flag to select write value (0 = 0, 1 = 0xFFF)
tone    bit     1       ; flag to indicate 440 Hz (0) or 880 Hz (1)
        org     0x00
        ljmp    start
        org     0x2B    ; TMR2 interrupt vector
        ljmp    T2ISR   ; jump to the actual TMR2 interrupt service routine
        org     0xAB
```

```
start:      mov     SP,#0x7F            ; establish stack pointer
            acall   sysInit             ; set up SYSCLK, assign I/O pins to peripheral functions
            acall   DAC0init
            acall   openTimer2          ; initialize Timer 2
            mov     RCAP2H,#lohi        ; start with low tone
            mov     RCAP2L,#lolo        ;     "
            mov     TMR2H,#lohi         ;     "
            mov     TMR2L,#lolo         ;     "
            mov     loophi,#F1hi        ; initialize count to low frequency loop count
            mov     looplo,#F1lo        ;     "
            setb    HiLo                ; next time to pull DAC0 pin to high (write 0xFFF)
            clr     tone                ; indicate that tone is low (440 Hz)
            mov     DAC0L,#0            ; start with the DAC0 pin low
            mov     DAC0H,#0            ;     "
            setb    TR2                 ; start TMR2
            mov     IE,#0xA0            ; enable Timer 2 interrupt, disable other interrupts
            setb    PT2                 ; set Timer 2 interrupt at high priority
forever:    nop                         ; stay here to perform other operations, the waveform
            ajmp    forever             ; will repeat forever

; ************************************************************************************************
; The following subroutine enables all peripheral functions and assigns port pins
; to peripheral functions. It configures SPI to 3-wire mode and switch to SFR page 0.
; ************************************************************************************************
sysInit:    mov     SFRPAGE,#0          ; switch to SFR page 0
            mov     SPI0CN,#01          ; enable and configure SPI to 3-wire mode
            mov     SFRPAGE,#0x0F       ; switch to SFR page F
            mov     WDTCN,#0xDE         ; disable watchdog timer (must be done)
            mov     WDTCN,#0xAD         ;     "
            mov     CLKSEL,#0           ; use external oscillator to generate SYSCLK
            mov     OSCXCN,#0x67        ; configure external oscillator control
            mov     R7,#255
againosc:   djnz    R7,againosc         ; wait for about 1 ms
chkstable:  mov     A,OSCXCN            ; wait until external crystal oscillator is stable
            anl     A,#0x80             ; before using it
            jz      chkstable           ;     "
            mov     CLKSEL,#1           ; use external crystal oscillator to generate SYSCLK
            mov     XBR0,#0F7H          ; route all peripheral function signals
            mov     XBR1,#0xFF          ; to I/O pins
            mov     XBR2,#0x5D          ; enable digital crossbar
            mov     XBR3,#0x8F          ;     "
            ret
; ************************************************************************************************
; The following function enables the DAC0 to start conversion by writing into DAC0H
; register, DAC0H:DAC0L right-justified.
; ************************************************************************************************
DAC0init:   mov     SFRPAGE,#DAC0_PAGE
            mov     DAC0CN,#0x80        ; enable DAC0, starts conversion by writing into
                                        ; the DAC0H register
            mov     REF0CN,#02          ; enable internal bias generator
            ret
```

```
; ************************************************************************************************
; The following function initializes Timer 2 to timer mode, auto reload on overflow,
; and uses SYSCLK as clock input.
; ************************************************************************************************
openTimer2:
            mov     SFRPAGE,#LEGACY_PAGE
            mov     TMR2CN,#0           ; TMR2 operates in timer mode, reload mode
            mov     TMR2CF,#0x0A        ; TMR2 counts up and uses SYSCLK as clock
            clr     TF2
            ret
; ************************************************************************************************
; The following is the actual TMR2 interrupt service routine.
; ************************************************************************************************
T2ISR:      clr     TF2                 ; clear TMR2 interrupt flag
            jnb     HiLo,islow          ; is HiLo flag equal to 0?
            mov     DAC0L,#0xFF         ; make DAC0 to output high
            mov     DAC0H,#0x0F         ;   "
            clr     HiLo                ; next time load 0 to DAC0H:DAC0L
            ajmp    chkloop
islow:      mov     DAC0L,#0            ; next time DAC0 outputs low
            mov     DAC0H,#0            ;   "
            setb    HiLo                ; next time load 0xFFF to DAC0H:DAC0L
chkloop:    jb      tone,HiFreq         ; jump if high tone (880 Hz)
            mov     A,looplo
            clr     C
            subb    A,#1                ; decrement loop count by 1
            mov     looplo,A            ;   "
            mov     A,loophi            ;   "
            subb    A,#0                ;   "
            mov     loophi,A            ;   "
            mov     A,looplo
            jnz     quit                ; no need to change setting yet
            mov     A,loophi
            jnz     quit                ; no need to change setting yet
            mov     RCAP2H,#hihi        ; switch to high frequency tone
            mov     RCAP2L,#hilo        ;   "
            mov     TMR2H,#hihi
            mov     TMR2L,#hilo
            mov     looplo,#F2lo        ; change loop count to high frequency (880 Hz)
            mov     loophi,#F2hi        ;   "
            setb    tone                ; switch to high tone
            reti
HiFreq:     mov     A,looplo
            clr     C
            subb    A,#1                ; decrement loop count by 1
            mov     looplo,A            ;   "
            mov     A,loophi            ;   "
            subb    A,#0                ;   "
            mov     loophi,A            ;   "
            mov     A,looplo
            jnz     quit                ; no need to change setting yet
```

```
         mov    A,loophi
         jnz    quit             ; no need to change setting yet
         mov    RCAP2H,#lohi     ; change TMR2 reload value to low frequency
         mov    RCAP2L,#lolo     ;   "
         mov    TMR2H,#lohi      ;   "
         mov    TMR2L,#lolo      ;   "
         mov    looplo,#F1lo     ; change loop count to low frequency (440 Hz)
         mov    loophi,#F1hi     ;   "
         clr    tone             ; switch to low tone from now on
quit:    reti
         end
```

The C language version of the program is straightforward and, hence, will be left for you as an exercise problem.

12.10 Chapter Summary

A data acquisition system consists of four major components: a transducer, a signal conditioning circuit, an A/D converter, and a computer. The transducer converts a nonelectric quantity into a voltage. The range of the transducer output may be too narrow and hence may cause significant errors in the A/D conversion process. The signal conditioning circuit shifts and scales the output from a transducer to a range that can take advantage of the full capacity of the A/D converter. However, signal conditioning circuit adds cost to the data acquisition system, users must be cautious in using it.

Because of the discrete nature of a digital system, the A/D conversion result has a quantization error. The accuracy of an A/D converter is dictated by the number of bits used to represent the analog quantity. The more bits are used, the smaller the quantization error will be.

There are four major A/D conversion algorithms:

1. Parallel (flash) A/D conversion method
2. Slope and double-slope conversion method
3. Sigma-delta A/D conversion method
4. Successive-approximation conversion method

The C8051F040 microcontroller uses the successive-approximation method to perform A/D conversion. The C8051F040 has two A/D converters: the 12-bit converter ADC0 and the 8-bit converter ADC2. The 12-bit ADC0 provides the capability to handle differential inputs and a programmable gain amplifier. When these features are used effectively, it often renders the external signal conditioning circuit unnecessary. When applications require high throughput but not high accuracy, the user may choose to use the 8-bit ADC2.

Both the ADC0 and ADC2 can use either an external or an internal reference voltage to perform the task of A/D conversion. The external-reference voltage provides better control flexibility. The conversion result of the ADC0 module can be programmed to be right-justified or left-justified. The highest clock rate for A/D conversion is 2.5 MHz. A sample of voltage takes 10 μs to convert.

The C8051F040 has an on-chip temperature sensor that measures the chip temperature. Since the chip temperature is not the same as the ambient temperature, the user must learn to use a compensation method when using this sensor to measure the ambient temperature. It would be easier to use external temperature sensor such as the TC1047A from Microchip to measure the ambient temperature.

The C8051F040 ADC0 and ADC2 also have a window detector that can detect whether the analog voltage input is within or outside a voltage range. This feature could be useful in some control applications.
The user has four choices of starting the A/D conversion:

1. Writing a 1 to the AD0BUSY bit of the ADC0CN register
2. A Timer 3 overflow
3. A rising edge detected on the external CNVSTR0 signal
4. A Timer 2 overflow

The operation and control of ADC2 are quite similar to those of the ADC0.

The TC1047A temperature sensor, the HIH-4000 humidity sensor, and MP3H6115A barometric-pressure sensor are used as examples to illustrate the A/D conversion process. The TC1047A can measure a temperature in the range from -40 to 125°C. The HIH-4000 can measure the relative humidity from 0 to 100 percent. The MP3H6115A can measure the barometric pressure from 15 to 115 kPa. Because of the high-resolution and programmable gain control of the ADC0, there is no need to use an external signal conditioning circuit.

There are applications that require much higher resolution than that provided by the C8051F040. For those applications, users can choose other SiLabs MCUs that provide higher A/D resolutions. For examples, the C8051F06x ($x = 0, \ldots, 7$) and C8051F352 provide 16-bit resolution, whereas the C8051F35y ($y = 0$ or 1) provides 24-bit resolution.

The C8051F040 has two identical 12-bit DACs. These two DACs can be used to generate digital waveforms of any shape. The output of DAC is from 0 V to V_{REFD}. There are four methods to start a D/A conversion:

1. Write the high byte of the value to be converted to the DACxH ($x = 0$ or 1) register: for this starting method, the user should write the low byte into DACxL first and then write the high byte into the DACxH register
2. Timer 3 overflow
3. Timer 4 overflow
4. Timer 2 overflow

12.11 Exercise Problems

E12.1 Design a circuit that can scale the voltage from the range of 0 ~ 100 mV to the range of 0 ~ 3 V.

E12.2 Design a circuit that can scale the voltage from the range of 0 ~ 75 mV to the range of 0 ~ 3 V.

E12.3 Design a circuit that can scale and shift the voltage from the range of −40 ~ 120 mV to the range of 0 ~ 3.0 V.

E12.4 Design a circuit that can scale and shift the voltage from the range of −50 ~ 75 mV to the range of 0 ~ 3.0 V.

E12.5 Design a circuit that can scale the voltage from the range of 2 ~ 2.2 V to the range of 0 ~ 3.0 V.

E12.6 Find the corresponding voltage values for the A/D conversion results 100, 300, 600, 1024, 2048, and 3072, assuming that you are using the ADC0 of the C8051F040 to perform A/D conversion with VREF0 set to 3 V.

E12.7 Find the corresponding voltage values for the A/D conversion results 20, 45, 80, 127, 192, assuming that you are using the ADC2 of the C8051F040 to perform A/D conversion with VREF2 set to 3 V.

E12.8 Assume that an unknown analog signal is connected to the AIN0.3 pin. The range of this voltage is from 0.2 to 1.5 V. Make an appropriate suggestion on how to configure the ADC0 so that this voltage can be measured correctly. Write an instruction sequence to carry out the suggested configuration, assuming that the C8051F040 is operating with a 24-MHz crystal oscillator.

E12.9 Assume that an unknown analog signal is connected to the AIN0.0 pin. The range of this voltage is from 0 to 15 V. Make an appropriate suggestion on how to configure the ADC0 so that this voltage can be measured correctly. Write an instruction sequence to carry out the suggested configuration, assuming that the C8051F040 is operating with a 24-MHz crystal oscillator.

E12.10 Assume that an unknown analog signal is connected to the AIN0.0 pin. The range of this voltage is from 0 to 1.0 V. Make an appropriate suggestion on how to configure the ADC0 so that this voltage can be measured correctly. Write an instruction sequence to carry out the suggested configuration, assuming that the C8051F040 is operating with a 24-MHz crystal oscillator.

E12.11 Write an instruction sequence to set up greater-than and less-than registers to interrupt the C8051F040 whenever the input is less than 1.0 V and higher than 2.3 V using the ADC2.

E12.12 Write an instruction sequence to set up greater-than and less-than registers to interrupt the C8051F040 whenever the input is between 1.5 and 2.5 V using the ADC0.

E12.13 Suppose you are assigned to measure an unknown voltage in the range from 2 to 12 V. Describe how you want to connect the circuit and the ADC0 configuration to measure the unknown voltage. Write a program to measure and display the voltage 10 times per second on the LCD.

E12.14 Suppose you are assigned to measure an unknown voltage in the range from 1 to 9 V. Describe how you want to connect the circuit and the ADC0 configuration to measure the unknown voltage. Write a program to measure and display the voltage 10 times per second on the LCD.

E12.15 Write an instruction sequence to configure ADC2 to operate with the following settings by assuming that the C8051F040 is operating with a 24-MHz external oscillator:

- Enable ADC2
- Convert the input AIN2.0 (P1.0) to digital value
- Set SAR2 clock rate to 6 MHz
- PGA gain set to 4
- Use Timer 3 overflow to start the A/D conversion
- Interrupt the MCU whenever the conversion is less than 0x30 or greater than 0x80
- Use VREF2 input as reference voltage

E12.16 Write a program to generate a triangular waveform using the on-chip DAC0 of the C8051F040. Use VREFD (connected to 3 V) as the voltage reference. Assume the MCU uses a 24-MHz crystal oscillator to generate SYSCLK. Increase V_{OUT} from 0 to 3 V in 100 steps and decrement to 0 also in 100 steps. Allow each step to take 10 μs to complete.

E12.17 Write a program to generate a triangular waveform using the on-chip DAC0 of the C8051F040. Use VREFD (connected to 3 V) as the voltage reference. Assume that the MCU use a 24-MHz crystal oscillator to generate SYSCLK. Increase the DAC0 pin voltage from 0 to 3 V in 4096 steps and decrement to 0 also in 4096 steps. Allow each step to take 10 μs to complete.

E12.18 Use the DAC0 to generate a sinusoidal waveform with a magnitude equal to 3 V and a period equal to 1.8 ms (180 steps with each step set to 10 μs).

12.12 Laboratory Exercise Problems and Assignments

L12.1 **Digital Voltmeter Experiment.** Follow the description of Example 12.6, make the program, and download the program (in hex file format) onto the SSE040 (or other demo board) demo board for execution. Turn the potentiometer that is connected to AIN0.0 and read the voltage value on the LCD.

L12.2 **Digital Thermometer.** Use the Centigrade temperature sensor LM35 from National Semiconductor to construct a digital thermometer. The pin assignment and circuit connection for converting temperature are shown in Figure L12.1. The temperature range to be converted is from 0 to 127°C. Describe the circuit connection between the LM35 and C8051F040 and write a program to display the temperature in three integral and one fractional digits on the LCD.

L12.3 **Play a Song using DAC0.** Connect the output of the DAC0 of the C8051F040 to a speaker or buzzer (available in SSE040 demo board). Write a program to play the United States national anthem (or any other song).

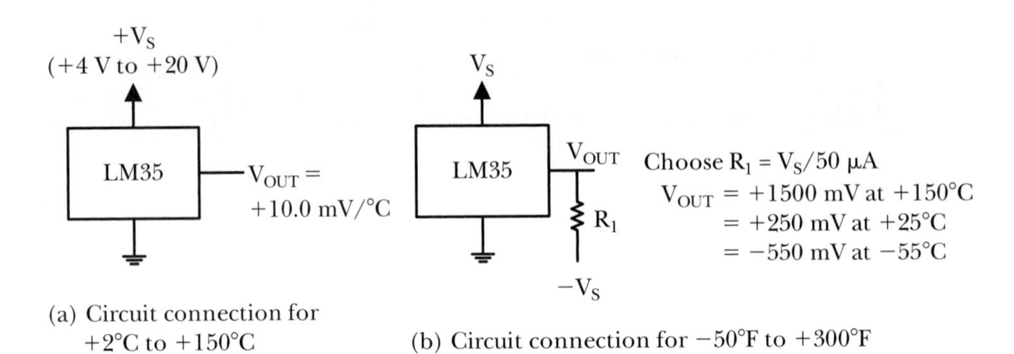

FIGURE L12.1
Circuit connection for the LM35.

CHAPTER 13

Controller Area Network (CAN)

13.1 Objectives

Upon successful completion of this chapter, you will be able to:

- Describe the layers of the CAN protocol
- Describe CAN's error-detection capability
- Describe the formats of CAN messages
- Describe CAN message handling
- Explain CAN error handling
- Describe CAN fault confinement
- Describe CAN bit timing
- Explain CAN synchronization issue and methods
- Describe the CAN message structures
- Compute timing parameters to meet the requirements of your application
- Write programs to configure the C8051F040 CAN module
- Write programs to transfer data over the CAN bus
- Design a CAN-based remote sensing system

13.2 Overview of Controller Area Network

The controller area network (CAN) is a two-wire bus initially created by German automotive system supplier Robert Bosch in the mid-1980s for automotive applications as a method for enabling robust serial communication. The goal was to make automobiles more reliable, safe, and fuel-efficient while at the same time to decrease wiring-harness weight and complexity. Since its inception, the CAN protocol has gained widespread use in industrial automation and automotive/truck applications. The description of CAN in this chapter is based on the CAN Specification 2.0 published in September 1991 by Bosch.

13.2.1 Layered Approach in CAN

The CAN protocol specifies the lowest two layers of the ISO seven-layer model: *data link* and *physical* layers. The data-link layer is further divided into two sublayers: logical link control (LLC) layer and medium access control (MAC) layer.

- The LLC sublayer deals with message-acceptance filtering, overload notification, and error-recovery management.
- The MAC sublayer presents incoming messages to the LLC sublayer and accepts messages to be transmitted that are forwarded by the LLC sublayer. The MAC sublayer is responsible for message framing, arbitration, acknowledgement, error detection, and signaling. The MAC sublayer is supervised by a self-checking mechanism, called *fault confinement*, which distinguishes short disturbances from permanent failures.

The physical layer defines how signals actually are transmitted and deals with the description of bit timing, bit encoding, and synchronization. CAN bus driver/receiver characteristics and the wiring and connectors are not specified in the CAN protocol. These two aspects are not specified so that the implementers can choose the most appropriate transmission medium and hence optimize signal-level implementations for their applications. The system designer can choose from multiple available media technologies, including twisted pair, single wire, optical fiber, radio frequency (RF), infrared (IR), and so on. The layered CAN protocol is shown in Figure 13.1.

13.2.2 General Characteristics of CAN

The CAN protocol is optimized for systems that need to transmit and receive relatively small amounts of information (as compared to Ethernet or USB, which are designed to move much larger blocks of data). The CAN protocol has the following features.

Carrier-Sense Multiple Access with Collision Detection (CSMA/CD)

The CAN protocol is a CSMA/CD protocol. Every node on the network must monitor the bus (carrier sense) for a period of no activity before trying to send a message on the bus. Once this period of no activity occurs, every node on the bus has an equal opportunity to transmit a message (multiple access). If two nodes happen to transmit at the same time, the nodes will detect the

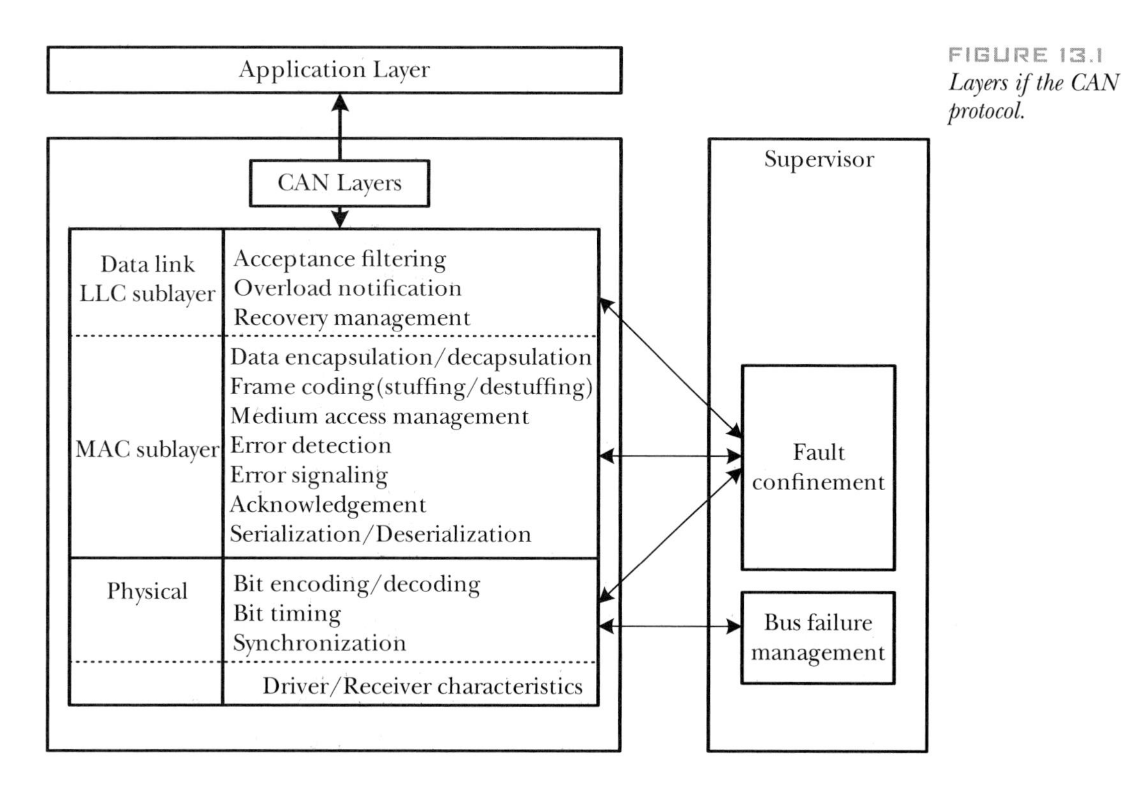

FIGURE 13.1
Layers if the CAN protocol.

collision and take the appropriate action. In the CAN protocol, a nondestructive bitwise arbitration method is utilized. Messages remain intact after arbitration is completed, even if collisions are detected. Message arbitration will not delay higher priority messages. To facilitate bus arbitration, the CAN protocol defines two bus states: *dominant* and *recessive.* The dominant state is represented by logic 0 (low voltage), whereas the recessive state is represented by logic 1 (high voltage). The dominant state will win over the recessive state.

Message-Based Communication

CAN is a message-based protocol not an address-based protocol. Embedded in each message is an *identifier.* This identifier allows messages to arbitrate the use of the CAN bus and also allows each node to decide whether to work on the message. The value of the identifier is used as the priority of the message. The lower the value, the higher the priority. Each node in the CAN system uses one or more filters to compare the identifier of the incoming message. Once the identifier passes the filter, the message will be worked on by the node. The CAN protocol also provides the mechanism for a node to request data transmission from another node. Since an address is not used in the CAN system, there is no need to reconfigure the system whenever a node is added to or deleted from a system. This capability allows the system to perform node-to-node or multicast communications.

Error Detection and Fault Confinement

The CAN protocol requires each sending node to monitor the CAN bus to find out if the bus value and the transmitted bit value are identical. For every message, a cyclic redundancy check is calculated and the checksum is appended to the message. CAN is an asynchronous protocol, and hence, clock information is embedded in the message rather than transmitted as a separate signal. A message with long sequence of identical bits could cause a synchronization problem. To resolve this problem, the CAN protocol requires the physical layer to use bit stuffing to avoid a long sequence of identical bit values. With these measures implemented, the residual probability for undetected corrupted messages in a CAN system is as low as

$$\textbf{Message error rate} \times 4.7 \times 10^{-11}$$

CAN nodes are able to distinguish short disturbances from permanent failures. Defective nodes are switched off from the CAN bus.

13.3 CAN Messages

The CAN protocol defines four different types of messages:

1. ***Data Frame.*** A data frame carries data from a transmitter to the receivers.
2. ***Remote Frame.*** A remote frame is transmitted by a node to request the transmission of the data frame with the same identifier.
3. ***Error Frame.*** An error frame is transmitted by a node on detecting a bus error.
4. ***Overload Frame.*** An overload frame is used to provide for an extra delay between the preceding and the succeeding data or remote frames.

Data frames and remote frames are separated from preceding frames by an interframe space. Applications do not need to send or handle error and overload frames.

13.3.1 Data Frame

As shown in Figure 13.2, a data frame consists of seven different bit fields: start-of-frame, arbitration, control, data, CRC, ACK, and end-of-frame.

Start-of-Frame Field

This field is a single, dominant bit that marks the beginning of a data frame. A node is allowed to start transmission only when the bus is idle. All nodes have to synchronize to the leading edge of this field of the node that starts transmission first.

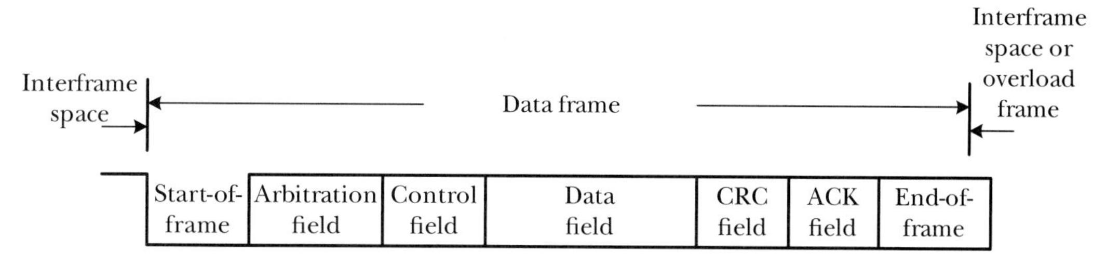

FIGURE 13.2
CAN Data frame.

Arbitration Field

The format of the arbitration field is different for *standard format* and *extended format* frames, as illustrated in Figure 13.3. The identifier's length is 11 bits for the standard format and 29 bits for the extended format.

The identifier of the standard format corresponds to the base ID in the extended format. These bits are transmitted as most-significant-bit first. The most significant 7 bits all cannot be recessive.

The identifier of the extended format comprises two sections: an 11-bit base ID and an 18-bit extended ID. Both the base ID and the extended ID are transmitted as most-significant-bit first. The base ID defines the base priority of the extended frame.

The *remote transmission request* (RTR) bit in data frames must be dominant. Within a remote frame, the RTR bit has to be recessive.

The *substitute remote request* (SRR) bit is a recessive bit. The SRR bit of an extended frame is transmitted at the position of the RTR bit in the standard frame and therefore substitutes for the RTR bit in the standard frame. As a consequence, collisions between a standard frame and an extended frame (where the base IDs of both frames are identical) are resolved in such a way that the standard frame prevails over the extended frame.

The *identifier extension* (IDE) bit belongs to the arbitration field for the extended format and the control field for the standard format. The IDE bit in the standard format is transmitted as dominant, whereas in the extended format, the IDE bit is recessive.

Control Field

The contents of this field are shown in Figure 13.4. The format of the control field is different for the standard format and the extended format. Frames in standard format include the data length code; the IDE bit, which is transmitted dominant; and the reserved bit **r0**. Frames in extended format include the data length code and two reserved bits, **r0** and **r1**. The reserved bits must be sent as dominant, but the receivers accept dominant and recessive bits in all combinations. The data length code specifies the number of bytes contained in the data field. Data length can be 0 to 8, as encoded in Table 13.1.

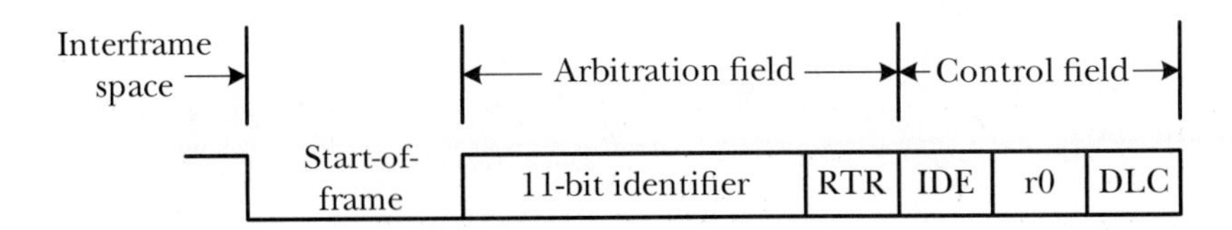

(a) Standard format

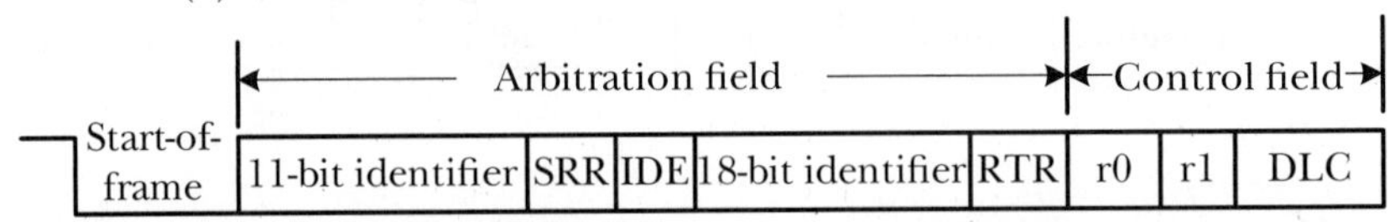

(b) Extended format

FIGURE 13.3 *Arbitration field.*

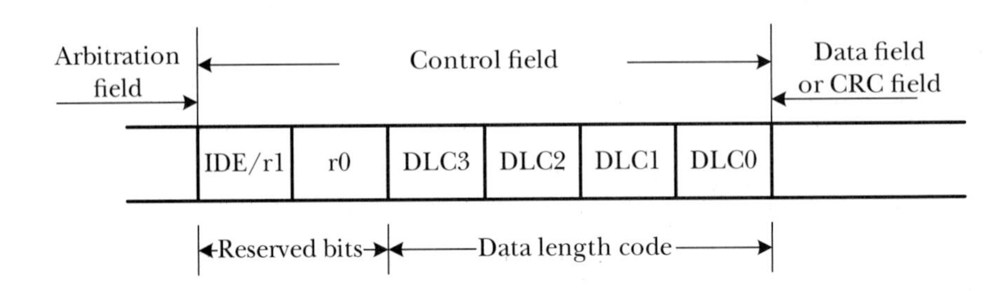

FIGURE 13.4 *Control field.*

TABLE 13.1 *CAN Data Length Coding*

DLC3	DLC2	DLC1	DLC0	Data Byte Count
d	d	d	d	0
d	d	d	r	1
d	d	r	d	2
d	d	r	r	3
d	r	d	d	4
d	r	d	r	5
d	r	r	d	6
d	r	r	r	7
r	d	d	d	8

d = dominant, *r* = recessive

Data Field

The data field consists of the data to be transmitted within a data frame. It may contain from 0 to 8 bytes, each of which contains 8 bits that are transferred as most-significant-bit first.

CRC Field

The CRC field contains the CRC sequence followed by a CRC delimiter, as shown in Figure 13.5.

The frame-check sequence is derived from a cyclic redundancy code (CRC) best suited to frames with bit counts less than 127. The CRC sequence is calculated by performing a polynomial division. The coefficients of the polynomial are given by the destuffed bit stream, consisting of the start of frame, arbitration field, control field, data field (if present), and 15 0s. This polynomial is divided (the coefficients are calculated using modulo-2 arithmetic) by the generator polynomial:

$$X^{15} + X^{14} + X^{10} + X^{8} + X^{7} + X^{4} + X^{3} + 1$$

The remainder of this polynomial division is the CRC sequence. In order to implement this function, a 15-bit shift register **CRC_RG (14:0)** is used. If **nxtbit** denotes the next bit of the bit stream given by the destuffed bit sequence from the start of frame until the end of the data field, the CRC sequence is calculated as

```
CRC_RG = 0;  // initialize shift register
do {
    crcnxt = nxtbit ^ CRC_RG(14);              // exclusive OR
    CRC_RG(14:1) = CRC_RG(13:0);               // shift left by one bit
    CRC_RG(0) = 0;
    if crcnxt
        CRC_RG(14:0) = CRC_RG(14:0) ^ 0x4599;
    } while (!(CRC SEQUENCE starts or there is an error condition));
```

After the transmission/reception of the last bit of the data field, **CRC_RG(14:0)** contains the CRC sequence. The *CRC delimiter* is a single recessive bit.

ACK Field

As shown in Figure 13.6, the ACK field is 2 bits long and contains the ACK slot and the ACK delimiter. A transmitting node sends two recessive bits in the ACK field. A receiver that has received a valid message reports this to the transmitter by sending a dominant bit in the ACK slot (i.e., it sends ACK). A node that has received the matching CRC sequence overwrites the recessive

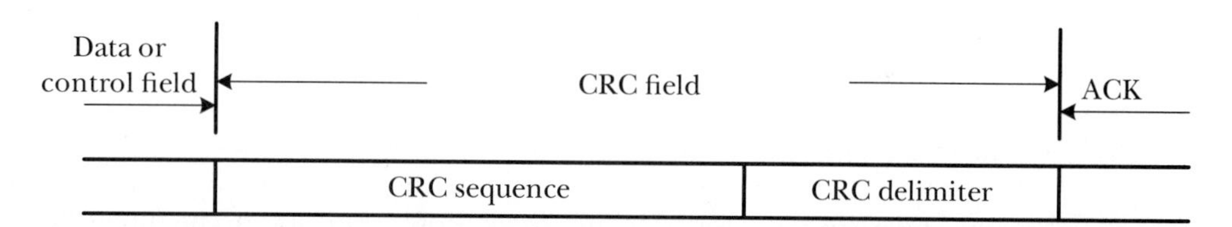

FIGURE 13.5
CRC field.

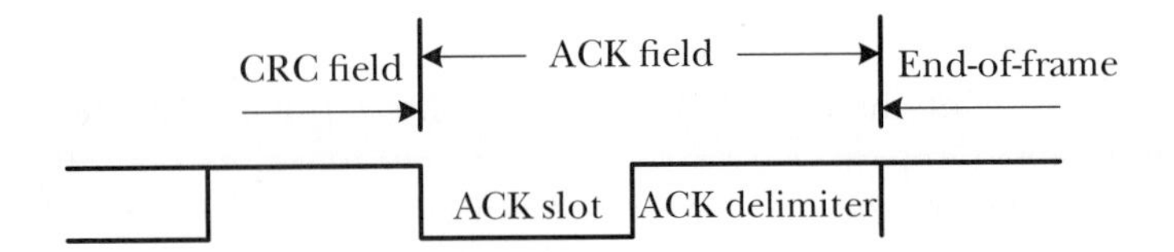

FIGURE 13.6
ACK field.

bit in the ACK slot with a dominant bit. This bit will be received by the data frame transmitter and learn that the previously transmitted data frame has been received correctly. The *ACK delimiter* has to be a recessive bit. As a consequence, the ACK slot is surrounded by two recessive bits (the CRC delimiter and the ACK delimiter).

End-of-Frame Field

Each data frame and remote frame is delimited by a flag sequence consisting of seven recessive bits. This 7-bit sequence is the *end-of-frame* sequence.

13.3.2 Remote Frame

A node that needs certain data can request the relevant source node to transmit the data by sending a remote frame. The format of a remote frame is shown in Figure 13.7. A remote frame consists of six fields: start-of-frame, arbitration, control, CRC, ACK, and end-of-frame. The polarity of the RTR bit in the arbitration field indicates whether a transmitted frame is a *data frame* (RTR-bit dominant) or a *remote frame* (RTR-bit recessive).

13.3.3 Error Frame

The error frame consists of two distinct fields. The first field is given by the superposition of error flags contributed from different nodes. The second field is the error delimiter. The format of the error frame is shown in Figure 13.8. In order to terminate an error frame correctly, an *error-passive node* may need the bus to be idle for at least three bit times (if there is a local error at an error-passive receiver). Therefore, the bus should not be loaded to 100 percent. An error-passive node has an error count greater than 127 but no more than 255. An *error-active node* has an error count less than 127. There are two forms of error flags:

- ***Active-Error Flag.*** This flag consists of six consecutive dominant bits.
- ***Passive-Error Flag.*** This flag consists of six consecutive recessive bits unless it is overwritten by dominant bits from other nodes.

An error-active node signals an error condition by transmitting an *active-error* flag. The error flag's form violates the law of bit stuffing (to be discussed shortly) and applies to all fields (from start-of-frame to CRC delimiter) or destroys the fixed-form ACK field or end-of-frame field. As a consequence, all

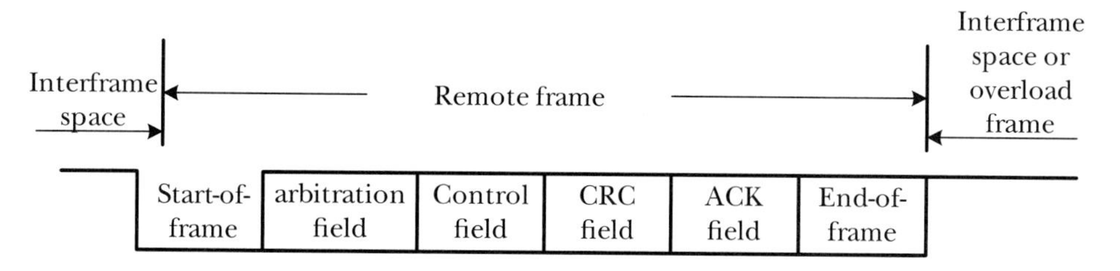

FIGURE 13.7
Remote frame.

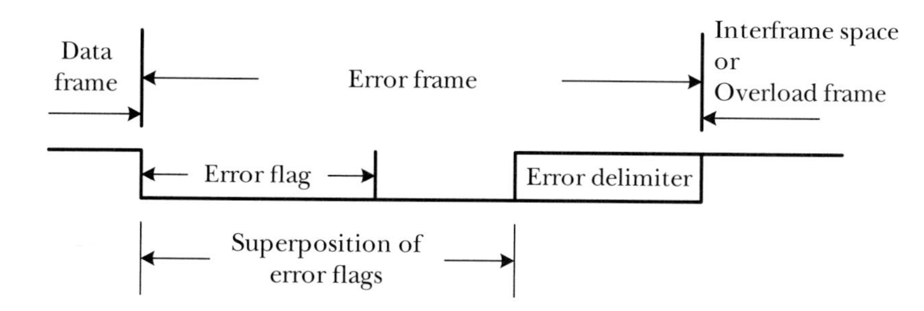

FIGURE 13.8
Error frame.

other nodes detect an error condition and each starts to transmit an error flag. Therefore, the sequence of dominant bits, which actually can be monitored on the bus, results from a superposition of different error flags transmitted by individual nodes. The total length of this sequence varies between a minimum of 6 and a maximum of 12 bits.

An error-passive node signals an error condition by transmitting a passive-error flag. The error-passive node waits for six consecutive bits of equal polarity, beginning at the start of the passive-error flag. The passive-error flag is complete when these equal bits have been detected.

The *error delimiter* consists of eight recessive bits. After transmission of an error flag, each node sends recessive bits and monitors the bus until it detects a recessive bit. Afterwards, it starts transmitting seven more recessive bits.

13.3.4 Overload Frame

The *overload frame* contains two bit fields: *overload flag* and *overload delimiter.* There are three different overload conditions that lead to the transmission of an overload frame.

1. The internal conditions of a receiver require a delay of the next data frame or remote frame.
2. At least one node detects a dominant bit during intermission.
3. A CAN node samples a dominant bit at the eighth bit (i.e., the last bit) of an error delimiter or overload delimiter. The error counters will not be incremented.

The format of an overload frame is shown in Figure 13.9. An overload frame resulting from condition 1 is only allowed to start at the first bit time of an expected intermission, whereas an overload frame resulting from overload conditions 2 and 3 starts one bit after detecting the dominant bit.

No more than two overload frames may be generated to delay the next data frame or remote frame. The overload flag consists of six dominant bits. The format of an overload frame is similar to that of the active-error flag. The overload flag's form destroys the fixed form of the *intermission field*. As a consequence, all other nodes also detect an overload condition, and each starts to transmit an overload flag. In the event that there is a dominant bit detected during the third bit of *intermission* locally at some node, it will interpret this bit as the start-of-frame.

The overload delimiter consists of eight recessive bits. The overload delimiter has the same form as the error delimiter. After the transmission of an overload flag, the node monitors the bus until it detects a transition from a dominant to a recessive bit. At this point in time, every bus node has finished sending its overload flag, and all nodes start transmission of seven more recessive bits in coincidence.

13.3.5 Interframe Space

Data frames and remote frames are separated from preceding frames by a field called *interframe space*. In contrast, overload frames and error frames are not preceded by an interframe space, and multiple overload frames are not separated by an interframe space. For nodes that are not error-passive or have been receivers of the previous message, the interframe space contains the bit fields of *intermission* and *bus idle*, as shown in Figure 13.10. The interframe space of an error-passive node consists of three subfields: *intermission, suspend transmission*, and *bus idle*, as shown in Figure 13.11.

The intermission subfield consists of three recessive bits. During intermission, no node is allowed to start transmission of the data frame or remote frame. The only action permitted is the signaling of an overload condition. The period of bus idle may be of arbitrary length. The bus is recognized to be

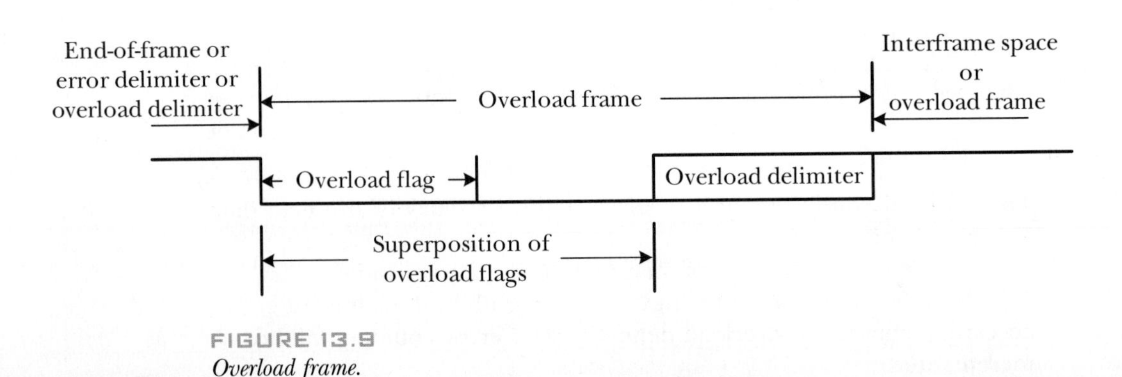

FIGURE 13.9
Overload frame.

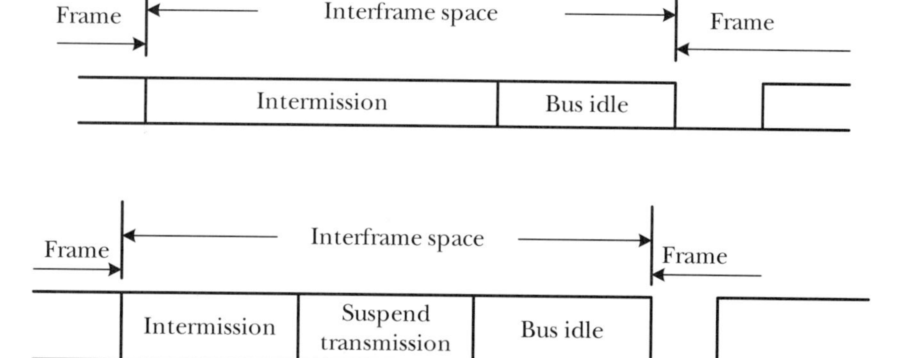

FIGURE 13.10 *Interframe space for non error-passive nodes or receiver of previous message.*

FIGURE 13.11 *Interframe space for error-passive nodes.*

free, and any node having something to transmit can access the bus. A message, pending during the transmission of another message, is started in the first bit following intermission. When the bus is idle, the detection of a dominant bit on the bus is interpreted as a start-of-frame. After an error-passive node has transmitted a frame, it sends eight recessive bits following an intermission before starting to transmit a new message or recognizing the bus as idle. Meanwhile, if a transmission (caused by another node) starts, the node will become the receiver of this message.

13.3.6 Message Filtering

A node uses filter(s) to decide whether to work on a specific message. Message filtering is applied to the whole identifier. A node optionally can implement mask registers that specify which bits in the identifier are examined with the filter. If mask registers are implemented, every bit of the mask registers must be programmable; in other words, they can be enabled or disabled for message filtering. The length of the mask register can comprise the whole identifier or only part of it.

13.3.7 Message Validation

The point in time at which a message is taken to be valid is different for the transmitters and receivers of the message. The message is valid for the transmitter if there is no error until the end-of-frame. If a message is corrupted, retransmission will follow automatically and according to the rules of prioritization. In order to be able to compete for bus access with other messages, retransmission has to start as soon as the bus is idle. The message is valid for the receiver if there is no error until the second-to-last bit of the end-of-frame field.

13.3.8 Bit Stream Encoding

The frame segments including start-of-frame, arbitration field, control field, data field, and CRC sequence are encoded by *bit stuffing*. Whenever a transmitter detects five consecutive bits of identical value in the bit stream to be transmitted, it automatically inserts a complementary bit in the actual transmitted bit stream. The remaining bit fields of the data frame or remote frame (CRC delimiter, ACK field, and end-of-frame) are of fixed form and not stuffed. The error frame and overload frame are also of fixed form and are not encoded by the method of bit stuffing.

The bit stream in a message is encoded using the *non-return-to-zero* (NRZ) method. This means that during the total bit time the generated bit level is either dominant or recessive.

13.4 Error Handling

There are five types of errors. These errors are not mutually exclusive.

13.4.1 Bit Error

A node that is sending a bit on the bus also monitors the bus. When the bit value monitored is different from the bit value being sent, the node interprets the situation as an error. There are two exceptions to this rule:

1. A node that sends a recessive bit during the stuffed bit stream of the arbitration field or during the ACK slot detects a dominant bit.
2. A transmitter that sends a passive-error flag detects a dominant bit.

13.4.2 Stuff Error

A stuff error is detected whenever six consecutive dominant or six consecutive recessive levels occur in a message field.

13.4.3 CRC Error

The CRC sequence consists of the result of the CRC calculation by the transmitter. The receiver calculates the CRC in the same way as the transmitter. A CRC error is detected if the calculated result is not the same as that received in the CRC sequence.

13.4.4 Form Error

A form error is detected when a fixed-form bit field contains one or more illegal bits. For a receiver, a dominant bit during the last bit of end-of-frame is not treated as a form error.

13.4.5 Acknowledgement Error

An acknowledgement error is detected whenever the transmitter does not monitor a dominant bit in the ACK slot.

13.4.6 Error Signaling

A node that detects an error condition signals the error by transmitting an error flag. An error-active node will transmit an *active-error* flag; an error-passive node will transmit a *passive-error* flag. Whenever a node detects a bit error, a stuff error, a form error, or an acknowledgement error, it will start transmission of an error flag at the next bit time. Whenever a CRC error is detected, transmission of an error flag will start at the bit following the ACK delimiter, unless an error flag for another error condition already has been started.

13.5 Fault Confinement

13.5.1 CAN Node Status

A node in error may be in one of three states: *error-active*, *error-passive*, or *bus-off*. An error-active node normally can take part in bus communication and sends an active-error flag when an error has been detected. An error-passive node must not send an active-error flag. It takes part in bus communication, but when an error has been detected only a passive-error flag is sent. After a transmission, an error-passive node will wait before initiating further transmission. A bus-off node is not allowed to have any influence on the bus.

13.5.2 Error Counts

The CAN protocol requires each node to implement *transmit error count* and *receive error count* to facilitate fault confinement. These two counts are updated according to 12 rules. These 12 rules can be found in the CAN specification. An error-count value greater than (roughly) 96 indicates a heavily disturbed bus. It may be advantageous to provide the means to test for this condition. If during system start-up only one node is on-line and if this node transmits some message, it will get no acknowledgement, detect an error, and repeat the message. It can become error-passive but not bus-off for this reason.

13.6 CAN Message Bit Timing

The setting of a bit time in a CAN system must allow a bit sent out by the transmitter to reach the far end of the CAN bus and allow the receiver to send back acknowledgement and reach the transmitter. In a CAN environment, the *nominal bit rate* is defined to be the number of bits transmitted per second in the absence of resynchronization by an ideal transmitter.

13.6.1 Nominal Bit Time

The inverse of the nominal bit rate is the *nominal bit time.* A nominal bit time can be divided into four nonoverlapping time segments, as shown in Figure 13.12.

The **sync_seg** segment is used to synchronize the various nodes on the bus. An edge is expected to lie within this segment. The **prop_seg** segment is used to compensate for the physical delay times within the network. It is twice the sum of the signal's propagation time on the bus line, the input comparator delay, and the output driver delay. The **phase_seg1** and **phase_seg2** segments are used to compensate for edge-phase errors (due to oscillator frequency variation). These segments can be lengthened or shortened by synchronization.

The *sample point* is the point in time at which the bus level is read and interpreted as the value of that respective bit. The sample point is located at the end of **phase_seg1**. A CAN controller may implement the three-samples-per-bit option in which the majority function is used to determine the bit value. Each sample is separated from the next sample by half a time quanta (CAN clock cycle). The *information processing time* is the time segment starting with the sample point reserved for calculation of the sample bit(s) level. The segments contained in a nominal bit time are represented in the unit of *time quantum.* The time quantum (t_Q) is a fixed unit of time that can be derived from the oscillator period (T_{OSC}). t_Q is expressed as a multiple of a *minimum time quantum.* This multiple is a programmable prescale factor. Thus, the time quantum can have the length of

$$\text{Time quantum} = M \times \text{minimum time quantum}$$

where M is the value of the prescaler.

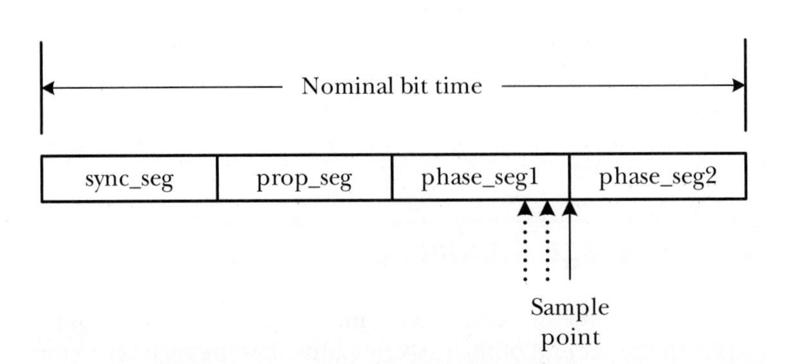

FIGURE 13.12
Nominal bit time.

13.6.2 Length of Time Segments

The segments of a nominal bit time can be expressed in the unit of time quantum as

- **sync_seg** is 1 time quantum long
- **prop_seg** is programmable to be 1, 2, . . . , 8 time quanta long
- **phase_seg1** is programmable to be 1, 2, . . . , 8 time quanta long
- **phase_seg2** is the maximum of **phase_seg1** and information processing time and hence will be programmable from 2 to 8 t_Qs
- The information processing time is equal to or less than 2 t_Q

The total number of time quanta in a bit time must be programmable over a range of at least 8 to 25.

13.7 Synchronization Issue

All CAN nodes must be synchronized while receiving a transmission, i.e., the beginning of each received bit must occur during each node's **sync_seg** segment. This is achieved by synchronization. Synchronization is required because of phase errors between nodes, which may arise because of nodes having slightly different oscillator frequencies or because of changes in propagation delay when a different node starts transmitting.

Two types of synchronization are defined: *hard synchronization* and *resynchronization.* Hard synchronization is performed only at the beginning of a message frame, when each CAN node aligns the **sync_seg** of its current bit time to the recessive-to-dominant edge of the transmitted start-of-frame. After a hard synchronization, the internal bit time is restarted with **sync_seg**. Resynchronization subsequently is performed during the remainder of the message frame whenever a change of bit value from recessive to dominant occurs outside of the expected **sync_seg** segment. Resynchronization is achieved by implementing a digital phase-lock loop (DPLL) function that compares the actual position of a recessive-to-dominant edge on the bus to the position of the expected edge.

13.7.1 Resynchronization Jump Width

There are three possibilities for the occurrence of the incoming recessive-to-dominant edge.

1. *After the* **sync_seg** *segment but before the sample point.* This situation is interpreted as a *late edge.* The node will attempt to resynchronize to the bit stream by increasing the duration of its **phase_seg1** segment of the

current bit by the number of time quanta by which the edge was late up to the resynchronization jump-width (SJW) limit.

2. *After the sample point but before the* **sync_seg** *segment of the next bit.* This situation is interpreted as an *early bit.* The node now will attempt to resynchronize to the bit stream by decreasing the duration of its **phase_seg2** segment of the current bit by the number of time quanta by which the edge was early by up to the resynchronization jump-width limit. Effectively, the **sync_seg** segment of the next bit begins immediately.
3. *Within the* **sync_seg** *segment of the current bit time.* This is interpreted as no synchronization error.

As a result of resynchronization, **phase_seg1** may be lengthened or **phase_seg2** may be shortened. The amount by which the phase buffer segments may be altered may not be greater than the *resynchronization jump width,* which is programmable to be between 1 and the smaller of 4 and **phase_seg1** time quanta.

Clocking information may be derived from transitions from one bit value to the other. The property that only a fixed maximum number of successive bits have the same value provides the possibility of resynchronizing a bus node to the bit stream during a frame.

The maximum length between two transitions that can be used for resynchronization is 29 bit times.

13.7.2 Phase Error of an Edge

The *phase error* of an edge is given by the position of the edge relative to **sync_seg**, measured in time quanta. The sign of phase error is defined as

$e < 0$ if the edge lies after the sample point of the previous bit
$e = 0$ if the edge lies within **sync_seg**
$e > 0$ if the edge lies before the sample point

13.8 The C8051F040 CAN Module

The C8051F040 includes a CAN controller that supports the CAN protocol version 2.0 part A and B. The data rate can be as high as 1 MBit/s. Two dedicated pins **CANTX** and **CANRX** are used by the CAN module to transmit and receive data. A typical CAN bus configuration is illustrated in Figure 13.13. The C8051F040 includes a CAN controller to handle the data transmission and reception to and from the CAN bus. The CAN bus could reach several hundred meters long, no commercial microcontroller can supply the current level required in CAN bus data transmission. A dedicated external CAN transceiver is needed to supply the current required in CAN bus data communica-

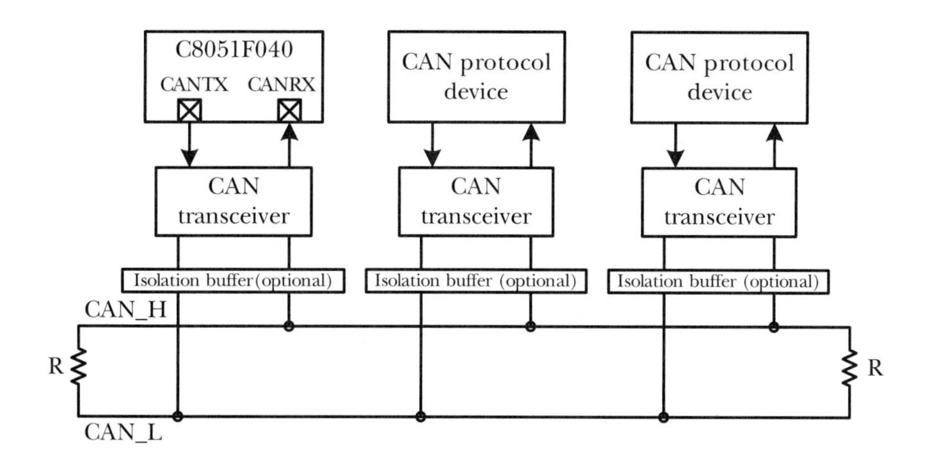

FIGURE 13.13
A typical CAN bus configuration.

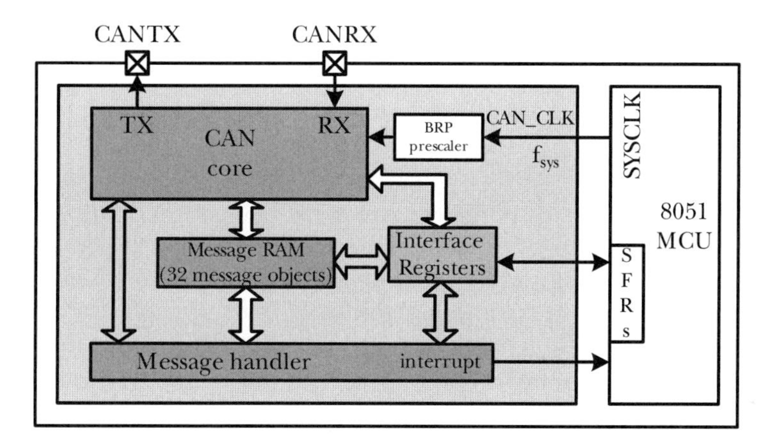

FIGURE 13.14
C8051F040 CAN controller diagram.

tion. The CAN bus must be terminated by a 120 Ω resistor on both ends to minimize the reflection interference.

As shown in Figure 13.14, the CAN controller consists of a CAN core, a message RAM (separate from MCU data memory), a message-handler state machine, interface registers, and control registers. The CAN controller does not provide physical layer drivers.

13.8.1 The Structure of the CAN Controller

The CAN core serializes the parallel message to be transmitted and shifts it out to the CAN bus. It also parallelizes the serial data received from the CAN bus and stores it in the message RAM.

The message RAM consists of 32 message objects. A message to be transmitted to the CAN bus must be framed and stored in one of the message objects in the message RAM.

A message frame received from the CAN bus will be stored in the message RAM temporarily before it is read by the MCU.

The C8051F040 system clock is divided by a prescaler before it is used to synchronize the data transmission and reception in the CAN core.

The block labeled as *interface registers* is divided into two sets of registers (IF1 and IF2). IF1 and IF2 control the CPU access to the message RAM. They buffer the data to be transferred to and from the RAM, avoiding conflicts between CPU accesses and message reception/transmission.

The message handler is responsible for generating control signals to transfer data back and forth between

- CAN core and message RAM
- Message RAM and interface registers

The message handler also may generate interrupts to the CPU in case of errors and data transfer completion.

13.8.2 Registers Related to CAN Module

A space of 256 bytes is allocated to the CAN module. The registers are organized as 16-bit registers, with the high byte at the odd address and the low byte at the even address. The two sets of registers IF1 and IF2 are also in this space. A summary of CAN registers is given in Table 13.2. The mnemonic of each CAN register index is in column 5 of this table. They are also made available to you in two files in the complementary CD: **canIndex.h** and **canIndex.inc**. Place one (or both) of these two files in your project directory, include them in your program, and you will be able to refer to them (used as index) by mnemonic names.

The CAN module registers are divided into four groups:

- ***CAN Core Registers:*** This group consists of CAN control register (CAN0CN), status register (CAN0STA), error counter, bit timing register, test register (CAN0TST), and BRP extension register
- ***Message Interface Register Set IF1:*** This set consists of 12 registers
- ***Message Interface Register Set IF2:*** This set consists of 12 registers
- ***Message Handling Registers:*** This group consists of five registers: interrupt register, transmission request 1 and 2 registers, new data 1 and 2 registers, interrupt pending 1 and 2 registers, message valid 1 and 2 registers

Among these registers, only CAN0CN, CAN0STA, and CAN0TST may be accessed directly by the CPU. The other registers must be accessed indirectly via the CAN address register (CAN0ADR) and CAN data registers (CAN0DATH and CAN0DATL). The CAN0CN, CAN0STA, and CAN0TST registers also can be accessed indirectly like other registers.

To write into a CAN register, the user first places the index of that CAN register in the CAN0ADR register and then writes the value into the

TABLE 13.2 *The C8051F040 CAN Register Summary*

Address	Index	Name	Reset value	Mnemonic name[4]
CAN base + 0x00	0x00	CAN control register	0x0001	
CAN base + 0x02	0x01	Status register	0x0000	
CAN base + 0x04	0x02	Error counter	0x0000	ERRCNT
CAN base + 0x06	0x03	Bit timing register	0x2301	BITREG
CAN base + 0x08	0x04	Interrupt register	0x0000	INTREG
CAN base + 0x0A	0x05	Test register	0x00 & 0br0000000[1)]	CANTSTR
CAN base + 0x0C	0x06	BRP extension register	0x0000	BRPEXT
CAN base + 0x0E		--reserved	---[3)]	
CAN base + 0x10	0x08	IF1 command register	0x0001	IF1CMDRQST
CAN base + 0x12	0x09	IF1 command mask	0x0000	IF1CMDMSK
CAN base + 0x14	0x0A	IF1 mask1	0xFFFF	IF1MSK1
CAN base + 0x16	0x0B	IF1 mask2	0xFFFF	IF1MSK2
CAN base + 0x18	0x0C	IF1 arbitration 1	0x0000	IF1ARB1
CAN base + 0x1A	0x0D	IF1 arbitration 2	0x0000	IF1ARB2
CAN base + 0x1C	0x0E	IF1 message control	0x0000	IF1MSGC
CAN base + 0x1E	0x0F	IF1 data A 1	0x0000	IF1DATA1
CAN base + 0x20	0x10	IF1 data A 2	0x0000	IF1DATA2
CAN base + 0x22	0x11	IF1 data B 1	0x0000	IF1DATB1
CAN base + 0x24	0x12	IF1 data B2	0x0000	IF1DATB2
CAN base + 0x28-0x3E		--reserved	---[3)]	
CAN base + 0x40	0x20	IF2 command register	0x0001[2)]	IF2CMDRQST
CAN base + 0x42	0x21	IF2 command mask	0x0000[2)]	IF2CMDMSK
CAN base + 0x44	0x22	IF2 mask1	0xFFFF[2)]	IF2MSK1
CAN base + 0x46	0x23	IF2 mask2	0xFFFF[2)]	IF2MSK2
CAN base + 0x48	0x24	IF2 arbitration 1	0x0000[2)]	IF2ARB1
CAN base + 0x4A	0x25	IF2 arbitration 2	0x0000[2)]	IF2ARB2
CAN base + 0x4C	0x26	IF2 message control	0x0000[2)]	IF2MSGC
CAN base + 0x4E	0x27	IF2 data A 1	0x0000[2)]	IF2DATA1
CAN base + 0x50	0x28	IF2 data A 2	0x0000[2)]	IF2DATA2
CAN base + 0x52	0x29	IF2 data B 1	0x0000[2)]	IF2DATB1
CAN base + 0x54	0x2A	IF2 data B2	0x0000[2)]	IF2DATB2
CAN base + 0x56-0x7E		--reserved	---[3)]	
CAN base + 0x80	0x40	Transmission request 1	0x0000	TRANSREQ1
CAN base + 0x82	0x41	Transmission request 2	0x0000	TRANSREQ2
CAN base + 0x84-0x8E		--reserved	---[3)]	
CAN base + 0x90	0x48	New data 1	0x0000	NEWDAT1
CAN base + 0x92	0x49	New data 2	0x0000	NEWDAT2
CAN base + 0x94-0x9E		--reserved	---[3)]	

(*continued*)

TABLE 13.2 *The C8051F040 CAN register summary (continued)*

Address	Index	Name	Reset value	Mnemonic name[4]
CAN base + 0xA0	0x50	Interrupt pending 1	0x0000	INTPEND1
CAN base + 0xA2	0x51	Interrupt pending 2	0x0000	INTPEND2
CAN base + 0xA4-0xAE		--reserved	---[3)]	
CAN base + 0xB0	0x58	Message valid 1	0x0000	MSGVAL1
CAN base + 0xB2	0x59	Message valid 2	0x0000	MSGVAL2
CAN base + 0xB4-0xBE		--reserved	---[3)]	

Note:
1. r signifies the actual value of the CAN_RX pin
2. The two sets of message interface registers – IF1 and IF2 —have identical functions
3. Reserved bits are read as '0' except for IF1 mask 2 register where they are read as '1'.
4. Users can use these names to refer to CAN registers in their program once they include the canIndex.h and/or canIndex.inc file in their C and assembly program.

CAN0DATH and CAN0DATL registers. For example, the following C statements will write the value 0x2304 into the bit timing register:

```
CAN0ADR  = BITREG;      // Load bit timing register's index
CAN0DATH = 0x23;        // Move the upper byte into data reg high byte
CAN0DATL = 0x04;        // Move the lower byte into data reg low byte
```

To read a CAN register, the user first places the index number of that register in the CAN0ADR register and then reads from CAN0DATH and CAN0DATL. For example, the following C statements read the data A1 registers of IF1:

```
CAN0ADR = IF1DATA;      // Load IF1 data A1's index
HiByte  = CAN0DATH;     // copy the high byte of IF1 data A1 to HiByte
LoByte  = CAN0DATL;     // copy the low byte of IF1 data A1 to LoByte
```

Please note that the CAN0DATL register must be accessed after the CAN0DATH register if both need to be accessed.

13.8.3 CAN0ADR Auto-incrementing Features

For ease of programming message objects, CAN0ADR provides auto-incrementing for the index ranges 0x08 to 0x12 (IF1) and 0x20 to 0x2A (IF2). When the CAN0ADR register has an index in these ranges, the CAN0ADR will auto-increment by 1 to point to the next CAN register 16-bit word on a read or write of the CAN0DATL register. This speeds up programming of frequently interfaced registers when configuring message objects.

13.8.4 The CAN Core Registers

The registers in this group control the operation modes and the configuration of the CAN bit timing and provide status information.

CAN Control Register

The contents of this register are shown in Figure 13.15. The CAN initialization is started by setting the **Init** bit. When the **Init** bit is set, all message transfers from and to the CAN bus are stopped, and the status of the CAN bus output **CAN_TX** is high (recessive). To initialize the CAN controller, the CPU has to set up the bit timing register and each message object (in message RAM). If a message object is not needed, it is sufficient to set its **MsgVal** bit to invalid. Otherwise, the whole message object has to be initialized.

15	14	13	12	11	10	9	8	7	6	5	4	3	2	1	0
res	res	res	res	res	res	res	res	Test	CCE	DAR	CANIF	EIE	SIE	IE	Init
r	r	r	r	r	r	r	r	rw	rw	rw	r	rw	rw	rw	rw

res: Unimplemented (reserved). **r** stands for read only. **rw** stands for read/writeable

Test: Test mode enable
- 0 = disable
- 1 = enable

CCE: Configuration change enable
- 0 = the CPU has no write access to the bit timing register (while init = 1)
- 1 = the CPU has write access to the bit timing register

DAR: Disable automatic retransmission
- 0 = automatic retransmission of disturbed messages enabled
- 1 = automatic retransmission disabled

CANIF: CAN interrupt flag
- 0 = CAN interrupt has not occurred
- 1 = CAN interrupt has occurred and is active

EIE: Error interrupt enable
- 0 = disabled–No error status interrupt will be generated
- 1 = enabled–A change in the bits Boff or Ewarn in the status register will generate an interrupt

SIE: Status change interrupt enable
- 0 = disabled–No status change interrupt will be generated.
- 1 = enabled–An interrupt will be generated when a message transfer is successful or a CAN error is detected.

IE: Module interrupt enable
- 0 = disabled–module interrupt output signal is always high
- 1 = enabled

Init: Initialization
- 0 = normal operation
- 1 = initialization is started

FIGURE 13.15
The CAN control register (CAN0CN, address 0x01 and 0x00).

Access to the bit timing register and to the BRP extension register for the configuration of the bit timing is enabled when both the **Init** and **CCE** bits in this register are set. Resetting the **Init** bit finishes the software initialization.

The initialization of message object (in the message RAM) is independent of the **Init** bit and can be done on the fly, but the message objects all should be configured to a particular identifier or set to not valid before starting the message transfer. The CAN specification requires the CAN implementation to provide automatic retransmission of frames that have lost arbitration or that have been disturbed by errors during transmission. This feature can be disabled by setting the **DAR** bit. This feature is useful to operate the CAN module within a time-triggered CAN (TTCAN) environment.

When the **DAR** bit is cleared and a message is not transmitted successfully due to arbitration failure or error, then the **NewDat** bit of the message object (Figure 13.32) will remain set. The user needs to set the **TxRqst** of the message to 1 to start retransmission.

The CAN module can interrupt the CPU when there is an error or when a message frame is transmitted or received successfully. Interrupt enabling and disabling can be done by setting or clearing the **IE**, **EIE**, and **SIE** bits of this register.

CAN Status Register

This register records the error status and also indicates whether a message has been received or transmitted successfully. A field of error code in this register indicates the type of error which just occurred. The contents of this register are shown in Figure 13.16.

Error Counter

This register holds the error counts that occurred during the transmission or reception of messages. The contents of this register are shown in Figure 13.17.

Bit Timing Register

This register sets the four segments of a bit time. The contents of this register are shown in Figure 13.18. The **TSeg1** field represents the sum of the required **phase_seg1** and **prop_seg** minus 1.

Test Register

This register allows the user to exercise the test functions provided by the CAN module. The contents of this register are shown in Figure 13.19. To write into this register, the user has to set the **Test** bit in the CAN control register to 1.

BRP Extension Register

This register allows the user to expand the prescaler to the CAN clock source. The contents of this register are shown in Figure 13.20. The BRPE field of this register is used as the most significant four bits of the baud rate prescaler.

15	14	13	12	11	10	9	8	7	6	5	4	3	2	1	0
res	res	res	res	res	res	res	res	Boff	EWarn	EPass	RxOk	TxOk	LEC		
r	r	r	r	r	r	r	r	r	r	r	rw	rw	rw		

res: Unimplemented (reserved), **r** stands for read only, **rw** stands for read/writeable
Boff: Busoff status
0 = the CAN module is not busoff
1 = the CAN module is in busoff state
EWarn: Warning status
0 = both error counters are below the error warning limit of 96
1 = at least one of the error counters has reached the error warning limit of 96
EPass: Error passive
0 = the CAN core is error active
1 = the CAN core is in the error passive state as defined in the CAN specification
RxOk: Receive a message successfully
0 = since this bit was last reset by the CPU, no message has been successfully received
1 = since this bit was last reset by the CPU, a message has been successfully received
TxOk: Transmitted message successfully
0 = since this bit was last reset by the CPU, no message has been successfully transmitted
1 = since this bit was last reset by the CPU, a message has been successfully transmitted
LEC: Last error code
000 = no error
001 = stuff error (more than five consecutive identical bits have occurred)
010 = form error: A fixed format part of a received frame has the wrong format
011 = AckError: The message this CAN core transmitted was not acknowledged by another node
100 = Bit1Error: When transmitting a 1 to the CAN, a dominant level was detected on the CAN bus
101 = Bit0Error: When transmitting a 0, the CAN module detected a recessive level on the CAN bus
110 = CRCError: The CRC check sum was incorrect
111 = unused

FIGURE 13.16
The CAN status register (CAN0STA, address 0x03 and 0x02).

15	14	13	12	11	10	9	8	7	6	5	4	3	2	1	0
RP	REC6-0							TEC7-0							
r	r							r							

r stands for read only. **rw** stands for read/writeable
RP: Receive error passive
0 = the receive error counter is below the error passive level
1 = the receive error counter has reached the error passive level
REC6-0: Receive error count
Actual state of the receive error counter. Values between 0 and 127
TEC7-0: Transmit error count
Actual state of the transmit error counter. Values between 0 and 255

FIGURE 13.17
Error counter (ERRCNT, address 0x05 and 0x04).

15	14	13	12	11	10	9	8	7	6	5	4	3	2	1	0
res	TSeg2			TSeg1				SJW		BRP					
r	rw			rw				rw		rw					

res: Unimplemented (reserved). **r** stands for read only. **rw** stands for read/writeable
TSeg1: The time segment before the sample point
0x01–0x0F: Valid value for the TSeg1 is from 1 to 15. The actual value used by the hardware is one plus this field
TSeg2: The time segment after the sample point
0x0–0x7: Valid value for TSeg2 is from 0 to 7. The actual value used by the hardware is one plus this field
SJW: Synchronization jump width
0x0–0x3: Valid value for SJW is from 0 to 3. The actual value used by the hardware is one plus this field
BRP: Baud rate prescaler
0x00–0x3F: Valid value for BRP is 00 to 63. The actual value used by the hardware is one plus this field. The oscillator frequency is divided by this value before being used as CAN clock signal

FIGURE 13.18
The CAN bit timing register (BITREG, addresses 0x07 and 0x06).

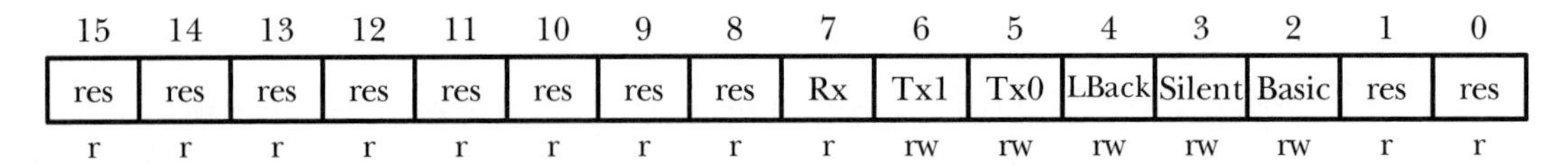

15	14	13	12	11	10	9	8	7	6	5	4	3	2	1	0
res	res	res	res	res	res	res	res	Rx	Tx1	Tx0	LBack	Silent	Basic	res	res
r	r	r	r	r	r	r	r	r	rw	rw	rw	rw	rw	r	r

res: Unimplemented (reserved). **r** stands for read only. **rw** stands for read/writeable
Rx: Monitors the actual value of the CANRX pin
0 = the CANRX pin level is dominant
1 = the CANRX pin level is recessive
TX1-TX0: Control of CANTX pin
00 = reset value. CANTX is controlled by the CAN core
01 = the sample point can be monitored at CANTX pin
10 = CANTX pin drives a dominant value (low)
11 = CANTX pin drives a recessive value (high)
Lback: Loopback mode
0 = loopback mode is disabled
1 = loopback mode is enabled
Silent: Silent mode
0 = normal operation
1 = the module is in silent mode
Basic: Basic mode
0 = basic mode is disabled
1 = IF1 registers used as Tx buffer, IF2 registers used as Rx buffer

FIGURE 13.19
CAN test register (CAN0TST, addresses 0x0B and 0x0A).

15	14	13	12	11	10	9	8	7	6	5	4	3	2	1	0
res	res	res	res	res	res	res	res	res	res	res	res	BRPE			
r	r	r	r	r	r	r	r	r	r	r	r	rw			

res : Unimplemented (reserved). **r** stands for read only. **rw** stands for read/writeable
BRPE: Baud rate prescaler extension
0x00–0x0F: Valid value for the BRPE is from 0 to 15. The actual value used by the hardware is one plus this field. This field allows the baud rate prescaler to be extended to values up to 1023

FIGURE 13.20
BRP extension register (BRPEXT, addresses 0x0D and 0x0C).

13.8.5 Message Interface Register Sets IF1 and IF2

These two sets of registers are used to control the CPU's access to the message RAM. The interface registers avoid conflicts between the CPU's access to the message RAM and CAN message reception and transmission by buffering the data to be transferred. A complete message object or parts of the message object may be transferred between the message RAM and the IFx (x = 0 or 1) message buffer registers in one single transfer.

The functions of the two interface register sets are identical (except in basic mode). They can be used in the way that one set of registers is used for data transfer to the message RAM while the other set of registers is used for the data transfer from the message RAM. A summary of the two interface register sets is given in Table 13.3.

Each set of interface registers consists of message buffer registers controlled by their own command registers. The command mask register specifies the direction of the data transfer and which parts of a message object will be transferred. The command request register selects a message object in the message RAM as the destination or source for the transfer and to start the action specified in the command mask register.

IFx Command Request Registers

A message transfer is started as soon as the CPU has written the message number to the command request register. With this write operation, the **Busy** bit automatically is set to 1, and the CAN module notifies the CPU that a transfer is in progress. After a period of 3 to 6 CAN_CLK cycles, the transfer between the interface registers and the message RAM has completed, which will reset the **Busy** bit to 0. The contents of the IF command request register are shown in Figure 13.21.

IFx Command Mask Register

The control bits of this register specify the transfer direction and select which of the IFx message buffer registers to be the source or destination of the data transfer. The contents of this register are shown in Figure 13.22.

TABLE 13.3 *IF1 and IF2 Message Interface Register Sets*

Address	IF1 Register Set	Address	IF2 Register Set
CAN base + 0x10	IF1 command request	CAN base + 0x40	IF2 command request
CAN base + 0x12	IF1 command mask	CAN base + 0x42	IF2 command mask
CAN base + 0x14	IF1 mask 1	CAN base + 0x44	IF2 mask 1
CAN base + 0x16	IF1 mask 2	CAN base + 0x46	IF2 mask 2
CAN base + 0x18	IF1 arbitration 1	CAN base + 0x48	IF2 arbitration 1
CAN base + 0x1A	IF1 arbitration 2	CAN base + 0x4A	IF2 arbitration 2
CAN base + 0x1C	IF1 message control	CAN base + 0x4C	IF2 message control
CAN base + 0x1E	IF1 data A 1	CAN base + 0x4E	IF2 data A 1
CAN base + 0x20	IF1 data A 2	CAN base + 0x50	IF2 data A 2
CAN base + 0x22	IF1 data B 1	CAN base + 0x52	IF2 data B 1
CAN base + 0x24	IF1 data B 2	CAN base + 0x54	IF2 data B 2

15	14	13	12	11	10	9	8	7	6	5	4	3	2	1	0
Busy	res	res	res	res	res	res	res	res	res	Message number					
r	r	r	r	r	r	r	r	r	r	rw					

res: Unimplemented (reserved). **r** stands for read only. **rw** stands for read/writeable
Busy: Busy flag
 0 = reset to 0 when read/write action to IFx has finished
 1 = set to 1 when writing into the IFx command request register
Message number:
 0x01-0x20: Valid message number, the message object in the message RAM with this number is selected for data transfer
 0x00: Not a valid message number, interpreted as 0x20
 0x21-0x3F: Not a valid message number, interpreted as 0x01-0x1F

FIGURE 13.21
IFx command request register (IF1CMDRQST, addresses 0x11 & 0x10 for IF1; and IF2CMDRQST, 0x41 & 0x40 for IF2).

- Bit 7 (**WR/RD**) sets the data transfer direction. When this bit is 1, the contents of the selected interface registers are written into the message object addressed by the command register. Otherwise, the contents of the message object addressed by the command register are transferred to the selected interface registers.
- Bit 6 (**Mask**) determines whether identifier mask and two other mask bits are to be changed. When this bit is 0, then mask information is not changed. Otherwise, mask information of the message frame will be updated (0 = no, 1 = yes).
- Bit 5 (**Arb**) determines whether the message identifier and the **Dir**, **Xtd**, and **MsgVal** bits are involved in the data transfer (0 = no, 1 = yes).

15	14	13	12	11	10	9	8	7	6	5	4	3	2	1	0
res								WR/RD	Mask	Arb	Control	ClrIntPnd	TxRqst/ NewDat	Data A	Data B
r	r	r	r	r	r	r	r	rw	rw	rw	rw	rw	rw	rw	rw

res: Unimplemented (reserved). **r** stands for read only. **rw** stands for read/writeable

WR/RD: Write/Read

0 = transfer data from the message object addressed by the command register into selected message interface registers

1 = transfer data from the selected interface registers to the message object addressed by the command request register

Direction = Write (RW/RD bit = 1)

Mask: Access mask bits

0 = mask bits unchanged

1 = transfer identifier mask + Mdir + MXtd (in IFx mask registers) to message object

Arb: Access arbitration bits

0 = arbitration bits unchanged

1 = transfer Identifier + Dir + Xtd + MsgVal (in IFx arbitration registers) to message object

Control: Access control bits

0 = control bits unchanged

1 = transfer control bits to message object

ClrIntPnd: Clear interrupt pending bit

This bit is ignored

TxRqst/NewDat: Access transmission request bit

0 = TxRqst bit unchanged

1 = set TxRqst bit

Data A: Access data bytes 0-3

0 = data bytes 0-3 unchanged

1 = transfer data bytes 0-3 to message object

Data B: Access data bytes 4-7

0 = data bytes 4-7 unchanged

1 = transfer data bytes 4-7 to message object

Direction = Read (RW/RD bit = 0)

Mask: Access mask bits

0 = mask bits unchanged

1 = transfer identifier mask + Mdir + MXtd (in message object) to IFx message interface registers

Arb: Access arbitration bits

0 = arbitration bits unchanged

1 = transfer Identifier + Dir + Xtd + MsgVal (in message object) to IFx message interface registers

Control: Access control bits

0 = control bits unchanged

1 = transfer control bits to IFx message interface register

ClrIntPnd: Clear interrupt pending bit

0 = IntPnd bits unchanged 1 = clear IntPnd bit in the message object

TxRqst/NewDat: Access transmission request bit

0 = TxRqst bit unchanged 1 = clear NewDat bit in the message object

Data A: Access data bytes 0-3

0 = data bytes 0-3 unchanged

1 = transfer data bytes 0-3 to IFx message interface registers

Data B: Access data bytes 4-7

0 = data bytes 4-7 unchanged

1 = transfer data bytes 4-7 to IFx message interface registers

FIGURE 13.22

IFx command mask registers (IF1CMDMSK 0x13 & 0x12, IF2CMDMSK, 0x43 & 0x42).

- Bit 4 (**Control**) determines whether the contents of the message control register are transferred during the message transfer (0 = no, 1 = yes).
- Bit 3 (**ClrIntPnd**) determines whether the interrupt pending bit in the message will be cleared during the message transfer (0 = no, 1 = yes).
- Bit 2 (**TxRqst/NewDat**) determines whether the TxRqst bit (when WR/RD bit = 1) or the NewDat bit (when WR/RD bit = 0) will be cleared to 0 or set to 1.
- Bit 1 (**Data A**) determines whether the first four data bytes of the message will be transferred (0 = no, 1 = yes).
- Bit 0 (**Data B**) determines whether the last four data bytes of the message will be transferred (0 = no, 1 = yes).

The user should set a mask bit to 1 only when the related information needs to be updated during a message transfer.

IFx Mask Registers

The contents of these registers mirror their equivalents in the message objects in the message RAM. These registers hold the identifier mask and control whether a specific identifier bit is used in matching the incoming messages. The contents of IFx mask 1 register and IFx mask 2 registers are shown in Figure 13.23.

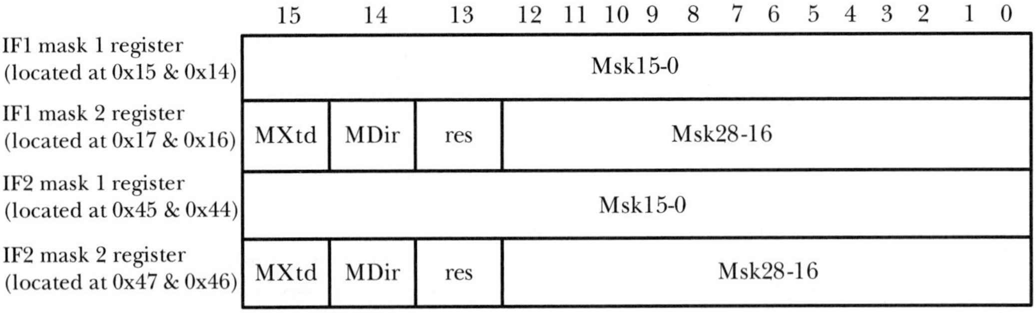

	15	14	13	12 11 10 9 8 7 6 5 4 3 2 1 0
IF1 mask 1 register (located at 0x15 & 0x14)	Msk15-0			
IF1 mask 2 register (located at 0x17 & 0x16)	MXtd	MDir	res	Msk28-16
IF2 mask 1 register (located at 0x45 & 0x44)	Msk15-0			
IF2 mask 2 register (located at 0x47 & 0x46)	MXtd	MDir	res	Msk28-16

MXtd: Mask extended identifier
  0 = the extended identifier bit (IDE) has no effect on the acceptance filtering
  1 = the extended identifier bit (IDE) is used for acceptance filtering
Mdir: Mask message direction
  0 = the message direction bit (Dir) has no effect on the acceptance filtering
  1 = the message direction bit (Dir) is used for acceptance filtering
Msk28-0: Identifier mask
  0 = the corresponding bit in the identifier of the message object cannot inhibit the match in the acceptance filtering
  1 = the corresponding identifier bit is used for acceptance filtering

FIGURE 13.23
IFx mask registers (IF1MSK1, IF1MSK2, IF2MSK1, IF2MSK2).

IFx Arbitration Registers

These registers hold the identifiers of the message in the IFx buffer and also indicate whether a message is valid. The contents of these registers are shown in Figure 13.24.

IFx Message Control Registers

These registers provide overall control to the message object to be transmitted and received. The contents of these registers are shown in Figure 13.25. Bits 15, . . . , 13 are status bits. The meanings of most bits are obvious except the **RmtEn** bit.

IFx Data A and Data B Registers

These registers hold the received data and data to be transmitted. The contents of these registers are shown in Figure 13.26.

Register	15	14	13	12–0
IF1 arbitration 1 register (located at 0x19 & 0x18)	ID15-0			
IF1 arbitration 2 register (located at 0x1B & 0x1A)	MsgVal	Xtd	Dir	ID28-16
IF2 arbitration 1 register (located at 0x49 & 0x48)	ID15-0			
IF2 arbitration 2 register (located at 0x4B & 0x4A)	MsgVal	Xtd	Dir	ID28-16

MsgVal: Message valid
  0 = the message object is ignored by the message handler
  1 = the message object is configured and should be considered by the message handler
Xtd: Extended identifier
  0 = the 11-bit identifier will be used for this message object
  1 = the 29-bit identifier will be used for this message object
Dir: Message direction
  0: direction = receive. On TxRqst, a remote frame with the identifier of this message object is transmitted. On reception of a data frame with matching identifier, that message is stored in this message object
  1: Direction = transmit. On TxRqst, the respective message object is transmitted as a data frame. On reception of a remote frame with matching identifier, the TxRqst bit of this message object is set (if RmtEn bit = 1)
ID28-0: Message identifier
  ID28–ID0: 29-bit identifier (extended frame)
  ID28–ID18: 11-bit identifier (standard frame)

FIGURE 13.24
IFx arbitration registers (IF1ARB1, IF1ARB2, IF2ARB1, IF2ARB2).

15	14	13	12	11	10	9	8	7	6	5	4	3 2 1 0
NewDat	MsgLst	IntPnd	UMask	TxIE	RxIE	RmtEn	TxRqst	EoB	res	res	res	DLC3-0
rw	rw	rw	rw	rw	rw	rw	rw	rw	r	r	r	rw

res: Unimplemented (reserved). **r** stands for read only. **rw** stands for read/writeable
NewDat: New data
0 = no new data has been written into the data portion of this message object since this bit was last cleared by CPU
1 = the message handler or CPU has written new data into the data portion of this message object
MsgLst: Message lost (only valid for message objects with direction = receive)
0 = no message lost since last time this bit was reset by the CPU
1 = the message handler stored a new message into this object when the NewDat bit was still set the CPU has lost a message
IntPnd: Interrupt pending
0 = this message is not the source of interrupt
1 = this message is the source of interrupt. The interrupt Identifier in the interrupt register will point to this message object if there is no other interrupt source with higher priority
UMask: Use acceptance mask
0 = mask ignored
1 = use mask (Msk28-0, MXtd, and Mdir) for acceptance filtering
TxIE: Transmit interrupt enable
0 = the IntPnd bit will be left unchanged after a successful transmission of a frame
1 = the IntPnd bit will be set after a successful transmission of a frame
RxIE: Receive interrupt enable
0 = the IntPnd bit will be left unchanged after a successful reception of a frame
1 = the IntPnd bit will be set after a successful reception of a frame
RmtEn: Remote enable
0 = at the reception of a remote frame, the TxRqst bit is left unchanged
1 = at the reception of a remote frame, the TxRqst bit is set
TxRqst: Transmit request
0 = this message is not waiting for transmission
1 = the transmission of this message object is requested but is not yet done
EoB: End of buffer
0 = message object belongs to a FIFO buffer and is not the last message object of that FIFO buffer
1 = single message object or last message object of a FIFO buffer
DLC3-DLC0: Data length code
0-8: data frame has 0-8 data bytes
9-15: data frame has 8 data bytes

FIGURE 13.25
IFx message control register (IF1MSG: at 0x1D & 0x1C, IF2MSG: at 0x4D & 0x4C for IF2).

13.8.6 Message Handling Registers

Registers in this group are read-only. Their contents are status information provided by the message handler finite-state machine. Registers in this group include interrupt register, transmission request registers, new data registers, interrupt pending registers, and message valid registers.

	15–8	7–0
IF1 data A1 (0x1F & 0x1E)	Data 1	Data 0
IF1 data A2 (0x21 & 0x20)	Data 3	Data 2
IF1 data B1 (0x23 & 0x22)	Data 5	Data 4
IF1 data B2 (0x25 & 0x24)	Data 7	Data 6
IF2 data A1 (0x4F & 0x4E)	Data 1	Data 0
IF2 data A2 (0x51 & 0x50)	Data 3	Data 2
IF2 data B1 (0x53 & 0x52)	Data 5	Data 4
IF2 data B2 (0x55 & 0x54)	Data 7	Data 6
	rw	rw

FIGURE 13.26
IFx data A and data B registers (IF1DATA1, IF1DATA2, IF2DATB1, IF2DATB2).

15–8	7–0
IntId15-8	IntId7-0
r	r

IntID15-0: Interrupt identifier (indicates source of interrupt)

0x0000:	No interrupt is pending
0x0001-0x0020:	Number of message object which caused the interrupt
0x0021-0x7FFF:	Unused
0x8000:	Status interrupt
0x8001-0xFFFF:	Unused

FIGURE 13.27
Interrupst register (INTREG at 0x08-0x09).

Interrupt Register

This register holds the identifier (message object number) of the interrupt source to be serviced by the CPU. If several interrupts are pending, this register points to the pending interrupt with the highest priority. An interrupt remains pending until the CPU has cleared it. The status interrupt has the highest priority. Among message object interrupts, the message object's interrupt priority decreases with increasing message number. A message interrupt is cleared by clearing the message object's **IntPnd** bit. The status interrupt is cleared by reading the status register. The contents of this register are shown in Figure 13.27.

Transmission Request Registers

These registers hold the **TxRqst** bits of the 32 message objects. By reading out the **TxRqst** bits, the CPU can check for which message object a transmission request is pending. The **TxRqst** bit of a specific message object can be set/reset by the CPU via the IFx message interface registers or by the message handler after reception of a remote frame or after a successful transmission. The contents of these registers are shown in Figure 13.28.

New Data Registers

These registers indicate whether the data portion of all 32 message objects in the message RAM hold new data. The contents of these registers are shown in Figure 13.29. The **NewDat** bit of a specific message object can be set/reset by the CPU via the IFx message interface registers or by the message handler after reception of a data frame or after a successful transmission.

Interrupt Pending Registers

These registers hold the **IntPnd** bits of the 32 message objects. By reading out the **IntPnd** bits, the CPU can check for which message object an interrupt is pending. The contents of these registers are shown in Figure 13.30.

Message Valid Registers

These registers hold the **MsgVal** bits of the 32 message objects. By reading out the **MsgVal** bits, the CPU can check for which message object is valid.

	15 14 13 12 11 10 9 8	7 6 5 4 3 2 1 0
Transmission request 1 register (addresses 0x81 & 0x80)	TxRqst16-9	TxRqst8-1
Transmission request 2 register (address 0x83 & 0x82)	TxRqst32-25	TxRqst24-17
	r	r

TxRqst32-1: Transmission request bits:
0 = this message is not waiting for transmission
1 = the transmission of this message is requested and is not yet done

FIGURE 13.28
Transmission request registers (TRANSREQ1, TRANSREQ2).

	15 14 13 12 11 10 9 8	7 6 5 4 3 2 1 0
New 1 register (addresses 0x91 & 0x90)	NewDat16-9	NewDat8-1
New 2 register (address 0x93 & 0x92)	NewDat32-25	NewDat24-17
	r	r

NewDat32-1: New data bits (of all message objects)
0 = no new data has been written into the data portion of this message object by the message handler or the CPU since last time this flag was cleared by the CPU
1 = the message handler or the CPU has written new data into the data portion of this message object

FIGURE 13.29
New data registers (NEWDAT1 & NEWDAT2).

13.8.7 Message Objects in the Message Memory

The C8051F040 does not access the 32 message objects in the message memory directly to avoid conflicting with CAN message reception and transmission. The C8051F040 CPU always accesses message objects through the interface registers.

The structure of a message object is shown in Figure 13.32. Each field has its counterpart in the message interface registers and also has the same function as its counterpart.

Register	15–8	7–0
Interrupt pending 1 register (addresses 0xA1 & 0xA0)	IntPnd16-9	IndPnd8-1
Interrupt pending 2 register (address 0xA3 & 0xA2)	IntPnd32-25	IntPnd24-17
	r	r

IntPnd32-1: Interrupt pending bits (of all message objects)
0 = this message is not the source of an interrupt
1 = this message object is the source of an interrupt

FIGURE 13.30
Interrupt pending registers (INTPEND1 & INTPEND2).

Register	15–8	7–0
Message valid 1 register (addresses 0xB1 & 0xB0)	MsgVal16-9	MsgVal8-1
Message valid 2 register (address 0xB3 & 0xB2)	MsgVal32-25	MsgVal24-17
	r	r

MsgVal32-1: Message valid bits (of all message objects)
0 = this message object is ignored by the message handler
1 = this message object is configured and should be considered by the message handler

FIGURE 13.31
Message valid registers (MSGVAL1 & MSGVAL2).

Message Object												
UMask	Msk28-0	MXtd	MDir	EoB	NewDat		MsgLst	RxIE	TxIE	IntPnd	RmtEn	TxRqst
MsgVal	ID28-0	Xtd	Dir	DLC3-0	Data 0	Data 1	Data 2	Data 3	Data 4	Data 5	Data 6	Data 7

FIGURE 13.32
Message structure in message RAM.

13.9 CAN Operation Modes

The CAN module can operate in one of the following modes depending on the purpose:

- Normal mode
- Basic mode
- Test mode
- Silent mode
- Loopback mode
- Loopback combined with silent mode

13.9.1 Normal Mode

In this mode, the CAN module carries out normal message transmission and reception. To send a message frame, the CPU writes the message frame into the interface registers and sets the **TxRqst** bit of the IFx command register, and the message handler takes care of the actual transmission. To read a received message, the CPU also reads from the interface registers.

13.9.2 Basic Mode

In this mode, the CAN module runs without the message RAM. The IF1 registers are used as the transmit buffer. The transmission of the contents of the IF1 registers is requested by writing the **Busy** bit of the IF1 command request register to 1. The IF1 registers are locked while the **Busy** bit is set. The **Busy** bit indicates that the transmission is pending.

As soon as the CAN bus is idle, the IF1 registers are loaded into the shift register of the CAN core and the transmission is started. When the transmission has completed, the **Busy** bit is reset and the locked IF1 registers are released.

A pending transmission can be aborted at any time by resetting the **Busy** bit in the IF1 command request register while the IF1 registers are locked. If the CPU has reset the **Busy** bit, a possible retransmission in case of lost arbitration or in case of an error is disabled.

The IF2 registers are used as receive buffers. After the reception of a message, the contents of the shift register is stored into the IF2 registers, without any acceptance filtering.

Additionally, the actual contents of the shift register can be monitored during the message transfer. Each time a read message object is initiated by setting the **Busy** bit of the IF2 command request register to 1, the contents of the shift register is stored into the IF2 registers.

In basic mode, the evaluation of all message object related control and status bits and of the control bits of the IFx command mask registers is turned off. The message number of the command request register is not evaluated. The **NewDat** and **MsgLst** bits of the IF2 message control register retain their function; DLC3-0 will show the received DLC, and the other control bits will be read as 0.

13.9.3 Test Mode

The test mode is entered by setting the **Test** bit in the CAN control register to 1. In test mode, the **Tx1**, **Tx0**, **LBack**, **Silent**, and **Basic** bits in the test register are writable. The **Rx** bit monitors the state of pin CANRX and therefore is only readable. All test-register functions are disabled when the **Test** bit is reset to 0. This mode allows the user to enter the silent and/or loopback mode to monitor the CAN bus traffic or test the CAN applications.

13.9.4 Silent Mode

The CAN core can enter the silent mode by setting the **Silent** bit of the CAN control register to 1. In silent mode, the CAN module is able to receive valid data frames and valid remote frames. However, it sends only recessive bits on the CAN bus, and it cannot start a transmission. If the CAN core is required to send a dominant bit (ACK bit, overload flag, active-error flag), the bit is rerouted internally so that the CAN core monitors this dominant bit, although the CAN bus may remain in recessive state. The silent mode can be used to analyze the traffic on a CAN bus without affecting it by the transmission of dominant bits (Acknowledge bits, error frames). Figure 13.33 shows the connection of signals CANTX and CANRX to the CAN core in silent mode.

13.9.5 LoopBack Mode

The loopback mode is entered by setting the **LBack** bit of the test register to 1. In loopback mode, the CAN core treats its own transmitted messages as received messages and stores them into a receive buffer. Figure 13.34 shows the

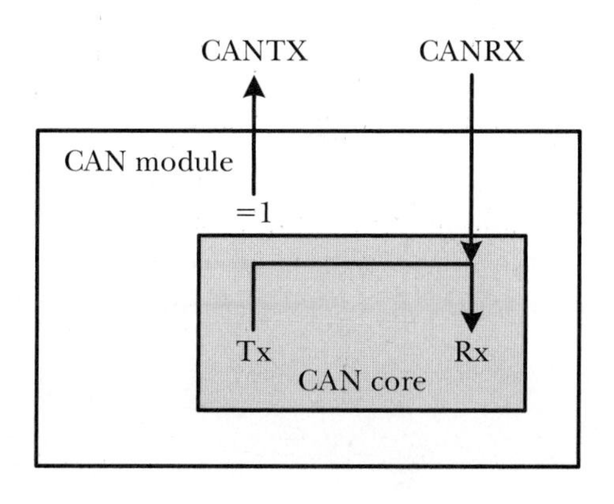

FIGURE 13.33 *CAN core in silent mode.*

connection of signals CANTX and CANRX to the CAN core in the loopback mode. This mode is provided for self-test functions. To be independent from external stimulation, the CAN core ignores acknowledgement errors (recessive bits sampled in the acknowledge slot of a data/remote frame) in the loopback mode. In this mode, the CAN core performs an internal feedback from its **Tx** output to its **Rx** input. The actual value of the CANRX input pin is disregarded by the CAN core. The transmitted messages can be monitored at the CANTX pin.

13.9.6 LoopBack Combined with Silent Mode

It is possible to combine loopback mode and silent mode by programming the bits **LBack** and **Silent** to 1 at the same time. This mode can be used for a "hot self-test," meaning that the CAN module can be tested without affecting a running CAN system connected to the pins CANTX and CANRX. In this mode, the CANRX pin is disconnected from the CAN core and the CANTX pin is held recessive. Figure 13.35 shows the connection of signals

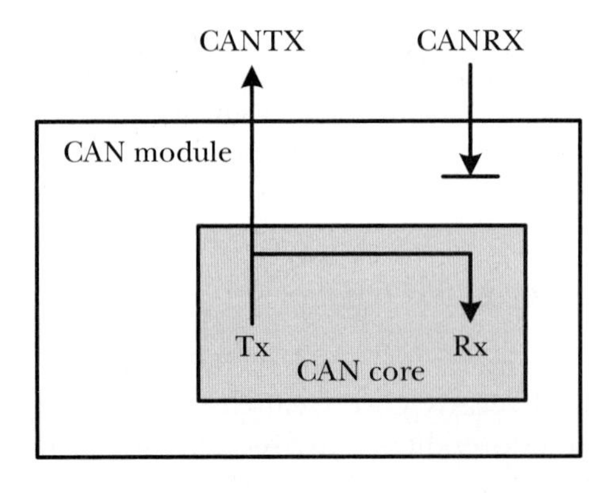

FIGURE 13.34 *CAN core in loopback mode.*

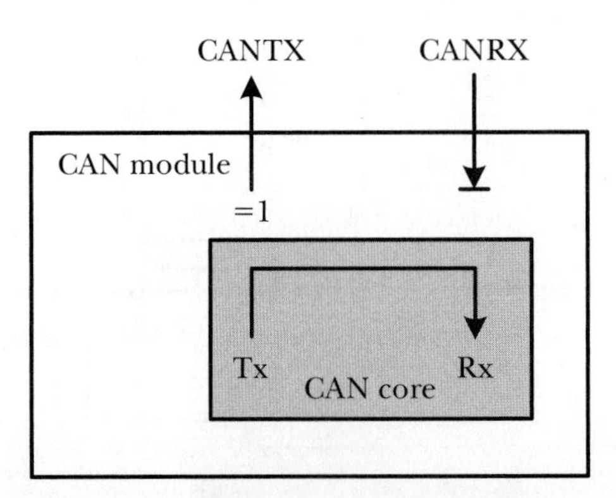

FIGURE 13.35 *CAN core in loopback combined with silent mode.*

CANTX and CANRX to the CAN core in case of the combination of loopback mode with silent mode.

13.10 CAN Module Operation

The detailed internal operation of the CAN module and the configuration of transmit and receive objects are described in this section.

13.10.1 Managing Message Objects

All of the message objects in the message RAM must be initialized by the CPU, or they must be set to invalid (with MsgVal bit = 0) and the bit timing must be configured before the CPU clears the **Init** bit in the CAN control register.

The configuration of message objects is done by programming the mask, arbitration, control, and data fields of one of the two sets of interface registers to the desired values. By writing the message object number into the corresponding IFx command request register (see Figure 13.21), the IFx message interface registers are loaded into the addressed message object in the message RAM.

When the **Init** bit of the CAN control register is cleared, the CAN core and the message handler take control of the CAN module's internal data flow. Received messages that pass the acceptance filtering are stored into the message RAM, while messages with pending transmission request are loaded into the CAN core's shift register and are transmitted via the CAN bus.

A simple way to initialize all message objects is to write the reset values (shown in Table 13.2) of IF registers into all message objects. The following example does just that.

Example 13.1 Write a function to clear all 32 message objects.

Solution: The assembly subroutine that performs the specified operation is as follows.

```
            $include    (canIndex.inc)
clrMsgObj:  push        B
            mov         B,#32
            mov         SFRPAGE,#CAN0_PAGE
            mov         CAN0ADR,#IF1CMDMSK      ; point to command mask register 1
            mov         CAN0DATL,#0xFF          ; set direction to write all IF registers to Msg Obj
wcloop:     mov         CAN0ADR,#IF1CMDRQST     ; write blank (reset) IF registers to each Msg Obj
            mov         CAN0DATL,B              ; select a message object to write
            djnz        B,wcloop
            pop         B
            ret
```

The C function that clears all message objects is as follows.

```
void clrMsgObj(void)
{
    char i;
    SFRPAGE = CAN0_PAGE;
    CAN0ADR = IF1CMDMSK;              // point to command mask register 1
    CAN0DATL = 0xFF;                  // set direction to WRITE all IF registers to Msg Obj
    for (i = 1; i < 33; i++){
        CAN0ADR  = IF1CMDRQST;        // write blank IF registers (reset values) to Msg obj
        CAN0DATL = i;                 // write to message object i
    }
}
```

In the simplest case, the CAN node has only one type of message to send. The user can set the identifier bits to 0s and use the same mask for all message frames. After the initial set up, the user need only fill in the data byte before the transmission of each data frame. The following example illustrates this situation.

Example 13.2 Write an assembly subroutine and C function to initialize the transmit message object so that the same mask and the same standard identifier are used in all transmissions. Set the identifier to 0 and the data length code to 1 for the message object (data length code can be another value).

Solution: The identifier with the highest priority is all 0s. The following assembly subroutine will perform as specified.

```
initMsgObjTx:                                     ; R0 specified the message object to set up
            mov     SFRPAGE,#CAN0_PAGE
            mov     CAN0ADR,#IF1CMDMSK            ; point to command mask register
            mov     CAN0DATH,#0x00                ; set to WRITE, alter all Msg Obj except
            mov     CAN0DATL,#0xB2                ; ID mask bits
            mov     CAN0ADR,#IF1ARB1              ; point to arbitration register
            mov     CAN0DATH,#0                   ; set arbitration ID to highest priority (0)
            mov     CAN0DATL,#0                   ;   "
            mov     CAN0DATH,#0xA0                ; autoincrement to arb2 high byte:
            mov     CAN0DATL,#0x00                ; set MsgVal bit, standard ID (= 0), Dir=WRITE
            mov     CAN0DATH,#0                   ; autoincrement to message control Msg Cntrl:
            mov     CAN0DATL,#0x81                ; disable interrupt, DLC = 1, disable remote frame
            mov     CAN0ADR,#IF1CMDRQST           ; point to command request register
            mov     CAN0DATH,#0                   ; select Msg Obj (passed in R0) passed
            mov     CAN0DATL,R0                   ; into this function
            ret
```

The C function that performs the set up is as follows.

```
sfr16 CAN0DAT = 0xD8;                // 16-bit SFR definition for CAN0DATH and CAN0DATL
void initMsgObjTx (char MsgNum)
{
    SFRPAGE  = CAN0_PAGE;
    CAN0ADR = IF1CMDMSK;
    CAN0DAT  = 0x00B2;               // set to WRITE, alter all Msg Obj except ID mask bits
    CAN0ADR = IF1ARB1;               // point to arbitration register
    CAN0DAT  = 0x0000;               // set arbitration ID (use standard ID)
    CAN0DAT  = 0xA000;               // Arb2 high byte: MsgVal = 1, Xtd = 0, Dir = 1 (Write)
    CAN0DAT  = 0x0081;               // Msg Ctrl: DLC=1, disable remote frame function
    CAN0ADR = IF1CMDRQST;            // point to command request reg.
    CAN0DAT  = MsgNum;               // select Msg Obj passed to this function parameter list
}
```

Example 13.3 Write an assembly subroutine and C function to initialize the receive message object that use the same mask for all reception, standard identifier, and set the identifier to 0.

Solution: The identifier with the highest priority is all 0s. The following assembly subroutine will perform as specified.

```
initMsgObjRx:                                      ; R0 specified the message object to set up
            mov     SFRPAGE,#CAN0_PAGE
            mov     CAN0ADR,#IF1CMDMSK             ; point to command mask register
            mov     CAN0DATH,#0x00                 ; set to WRITE, alter all Msg Obj except
            mov     CAN0DATL,#0xB8                 ; ID mask bits
            mov     CAN0ADR,#IF1ARB1               ; point to arbitration register
            mov     CAN0DATH,#0                    ; set arbitration ID to highest priority
            mov     CAN0DATL,#0                    ;      "
            mov     CAN0DATH,#0x80                 ; autoincrement to arbitration 2 register high byte:
            mov     CAN0DATL,#0x00                 ; MsgVal = 1, Xtd = 0, Dir = 0 (Read)
            mov     CAN0DATH,#0x04                 ; autoincrement to message control Msg Cntrl:
            mov     CAN0DATL,#0x80                 ; set RXIE, disable remote frame
            mov     CAN0ADR,#IF1CMDRQST            ; point to command request register
            mov     CAN0DATH,#0                    ; select Msg Obj passed into the function
            mov     CAN0DATL,R0                    ;      "
            ret
```

The C function that performs the set up is as follows.

```
sfr16 CAN0DAT = 0xD8;
void initMsgObjRx (char MsgNum)
{
    SFRPAGE  = CAN0_PAGE;
    CAN0ADR = IF1CMDMSK;             // point to command mask
    CAN0DAT  = 0x00B8;               // set to WRITE, alter all Msg Obj except ID mask & data bits
    CAN0ADR = IF1ARB1;               // point to arbitration register
```

```
    CAN0DAT = 0x0000;          // set arbitration ID (use standard ID)
    CAN0DAT = 0x8000;          // Arb2 high byte:set MsgVal bit, standard ID (=0), Dir = Read
    CAN0DAT = 0x0480;          // Msg Ctrl: set RXIE. remote frame not enabled
    CAN0ADR = IF1CMDRQST;      // point to command request reg.
    CAN0DAT = MsgNum;          // select Msg Obj passed to this function parameter list
}
```

13.10.2 Data Transfer to/from Message RAM

When the CPU initiates a data transfer between the IFx registers and message RAM, the message handler sets the **Busy** bit in the respective command request register to 1. After transfer is completed, the **Busy** bit is set back to 0.

The respective command mask register specifies whether a complete message object or only part of it will be transferred. Due to the structure of the message RAM, it is not possible to write a few bits/bytes of one message object; it is always necessary to write a complete message object into the message RAM. Therefore, the data transfer from the IFx registers to the message RAM requires a **read-modify-write** cycle. First, the parts of the message object that are not to be changed are read from the message RAM. Then, the complete contents of the message interface registers are written into the message object.

After the partial write of a message object, the message interface registers that are not selected in the command mask register will be set to the actual contents of the selected message object. After the partial read of a message object, the message interface registers that are not selected in the command mask register will be left unchanged.

13.10.3 Transmission of Messages

If the shift register of the CAN core is ready for loading and if there is no data transfer between the IFx registers and message RAM, the **MsgVal** bits in the message valid register and the **TxRqst** bits in the transmission request register are evaluated. The valid message object with the highest priority pending transmission request is loaded into the shift register by the message handler, and the transmission is started. The message object's **NewDat** bit is reset.

After a successful transmission and if no new data was written to the message object (NewDat bit = 0) since the start of the transmission, the **TxRqst** bit will be reset. If the **TxIE** bit is set, the **IntPnd** bit will be set after a successful transmission. If the CAN module lost the arbitration or if an error occurred during the transmission, the message will be retransmitted as soon as the CAN bus is free again.

13.10.4 Sending Remote Frame

A CAN node can request other node to send back a data frame by sending a remote frame. The procedure for sending a remote frame is as follows.

Step 1.
Select a message buffer as a receive buffer.

Step 2.
Load the message identifier of the receive buffer with the ID of the remote frame whose data you want to receive.

Step 3.
Send the remote frame by setting the **TxRqst** bit of the buffer selected as a receive one.

After this, the message object selected as a receive buffer will send a remote frame. The node which has a buffer in transmit mode with a matching ID in turn automatically will set its **TxRqst** bit (if RmtEn = 1) and will reply with the message with the requested data.

13.10.5 Auto Reply to Remote Frames

The **RmtEn** bit of the message control register enables an **auto reply** to the remote frame. To utilize this feature, the user needs to take the following actions.

- Prepare a message object with the identifier identical to that of the incoming remote frame
- Set the **Dir** bit of the IFx arbitration register to **1**
- Set the **RmtEn** bit of the IFx message control register to **1**
- Clear the **TxRqst** bit (in the message control register) of the message object to **0**

When a matched remote request frame arrived, the **TxRqst** bit will be set to 1, and this message object will be transmitted as a reply to the remote frame.

13.10.6 Acceptance Filtering of Received Messages

When the arbitration and control fields (identifier + IDE + RTR + DLC) of an incoming message is shifted completely into the Rx/Tx shift register of the CAN core, the message handler finite-state machine starts scanning the message RAM for a matching, valid message object.

To scan the message RAM for a matching message object, the acceptance filtering unit is loaded with the arbitration bits from the CAN-core shift register. Then the arbitration and mask fields (including **MsgVal**, **UMask**, **NewDat**, and **EoB**) of message object 1 are loaded into the acceptance filtering unit and compared with the arbitration field from the shift register. This is repeated with each following message object until either a matching message object is found or the end of the message RAM is reached.

If a match occurs, the scanning is stopped, and the message handler FSM proceeds, depending on the type of frame (data frame or remote frame) received.

13.10.7 Reception of Data Frame

The message handler FSM stores the message from the CAN-core shift register into the respective message object in the message RAM. Not only the data bytes, but all of the arbitration bits and the data length code are stored in the corresponding message object. This is implemented to keep the data bytes connected with the identifier, even if arbitration mask registers are used.

The **NewDat** bit is set to indicate that new data has been received. The CPU should reset the **NewDat** bit when it reads the message object. If at the time of reception the **NewDat** bit was already set, the **MsgLst** bit is set to indicate that the previous data is lost. If the RxIE bit is set, the **IntPnd** bit will be set after the new message is received, causing the interrupt register to point to this message object.

The **TxRqst** bit of this message object is reset to prevent the transmission of a remote frame, while the requested data frame has just been received.

13.10.8 Reception of Remote Frame

When a remote frame is received, three different configurations of the matching message objects have to be considered.

1. Dir = 1 (direction = transmit), RmtEn = 1, and UMask = 0 or 1
   At the reception of a matching remote frame, the **TxRqst** bit of this message object is set. The rest of the message object remains unchanged. After this, the message object that matches the remote frame will be transmitted according to its priority.
2. Dir = 1, RmtEn = 0, and UMask = 0
   At the reception of a matching remote frame, the **TxRqst** bit of this message object remains unchanged; the remote frame is ignored.
3. Dir = 1, RmtEn = 0, and UMask = 1
   At the reception of a matching remote frame, the **TxRqst** bit of this message object is reset. The arbitration and control field (Identifier + IDE + RTR + DLC) from the shift register is stored in the message object in the message RAM and the **NewDat** bit of this message object is set. The data field of the message object remains unchanged; the remote frame is treated similar to a received data frame.

13.10.9 Receive/Transmit Priority

The receive/transmit priority for message objects is set by the message number. Message object 1 has the highest priority, while message object 32 has the lowest priority. If more than one transmission request is pending, they are serviced according to the priorities of message objects.

13.11 Configuration of Message Objects

Message objects to be transmitted and the type of message objects to be received need to be configured beforehand.

Example 13.4 Write an instruction sequence to prepare a message object for receiving data frames with standard identifiers that start with the ASCII code of the letter '**V**'.

Solution: The following instruction sequence will prepare message object 1 to receive data frames with standard identifiers starting with the letter 'V'.

```
mov    SFRPAGE,#CAN0_PAGE
mov    CAN0ADR,#IF1CMDMSK
mov    CAN0DATH,#0x00      ; transfer mask, arbitration bits, and control from
mov    CAN0DATL,#0xF0      ; IF1 to message RAM
mov    CAN0DATH,#0x00      ; set mask bits to compare the highest 8 bits of
mov    CAN0DATL,#0x00      ; identifier (in identifier 2 register)
mov    CAN0DATH,#0xDF      ;     "
mov    CAN0DATL,#0xE0      ;     "
mov    CAN0DATH,#0         ; set identifier to 0 (arbitration 1)
mov    CAN0DATL,#0         ;     "
mov    CAN0DATH,#0x8A      ; MsgVal = 1, Xtd = 0, Dir = 0 (arbitration 2)
mov    CAN0DATL,#0xC0      ; bit 28-18 contains 01010110000b (ASCII of V)
mov    CAN0DATH,#0x10      ; use mask, TxRqst = 0, EoB = 1, DLC = 0
mov    CAN0DATL,#0x80      ; (message control register)
mov    CAN0ADR,#IF1CMDRQST ; point to command request register
mov    CAN0DATH,#0         ; select message object 1 for receiving incoming
mov    CAN0DATL,#1         ; data frames
```

The following C statements will prepare message object 1 for receiving data frames with standard identifiers that start with the letter 'V'.

```
sfr16 CAN0DAT = 0xD8;
SFRPAGE  = CAN0_PAGE;
CAN0ADR = IF1CMDMSK;     // set index to IF1 command mask register
CAN0DAT = 0x00F0;        // transfer mask, arbitration bits, and control
CAN0DAT = 0x0000;        // mask set to compare the highest 8 bits of identifier
CAN0DAT = 0xDFE0;        //     "
CAN0DAT = 0x0000;        // identifier set to 0 (arbitration 1 register)
CAN0DAT = 0x8AC0;        // MsgVal = 1, Xtd = 0, Dir = 0 (arbitration 2 register)
CAN0DAT = 0x1080;        // TxRqst = 0,EoB = 1, DLC = 0 (message control register)
CAN0ADR = IF1CMDRQST;    // point to command request register
CAN0DAT = 1;             // select message object 1 for matching incoming data frames
```

FIGURE 13.36
Initialization of a message object to be transmitted.

MsgVal	Arb	Data	Mask	EoB	Dir	NewDat	MsgLst	RxIE	TxIE	IntPnd	RmtEN	TxRqst
1	appl.	appl.	appl.	1	1	0	0	0	appl.	0	appl.	0

13.11.1 Configuration of a Message Object to be Transmitted

A message object to be transmitted should be configured as shown in Figure 13.36.

The values of arbitration registers (ID28-ID0 and Xtd bit) are given by the application. They define the identifier and type of the outgoing message. If an 11-bit identifier (standard frame) is used, it is programmed to ID28-ID18, and ID17-ID0 can be ignored.

If the TxIE bit is set, the **IntPnd** bit will be set after a successful transmission of the message object. If the **RmtEn** bit is set, a matching received remote frame will cause the **TxRqst** bit to be set; the remote frame automatically will be answered by a data frame.

The data registers (Data7-0 and DLC3-0) are given by the application. The **TxRqst** and **RmtEn** bits may not be set before the data is valid. The mask registers (Msk28-0, UMask, MXtd, and MDir bits) may be used (UMask = 1) to allow group of remote frames with similar identifiers to set the **TxRqst** bit.

Example 13.5 Write an instruction sequence to prepare a data frame to be transmitted. The data frame uses standard identifier that starts with letter 'V' and holds the string **3.5V**.

Solution: The following instruction sequence will prepare a data frame that uses standard identifier starting with letter 'V' and holds the string **3.5V** to be transmitted.

```
mov   SFRPAGE,#CAN0_PAGE
mov   CAN0ADR,#IF1CMDMSK
mov   CAN0DATH,#0x00       ; transfer arbitration bits, control, and data
mov   CAN0DATL,#0xB7       ; set TxRqst bit
mov   CAN0ADR,#IF1ARB1     ; set index to point to arbitration 1
mov   CAN0DATH,#0          ; set identifier to 'V' (arbitration 1)
mov   CAN0DATL,#0          ;    "
mov   CAN0DATH,#0xAA       ; MsgVal = 1, Xtd = 0, Dir = 1 (arbitration 2)
mov   CAN0DATL,#0xC0       ; bits 28 ~ 16 contains ASCII code of 'V' & 00000
mov   CAN0DATH,#0x81       ; RmtEn = 0, TxRqst = 1, EoB = 1, DLC = 4
mov   CAN0DATL,#0x84       ; (message control register)
mov   CAN0DATH,#'3'        ; string to be sent in response to remote frame
mov   CAN0DATL,#'.'        ; request
mov   CAN0DATH,#'5'        ;    "
mov   CAN0DATL,#'V'        ;
mov   CAN0ADR,#IF1CMDRQST  ; point to command request register
mov   CAN0DATH,#0          ; select message object 2 to hold data frame
mov   CAN0DATL,#2          ; to be transmitted
```

The following C statements will perform the desired operation.

```
sfr16 CAN0DAT = 0xD8;
SFRPAGE  = CAN0_PAGE;
CAN0ADR = IF1CMDMSK;
CAN0DAT = 0x00B7;          // transfer arbitration bits, control, and data
CAN0ADR = IF1ARB1;         // set index to point to arbitration 1
CAN0DAT = 0x0000;          // set identifier to start with 'V'
CAN0DAT = 0xAAC0;          // MsgVal = 1, Xtd = 0, Dir = 1, ASCII of 'V' & 000
CAN0DAT = 0x8184;          // RmtEn = 0, TxRqst = 1, EoB = 1, DLC = 4
CAN0DAT = 0x332E;          // "3."
CAN0DAT = 0x3556;          // "5V"
CAN0ADR = IF1CMDRQST;      // set index to point command request reqister
CAN0DAT = 2;               // select message object 2 to hold this data frame
```

13.11.2 Updating a Transmit Object

The CPU may update the data bytes of a transmit object any time via the IFx interface registers. Neither **MsgVal** nor **TxRqst** have to be reset before the update.

Even if only a part of the data bytes are to be updated, all four bytes of the corresponding IFx data A register or IFx data B register have to be valid before the content of that register is transferred to the message object. Either the CPU has to write all four bytes into the IFx data registers or the message object is transferred to the IFx data register before the CPU writes the new data bytes.

When only data bytes need to be updated, the user first should write the value of 0x0087 into the command mask register and then place the index number of the message object in the command request register. The value 0x0087 causes data bytes to be updated while at the same time sets the **TxRqst** bit.

13.11.3 Configuration of a Receive Object

The user needs to initialize the type of message to be received as shown in Figure 13.37.

The arbitration registers (ID28-0 and **Xtd** bit) are given by the application. They define the identifier and type of accepted received messages. If an 11-bit identifier is used, it is programmed to ID28-ID18. ID17-ID0 can then be

MsgVal	Arb	Data	Mask	EoB	Dir	NewDat	MsgLst	RxIE	TxIE	IntPnd	RmtEN	TxRqst
1	appl.	appl.	appl.	1	0	0	0	appl.	0	0	0	0

FIGURE 13.37
Initialization of a receive message object.

disregarded. When a data frame with an 11-bit identifier is received, ID17-ID0 will be set to 0.

If the RxIE bit is set, the **IntPnd** bit will be set when a received data frame is accepted and stored in the message object.

The data-length code is given by the application. When the message handler stores a data frame in the message object, it will store the received data-length code and eight data bytes. If the data-length code is less than 8, the remaining bytes of the message object will be overwritten by nonspecified values.

The mask registers (Msk28-0, UMask, MXtd, and MDir bits) may be used (UMask = 1) to allow groups of data frames with similar identifiers to be accepted.

Example 13.6 Write a sequence of instructions and C statements to send a remote frame with the most significant 8 bits of the identifier (using standard identifier) set to be the ASCII code of the letter '**T**'.

Solution: The following instruction sequence sets up message object 5 to send a remote frame.

```
mov     SFRPAGE,#CAN0_PAGE
mov     CAN0ADR,#IF1CMDMSK
mov     CAN0DATH,#0x00      ; transfer mask, arbitration bits, and control from
mov     CAN0DATL,#0xF0      ; IF1 to message RAM
mov     CAN0DATH,#0x00      ; set mask bits to compare the highest 8 bits of
mov     CAN0DATL,#0x00      ; identifier
mov     CAN0DATH,#0xDF      ;   "
mov     CAN0DATL,#0xE0      ;   "
mov     CAN0DATH,#0         ; set identifier to 'T' (arbitration 1)
mov     CAN0DATL,#0         ;   "
mov     CAN0DATH,#0x8A      ; MsgVal = 1, Xtd = 0, Dir = 0 (arbitration 2)
mov     CAN0DATL,#0x80      ; ASCII of 'T' & 00000 (bits 28-16 in arbitration field)
mov     CAN0DATH,#0x01      ; TxRqst = 1, EoB = 1, DLC = 0 (message control
mov     CAN0DATL,#0x80      ; register)
mov     CAN0ADR,#IF1CMDRQST ; point to command request register
mov     CAN0DATH,#0         ; select message object 5 to deliver remote request
mov     CAN0DATL,#5         ; frame
```

The C statements that requests the specified remote frame transmission is as follows.

```
sfr16 CAN0DAT = 0xD8;
SFRPAGE = CAN0_PAGE;
CAN0ADR = IF1CMDMSK;        // set index to IF1 command mask register
CAN0DAT = 0x00F0;           // transfer mask, arbitration bits, and control
CAN0DAT = 0x0000;           // mask set to compare the highest 8 bits of identifier
CAN0DAT = 0xDFE0;           //    "
CAN0DAT = 0x0000;           // arbitration bits 15-0 set to 0
CAN0DAT = 0x8A80;           // MsgVal = 1, Xtd = 0, Dir = 0, ASCII of 'T'& 00000
```

```
CANODAT  = 0x0180;          // TxRqst = 1,EoB = 1, DLC = 0 (message control register)
CANOADR  = IF1CMDRQST;      // point to command request register
CANODAT  = 5;               // select message object 5 to carry remote request frame
```

Example 13.7 Write a sequence of instructions and C statements to prepare a message object to reply to a remote frame with standard identifier starting with the ASCII code of letter 'T'. The data field to be sent is the string **50.6°C**.

Solution: The following instruction sequence sets up message object 6 to reply to the specified remote-request frame automatically.

```
mov     SFRPAGE,#CANO_PAGE
mov     CANOADR,#IF1CMDMSK
mov     CANODATH,#0x00          ; transfer mask, arbitration bits, control, and data
mov     CANODATL,#0xF3          ;    "
mov     CANODATH,#0x00          ; set mask bits to compare the highest 8 bits of
mov     CANODATL,#0x00          ; identifier
mov     CANODATH,#0xDF          ;    "
mov     CANODATL,#0xE0          ;    "
mov     CANODATH,#0             ; set arbitration bits 15-0 to 0
mov     CANODATL,#0             ;    "
mov     CANODATH,#0xAA          ; MsgVal = 1, Xtd = 0, Dir = 1, identifier bits
mov     CANODATL,#0x80          ; 28 to 21 set to ASCII code of 'T' (0x54)
mov     CANODATH,#0x92          ; RmtEn = 1, TxRqst = 0, EoB = 1, DLC = 6
mov     CANODATL,#0x86          ; (message control register)
mov     CANODATH,#'5'           ; string to be sent in response to remote frame
mov     CANODATL,#'0'           ; request
mov     CANODATH,#'.'           ;    "
mov     CANODATL,#'6'           ;
mov     CANODATH,#223           ; degree character
mov     CANODATL,#'C'           ;
mov     CANOADR,#IF1CMDRQST     ; point to command request register
mov     CANODATH,#0             ; select message object 6 to deliver remote request
mov     CANODATL,#6             ; frame
```

The following C statements will perform the same function.

```
sfr16 CANODAT = 0xD8;
SFRPAGE   = CANO_PAGE;
CANOADR   = IF1CMDMSK;      // set index to IF1 command mask register
CANODAT   = 0x00F3;         // transfer mask, arbitration bits, and control
CANODAT   = 0x0000;         // mask set to compare the highest 8 bits of identifier
CANODAT   = 0xDFE0;         //    "
CANODAT   = 0x0000;         // identifier bits 15-0 set to 0
CANODAT   = 0xAA80;         // MsgVal=1,Xtd=0, Dir = 1, identifier bits 28-21 sets to ASCII
                            // code of 'T' (0x54)
CANODAT   = 0x9286;         // TxRqst=0,EoB=1, DLC=6, RmtEn=1 (message control reg)
```

```
CANODATH = '5';           // data string to be sent in response to remote request frame
CANODATL = '0';           //    "
CANODATH = '.';           //    "
CANODATL = '6';           //    "
CANODATH = 223;           // degree character
CANODATL = 'C';           // Celsius character
CANOADR  = IF1CMDRQST;    // point to command request register
CANODAT  = 6;             // select message object 6 to carry remote request frame
```

13.12 Handling of Received Messages

The CPU can read a received message any time via the IFx interface registers; the data consistency is guaranteed by the message-handler state machine. Typically, the CPU will write first 0x007F to the command mask register, and then the number of the message object to the command request register. That combination will transfer the whole received message from the message RAM into the message interface registers and clear the **NewDat** and **IntPnd** bits in the message RAM.

If the message object uses masks for acceptance filtering, the arbitration bits show which of the matching messages has been received. The actual value of the **NewDat** bit shows whether a new message has been received since the last time a message object was read. The actual value of **MsgLst** shows whether more than one message has been received since last time this message object was read. The **MsgLst** bit will not be reset automatically.

13.13 Configuration of a FIFO Buffer

By programming the identifiers and masks of two or more message objects to matching values, a FIFO is formed. The message object (in a FIFO) with the lowest number is the first message object in a FIFO. The **EoB** bits of all message objects of a FIFO buffer (except the last) have to be programmed to zero. The **EoB** bit of the last message object of a FIFO buffer is set to one, configuring it as the end of the block.

Received messages with identifiers matching to a FIFO buffer are stored into a message object of this FIFO buffer starting with the message object with the lowest message number. When a message is stored into a message object of a FIFO buffer, the **NewDat** bit of this message object is set. By setting the **NewDat** bit while the **EoB** bit is 0, the message object is locked for further write accesses by the message handler until the CPU has written the **NewDat** bit back to zero.

Messages are stored into a FIFO buffer until the last message object of this FIFO buffer is reached. If none of the preceding message objects is released by writing **NewDat** to zero, all further messages for this FIFO buffer will be written into the last message object of the FIFO buffer and therefore overwritting previous messages.

When the CPU transfers the contents of a message object to the IFx registers by writing its number to the IFx command request register, the corresponding command mask register should be programmed the way that bits **NewDat** and **IntPnd** are reset to zero (TxRqst/NewDat = 1 and ClrIntPnd = 1). The values of these bits in the message control register always reflect the status before resetting the bits.

To assure the correct function of a FIFO buffer, the CPU should read out the message objects starting at the FIFO object with the lowest message number. Figure 13.38 shows how a set of message objects that are concatenated to a FIFO buffer can be handled by the CPU.

When multiple data frames with the same identifier from multiple CAN nodes arrive at about the same time, there is a danger of losing data frames.

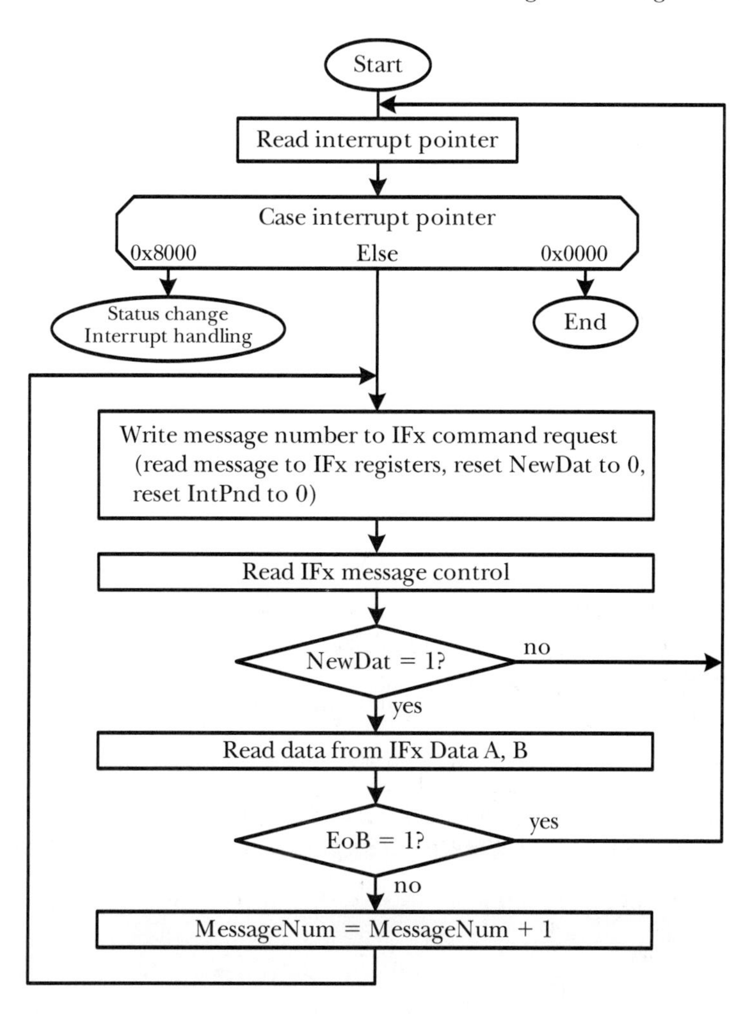

FIGURE 13.38 *C8051F040 CPU handling of a CAN FIFO buffer.*

Using FIFO may resolve this problem. The following example illustrates how to set up a FIFO to receive data frames with the same identifier.

Example 13.8 Write a sequence of instructions and C statements to set up a FIFO of eight message objects for receiving data frames with standard identifier that starts with the ASCII code of letter 'V'.

Solution: The following instruction sequence will set up a FIFO of eight data objects accordingly (starting from message object 2 to message object 9):

```
; all message objects should have the same setup except that the last message object in the
; FIFO should have its EoB bit set to 1. All other objects should have the EoB bit set to 0

          mov     R7,#2                  ; the number of the first message object in the FIFO
          mov     B,#7                   ; used B as loop count
          mov     SFRPAGE,#CAN0_PAGE
          mov     CAN0ADR,#IF1CMDMSK
          mov     CAN0DATH,#0x00         ; transfer mask, arbitration bits, and control from
          mov     CAN0DATL,#0xF0         ; IF1 to message RAM
          mov     CAN0DATH,#0x00         ; set mask bits to compare the first 8 bits of identifier
          mov     CAN0DATL,#0x00         ;    "
          mov     CAN0DATH,#0xDF         ;    "
          mov     CAN0DATL,#0xE0         ;    "
          mov     CAN0DATH,#0            ; set identifier to 'V' (arbitration 1)
          mov     CAN0DATL,#0            ;    "
          mov     CAN0DATH,#0x8A         ; MsgVal = 1, Xtd = 0, Dir = 0 (arbitration 2)
          mov     CAN0DATL,#0xC0         ; bit 28-18 contains ASCII of 'V' (0x56)
          mov     CAN0DATH,#0x14         ; use mask, TxRqst = 0, EoB = 0, DLC = 0, RxIE=1
          mov     CAN0DATL,#0x00         ; (message control register)
FIFOlp:   mov     CAN0ADR,#IF1CMDRQST    ; point to command request register
          mov     CAN0DATH,#0            ; select message object specified in R7 for matching
          mov     CAN0DATL,R7            ; incoming data frames
          inc     R7
          djnz    B,FIFOlp
; set the EoB bit of the last object in FIFO to 1
          mov     CAN0ADR,#IF1MSGC       ; point to message control register
          mov     CAN0DATH,#0x14         ; use mask, TxRqst = 0, EoB = 1, DLC = 0,RxIE=1
          mov     CAN0DATL,#0x80         ;    "
          mov     CAN0ADR,#IF1CMDRQST    ; point to command request register
          mov     CAN0DATH,#0
          mov     CAN0DATL,#9            ; set up message object 9 as the last object in the FIFO
```

The sequence of C statements that sets up the specified FIFO is as follows:

```
sfr16 CAN0DAT = 0xD8;
char i;
sysinit();
```

```
SFRPAGE  = CAN0_PAGE;
CAN0ADR = IF1CMDMSK;            // set index to IF1 command mask register
CAN0DAT = 0x00F0;               // transfer mask, arbitration bits, and control
CAN0DAT = 0x0000;               // mask set to compare the first 8 bits of identifier
CAN0DAT = 0xDFE0;               //        "
CAN0DAT = 0x0000;               // identifier set to 'V'
CAN0DAT = 0x8AC0;               // MsgVal = 1, Xtd = 0, Dir = 0, ASCII of V
CAN0DAT = 0x1400;               // TxRqst = 0, RxIE=1, EoB = 0, DLC = 0 (message control
                                // register)
for (i = 2; i < 9; i++){ // set up message object 2 to 8 of FIFO
    CAN0ADR = IF1CMDRQST;       // point to command request register
    CAN0DAT = i;                // select message object i for matching incoming data
                                // frames
    }
    CAN0ADR = IF1MSGC;
    CAN0DAT = 0x1480;           // TxRqst = 0, RxIE=1, EoB = 1, DLC = 0
    CAN0ADR = IF1CMDRQST;       // set index to IF1 command request register
    CAN0DAT = 9;                // select message object 9
```

13.14 Handling of CAN Interrupts

If several interrupts are pending, the CAN interrupt register will point to the pending interrupt with the highest priority, disregarding their chronological order. An interrupt remains pending until the CPU has cleared it.

The status interrupt has the highest priority. Among the message interrupts, the message object's interrupt priority decreases with increasing message numbers. A message interrupt is cleared by clearing the message object's **IntPnd** bit. The status interrupt is cleared by reading the status register.

The interrupt identifier **IntId** in the interrupt register indicates the cause of the interrupt. When no interrupt is pending, the register will hold the value 0. If the value of the interrupt register is not 0, then there is an interrupt pending, and if the IE bit (of CAN control register) is 1, the CPU will be interrupted. The interrupt remains active until the interrupt register is back to 0 or until the IE bit is cleared.

The value 0x8000 indicates that status interrupt is pending, because the CAN core has updated the status register (due to errors or status change). This interrupt has the highest priority.

All other values indicate that the source of the interrupt is one of the message objects. The **IntId** value points to the pending-message interrupt with the highest interrupt priority. The status change and error interrupts are enabled and disabled by the CAN control register (**SIE**, **EIE**, and **IE** bits). There are two ways to identify the source interrupts. The first method is to check the **IntId** in the interrupt register. The second method is to poll the interrupt pending register. An interrupt service routine reading the message that is the source of interrupt also may reset the message object's **IntPnd** bit at the same time (by setting the **ClrIntPnd** bit of the IFx command mask register). When the **IntPnd** bit is cleared, the interrupt register will point to the next message object with a pending interrupt.

13.15 Setting the CAN Bit Timing Parameters

All devices on the CAN bus must use the same bit rate. However, all devices are not required to have the same master oscillator clock frequency. For the different clock frequencies of the individual devices, the bit rate has to be adjusted by appropriately setting the baud rate prescaler and the number of time quanta in each time segment.

The nominal bit rate is the number of bits transmitted per second, assuming an ideal transmitter with an ideal oscillator and the absence of resynchronization. The maximum nominal bit rate is 1 Mb/s. The nominal bit time is defined as

$$T_{BIT} = 1/(\text{nominal bit rate}) \tag{13.1}$$

According to the CAN specification, the bit time is divided into four non-overlapping time segments (see Figure 13.12): **sync_seg**, **prop_seg**, **phase_seg1**, and **phase_seg2**. Each segment consists of a specific, programmable number of time quanta (see Section 13.6.2). The length of time quantum (t_q), which is the basic unit of the bit time, is derived from the CAN controller system clock f_{sys} and the baud rate prescaler (BRP) as follows:

$$t_q = \text{BRP} / f_{sys} \tag{13.2}$$

The purpose of the **sync_seg** is to inform all nodes that a bit time is just started. The existence of the propagation-delay segment **prop_seg** is due to the fact that the CAN protocol allows for nondestructive arbitration between nodes contending for access to the bus and the requirement for *in-frame acknowledgement.* In the case of nondestructive arbitration, more than one node may be transmitting during the arbitration period. Each transmitting node samples data from the bus in order to determine whether it has won or lost the arbitration and also to receive the arbitration field in case it loses arbitration. When each node samples a bit, the value sampled must be the logical superposition of the bit values transmitted by each of the nodes arbitrating for bus access. In the case of the acknowledge field, the transmitting node transmits a recessive bit but expects to receive a dominant bit; in other words, a dominant value must be sampled at the sample time. The length of the **prop_seg** segment must be selected for the earliest possible sample of the bit by a node until the transmitted bit values from all of the transmitting nodes have reached all of the nodes. When multiple nodes are arbitrating for the control of the CAN bus, the node that transmitted the earliest must wait for the node that transmitted the latest in order to find out if it won or lost. That is, the value for t_{PROP_SEG} is given by

$$t_{PROP\_SEG} = t_{PROP(A,B)} + t_{PROP(B,A)} \tag{13.3}$$

where $t_{PROP(A,B)}$ and $t_{PROP(B,A)}$ are the propagation delays from node A to node B and node B to node A, respectively. In the worst case, node A and node B are at the two ends of the CAN bus. The propagation delay from node A to node B is given by

$$t_{PROP(A,B)} = t_{BUS} + t_{Tx} + t_{Rx} \tag{13.4}$$

where, t_{BUS}, t_{Tx}, and t_{Rx} are data traveling time on the bus, transmitter propagation delay, and receiver propagation delay, respectively.

When node A and node B are two ends at opposite ends of the CAN bus, then the value for t_{PROP_SEG} is

$$t_{PROP\_SEG} = 2 \times (t_{BUS} + t_{Tx} + t_{Rx}) \tag{13.5}$$

The minimum number of time quantum (t_q) that must be allocated to the **prop_seg** segment is therefore

$$\text{prop\_seg} = \text{round\_up}(t_{PROP\_SEG} \div t_q) \tag{13.6}$$

where the **round_up()** function returns a value that equals the argument rounded up to the next integer value.

In the absence of bus errors, bit staffing guarantees a maximum of 10-bit periods between resynchronization edges (5 dominant bits followed by 5 recessive bits and then by a dominant bit). This represents the worst-case condition for the accumulation of phase error during normal communication. The accumulated phase error must be compensated for by resynchronization and, therefore, must be less than the programmed resynchronization jump width (t_{RJW}). The accumulated phase error is due to the tolerance in the CAN system clock, and this requirement can be expressed as

$$(2 \times \Delta f) \times 10 \times t_{NBT} < t_{RJW} \tag{13.7}$$

where Δf is the largest crystal-oscillator frequency variation (in percentage) of all CAN nodes in the network. The transmitter may have a frequency error of $+\Delta f$ whereas the receiver may have the frequency error of $-\Delta f$ in the worst case. This explains the factor of 2 in Equation 13.7.

Real systems must operate in the presence of electrical noise, which may induce errors on the CAN bus. A node transmits an error flag after it detects an error. In the case of a local error, only the node that detects the error will transmit the error flag. All other nodes receive the error flag and transmit their own error flags as an echo. If the error is global, all nodes will detect it within the same bit time and will transmit error flags simultaneously. Therefore, node can differentiate between a local error and a global error by detecting whether there is an echo after its error flag. This requires that a node can sample the first bit correctly after transmitting its error flag.

An error flag from an error-active node consists of 6 dominant bits, and there could be up to 6 dominant bits before the error flag (for example, if the error was a stuff error). A node therefore must sample the thirteenth bit correctly after the last resynchronization. This can be expressed as

$$(2 \times \Delta f) \times (13 \times t_{NBT} - t_{PHASE\_SEG2}) < \text{MIN}(t_{PHASE\_SEG1}, t_{PHASE\_SEG2}) \qquad (13.8)$$

where the function **MIN(arg1,arg2)** returns the smaller of the two arguments.

Thus, there are two clock tolerance requirements that must be satisfied. The selection of bit timing values involves consideration of various fundamental system parameters. The requirement of the **prop_seg** value imposes a trade off between the maximum achievable bit rate and the maximum propagation delay due to the bus length and the characteristics of the bus driver circuit. The highest bit rate only can be achieved with a short bus length, a fast bus driver circuit, and a high-frequency CAN clock source with high tolerance.

The time after the sample point that is needed to calculate the next bit to be sent (e.g., data bit, CRC bit, stuff bit, error flag, or idle) is called *information processing time* (IPT). The IPT is application specific but may not be longer than $2t_q$. The C8051F040 CAN module's IPT is zero.

The procedure for determining the optimum bit-timing parameters that satisfy the requirements for proper bit sampling is as follows.

Step 1
Compute the nominal bit time (t_{NBT}) (= 1 / (bit rate)) and determine the minimum permissible time (t_{PROP_SEG}) for the **prop_seg** segment using Equation 13.4. Tentatively, set the prescaler to 1.

Step 2
Calculate the CAN clock period (t_q) which is t_{sys} **× prescaler**.

Step 3
Compute the nominal bit time (**NBT**) (in units of t_q) to be $t_{NBT} \div t_q$.

Step 4
Calculate the value for **prop_seg** using Equation 13.6.

Step 5
Determine the values for **phase_seg1** and **phase_2**. The sum of **phase_seg1** and **phase_seg2** is **NBT – prop_seg – 1**. Divide this value evenly between **phase_seg1** and **phase_seg2**. If the sum of **phase_seg1** and **phase_seg2** is odd, then make **phase_seg2** equal to **phase_seg1** plus 1.

Step 6
Compute the sum of **prop_seg** and **phase_seg1**. If the sum is larger than 16, then increase the CAN prescaler by 1 and return to Step 2. If the sum is less than 2, then the specified bit rate cannot be achieved. Lower the bit rate and return to Step 1.

Step 7
Determine the resynchronization jump width (RJW). The RJW is the smaller one of 4 and **phase_seg1**.

Step 8
Calculate the required oscillator tolerance from Equations 13.7 and 13.8.

Example 13.9 Calculate the CAN bit-timing segments for the constraints:

- Bit rate = 100 Kb/s
- Bus length = 10 m
- Bus propagation delay = 5×10^{-9} s/m
- CAN transceiver plus receiver propagation delay = 150 ns at 75°C

Solution: Physical delay of the CAN bus is 5 ns/m × 10 m = 50 ns.

$t_{PROP\_SEG} = 2 \times (50 + 150) = 400$ ns

Set CAN prescaler to 24.

$t_q = 24/24$ MHz $= 1$ μs

prop_seg = roundup (400 / 1000) = roundup (0.4) ≈ 1

$t_{NBT} = 10$ μs

NBT $= t_{NBT} / t_q = 10$

phase_seg1 + phase_seg2 = 10 − 1 − 1 = 8

Set **phase_seg1** and **phase_seg2** to 4 and 4, respectively.

RJW = Min(4, phase_seg1) = 4

Frequency tolerance using equation 13.7 is 3 / (20 × 10) = 2.0%

Frequency tolerance using equation 13.8 is min(4, 4) / 2(13 × 10 − 4) = 1.56%

Therefore, the frequency tolerance for the oscillator is 1.56%.

The value to be written into the BTR register is 0011 0100 1101 0111 = 0x34D7.

Example 13.10 Calculate the CAN bit-timing segments for the constraints:

- Bit rate = 500 kb/s
- Bus length = 80 m
- Bus propagation delay = 5×10^{-9} s/m
- CAN transceiver plus receiver propagation delay = 110 ns at 25°C

Solution: Physical delay of CAN bus is 5 ns/m × 80 m = 400 ns.

$t_{PROP\_SEG} = 2 \times (400 + 110) = 1020$ ns

Set CAN prescaler to 4 (2 or 3 does not work).

$t_q = 4/24$ MHz $= 166.67$ ns

prop_seg = roundup(1020 / 166.67) = 7

$t_{NBT} = 2\ \mu s$

NBT $= t_{NBT}/t_q = 12$

phase_seg1 + phase_seg2 = NBT − prop_seg − 1 = 12 − 7 − 1 = 4

Set **phase_seg1** and **phase_seg2** to 2 and 2, respectively.

RJW = Min(4, phase_seg1) = 2

Frequency tolerance using equation 13.7 is 2 / (20 × 16) = 0.63%

Frequency tolerance using equation 13.8 is min(2, 2) / 2(13×16 − 3) = 0.73%

Therefore, the frequency tolerance for the oscillator is 0.49%.

The value to be written into the BTR register is 0 010 1011 10 000010 = 0x1843.

13.16 Physical CAN Bus Connection

As shown in Figure 13.13, a CAN node is connected to the CAN bus through a transceiver. The CAN transceiver is connected to the CAN bus, which provides differential receiver and transceiver capabilities. The CAN bus levels for a 3.3-V power supply are shown in Figure 13.39.

Many semiconductor manufacturers produce CAN bus transceivers. Most of them require a 5-V power supply to operate. With the newer microcontrollers using a lower power supply, CAN bus transceivers using a 3.3-V or lower power supply begin to appear. The TI SN65HVD230 is a CAN bus transceiver that operates with a 3.3-V power supply.

13.16.1 The SN65HVD230 CAN Bus Transceiver

The SN65HVD230 is designed to provide differential transmiting capability to the bus and differential capability to a CAN controller at speeds up to 1 Mbps. The SN65HVD230 converts the digital signals generated by a CAN controller

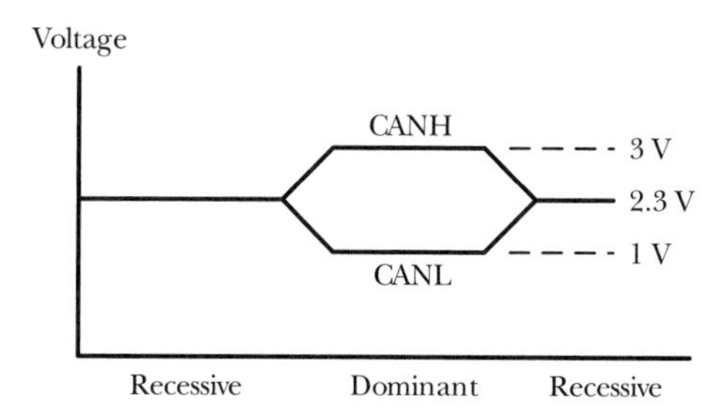

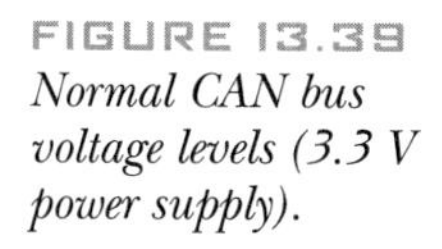
FIGURE 13.39
Normal CAN bus voltage levels (3.3 V power supply).

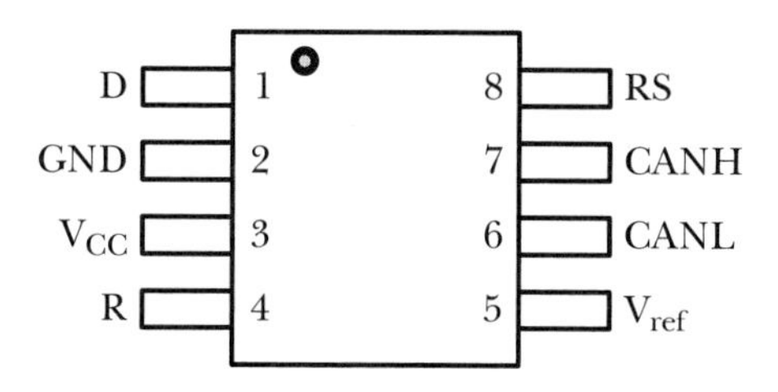

FIGURE 13.40
The SNHVD230 CAN Transceiver.

to signals suitable for transmission over the CAN bus cabling. It operates over a −2V to 7V common-mode range on the bus, and it can withstand common-mode transients of ±25V. The pin assignment of the SN65HVD230 is shown in Figure 13.40.

The D pin is the driver input, which should be connected to the MCU's CANTX pin. The R pin is the receiver output and should be connected to the MCU's CANRX pin. The V_{ref} pin provides a $V_{CC}/2$ output reference. The CANL pin is the low bus output, whereas the CANH pin provides the high bus output.

The RS input provides three different modes of operation: **high-speed**, **slope control**, and **low-power** modes. The high-speed mode of operation is selected by connecting pin 8 to ground, allowing the transmitter output transistors (of CANH and CANL pins) to switch as fast as possible with no limitation on the rise and fall slopes. The rise and fall slopes can be adjusted by connecting a resistor to ground at pin 8, since the slope is proportional to the pin's output current. This slope control is implemented with external-resistor values of 10 kΩ to achieve a 15−V/μs slew rate, and 100 kΩ to achieve a 2−V/μs slew rate. The resistance versus slew rate relationship is shown in Figure 13.41.

If a logic high (>0.75 V_{CC}) is applied to the RS pin, the circuit of the SN65HVD230 enters a low-current, listen-only standby mode, in which the driver is switched off and the receiver remains active. In the listen-only state, the transceiver is completely passive to the bus. The RS pin usually is driven by the MCU in this mode. When detecting the CAN bus activity, the MCU can lower the voltage level on the RS pin and put it in normal high-speed mode.

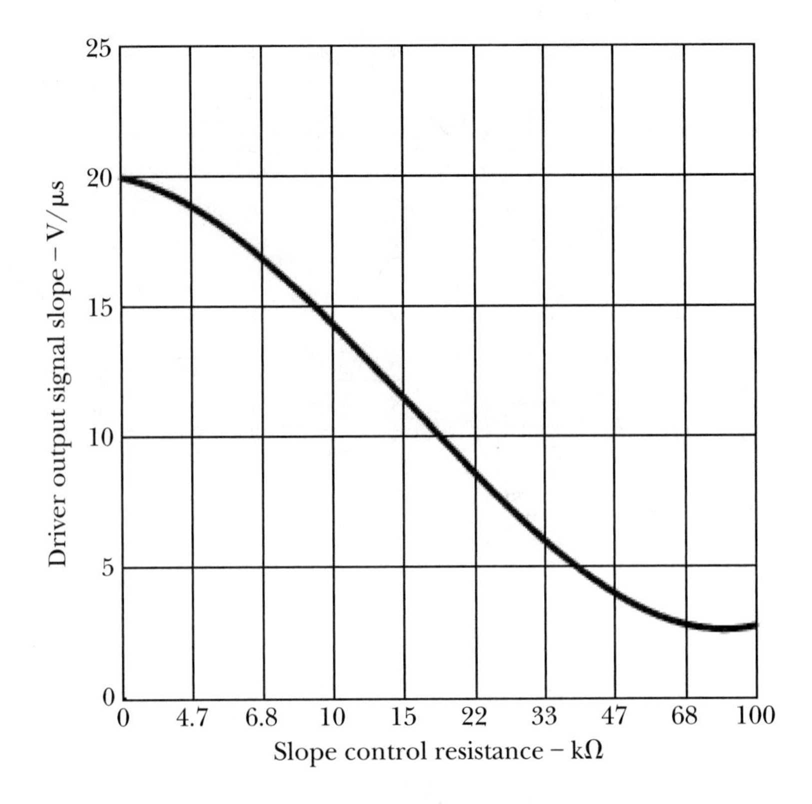

FIGURE 13.41 *SN65HVD230 driver-output signal slope versus slope-control resistance value.*

Transceiver loop delay is a measure of the overall device propagation delay, consisting of the delay from the driver input to the differential outputs, plus the delay from the receiver inputs to its output. When no slope control is used (RS pin connected to ground), the loop delay is 70.7 ns. The loop delay becomes 100 ns when employing slope control with a 10−kΩ resistor and ≈500 ns with a 100−kΩ resistor. The ISO11898 standard requires that the loop delay to be no larger than 150 ns.

Up to 120 CAN nodes can connect to the same bus using the SN65HVD230 transceiver to interface with the CAN bus.

13.16.2 Interfacing the SN65HVD230 to the C8051F040

A typical method of interfacing the SN65HVD230 transceiver to the C8051F040 CAN module is shown in Figure 13.42.

The maximum achievable bus length in a CAN bus network is determined by the following physical factors.

1. The loop delay of the connected bus nodes (CAN controller and transceiver) and the delay of the bus line

2. The differences in bit-time quantum length due to the relative oscillator tolerance between nodes
3. The signal amplitude drop due to the series resistance of the bus cable and the input resistance of bus node

The resultant equation after taking these three factors into account would be very complicated. The bus length that can be achieved as a function of the bit rate in the high-speed mode and with CAN bit timing parameters being optimized for maximum propagation delay is shown in Table 13.4. The types and the cross sections suitable for the CAN bus trunk cable that has more than 32 nodes connected and spans more than 100 meters are listed in Table 13.5.

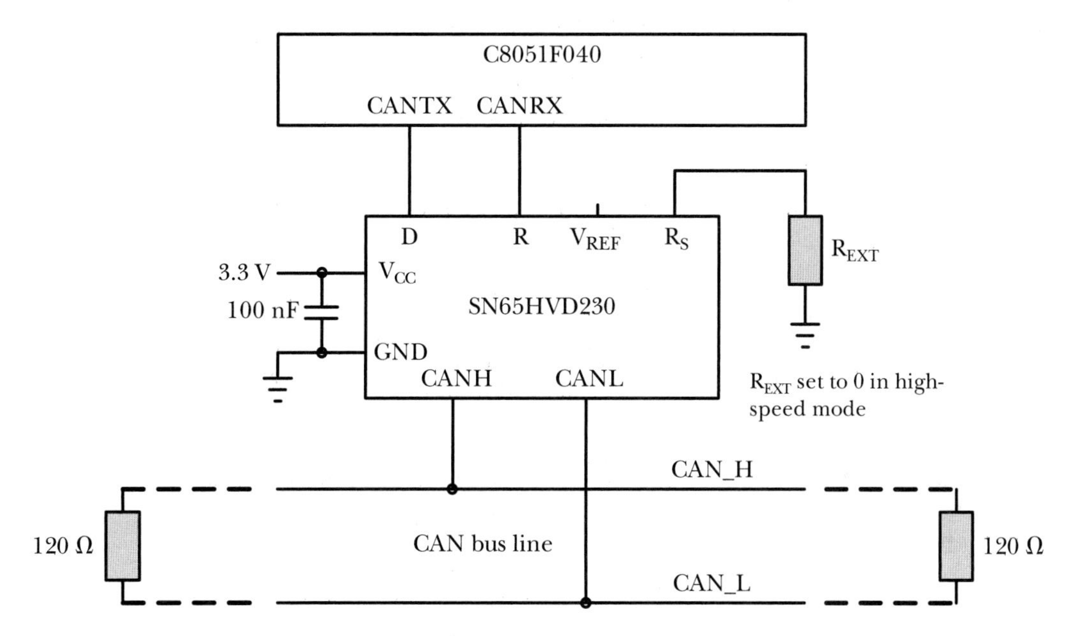

FIGURE 13.42
Interfacing the SN65HVD230 with the C8051F040.

TABLE 13.4 *CAN Bus Bit Rate/Bus Length Relation*

Bit Rate (Kbps)	Bus Length
1000	40
500	100
250	250
125	500
62.5	1000

TABLE 13.5 *Minimum Recommended Bus Wire Cross Section for the Trunk Cable*

Bus Length/ Number of Nodes	32	64	100
100 m	0.25 mm^2 or AWG 24	0.25 mm^2 or AWG 24	0.25 mm^2 or AWG 24
250 m	0.34 mm^2 or AWG 22	0.50 mm^2 or AWG 20	0.50 mm^2 or AWG 20
500 m	0.75 mm^2 or AWG 18	0.75 mm^2 or AWG 18	0.75 mm^2 or AWG 18

Example 13.11 Assume that two C8051F040 demo boards are attached to the same CAN bus. Write a program to configure one demo board to send LED patterns specified in Lab Assignment L5.1 one byte every 200 ms to another demo board. The demo board that receives the data simply outputs the single data byte to P5. Write another program to configure the second demo board to increment a counter every 200 ms and send it to the first demo board using a standard identifier. The first demo board also outputs the single-byte data to P5. The CAN cable is 10-m long, and both demo boards are running with a 24-MHz crystal oscillator.

Solution: These two C8051F040 demo boards can use the values computed in Example 13.9 to set the bit timing register. Message reception should be interrupt-driven due to the unpredictability of arrival of messages. The first demo board sets the identifier to 1 for all the messages to be transmitted and uses 2 as the identifier of all the messages to be received. The second board sets the identifier to 2 for all the messages to be transmitted and uses 1 as the identifier of all the messages to be received.

Program 1

```
#include <c8051F040.h>
#include <delays.h>
#include <canIndex.h>
#define LED P5
char MsgNum;
char status, temp;

sfr16 CAN0DAT = 0xD8;
void clrMsgObj(void);
void initMsgObjTx(char MsgNum);
void initMsgObjRx(char MsgNum);
void startCAN(void);
void sndLEDpattern(char MsgNum, char pattern);
void rcvData(char MsgNum);
void sysInit(void);
```

```
void CANISR(void);
code unsigned char ledPat[]= {0x00,0xFF,0x00,0xFF,0x00,0xFF,0x00,0xFF,
                              0x80,0x40,0x20,0x10,0x08,0x04,0x02,0x01,
                              0x01,0x02,0x04,0x08,0x10,0x20,0x40,0x80,
                              0x80,0x40,0x20,0x10,0x08,0x04,0x02,0x01,
                              0x01,0x02,0x04,0x08,0x10,0x20,0x40,0x80,
                              0x80,0x40,0x20,0x10,0x08,0x04,0x02,0x01,
                              0x01,0x02,0x04,0x08,0x10,0x20,0x40,0x80,
                              0x80,0x40,0x20,0x10,0x08,0x04,0x02,0x01,
                              0x01,0x02,0x04,0x08,0x10,0x20,0x40,0x80,
                              0x81,0x42,0x24,0x18,0x18,0x24,0x42,0x81,
                              0x81,0x00,0x81,0x00,0x81,0x00,0x81,0x00,
                              0x42,0x00,0x42,0x00,0x42,0x00,0x42,0x00,
                              0x24,0x00,0x24,0x00,0x24,0x00,0x24,0x00,
                              0x18,0x00,0x18,0x00,0x18,0x00,0x18,0x00,
                              0x18,0x00,0x18,0x00,0x18,0x00,0x18,0x00,
                              0x24,0x00,0x24,0x00,0x24,0x00,0x24,0x00,
                              0x42,0x00,0x42,0x00,0x42,0x00,0x42,0x00,
                              0x81,0x00,0x81,0x00,0x81,0x00,0x81,0x00};
void main (void)
{
    int   ix;
    sysInit();
    clrMsgObj();
    initMsgObjTx(0x02);
    initMsgObjRx(0x01);
    EIE2 = 0x20;                    // enable CAN interrupts
    startCAN();                     // configure CAN module
    EA = 1;                         // globally enable CAN interrupt
    while(1){
        for(ix = 0; ix < 144; ix++){
            sndLEDpattern(2,ledPat[ix]);
            delayby100ms(2);
        }
    }
}
void sysInit(void)
{
    unsigned char n;
    SFRPAGE  = CONFIG_PAGE;
    WDTCN    = 0xDE;
    WDTCN    = 0xAD;
    OSCXCN   = 0x67;
    for (n = 0; n < 255; n++);
    while(!(OSCXCN & 0x80));        // wait until crystal oscillator becomes stable
    CLKSEL   = 1;                   // switch to external crystal oscillator
    XBR3     = 0x80;                // configure CAN TX pin (CTX) as push-pull digital output
    P5MDOUT = 0xFF;                 // configure P5 to be push pull
    XBR2     = 0x40;                // enable crossbar decoder
}
```

```
//*************************************************************************************************
// The following function clears all 32 message objects.
//*************************************************************************************************
void clrMsgObj(void)
{
    char i;
    SFRPAGE   = CAN0_PAGE;
    CAN0ADR   = IF1CMDMSK;               // point to command mask register 1
    CAN0DATL = 0xFF;                     // set direction to WRITE all IF registers to Msg Obj
    for (i = 1; i < 33; i++){
        CAN0ADR   = IF1CMDRQST;          // write blank IF registers to Msg obj
        CAN0DATL = i;
    }
}
//*************************************************************************************************
// The following function initializes message object for reception
//*************************************************************************************************
void initMsgObjRx(char MsgNum)
{
    SFRPAGE   = CAN0_PAGE;
    CAN0ADR   = IF1CMDMSK                // point to command register 1
    CAN0DAT   = 0x00B8;                  // set to Write, alter all Msg Obj except ID mask and data bits
    CAN0ADR   = IF1ARB1;                 // point to arbitration register
    CAN0DAT   = 0x0000;                  // set arbitration ID to "0"
    CAN0DAT   = 0x8000;                  // Arb2 high byte:set MsgVal bit, no extended ID, Dir=receive
    CAN0DAT   = 0x0480;                  // Msg ctrl: set RXIE, remote frame function disabled
    CAN0ADR   = IF1CMDRQST;              // point to command request reg.
    CAN0DATL = MsgNum;                   // select Msg Obj passed into function parameter list
}
//*************************************************************************************************
// The following function initializes message object for transmission.
//*************************************************************************************************
void initMsgObjTx(char MsgNum)
{
    SFRPAGE  = CAN0_PAGE;
    CAN0ADR = IF1CMDMSK;
    CAN0DAT = 0x00B2;                    // set to WRITE, alter all Msg Obj except ID mask bits
    CAN0ADR = IF1ARB1;                   // point to arbitration 1 register
    CAN0DAT = 0x0000;                    // set arbitration ID to highest priority
    CAN0DAT = 0xA000;                    // Arb2 high byte:set MsgVal bit, no extended ID, Dir = Write
    CAN0DAT = 0x0081;                    // Msg Ctrl: DLC=1, remote frame function not enabled
    CAN0ADR = IF1CMDRQST;                // point to command request reg.
    CAN0DAT = MsgNum;                    // select Lsg Obj passed to this function parameter list
}
//*************************************************************************************************
// This function sets CAN timing, configure transmission and reception.
//*************************************************************************************************
void startCAN(void)
{
    SFRPAGE = CAN0_PAGE;
```

```
    CAN0CN   = CAN0CN | 0x41;          // configuration change enable CCE and INIT
    CAN0ADR  = BITREG;                 // point to bit timing register
    CAN0DAT  = 0x34D7;                 // set CAN timing
    CAN0ADR  = IF1CMDMSK;              // point to command mask register 1
    CAN0DAT  = 0x0087;                 // Config for TX: WRITE to CAN RAM, write data bytes,
                                       // set TXrqst/NewDat (actual transfer will occur later)
                                       //RX-IF2 operation may interrupt TX-IF1 operation
    CAN0ADR  = IF2CMDMSK;              //point to command mask 2
    CAN0DATL = 0x1F;                   // Config for RX: read CAN RAM, read data bytes,
                                       // clear NewDat and IntPnd (actual read will occur later)
    CAN0CN  |= 0x06;                   // global interrupt enable IE and SIE
    CAN0CN  &= ~0x41;                  // clear CCE and INIT bits, enter normal mode
}
//************************************************************************************************************
// This function transmits LED pattern to be used by other node
//************************************************************************************************************
void sndLEDpattern(char MsgNum, char pattern)
{
    SFRPAGE  = CAN0_PAGE;
    CAN0ADR  = IF1CMDMSK;              // point to command mask 1
    CAN0DAT  = 0x0087;                 // Config to Write to CAN RAM, write data bytes, set
                                       // TXrqst/NewDat
    CAN0ADR  = IF1DATA1;               // point to 1st byte of data field
    CAN0DATL = pattern;                // this pattern is to be output to P5
    CAN0ADR  = IF1CMDRQST;             // point to command register reg.
    CAN0DATL = MsgNum;                 // move new data for TX to Msg Obj "MsgNum"
}
//************************************************************************************************************
// This function receives data from IF2 buffer
//************************************************************************************************************
void rcvData(char MsgNum)
{
    SFRPAGE  = CAN0_PAGE;
    CAN0ADR  = IF2CMDRQST;             // point to command request reg.
    CAN0DATL = MsgNum;                 // move new data for RX from Msg Obj "MsgNum"
    CAN0ADR  = IF2DATA1;               // point to 1st byte of Data Field
    temp     = CAN0DATL;               // make a copy of CAN0DATL
    SFRPAGE  = 0x0F;                   // switch LED page
    LED      = temp;                   // use received data to drive LEDs
}
//************************************************************************************************************
//
//************************************************************************************************************
void CANISR(void) interrupt 19
{
    status = CAN0STA;
    if((status & 0x10) != 0){ // RxOk is set, interrupt caused by reception
        CAN0STA = (CAN0STA & 0xEF) | 0x07; //reset RxOk, set LEC to no change
        rcvData(0x01);
    }
```

```
    if((status & 0x08) != 0){ //TxOk is set, interrupt caused by transmission
        CANOSTA = (CANOSTA & 0xF7) | 0x07;      // reset TxOk, set LEC to no change
    }
    if(((status & 0x07)!= 0) && ((status & 0x07) != 7)){//Error interrupt, LEC changed
        // error handling
        CANOSTA = CANOSTA | 0x07; // set LEC to no change
    }
}
```

Program 2

```
#include <c8051F040.h>
#include <delays.h>
#include <canIndex.h>
#define  LED  P5
char MsgNum;
char status, temp, cnt;

sfr16  CANODAT = 0xD8;
void clrMsgObj(void);
void initMsgObjTx(char MsgNum);
void initMsgObjRx(char MsgNum);
void startCAN(void);
void sndLEDpattern(char MsgNum, char pattern);
void rcvData(char MsgNum);
void sysInit(void);
void CANISR(void);

void main (void)
{
    unsigned char ix;
    sysInit();
    clrMsgObj();
    //initialize message object to transmit data    initMsgObjTx(0x02);
    //initialize message object to receive data     initMsgObjRx(0x02);
    EIE2 = 0x20;                                    // enable CAN interrupts
    startCAN();                                     // configure CAN module
    EA  = 1;                                        // globally enable CAN interrupt
    cnt  = 0;
    while(1){
        for(ix = 0; ix < 144; ix++){
            sndLEDpattern(2,cnt++);
            delayby100ms(2);
        }
    }
}
void sysInit(void)
{
    unsigned char n;
```

```
    SFRPAGE  = CONFIG_PAGE;
    WDTCN    = 0xDE;
    WDTCN    = 0xAD;
    OSCXCN   = 0x67;
    for (n = 0; n < 255; n++);
    while(!(OSCXCN & 0x80));        // wait until crystal oscillator becomes stable
    CLKSEL   = 1;                   // switch to external crystal oscillator
    XBR3     = 0x80;                // configure CAN TX pin (CTX) as push-pull digital output
    P5MDOUT = 0xFF;                 // configure P5 to be push pull
    XBR2     = 0x40;                // enable crossbar decoder
}
//*****************************************************************************************
// The following function clears all 32 message objects.
//*****************************************************************************************
void clrMsgObj(void)
{
    char i;
    SFRPAGE  = CAN0_PAGE;
    CAN0ADR = IF1CMDMSK;            // point to command mask register 1
    CAN0DATL = 0xFF;                // set direction to WRITE all IF registers to Msg Obj
    for (i = 1; i < 33; i++){
        CAN0ADR = IF1CMDRQST;       // write blank IF registers (reset values) to Msg obj
        CAN0DATL= i;
    }
}
//*****************************************************************************************
// The following function initializes message object for receive
//*****************************************************************************************
void initMsgObjRx(char MsgNum)
{
    SFRPAGE  = CAN0_PAGE;
    CAN0ADR  = IF1CMDMSK;           // point to command register 1
    CAN0DAT  = 0x00B8;              // set to Write, alter all Msg Obj except ID mask and data bits
    CAN0ADR  = IF1ARB1;             // point to arbitration register
    CAN0DAT  = 0x0000;              // set arbitration ID to "0"
    CAN0DAT  = 0x8000;              // Arb2 high byte: MsgVal = 0, Xtd = 0, Dir = 0 (read)
    CAN0DAT  = 0x0480;              // Msg ctrl: set RXIE, remote frame function disabled
    CAN0ADR  = IF1CMDRQST;          // point to command request reg.
    CAN0DATL = MsgNum;              // select Msg Obj passed into function parameter list
}
//*****************************************************************************************
// The following function initializes message object for transmission.
//*****************************************************************************************
void initMsgObjTx(char MsgNum)
{
    SFRPAGE  = CAN0_PAGE;
    CAN0ADR = IF1CMDMSK;
    CAN0DAT = 0x00B2;               // set to WRITE, alter all Msg Obj except ID mask bits
    CAN0ADR = IF1ARB1;              // point to arbitration register
    CAN0DAT = 0x0000;               // set arbitration ID to highest priority
```

```
    CANODAT = 0xA000;              // Arb2 high byte: MsgVal = 1, Xtd = 0, Dir = 1 (Write)
    CANODAT = 0x0081;              // Msg Ctrl: DLC=1, remote frame function not enabled
    CANOADR = IF1CMDRQST;          // point to command request reg.
    CANODAT = MsgNum;              // select Lsg Obj passed to this function parameter list
}
//*****************************************************************************************
// include void startCAN(void) from Program 1
//*****************************************************************************************
//*****************************************************************************************
// include void sndLEDpattern(char MsgNum, unsigned char pattern) from Program 1
//*****************************************************************************************
//*****************************************************************************************
// include void rcvData(char MsgNum) from Program 1
//*****************************************************************************************
//*****************************************************************************************
// include void CANISR(void) interrupt 19 from Program 1
//*****************************************************************************************
```

13.17 Chapter Summary

The CAN bus specification initially was proposed as a data-communication protocol for automotive applications. However, it also can fulfill the data-communication needs of a wide range of applications, from high-speed networks to low-cost multiplexed wiring.

The CAN protocol has gone through several revisions. The latest revision is 2.0A/B. The CAN 2.0A uses standard identifiers, whereas the CAN 2.0B specification accepts extended identifiers. The C8051F040 CAN module supports the CAN 2.0A and 2.0B specifications.

The CAN protocol supports four types of messages: data frame, remote frame, error frame, and overload frame. Users need to transfer only data frames and remote frames. The other two types of frames only are used by the CAN controller to control data transmission and reception. Data frames are used to carry normal data; remote frames are used to request other nodes to send messages with the same identifier as in the remote frame.

The CAN protocol allows all nodes on the bus to transmit simultaneously. When there are multiple transmitters on the bus, they arbitrate, and the CAN node transmitting a message with the highest priority wins. The simultaneous transmission of multiple nodes will not cause any damage to the CAN bus. CAN data frames are acknowledged in-frame; that is, a receiving node sets a bit only in the acknowledge field of the incoming frame. There is no need to send a separate acknowledge frame.

The CAN bus has two states: *dominant* and *recessive*. The dominant state represents logic 0, and the recessive state represents logic 1 for most CAN implementations. When one node drives the dominant voltage to the bus while other nodes drive the recessive level to the bus, the resultant CAN bus state will be the dominant state.

Synchronization is a critical issue in the CAN bus. Each bit time is divided into four segments: **sync_seg**, **prop_seg**, **phase_seg1**, and **phase_seg2**. The **sync_seg** segment signals the start of a bit time. The sample point is between **phase_seg1** and **phase_seg2**. At the beginning of each frame, every node performs a hard synchronization to align its **sync_seg** segment of its current bit time to the recessive-to-dominant edge of the transmitted start-of-frame. Resynchronization is performed during the remainder of the frame whenever a change of bit value from recessive to dominant occurs outside of the expected **sync_seg** segment.

A CAN module (also called the CAN controller) requires a CAN bus transceiver, such as the Philips PCA82C250 or the Microchip MCP2551, to interface with the CAN bus. Most CAN controllers allow more than 100 nodes to connect to the CAN bus. The CAN trunk cable could be a shielded cable, unshielded twisted pair, or simply a pair of insulated wires. It is recommended that shielded cable be used for high-speed transfer when there is a radio frequency interference problem. Up to a 1-Mbps data rate is achievable over a distance of 40 meters. The CAN module of the C8051F040 needs to use a CAN transceiver operating at 3.3-V power supply (for example, TI's SN65HVD230) to form a CAN node.

The CAN module of the C8051F040 MCU has six major modes of operations.

- Basic mode
- Test mode
- Normal operation mode
- Silent mode
- Loopback mode
- Loopback combined with silent mode

Most of the CAN configuration operations can be performed without stopping the CAN module. The only exception is changing bit timing. To change bit timing, the user has to set the CCE bit and the **Init** bit of the CAN0CN register. Other configuration can be performed on-the-fly. A CAN application usually has several types of message objects to be transmitted and received. For a certain type of message objects, the user can initialize the mask and arbitration information beforehand and write it into the message RAM. When transmitting a message object of that type, the user simply fills in the data bytes and writes them into the message RAM via the interface registers. Message reception can be performed in the similar manner.

The C8051F040 use message RAM to buffer message objects to be transmitted or received. Each transmit and receive object has a 16-bit command request register, a 16-bit command mask register, four 16-bit data registers to hold data, two 16-bit arbitration registers for identification and arbitration purposes, two 16-bit mask registers, and a 16-bit message-control register. Data transmission involves copying data into identifier registers, data registers, and the message-control register, and setting the **TxRqst** bit to mark the buffer ready for transmission. Data reception in the CAN bus is often interrupt-driven due to the unpredictability of message arrival times.

13.18 Exercise Problems

E13.1 Calculate the bit segments for the following system constraints assuming that the TI SN65HVD230 transceiver is used.
Bit rate = 400 Kbps
Bus length = 100 m
Bus propagation delay = 5×10^{-9} s/m
TI SN65HVD230 transceiver loop delay = 75 ns at 25°C
Oscillator frequency = system clock frequency = 24 MHz

E13.2 Calculate the bit segments for the following system constraints assuming that the TI SN65HVD230 transceiver is used.
Bit rate = 250 kbps
Bus length = 120 m
Bus propagation delay = 5×10^{-9} s/m
TI SN65HVD230 transceiver loop delay = 150 ns at 85°C
Oscillator frequency = system clock frequency = 24 MHz

E13.3 Calculate the bit segments for the following system constraints assuming that the TI SN65HVD230 transceiver is used.
Bit rate = 1 Mbps
Bus length = 20 m
Bus propagation delay = 5×10^{-9} s/m
SN65HVD230 transceiver loop delay = 75 ns at 25°C
Oscillator frequency = system clock frequency = 24 MHz

E13.4 Calculate the bit segments for the following system constraints assuming that the TI SN65HVD230 transceiver is used.
Bit rate = 200 kbps
Bus length = 200 m
Bus propagation delay = 5×10^{-9} s/m
SN65HVD230 transceiver and receiver
propagation delay = 75 ns at 25°C
Oscillator frequency = system clock frequency = 24-MHz

E13.5 Write a subroutine to configure the CAN module of the C8051F040 with the bit timing parameters computed in Problem E13.2. Enable the receive interrupt but disable the transmit interrupt. Configure the CAN so that it accepts only messages with the extended or standard identifiers starting with '**H**', '**P**', or '**T**'.

E13.6 Write a C function to configure CAN bit timing using the same setting as was computed in Problem E13.4.

E13.7 Provide two different ways of accepting messages that have the letter '*K*' as the first letter of the identifier.

E13.8 Write an assembly program to send out the string **Monday** using '**WD**' as the first two characters of the extended identifier. The remaining bits of the identifier are cleared to 0.

E13.9 Write a C program to send out the string **too high** using '**ans**' as the first three letters of the extended identifier. The remaining bits of the extended identifier are cleared to 0.

E13.10 Write a C program to send out the seven-segment display patterns and digit-select values to the CAN bus so that the receiving CAN node can use it to drive seven-segment displays using the time-multiplexing technique. The sending node sends out the values for driving eight seven-segment displays in two transfers. Also, write a program to be run on another C8051F040 that receives information from the CAN bus and uses the received values to drive the eight seven-segment displays. The sending node sends out new values to be displayed by another node every 5 seconds.

E13.11 Write a C program to be run by a C8051F040 demo board which will measures the voltage output of a potentiometer attached to an AIN0.0 pin once every 100 ms and saves it in a data frame to reply to a remote frame. Write another program to send out a remote frame every 100 ms. Both programs use the standard identifier that starts with the ASCII code of letter '**V**'. The node that sends out the remote frame displays the returned data on the LCD screen.

13.19 Laboratory Exercise Problems and Assignments

L13.1 Practice data transfer over the CAN bus using the following procedures.

Step 1
Use a pair of insulated wires about 10-m long. Connect the CAN_H pins of two demo boards (e.g., the SSE040 or C8051F040-DK demo boards) with one wire and connect the CAN_L pins of both demo boards with another wire. Connect both ends of these two wires together with 120-Ω resistors (both the SSE040 and C8051F040-DK have terminating resistors on board). The circuit connection is shown in Figure L13.1.

Step 2
Write a program to be downloaded onto one of the demo boards (called board A) that performs A/D conversion 10 times every second. Use the on-board potentiometer to generate a voltage to be converted. Send out the A/D conversion result over the CAN bus every

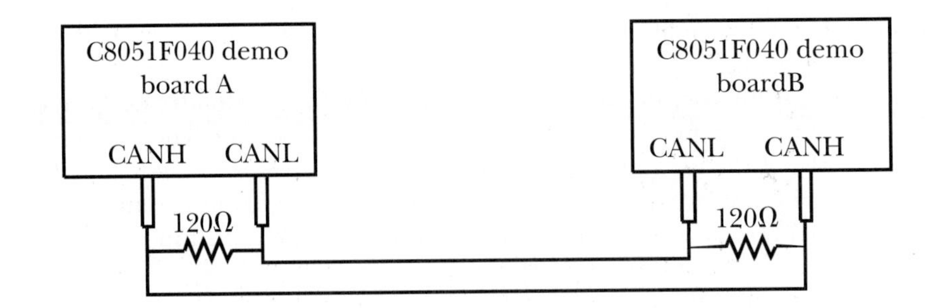

FIGURE L13.1
CAN circuit connection for L13.1.

100 ms. Use the letter '**W**' as the identifier of the data frame. (Use the timing parameters computed in Problem E13.9 for the SSE040 demo board.)

Step 3
Write a program to be downloaded onto another demo board (called board B). This program will send out the number of data frames received so far over the CAN bus. This program will use the letter '**R**' as the identifier. After the number reaches 99, the program will reset the number to 0 and start over again. Each digit of the decimal number is encoded in ASCII code.

Step 4
Board A will display the number received over the CAN bus on eight LEDS in BCD format.

Step 5
Board B will display the received A/D result in an LCD display.

L13.2 This assignment requires three students to wire three SSE040 demo boards together using the CAN bus. Assign each of these three demo boards with a number 0, 1, and 2, respectively. Write a program to be run on each demo board and perform the following procedure.

Step 1
Use two potentiometers to emulate the temperature-sensor and barometric-sensor outputs (0 to 3.3 V). Connect the demo boards as shown in Figure L13.2. The temperature-sensor output represents the temperature range from −40 to 125°C, whereas the barometric-sensor output represents the pressure range from 948 to 1083.8 mbar.

Step 2
Each demo board performs temperature and barometric-pressure measurements once per second and stores the result in separate buffers formatted to have the same structure as a transmit buffer. When storing the measurement data, use Ti (for node i) as the identifier for the temperature value and use Pi (for node i) as the identifier for the pressure value.

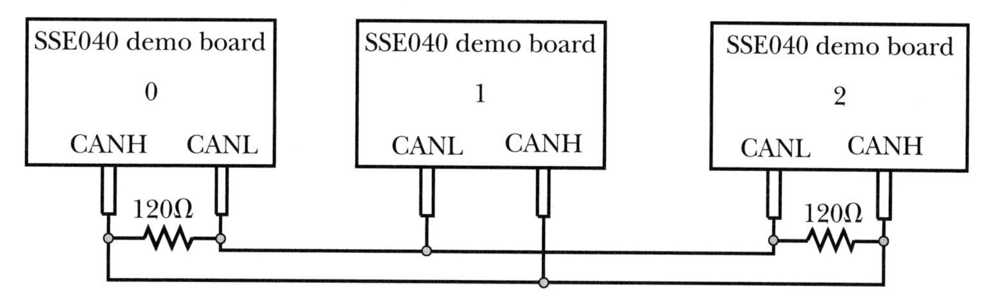

FIGURE L13.2
CAN circuit connection for L13.2.

Step 3
After performing temperature and pressure measurements and storing data in buffers, each demo board uses a random number generator to generate a number ***k*** in the range of 0 to 2 and waits for that amount of time (seconds) to send out a remote transmission request to request node ***k*** to send in temperature or pressure data and display them on the LCD. The received data should be displayed as follows.

received Tk = xxx.y

or

received Pk = xxxx.y

The demo board that transmits the data should display the message:

send Tk = xxx.y

or

send Pk = xxxx.y

CHAPTER 14

Memory Subsystem

14.1 Objectives

Upon successful completion of this chapter, you will be able to:

- Understand the overall C8051F040 memory organization
- Erase and program the on-chip flash memory
- Control the operation, programming, and protection of the on-chip flash memory
- Understand the external memory expansion issue
- Make memory space assignment
- Perform timing analysis for the memory system

14.2 Introduction

Memory technologies and the 8051 memory organization have been introduced in Chapter 1. DRAM has the lowest per-bit cost among all the memory technologies. However, the circuit design of DRAM requires a periodic refresh operation to maintain its contents. Because of that, interfacing DRAM to microprocessors is more complicated than any other memory technologies. Since 8-bit microprocessors and microcontrollers have small memory spaces (mostly 64 kB), their memory needs easily can be satisfied by one memory chip. There is no reason to use DRAM in 8-bit microprocessor or microcontroller applications.

When the 8051 was first introduced, it has only 4 kB of on-chip program memory. It often was necessary to add external program memory to satisfy the memory needs of applications. Today, you can always find an 8051 variant with

64-kB program memory from any vendor. This eliminates the need to add external program memory.

It is not so common for an 8051 variant to have large on-chip data memory. When an application needs large amount of data memory, the user has the following options:

- Add external parallel data memory
- Add external serial memory

The parallel data memory is limited to 64 kB in size. However, there is no such size limit if serial memory is used. Unless the application requires high-speed access to data, most of the data memory needs can be satisfied by using serial memory. Serial memory chips with SPI or I^2C interface have been discussed in Chapters 10 and 11.

Serial memories have one great advantage over the parallel memories: they use fewer pins than parallel memory (only two to four pins) to interface with the microcontroller. This allows more I/O pins available for applications. I/O pins are a premium for 8-bit microprocessors and microcontrollers.

The C8051F040 has 64-kB on-chip flash memory for program code and 4-kB SRAM for data memory (referred to as XRAM). Since the flash memory is already 64 kB, there is no room for program memory expansion. However, it is possible to add up to 60-kB data memory to the C8051F040. Adding external memory to a microcontroller involves the use of buses. We will introduce the basic concepts related to buses before discussing the external memory expansion issue.

14.3 Basic Concepts of Buses

A bus is a group of conducting wires interconnecting the processor, memory, and I/O devices. There are three types of buses: address, data, and control. A microprocessor bus can be *active* or *passive.* As illustrated in Figure 14.1, an active bus has a pull-up device to pull its voltage to high, whereas a passive does not. The pull-up device can be a resistor or a transistor. A bus is often in the form of conductors in a printed circuit board. Each bus conductor and the ground plane of the printed circuit board form a capacitor.

For an active bus, the microcontroller will need to configure its address or data pins to be open-drain in order to drive the bus. As shown in Figure 14.1, the voltage of an active bus will be low when one or more devices attached to the bus turn on the transistors connected to the bus. When no device turns on the driver to the bus line, the bus voltage will be pulled up to high (V_{CC}). When a bus line is not driven, its equivalent circuit is like an RC circuit connected to a power supply V_{CC}, as shown in Figure 14.2. The time constant for the bus to be pulled up to 63 percent of V_{CC} is $R_{PU}C_{BUS}$, which determines the propagation delay of the pull-up bus.

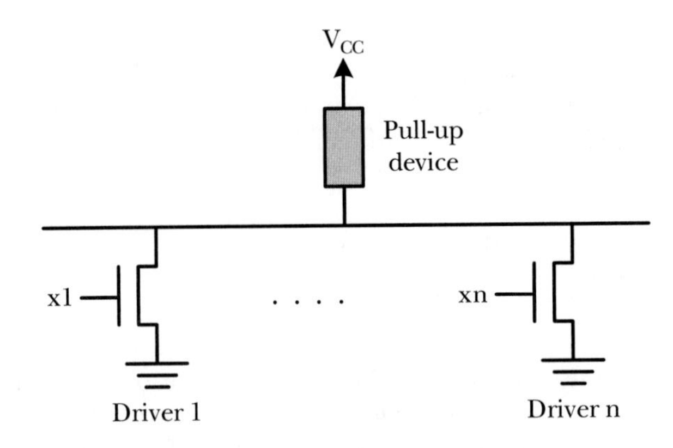

FIGURE 14.1
Circuit of an active bus.

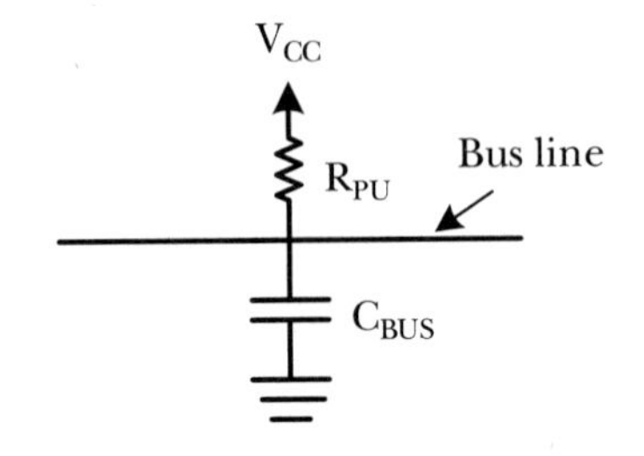

FIGURE 14.2
Equivalent circuit for an undriven bus.

An open-drain MOS transistor can be used to drive an active bus. When the open-drain MOS transistor is turned on, its drain terminal will pull the bus line to the ground level. Otherwise, the bus line will be pulled up to V_{CC} by the pull-up device. When the output transistor of a device with open-drain is turned off, the device essentially is disconnected from the bus. This property allows many devices with open-drain output transistors to be connected to an active bus. Since the device intending to drive the bus line to high turns off its output transistor, the bus contention issue does not exist in an active bus. *Bus contention* is the situation in which one device intends to drive the bus line high, whereas there is another device which intends to drive the bus line low.

A passive bus requires the devices attached to the bus to drive the bus to high or low. To write data to the bus, a *bus driver* is needed. To read data from the bus, a *bus receiver* is needed. The bus driver and bus receiver usually are combined to form a *bus transceiver.* The system designer should design the passive bus circuit so that it is driven by no more than one device at any time. Otherwise, bus contention may occur.

14.4 Representing Bus Signals

A digital signal has two states, high and low, as shown in Figure 14.3a. Ideally, the transition from high to low and from low to high is instantaneous. In reality, however, signal transition takes time. The time needed for a signal to rise from 10 to 90 percent of its peak value (V_{DD}) is referred to as the *rise time* (t_r), whereas the time needed for a signal to fall from 90 to 10 percent of its peak value (V_{DD}) is called the *fall time* (t_f). An exaggerated representation of a real signal is shown in Figure 14.3b. Not all companies follow this definition.

A single signal is drawn as a set of connected line segments, whereas multiple signals of the same nature (for example, address and data) are grouped together and represented as parallel lines with crossovers, as illustrated in Figures 14.4 and 14.5, respectively. A crossover in Figure 14.5 represents the interval during which one or multiple signals change values.

There are times that a signal is changing its value, and hence, its value cannot be determined. The convention for representing changing signals is by adding hatched areas to the signal waveforms. There are other times that one or multiple signals are not driven by any devices and, hence, cannot be received. A signal that is not driven by any device is *floating*. The convention for representing a floating signal is to draw it at a level between logic-high and logic-low. Figure 14.6 illustrates a single signal waveform with unknown and floating intervals, whereas Figure 14.7 shows a multiple-signal waveform with unknown and floating intervals.

In a microprocessor or microcontroller system, all address signals are carried by the *address bus*, data signals are transferred on the *data bus*, and control signals travel on the *control bus*.

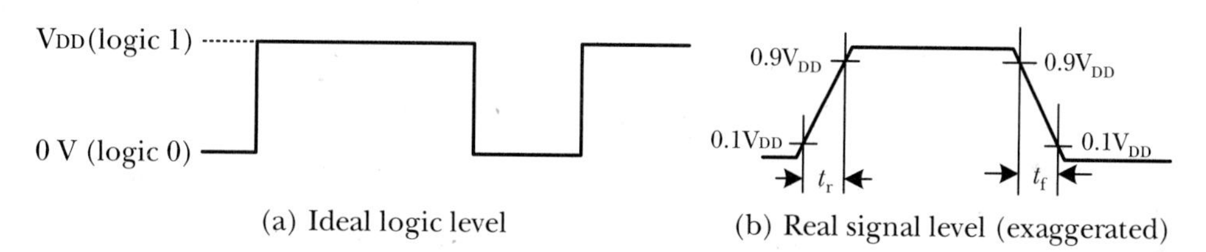

FIGURE 14.3
Digital signal representation.

FIGURE 14.4
Single-signal representation.

FIGURE 14.5
Multiple-signal representation.

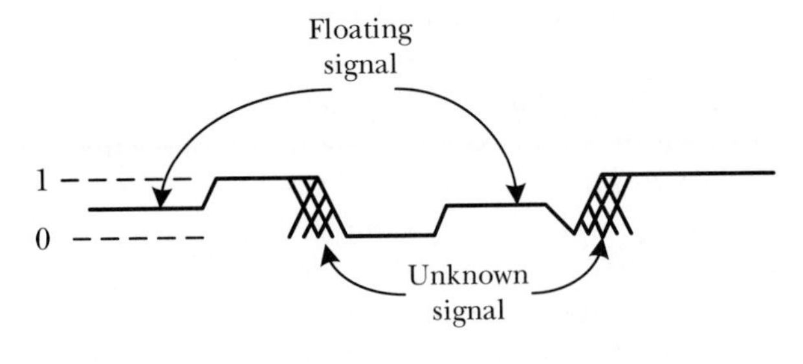

FIGURE 14.6
Single signal with unknown and floating intervals.

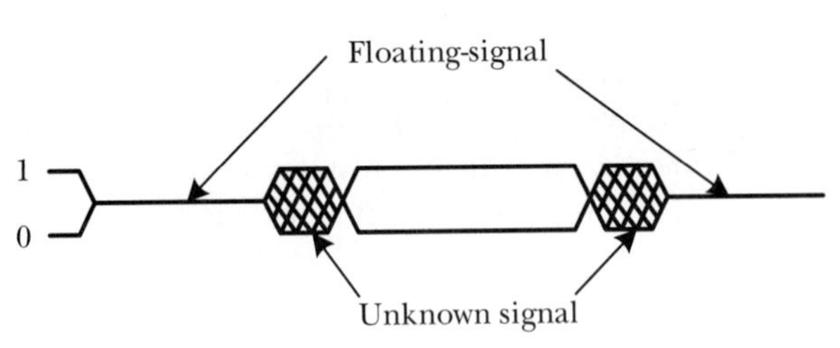

FIGURE 14.7
Multiple signal waveform with unknown and floating intervals.

14.5 Bus Transactions

The activities that a microcontroller or a microprocessor uses to read data and write data through the bus are referred to as *bus transactions.* A bus transaction also is called a *bus cycle.* A bus cycle requires an initiator and a responder. The bus cycle initiator is referred to as *bus master,* whereas the responder is called a *bus slave.* A bus master is usually a processor, whereas a bus slave is usually a memory module or an I/O device. Synchronization is required for a bus cycle to be successful. A bus cycle that uses a clock signal to synchronize the bus master and the bus slave is called a *synchronous bus cycle.* A bus cycle that does not use a clock signal to synchronize the bus master and bus slave is referred to as an *asynchronous bus cycle.*

A *read bus cycle* allows the processor to fetch data from the memory module, whereas the processor writes data into the memory module in a *write bus cycle.* An example of a synchronous write bus cycle is illustrated in Figure 14.8, whereas an asynchronous read bus cycle is shown in Figure 14.9.

In Figure 14.8, a bus cycle takes one cycle to complete if the involved memory module is fast enough. However, the bus cycle also provides a $\overline{\text{RDY}}$ signal to extend the bus cycle to multiple clock cycles. This feature allows slower memory components to be used in the memory system. The memory system may extend the bus cycle by not pulling the $\overline{\text{RDY}}$ signal to low at the end of the clock cycle. An asynchronous bus cycle needs a handshake signal to complete the bus transaction. In Figure 14.9, the bus master uses the REQ signal to request a new bus cycle, whereas the slave uses the ACK signal to indicate its readiness for a bus transaction. Both synchronous and asynchronous bus cycles use the R/$\overline{\text{W}}$ signal to specify the data-transfer direction. When this signal is low, a write cycle is requested. Otherwise, a read cycle is started.

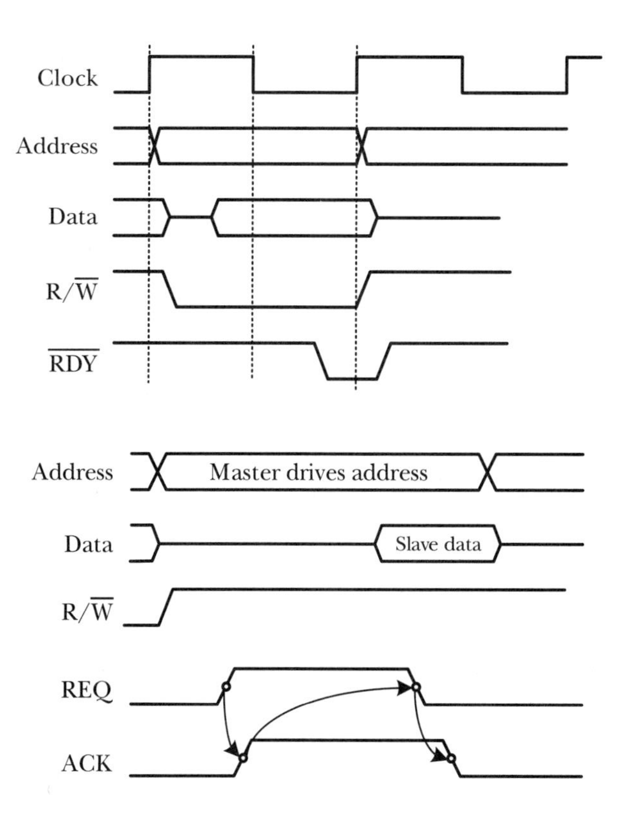

FIGURE 14.8 *An example of a synchronous bus write cycle.*

FIGURE 14.9 *An example of an asynchronous read bus cycle.*

14.6 Bus Multiplexing

Designers of microcontrollers prefer to minimize the number of signal pins because they will make the chip less expensive to manufacture and use. By multiplexing the address and data buses, some pins can be used for other purposes than carrying address information. The drawback of bus multiplexing is that the achievable bus transaction performance is compromised. Most 8-bit and many 16-bit microcontrollers multiplex their address and data buses.

For any bus transaction, the address signal input to the memory chips must be stable throughout the whole bus transaction. The memory system will need to use a circuit to make a copy of the address signals so that they stay valid throughout the whole bus cycle. In a microcontroller that multiplexes the address and data signals, the address signals are placed on the multiplexed bus first, along with certain control signals to indicate that the address signals are valid. After the address signals are on the bus long enough so that the external logic has time to latch them, the microcontroller stops driving the address signals and either waits for the memory devices to place data on the multiplexed bus (in a read bus cycle) or places data on the multiplexed bus (in a write cycle).

14.7 The C8051F040 Flash Program Memory

The C8051F040 includes 64 kB + 128 bytes of on-chip, reprogrammable flash memory for program code and non-volatile data storage. The flash memory can be programmed in a system, one byte at a time, through the JTAG interface or by software using the **MOVX** write instructions. Once cleared to logic 0, a flash bit must be erased to set it back to logic 1. The bytes typically would be erased (set to 0xFF) before being reprogrammed. Flash write and erase operations are timed automatically by hardware for proper execution; data polling to determine the end of the write/erase operation is not required. The CPU is stalled during write/erase operations while the device peripherals remain active. Interrupts that occur during flash write/erase operations are held and are then serviced in their priority order once the flash operation has completed.

14.7.1 Programming the Flash Memory

The programming and erasure of flash memory are controlled by three registers: the *flash access limit register* (FLACL), the *flash memory control register* (FLSCL), and the *program store read/write control register* (PSCTL). The contents of these three registers are shown in Figures 14.10, 14.11, and 14.12, respectively.

The simplest means of programming the flash memory is through the JTAG interface using programming tools provided by SiLabs or a third-party vendor. This is the only means for programming a non-initialized device.

Flash memory also can be programmed by software using the **MOVX** instruction with the address and data to be programmed provided as normal operands. Before writing to flash memory using **MOVX**, flash write operations must be enabled by setting the *program store write enable* bit (the **PSWE** bit of the PSCTL register) to logic 1. This directs **MOVX** to write to flash memory instead of to XRAM, which is the default target. The **PSWE** bit remains set until cleared by software. To avoid errant flash writes, it is recommended that interrupts be disabled while the PSWE bit is in logic 1.

FIGURE 14.10
Flash access limit register (FLACL).

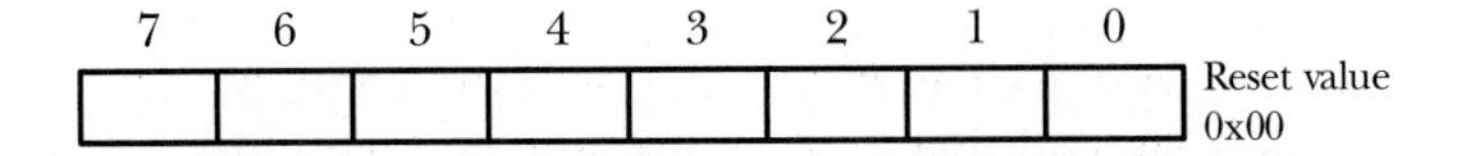

FLACL: Flash access limit
This register holds the high byte of the 16-bit program memory read/write/erase limit address. The entire 16-bit access limit address value is calculated as 0xNN00 where NN is replaced by contents of FLACL

7	6	5	4	3	2	1	0	
FOSE	FRAE	res	res	res	res	res	FLWE	Reset value 0x80

FOSE: Flash one-shot timer enable
0 = flash one-shot timer disabled
1 = flash one-shot timer enabled (recommended setting)
FRAE: Flash read always enable
0 = flash reads occur as necessary (recommended setting)
1 = flash reads occur every system clock cycle
FLWE: Flash write/erase enable
0 = flash writes/erases disabled
1 = flash writes/erases enabled

FIGURE 14.11 *Flash memory control register (FLSCL).*

7	6	5	4	3	2	1	0	
--	--	--	--	--	SFLE	PSEE	PSWE	Reset value 0x00

SFLE: Scratchpad flash memory access enable
0 = flash access from user software directed to the program/data flash sector
1 = flash access from user software directed to the 128 bytes scratchpad sector
PSEE: Program store erase enable
0 = flash program memory erasure disabled
1 = flash program memory erasure enabled
PSWE: Program store write enable
0 = write to flash program memory disabled. MOVX write operations target external RAM
1 = write to the flash program memory enabled. MOVX write operations target flash memory

FIGURE 14.12 *Program store read/write contol register (PSCTL).*

Flash memory is read using the **MOVC** instruction. **MOVX** reads are always directed to XRAM, regardless of the state of the PSWE bit. To ensure the integrity of flash memory contents, it is strongly recommended that the on-chip V_{DD} monitor be enabled by connecting the V_{DD} monitor enable pin (MONEN) to V_{DD} in any system that executes code that writes and/or erases flash memory from software.

Due to the nature of flash memory, a write operation can only clear (but not set) bits in flash memory; only erase operation can set bits in flash memory. A location to be programmed must be erased before a new value can be written. The C8051F040 flash memory is organized into 512-byte pages. The erase operation applies to an entire page. The procedure for programming (by user software) flash memory is as follows.

Step 1
Disable interrupts.

Step 2
Set the **FLWE** bit (bit 0 of the FLSCL register) to enable flash writes/erases via user software.

Step 3
Set the **PSEE** bit (bit 1 of the PSCTL register) to enable flash erase.

Step 4
Set the **PSWE** bit (bit 0 of the PSCTL register) to redirect **MOVX** writes to flash memory.

Step 5
Use the **MOVX** instruction to write a data byte to any location within the 512-byte page to be erased.

Step 6
Clear the **PSEE** bit to disable flash erases.

Step 7
Use the **MOVX** instruction to write a data byte to the desired location within the erased 512-byte page. Repeat this step until all of the desired bytes are written.

Step 8
Clear the **PSWE** bit to redirect the **MOVX** instruction to the XRAM data space.

Step 9
Re-enable interrupts.

Write/Erase timing is controlled automatically by hardware. Instruction execution is stalled in the MCU while the flash memory is being programmed or erased. The 512 bytes above 0xFE00 for the C8051F040 are reserved. Flash writes and erases targeting the reserved area should be avoided.

Example 14.1 Write a program to erase the page starting at 0x4000, to write the value i into memory location 0x4000+i for i from 0 to 255, and then, to write i to memory location 0x4100+i for i from 0 to 255.

Solution: This program can be written by following the procedure described previously.

```
            $nomod51
            $include        (c8051F040.inc)
cnt         equ             30
            cseg at         0x00
            ajmp            start
            org             0xAB
start:      acall           sysinit
            mov             cnt,#0
            mov             SFRPAGE,#0
```

```
            mov       FLSCL,#0x81          ; enable flash memory erase
            orl       PSCTL,#0x03          ; enable erase and write to flash program memory
            mov       DPTR,#0x4000         ; set DPTR to program memory at 0x4000
            mov       A,#0
            movx      @DPTR,A              ; erase the program memory page starting at 0x4000
            anl       PSCTL,#0xFD          ; disable flash erase
            mov       DPTR,#0x4000         ; write value to code memory starting from 0x4000
            mov       B,#2                 ; up to 0x41FF
loope:      mov       cnt,#0               ; cnt starts with 256
            mov       A,#0                 ; write i to memory location at 0x4000+i for i from 0
loopi:      movx      @DPTR,A              ; to 255 and write i to memory location 0x4100+i for i
            inc       DPTR                 ; start from 0 to 255
            inc       A                    ;     "
            djnz      cnt,loopi            ;     "
            djnz      B,loope              ;     "
            anl       PSCTL,#0xFE          ; disable write to flash program memory
            mov       SFRPAGE,#0x0F        ; switch to SFR page F
            mov       DPTR,#0x4000         ; read out the page that was just written into
            mov       R1,#0                ; and display the contents of every location
            mov       A,R1                 ; in the page for 200 ms on LEDs driven by P5
loopi2:     movc      A,@A+DPTR            ;     "
            mov       P5,A                 ;     "
            mov       R0,#2                ;     "
            acall     delayby100ms         ;     "
            inc       R1                   ;     "
            mov       A,R1                 ;     "
            cjne      R1,#0,loopi2         ;     "
            mov       DPTR,#0x4100         ;     "
            mov       R1,#0                ;     "
            mov       A,R1                 ;     "
loopi3:     movc      A,@A+DPTR            ;     "
            mov       P5,A                 ;     "
            mov       R0,#2                ;     "
            acall     delayby100ms         ;     "
            inc       R1                   ;     "
            mov       A,R1                 ;     "
            cjne      R1,#0,loopi3         ;     "
            ajmp      $                    ; force program to stay here

sysinit:    mov       SFRPAGE,#CONFIG_PAGE
            mov       WDTCN,#0xDE          ; disable watchdog timer
            mov       WDTCN,#0xAD          ;     "
            mov       CLKSEL,#0            ; select internal oscillator to generate
            mov       OSCICN,#0x83         ; SYSCLK
            mov       P5MDOUT,#0xFF        ; configure P5 to push pull
            mov       P5,#0                ; turn off LEDs
            ret
            $include  (delays.a51)
            end
```

14.7.2 Non-Volatile Data Storage

The flash memory can be used for non-volatile data storage as well as program code. This allows data (such as calibration coefficients) to be calculated and stored at run time. Data is written using the **MOVX** write instruction and read using the **MOVC** instruction.

An additional 128-byte sector of flash memory is included for non-volatile data storage. This 128-byte sector is mapped to the address range from 0x00 to 0x7F. To access this sector, the user needs to set the **SFLE** bit of the PSCTL register. Code execution from this sector is not permitted.

Example 14.2 Write a program to erase the whole scratchpad sector and write the value i into memory location i for $i = 0$ to 127.

Solution: This program is a minor modification to the program in Example 14.1.

```
            $nomod51
            $include    (c8051F040.inc)
cnt         equ         30
            cseg at     0x00
            ajmp        start
            org         0xAB
start:      acall       sysinit
            mov         cnt,#0
            mov         SFRPAGE,#0
            mov         FLSCL,#0x81         ; enable flash memory erase
            orl         PSCTL,#0x07         ; enable erase and write to the scratchpad sector
            mov         DPTR,#0x00          ; set DPTR to scratchpad sector at 0x00
            mov         A,#0
            movx        @DPTR,A             ; erase the scratchpad sector starting at 0x00
            anl         PSCTL,#0xFD         ; disable flash erase
            mov         DPTR,#0x00          ; write value to scratchpad sector starting from 0x00

            mov         cnt,#128
            mov         A,#0
loopi:      movx        @DPTR,A
            inc         DPTR
            inc         A
            djnz        cnt,loopi
; ***************************************************************************************************
; Read out the scratchpad sector one byte at a time and display its contents on P5 for 200 ms.
; ***************************************************************************************************
            anl         PSCTL,#0xFE         ; disable write to scratchpad sector
            mov         SFRPAGE,#0x0F       ; switch to SFR page F
            mov         DPTR,#0x00
            mov         R1,#0
            mov         A,R1
```

```
loopi2:     movc     A,@A+DPTR          ; read a location from the scratchpad sector
            mov      P5,A               ; output memory contents to P5
            mov      R0,#2
            acall    delayby100ms       ; wait for 200 ms
            inc      R1
            mov      A,R1
            cjne     R1,#128,loopi2
            ajmp     $
sysinit:    mov      SFRPAGE,#CONFIG_PAGE
            mov      WDTCN,#0xDE
            mov      WDTCN,#0xAD
            mov      CLKSEL,#0          ; use internal oscillator to generate SYSCLK
            mov      OSCICN,#0x83       ;     "
            mov      P5MDOUT,#0xFF
            mov      P5,#0              ; turn off LEDs
            ret
            $include (delays.a51)
            end
```

14.7.3 Security Options

The C8051F040 provides security options to protect the flash memory from inadvertent modification by software and also protect proprietary programs from being copied. The *program store write enable* and the *program store erase enable* bits of the PSCTL register protect the flash memory from accidental modification by software. These bits must be explicitly set to logic 1 before software can write or erase the flash memory.

The C8051F040 includes a set of security lock bytes stored at 0xFDFE and 0xFDFF to protect the flash program memory from being read or altered across the JTAG interface. Each bit in a security lock-byte protects one 8-kB block of memory. Clearing a bit to logic 0 in a *read lock byte* prevents the corresponding block of flash memory from being read via the JTAG interface. Clearing a bit in the *write/erase lock byte* protects the block from being erased or written via the JTAG interface.

The *read lock byte* is located at 0xFDFF whereas the *write/erase lock byte* is located at 0xFDFE. The 512-byte sector that contains the lock bytes can be written into but not erased by software. An attempted read of a read-locked byte returns undefined data. Debugging code in a read-locked sector through the JTAG interface is not possible.

The lock bits always can be read and cleared to 0 regardless of the security setting applied to the block containing the security bytes. This allows additional blocks to be protected after the block containing the security bytes has been locked. The only means of removing a lock bit that has been set is to erase the entire program memory space by performing a JTAG erase operation (i.e., cannot be done in user firmware). Performing a JTAG erasure operation with the address set to one of the security bytes will erase the entire program memory space. If a non-security byte in the 0xFC00-0xFDFF page is addressed during the JTAG erasure, only that page (including the security bytes) will bc crascd.

The C8051F040 provides a *flash access limit register* (FLACL) to further enhance the security of application software. The FLACL register holds the upper byte of the software access limit address (lower byte is 0x00). Software running at above this address is prohibited from using the **MOVC** or **MOVX** instructions to read, write, or erase flash locations below this address. Any attempt to read locations below this limit will return the value 0x00. Software in the upper partition (above the access limit) can execute (by calling or jumping) code in the lower partition. Software running in the lower partition can access locations in both the upper and lower partition without restriction.

The page containing the security bytes cannot be erased by programs located below or above the software access limit. However, if the page containing the security bytes is unlocked, it can be erased directly via the JTAG interface. Doing so will reset the security bytes and unlock all pages of flash. If the page containing the security bytes is locked, it cannot be erased directly. To unlock the page containing the security bytes, a full JTAG device erase is required, which will erase all flash pages.

14.8 External Data Memory Interface and On-Chip XRAM

In addition to having 4-kB on-chip data memory, the C8051F040 includes an external data memory interface (EMI) that can be used to access off-chip memories and memory-mapped devices connected to the GPIO ports. The external data memory space may be accessed using the **MOVX** instruction along with the DPTR register, or using the **MOVX** instruction along with R0 or R1 register. If the **MOVX** instruction is used with an 8-bit address operand (such as @R0), then the high byte of the 16-bit address is provided by the *external memory interface control register* (EMI0CN). The contents of the EMI0CN register (shown in Figure 14.13) select an XRAM page to access.

FIGURE 14.13 *External memory interface control register (EMI0CN).*

7	6	5	4	3	2	1	0	
PGSEL7	PGSEL6	PGSEL5	PGSEL4	PGSEL3	PGSEL2	PGSEL1	PGSEL0	Reset value 0x00

PGSEL[7:0]: XRAM page select bits

The XRAM page select bits provide the high byte of the 16-bit external data memory address when using an 8-bit MOVX command

0x00: 0x0000 to 0x00FF

0x01: 0x0100 to 0x01FF

....

0xFE: 0xFE00 to 0xFEFF

0xFF: 0xFF00 to 0xFFFF

For example, this instruction sequence loads the contents of the data memory location at 0x2020 into accumulator A:

```
mov     EMI0CN,#0x20
mov     R0,#0x20
movx    A,@R0          ; load the contents of 0x2020 into A
```

14.9 Configuring the External Memory Interface

The C8051F040 external memory interface (EMI) has the following options:

1. EMI can be configured to multiplex with lower ports (P0 through P3) or upper ports (P4 through P7)
2. I/O ports associated with EMI can be configured to be open-drain or push-pull
3. EMI lower address signals and data pins can be multiplexed to the same I/O port or be separate
4. Four memory modes: *on-chip only, split mode without bank select, split mode with bank select,* or *off-chip only*
5. Programmable timing to work with slower memory chip or peripheral devices

Each of these five options will be discussed in detail in the following sections.

14.9.1 Port Selection and Configuration

The EMI can appear on Ports P3, P2, P1, and P0 or on Ports P7, P6, P5, and P4, depending on the state of the **PRTSEL** bit (bit 5 of the EMI0CF register). The contents of the EMI0CF register are shown in Figure 14.14. If the lower ports are selected, the **EMIFLE** bit of the XBR2 register must be set to 1 so that the crossbar will skip over P0.7 (/WR), P0.6 (/RD), and (if multiplexed mode is selected) P0.5 (ALE).

The external memory interface claims the associated port pins for memory operations only during the execution of an off-chip **MOVX** instruction. Once the **MOVX** instruction has completed, control of the port pins reverts to the port latches or to the crossbar (on Ports P3, P2, P1, and P0).

During the execution of the **MOVX** instruction, the external memory interface will explicitly disable the drivers on all port pins that are acting as inputs (for example, data[7:0] during a read operation). The output mode of the port pins is unaffected by the external memory interface operation and remains controlled by the PnMDOUT registers. In most cases, the output modes of all **EMIF** pins should be configured for push-pull mode.

FIGURE 14.14
External memory configuration register (EMI0CF).

7	6	5	4	3	2	1	0	
--	--	PRTSEL	EMD2	EMD1	EMD0	EALE1	EALE0	Reset value 0x00

PRTSEL: EMIF port select
 0: EMIF active on P0-P3
 1: EMIF active on P4-P7
EMD2: EMIF multiplex mode select
 0: EMIF operates in multiplexed address/data mode
 1: EMIF operates in non-multiplexed mode (separate address and data pins
EMD1-EMD0: EMIF operating mode select
 00: Internal only: MOVX accesses on-chip XRAM only
 01: Split mode without bank select. Accesses below 4k boundary are directed on-chip. Accesses above the 4k boundary are directed off-chip 8-bit off-chip MOVX operations use the current contents of the Address High port latches to resolve upper address byte
 10: Split mode with bank select: Accesses below 4k boundary are directed on-chip. Accesses above 4k boundary are directed off-chip 8-bit off-chip MOVX operations use the contents of EMI0CN to determine the high-byte of the address
 11: External only: MOVX accesses off-chip XRAM only
EALE1-EALE0: ALE pulse width select bits (has effect when EMD2=1)
 00: ALE high and ALE low pulse width = 1 SYSCLK cycle
 01: ALE high and ALE low pulse width = 2 SYSCLK cycle
 10: ALE high and ALE low pulse width = 3 SYSCLK cycle
 11: ALE high and ALE low pulse width = 4 SYSCLK cycle

14.9.2 Multiplexed and Non-Multiplexed Selection

The C8051F040 external memory interface is capable of acting in a multiplexed mode or a non-multiplexed mode, depending on the state of the **EMD2** bit of the EMI0CF register.

In multiplexed mode, the data bus and the lower 8 bits of the address bus share the same port pins: AD[7:0]. In this mode, an external latch is needed to hold the lower address bits A7:A0, because the external memory chip requires the address signal to be stable throughout the access cycle. The external latch is controlled by the address latch enable (ALE) signal, which is driven by the external memory interface logic. An example of a multiplexed configuration is shown in Figure 14.15.

In multiplexed mode, the external **MOVX** operation can be broken into two phases delineated by the state of the ALE signal. During the first phase, ALE is high and the lower 8 address bits are presented to AD[7:0]. During this phase, the address latch is configured such that the Q outputs reflect the states of the 'D' inputs. When ALE falls, signaling the beginning of the second

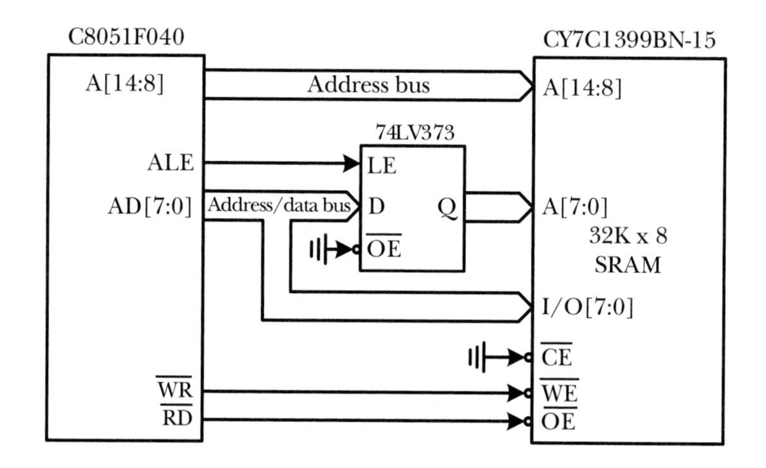

FIGURE 14.15 *A multiplexed memory configuration Example.*

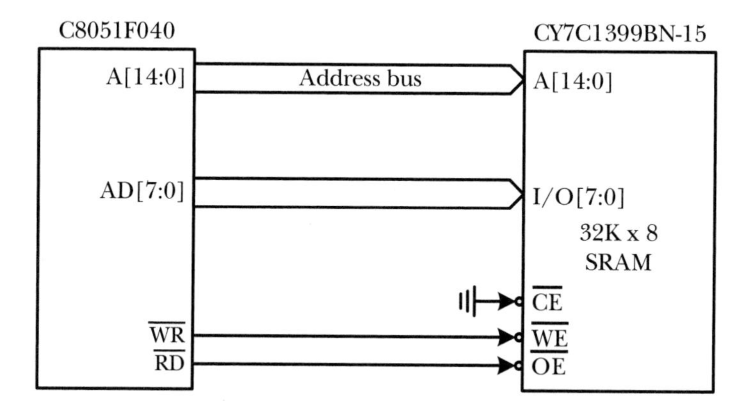

FIGURE 14.16 *A non-multiplexed memory configuration Example.*

phase, the address latch outputs remain fixed and are no longer dependent on the latch inputs. Later in the second phase, the data bus controls the state of the AD[7:0] port at the time /RD or /WR is asserted.

In non-multiplexed mode, the data bus and the address bus pins are not shared. An example of a non-multiplexed configuration is shown in Figure 14.16.

14.9.3 Memory Mode Selection

The external data memory space can be configured in one of four modes, shown in Figure 14.17, based on the **EMIF** mode bits in the EMI0CF register. These modes are summarized below.

Internal XRAM Only

When EMI0CF[3:2] are set to 00, all **MOVX** instructions will target the internal XRAM space on the device. Memory accesses to addresses beyond the populated space will wrap on 4-k boundaries. For example, the addresses

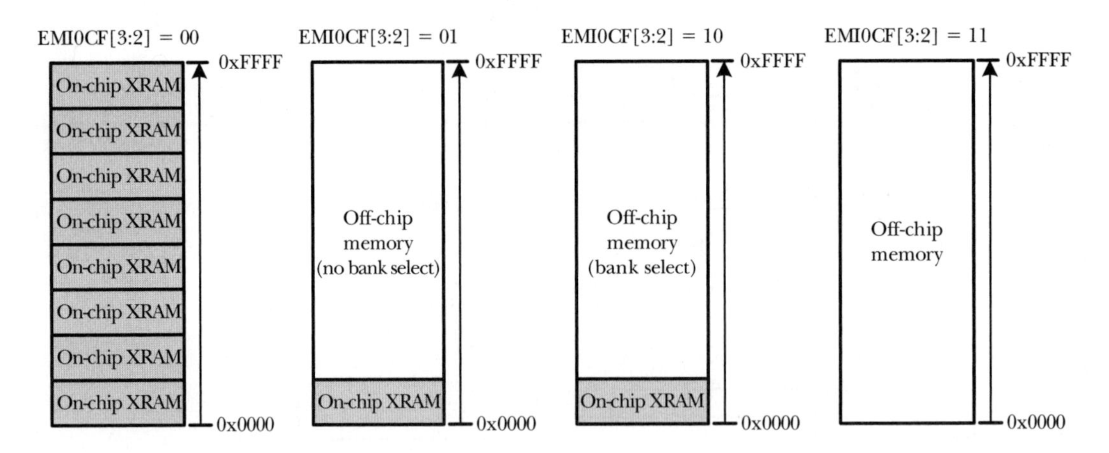

FIGURE 14.17
EMIF Operation Modes.

0x1000 and 0x2000 both evaluate to address 0x0000 in on-chip XRAM space. This is the default memory mode after reset.

Split Mode without Bank Select

When EMI0CF[3:2] are set to 01, the XRAM memory map is split into two areas, on-chip space and off-chip space. The address below the 4-k boundary will access on-chip XRAM space. The address above the 4-k boundary will access off-chip space.

The 8-bit **MOVX** operation uses the contents of EMI0CN to determine whether the memory access is on-chip or off-chip. However, in the *no bank select* mode, an 8-bit **MOVX** operation will not drive the upper 8 bits (A[15:8]) of the address bus during an off-chip access. This allows the user to manipulate the upper address bits at will by setting the port state directly via the port latches. This behavior is in contrast with *split mode with bank select.* The lower 8 bits of the address bus A[7:0] are driven, determined by R0 or R1.

The 16-bit **MOVX** operation uses the contents of DPTR to determine whether the memory access is on-chip or off-chip, and unlike 8-bit **MOVX** operations, the full 16 bits of the address bus are driven during the off-chip transaction.

14.9.4 Split Mode with Bank Select

In this mode, the XRAM memory map is split into two areas, on-chip space and off-chip space. An address below the 4-k boundary accesses on-chip XRAM, whereas the address above the 4-k boundary accesses off-chip XRAM. An 8-bit **MOVX** operation uses the contents of EMI0CN to determine whether the memory access is on-chip or off-chip. The upper 8 bits of the address bus A[15:8] are determined by EMI0CN, whereas the lower 8 address bits are supplied by R1 or R0.

A 16-bit MOVX operation uses the contents of the DPTR to determine whether the memory access is on-chip or off-chip, and the full 16 bits of the address bus A[15:0] are driven during the off-chip transaction.

External Only

In this mode, all **MOVX** operations are directed to off-chip space. On-chip XRAM is invisible to the CPU. This mode is useful for accessing off-chip memory located between 0x0000 and the 4-k boundary (0x0FFF).

Eight-bit **MOVX** operations ignore the contents of EMI0CN (same as split mode without bank select). The upper address bits are not driven. This allows the user to manipulate the upper address bits at will by setting the port state directly. The lower 8 bits of the address are supplied by R0 or R1.

Sixteen-bit **MOVX** operations use the contents of the DPTR register to determine the effective address A[15:0]. The full 16-bit address bus is driven during the off-chip transaction.

14.9.5 Timing

Timing is critical for external memory to work with the microprocessor/microcontroller. There are 16 different combinations of bus cycles for the C8051F040. These bus cycles often are represented as timing diagrams of a set of involved signals. For example, the 16-bit address, non-multiplexed read and write bus cycles, and the 16-bit address, multiplexed read and write bus cycles are illustrated in Figures 14.18 through 14.21. The timing diagram of 8-bit **MOVX** multiplexed read cycle with bank select is identical to that of the 16-bit multiplexed **MOVX** read cycle—with the exception that A[15:8] be supplied by EMI0CN register. The timing diagram of 8-bit **MOVX** multiplexed write cycle with bank select is identical to that of the 16-bit multiplexed MOVX write cycle—with the exception that A[15:8] be supplied by EMI0CN register.

There are several timing parameters that must be met in order for a read or write bus cycle to be successful. The minimum and maximum values of

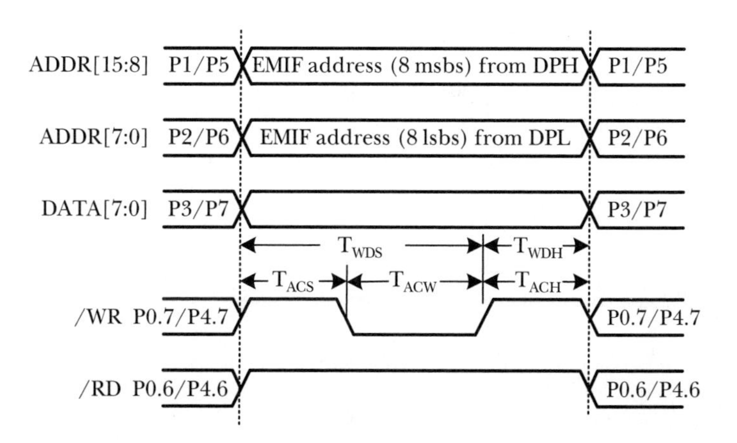

FIGURE 14.18 *16-bit MOVX non-multiplexed mode bus write cycle.*

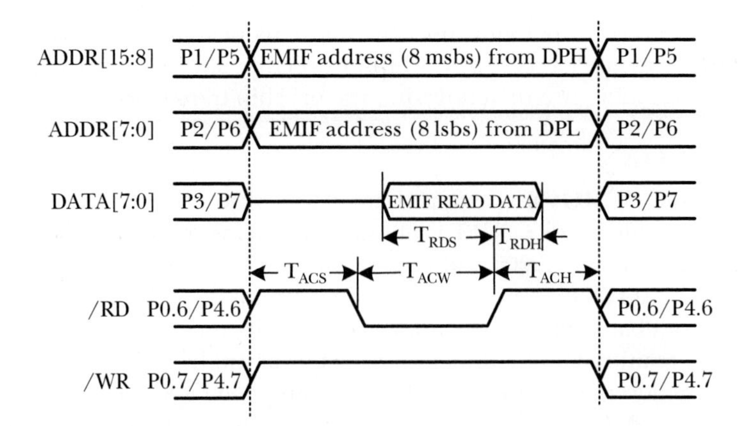

FIGURE 14.19
16-bit MOVX non-multiplexed mode bus read cycle.

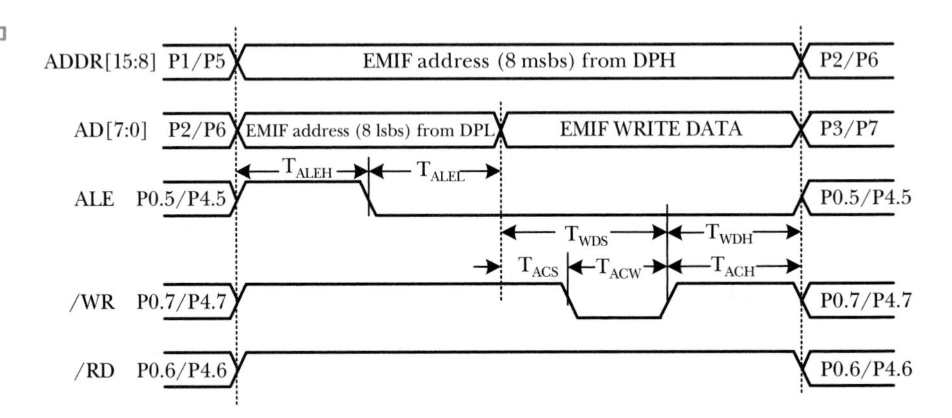

FIGURE 14.20
16-bit MOVX multiplexed mode bus write cycle.

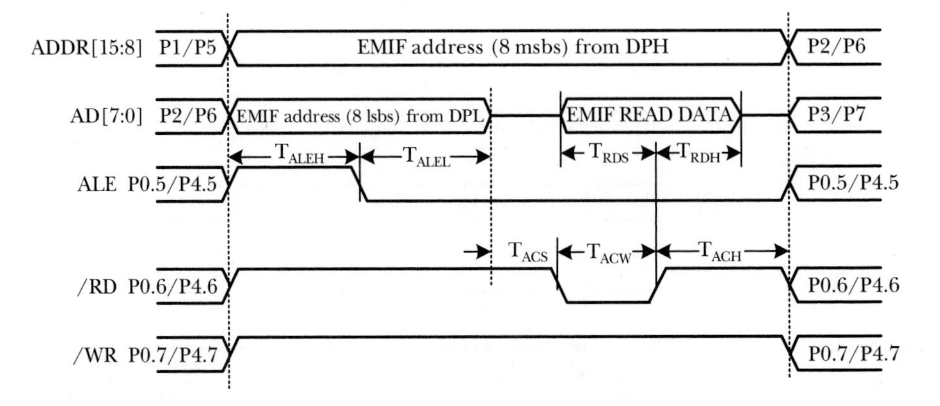

FIGURE 14.21
16-bit MOVX multiplexed mode bus read cycle.

these parameters are listed in Table 14.1. The meanings of these parameters are as follows.

- T_{ACS} (address/control setup time): the amount of time that the address signal will be valid before the control signal becomes valid (/WR or /RD signal)
- T_{ACW} (address/control pulse width): the minimum pulse width of the control signal (/WR or /RD)
- T_{ACH} (address/control hold time): the amount of time that the address signals will stay valid after the control becomes invalid (/WR or /RD)
- T_{ALEH} (address latch enable high time): the amount of time that the ALE signal stays at high
- T_{ALEL} (address latch enable low time): the amount of time that address signals will stay valid after the ALE signal becomes invalid (low)
- T_{WDS} (write data setup time): the amount of time that the write data will become valid before the /WR signal goes high
- T_{WDH} (write data hold time): the amount of time that the write data will stay valid after the /WR signal becomes invalid
- T_{RDS} (read data setup time): the minimum amount of time that the read data must be valid before the /RD signal goes high in order for the read operation to be successful
- T_{RDH} (read data hold time): the minimum amount of time that the read data must stay valid after /RD goes high in order for the read operation to be successful

Many of the timing parameters in Table 14.1 are programmable via the EMI0TC register. The contents of this register are shown in Figure 14.22. These parameters must be programmed to meet the requirements of the memory chip.

TABLE 14.1 *AC Parameters for External Memory Interface*

Parameter	Description	Min	Max	Unit
T_{SYSCLK}	System clock period	40	--	ns
T_{ACS}	Address/control setup time	0	$3 \times T_{SYSCLK}$	ns
T_{ACW}	Address/control pulse width	$1 \times T_{SYSCLK}$	$16 \times T_{SYSCLK}$	ns
T_{ACH}	Address/control hold time	0	$3 \times T_{SYSCLK}$	ns
T_{ALEH}	Address latch enable high time	$1 \times T_{SYSCLK}$	$4 \times T_{SYSCLK}$	ns
T_{ALEL}	Address latch enable low time	$1 \times T_{SYSCLK}$	$4 \times T_{SYSCLK}$	ns
T_{WDS}	Write data setup time	$1 \times T_{SYSCLK}$	$19 \times T_{SYSCLK}$	ns
T_{WDH}	Write data hold time	0	$3 \times T_{SYSCLK}$	ns
T_{RDS}	Read data setup time	20	--	ns
T_{RDH}	Read data hold time	0	--	ns

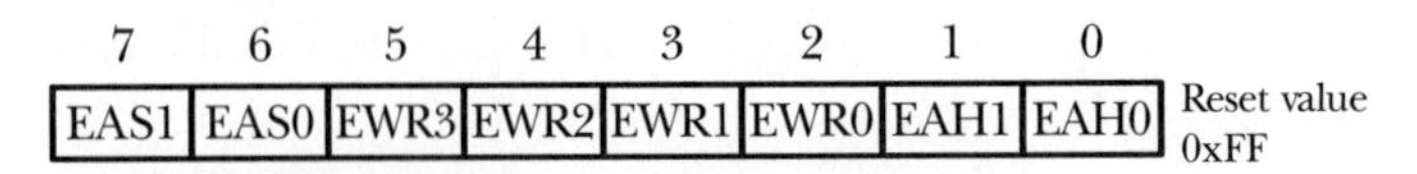

FIGURE 14.22 *External memory timing control register (EMI0TC).*

EAS1~EAS0: EMIF address set up time bits
  00: address set up time = 0 SYSCLK cycles
  01: address set up time = 1 SYSCLK cycles
  10: address set up time = 2 SYSCLK cycles
  11: address set up time = 3 SYSCLK cycles
EWR3~EWR0: EMIF /WR and /RD pulse-width control bits
  0000: /WR and /RD pulse width = 1 SYSCLK cycle
  0001: /WR and /RD pulse width = 2 SYSCLK cycle
  ....
  1111: /WR and /RD pulse width = 16 SYSCLK cycle
EAH1~EAH0: EMIF address hold time bits
  00: address hold time = 0 SYSCLK cycles
  01: address hold time = 1 SYSCLK cycles
  10: address hold time = 2 SYSCLK cycles
  11: address hold time = 3 SYSCLK cycles

14.10 Issues Related to Adding External Memory

Adding external memory is necessary for the C8051F040 only when the application needs more SRAM to hold dynamic data. When adding external memory, there are three issues that need to be considered:

1. Memory space assignment
2. Address decoder and control circuitry design
3. Timing verification

14.10.1 Memory Space Assignment

Any space unoccupied by the on-chip XRAM can be assigned to external memory devices. When making memory space assignment, the designer has two options to choose from.

1. ***Equal Size Assignment.*** In this method, the available memory space is divided into blocks of equal size, and then each block is assigned to a memory device without regard for the actual size of each memory device. A memory-mapped device could be a memory chip or a peripheral device. Memory space tends to be wasted using this approach, because most memory-mapped peripheral chips need only a few bytes to be assigned to their internal registers.
2. ***Demand Assignment.*** In this approach, the designer assigns the memory space according to the size of memory devices.

Example 14.3 Suppose that the user is assigned to design a C8051F040-based embedded product that requires 32 KB of SRAM and 32 KB of EEPROM. The only SRAM available to this user is the 32 K × 8 SRAM chip, whereas the only available EEPROM is the 32 K × 8 EEPROM chip. Suggest a workable memory space assignment.

Solution: This user is assigned to design an 8-bit memory system using the 8-bit wide SRAM and EEPROM chips. Since the C8051F040 has 4-KB on-chip XRAM, an easy solution is to assign the lower 32 KB (from 0x0000 to 0x7FFF) and upper 32 KB (0x8000 to 0xFFFF) of the data memory space to the SRAM and EEPROM chip, respectively. If the user selects the split mode for data memory, then the first 4 kB of external SRAM won't be available for access.

14.10.2 Address Decoder Design

The function of an address decoder is to make sure that there is no more than one memory device enabled to drive the data bus at a time. All memory devices or peripheral devices have control signals—such as chip-enable (CE), chip-select (CS), or output-enable (OE)—to control their read and write operations. These signals often are asserted low. When these signals are high, the memory-device data pins are in a high-impedance state and are isolated from the data bus. The address decoder outputs will be used as the chip-select or chip-enable signals of external memory devices.

Two address-decoding schemes have been used: *full decoding* and *partial decoding*. In the full decoding scheme, each addressable location of the memory chip responds to one and only one address on the address bus. In the partial decoding scheme, each location of the memory chip may respond to more than one address on the address bus. Memory components (such as SRAM and EEPROM) use the full decoding scheme more often, whereas peripheral chips use the partial decoding scheme more often.

Address decoder design is related closely to memory space assignment. For the memory space assignment made in Example 14.3, the decoder design is trivial. One needs only to use the address signal A15 to select one memory chip and use the complement (inverted by an inverter chip) of A15 to select another memory chip.

14.10.3 Timing Verification

When designing a memory system, the designer needs to make sure that timing requirements for both the microcontroller and the memory system are satisfied. In a read cycle, the most critical timing requirements are the data setup time and data hold time required by the C8051F040. In addition, the designer must make sure that the address setup time and hold time requirements for the memory devices are met. The control signals needed by memory devices during a read cycle must be asserted at the appropriate times.

In a write cycle, the most critical timing requirements are the write-data setup time and write-data hold time required by the memory devices. As in a read cycle, the address setup time and address hold time also must be satisfied. Control signals required during a write cycle also must be generated at proper times.

14.11 Memory Devices

The control circuit designs for interfacing the SRAM, the EEPROM, the EPROM, and the flash memory are quite similar. The following sections illustrate how to add SRAM and EEPROM chips with the 32 K × 8 organization to the C8051F040 microcontroller.

14.11.1 The CY7C1399BN SRAM

The CY7C1399BN is a 32 K × 8 asynchronous SRAM chip made by Cypress. This device uses 3.3-V power supply and has a short access time (12 ns, or 15 ns, or 20 ns) and three-state output. This device also has an automatic power-down feature, reducing the power consumption by more than 95 percent when it is not selected.

The active-low $\overline{WE}$ signal controls the writing/reading operation of the memory. When $\overline{CE}$ and $\overline{WE}$ are both low, the data on the eight data input/output pins (I/O_0 through I/O_7) is written into the memory location addressed by the address present on the address pins (A0 through A14). Reading the device is accomplished by selecting the device and enabling the outputs. This is achieved by pulling $\overline{CE}$ and $\overline{OE}$ to low, while at the same time pulling the $\overline{WE}$ signal to high. The contents of the memory location addressed by the address-input pins appear on the eight data input/output pins. The pin assignment of the CY7C1399BN is shown in Figure 14.23.

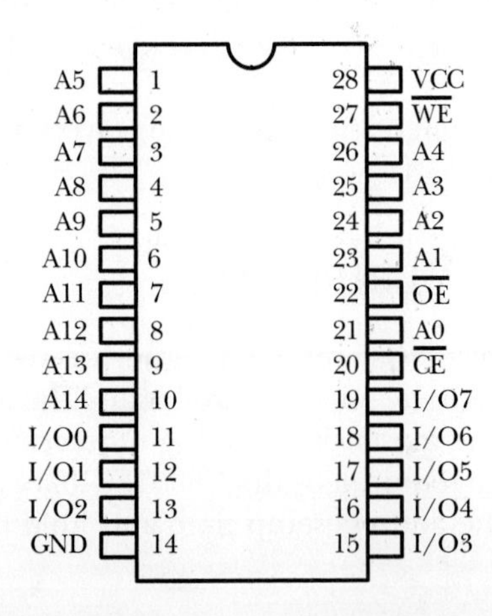

FIGURE 14.23 *CY7C1399BN SRAM.*

Depending on the timing of the control signals, there are two timing diagrams for the read cycle, whereas there are three timing diagrams for the write cycle. Figure 14.24 is the read timing diagram under the situation that the device is continuously selected (i.e., $\overline{CE}$ and $\overline{OE}$ are both low). Figure 14.25 is the read timing diagram in which $\overline{CE}$ and $\overline{OE}$ are not always low and $\overline{WE}$ is high. Figure 14.26 is the $\overline{WE}$-controlled write timing diagram. Figure 14.27 is the $\overline{CE}$-controlled write timing diagram. Figure 14.28 is the $\overline{WE}$-controlled write timing diagram with the $\overline{OE}$ pin tied to low. The values of timing parameter of the CY7C1399BN are listed in Table 14.2.

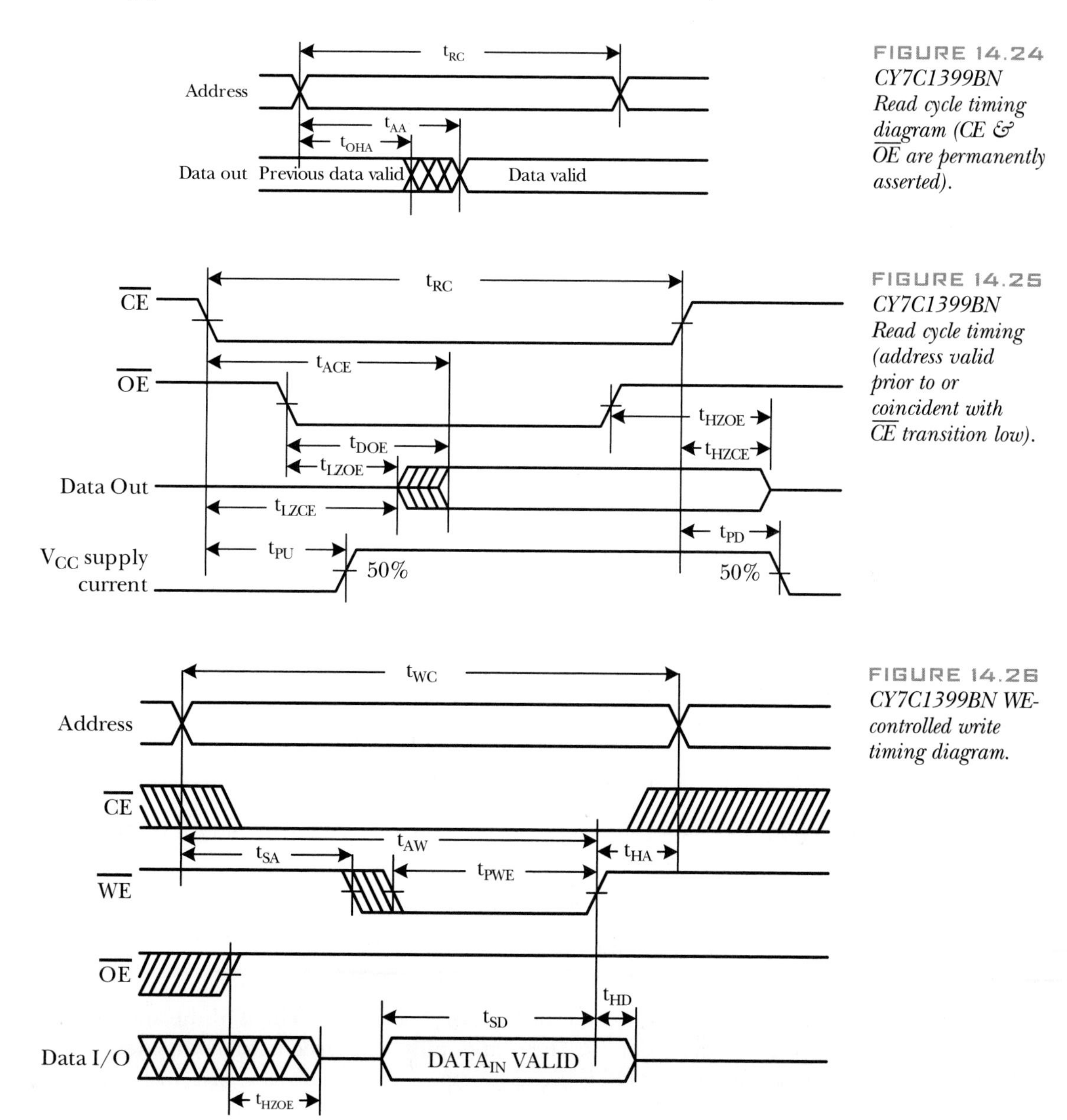

FIGURE 14.24 *CY7C1399BN Read cycle timing diagram ($\overline{CE}$ & $\overline{OE}$ are permanently asserted).*

FIGURE 14.25 *CY7C1399BN Read cycle timing (address valid prior to or coincident with $\overline{CE}$ transition low).*

FIGURE 14.26 *CY7C1399BN WE-controlled write timing diagram.*

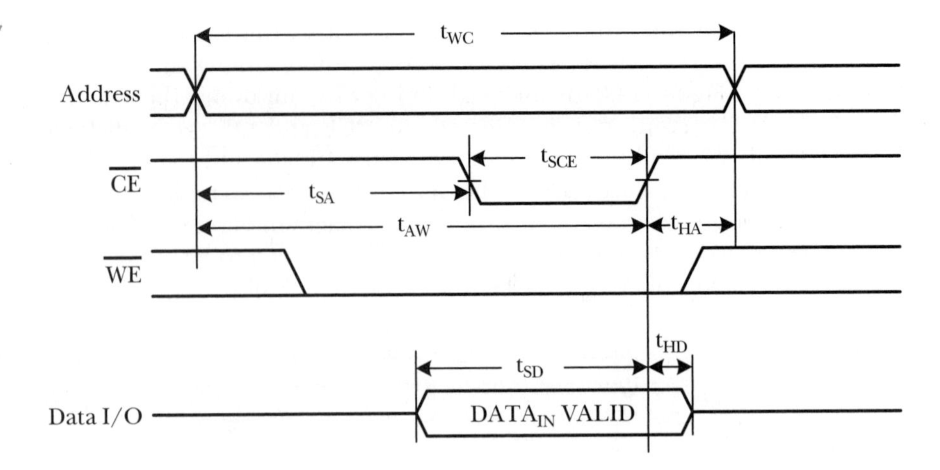

FIGURE 14.27 *CY7C1399BN $\overline{CE}$-controlled write timing diagram.*

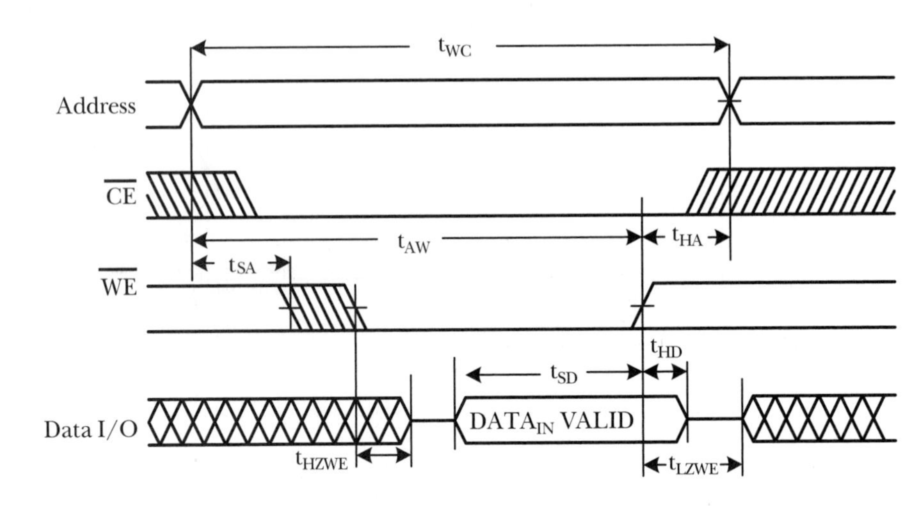

FIGURE 14.28 *CY7C1399BN WE-controlled write timing ($\overline{OE}$ low).*

14.11.2 AT28BV256 EEPROM

The AT28BV256 is a 32 K × 8 EEPROM with 200-ns access time and operates with a power supply ranging from 2.7 to 3.6 V. The AT28BV256 is accessed like a SRAM for the read or write cycle without the need for external components. The device contains a 64-byte page register to allow writing up to 64 bytes before starting the internal write operation. The pin assignment of the AT28BV256 is shown in Figure 14.29.

Read Operation

The AT28BV256 is accessed like a SRAM. When $\overline{CE}$ and $\overline{OE}$ are low and $\overline{WE}$ is high, data stored at the memory location specified by the address pins appears on the I/O pins. The output pins are placed in high-impedance state when either $\overline{CE}$ or $\overline{OE}$ is high.

TABLE 14.2 *CY7C1399BN Timing Parameter Values at Two Speed Grades*

		−12		−15		
Parameter	**Description**	**Min**	**Max**	**Min**	**Max**	**Unit**
Read Cycle						
t_{RC}	Read cycle time	12		15		ns
t_{AA}	Address to data valid		12		15	ns
t_{OHA}	Data hold from address change	3		3		ns
t_{ACE}	CE low to data valid		12		15	ns
t_{DOE}	OE low to data valid		5		6	ns
t_{LZOE}	OE low to low Z	0		0		ns
t_{HZOE}	OE high to high Z		5		6	ns
t_{LZCE}	CE low to low Z	3		3		ns
t_{HZCE}	CE high to high Z		6		7	ns
t_{PU}	CE low to power up	0		0		ns
t_{PD}	CE high to power down		12		15	ns
Write Cycle						
t_{WC}	Write cycle time	12		15		ns
t_{SCE}	CE low to write end	8		10		ns
t_{AW}	Address setup to write end	8		10		ns
t_{HA}	Address hold from write end	0		0		ns
t_{SA}	Address setup to write start	0		0		ns
t_{PWE}	WE pulse width	8		10		ns
t_{SD}	Data setup to write end	7		8		ns
t_{HD}	Data hold after write end	0		0		ns
t_{HZWE}	WE low to high Z		7		7	ns
t_{LZWE}	WE high to low Z	3		3		ns

Byte Write Operation

A low pulse on the WE (or $\overline{CE}$) pin with $\overline{CE}$ (or $\overline{WE}$) low and $\overline{OE}$ high initiates a write cycle. The address is latched on the falling edge of the $\overline{CE}$ or $\overline{WE}$ signal, whichever occurs last. The data is latched by the first rising edge of the $\overline{CE}$ or $\overline{WE}$ signal, whichever occurs first. Once a byte write has been started, it will automatically time itself to completion. Once a programming operation has been initiated, and for the duration of t_{WC}, a read operation effectively will be a polling operation.

Page Write Operation

The page write operation of the AT28BV256 allows 1 to 64 bytes of data to be written into the device during a single internal programming period. A page write operation is initiated in the same manner as a byte write; the first

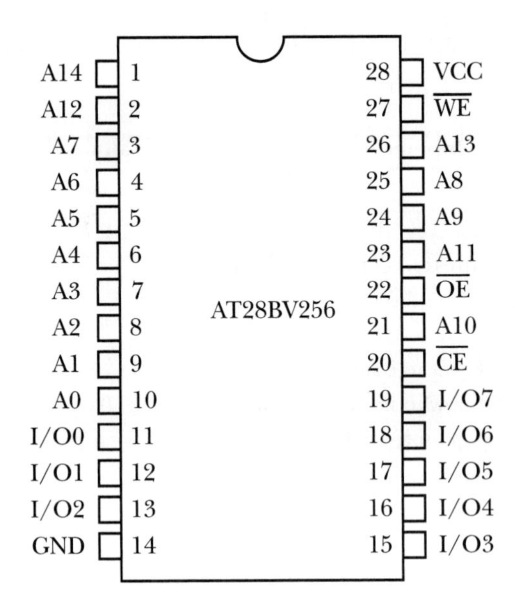

FIGURE 14.29
The AT28BV256 pin assignment.

byte written then can be followed by 1 to 63 additional bytes. Each successive byte must be written within 150 μs (t_{BLC}) of the previous byte. If the t_{BLC} limit is exceeded, the AT28BV256 will cease accepting data and commence the internal programming operation. All bytes during a page write operation must reside on the same page, as defined by the state of the A6-A14 inputs. For each WE high-to-low transition during the page write operation, A6-A14 must be the same. The A0 to A5 inputs are used to specify which bytes within the page are to be written. The bytes may be loaded in any order and may be altered within the same load period. Only bytes which are specified for writing will be written; unnecessary cycling of other bytes within the same page does not occur.

Data Polling

The AT28BV256 allows the user to poll the I/O7 pin to find out if the internal write operation has been completed. The complement of bit 7 of the last written data byte will appear on the I/O7 pin if the internal write operation is not completed yet.

Toggle Bit

In addition to data polling, the AT28BV256 provides another method for determining the end of a write cycle. During the write operation, successive attempts to read data from the device will result in the I/O6 signal toggling between one and zero. Once the write operation has completed, the I/O6 signal will stop toggling, and valid data will be read. Reading the toggle bit may begin at any time during the write cycle.

Data Protection

The AT28BV256 provides both hardware and software methods to protect the internal data.

Whenever power is turned on to the EEPROM chip and when it rises to 1.8 V, the device automatically will timeout for 10 ms before allowing a write. Holding $\overline{OE}$ low, $\overline{CE}$ high, or $\overline{WE}$ high will not initiate a write cycle.

The AT28BV256 also provides a software data-protection scheme to provide additional protection to the internal data. A series of three write commands to specific addresses with specific data must be presented to the device before writing in the byte or page mode. The same three write commands must begin each write operation. All software write commands must obey the page mode write timing specifications. The data in the 3-byte command sequence is not written to the device; the address in the command sequence can be utilized just like any other location in the device.

Any attempt to write to the device without the 3-byte sequence will start the internal write timers. No data will be written to the device; however, for the duration of t_{WC}, read operations effectively will be polling operations.

The procedure for writing data to any location(s) is as follows.

Step 1
Write the value **0xAA** to the memory location at **0x5555**.

Step 2
Write the value **0x55** to the memory location at **0x2AAA**.

Step 3
Write the value **0xA0** to the memory location at **0x5555**.

Step 4
Write 1 to 64 bytes of data to the desired memory locations.

Read Timing

To read from this device, both $\overline{CE}$ and $\overline{OE}$ must be pulled low, while at the same time, the $\overline{WE}$ signal must be high. The timing diagram of the read cycle is shown in Figure 14.30. The read access time is 200 ns. The values of read timing parameters are listed in Table 14.3.

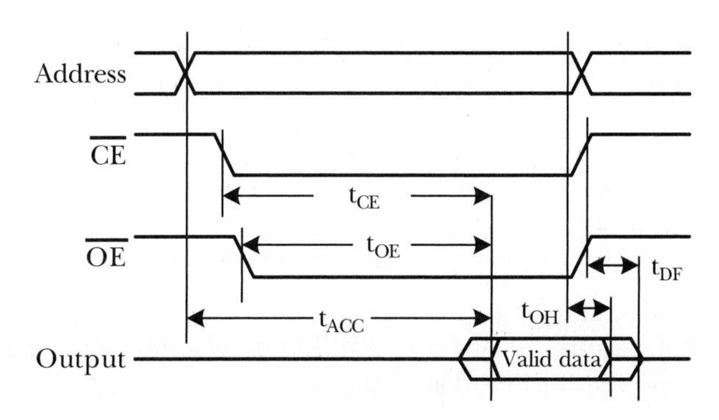

FIGURE 14.30 *Read timing of AT28BV256.*

TABLE 14.3 *AT28BV256 Timing Parameter Values*

Parameter	Description	AT28BV256 Min	AT28BV256 Max	Unit
Read Cycle				
t_{ACC}	Address to output delay		200	ns
t_{CE}	$\overline{CE}$ to output delay		200	ns
t_{OE}	OE to output delay	0	80	ns
t_{DF}	$\overline{CE}$ or $\overline{OE}$ to output float	0	55	ns
t_{OH}	Output hold from OE, CE or	0		ns
	Address whichever occurred			ns
	first			ns
Write Cycle				
t_{AS}, t_{OES}	Address, OE setup time	0		ns
t_{AH}	Address hold time	50		ns
t_{CS}	Chip select set-up time	0		ns
t_{CH}	Chip select hold time	0		ns
t_{WP}	Write pulse width (WE or CE)	200		ns
t_{DS}	Data setup time	50		ns
t_{DH}, t_{OEH}	Data, OE hold time	0		ns
t_{DV}	Time to data valid	NR[1]		ns

Note:
1. NR = No restriction.

Write Timing

During a write cycle, the $\overline{OE}$ signal should be pulled high. Depending on the assertion times of $\overline{CE}$ and $\overline{WE}$, there are two versions of the write-cycle timing diagram, as shown in Figures 14.31 and 14.32. The values of write timing parameters are listed in Table 14.3.

In a $\overline{WE}$-controlled write cycle, the $\overline{CE}$ signal is asserted (goes low) earlier than the $\overline{WE}$ signal and becomes inactive later than the $\overline{WE}$ signal. In a $\overline{CE}$-controlled write cycle, the $\overline{WE}$ signal goes low earlier and returns to high later than the $\overline{CE}$ signal. In Figures 14.31 and 14.32, the control signal which is asserted the last must have a minimal pulse width of 200 ns. Address signals must be valid earlier than the control signal (that is asserted the latest) and must remain valid for at least 50 ns after assertion of the same control signal.

The write timing waveform illustrates only how the CPU writes data into the EEPROM. The EEPROM still needs to initiate an internal programming process to actually write data into the special location. The CPU can find out whether the internal programming process has been completed by using the polling or the toggle-bit polling method.

In the following, we assume that you have an application that justifies the use of external parallel SRAM and EEPROM. We will use the addition of 32-kB SRAM and 32-kB EEPROM to the C8051F040 to illustrate the memory expansion process.

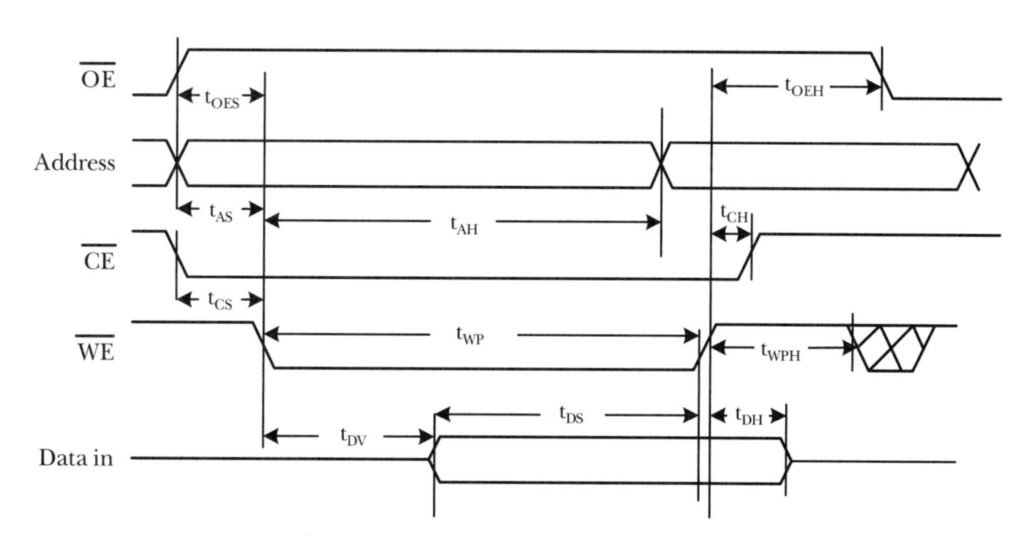

FIGURE 14.31
AT28BV256 WE-controlled write-cycle timing diagram.

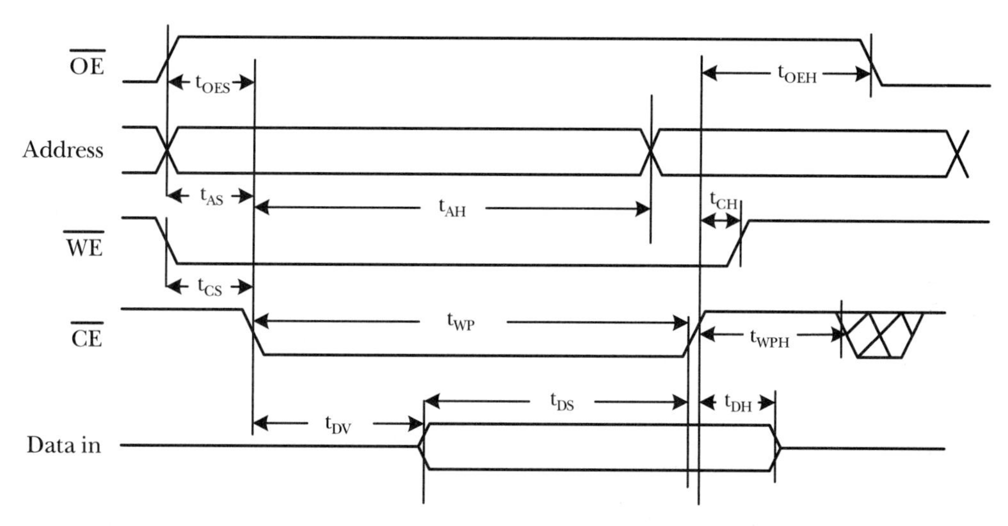

FIGURE 14.32
AT28BV256 $\overline{CE}$-controlled write-cycle timing diagram.

14.12 Example of External Memory Expansion for the C8051F040

To add a 32-kB SRAM (using the version with 15-ns access time) and 32-kB EEPROM to the C8051F040, one needs an address latch and an inverter. The inverter serves as the address decoder.

14.12.1 Memory Space Assignment

Memory space assignment is very simple in this example. We assign 0x0000~0x7FFF to the SRAM and 0x8000 ~ 0xFFFF to the EEPROM.

14.12.2 Address Latch

Multiplexed mode is selected to enable the lower eight address signals and the data signals to share the pins. The 8-bit latch 74LV373 will be used to hold A7 through A0 valid throughout a bus cycle. The pin assignments of the 74LV373 and the address latch circuit for the C8051F040 are shown in Figures 14.33 and 14.34, respectively. The 74LV373 uses the falling edge of LE to latch the data inputs (D7~D0) to Q7~Q0. The propagation delay from D_n to Q_n is 10 ns. The propagation from LE to Q is 12 ns. When LE is high, the 74LV373 is transparent, i.e., any change of value in D_n will be available at Q_n after 10 ns.

14.12.3 Address Decoder Design

In this memory design example, the 64-kB data memory space is divided into two 32-kB modules. The highest address signal A15 and its complement are used as the chip-enable signals for SRAM and EEPROM. The address decoder circuit is shown in Figure 14.35.

When A15 = 0, SRAM chip is selected. Otherwise, EEPROM is selected. The 74LV04 inverter has a 6-ns propagation delay.

The $\overline{\text{WR}}$ signal from the C8051F040 should be connected to the $\overline{\text{WE}}$ input of SRAM and EEPROM, whereas the $\overline{\text{RD}}$ signal from the C8051F040 should be connected to the $\overline{\text{OE}}$ input of both memory devices. The complete C8051F040 external memory expansion system is shown in Figure 14.36.

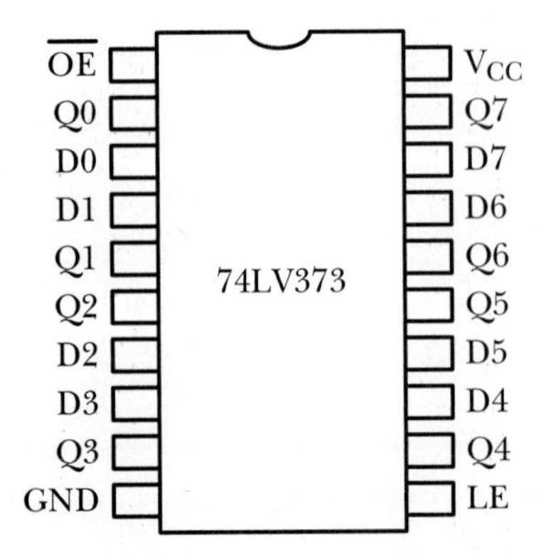

FIGURE 14.33
Pin assignment of 74LV373.

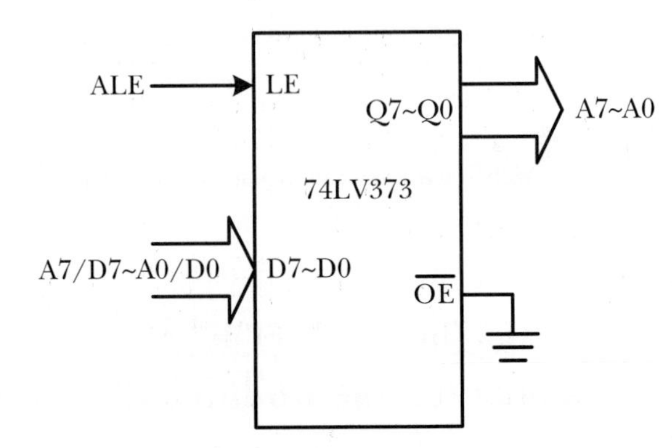

FIGURE 14.34
Address latch circuit for the C8051F040.

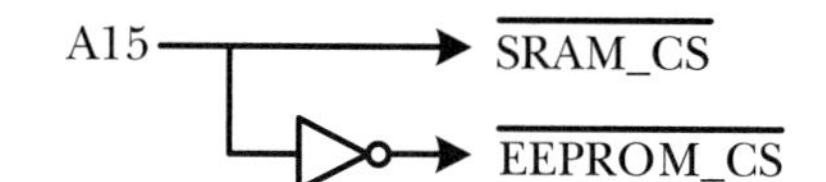

FIGURE 14.35 *Address decoder example circuit.*

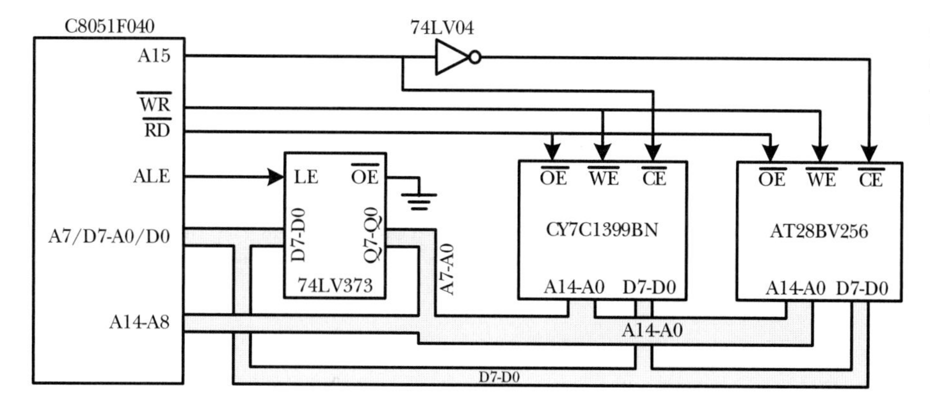

FIGURE 14.36 *A C8051F040 External memory system example.*

14.12.4 Timing Verification

The last step in the memory-system design is to verify that all timing parameter requirements have been met. In the following example, we need to figure out the value to be programmed into the EMI0CF and EMI0TC registers.

Example 14.4: Suggest values to be written into the EMI0CF and the EMI0TC registers and verify that all timing requirements are met, assuming that f_{SYSCLK} = 24 MHz and the multiplexed mode is selected.

Solution:

$T_{SYSCLK} = 1 / 24{,}000{,}000 = 41.67$ ns

The AT28BV256 requires the WE and RD pulse widths to be at least 200 ns (less than 5 T_{SYSCLK} but greater than 4 T_{SYSCLK}).

The following settings are suggested for this memory system:

- ALE high and low pulse set to 1 SYSCLK cycle
- Address setup time set to 0 SYSCLK cycle
- Address hold time set to 0 SYSCLK cycle
- /WR and /RD pulse width set to 5 SYSCLK cycles

Address Timing A15-A0 are driven at the beginning of a read and write cycle. The values of A7 through A0 will become valid to the SRAM and EEPROM 12 ns after the start of a read or write cycle, because they pass through the 74LV373. The CE input to the CY7C1399BN becomes valid from the start of a read or write cycle, whereas the $\overline{CE}$ input to the AT28BV256 is valid 6 ns after the start of a read or write cycle.

Read Cycle Timing The $\overline{RD}$ signal is driven low 2 T_{SYSCLK} after the start of the read cycle and will stay low for 5 T_{SYSCLK} cycles. The data from the CY7C1399BN will be available $2 \times T_{SYSCLK} + t_{DOE} = 89.6$ ns after the start of the read cycle or $5T_{SYSCLK} - 15$ ns before the rising edge of the $\overline{OE}$ signal. This definitely satisfies the read-data setup time (20 ns) of the C8051F040. The data from CY7C1399BN will stay valid for 6 ns after the $\overline{OE}$ signal goes high and hence also wil meet the read-data hold-time requirement for the C8051F040.

For the AT28BV256, the read data becomes valid at the time that is the longer one of the following two expressions after the start of the read cycle:

- 2 T_{SYSCLK} + 80 ns = 163.33 ns (calculated using the $\overline{OE}$ signal)
- Inverter delay (6 ns) + t_{CE} (200 ns) = 206 ns (calculated using the $\overline{CE}$ signal)

Obviously, the larger value is the second expression, and therefore, the read data from the AT28BN256 is valid 206 ns after the start of a read cycle. Since the read data is expected to be valid 20 ns before the rising edge of the $\overline{RD}$ signal ($7T_{SYSCLK} - 20$ ns), it also is satisfied. The AT28BV256 provides a 0-ns read-data hold time and also meets the requirement.

Write Cycle Timing Since the C8051F040 drives the $\overline{WR}$ signal 2 T_{SYSCLK} cycles after the start of a write cycle in our suggested timing parameters, both the write cycles of the CY7C1399BN and the AT28BV256 are $\overline{WE}$-controlled. Since T_{ACS} is set to 0 and T_{ACW} is set to 5 T_{SYSCLK}, the write data setup time is $5T_{SYSCLK}$ and is much longer than the write-data setup time requirements for both the CY7C1399BN (8 ns) and the AT28BV256 (50 ns).

The write-data hold time is set to 0 SYSCLK cycle and also satisfies the requirement for both memory chips (0 ns).

From this analysis, we conclude that the timing requirements for both the C8051F040 and memory chips are satisfied completely, and hence, both the read and write operations should be performed successfully.

14.13 Chapter Summary

The C8051F040 has on-chip memory resources, including special-function registers, data memory, XRAM, and flash memory. Special-function registers and the on-chip data memory do not occupy the 64-kB data memory space.

Since the on-chip flash program memory has reached the 64-kB size limit, the C8051F040 does not allow the user to add external program memory. The on-chip XRAM is 4 kB in size, and hence, the C8051F040 allows the user to expand external data memory to the 64-kB size limit.

The flash memory needs to be erased before it can be programmed. The C8051F040 uses three registers to control the erasure and programming of the flash memory. The flash memory is organized into 512-byte pages. Each erasure operation erases one page of flash memory. To erase and program the flash memory, the user needs to use the **MOVX** instruction with the address and data to be programmed provided as normal operands. Before writing to flash memory using the **MOVX** instruction, flash write operations must be enabled by setting the **PSWE** bit of the PSCTL register to logic 1. This directs the **MOVX** instruction to the flash memory instead of XRAM.

In addition to being used to hold program instructions, the on-chip flash memory also can be used to store dynamic data just like the on-chip XRAM.

Some applications may need more data memory than the amount provided by the C8051F040. The user has two options: the first option is to use serial memory with SPI or I^2C interface; the second option is to add external parallel memory. Cost and performance are two major considerations when making the decision. The low-capacity parallel-memory chips are quite rare and expensive these days due to the low demand.

The C8051F040 allows the user to use the **MOVX** instruction together with the DPTR register to access external memory. It also allows the user to use either R0 or R1 as the pointer to specify the lower 8 bits of the address and the EMI0CN register to provide the page address. The user can choose to multiplex the lower 8 address bits with the data bits or to use separate pins to carry these two groups of signals (non-multiplexed mode). Using non-multiplexed mode requires the use of eight extra signal pins.

External memory expansion involves three issues: memory space assignment, address decoder design, and timing verification.

14.14 Exercise Problems

E14.1 Write an instruction sequence to store the value 59 in data memory at 0x4050, using DPTR as the pointer.

E14.2 Write an instruction sequence to store the value 59 in data memory at 0x4050, using R1 as the pointer. Place the upper eight address bits in the EMI0CN register.

E14.3 Give an instruction to configure EMI as
- EMIF active on P4-P7
- EMIF operates in multiplexed mode
- Split mode with bank select
- Set ALE high and low pulse width to 1 SYSCLK

E14.4 Give an instruction to configure EMI as
- EMIF active on P0-P3

- EMIF operates in multiplexed mode
- Split mode with no bank select
- Set ALE high and low pulse width to 2 SYSCLK

E14.5 Write a C program to perform the operation specified in Example 14.1.

E14.6 Write a C program to perform the operation specified in Example 14.2.

E14.7 Write an instruction sequence to configure EMI timing as
- EMIF active on P4-P7
- EMIF operates in multiplexed mode
- Split mode with bank select
- Set ALE high and low pulse width to 1 SYSCLK
- Address setup time = 1 SYSCLK
- /WR and /RD pulse width = 4 SYSCLK
- Address hold time = 1 SYSCLK

E14.8 Write an instruction sequence to configure EMI timing as
- EMIF active on P4-P7
- EMIF operates in non-multiplexed mode
- Split mode with bank select
- Set ALE high and low pulse width to 2 SYSCLK
- Address setup time = 0 SYSCLK
- /WR and /RD pulse width = 3 SYSCLK
- Address hold time = 0 SYSCLK

E14.9 For the circuit in Figure 14.36, write an instruction sequence to write the values 0 to 15 to the first 16 bytes of the AT28BV256.

E14.10 For the circuit in Figure 14.36, write an instruction sequence to write the values 2, 4, 6, and so on to the first 16 bytes of the CY7C1399BN.

14.15 Laboratory Exercise Problems and Assignments

L14.1 Write a program to erase the page located at 0x4000~0x41FF, store the string **The 8051 architecture has new implementations which prevents it from becoming obsolete in the 8-bit microcontroller arena** in this page, and then read it out and display the string on the terminal window (using the **HyperTerminal** program).

APPENDIX A

The 8051 Instruction Execution Times (Courtesy of Intel and Silabs)

Instruction		Operation	# of Bytes	# of Machine Cycles	
Arithmetic Operations				**Original 8051**	**C8051F040**
ADD	A,Rn	Add register to accumulator	1	1	1
ADD	A,direct	Add direct byte to accumulator	2	1	2
ADD	A,@Ri	Add indirect RAM to accumulator	1	1	2
ADD	A,#data	Add immediate to accumulator	2	1	2
ADDC	A,Rn	Add register to accumulator with carry	1	1	1
ADDC	A,direct	Add direct byte to accumulator with carry	2	1	2
ADDC	A,@Ri	Add indirect RAM to accumulator with carry	1	1	2
ADDC	A,#data	Add immediate to accumulator with carry	2	1	2
SUBB	A,Rn	Subtract register from accumulator with borrow	1	1	1
SUBB	A,direct	Subtract direct byte from accumulator with borrow	2	1	2
SUBB	A,@Ri	Subtract indirect RAM from accumulator with borrow	1	1	2
SUBB	A,#data	Subtract immediate from accumulator with borrow	2	1	2
INC	A	Increment accumulator	1	1	1
INC	Rn	Increment register	1	1	1
INC	direct	Increment direct byte	2	1	2
INC	@Ri	Increment indirect RAM	1	1	2
INC	DPTR	Increment data pointer	1	1	1
DEC	A	Decrement accumulator	1	1	1
DEC	Rn	Decrement register	1	1	2
DEC	direct	Decrement direct byte	2	1	2
DEC	@Ri	Decrement indirect RAM	1	2	1
MUL	AB	Multiply A and B	1	4	4
DIV	AB	Divide A by B	1	4	8
DA	A	Decimal adjust accumulator	1	1	1

Notes:
1. One machine cycle consists of 12 oscillator cycles in the original 8051.
2. One machine cycle consists of 1 oscillator cycle in the SiLabs C8051F040.
3. Ri refers to R0 or R1.
4. Rn refers to R0 to R7.

Instruction		Operation	# of Bytes	# of Machine Cycles	
				Original 8051	C8051F040
Logical Operations					
ANL	A,Rn	AND register to accumulator	1	1	1
ANL	A,direct	AND direct byte to accumulator	2	1	2
ANL	A,@Ri	AND indirect RAM to accumulator	1	1	2
ANL	A,#data	AND immediate data to accumulator	2	1	2
ANL	direct,A	AND accumulator to direct byte	2	1	2
ANL	direct,#data	AND immediate data to direct byte	3	2	3
ORL	A,Rn	OR register to accumulator	1	1	1
ORL	A,direct	OR direct byte to accumulator	2	1	2
ORL	A,@Ri	OR indirect RAM to accumulator	1	1	2
ORL	A,#data	OR immediate data to accumulator	2	1	2
ORL	direct,A	OR accumulator to direct byte	2	1	2
ORL	direct,#data	OR immediate data to direct byte	3	2	3
XRL	A,Rn	Exclusive OR register to accumulator	1	1	1
XRL	A,direct	Exclusive OR direct byte to accumulator	2	1	2
XRL	A,@Ri	Exclusive OR indirect RAM to accumulator	1	1	2
XRL	A,#data	Exclusive OR immediate to accumulator	2	1	2
XRL	direct,A	Exclusive OR accumulator to direct byte	2	1	2
XRL	direct,#data	Exclusive OR immediate to direct byte	3	2	3
CLR	A	Clear accumulator	1	1	1
CPL	A	Complement accumulator	1	1	1
RL	A	Rotate accumulator left	1	1	1
RLC	A	Rotate accumulator left through carry	1	1	1
RR	A	Rotate accumulator right	1	1	1
RRC	A	Rotate accumulator right through carry	1	1	1
SWAP	A	Swap nibbles within accumulator	1	1	1
Data Transfer					
MOV	A,Rn	Move register to accumulator	1	1	1
MOV	A,direct	Move direct byte to accumulator	2	1	2
MOV	A,@Ri	Move indirect RAM to accumulator	1	1	2
MOV	A,#data	Move immediate data to accumulator	2	1	2
MOV	Rn,A	Move accumulator to register	1	1	1
MOV	Rn,direct	Move direct byte to register	2	2	2
MOV	Rn,#data	Move immediate data to register	2	1	2
MOV	direct,A	Move accumulator to direct byte	2	1	2
MOV	direct,Rn	Move register to direct byte	2	2	2
MOV	direct,direct	Move direct byte to direct byte	3	2	3

(*continued*)

Instruction		Operation	# of Bytes	# of Machine Cycles	
Logical Operations				Original 8051	C8051F040
MOV	direct,@Ri	Move indirect RAM to direct byte	2	2	2
MOV	direct,#data	Move immediate data to direct byte	3	2	3
MOV	@Ri,A	Move A to indirect RAM	1	1	2
MOV	@Ri,direct	Move direct byte to indirect RAM	2	2	2
MOV	@Ri,#data	Move immediate to indirect RAM	2	1	2
MOV	DPTR,#data16	Load DPTR with 16-bit constant	3	2	3

Notes:
1. One machine cycle consists of 12 oscillator cycles in the original 8051.
2. One machine cycle consists of 1 oscillator cycle in the SiLabs C8051F040.
3. Ri refers to R0 or R1.
4. Rn refers to R0 to R7.

Instruction		Operation	# of Bytes	# of Machine Cycles	
				Original 8051	C8051F040
Data Transfer					
MOVC	A,@A+DPTR	Move code byte relative to DPTR to A	1	2	3
MOVC	A,@A+PC	Move code byte relative to PC to A	1	2	3
MOVX	A,@Ri	Move external RAM byte (8-bit addr.) to A	1	2	3
MOVX	A,@DPTR	Move external RAM byte (16-bit addr.) to A	1	2	3
MOVX	@Ri,A	Move A to external RAM (8-bite addr.)	1	2	3
MOVX	@DPTR,A	Move A to external RAM (16-bit addr.)	1	2	3
PUSH	direct	Push direct byte to stack	2	2	2
POP	direct	Pop direct byte from stack	2	2	2
XCH	A,Rn	Exchange register with A	1	1	1
XCH	A,direct	Exchange direct byte with A	2	1	2
XCH	A,@Ri	Exchange indirect RAM with A	1	1	2
XCHD	A,@Ri	Exchange low order digit of indirect RAM with A	1	1	2
Boolean Variable Manipulation					
CLR	C	Clear carry	1	1	1
CLR	bit	Clear direct bit	2	1	2
SETB	C	Set carry	1	1	1
SETB	bit	Set direct bit	2	1	2
CPL	C	Complement carry	1	1	1
CPL	bit	Complement direct bit	2	1	2
ANL	C,bit	AND direct bit with carry	2	2	2
ANL	C,/bit	AND complement of direct bit with carry	2	2	2
ORL	C,bit	OR direct bit with carry	2	2	2
ORL	C,/bit	OR complement of direct bit with carry	2	2	2
MOV	C,bit	Move direct bit to carry	2	1	2
MOV	bit,C	Move carry to direct bit	2	2	2
JC	rel	Jump if carry is set	2	2	2/3[5]
JNC	rel	Jump if carry is not set	2	2	2/3[5]
JB	bit,rel	Jump if direct bit is set	3	2	3/4[5]
JNB	bit,rel	Jump if direct bit is not set	3	2	3/4[5]
JBC	bit,rel	Jump if direct bit is set and clear it	3	2	3/4[5]

Notes:

1. One machine cycle consists of 12 oscillator cycles in the original 8051.
2. One machine cycle consists of 1 oscillator cycle in the SiLabs C8051F040.
3. Ri refers to R0 or R1.
4. Rn refers to R0 to R7.
5. The execution time for a untaken jump is to the left of the slash(/). The execution time for a taken jump is to the right of the slash.

Instruction		Operation	# of Bytes	# of Machine Cycles	
Program Branching				**Original 8051**	**C8051F040**
ACALL	Addr11	Absolute subroutine call	2	2	3
LCALL	Addr16	Long subroutine call	3	2	4
RET		Return from subroutine	1	2	5
RETI		Return from interrupt	1	2	5
AJMP	Addr11	Absolute jump	2	2	3
LJMP	Addr16	Long jump	3	2	4
SJMP	rel	Short jump (relative address)	2	2	3
JMP	@A+DPTR	Jump indirect relative to DPTR	1	2	3
JZ	rel	Jump if accumulator is zero	2	2	2/3[5]
JNZ	rel	Jump if accumulator is not zero	2	2	2/3[5]
CJNE	A,direct,rel	Compare direct byte to A and jump if not equal	3	2	3/4[5]
CJNE	A,#data,rel	Compare immediate to A and jump if not equal	3	2	3/4[5]
CJNE	Rn,#data,rel	Compare immediate to Rn and jump if not equal	3	2	3/4[5]
CJNE	@Ri,#data,rel	Compare immediate to indirect RAM and jump if not equal	3	2	4/5[5]
DJNZ	Rn,rel	Decrement register and jump if not equal to zero	2	2	2/3[5]
DJNZ	Direct,rel	Decrement direct byte and jump if not equal to zero	3	2	3/4[5]
NOP		No operation	1	1	1

Notes:

1. One machine cycle consists of 12 oscillator cycles in the original 8051.
2. One machine cycle consists of 1 oscillator cycle in the Silabs C8051F040.
3. addr11 is a11-bit address.
4. addr16 is a 16-bit address.
5. The execution time for a untaken jump is to the left of the slash(/). The execution time for a taken jump is to the right of the slash.

APPENDIX B

Tutorial for Using the Keil's μVision IDE

Keil's μVision IDE is one of the most popular commercial IDEs for the 8051 variants. Like many other IDEs, μVision uses a project as the unit for managing software programs. We will illustrate the procedure for building (including compiling, assembling, and linking) a project using both the assembly and C programs in this tutorial.

B.1 Procedure for Assembly Program Development

We will use the bubble sort subroutine and its test program from Example 4.14 to illustrate the process. The procedure is as follows.

Step 1

Start the μVision IDE by clicking on its icon. After the μVision is started, a window similar to that shown in Figure B.1 will appear.

The project workspace is used to display the project name and the names of files contained in the project. The output window is used to output the messages related to compiling (or assembling) and linking processes. It also allows the user to enter commands to the μVision IDE. The Editing workspace is used by the editor for entering programs.

Step 2

Create a New Project. This is done by pressing the **Project** menu and select **New Project . . .** (shown in Figure B.2). A popup dialog box similar to that shown in Figure B.3 will appear. Enter the project name of your choice (let's use bubble as the project name) then click on **Save**, and the dialog box for the new project will disappear. Another new popup dialog box (as shown in Figure B.4)

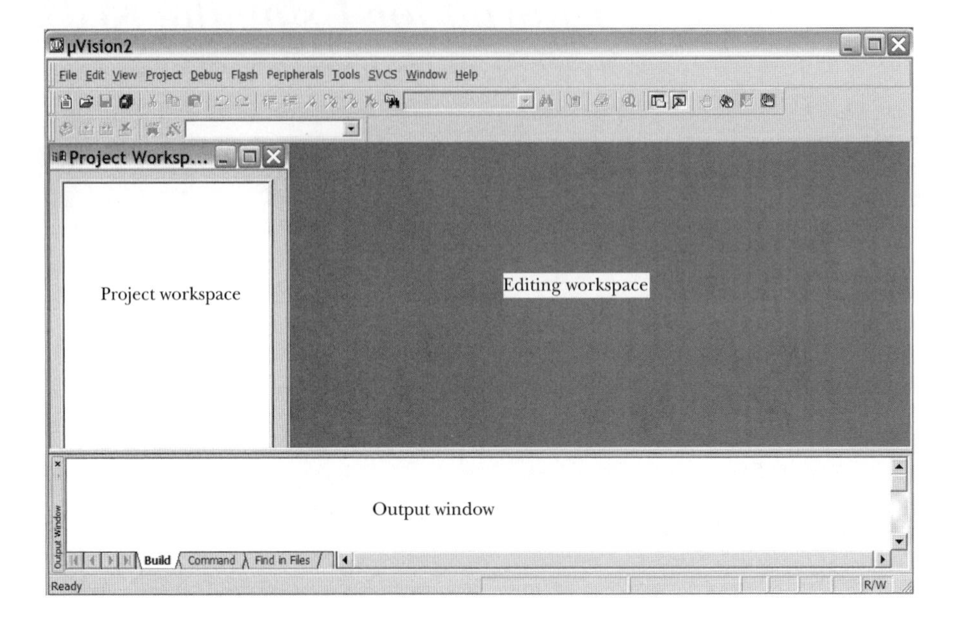

FIGURE B.1 µVision window screen.

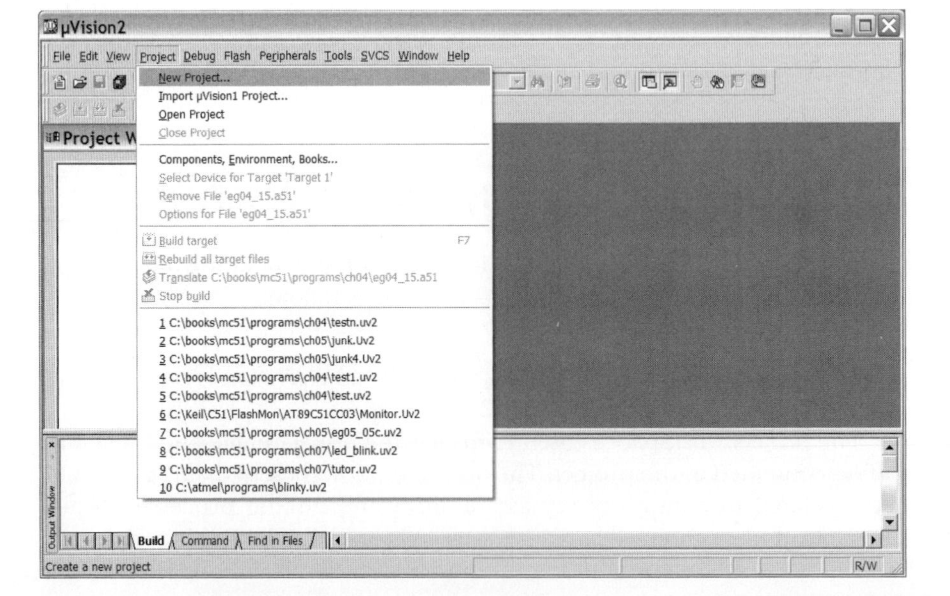

FIGURE B.2 Menu selection for creating a new project.

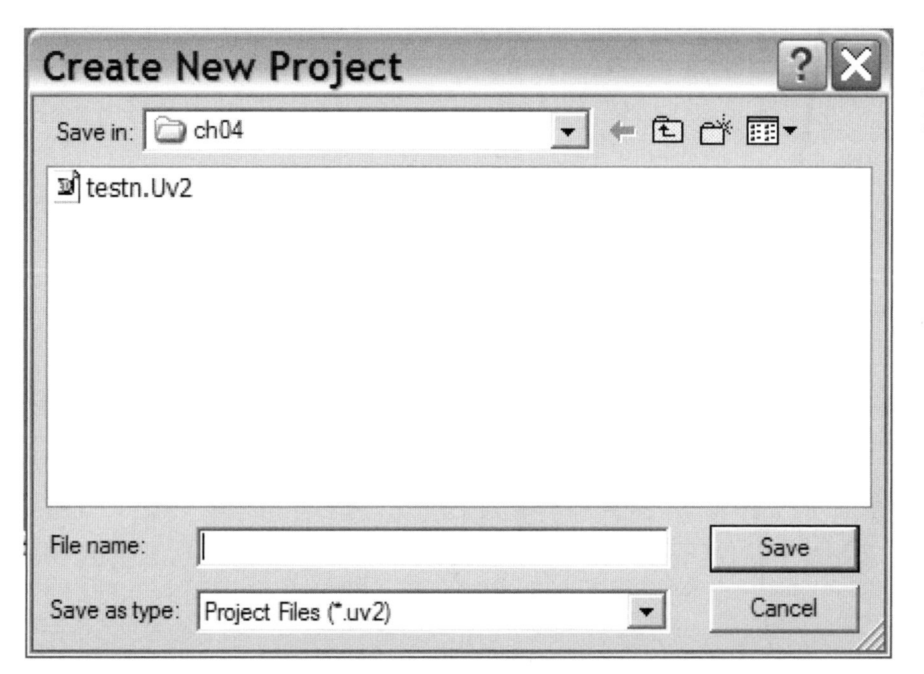

FIGURE B.3 Dialog box for creating a new project.

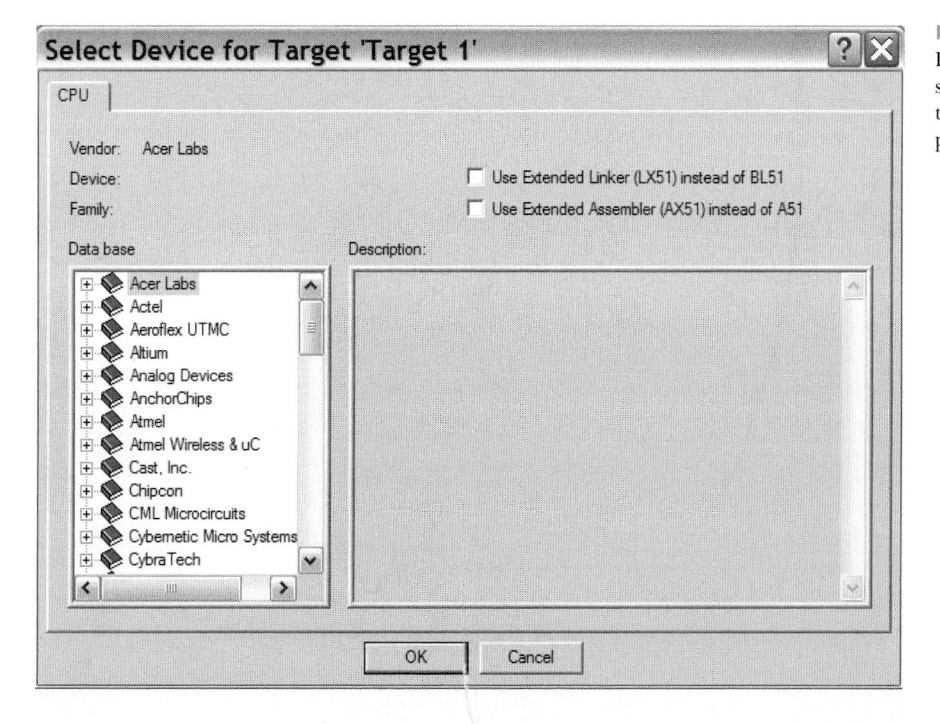

FIGURE B.4 Dialog box for selecting a device to be used in the project.

will appear for you to select the device for the project. Select **Atmel** and then select **T89C51AC2** (or other company's device) as the target device (as shown in Figure B.5). After selecting the target device, a brief description about the chosen device will appear in the Description window, as shown in Figure B.5. Click on **OK**. The μVision IDE will then ask you if you want to add a startup code (shown in Figure B.6) to your project. Since this is an assembly program, click on **No** and the dialog box disappears.

Step 3

Add the Assembly Program to the Project. There are two possibilities here. The first possibility is that the program does not exist and must be entered as a new file. The second possibility is that the program already existed and needs to be added to the project. To enter a new program, press the **File** menu and select

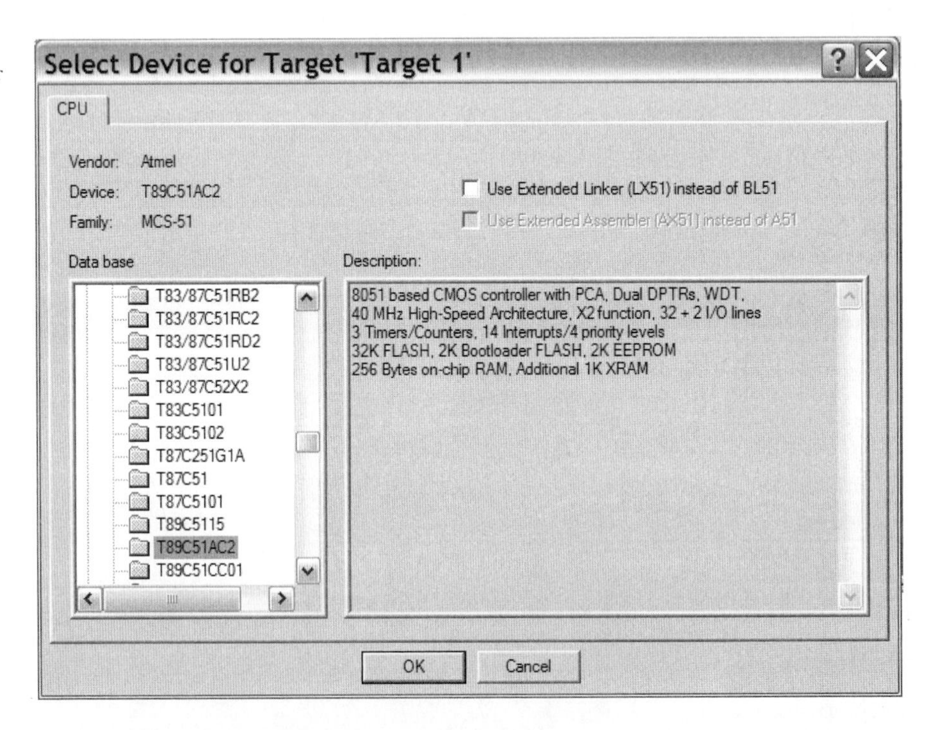

FIGURE B.5 Selecting a device for implementing the project.

FIGURE B.6 Dialog for adding the startup code to the project.

New. The resultant screen is shown in Figure B.7. You may want to resize the **Text2** window before entering the program. The screen after entering and saving the **bubble.a51** (or **bubble.asm**) program is shown in Figure B.8.

Adding **bubble.a51** or an existing assembly program to the project is achieved by clicking on **Source Group 1** in the Project Worksp . . . window and then press the right mouse button to select **Add Files to 'Source Group 1'**.

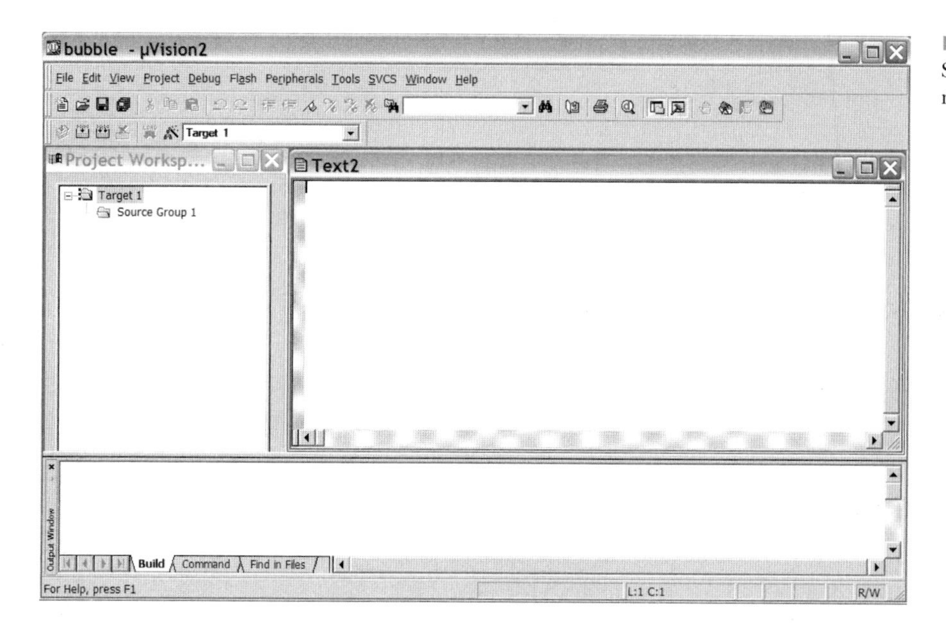

FIGURE B.7 Screen for entering a new program file.

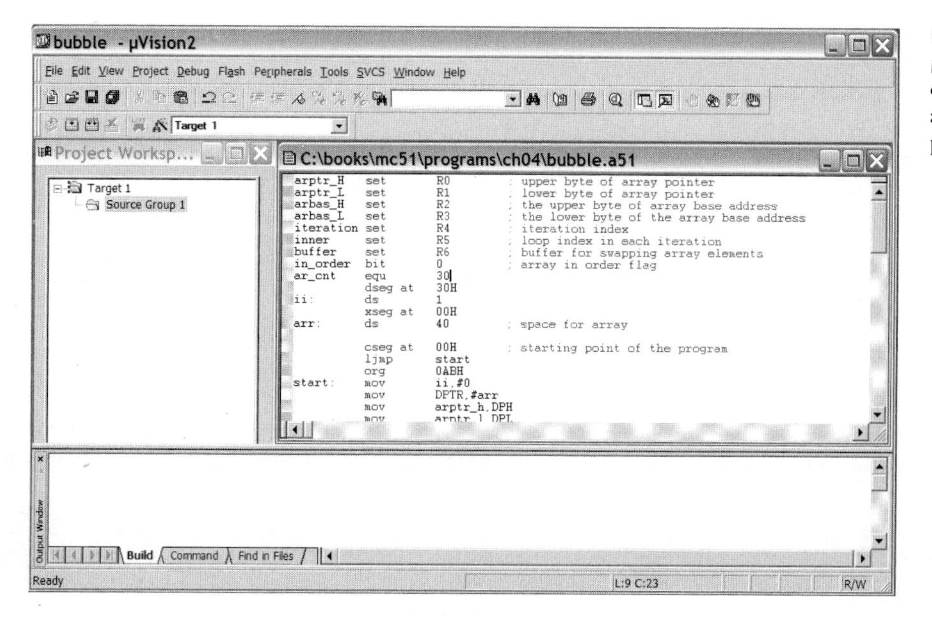

FIGURE B.8 µVision screen after entering and saving a new assembly program.

After this, a dialog box as shown in Figure B.9 will appear for you to enter the file name to be added. Enter **bubble.a51** in the blank and click on **Add** and then **Close** and the screen will be changed one similar to Figure B.10.

Sometimes, you may want to delete a file from the project. This can be achieved by pressing the right mouse button on the file name (inside the

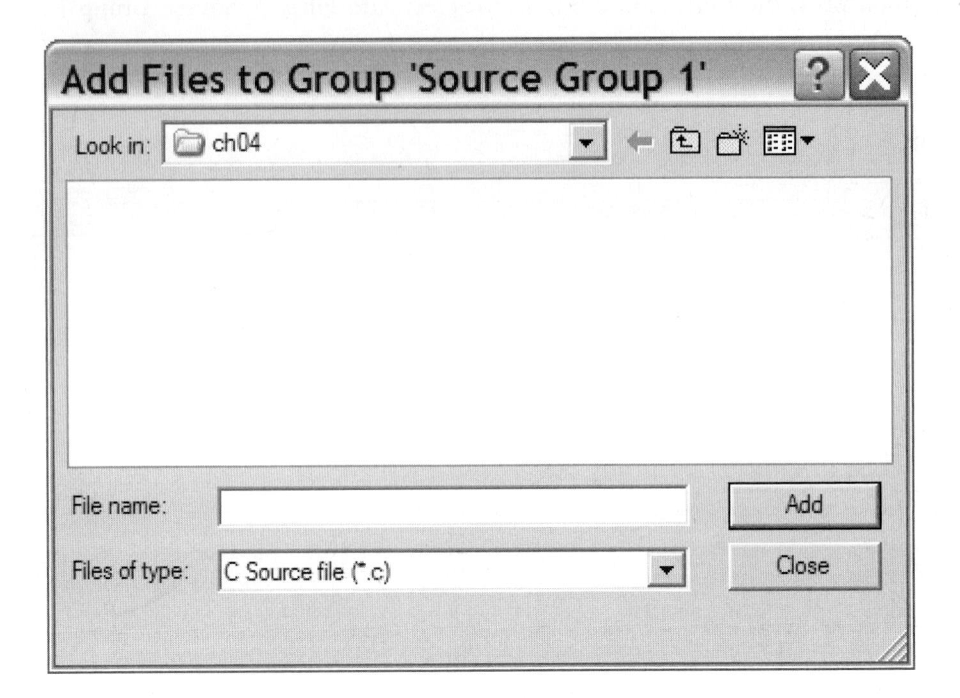

FIGURE B.9 Dialog box for adding files to a project.

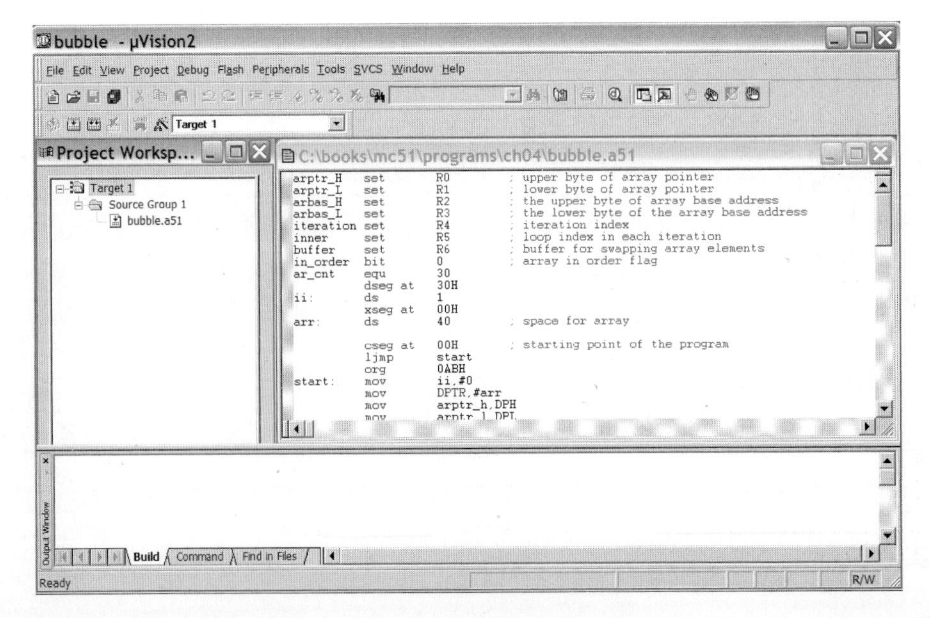

FIGURE B.10 µVision screen after adding a file to the project.

project workspace) to be deleted from the project (shown in Figure B.11) and select **Remove File 'bubble.a51'** and the file will be deleted from the project. Of course, you don't want to delete bubble.a51 from this project.

Step 4

Compile and Link the Program. This step is called **build the project** and is done by pressing on the **Project** menu and select **Build target** (as shown in Figure B.12). The result of **Build target** will be displayed in the **Output window**. The bubble.a51 program has no syntax error, and the output message shows it in Figure B.13. The Build result message also indicates the amount (in bytes) of data memory, XDATA memory, and program memory used by the project.

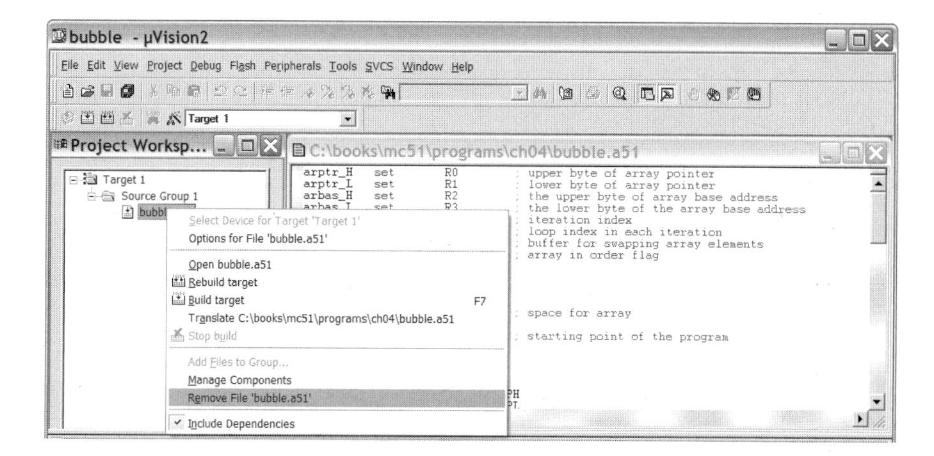

FIGURE B.11 Delete a file from the project (press the right mouse button on the file name).

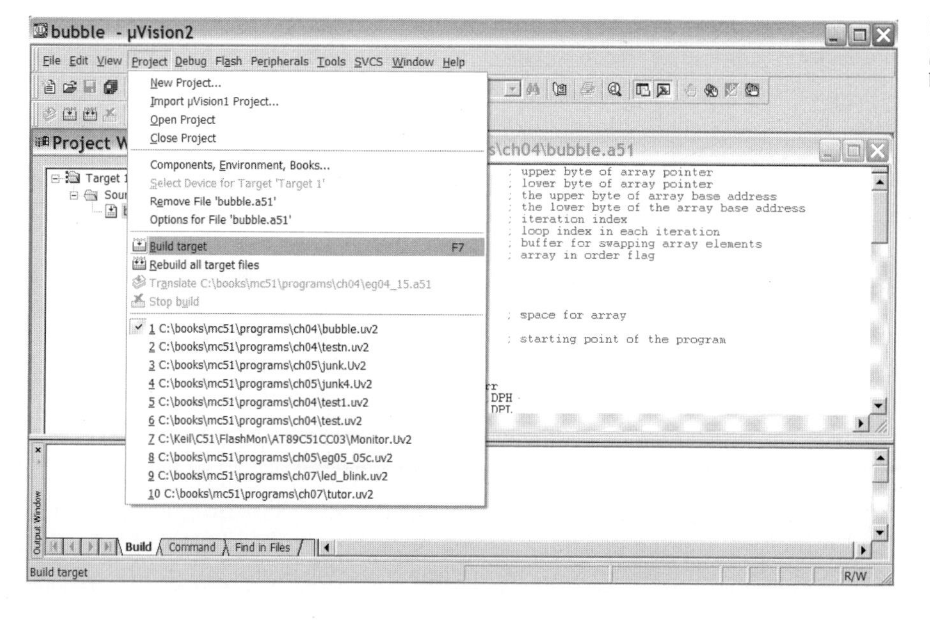

FIGURE B.12 μVision screen for building a project.

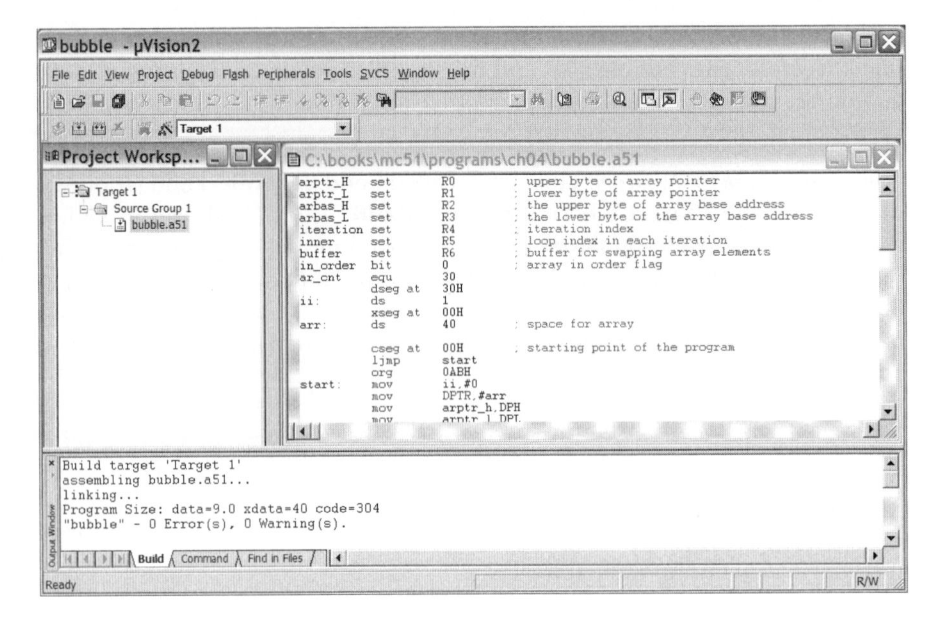

FIGURE B.13 Build target result message for a build without errors.

If there are any syntax errors, then you will need to fix them before the program can be tested.

Step 5

Debug the Program. The **bubble.a51** program is written to test the bubble sort subroutine. The bubble sort subroutine sorts an array of N 8-bit numbers. The main program first copies the array from the program memory to the XDATA memory and leaves the sorted array in the XDATA memory. The loop for copying the array starts from the instruction with the label **scopylp** and ends with the **cjne A,#ar_cnt,scopylp** instruction.

Before debugging the program, you need to start the debug session. The debug session is started by pressing the **Debug** menu and selecting **Start/Stop Debug Session** from the menu. After this, the screen changes to that in Figure B.14.

We need to display the values of certain related MCU resources in order to determine whether the program execution results are correct. The MCU resources that can be displayed are listed in the View menu, as shown in Figure B.15. To display any resource, select it (by double clicking on the resource name) from this menu. You may want to resize the µVision to get a better view. The Watch window and Memory window are brought out in Figure B.15.

Enter Program Variables in the Watch Window

The Watch window allows the user to enter the program variables and watch the change of their values after program execution or at a breakpoint. Two Watch windows are available for displaying the values of program variables. This is useful when the program has many variables to be displayed.

To place a program variable in the watch list, click the left mouse button on any character of the variable and press the right mouse button, and a

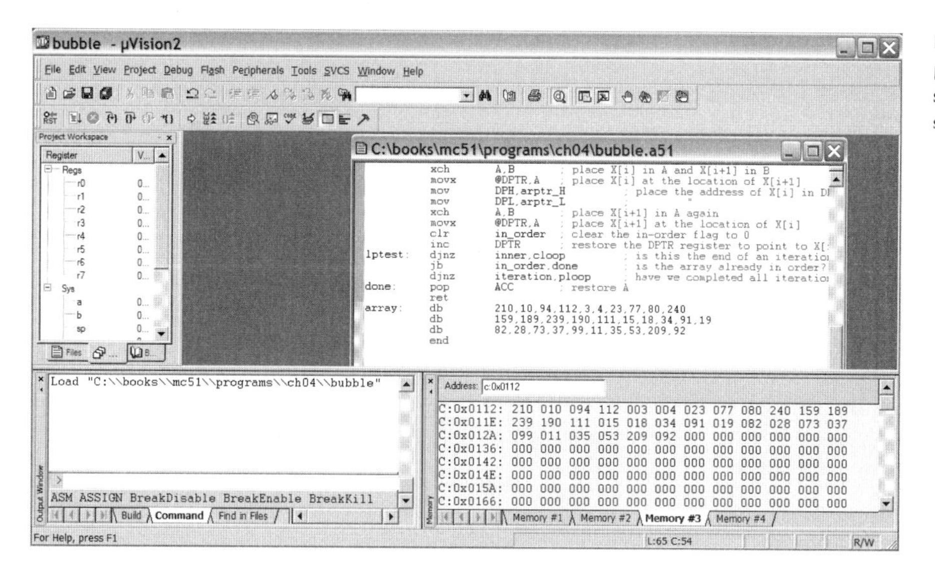

FIGURE B.14
μVision screen after starting a debug session.

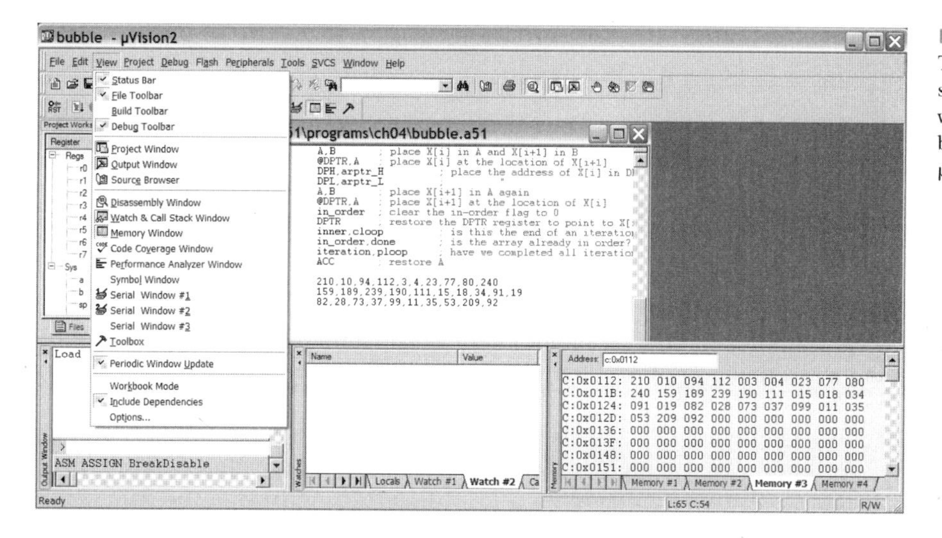

FIGURE B.15
The View menu shows all the windows that can be displayed in a μVision screen.

popup menu will be brought up for you to select the Watch window to enter the program variable. For example, Figure B.16 illustrates the procedure for entering variable **ii** into Watch window #1. Since there are two Watch windows, you need to click on the one that you want to examine. After entering variable **ii** into Watch window #1 and making it active by clicking on **Watch #1**, the Watch window changes to that in Figure B.17.

Configure Memory Window

The Memory window can be used to display four different memory areas. Since we have four different types of memory (data, XDATA, program, and indirect data), it is logical to use one memory area to display the contents of

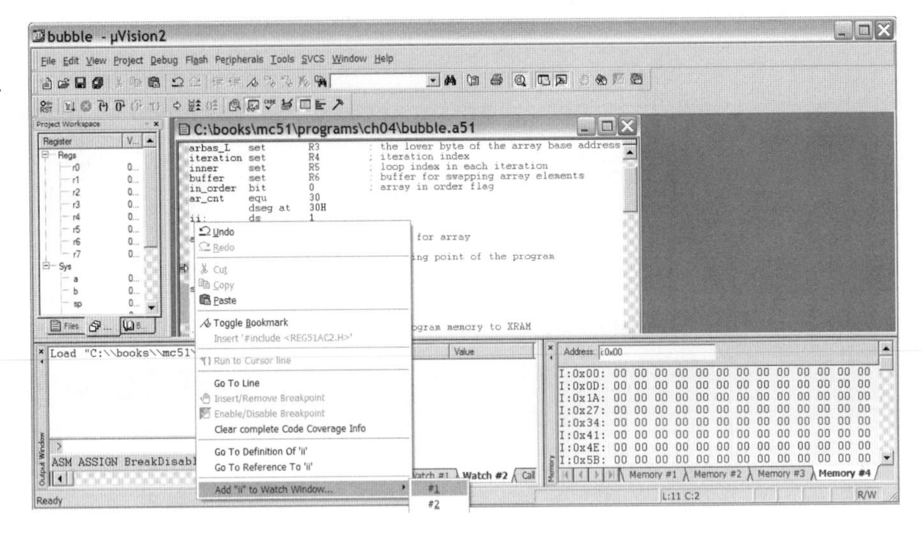

FIGURE B.16 Enter variable ii to the Watch window #1.

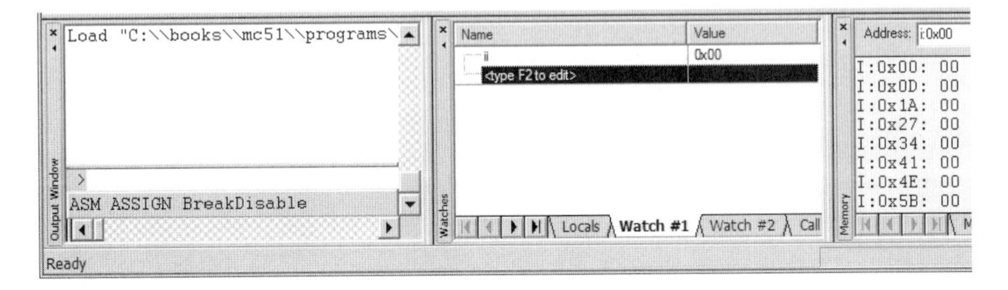

FIGURE B.17 The contents of Watch window #1 after entering variable ii.

one type of memory. To use Memory #1, #2, #3, and #4 to display the contents of data memory, XDATA memory, program memory, and indirect data memory, respectively, take the following actions.

- Click on Memory #1 and enter D:0x00 in the space to the right of Address on the top of the Memory window. The letter D specifies data memory whereas the number 0x00 specifies the starting address of the data memory area to be displayed. The starting address need not be 0x00.
- Click on Memory #2 and enter X:0x0000 in the space to the right of Address on top of the Memory window. The letter X specifies XDATA memory and external data memory, whereas the number 0x0000 specifies the starting address of the XDATA memory area to be displayed. The starting address can be any other value.
- Click on Memory #3 and enter C:0x0000 in the space to the right of Address on top of the Memory window. The letter C selects the program memory, whereas the value 0x0000 sets the starting address of the program memory area to be displayed. For this example, the starting address of **array** can be found by looking at the Symbols window. The Symbols window can be displayed by pressing on the View menu and select Symbol

window. The Symbols window is shown in Figure B.18. The starting address of every symbol (variable) is displayed in this window. The array to be sorted is found to be stored in program memory starting from 0x0112. It would be more convenient to set the starting address of the program memory to be 0x0112 instead of 0x0000. The Symbols window can be set to other modes, as shown in Figure B.19. The Lines mode tells you the starting address of each line or statement (in C language).

- Click on Memory #4 and enter I:0x00 in the space to the right of Address on top of the Memory window. The letter I specifies the indirect data memory, whereas the value 0x00 sets the starting address of the indirect data memory area to be displayed.

To make the contents of a specific memory area appear on the memory window, click on the corresponding memory number (from 1 to 4).

Set a Breakpoint

Breakpoints should be set at those instructions that you want to check the program execution result. For this tutorial, we choose to set a breakpoint (at the mov DPTR,#arr instruction) after the array has been copied from program memory (at 0x112) to XDATA memory (starting from 0x0000) and a breakpoint after the bubble subroutine has been called (the jmp $ instruction). A breakpoint can be set by clicking (left mouse button) on the breakpoint instruction, then pressing the right mouse button, and then selecting Insert/Remove Breakpoint, as shown in Figure B.20. After this, a red rectangle will be attached to the left of the breakpoint instruction.

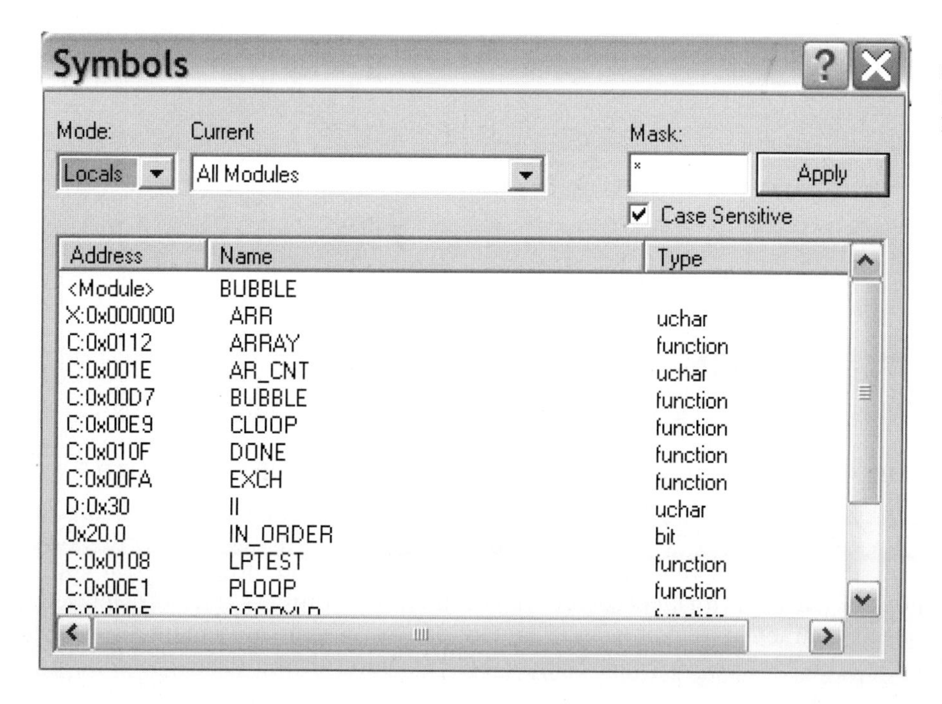

FIGURE B.18
Symbols window with locals mode set to **Locals**.

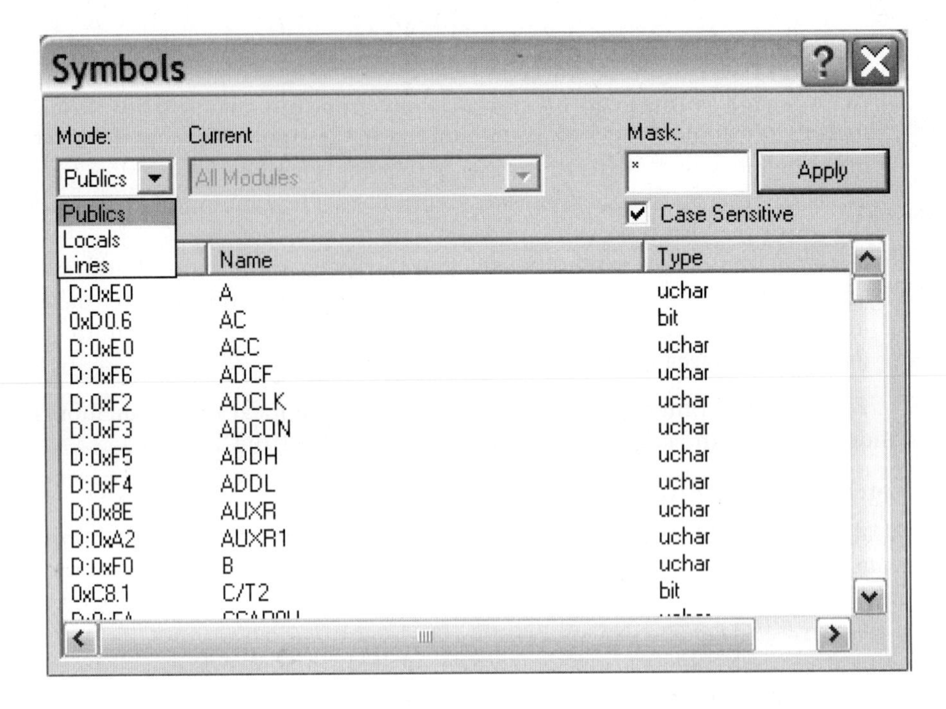

FIGURE B.19 The modes of the Symbols window.

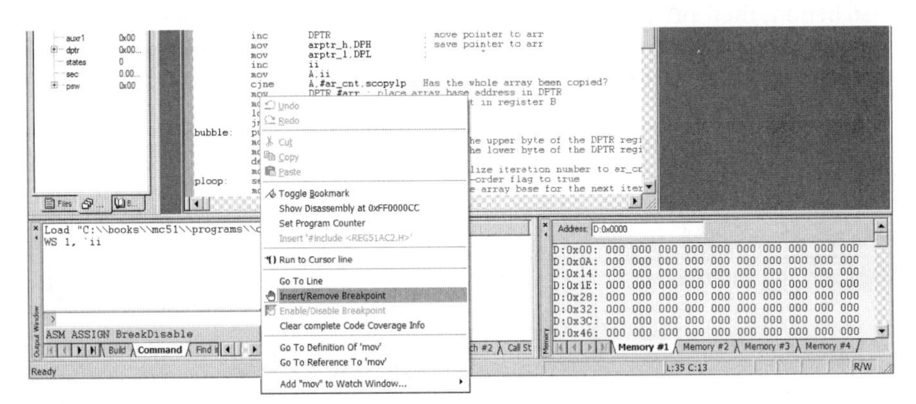

FIGURE B.20 Setting a breakpoint at the instruction **"mov DPTR, # arr"**.

Execute the Program

When first executing a program, you want to make sure that the program is run from the start of the program. This can be achieved by resetting the MCU. The MCU is reset by pressing the **RST** button on the toolbar, as shown in Figure B.21. To execute the program, press the function key F5 or the button next to the RST button. The program will stop at the first breakpoint. You can then examine the XDATA memory contents to find out if the array copy operation is performed correctly. The memory contents can be displayed using characters (8 bits) or integers (16 bits) as the unit and can be displayed in decimal or hexadecimal format. The selection of the format is done by

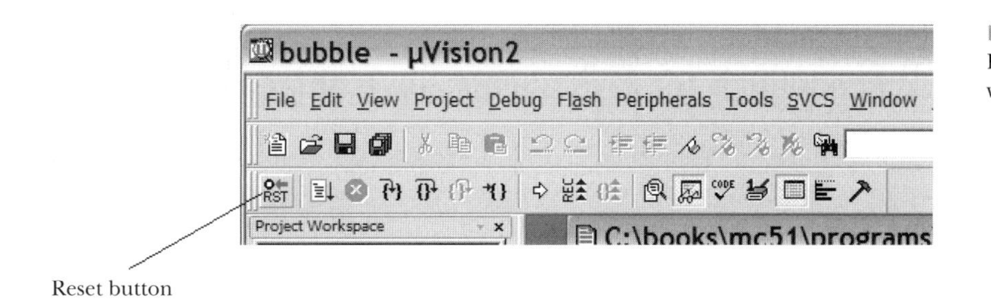

FIGURE B.21
Press the RST button will reset the MCU.

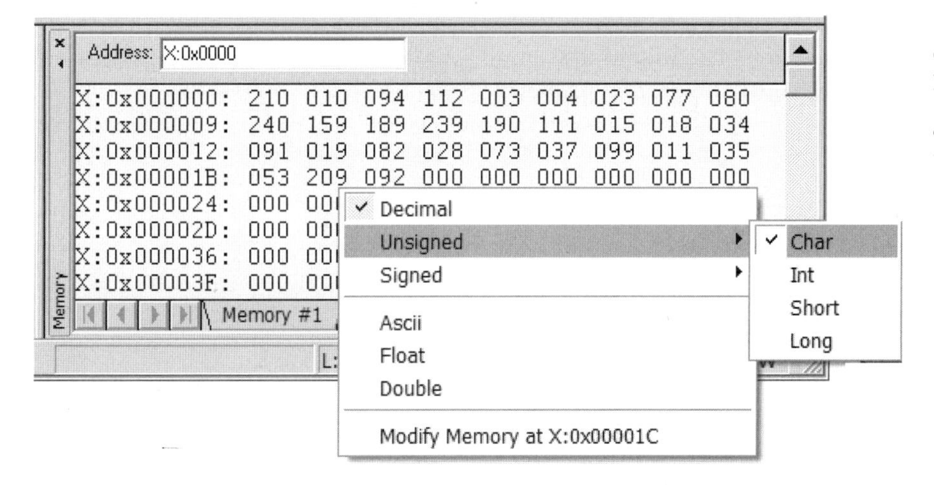

FIGURE B.22
Selecting the appropriate display format (one character and decimal format in this example).

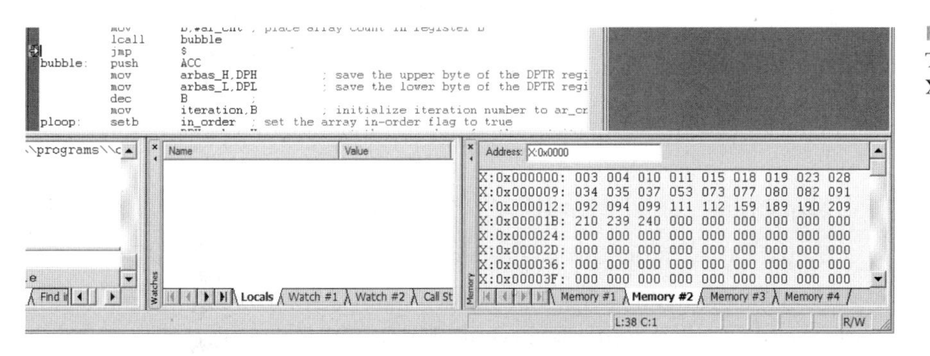

FIGURE B.23
The sorted array in XDATA memory.

pressing the right mouse button inside the Memory window and then choosing the appropriate items, as shown in Figure B.22. It is easy to verify that the copy operation is successful.

Set another breakpoint at the **jmp $** instruction, rerun the program, and check the XDATA memory. The resultant contents of the XDATA memory are shown in Figure B.23 and are correct.

This concludes the tutorial for assembly program development.

B.2 Procedure for C Program Development

The procedure for debugging C programs using the Keil μVision is similar to that for assembly programs. We will use one example to demonstrate the process for C program development.

Example B.1 Write a C program that finds three four-digit integers with the following property:

The square of the sum of the first two digits and the last two digits equals the original number.

Solution: The following C program will find three four-digit integers that have the desired property.

```
int main (void)
{
        char cnt;
        unsigned int ix,iy,iz,arr[10];

        cnt = 0;
        for (ix = 1000; cnt < 3; ix++) {
            iy = ix / 100;
            iz = ix % 100;
            iy += iz;
            if (iy * iy == ix)
                arr[cnt++] = ix;
        }
        return 0;
}
```

The procedure for debugging this program using μVision IDE is as follows.

Step 1 Start the μVision IDE.

Step 2 Create a new project called appF01.

Step 3 Create a new file with the file name appF01.c to hold the C program and add it to the project appF01.

Step 4 Build the project. If there are any syntax errors, fix them. When all the syntax errors have been corrected, the resultant μVision screen should be as shown in Figure B.24.

Step 5 Start the debug session by selection the menu item **Debug->Start/Stop Debug Session**. The resultant screen is shown in Figure B.25. The Watch window and Memory window have been shown in the screen. We also activate the local variables in the Watch window. Since we use a local variable array (arr[]) to hold the result, it is dis-

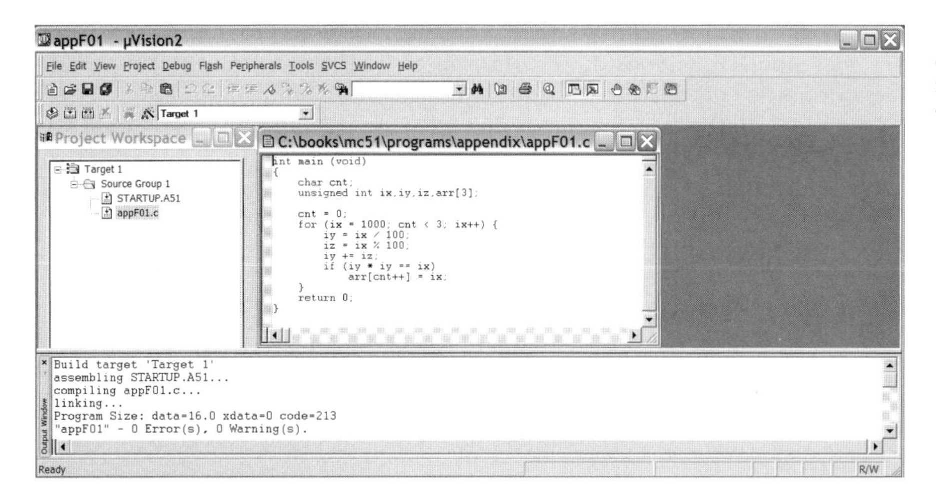

FIGURE B.24 Screen for a correctly compiled C program **(appF01.c)**.

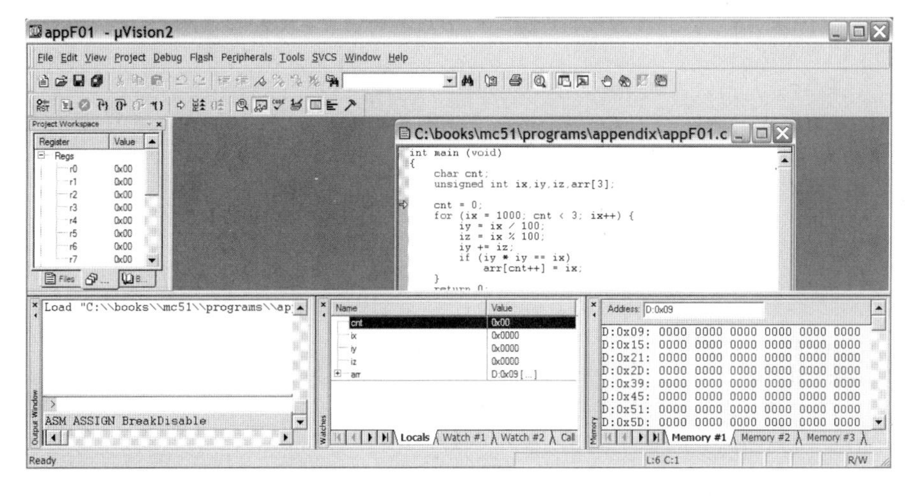

FIGURE B.25 The resultant screen after starting the debug session for the project **appF01**.

played in this window. Figure B.25 shows that the array arr[] is allocated in data memory and its base address is 0x09. We use Memory #1 to hold the contents of data memory and set the starting address of this area to 0x09, same as the starting address of the arr[] array. The display format for Memory #1 is set to decimal format with its unit set to integer (INT).

Step 6 Set a breakpoint at the statement **return 0** by pressing the right mouse button on this statement and select **Insert/Remove Breakpoint**. The resultant screen is shown in Figure B.26. Run the program, and the program control will stop at the breakpoint. The first three integers (2025, 3025, and 9801) in the Memory window satisfy the requirement.

As demonstrated in this example, local variables (declared inside **main** ()) are added automatically into the Watch window. You don't need to insert them. However, global variables are not added to the

Watch window, and you will need to add them manually. The Keil µVision IDE cannot display the contents of an array in the Watch window; you will need to display them using the Memory window or insert them into the Watch window one element at a time. In order to display an array in the Memory window, you will need to find out the array base address. The array base address can be found in the Symbols window (shown in Figure B.27). This can be done by press the View menu and select **Symbol Window**, as shown in Figures B.18 and B.19. To display

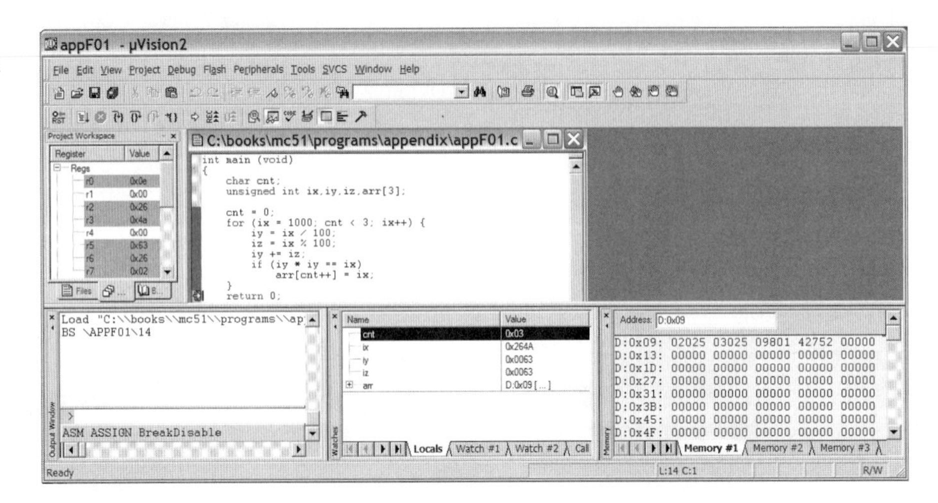

FIGURE B.26
The resultant screen after executing the appF01 program.

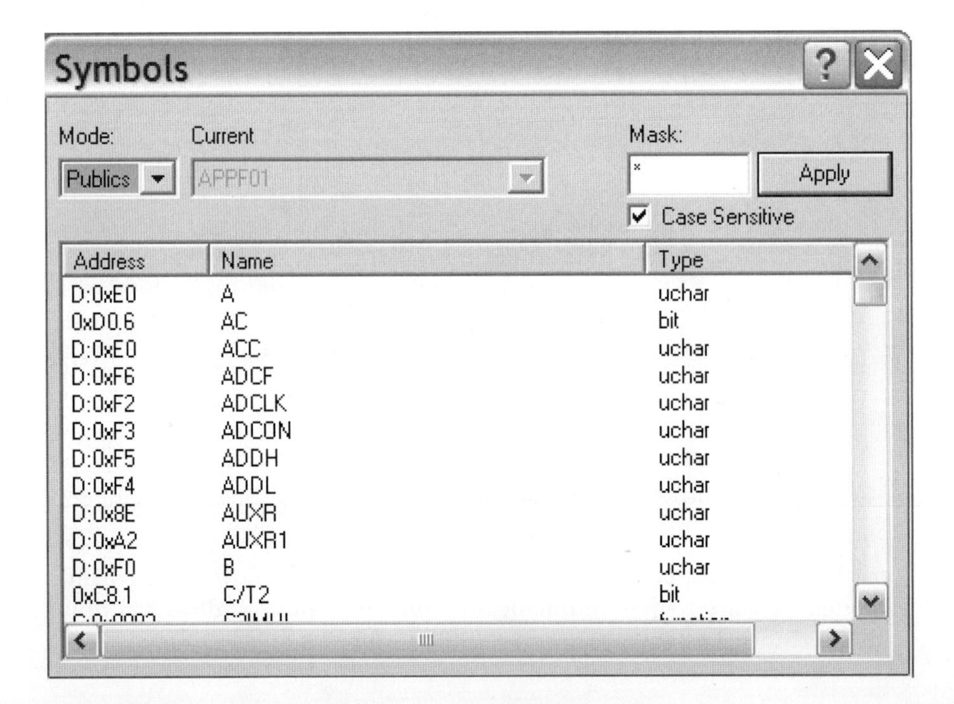

FIGURE B.27
Symbols window.

global variables, you need to set the mode to **Public** and select the project name (APPF01) in the box under **Current**. The names of global variables can be found by scrolling the scroll bar on the right side of the Symbols window. Since this program does not have any global variables, you don't need to do this step.

This completes the tutorial for using μVision to debug a C program.

B.3 Using μVision and SiLabs EC2 Adapter to Perform Source Debugging on SSE040 and SiLabs Target Kits

SiLabs provides a driver of EC2 for μVision 2 and μVision 3, which allows the user to perform source level debugging using μVision on any demo board that incorporates the 8051 MCU from SiLabs. The procedure to use this driver is as follows:

Step 1
Download and install the EC2 driver for μVision 2 or μVision 3 depending on which version you have and install it. You will be asked where to install during this process. Enter the installed directory of μVision 3 (for example, **c:\keil**). This step needs to be done only once.

Whenever you are creating a new project, take the following steps.

Step 2
After adding program files into the project, select **Project=>Options for File** 'file name' (as shown in Figure B.28) and a screen as shown in Figure B.29 appears. Click on **OK** to dismiss this screen.

Step 3
Select **Project=>Options for target 'Target 1'** to bring up the screen shown in Figure B.30. Click on **Debug**, and the screen will change to that shown in Figure B.31.

Step 4
Select **Silicon Laboratories C8051Fxxx μVision Driver** by pressing the downward arrow and release it (shown in Figure B.32). Click on the circular space to the left of **Use**.

Step 5
Click on **Setting**, and a popup dialog similar to that shown in Figure B.33 will appear. Select either **RS232 Serial Adapter** or **USB Debug Adapter** (depending on the type of debug adapter you are using) and click **OK** twice to dismiss all popup dialogs.

After this, you can start to debug your program directly on the target hardware. The process is similar to the simulator approach.

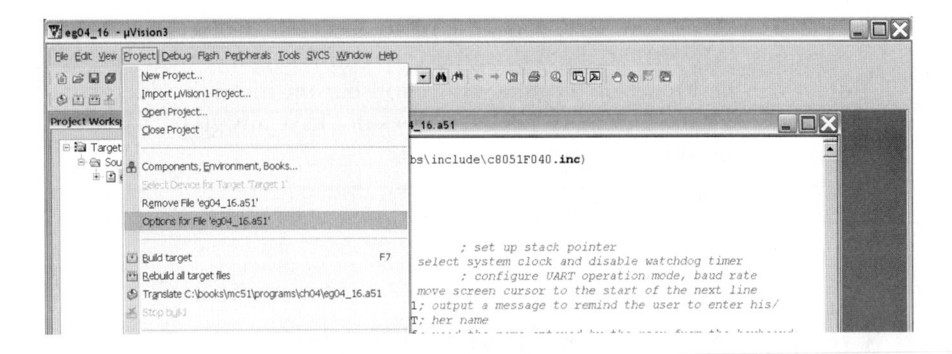

FIGURE B.28 Select project options for project.

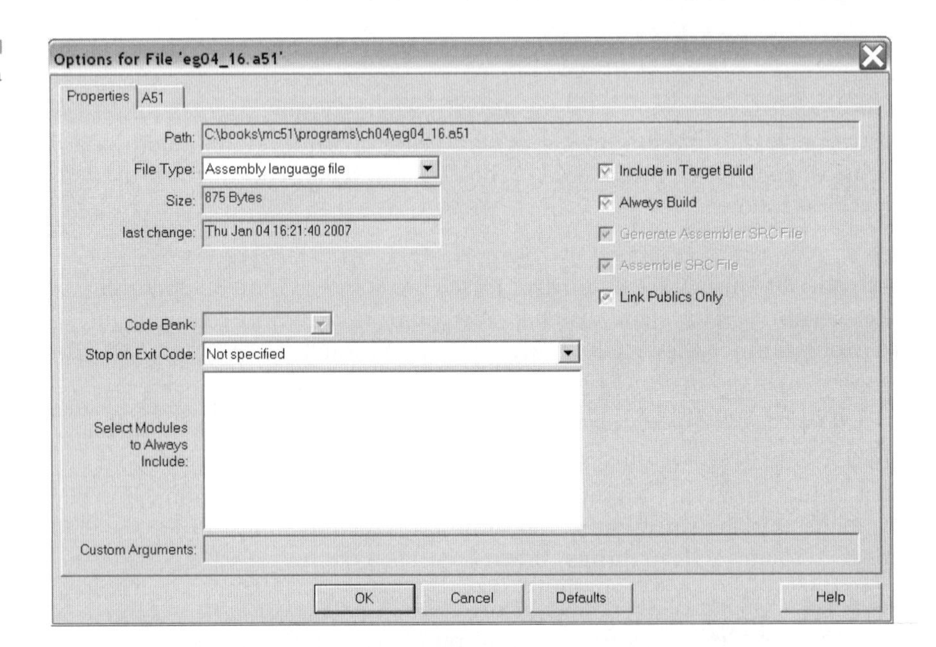

FIGURE B.29 Option screen for a new project.

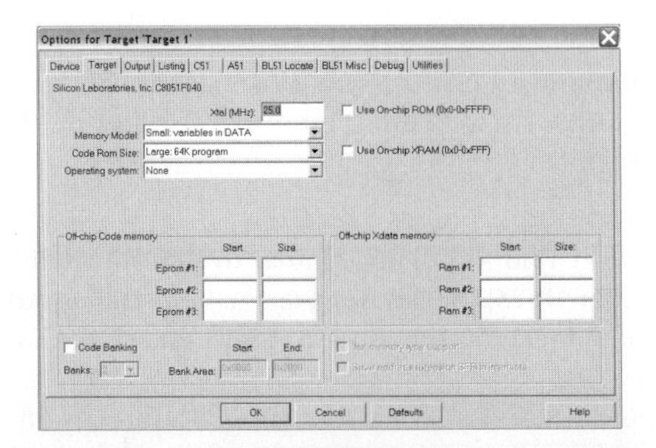

FIGURE B.30 µVision screen for setting target options for the project.

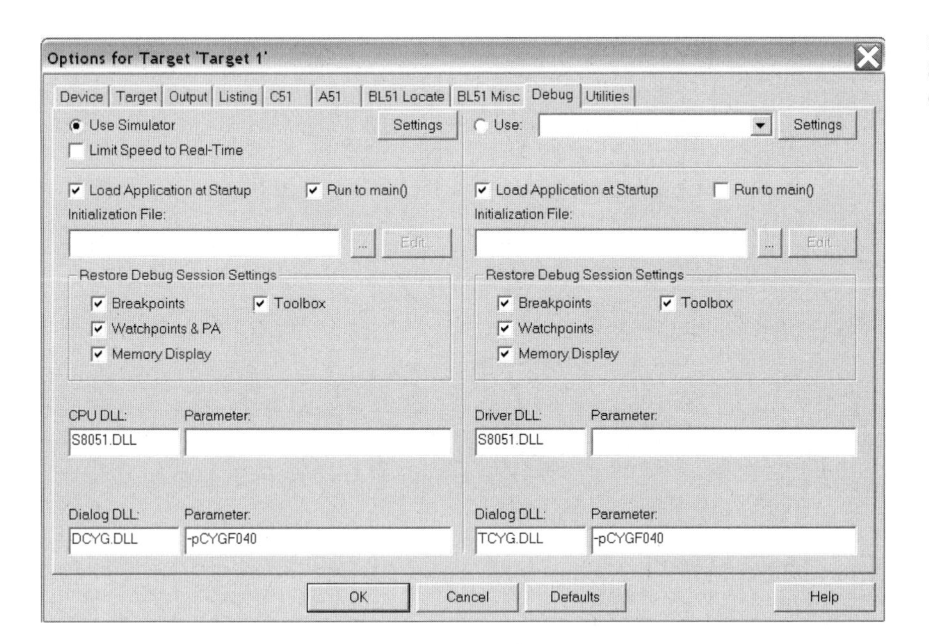

FIGURE B.31 Screen for setting debug options.

FIGURE B.32 Select Silicon Laboratories C8051Fxxx µVision Driver.

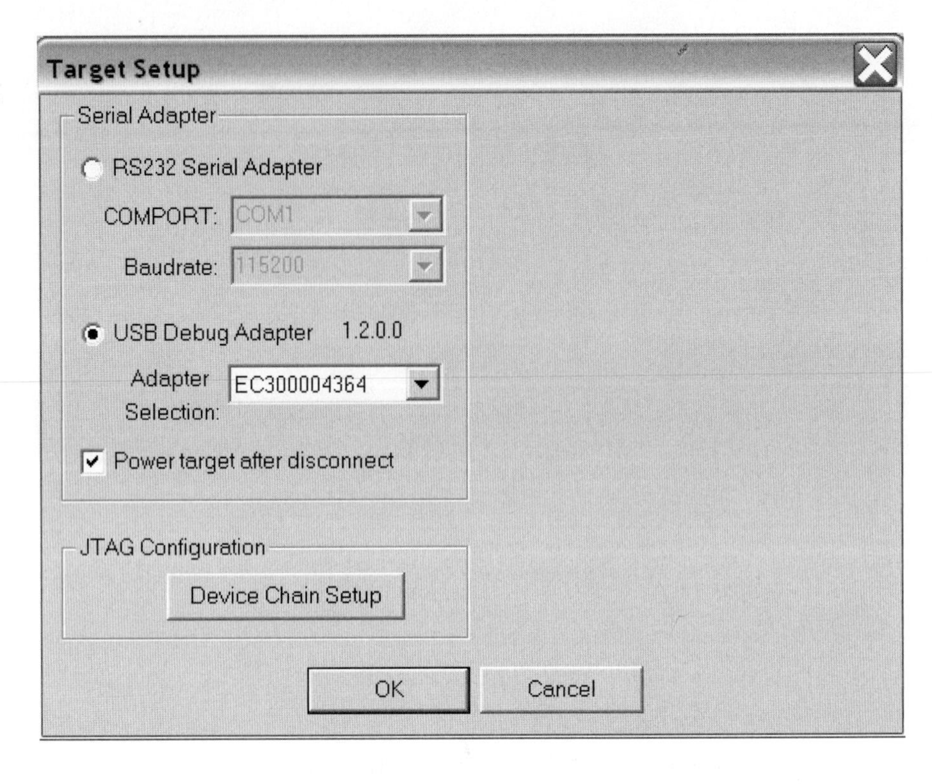

FIGURE B.33 Select **USB Debug Adapter** for µVision.

APPENDIX C

Tutorial for Using the Raisonance's Ride IDE

Raisonance's RIDE is another popular IDE for the 8051 variants. Like many other IDEs, RIDE uses a project as the unit for managing software programs. We will illustrate the procedure for building (including compiling, assembling and linking) a project using both the assembly and C programs in this tutorial?

C.1 Procedure for Assembly Program Development

We will use the bubble sort subroutine and its test program to illustrate the process. The procedure is as follows.

Step 1

Start the RIDE by clicking on its icon. After the RIDE is started, a window will appear as shown in Figure C.1.

The project window is used to display the project name and the names of files contained in the project. The Output window is used to output the messages related to compiling, assembling, and linking processes. The user also can enter commands to RIDE. However, using commands to control simulation is beyond the scope of this tutorial. The Editor window is used for entering programs.

Step 2

Create a New Project. This is done by pressing the **Project** menu and selecting **New . . .** (shown in Figure C.2). A popup dialog box as shown in Figure C.3 will appear. Enter the project name of your choice (let's use bubble as the project name), select the appropriate directory then click on **Next**, and another dialog box as shown in Figure C.4 will appear for you to select the device for the project. Select **Atmel's T89C51AC2** (or **SILabs' C8051F040** as you wish) as the target device and click on **Finish**, and the screen will change to that in Figure C.5.

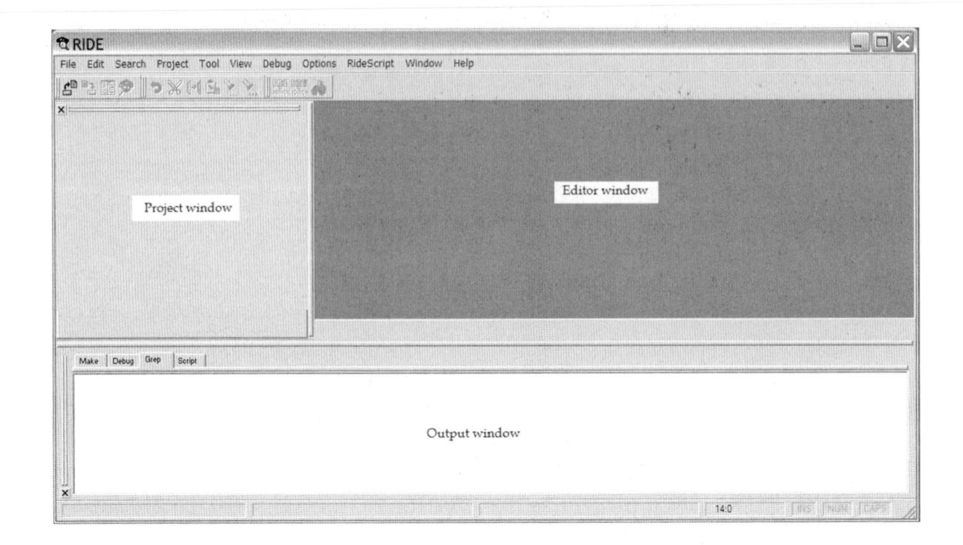

FIGURE C.1 The startup screen for Raisonance's RIDE.

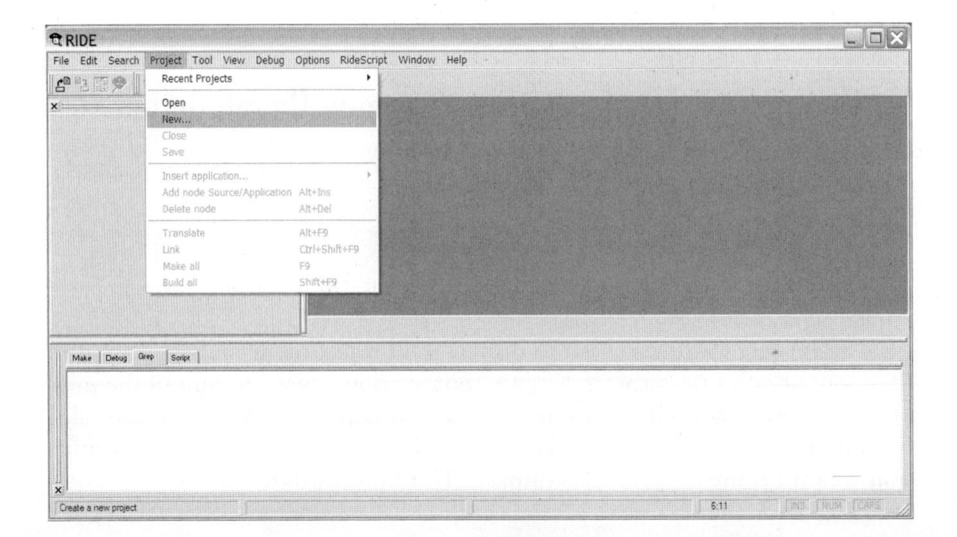

FIGURE C.2 Menu selection for creating a new project.

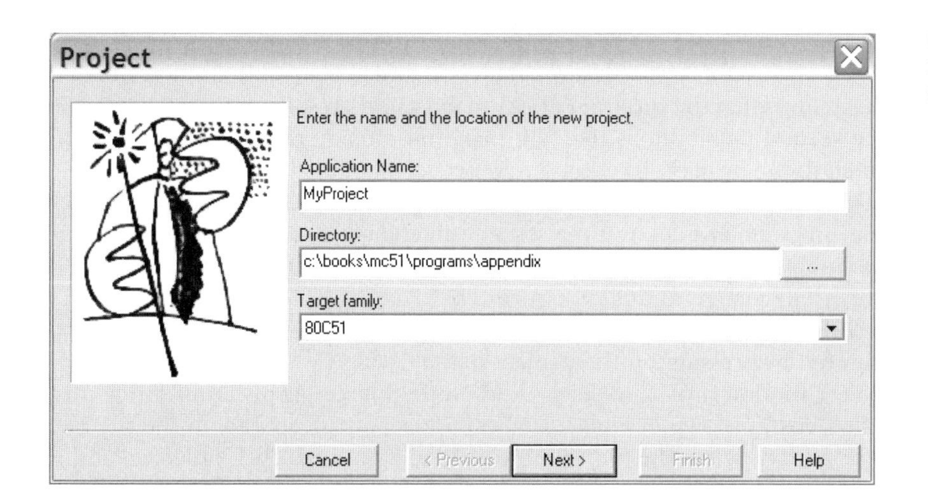

FIGURE C.3 New project dialog for RIDE.

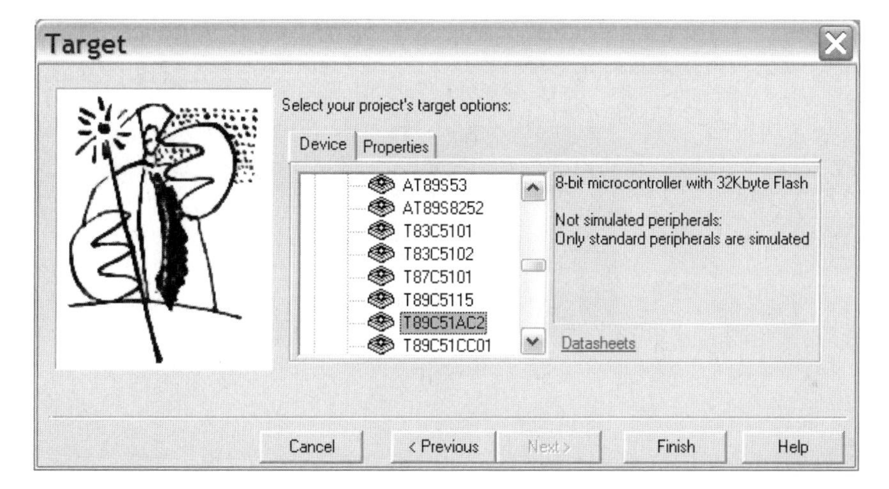

FIGURE C.4 Selecting the target device for the project.

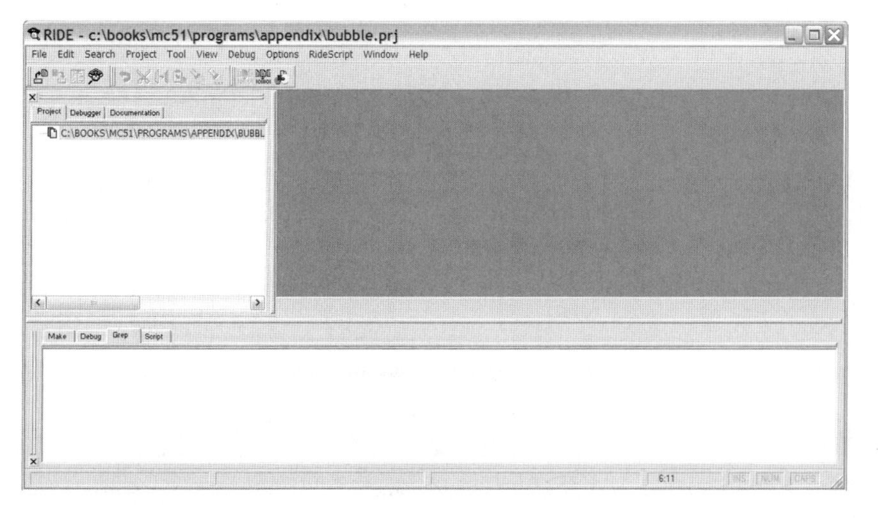

FIGURE C.5 RIDE screen after creating the new project bubble.

Step 3

Add the Assembly Program to the Project. There are two possibilities here. The first possibility is that the program does not exist and must be entered as a new file. The second possibility is that the program already existed and needs to be added to the project. To enter a new program, press the **File** menu and select **New**. After that, you will be asked to select the file type, as shown in Figure C.6. Select **Assembler Files** and the screen will change to that of Figure C.7. Enter the bubble sort program and save it with the filename **bubble.a51**, and the screen will change to that of Figure C.8.

Adding **bubble.a51** or an existing assembly program to the project is achieved by pressing on the project name (using right mouse button) in the Project window and then select **Add Node Source/Application**. After this, a dialog box as shown in Figure C.9 will appear to ask you about the file name to be added. Select the file (**bubble.a51**) to be added, click on **Open** and the screen will be changed to that in Figure C.10.

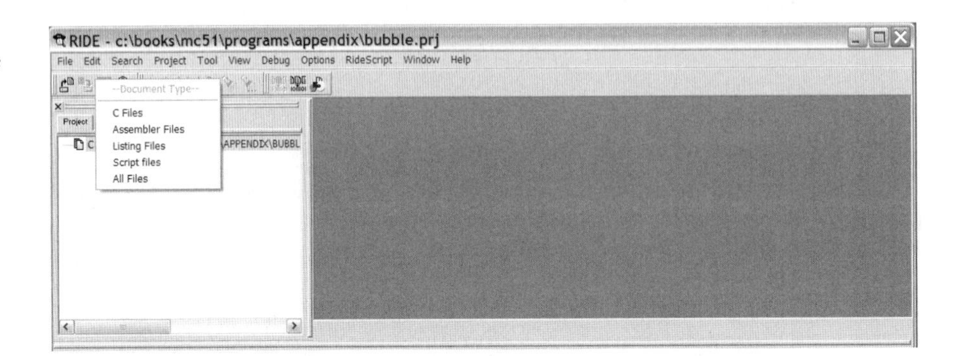

FIGURE C.6
Setting the file type for the new file.

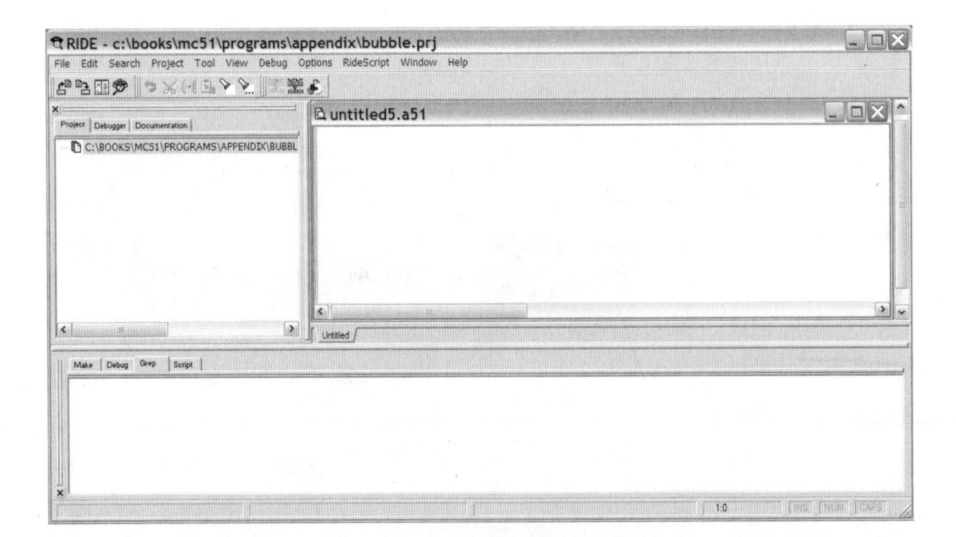

FIGURE C.7
The RIDE screen before entering a new assembler file.

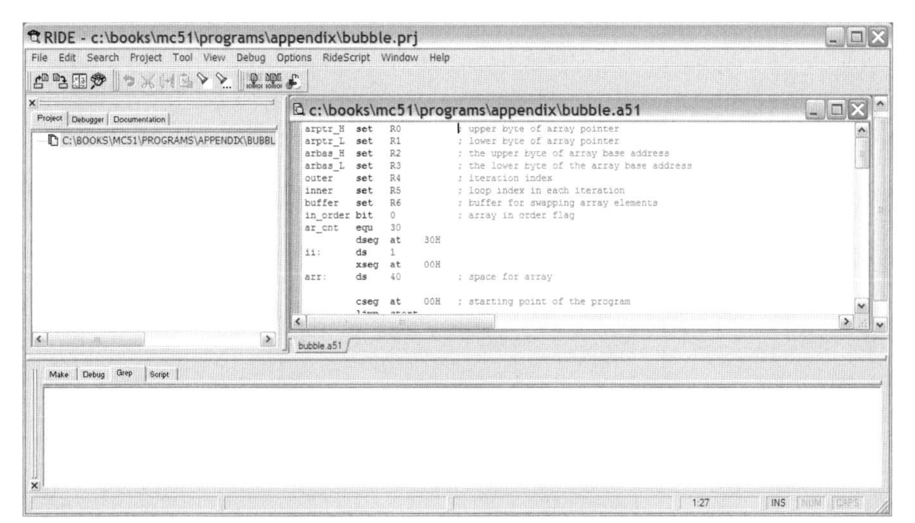

FIGURE C.8 RIDE screen after entering and saving a new file with the filename bubble. a51.

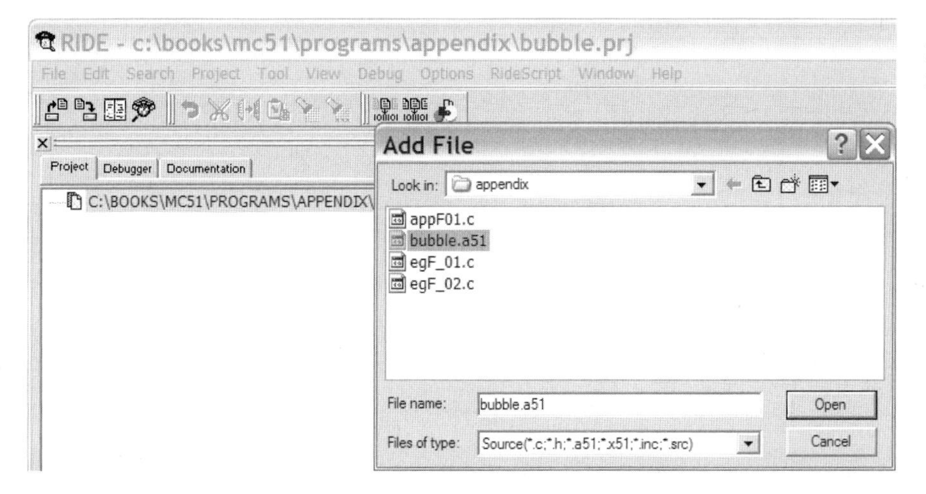

FIGURE C.9 Dialog for adding a file to the project.

Sometimes, you may want to delete a file from the project. This can be achieved by pressing the right mouse button on the name of the file (inside the project workspace) to be deleted from the project (shown in Figure C.11) and select **Delete node** and the file will be deleted from the project. For this tutorial you don't want to delete bubble.a51.

Step 4

Compile and Link the Program. This step is called **make the project** and is done by pressing on the **Project** menu and select **Make all** (as shown in Figure C.12). You can also select **Build all**. The result of **Make all** will be displayed in the **Output window**. The bubble.a51 program has no syntax error, and the output

FIGURE C.10
Screen after adding a file to the project.

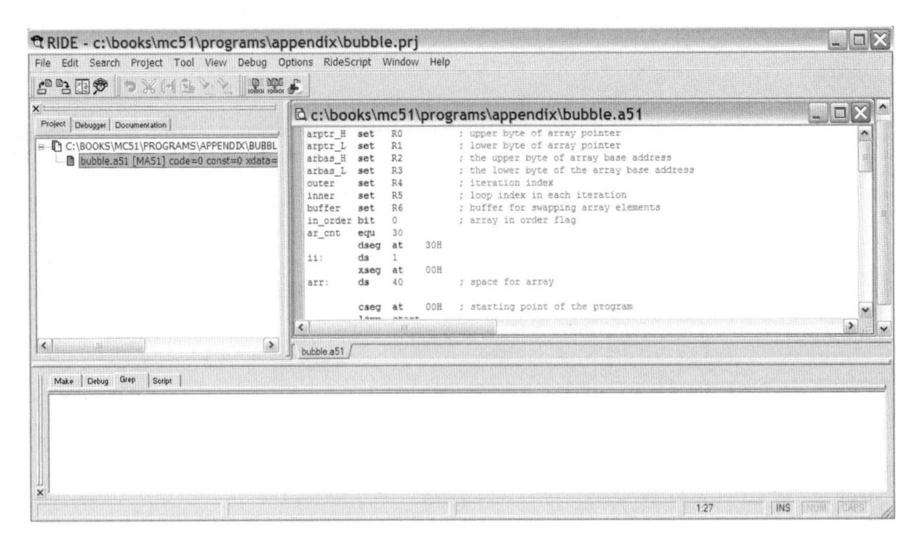

FIGURE C.11
Screen for deleting the bubble.a51 from the project.

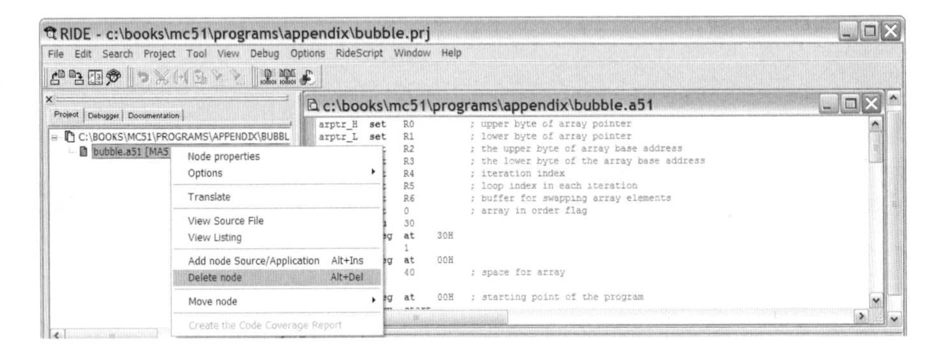

FIGURE C.12
Screen for making a project.

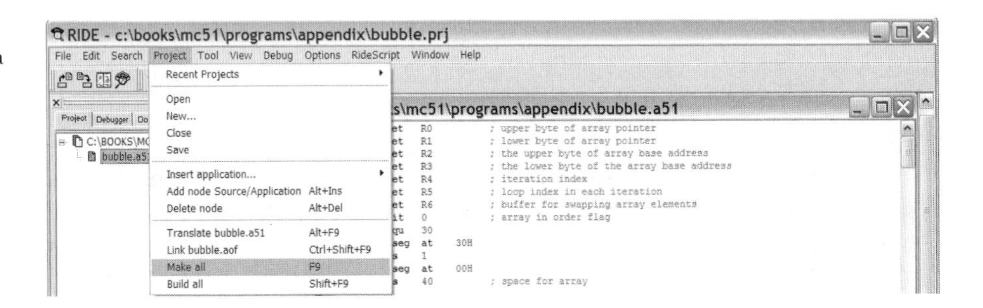

message shows it in Figure C.13. If there are any syntax errors, then you will need to fix them before the program can be tested.

Step 5

Debug the Program. The bubble.a51 program is written to test the bubble sort subroutine. The bubble sort subroutine sorts an array of 8-bit numbers in ascending order. The main program copies the array from the program memory to the XDATA memory and then calls the bubble subroutine. The sorted array is left in the XDATA memory. The loop for copying the array starts from the instruction with the label **scopylp** and ends with the **cjne A,#ar_cnt,scopylp** instruction.

Before debugging the program, you need to start the debug session. The debug session is started by pressing **Debug** menu and select **Start bubble.aof** from the menu (as shown in Figure C.14). After this, a popup dialog box (as shown in Figure C.15) will appear to ask you to enter the debug options. Click on **OK** and another dialog box (for Advanced Options) will appear. Also click on **OK** and the screen will be changed to Figure C.16.

We need to display the values of certain related MCU resources in order to determine whether the program execution results are correct. The MCU

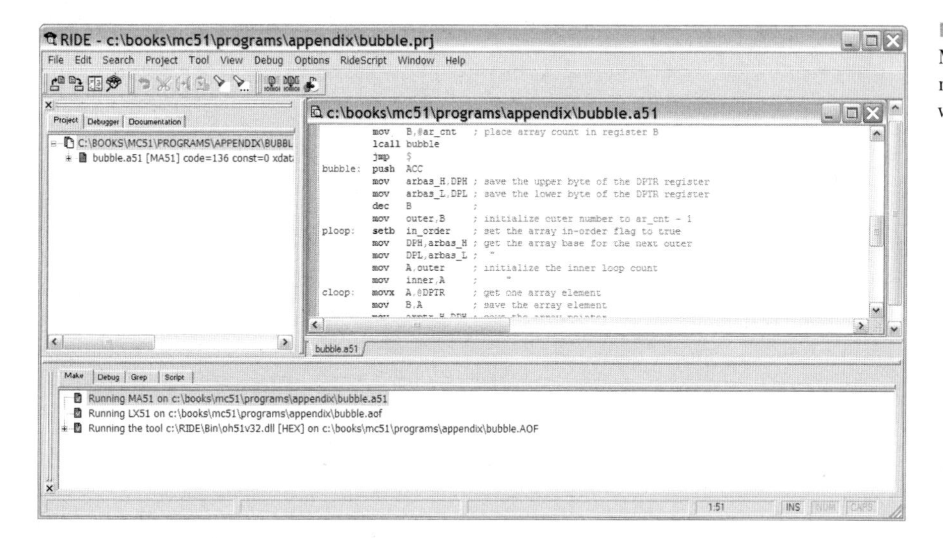

FIGURE C.13
Make all result message for a project without an error.

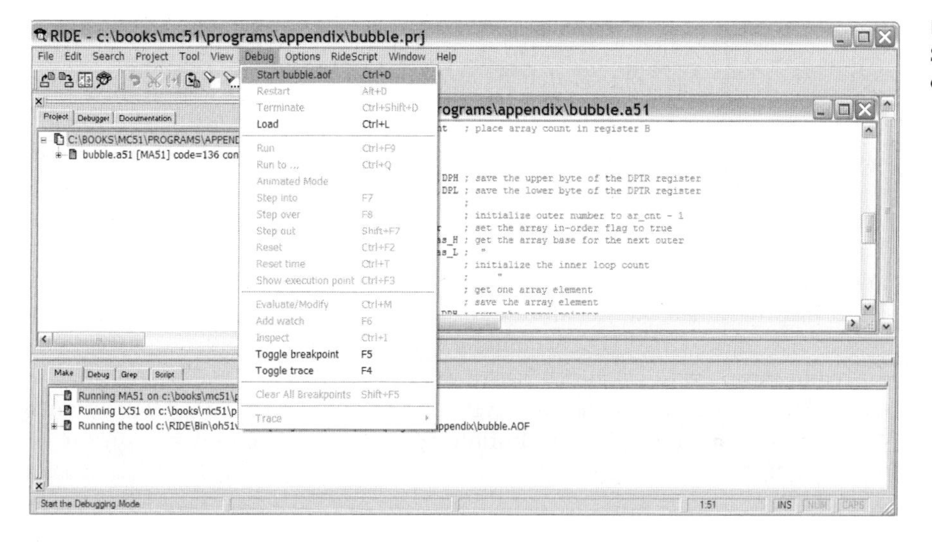

FIGURE C.14
Screen for starting debugging project.

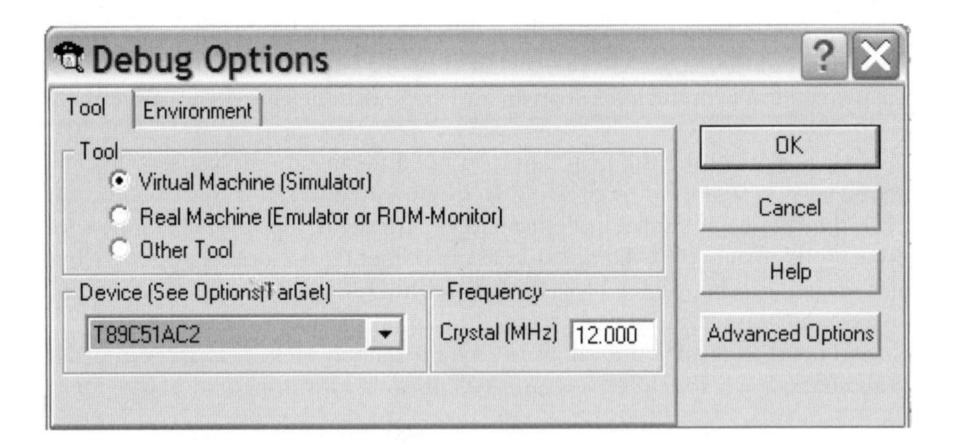

FIGURE C.15
Dialog for setting debug options.

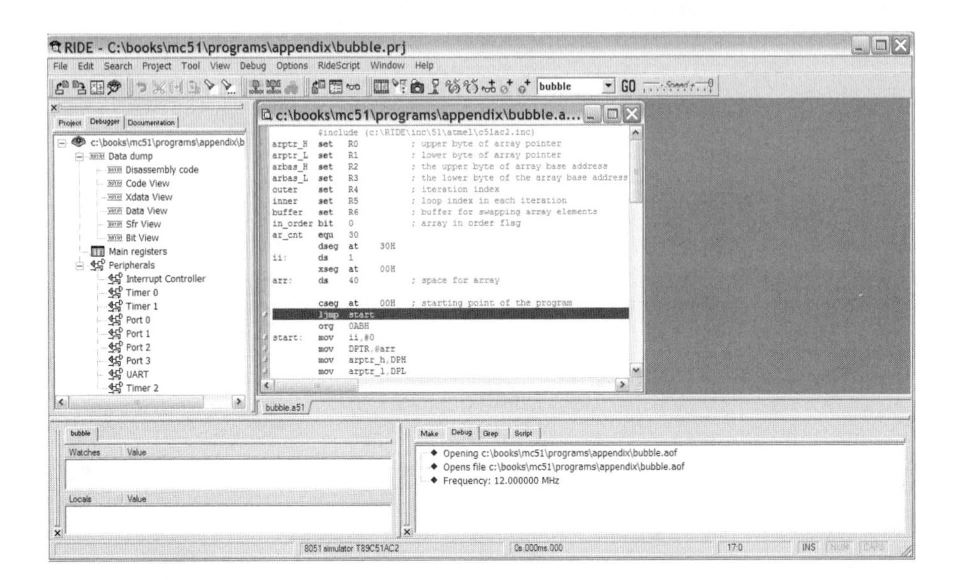

FIGURE C.16
RIDE screen after starting debugging operation.

resources that can be displayed are listed in the View menu, as shown in Figure C.17. To display any resource, select it from this menu. You may want to resize the RIDE screen for a better view. Program variables can be found in the Symbol table (select View−>Symbols) shown in Figure C.18. The variables of interest to us in this project are **array** and **arr**. Their starting addresses can be found by scrolling the scrollbar in Figure C.18. The unsorted array is stored in program (code) memory starting from 0x112, whereas the sorted array is to be stored in XDATA memory starting from 0x0000.

To check program execution result, you need to display the XDATA memory, which can be done by selecting **XDATA View** in Figure C.17. The contents of XDATA memory before the bubble program is executed are shown in Figure C.19.

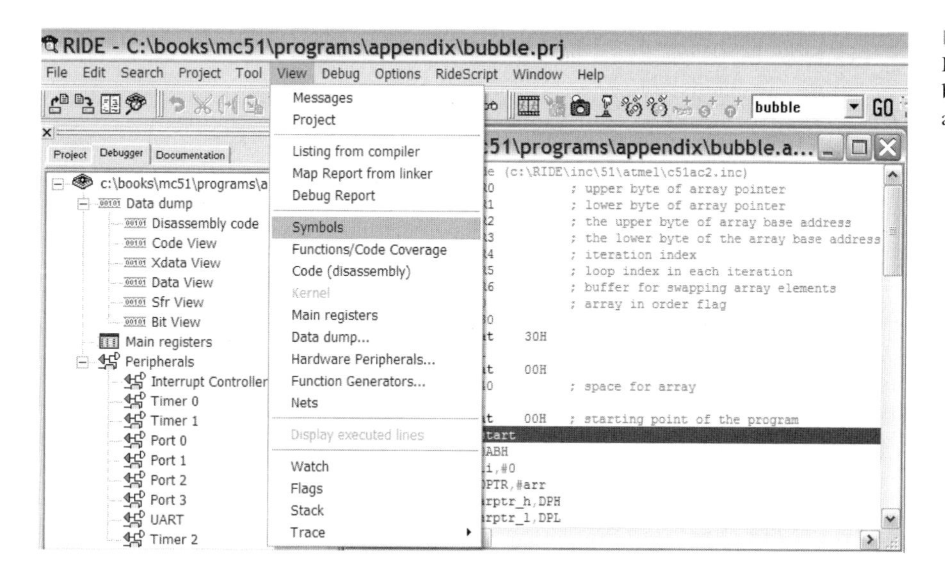

FIGURE C.17
Resources that can be displayed during a debugging session.

Global Symbols (bubble)

Name	Space	Address	Value	Type	File
ADCLK	SFR	F2	0x0		c:\book
ADCON	SFR	F3	0x0		c:\book
ADDH	SFR	F5	0x0		c:\book
ADDL	SFR	F4	0x0		c:\book
ARR	XDATA	0000	0xFF		c:\book
ARRAY	CODE	0112	0xD2		c:\book
AUXR	SFR	8E	0x0		c:\book
AUXR1	SFR	A2	0x0		c:\book
B	SFR	F0	0x0		
B.0	BIT	F0	FALSE		
B.1	BIT	F1	FALSE		
B.2	BIT	F2	FALSE		
B.3	BIT	F3	FALSE		
B.4	BIT	F4	FALSE		

Filters:
- [x] Code
- [x] Xdata
- [x] Data
- [x] Sfr
- [x] Bit

Search:

bubble.a51 | Xdata (bubble) | Global Symbols (bubble)

FIGURE C.18
Symbol table for the bubble project.

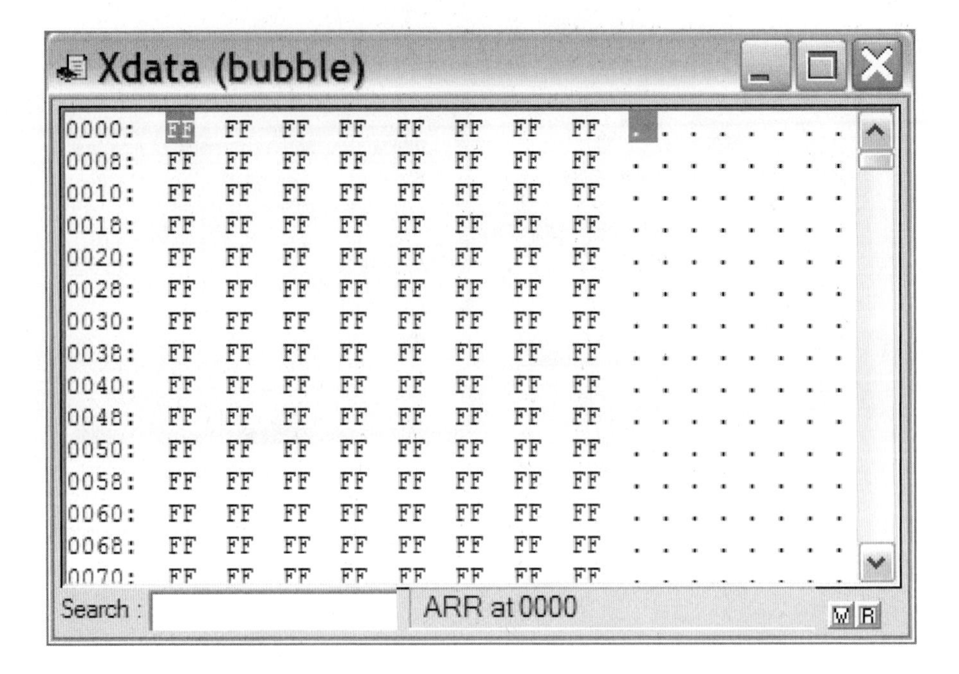

FIGURE C.19 XDATA contents before running the program.

Set a Breakpoint

Breakpoints should be set at any instruction that you want to check the program execution result. For this tutorial, we choose to set a breakpoint (at the mov DPTR,#arr instruction) after the array has been copied from program memory (at 0x112) to XDATA memory (starting from 0x0000) and a breakpoint after the bubble subroutine has been called (the jmp $ instruction). A breakpoint can be set by clicking (left mouse button) on the breakpoint instruction, then pressing the right mouse button, and then selecting **Toggle breakpoint**, as shown in Figure C.20. After this, the breakpoint instruction will be highlighted by the red color and there will be a letter **S** on its left.

Execute the Program

When executing a program for the first time, you want to make sure that the program is run from the start of the program. This can be achieved by resetting the MCU. The MCU is reset by pressing the **reset** button on the toolbar, as shown in Figure C.21. To execute the program, press the **GO** button on the toolbar. The program will stop at the first breakpoint. You can examine the XDATA memory contents (shown in Figure C.22) to find out if the memory copy operation is performed correctly. It is easy to verify that the copy operation is successful.

Set another breakpoint at the **jmp $** instruction, rerun the program, and check the XDATA memory. The resultant contents of the XDATA memory are shown in Figure C.23 and are correct.

This concludes the tutorial for assembly program development.

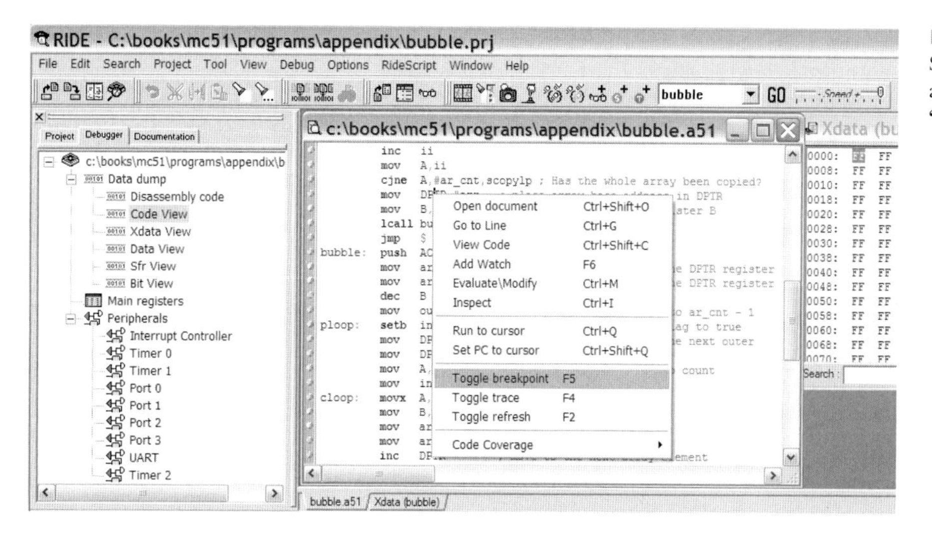

FIGURE C.20 Setting a breakpoint at the instruction **"mov DPTR, #arr"**.

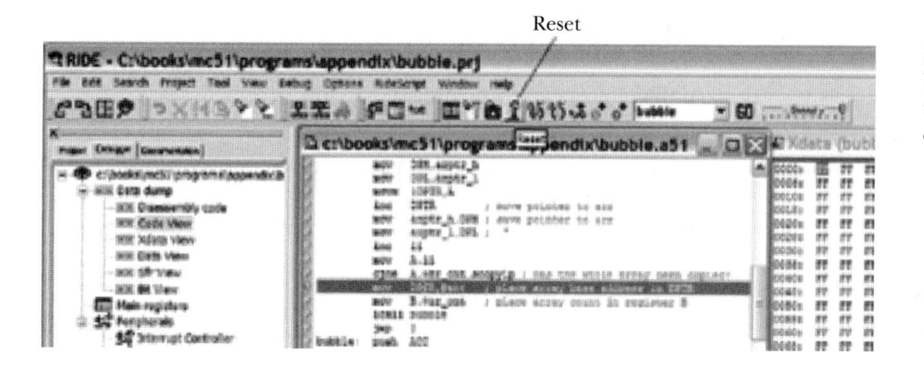

FIGURE C.21 Use a toolbar button to activate a RIDE command.

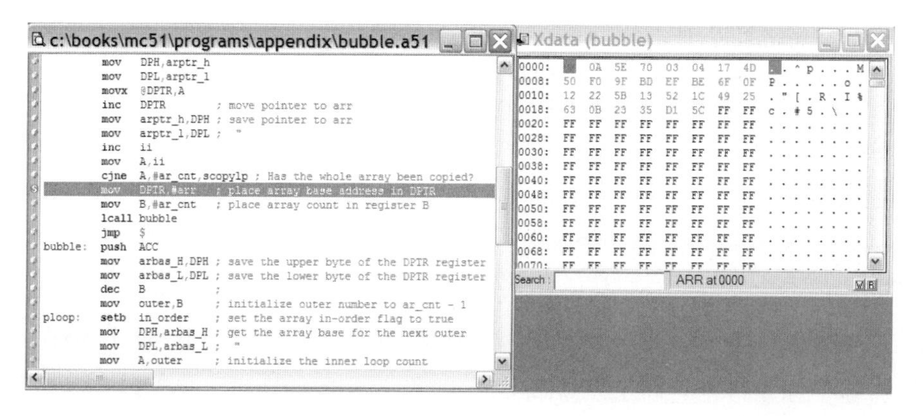

FIGURE C.22 The contents of the XDATA at the first breakpoint of the project.

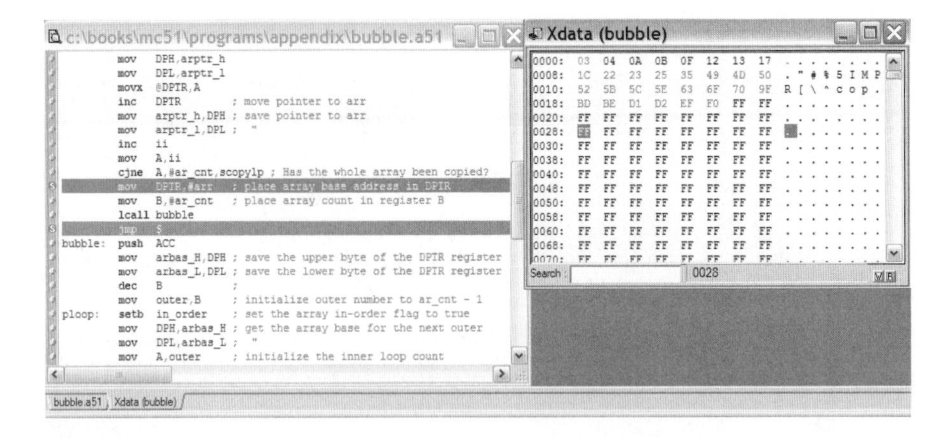

FIGURE C.23 The contents of XDATA memory (array arr) after calling the bubble subroutine.

C.2 Procedure for C Program Development

The procedure for debugging C programs using the RIDE is identical to that for assembly programs. We will use one example to demonstrate the process for C program development.

Example C.1 Write a C program that finds three four-digit integers with the following property:

The square of the sum of the first two digits and the last two digits equals the original number.

Solution: The following C program will find three four-digit integers that have the desired property.

```
int main (void)
{
        char cnt;
        unsigned int ix, iy, iz, arr[3];

        cnt = 0;
        for (ix = 1000; cnt < 3; ix++) {
            iy = ix / 100;
            iz = ix % 100;
            iy += iz;
            if (iy * iy == ix)
                arr[cnt++] = ix;
    }
        return 0;
}
```

The procedure for debugging this program using RIDE is as follows.

Step 1 Start the RIDE.

Step 2 Create a new project called appG01.

Step 3 Create a new file with the file name appG01.c to hold the C program and add it to the project appG01.

Step 4 Build the project. If there are any syntax errors, fix them. When all of the syntax errors have been corrected, the resultant screen should be as shown in Figure C.24.

Step 5 Start the debug session by selecting the menu item **Debug−>Start appg01.aof**. Click on **OK** for the Debug options (there are two popup buttons to be dealt with). The resultant screen is shown in Figure C.25. The Watch window and Locals window have been shown

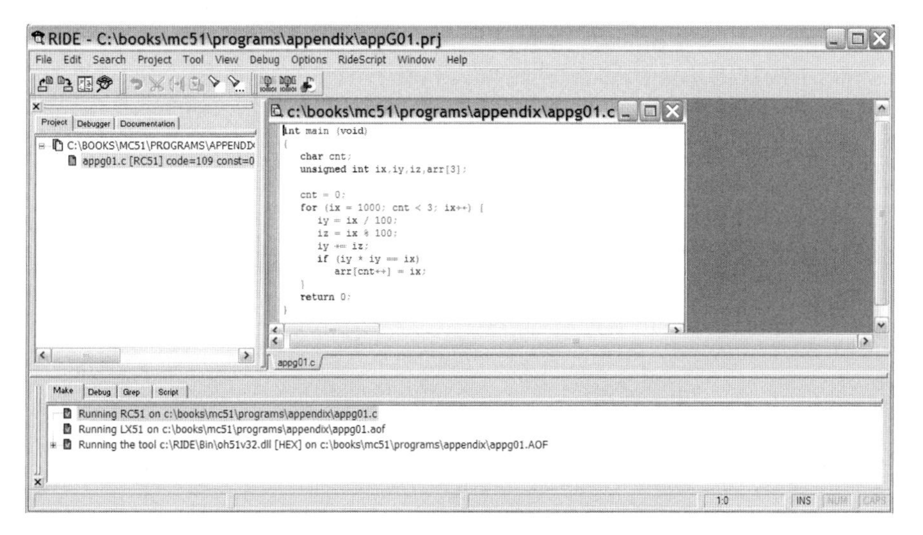

FIGURE C.24 The RIDE screen after the C program is correctly compiled.

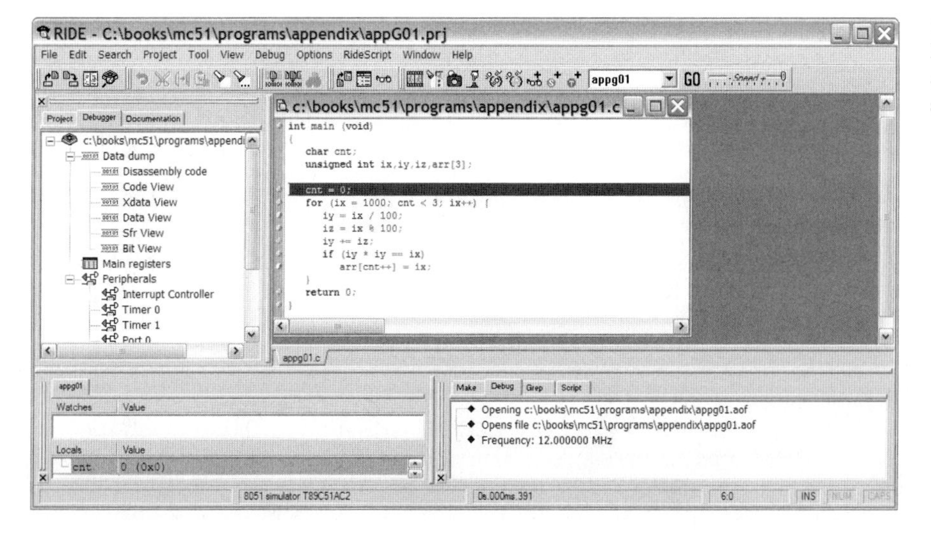

FIGURE C.25 RIDE screen after starting a debug session for a project.

in the screen (on the lower left corner). We are interested in **arr[0]**, **arr[1]**, and **arr[2]** for this project. We can either add these three variables in the watch list or simply examine their values inside the Locals window. The screen for adding a variable into the Watch window is shown in Figure C.26. After clicking on any character of the **arr** symbol and pressing the right mouse button, a dialog box (shown in Figure C.27) will then appear for you to enter the variable to be inserted. You can only enter one variable at a time. Enter **arr[0]** in the dialog box. Repeat the same operation for **arr[1]** and **arr[2]**, and the resultant RIDE screen is shown in Figure C.28.

You can choose the desired display format for the Watch window. Press the right mouse button inside the Watch window and a popup menu (as shown in Figure C.29) will allow you to select the desired format. A dialog box (as shown in Figure C.30) allows you to make the selection. Select **Dec. and Hexa.** for this tutorial.

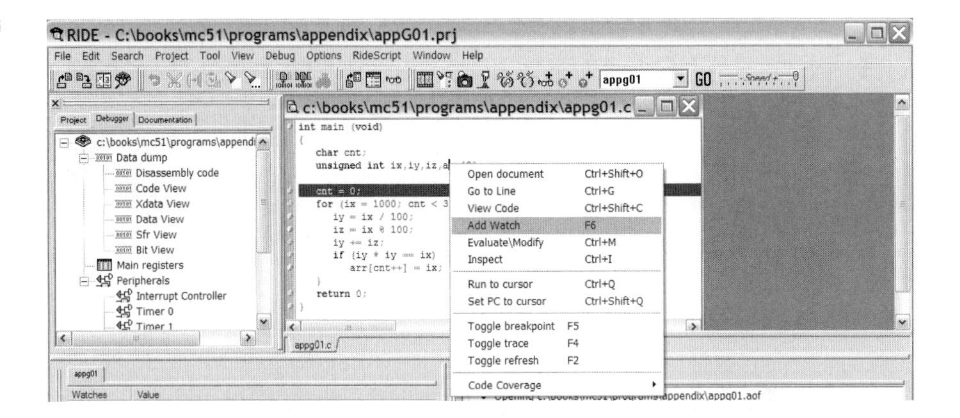

FIGURE C.26
RIDE screed for adding arr[] into watch list.

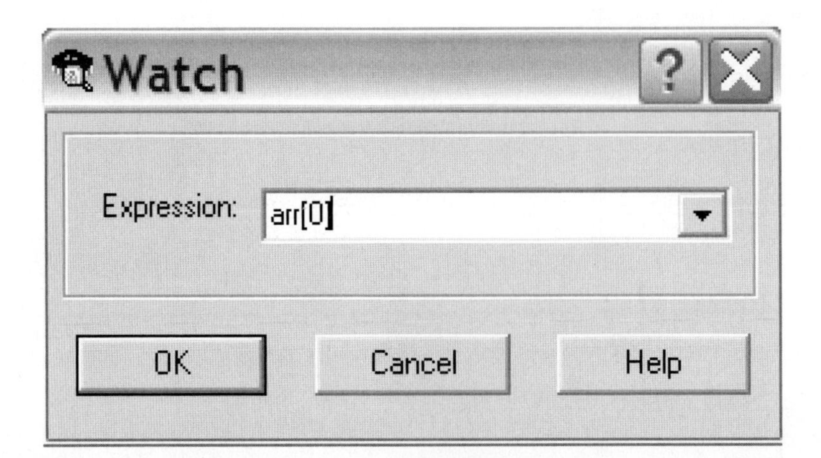

FIGURE C.27
Dialog for adding a variable into a Watch list.

Step 6

Set a breakpoint at the statement **return 0** by pressing the right mouse button on this statement and select **Toggle breakpoint**. The resultant Watch window screen is shown in Figure C.31. Run the program until the breakpoint and the values of arr[0], arr[1], and arr[2] in the Watch window are changed to 2025, 3025, and 9801, respectively. You can verify that these three numbers have the required property.

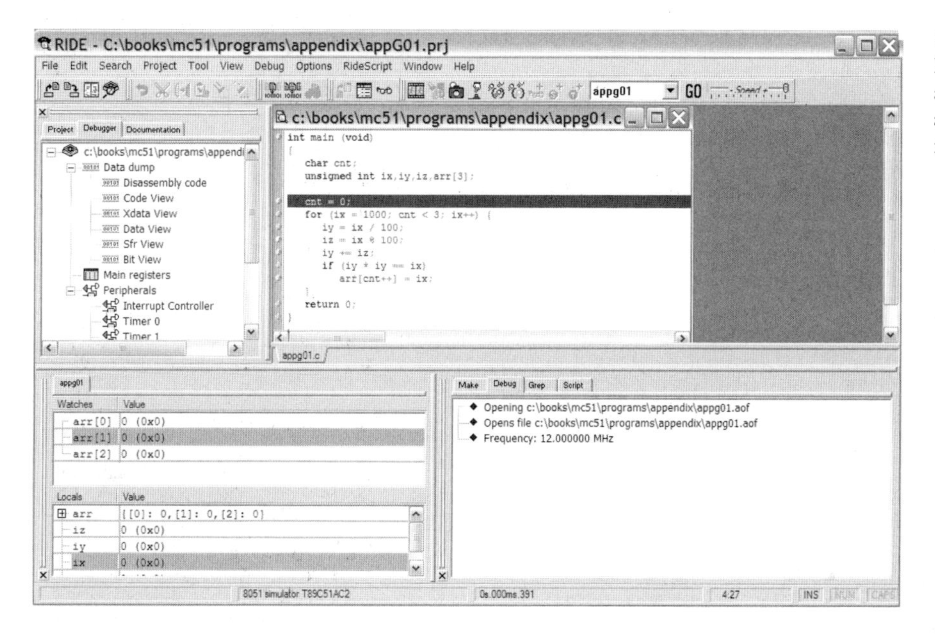

FIGURE C.28
RIDE screen after adding arr[0], arr[1], and arr[2] into the Watch list.

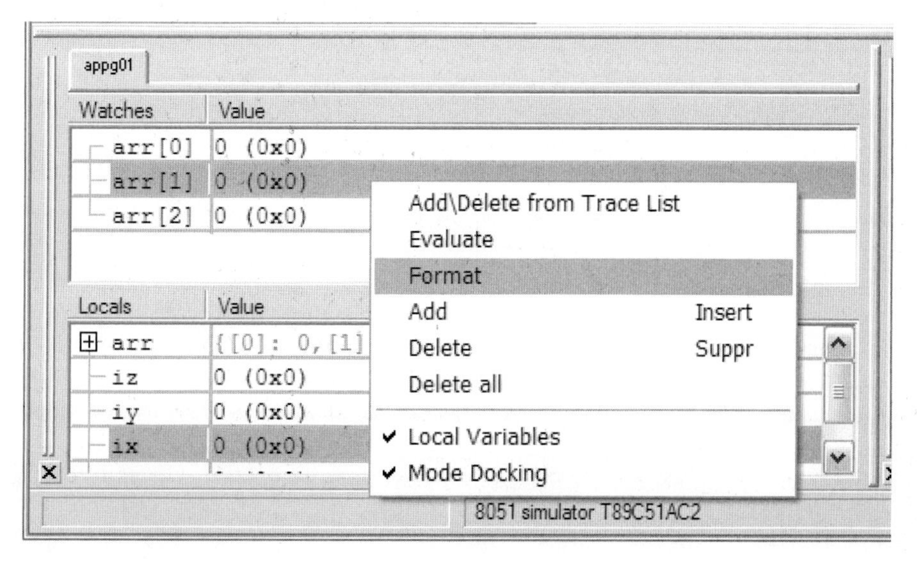

FIGURE C.29
Select display format for the Watch window.

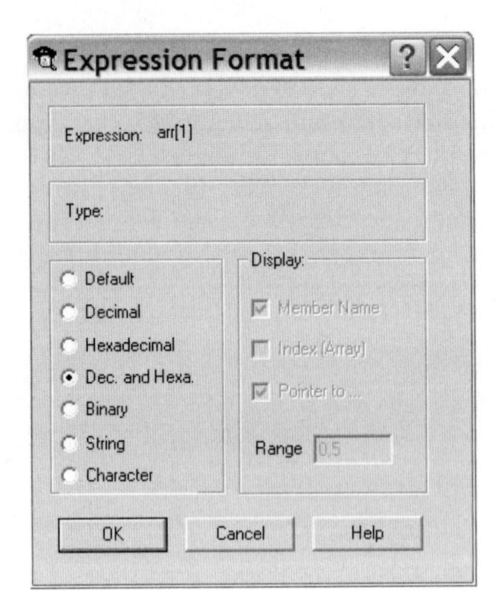

FIGURE C.30 RIDE screen for selecting Expression Format.

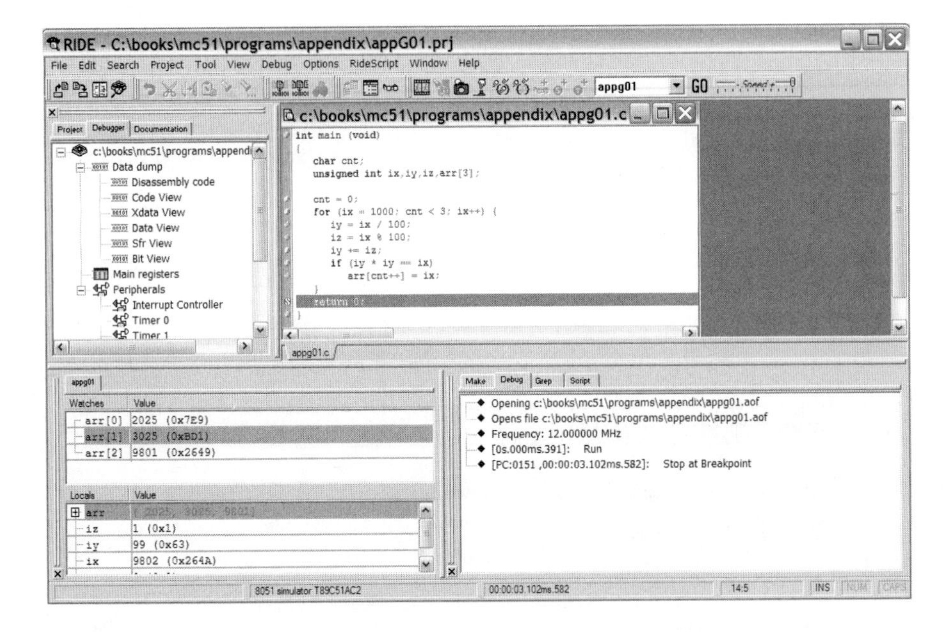

FIGURE C.31 RIDE screen after execution to a breakpoint.

This completes the tutorial for using RIDE to debug a C program.

APPENDIX D

C8051F040 Special Function Registers

Register	Address	SFR Page	Description
ACC	0xE0	All pages	Accumulator
ADC0CF	0xBC	0	ADC0 configuration
ADC0CN	0xE8	0	ADC0 control
ADC0GTH	0xC5	0	ADC0 greater-than high
ADC0GTL	0xC4	0	ADC0 greater-than low
ADC0H	0xBF	0	ADC0 data word high
ADC0L	0xBE	0	ADC0 data word low
ADC0LTH	0xC7	0	ADC0 less-than high
ADC0LTL	0xC6	0	ADC0 less-than low
ADC2	0xBE	2	ADC2 control
ADC2CF	0xBC	2	ADC2 analog multiplexer configuration
ADC2CN	0xE8	2	ADC2 control
ADC2GT	0xC4	2	ADC2 window comparator greater-than
ADC2LT	0xC6	2	ADC2 window comparator less-than
AMX0CF	0xBA	0	ADC0 multiplexer configuration
AMX0PRT	0xBD	0	ADC0 port 3 I/O pin select
AMX0SL	0xBB	0	ADC0 multiplexer channel select
AMX2CF	0xBA	2	ADC2 multiplexer configuration
AMX2SL	0xBB	2	ADC2 multiplexer channel select
B	0xF0	All pages	B register
CAN0ADR	0xDA	1	CAN0 address
CAN0CN	0xF8	1	CAN0 control
CAN0DATH	0xD9	1	CAN0 data register high
CAN0DATL	0xD8	1	CAN0 data register low
CAN0STA	0xC0	1	CAN0 status
CAN0TST	0xDB	1	CAN0 test register
CKCON	0x8E	0	Clock control
CLKSEL	0x97	F	Oscillator clock selection register
CPT0MD	0x89	1	Comparator 0 mode selection
CPT1MD	0x89	2	Comparator 1 mode selection
CPT2MD	0x89	3	Comparator 2 mode selection
CPT0CN	0x88	1	Comparator 0 control
CPT1CN	0x88	2	Comparator 1 control
CPT2CN	0x88	3	Comparator 2 control
DAC0CN	0xD4	0	DAC0 control
DAC0H	0xD3	0	DAC0 high
DAC0L	0xD2	0	DAC0 low
DAC1CN	0xD4	1	DAC1 control

(*continued*)

Register	Address	SFR Page	Description
DAC1H	0xD3	1	DAC1 high byte
DAC1L	0xD2	1	DAC1 low byte
DPH	0x83	All pages	Data pointer high
DPL	0x82	All pages	Data pointer low
EIE1	0xE6	All pages	Extended interrupt enable 1
EIE2	0xE7	All pages	Extended interrupt enable 2
EIP1	0xF6	All pages	Extended interrupt priority 1
EIP2	0xF7	All pages	Extended interrupt priority 2
EMI0CF	0xA3	0	EMIF configurration
EMI0CN	0xA2	0	External memory interface control
EMI0TC	0xA1	0	EMIF timing control
FLACL	0xB7	F	Flash access limit
FLSCL	0xB7	0	Flash scale
HVA0CN	0xD6	0	High voltage differential amp control
IE	0xA8	All pages	Interrupt enable
IP	0xB8	All pages	Interrupt priority
OSCICL	0x8B	F	Internal oscillator calibration
OSCICN	0x8A	F	Internal oscillator control
OSCXCN	0x8C	F	External oscillator control
P0	0x80	All pages	Port 0 latch
P0MDOUT	0xA4	F	Port 0 output mode configuration
P1	0x90	All pages	Port 1 latch
P1MDIN	0xAD	F	Port 1 input mode configuration
P1MDOUT	0xA5	F	Port 1 output mode configuration
P2	0xA0	All pages	Port 2 latch
P2MDIN	0xAE	F	Port 2 input mode configuration
P2MDOUT	0xA6	F	Port 2 output mode configuration
P3	0xB0	All pages	Port 3 latch
P3MDIN	0xAF	F	Port 3 input mode configuration
P3MDOUT	0xA7	F	Port 3 output mode configuration
P4	0xC8	F	Port 4 latch
P4MDOUT	0x9C	F	Port 4 output mode configuration
P5	0xD8	F	Port 5 latch
P5MDOUT	0x9D	F	Port 5 output mode configuration
P6	0xE8	F	Port 6 latch
P6MDOUT	0x9E	F	Port 6 output mode configuration
P7	0xF8	F	Port 7 latch
P7MDOUT	0x9F	F	Port 7 output mode configuration

(*continued*)

Register	Address	SFR Page	Description
PCA0CN	0xD8	0	PCA control
PCA0CPH0	0xFC	0	PCA capture 0 high
PCA0CPH1	0xFE	0	PCA capture 1 high
PCA0CPH2	0xEA	0	PCA capture 2 high
PCA0CPH3	0xEC	0	PCA capture 3 high
PCA0CPH4	0xEE	0	PCA capture 4 high
PCA0CPH5	0xE2	0	PCA capture 5 high
PCA0CPL0	0xFB	0	PCA capture 0 low
PCA0CPL1	0xFD	0	PCA capture 1 low
PCA0CPL2	0xE9	0	PCA capture 2 low
PCA0CPL3	0xEB	0	PCA capture 3 low
PCA0CPL4	0xED	0	PCA capture 4 low
PCA0CPL5	0xE1	0	PCA capture 5 low
PCA0CPM0	0xDA	0	PCA module 0 mode register
PCA0CPM1	0xDB	0	PCA module 1 mode register
PCA0CPM2	0xDC	0	PCA module 2 mode register
PCA0CPM3	0xDD	0	PCA module 3 mode register
PCA0CPM4	0xDE	0	PCA module 4 mode register
PCA0CPM5	0xDF	0	PCA module 5 mode register
PCA0H	0xFA	0	PCA counter high
PCA0L	0xF9	0	PCA counter low
PCA0MD	0xD9	0	PCA mode
PCON	0x87	All pages	Power control
PSCTL	0x8F	0	Program store R/W control
PSW	0xD0	All pages	Program status word
RCAP2H	0xCB	0	Timer/counter 2 capture/reload high
RCAP2L	0xCA	0	Timer/counter 2 capture/reload low
RCAP3H	0xCB	1	Timer/counter 3 capture/reload high
RCAP3L	0xCA	1	Timer/counter 3 capture/reload low
RCAP4H	0xCB	2	Timer/counter 4 capture/reload high
RCAP4L	0xCA	2	Timer/counter 4 capture/reload low
REF0CN	0xD1	0	Programmable voltage reference control
RSTSRC	0xEF	0	Reset source register
SADDR0	0xA9	0	UART0 slave address
SADEN0	0xB9	0	UART0 slave address enable
SBUF0	0x99	0	UART0 data buffer
SBUF1	0x99	1	UART1 data buffer
SCON0	0x98	0	UART0 control
SCON1	0x98	1	UART1 control

(*continued*)

Register	Address	SFR Page	Description
SFRPAGE	0x84	All pages	SFR page register
SFRPGCN	0x96	F	SFR page control register
SFRNEXT	0x85	All pages	SFR next page stack access register
SFRLAST	0x86	All pages	SFR last page stack access register
SMB0ADR	0xC3	0	SMBus slave address
SMB0CN	0xC0	0	SMBus control
SMB0CR	0xCF	0	SMBus clock rate
SMB0DAT	0xC2	0	SMBus data
SMB0STA	0xC1	0	SMbus status
SP	0x81	All pages	Stack pointer
SPI0CFG	0x9A	0	SPI configuration
SPI0CKR	0x9D	0	SPI clock rate control
SPI0CN	0xF8	0	SPI control
SPI0DAT	0x9B	0	SPI data
SSTA0	0x91	0	UART0 status and clock selection
TCON	0x88	0	Timer/counter control
TH0	0x8C	0	Timer/counter 0 high
TH1	0x8D	0	Timer/counter 1 high
TL0	0x8A	0	Timer/counter 0 low
TL1	0x8B	0	Timer/counter 1 low
TMOD	0x89	0	Timer/counter mode
TMR2CF	0xC9	0	Timer/counter 2 configuration
TMR2CN	0xC8	0	Timer/counter 2 control
TMR2H	0xCD	0	Timer/counter 2 high
TMR2L	0xCC	0	Timer/counter 2 low
TMR3CF	0xC9	1	Timer/counter 3 configuration
TMR3CN	0xC8	1	Timer/counter 3 control
TMR3H	0xCD	1	Timer/counter 3 high
TMR3L	0xCC	1	Timer/counter 3 low
TMR4CF	0xC9	2	Timer/counter 4 configuration
TMR4CN	0xC8	2	Timer/counter 4 control
TMR4H	0xCD	2	Timer/counter 4 high
TMR4L	0xCC	2	Timer/counter 4 low
WDTCN	0xFF	All pages	Watchdog timer control
XBR0	0xE1	F	Port I/O crossbar control 0
XBR1	0xE2	F	Port I/O crossbar control 1
XBR2	0xE3	F	Port I/O crossbar control 2
XBR3	0xE4	F	Port I/O crossbar control 3
0x97, 0xA2, 0xB3, 0xB4, 0xCE, 0xDF			Reserved

APPENDIX E

C8051F040 SFR PAGE Definition (Keil and Raisonance)

In Assembly Language

```
CONFIG_PAGE     EQU     0FH     ; system and port configuration page
LEGACY_PAGE     EQU     00H     ; legacy SFR page
TIMER01_PAGE    EQU     00H     ; Timer 0 and Timer 1
CPT0_PAGE       EQU     01H     ; comparator 0
CPT1_PAGE       EQU     02H     ; comparator 1
CPT2_PAGE       EQU     03H     ; comparator 2
UART0_PAGE      EQU     00H     ; UART 0
UART1_PAGE      EQU     01H     ; UART 1
SPI0_PAGE       EQU     00H     ; SPI 0
EMI0_PAGE       EQU     00H     ; External memory interface
ADC0_PAGE       EQU     00H     ; ADC 0
ADC2_PAGE       EQU     02H     ; ADC 2
SMB0_PAGE       EQU     00H     ; SMBUS 0
TMR2_PAGE       EQU     00H     ; TIMER 2
TMR3_PAGE       EQU     01H     ; TIMER 3
TMR4_PAGE       EQU     02H     ; TIMER 4
DAC0_PAGE       EQU     00H     ; DAC 0
DAC1_PAGE       EQU     01H     ; DAC 1
PCA0_PAGE       EQU     00H     ; PCA 0
CAN0_PAGE       EQU     01H     ; CAN 0
```

In C Language

```
#define    CONFIG_PAGE     0x0F    // system and port configuration page
#define    LEGACY_PAGE     0x00    // legacy SFR page
#define    TIMER01_PAGE    0x00    // Timer 0 and Timer 1
#define    CPT0_PAGE       0x01    // comparator 0
#define    CPT1_PAGE       0x02    // comparator 1
#define    CPT2_PAGE       0x03    // comparator 2
#define    UART0_PAGE      0x00    // UART 0
#define    UART1_PAGE      0x01    // UART 1
#define    SPI0_PAGE       0x00    // SPI 0
#define    EMI0_PAGE       0x00    // External memory interface
#define    ADC0_PAGE       0x00    // ADC 0
#define    ADC2_PAGE       0x02    // ADC 2
#define    SMB0_PAGE       0x00    // SMBUS 0
#define    TMR2_PAGE       0x00    // TIMER 2
#define    TMR3_PAGE       0x01    // TIMER 3
#define    TMR4_PAGE       0x02    // TIMER 4
#define    DAC0_PAGE       0x00    // DAC 0
#define    DAC1_PAGE       0x01    // DAC 1
#define    PCA0_PAGE       0x00    // PCA 0
#define    CAN0_PAGE       0x01    // CAN 0
```

APPENDIX F

Procedure for Setting up HyperTerminal

Step 1

Start the HyperTerminal program by selecting **Start=>All Programs=> Accessories=>Communications=>HyperTerminal**. After starting the **HyperTerminal** program, a screen as shown in Figure F.1 will appear in the PC monitor screen.

Step 2

The screen shown in Figure F.1 expects you to enter a name for the connection. Enter a name that you like and click on **OK** and the screen will change to that in Figure F.2. In the **Connect Using** field, select **COM1** or **COM3** depending on the COM port that you are using (as shown in Figure F.3) and click on **OK**, and a dialog screen as shown in Figure F.4 will appear.

Step 3

You need to enter the properties of the chosen COM port. Enter 19200 in the **Bits per second** field. Set the **Flow control** field to none. The default values of the remaining fields are acceptable. The resultant screen is shown in Figure F.5. Click on **OK**, and the dialog screen will disappear. The resultant screen is shown in Figure F.6.

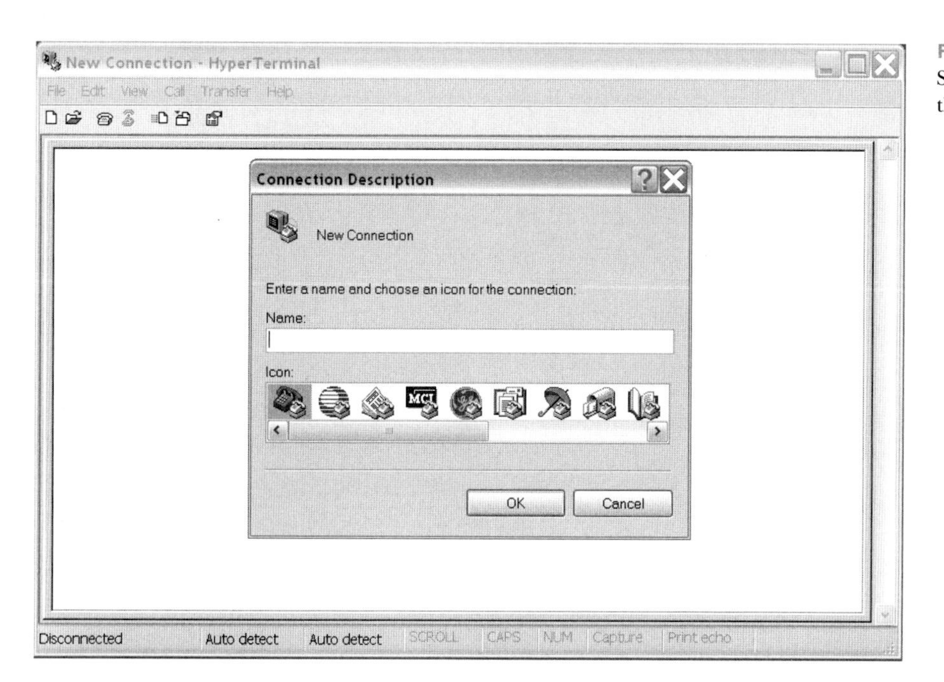

FIGURE F.1 Start up screen of the HyperTerminal.

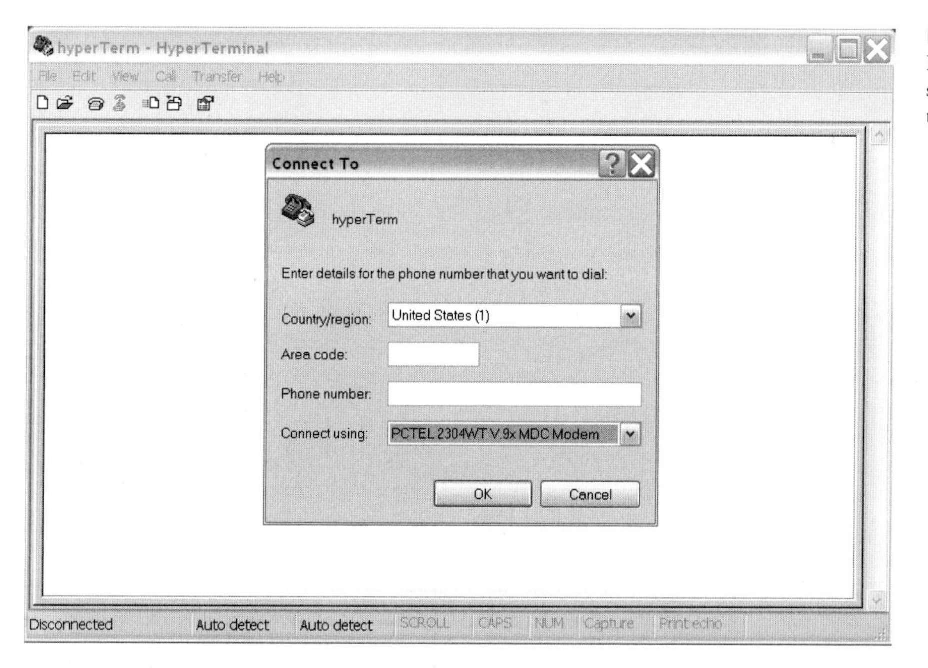

FIGURE F.2 HyperTerminal screen after entering the connection name.

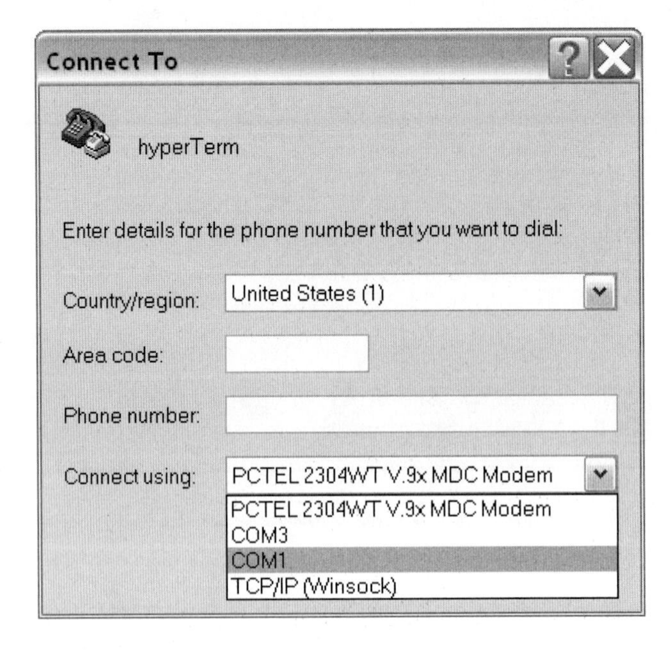

FIGURE F.3
Dialog for selecting connection method.

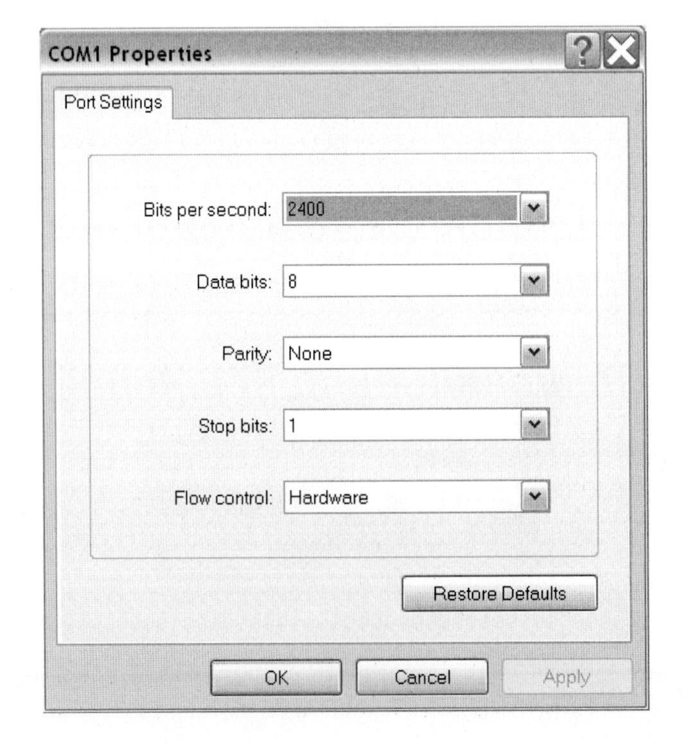

FIGURE F.4
Dialog for setting up COM port property.

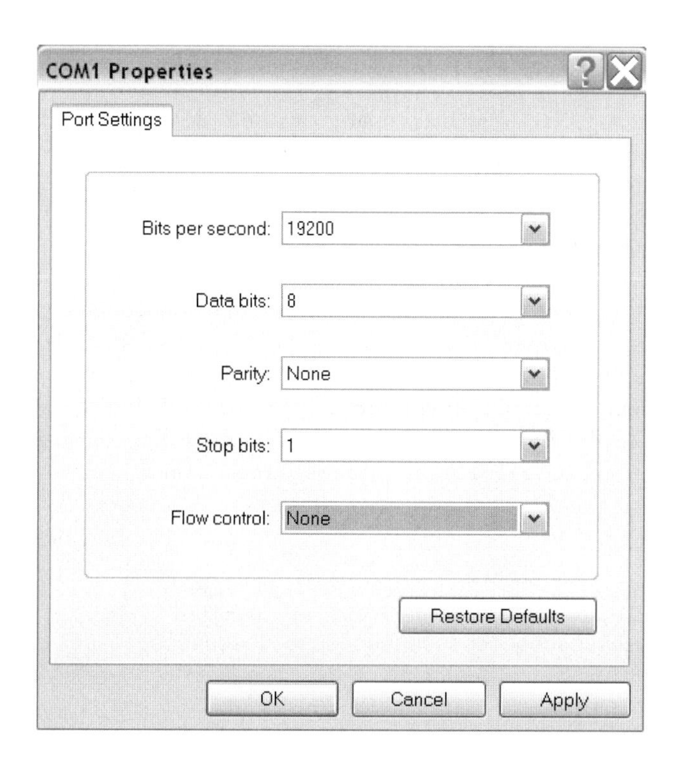

FIGURE F.5
COM1 properties that can work with the demo board settings.

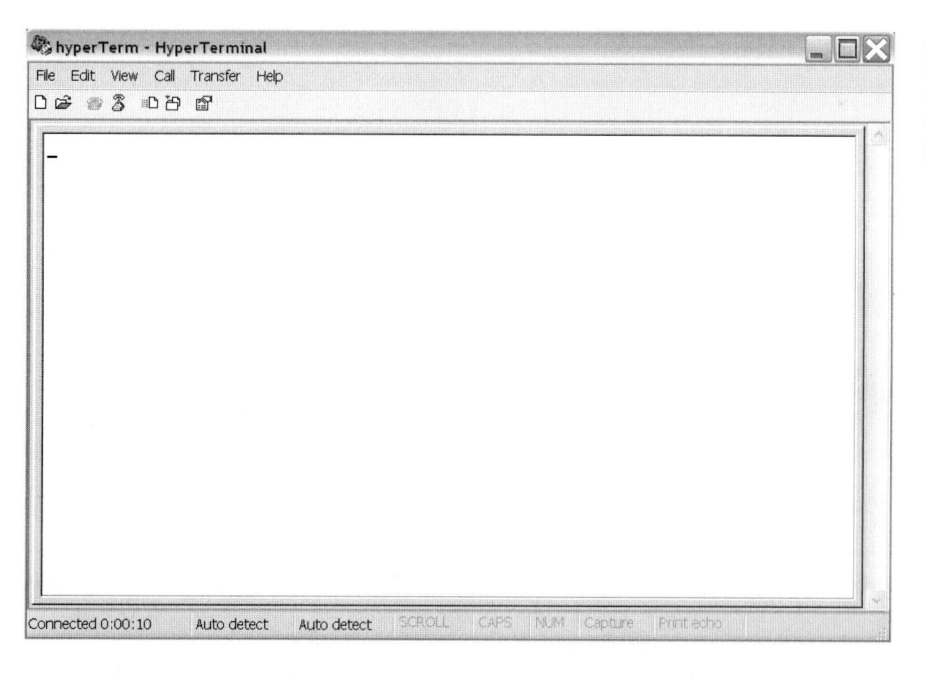

FIGURE F.6
Hyper Terminal screen after setting communciation properties.

Step 4

Start to communicate. Make sure that the **HyperTerminal** is ready to communicate. This can be verified by pressing on the Call menu. If all except the **Disconnect** menu item are gray, then the **HyperTerminal** is ready to communicate. In this situation, if your demo board sends a string to the UART port, then it will appear on the **HyperTerminal** screen.

This is the end of the setup procedure of **HyperTerminal**.

HyperTerminal needs only be set up once. You can follow the path of **Start=>All Programs=>Accessories=>Communications=>HyperTerminal** and see the name of **hyperTerm** to the right of **HyperTerminal**. **HyperTerm** is the name that you enter as the connection name. You can select it and the **HyperTerminal** will start, and you are ready to use it. You also can bring this connection to the desktop by using the mouse to drag it to the monitor screen, and it will stay there waiting for you to click on it.

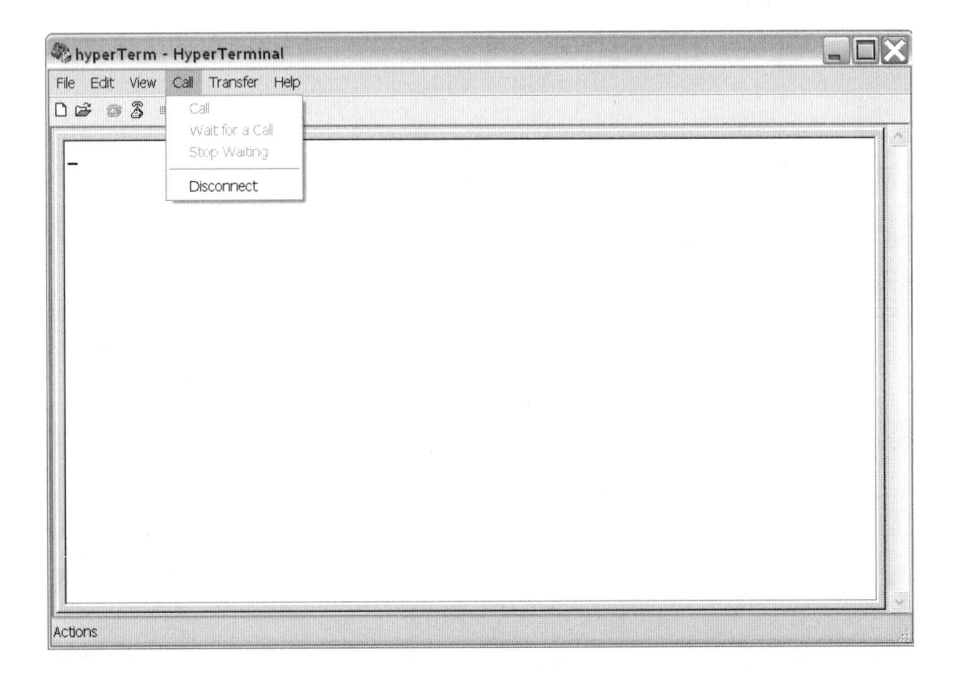

FIGURE F.7 A ready-to-communicate HyperTerminal screen.

APPENDIX G

Keil C Library Functions

These library functions are also available in Raisonance C and SDCC.

TABLE G.1 *Buffer Manipulation Functions*

Routine	Attributes	Description
memccpy		Copies data bytes from one buffer to another until a specified character or specified number of characters has been copied.
memchr	reentrant	Returns a pointer to the first occurrence of a specified character in a buffer.
memcmp	reentrant	Compares a given number of characters from two different buffers.
memcpy	reentrant	Copies a specified number of data bytes from one buffer to another.
memmove	reentrant	Copies a specified number of data bytes from one buffer to another.
memset	reentrant	Initializes a specified number of data bytes in a buffer to a specified character value.

Notes:
One needs to include the **string.h** header file in his/her program to use these functions.

TABLE G.2 *Character Conversion and Classification Functions*

Routine	Attributes	Description
isalnum	reentrant	Tests for an alphanumeric character.
isalpha	reentrant	Tests for an alphabetic character.
iscntrl	reentrant	Tests for a control character.
isdigit	reentrant	Tests for a decimal digit.
isgraph	reentrant	Tests for a printable character with the exception of space.
islower	reentrant	Tests for a lowercase alphabetic character.
isprint	reentrant	Tests for a printable character.
ispunct	reentrant	Tests for a punctuation character.
isspace	reentrant	Tests for a whitespace character.
isupper	reentrant	Tests for an uppercase alphabetic character.
isxdigit	reentrant	Tests for a hexadecimal digit.
toascii	reentrant	Converts a character to an ASCII code.
toint	reentrant	Converts a hexadecimal digit to a decimal value.
tolower	reentrant	Tests a character and converts it to lowercase if it is uppercase.
_tolower	reentrant	Unconditionally converts a character to lowercase.
toupper	reentrant	Tests a character and converts it to uppercase if it is lowercase.
_toupper	reentrant	Unconditionally converts a character to uppercase.

Notes:
1. One needs to include the **ctype.h** header file in his/her program to use these functions.
2. The **_tolower**, **_toupper**, and **toascii** routines are implemented as macros.

TABLE G.3 *Data Conversion Functions*

Routine	Attributes	Description
abs	reentrant	Generates the absolute value of an integer type.
atof		Converts a string to a float.
atoi		Converts a string to an int.
atol		Converts a string to a long.
cabs	reentrant	Generates the absolute value of a character type.
labs	reentrant	Generates the absolute value of a long type.
strtod		Converts a string to a float.
strtol		Converts a string to a long.
strtoul		Converts a string to an unsigned long.

Notes:
1. One needs to include the **math.h** file in his/her program to use abs, cabs, and labs.
2. One needs to include the **stdlib.h** file in his/her program to use the other functions in the table.

TABLE G.4 *Math Functions*

Routine	Attributes	Description
acos		calculates the arc cosine of a specified number.
asin		calculates the arc sine of a specified number.
atan		calculates the arc tangent of a specified number.
atan2		calculates the arc tangent of a fraction.
ceil		Find the integer ceiling of a specified number.
cos		Calculates the cosine of a specified number.
cosh		Calculates the hyperbolic cosine of a specified number.
exp		Calculates the exponential function of a specified number.
fabs	reentrant	Finds the absolute value of a specified number.
floor		Finds the largest integer less than or equal to a specified number.
fmod		Calculates the floating-point remainder.
log		Calculates the natural logarithm of a specified number.
log10		Calculates the common logarithm of a specified number.
modf		Generates integer and fractional components of a specified number.
pow		Calculates a value raised to a power.
rand	reentrant	Generates a pseudo random number.
sin		Calculates the sine of a specified number.
sinh		Calculates the hyperbolic sine of a specified number.
srand		Initializes the pseudo random number generator.
sqrt		Calculates the square root of a specified number.
tan		Calculates the tangent of a specified number.
tanh		Calculates the hyperbolic tangent of a specified number.
_chkfloat	intrinsic, reentrant	Checks the status of float numbers.
_crol	intrinsic, reentrant	Rotates an unsigned char left a specified number of bits.
_cror	intrinsic, reentrant	Rotates an unsigned char right a specified number of bits.
_irol	intrinsic, reentrant	Rotates an unsigned integer left a specified number of bits.
_iror	intrinsic, reentrant	Rotates an unsigned integer right a specified number of bits.
_lrol	intrinsic, reentrant	Rotates an unsigned long left a specified number of bits.
_lror	intrinsic, reentrant	Rotates an unsigned long right a specified number of bits.

Notes:
1. Most of these functions are prototyped in **math.h**.
2. The **rand** and **srand** routines are prototyped in **stdlib.h**.
3. The **_chkfloat_**, **_crol_**, **_cror_**, **_irol_**, **_iror_**, **_lrol_**, and **_lrol_** functions are prototyped in **intrins.h**.

TABLE G.5 *Memory Allocation Functions*

Routine	Attributes	Description
calloc		Allocates storage for an array from the memory pool.
free		Frees a memory block that was allocated using calloc, malloc, or realloc.
init_mempool		Initializes the memory location and size of the memory pool.
malloc		Allocates a block from the memory pool.
realloc		Reallocates a block from the memory pool.

Notes:
1. These functions are prototyped in **stdlib.h**.
2. Before using any of these functions to allocate memory, you must first specify, using the **init_mempool**, the location and size of a memory pool from which subsequent memory requests are satisfied.

TABLE G.6 *Stream Input and Output Functions*

Routine	Attributes	Description
getchar	reentrant	Reads and echoes a character using the _getkey and putchar routine.
_getkey		Reads a character using the 8051 serial interface.
gets		Reads and echoes a character string using the getchar routine.
printf		Writes formatted data using the putchar routine.
putchar		Writes a character using the 8051 serial interface.
puts	reentrant	Writes a character string and newline ("\n") character using putchar routine.
scanf		Reads formatted data using the getchar routine.
sprintf		Writes formatted data to a string.
sscanf		Reads formatted data from a string.
ungetchar		Places a character back into the getchar input buffer.
vprintf		Writes formatted data using the putchar function.
vsprintf		Writes formatted data to a string.

Notes:
1. One needs to include the **stdio.h** file in his/her program to use these functions.
2. The default **_getkey** and **putchar** functions read and write characters using the 8051 serial interface. You can change these two functions and substitute them for the library routines.
3. The source for **_getkey** and **putchar** is in the **c:\keil\lib** directory.

TABLE G.7 *String Manipulation Functions*

Routine	Attributes	Description
strcat		Concatenates two strings.
strchr	reentrant	Returns a pointer to the first occurrence of a specified character in a string.
strcmp	reentrant	Compares two strings.
strcpy		Copies one string to another.
strcspn		Returns the index of the first character in a string that matches any character in a second string.
strlen	reentrant	Returns the length of a string.
strncat		Concatenates up to a specified number of characters from one string to another.
strncmp		Compares two strings up to a specified number of characters.
strncpy		Copies up to a specified number of characters from one string to another.
strpbrk		Returns a pointer to the first character in a string that matches any character in a second string.
strpos	reentrant	Returns the index of the first occurrence of a specified character in a string.
strrchr	reentrant	Returns a pointer to the last occurrence of a specified character in a string.
strrpbrk		Returns a pointer to the last character in a string that matches any character in a second string.
strrpos	reentrant	Returns the index of the last occurrence of a specified character in a string.
strspn		Returns the index of the first character in a string that does not match any character in a second string.
strstr		Returns a pointer in a string that is identical to a second sub-string.

Notes:
1. These functions are prototyped in the **string.h** header file.
2. All string functions operate on NULL-terminated character strings.

APPENDIX H

Keil C Library Function Prototypes

The prototypes of Keil C library functions are listed in alphabetic order.

int **abs**(int *val*);	/* returns the absolute value of value **val**. */
float **acos**(float *x*);	/* returns the arc cosine of value x. */
float **asin**(float *x*);	/* returns the arc sine of value x. */
void **assert**(*expression*);	/* tests *expression* and prints a diagnostic message using the **printf** library routine if it is false. */
float **atan** (float *x*);	/* returns the arc tangent value of x. */
float **atan2**(float *y*, float *x*);	/* returns the arc tangent of the ratio **y/x**. */
float **atof** (void **string*);	/* converts *string* into a floating-point value. */
int **atoi** (void **string*);	/* converts *string* into an integer value. */
long **atol** (void **string*);	/* converts *string* into a long integer value. */
char **cabs** (char *val*);	/* returns the absolute value of *val.* */
void ***calloc** (unsigned int *num*, unsigned int *len*);	/* allocates memory for an array of *num* elements. Each element occupies *len* bytes and initialized to 0. */
float **ceil** (float *val*);	/* returns the smallest integer that is greater than or equal to *val.* */
unsigned char **_chkfloat_** (float *val*);	/* checks the status of a floating-point number and returns one of the five possible values: 0: standard floating-point numbers 1: floating-point value 0 2: **+INF** (positive overflow) 3: **−INF** (negative overflow) 4: **NaN** (not a number) error status

float **cos** (float *x*);	/* returns the cosine of floating-point number x. */
float **cosh** (float *x*);	/* returns the hyperbolic cosine of x. */
unsigned char **_crol_** (unsigned char *c*, unsigned char *b*);	/* rotates character c left by b bits. */
unsigned char **_cror_** (unsigned char *c*, unsigned char *b*);	/* rotates character c right by b bits. */
float **exp**(float *x*);	/* computes e^x. */
float **fabs** (float *val*);	/* returns the absolute value of *val.* */
float **floor** (float *val*);	/* returns the largest integer value that is less than or equal to *val.*/
float **fmod** (float *x*, float *y*);	/* returns a value *f* such that *f* has the same sign as x, the absolute value of *f* is less than the absolute value of y, and there is an integer k such that k*y + f equals x. */
void **free** (void xdata **p*);	/* returns the memory block pointed to by *p* to the memory pool. */
char **getchar** (void);	/* reads a single character from the input stream using the **_getkey** function. The character read is echoed. */
char **_getkey**(void);	/* waits for a character from the serial port. */
char ***gets** (char **string*, int *len*)	/* reads a line of characters terminated by newline character into the buffer pointed to by string. The newline character is replaced by a NULL. */
void **init_mempool** (void xdata **p*, unsigned int *size*);	/* initializes the memory management routines and provides the starting address and size of the memory pool. */
unsigned int **_irol_** (unsigned int *i*, unsigned char *b*);	/* rotates the bit pattern for the integer i left b bits. */
unsigned int **_iror_** (unsigned int *i*, unsigned char *b*);	/* rotates the bit pattern for the integer *i* right *b* bits. */
bit **isalnum** (char *c*);	/* tests *c* to determine if it is an alphanumeric character. */
bit **isalpha**(char *c*);	/* tests *c* to determine if it is an alphabetic character. */
bit **iscntrl**(char *c*);	/* tests *c* to determine if it is a control character (0x00~0x1F or 0x7F. */
bit **isdigit** (char *c*);	/* tests *c* to determine if it is a decimal digit. */
bit **isgraph** (char *c*);	/* tests *c* to determine if it is a printable character (between 0x21 and 0x7E. */
bit **islower** (char *c*);	/* tests *c* to determine if it is a lowercase character. */
bit **isprint** (char *c*);	/* tests *c* to determine if it is a printable character (between 0x20 and 0x7E). */
bit **ispunct** (char *c*);	/* tests *c* to determine if it is a punctuation character. */
bit **isspace** (char *c*);	/* tests *c* to determine if it is a whitespace character. */
bit **isupper** (char *c*);	/* tests *c* to determine if it is an uppercase alphabetic character. */
bit **isxdigit** (char *c*);	/* tests *c* to determine if it is a hexadecimal digit. */
long **labs** (long *val*);	/* determines the absolute value of the long integer *val.* */
float **log** (float *val*);	/* computes the natural logarithm for the floating-point number *val.*/
float **log10** (float val);	/* computes the logarithm for the floating-point number *val.* */
void **longjmp** (jmp_buf *env*, int *retval*);	/* restores the state which was previously stored in *env* by the **setjmp** function. The *retval* argument specifies the value to return from the **setjmp** function call. */
unsigned **long _lrol_** (unsigned long *l*, unsigned char *b*);	/* rotates the bit pattern for the long integer *l* left *b* bits. */
unsigned **long _lror_** (unsigned long *l*, unsigned char *b*);	/* rotates the bit pattern for the long integer *l* right *b* bits. */
void ***malloc** (unsigned int *size*);	/* allocates a memory block from memory pool of *size* bytes in length. The **malloc** function returns a pointer to the allocated block or a null pointer if there was not enough memory to satisfy the allocation request. */
void *memccpy (void **dest*, *void **src*, char *c*, int *len*);	/* copies 0 or more characters from *src* to *dest.* Characters are copied until the character c is copied or until *len* bytes have been copied, whichever comes first. The **memccpy** function returns a pointer to the byte in *dest* that follows the last character copied or a null pointer if the last character copied was c. */
void ***memchr** (void **buf*, char *c*, int *len*);	/* scans *buf* for the character c in the first *len* bytes in the buffer. The **memchr** function returns a pointer to the character c in *buf* or a null pointer if the character was not found. */

char **memcmp** (void **buf1*, void **buf2*, int *len*);
/* compares two buffers *buf1* and *buf2* for *len* bytes and returns a value indicating their relationship as follows:
< 0: *buf1* less than *buf2*
= 0: *buf1* equal to *buf2*
> 0: *buf1* greater than *buf2*

void ***memcpy** (void **dest*, void **src*, int *len*);
/* copies *len* bytes from *src* to *dest*. The **memcpy** function returns *dest*. */

void ***memmove** (void **dest*, void **src*, int *len*);
/* copies *len* bytes from *src* to *dest*. The **memmove** function returns *dest*. */

void ***memset** (void **buf*, char *c*, int *len*);
/* sets the first *len* bytes in *buf* to c. The **memset** function returns *dest*. */

float **modf** (float *val*, float **ip*);
/* splits the floating-point number *val* into integer and fractional components. The fractional part of *val* is returned as a signed floating-point number. The integer part is stored as a floating-point number at *ip*. */

void **_nop_** (void);
/* inserts an NOP instruction into the program. */

int **offsetof** (*structure*, *member*);
/* calculates the offset of the *member* structure element from the beginning of the structure. The *structure* argument must specify the name of a structure. The *member* argument must specify the name of a member of the structure. The **offsetof** macro returns the offset, in bytes, of the *member* element from the beginning of **struct** *structure*. */

float **pow** (float *x*, float *y*);
/* calculates and returns x^y. */

int **printf** (const char **fmtstr* [, *arguments*] . . .);
/* formats a series of strings and numeric values and builds a string to write to the output stream using the **putchar** function. The *fmtstr* argument is a format string as described in Section 5.5. */

char **putchar** (char *c*);
/* transmits the character *c* using the 8051 serial port. */

int **puts** (const char **string*);
/* writes *string* followed by a newline character to the output stream using the putchar function. The **puts** function returns **EOF** if an error occurred and a value of 0 if no errors were encountered. */

void ***realloc** (void xdata **p*, unsigned int *size*);
/* changes the size of a previously allocated memory block. The *p* argument points to the allocated block and the *size* argument specifies the new size for the block. The contents of the existing block are copied to the new block. This function returns a pointer to the new block. */

int **scanf** (const char **fmstr*[, *argument*] . . .);
/* reads data using the **getchar** routine. Data inputs are stored in the locations specified by argument according to the format string *fmstr*. The **scanf** function returns the number of input fields that were successfully converted. An **EOF** is returned if an error is encountered. */

int **setjmp** (jmp_buf *env*);
/* saves the current instruction address as well as other CPU registers. A subsequent call to the **longjmp** function restores the instruction pointer and registers, and execution resumes at the point just after the **setjmp** call. Local variables and function arguments are restored only if declared with the **volatile** attribute. */

float **sin** (float *x*);
/* computes the sine of the floating-point value *x*. The value of *x* must be in the −65535 to +65535 range or an **NaN** error value is generated. */

float **sinh** (float *val*);
/* calculates the hyperbolic sine of the floating-pointvalue *x*. The value of *x* must be in the −65535 to +65535 range or an **NaN** error value is generated. */

int **sprintf** (char **buffer*, const char **fmtstr* [,*argument*] . . .);
/* formats a series of strings and numeric values and stores the resulting string in *buffer*. The *fmtstr* argument is a format string and has the same requirements as specified for the **printf** function. The **sprintf** function returns the number of characters actually written to *buffer*. */

float **sqrt** (float *x*);
/* calculates and returns the square root of *x*. */

int **sscanf** (char **buffer*, const char **fmtstr* [, *argument*] . . .);
/* reads data from the string *buffer*. Data inputs are stored in the locations specified by *argument* according to the format string *fmtstr*. Each argument

must be a pointer to a variable that corresponds to the type defined in *fmtstr* which controls the interpretation of the input data. *fmtstr* specified a format string as described in Section 5.5. This function returns the number of input fields that were successfully converted. An **EOF** is returned if an error is encountered. */

char ***strcat** (char **dest*, char **src*);
/* concatenates or appends *src* to *dest* and terminates *dest* with a null character. A pointer to *dest* is returned. */

char ***strchr** (const char **string*, char *c*);
/* searches string for the first occurrence of *c*. The **strchr** function returns a pointer to the character c found in *string* or a null pointer if no matching character is found. */

char **strcmp** (char **string1*, char **string2*);
/* compares the contents of *string1* and *string2* and returns a value indicates their relationship.

Value	Meaning
< 0:	*string1* less than *string2*
= 0:	*string1* equal to *string2*
> 0:	*string1* greater than *string2* */

char ***strcpy** (char **dest*, char **src*);
/* copies *src* to *dest* and appends a null character to the end of *dest*. The value of *dest* is returned. */

int **strcspn** (char **src*, char **set*);
/* searches the *src* string for any of the characters in the *set* string. If the first character in *src* matches a character in *set*, a value 0 is returned. If there are no matching characters in *src*, the length of the string is returned. */

int **strlen** (char **src*);
/* calculates the length, in bytes, of *src*. The null character is included in counting. */

char ***strncat** (char **dest*, char **src*, int *len*);
/* appends at most *len* characters from *src* to *dest* and terminates *dest* with a null character. The function returns *dest* to the caller. */

char **strncmp** (char **string1*, char **string2*, int *len*);
/* compares the first *len* bytes of string1 and string2 and returns a value indicating their relationship:

Value	Meaning
< 0:	*string1* less than *string2*
= 0:	*string1* equal to *string2*
> 0:	*string1* greater than *string2* */

char ***strncpy** (char **dest*, char **src*, int *len*);
/* copies at most *len* characters from *src* to *dest*. The function returns *dst* to the caller. */

char ***strpbrk** (char **string*, char **set*);
/* searches *string* for the first occurrence of any character from *set*. The null terminator is not included in the search. The function returns a pointer to the matching character in *string*. If *string* contains no characters from *set*, a null pointer is returned. */

int **strpos** (const char **string*, char *c*);
/* searches *string* for the first occurrence of c. The null character is included in the search. The function returns the index of the character matching *c* in *string* or a value of −1 if no matching character was found. */

char ***strrchr** (const char **string*, char *c*);
/* searches *string* for the last occurrence of *c*. The null character is included in the search. The function returns a pointer to the last character c found in *string* or null pointer if no matching character was found. */

char ***strrpbrk** (char **string*, char **set*);
/* searches *string* for the last occurrence of any character from *set*. The null character is not included in the search. The function returns a pointer to the last matching character in *string*. */

int **strrpos** (const char **string*, char *c*);
/* searches *string* for the last occurrence of *c*. The null character terminating *string* is included in the search. The function returns the index of the last character matching *c* in *string* or a value of −1 if no matching character was found. */

int **strspn** (char **src*, char **set*);

/* searches the *src* string for characters not found in the *set* string. The function returns the index of the first character located in *src* that does not match a character in *set*. If the first character in *src* does not march any character in *set*, a 0 is returned. If all characters in *src* are found in *set*, the length of *src* is returned. */

char ***strstr** (const char **src*, char **sub*);

/* locates the first occurrence of the string *sub* in the string *src* and returns a pointer to the beginning of the first occurrence. If no such *sub* string exists in *src* a null pointer is returned. */

float **strtod** (const char **string*, char ***ptr*);

/* converts *string* into a floating- point value. The input *string* is a sequence of characters that can be interpreted as a floating-point number. Whitespace characters at the beginning of string are skipped. The **strtod** function requires the string to have the following format:

[{+ | −}] digits [.digits] [{e | E} [{+ | −}] digits]

where:

digits may be one or more decimal digits.

The value of *ptr* is set to point to the first character in the *string* immediately following the converted part of the *string*. If no conversion is possible, then *ptr* is set to the value of *string* and the value 0 is returned by **strtod**. */

long **strtol** (const char **string*, char ***ptr*, unsigned char *base*);

/* converts *string* into a long value. The input *string* is a sequence of characters that can be interpreted as an integer. Whitespace characters at the beginning of *string* are skipped. The **strtol** function requires *string* to have the following format:

[whitespace] [{+ | −}] digits */

unsigned long **strtoul** (const char **string*, char ***ptr*, unsigned char *base*);

/* converts *string* into an unsigned long value. The input *string* is a sequence of characters that can be interpreted as an integer number. Whitespace characters at the beginning of string are skipped. */

float **tan** (float *x*);

/* calculates the tangent of the floating-point value of x. The value of x must be in the −65535 to +65535 range or an **NaN** error value is generated. */

float **tanh** (float *x*);

/* calculates the hyperbolic tangent for the floating-point value x. */

bit **_testbit_** (bit *b*);

/* produces a JBC instruction in the generated program code to simultaneously test the bit b and clear it to 0. This routine can be used only on directly addressable bit variables and is invalid on any type of expression.*/

char **toascii** (char *c*);

/* converts c to a 7-bit ASCII character. This macro clears the bit 7 of character *c*. */

char **toint** (char *c*);

/* interprets *c* as a hexadecimal value. ASCII characters '0' through '9' generate values of 0 to 9. ASCII characters 'A' through 'F' and 'a' through 'f' generates values of 10 to 15. If the value of c is not a hexadecimal digit, the function returns −1. */

char **tolower** (char *c*);

/* converts *c* to a lowercase character. If *c* is not an alphabetic letter, the **tolower** function has no effect. */

char **_tolower** (char *c*);

/* is a macro version of **tolower** that can be used when *c* is known to be an uppercase character. */

char **toupper** (char *c*);

/* converts *c* to a uppercase character. If *c* is not an alphabetic letter, the **toupper** function has no effect. */

char **_toupper** (char *c*);

/* is a macro version of **toupper** that can be used when *c* is known to be an lowercase character. */

char **ungetchar** (char *c*);

/* stores the character *c* back into the input stream. Subsequent calls to **getchar** and other stream input functions return *c*. Only one character may be passed to **ungetchar** between calls to **getchar**. */

type **va_arg** (*argptr, type*);	/* is used to extract subsequent arguments from a variable-length argument list referenced by *argptr*. The *type* argument specifies the data type of the argument to extract. This macro may be called only once for each argument and must be called in the order of the parameters in the argument list. The first call to this function returns the first argument after the *prevparm* argument specified in the **va_start** macro. Subsequent calls to **va_arg** return the remaining arguments in succession. */
void **va_end** (*argptr*);	/* is used to terminate use of the variable-length argument list pointer *argptr* that was initialized using the **va_start** macro. */
void **va_start** (*argptr, prevparm*);	/* when used in a function with a variable-length argument list, initializes *argptr* for subsequent use by the **va_arg** and **va_end** macros. The *prevparm* argument must be the name of the function argument immediately preceding the optional arguments specified by an ellipses (. . .). This function must be called to initialize a variable-length argument list pointer before any access using the **va_arg** macro is made. */
void **vprintf** (const char **fmtstr*, char **argptr*);	/* formats a series of strings and numeric values and builds a string to write to the output stream using the **putchar** function. The function is similar to the counterpart **printf**, but it accepts a pointer to a list of arguments instead of an argument list. */
void **vsprintf** (char **buffer*, const char **fmtstr*, char **argptr*);	/* formats a series of strings and numeric values and stores the string in *buffer*. The function is similar to the counterpart *sprintf*, but it accepts a pointer to a list of argument instead of an argument list. */

APPENDIX I

Music Note Frequencies

A_4 = 440 Hz
Speed of sound = 345 m/s
(Middle C is C_4)

APPENDIX I *Complete List of Frequencies of Music Notes*

Note	Frequency (Hz)	Wavelength (cm)	Note	Frequency (Hz)	Wavelength (cm)
C_0	16.35	2100	$A^{\#}_0/B^b_0$	29.14	1180
$C^{\#}_0/D^b_0$	17.32	1990	B_0	30.87	1110
D_0	18.35	1870	C_1	32.70	1050
$D^{\#}_0/E^b_0$	19.45	1770	$C^{\#}_1/D^b_1$	34.65	996
E_0	20.60	1670	D_1	36.71	940
F_0	21.83	1580	$D^{\#}_1/E^b_1$	38.89	887
$F^{\#}_0/G^b_0$	23.12	1490	E_1	41.20	837
G_0	24.50	1400	F_1	43.65	790
$G^{\#}_0/A^b_0$	25.96	1320	$F^{\#}_1/G^b_1$	46.25	746
A_0	27.50	1250	G_1	49.00	704

(*continued*)

APPENDIX I *Complete List of Frequencies of Music Notes (continued)*

Note	Frequency (Hz)	Wavelength (cm)	Note	Frequency (Hz)	Wavelength (cm)
$G^\#_1/A^b_1$	51.91	665	C_5	523.25	65.9
A_1	55.00	627	$C^\#_5/D^b_5$	554.37	62.2
$A^\#_1/B^b_1$	58.27	592	D_5	587.33	58.7
B_1	61.74	559	$D^\#_5/E^b_5$	622.25	55.4
C_2	65.41	527	E_5	659.26	52.3
$C^\#_2/D^b_2$	69.30	498	F_5	698.46	49.4
D_2	73.42	470	$F^\#_5/G^b_5$	739.99	46.6
$D^\#_2/E^b_2$	77.78	444	G_5	783.99	44.0
E_2	82.41	419	$G^\#_5/A^b_5$	830.61	41.5
F_2	87.31	395	A_5	880.00	39.2
$F^\#_2/G^b_2$	92.50	373	$A^\#_5/B^b_5$	932.33	37.0
G_2	98.00	352	B_5	987.77	34.9
$G^\#_2/A^b_2$	103.83	332	C_6	1046.50	33.0
A_2	110.00	314	$C^\#_6/D^b_6$	1108.73	31.1
$A^\#_2/B^b_2$	116.54	296	D_6	1174.66	29.4
B_2	123.47	279	$D^\#_6/E^b_6$	1244.51	27.7
C_3	130.81	264	E_6	1318.51	26.2
$C^\#_3/D^b_3$	138.59	249	F_6	1396.91	24.7
D_3	146.83	235	$F^\#_6/G^b_6$	1479.98	23.3
$D^\#_3/E^b_3$	155.56	222	G_6	1567.98	22.0
E_3	164.81	209	$G^\#_6/A^b_6$	1661.22	20.8
F_3	174.61	198	A_6	1760.00	19.6
$F^\#_3/G^b_3$	185.00	186	$A^\#_6/B^b_6$	1864.66	18.5
G_3	196.00	176	B_6	1975.53	17.5
$G^\#_3/A^b_3$	207.65	166	C_7	2093.00	16.5
A_3	220.00	157	$C^\#_7/D^b_7$	2217.46	15.6
$A^\#_3/B^b_3$	233.08	148	D_7	2349.32	14.7
B_3	246.94	140	$D^\#_7/E^b_7$	2489.02	13.9
C_4	261.63	132	E_7	2637.02	13.1
$C^\#_4/D^b_4$	277.18	124	F_7	2793.83	12.3
D_4	293.66	117	$F^\#_7/G^b_7$	2959.96	11.7
$D^\#_4/E^b_4$	311.13	111	G_7	3135.96	11.0
E_4	329.63	105	$G^\#_7/A^b_7$	3322.44	10.4
F_4	349.23	98.8	A_7	3520.00	9.8
$F^\#_4/G^b_4$	369.99	93.2	$A^\#_7/B^b_7$	3729.31	9.3
G_4	392.00	88.0	B_7	3951.07	8.7
$G^\#_4/A^b_4$	415.30	83.1	C_8	4186.01	8.2
A_4	440.00	78.4	$C^\#_8/D^b_8$	4434.92	7.8
$A^\#_4/B^b_4$	466.16	74.0	D_8	4698.64	7.3
B_4	493.88	69.9	$D^\#_8/E^b_8$	4978.03	6.9

REFERENCES

1. "AN10216-01 I²C Manual," Philips Semiconductor, March 24, 2003.
2. "The I²C-bus Specification Version 2.1," January 2000.
3. "System Management Bus Specification—Version 1.1," SBS Implementers Forum, December 11, 1998.
4. "System Management Bus Specification—Version 2.0," SBS Implementers Forum. August 3, 2000.
5. "C8051F04x Datasheet," Rev. 1.5, Silicon Laboratory, December 2005.
6. "C8051F04x-Dk Development Kit User's Guide", Rev. 0.5, Silicon Laboratory, May 2005.
7. "MAX6952 4-Digit 5 x 7, Matrix LED Display User's guide", Rev. 1, Maxim-IC, October 2002.
8. *8051 Microcontroller,* Scott MacKenzie, Prentice Hall, July 2006.
9. *8051 Microcontroller,* Kenneth Ayala, Cengage Delmar Learning, June 2004.
10. *8051 and Embedded System Design,* Mazidi & Mazidi, Prentice Hall, July 2007.
11. "The HCS12/9S12—An Introduction", Han-Way Huang, Cengage Delmar Learning, July 2005.
12. "PIC Microcontroller—An Introduction", Han-Way Huang, Cengage Delmar Learning, July 2004.

INDEX

D

H

I

J

T

U

V